Lecture Notes in Physics

350

G. Tenorio-Tagle M. Moles
J. Melnick (Eds.)

Structure and Dynamics of the Interstellar Medium

Proceedings of IAU Colloquium No. 120
Held on the Occasion of Guido's Jubilee
in Granada, Spain, April 17–21, 1989

Springer-Verlag
Berlin Heidelberg New York London Paris Tokyo Hong Kong

Editors

Guillermo Tenorio-Tagle
Max-Planck-Institut für Physik und Astrophysik
Karl-Schwarzschild-Straße 1, D-8046 Garching, FRG

Mariano Moles
Instituto de Astrofisica de Andalucia
Apartado 2144, E-18080 Granada, Spain

Jorge Melnick
European Southern Observatory
La Silla, Chile

ISBN 3-540-51956-4 Springer-Verlag Berlin Heidelberg New York
ISBN 0-387-51956-4 Springer-Verlag New York Berlin Heidelberg

Printed in Germany

Printing: Druckhaus Beltz, Hemsbach/Bergstr.
Bookbinding: J. Schäffer GmbH & Co. KG., Grünstadt
2158/3140-543210 – Printed on acid-free paper

FOREWORD

IAU colloquium no. 120 was hosted by the Instituto de Astrofísica de Andalucía to celebrate the jubilee of Guido Münch, who pioneered many aspects of the study of interstellar matter, and whose multiple fundamental contributions make up the backbone of our present understanding of the structure and dynamics of the interstellar medium.

The study of interstellar matter has traditionally been divided into a number of areas that have followed a more or less parallel and independent development. As the field approaches maturity, however, it is increasingly clear that further development in it requires a basic understanding of the primary mechanisms that govern the interactions between the various components.

Because of the diversity and complexity of this subject, the meeting was organized around a number of invited talks (22 in total), in which the speakers were requested by the Scientific Organizing Committee to present the "state of the art" of each subject. These talks were complemented by 50 selected oral contributions and 40 poster papers.

The Local Organizing Committee accepted the double challenge of arranging an incredibly demanding scientific programme and at the same time a celebration of Guido's jubilee worthy of Guido's legend.

The magnificent city of Granada, situated in the foothills of the Sierra Nevada, provided an ideal setting for the conference, and the spring in Granada was enjoyed by 150 participants from 21 countries. The renowned hospitality of the inhabitants of Granada, and of the staff of the Instituto de Astrofísica de Andalucía, in particular, was apparent when they helped the participants to survive the loaded scientific programme, which took all day, and the celebrations, which took a good part of every night!

This book contains most of the papers that were presented at the conference and the abstracts of all posters. The order in which the material is presented here differs slightly from the order in which the papers were read at the conference since the editors felt it appropriate to rearrange some contributions to preserve the thematic unity of the book.

The scientific meetings took place in a very friendly atmosphere that stimulated many lively discussions and a "duel" between theorists and observers. The Local Organizing Committee made an effort to record these discussions, most of which are reproduced in this book.

The girls at the conference desk, Milagros Estepa, Susana Gómez, and Josefina Molina, solved every problem and were instrumental to the success of the meeting. Working mostly behind the scenes, Dr. Victor Costa made sure that everything ran smoothly, and was there whenever needed. The editors are particularly grateful to Susana Gómez and Josefina Molina from the IAA, and to Petra Berkemeyer from the MPI für Astrophysik for their invaluable help preparing the manuscripts for publication.

Finally, we thank all the participants of the meeting for their excellent science and for their songs, dances, and cheerfulness during the celebrations.

Granada, September 1989.

Guillermo Tenorio-Tagle
Mariano Moles
Jorge Melnick

SCIENTIFIC ORGANIZING COMMITTEE

LOCAL ORGANIZING COMMITTEE

SPONSORS

International Astronomical Union (IAU).
Consejo Superior de Investigaciones Científicas (CSIC).
Universidad de Granada.
Junta de Andalucía.
Comisión Interministerial de Ciencia y Tecnología (CICYT).
Max-Planck-Institut für Astronomie.
Max-Planck-Institut für Astrophysik.

CONTENTS

I. Molecular Clouds, Star Formation and HII Regions

○ = Invited Review, • = Oral Contribution, ◦ = Poster Contribution

II. Mechanical Energy Sources

III. Discs, Outflows, Jets and HH Objects

IV. The Orion Nebula

V. The Extragalactic ISM

VI. Interstellar Matter at High Galactic Latitudes

VII. The Structure of the Interstellar Medium

List of Participants

A. ALBERDI, Instituto de Astrofísica de Andalucía, Granada, Spain.
C.E. ALBERT, Physics Deparment, U.S. Naval Academy, U.S.A.
E. ALFARO, Instituto de Astrofísica de Andalucía, Granada, Spain.
P. ANDRE, Institut d'Radio-Astronomie Millimetrique, Granada, Spain.
G. ANGLADA, Departamento de Astronomía, Univer. Barcelona, Spain.
R. BACHILLER, Centro Astronómico de Yebes, Guadalajara, Spain.
J. BALLET, Service d'Astrophysique CEA Saclay, France.
J. BALLY, Department Bell Laboratories, Holmdel, U.S.A.
B. BARSELLA, Instituto di Astronomia, Pisa, Italy.
M. BARSONY, Deparment of Astronomy, University of California, U.S.A.
E. BATTANER, Departamento Física Moderna, Univer. Granada, Spain.
L. BINETTE, Canadian Inst. for Theoretical Astrophysics, Toronto, Canada.
K. BIRKLE, Centro Astronómico Hispano-Alemán, Calar Alto, Spain.
J.C. BLADES, Space Telescope Science Institute, Baltimore, U.S.A.
H. BLOEMEN, Leiden University Observatory, Leiden, Holland.
K.H. BÖHM, Astronomy Department, University of Washington, Seattle, U.S.A.
P.W.J.L. BRAND, Royal Observatory, University of Edinburgh, U.K.
D. BREITSCHWERDT, Max-Planck-Institut für Kernphysik, Heidelberg, F.R.G.
E. BRINKS, Royal Greenwich Observatory, Hailsham, U.K.
J. BUJ, Departamento de Astronomía, Univer. Barcelona, Spain.
M. CAMERON, Max-Planck-Institut für Extraterres. Physik, Garching, F.R.G.
A. CAMPOS, Instituto de Astrofísica de Andalucía, Granada, Spain.
H. CASTAÑEDA, Instituto de Astrofísica de Canarias, La Laguna, Spain.
M. CENTURION, Osservatorio Astronomico di Trieste, Italy.
J. CERNICHARO, Institut d'Radio-Astronomie Millimetrique, Granada, Spain.
A.D. CESARSKY, C.N.R.S., Institute d'Astrophysique, Paris, France.
C.J. CESARSKY, Cen-Saclay, Service d'Astrophysique, France.
S. CHANDRASEKHAR, The Enrico Fermi Institute, University of Chicago, U.S.A.
R. CHEVALIER, Department of Astronomy, University of Virginia, Charlottesville, U.S.A.
A. CLARET DOS SANTOS, Instituto de Astrofísica de Andalucía, Granada, Spain.
V. COSTA, Instituto de Astrofísica de Andalucía, Granada, Spain.
G. COURTES, Observatoire de Marseille, Marseille, France.
D. COX, Department of Physics, UW Madison, U.S.A.
A. DALGARNO, Harvard-Smithsonian Center for Astrophysics, Cambridge, U.S.A.
L. DANLY, Space Telescope Institute, Baltimore, U.S.A.
K.S. de BOER, Sternwarte der Universität Bonn, F.R.G.
A. DELGADO, Instituto de Astrofísica de Andalucía, Granada, Spain.
A. DIAZ, Departamento de Física Teórica, Universidad Autónoma de Madrid, Spain.
J.M. DICKEY, Sterrewacht Leiden, Holland.
D. DULTZIN-HACYAN, Instituto de Astronomía, UNAM, México.
J. DYSON, Department of Astronomy, The University of Manchester, U.K.
C. EIROA, Oservatorio Astronómico Nacional, Madrid, Spain.
P. ELOSEGUI, Instituto de Astrofísica de Andalucía, Granada, Spain.

H. ELSAESSER, Max-Planck- Institut für Astronomie, Heidelberg, R.F.A.
R. ESTALELLA, Departamento de Astronomía, Universidad de Barcelona, Spain.
N.J. EVANS, Department of Astronomy, University of Texas, Austin, U.S.A.
E. FALGARONE, Caltech, George Downs Laboratory, Pasadena, U.S.A.
S. FALLE, Department of Applied Maths., University of Leeds, U.K.
M. FELLI, Instituto di Astronomia di Arcetri, Florenz, Italy.
M. FERNANDEZ, Observatorio Astronómico Nacional, Madrid, Spain.
A. FERRARA, Instituto di Astronomia, Pisa, Italy.
F. FERRINI, Instituto di Astronomia, Pisa, Italy.
E. FLORIDO, Departamento de Física Moderna, Universidad de Granada, Spain.
G.A.P. FRANCO, University Observatory, Kopenhagen, Denmark.
J. FRANCO, Instituto de Astronomía, UNAM, México.
G. GARAY, Departamento de Astronomía, Universidad de Santiago, Santiago de Chile, Chile.
A.I. GOMEZ DE CASTRO, Observatorio Astronómico Nacional, Madrid, Spain.
J.F. GOMEZ, Instituto de Astrofísica de Andalucía, Granada, Spain.
R. GREDEL, European Southern Observatory, Santiago de Chile, Chile.
HARI OM VATS, Physical Research Laboratory, Ahmedabad, India.
T. HASEGAWA, Institute of Astronomy, University of Tokyo, Tokyo, Japan.
S.S. HAYASHI, Joint Astronomy Center, Hilo, Hawaii, U.S.A.
C. HEILES, Department of Astronomy, University of California, Berkeley, U.S.A.
U. HERBSTMEIER, Radioastronomisches, Institut der Universität, Bonn, R.F.A.
H. HIPPELEIN, Max-Planck-Institut fuer Kernphysik, Heidelberg, R.F.A.
N. HIRANO, Nobeyama Radio Observatory, Nagano, Japan.
V.A. HUGHES, Astron. Group Depart. of Physics, Queen's Univer., Ontario, Canada.
J. IBAÑEZ, Universidad de las Islas Baleares, Palma de Mallorca, Spain.
G. JONCAS, Department of Physics, University Laval, Quebec, Canada.
F.D. KHAN, Department of Astronomy, University of Manchester, U.K.
N. KAIFU, Nobeyama Radio Observatory, Nagano, Japan.
R. KAWAVE, Nobeyama Radio Observatory, Nagano, Japan.
E. KETO, Lawrence Livermore National Laboratory, Livermore, U.S.A.
M. KUN, Konkoly Observatory, Budapest, Hungary.
R. KUNZE, Institut für Astronomie und Astrophysik, Würzburg, R.F.A.
R.B. LARSON, Yale Astronomi Department, New Haven, U.S.A.
J. LATTANZIO, Institute of Geophysics, Planetary Physics, Livermore, U.S.A.
A. LAVAL, Observatoire de Marseille, Marseille, France.
M. LOPEZ ARROYO, Observatorio Astronómico Nacional, Madrid, Spain.
L. LUCY, European Southern Observatory, Garching, R.F.A.
A. MANCHADO, Instituto de Astrofísica de Canarias, La Laguna, Spain.
J.M. MARCAIDE, Instituto de Astrofísica de Andalucía, Granada, Spain.
J. MARTIN PINTADO, Centro Astronómico de Yebes, Guadalajara, Spain.
J. MASEGOSA, Instituto de Astrofísica de Andalucía, Granada, Spain.
U. MEBOLD, Radioastronomisches Institute der Universität Bonn, R.F.A.
J. MELNICK, European Southern Observatory, La Silla, Chile.
E. MENDOZA, Instituto de Astronomía, UNAM, México.

F. MIRABEL, Departamento de Física, Universidad de Puerto Rico, Puerto Rico.
P. MOLARO, Osservatorio Astronomico de Trieste, Trieste, Italy.
M. MOLES, Instituto de Astrofísica de Andalucía, Granada, Spain.
G. MÜNCH, Centro Astronómico Hispano-Alemán, Calar Alto, Spain.
Y. MURATA, Nobeyama Radio Observatory, Nagano, Japan.
A. NATTA, Osservatorio di Arcetri, Firenze, Italy.
C. NORMAN, Space Telescope Institute, Baltimore, U.S.A.
A. DEL OLMO, Instituto de Astrofísica de Andalucía, Granada, Spain.
D.E. OSTERBROCK, Lick Observatory, University of California, U.S.A.
B.E.J. PAGEL, Royal Greenwich Observatory, Hailsham, U.K.
F. PALLA, Osservartorio Astrofisico di Arcetri, Firenze, Italy.
J. PALOUS, Astronomical Institute of Czecheoslovak, Praga, Czechoslovakia.
P. PAOLICCI, Instituto di Astronomia, Pisa, Italy.
F. PARESCE, Space Telescope Science Institute, Baltimore, U.S.A.
J. PASTOR, Departamento de Astronomía, Universidad de Barcelona, Spain.
J.C. PECKER, College de France, Paris, France.
J. PEREA, Instituto de Astrofísica de Andalucía, Granada, Spain.
J.P. PHILLIPS, Instituto de Astrofísica de Canarias, La Laguna, Spain.
E. PITZ, Max-Planck-Institut für Astronomie, Heidelberg, F.R.G.
P. PLANESAS, Owens Valley Radio Observatory, Pasadena, U.S.A.
M. PRIETO, Instituto de Astrofísica de Canarias, La Laguna, Spain.
M.A. PRIETO, IUE Observatory VILSPA, Madrid, Spain.
J.M. QUINTANA, Instituto de Astrofísica de Andalucía, Granada, Spain.
A. RAGA, Canadian Institute for Theoretical Astrophysics, Toronto, Canada.
J. RICHER, Cavendish Laboratory, Cambridge, U.K.
A. RIERA, Instituto de Astrofísica de Canarias, La Laguna, Spain.
M.J. RIOJA, Instituto de Astrofísica de Andalucía, Granada, Spain.
L.F. RODRIGUEZ, Harvard-Smithsonian Center for Astrophysics, Cambridge, U.S.A.
J.R. ROY, Dept. of Physics, Faculty of Sciences, Quebec, Canada.
M. ROZYCZKA, Warsaw University Observatory, Warszawa, Poland.
R. RUBIN, NASA Ames Research Center, California, U.S.A.
N.S. SABALISCK, Instituto Astronomico e Geofisico, Sao Paulo, Brazil.
F. SANCHEZ, Instituto de Astrofísica de Canarias, La Laguna, Spain.
M.L. SANCHEZ, Departamento de Física Moderna, Universidad de Granada, Spain.
G. SANDELL, Joint Astronomy Centre, Hilo, Hawaii, U.S.A.
A. SARGENT, Downs Laboratory, Pasadena, U.S.A.
J.L SAUVAGEOT, Service d'Astrophysique, CEN de Saclay, France.
A. SCHULZ, Max-Planck Institut für Radioastronomie, Bonn, R.F.A.
R. SNELL, Five College R.A.O., University of Massachussetts, Amherst, U.S.A.
S. SNOWDEN, Max-Planck Inst. Extraterr. Physik., Garching, F.R.G.
P. SOLOMON, Groupe de Radioastronomie, Laboratoire de Physique, Paris, France.
A. TALAVERA, IUE Observatory - VILSPA, Madrid, Spain.
D. TAYLOR, The University of Manchester, Manchester, U.K.
G. TENORIO-TAGLE, Max Planck Institut für Physik und Astrophysik, Garching, F.R.G.

E. TERLEVICH, Royal Greenwich Observatory, Hailsham, U.K.
R. TERLEVICH, Royal Greenwich Observatory, Hailsham, U.K.
J.M. TORRELLES, Instituto de Astrofísica de Andalucía, Granada, Spain.
V. TRIMBLE, Dept. of Physics, University of California, U.S.A.
H. VAN WOERDEN, Kapteyn Astron. Institute, University of Groningen, Holland.
L. VERDES-MONTENEGRO, Instituto de Astrofísica de Andalucía, Granada, Spain.
J.M. VILCHEZ, Instituto de Astrofísica de Canarias, La Laguna, Spain.
T. VIVES, Centro Astronómico Hispano-Alemán, Calar Alto, Spain.
G. VLADILO, Osservatorio Astronomico di Trieste, Trieste, Italy.
H.J. VÖLK, Max-Planck-Institut für Kernphysik, Heidelberg, F.R.G.
B. WAKKER, Kapteyn Astronomical Institute, Groningen, Holland.
Y.A. WALTER, Institute for Theoretical Astrophysics, Frankfurt, F.R.G.
J. WAMPLER, European Southern Observatory, Garching, F.R.G.
J.S. WANG, Yunnan Observatory, Kunming, China.
A. WOOTTEN, National Radioastronomy Observatory, Charlottesville, U.S.A.
H. YORKE, Universitat Sternwarte, Gottingen, F.R.G.
H. ZINNECKER, Max-Planck-Institute für Extraterrestrische Physik, Garching, F.R.G.

GUIDO MÜNCH: AN APPRECIATION

Donald E. Osterbrock

Professor GUIDO MÜNCH

Guido Münch was born in the state of Chiapas, Mexico on June 9, 1921. He received his undergraduate education at the Universidad Nacional Autonoma de Mexico, where he began in civil engineering, but then switched to mathematics. He earned his B. S. in 1938. Two of his early heroes were mathematicians George D. Birkhoff and Solomon Lefschetz, who visited Mexico in his student days. Guido continued at UNAM as a graduate student, and two of his teachers who influenced him greatly were Spanish refugees who had fled to Mexico after the fall of the Republic, Pedro Carrasco, the former director of the Madrid Observatory, who taught astronomy, and Blas Cabrera, who taught atomic physics. Guido was also inspired by the group of American astronomers, including Walter S. Adams, Henry Norris Russell, Joel Stebbins, Harlow Shapley, Otto Struve, Donald H. Menzel, Cecilia Payne-Gaposchkin, W. W. Morgan, and Nicholas U. Mayall, who came to Mexico in early 1942 to participate in the inauguration of the Tonantzintla Observatory. As a result of this meeting Struve later offered Guido a position as a night assistant at McDonald Observatory. He accepted and after one month of training at Yerkes Observatory (April 1943), went to McDonald where he worked until the fall. Then he returned to Yerkes as a graduate student of the University of Chicago, of which the observatory is part.

Guido was a graduate student at Yerkes for three years, and earned his Ph.D. in 1946. He worked on spectroscopy of stars with Struve and Morgan, but chiefly on radiative transfer and the theory of stellar atmospheres with S. Chandrasekhar, his thesis adviser, applying Chandra's new theoretical methods to the Sun and stars, and extending them as well.

After receiving his Ph.D. Guido returned to a faculty appointment at UNAM, but came back to Yerkes as an instructor in the fall of 1947. There he continued very creatively his stellar atmospheres research. With Chandra he also undertook a theoretical discussion of the statistics of stellar rotation velocities, and their joint paper on this subject was an important one. Guido was promoted to assistant professor in 1949. Yerkes Observatory was a center of research on interstellar matter, and Guido became interested in the cloud structure of interstellar matter as revealed by the fluctuations in brightness in the Milky Way. He and Chandra developed the theory of these fluctuations, and applied them to deriving parameters of the cloud structure, in an important series of papers.

Guido went to the California Institute of Technology in 1951 as an assistant professor and staff member of the Mount Wilson and Palomar Observatories. In 1948 Jesse L. Greenstein had been the first to move there from Yerkes, to start the Caltech Astrophysics Department in its modern form; he brought Guido to join him three years later, myself two years after that, and Arthur D. Code three years still later, all of us Yerkes Ph.D.'s. At Caltech Guido was promoted to associate professor in 1954, and to full professor in 1959. He became a U.S. citizen in 1957.

In research he continued his work on stellar atmospheres, especially on hot stars, and most particularly on hot population II "subdwarfs" (O stars). With the coudé spectrograph of the 200–inch Hale telescope Guido opened a new era in the study of interstellar absorption lines, going beyond the outstanding work Adams had done with the 100–inch. Guido, with the telescope necessary to observe the entire Milky Way, obtained high-dispersion spectrograms which not only clearly resolved the absorption lines into "cloud" components, but which also showed groups of them in the outer Perseus spiral arm, well separated from the clouds in our (Orion) arm. It was widely recognized as a very important direct confirmation of the nature and location of the spiral arms in our Galaxy. Guido's papers and review articles on the interstellar medium are classics. He also used O stars in high galactic latitudes to study interstellar gas far from the plane. Everyone knew "it should not be there," but Guido showed that it is, and that it is *infalling*.

He also carried out very high-dispersion emission-line studies, especially of the Orion Nebula. He began this work with Olin C. Wilson, who had been applying these methods to planetary nebulae. Guido was always very interested in instruments, especially in pushing the limits of *wavelength resolution*. Beginning with this work on Orion, he became more and more active in conceiving new methods, improvements in existing instruments, finally whole *new* instruments, working closely with students, postdocs, engineers and technicians to bring them to reality and practice. He made especially important studies of the velocity and density fields within the Orion nebula, and of the *neutral gas* in it.

Guido also produced many new results on the physics of planets. He was one of the first well known U. S. observational astronomers to take the space program seriously and to get actively involved in it himself. With Hyron Spinrad and Lewis D. Kaplan he made the first detection of H_2O vapor on Mars, by high-dispersion near-infrared spectroscopy. Their spectroscopic measurement of the density of Mars's atmosphere (from the column density of CO_2 they determined from their spectra) was crucial in planning the first Martian landers. Guido also pioneered in space astrophysics in applying infrared measurements, especially to Jupiter and its atmosphere, Mars, and Venus (the Mariner missions). He worked with Gerry Neugebauer, Lawrence Trafton, and many other former Caltech graduate students on these subjects.

With the 200–inch Hale telescope Guido also worked on interstellar matter in and near the nucleus of M 31, and with his brother Luis on the rotational velocities in the inner part of our Galaxy (from stellar spectra). Guido's pioneering theoretical paper on Lα in interstellar matter, and how it is resonance-scattered by H atoms and destroyed by absorption by dust, is an extremely important one. He has always tried to use instruments to their limits, and beyond. He has thought creatively of problems that needed to be solved, of how they could be solved with high spectral resolution, and then he has gone out and solved them. He had the theoretical insights, the mathematical facility, and the instrumental skills to do so.

In 1977 Guido left Caltech to accept the position of co-director of the Max-Planck-Institut für Astronomie in Heidelberg. There, with Hans Elsässer, he created a school of high-resolution spectroscopy. The work of this whole group is well known from their many important papers in *Astronomy and Astrophysics*.

Guido has always kept close connections with his native land and with Mexican astronomy. He returned to Mexico often, and advised many Mexican students informally about astronomical opportunities in the United States.

The astrophysics graduate students who did their Ph.D. theses under Guido's supervision at Caltech form an imposing list: the late John C. Stewart, Manuel Mendez, Robert Norton, Robert A. R. Parker, Lawrence Trafton, James G. Gunn, Richard Larson, Christopher M. Anderson, Jeffrey Scargle, Bruce A. Peterson, Vicky A. Peterson, Virginia Trimble, Jay Frogel, S. Eric Persson, Judith Cohen and Jorge Melnick. He also had Robert Brinkman as a Ph.D. thesis student in aeronomy. In addition, he has had two more doctoral thesis students at Heidelberg, P. Gomez-Garrido and Roland Gredel.

Guido was elected to the American Academy of Arts and Sciences in 1962, and to the U. S. National Academy of Sciences in 1967. He was awarded the NASA Medal for Exceptional Scientific Achievement in 1974, and Spain's Prince of Asturias Prize for Scientific and Technical Research in 1989. He was elected to the Third World Academy of Sciences in 1984. He is an outstanding research scientist, and I am proud to have been his student, colleague, friend and admirer for forty years.

I am very grateful to S. Chandrasekhar, J. L. Greenstein, A. D. Code and E. E. Mendoza for their memories and thoughts, which I have incorporated with my own in this appreciation.

MOLECULAR CLOUDS, STAR FORMATION AND HII REGIONS

CHEMISTRY OF MOLECULAR CLOUDS

A. Dalgarno
Harvard-Smithsonian Center for Astrophysics
60 Garden Street, Cambridge
Massachusetts 02138 USA

Abstract

A discussion is presented of the chemistry of quiescent molecular clouds, and the effects of the presence of polycyclic aromatic hydrocarbon molecules and of cosmic-ray induced ultraviolet photons are examined. A comparison is made with the chemistry occurring in molecular clouds that are subjected to shocks and the differences between dissociative and non-dissociative shocks are described. The changes in composition caused by intense cosmic ray fluxes or intense ultraviolet radiation fields are explored.

1. Introduction

Molecular clouds in the interstellar medium are characterized as diffuse, translucent or dense, depending on the column density of material contained in the clouds. Diffuse clouds correspond to visual extinctions A_V less than about one magnitude, equivalent to a column density N_H of hydrogen nuclei less than 1.9×10^{21} cm^{-2}. Translucent clouds are more extended with visual extinctions between one and five magnitudes. Dense clouds have visual extinctions greater than five magnitudes. Dense clouds include clouds in which active star formation is in progress and dark clouds which appear to be quiescent.

Diffuse and translucent molecular clouds only partly obscure the light from the stars that lie behind them and they have been studied mostly through observations of the absorption lines in the ultraviolet, visible and infrared spectrum to which they give rise. Some molecules in diffuse, and more so in translucent clouds, have been detected also through the appearance of emission lines in the millimetre region. The penetration of the interstellar radiation field diminishes exponentially with A_V and molecules in dense molecular clouds are detectable through their emission and absorption lines in the millimetre and radio regions.

The densities of molecular clouds range from about 10^2 cm^{-3} characteristic of diffuse and translucent clouds to 10^3 cm^{-3} or 10^4 cm^{-3} characteristic of dense clouds though clumps of much higher density also occur. The temperatures range from about 70K or more in diffuse clouds to 10K in dark clouds. Temperatures may be higher in localized regions which have been disturbed by processes associated with star formation and stellar evolution.

A diverse array of molecules has been discovered in dense clouds. The list contains over seventy-five distinct species and includes many complex organic molecules and several molecular ions. Interstellar molecules have been detected containing the elements hydrogen, carbon, oxygen, nitrogen, sulphur, silicon, chlorine and phosphorous. In dense clouds where the complex molecules are found, carbon monoxide has an abundance relative to hydrogen of about 10^{-4} and the other molecules have fractional abundances between 10^{-7} and 10^{-10} (cf. Irving, Goldsmith and Hjalmarson 1987). In diffuse and translucent clouds, only diatomic molecules have been observed, though C_3H_2 has been detected in an apparently diffuse region towards Cas A. There is also indirect evidence for the presence of very large molecules containing more than twenty carbon atoms in the diffuse interstellar medium and in diffuse molecular clouds.

The chemical composition in molecular clouds and its response to intense radiation fields, enhanced ionization sources, dissociative and non-dissociative shocks and to the interactions of outflowing material from protostellar objects, evolving stars, novae and supernovae is a potentially significant source of information about the structure and evolution of interstellar clouds and of the mechanism by which stars are formed in the Milky Way galaxy and external galaxies. The chemical composition affects the cloud evolution through its influence on the thermal and ionization balance and may yet serve as a chemical clock for determining the age of clouds.

2. Chemistry

The chemistry of molecular clouds is broadly the chemistry of a gas of molecular and atomic hydrogen and atomic helium with a small admixture of heavy elements. The gas contains solid refractory dust grains. The grains play a crucial role in excluding the photons of the interstellar radiation field from the interiors of large clouds and thereby shielding the molecules from the destructive effects of photodissociation. They are important also as sites for the formation of molecular hydrogen and possibly of other molecules and as sinks for the removal of heavy atoms and molecules from the gas phase.

In diffuse and translucent clouds and in the outer envelopes of dense clouds, photodissociation and photoionization are significant events in the chemistry. For clouds near to external intense sources of ultraviolet radiation, processes initiated by the absorption of the photons exert a major influence on the chemistry for distances into the clouds up to and beyond $A_V = 5$. Those molecular envelopes subjected to intense ultraviolet radiation have been called photodissociation regions (Tielens and Hollenbach 1985) or photochemical regions (van Dishoeck 1988).

In addition to the interstellar radiation field, the clouds are bombarded by cosmic rays. Ionization caused by the cosmic rays leads to the formation of molecular ions whose presence in the interiors of interstellar clouds has been established observationally. Because many neutral particle reactions have activation barriers and are slow at the temperatures prevailing in molecular clouds, the chemistry is mainly an ion-molecule chemistry driven by the interstellar radiation field in the outer envelopes and by cosmic rays in the interiors. Most ion-molecule reactions remain rapid at low temperatures and for heteronuclear molecules may increase in efficiency.

In the envelope, carbon is ionized by interstellar photons and the carbon chemistry is initiated by the radiative association

$$C^+ + H_2 \rightarrow CH_2^+ + h\nu \ . \tag{1}$$

In the interior, reaction(1) occurs but the major source of C^+ is the charge transfer reaction

$$He^+ + CO \rightarrow C^+ + O \ + He \ , \tag{2}$$

the He^+ being a product of cosmic ray ionization. Following (1), the abstraction reaction

$$CH_2^+ + H_2 \rightarrow CH_3^+ + H \tag{3}$$

occurs. The molecular ion CH_3^+ may be removed by dissociative recombination

$$CH_3^+ + e \rightarrow CH_2 + H \tag{4a}$$

$$\rightarrow CH + H_2 \tag{4b}$$

but in dense clouds where the electron density is low, radiative association with H_2 to form CH_5^+,

$$CH_3^+ + H_2 \rightarrow CH_5^+ + h\nu \ , \tag{5}$$

may occur more rapidly. In diffuse regions, CH_2 formed in process (4a) is photodissociated to yield CH,

$$CH_2 \ + h\nu \rightarrow CH + H \ . \tag{6}$$

Methane can be formed from CH_5^+ by reaction with CO,

$$CH_5^+ + CO \rightarrow CH_4 + HCO^+ \ . \tag{7}$$

Complex hydrocarbons can then be built by insertion reactions. An example is the reaction

$$C^+ + CH_4 \rightarrow C_2 \ H_3^+ + H \ . \tag{8}$$

Dissociative recombination,

$$C_2H_3^+ + e \rightarrow C_2 \ H_2 + H \ , \tag{9}$$

then leads to acetylene. Condensation reactions such as

$$CH_3^+ + CH_4 \rightarrow C_2H_5^+ + H_2 \tag{10}$$

are also effective.

Reactions of the hydrocarbon ions with neutral heavy atoms lead to the formation of molecules incorporating oxygen, nitrogen, sulphur and other elements. Thus

$$C_2H_4^+ + O \rightarrow C_2H_3O^+ + H \qquad (11)$$

$$C_2H_3O^+ + e \rightarrow CH_2CO + H \qquad (12)$$

is a source of ketene and

$$C_2H_4^+ + N \rightarrow C_2H_3N^+ + H \qquad (13)$$

$$C_2H_3N^+ + e \rightarrow CH_2CN + H \qquad (14)$$

is a source of the cyanomethyl radical CH_2CN.

Cosmic ray ionization of H_2 to produce H_2^+ leads to H_3^+,

$$H_2^+ + H_2 \rightarrow H_3^+ + H \quad . \qquad (15)$$

The molecular ion H_3^+ undergoes proton transfer

$$H_3^+ + X \rightarrow HX^+ + H_2 \qquad (16)$$

with many of the neutral systems X. The ion HX^+ then initiates a sequence of abstraction reactions. Thus

$$H_3^+ + C \rightarrow CH^+ + H_2 \qquad (17)$$

leads to CH_3^+ and

$$H_3^+ + O \rightarrow OH^+ + H_2 \qquad (18)$$

leads to H_3O^+. Dissociative recombination of H_3O^+ produces H_2O and possibly OH. In diffuse and translucent clouds the oxygen sequence can also be by H^+ ions, produced by cosmic ray ionization of H, which undergo charge transfer

$$H^+ + O \rightarrow H + O^+ \qquad (19)$$

followed by

$$O^+ + H_2 \rightarrow OH^+ + H \quad . \qquad (20)$$

In both diffuse and dense clouds the production of OH is proportional to the cosmic ray ionization rate and the abundance of OH is a measure of the flux of cosmic rays penetrating the clouds.

Radiative association reactions are effective in building large molecules and in mixing the different element chemistries. Methanol can be formed by

$$CH_3^+ + H_2O \rightarrow CH_3OH_2^+ + h\nu \qquad (19)$$

$$CH_3OH_2^+ + e \rightarrow CH_3OH + H \qquad (20)$$

and ethanol by

$$H_3O^+ + C_2H_4 \rightarrow CH_3CH_2OH_2^+ + h\nu \qquad (21)$$

$$CH_3CH_2OH_2^+ + e \rightarrow CH_3CH_2OH + H \quad . \qquad (22)$$

Comprehensive accounts of the chemistry of complex molecules and the results of detailed calculations of the molecular abundances have been presented by Graedel and Langer (1989) and Herbst and Leung (1989). Their models do not take full account of the effects of the ultraviolet photons that are generated internally by the cosmic rays (Prasad and Tarafdar 1983). The secondary electrons accompanying ionization by cosmic rays excite molecular hydrogen and the excited states decay by spontaneous emission in the ultraviolet, producing photons that are energetically capable of photodissociating many and photoionizing some of the interstellar molecules (Sternberg, Dalgarno and Lepp 1987). The destructive effects of the cosmic ray-induced photons amplify along a chemical chain as molecules are built by the addition of carbon atoms. The effects are mitigated by the photodissociation of CO

$$CO + h\nu \rightarrow C + O \quad . \qquad (23)$$

It happens that many of the emission lines of H_2 overlap the absorption lines of CO that lead to dissociation of CO (Gredel, Lepp and Dalgarno 1987).

Photodissociation of CO by the cosmic ray-induced photons is a major source of neutral carbon atoms which counteracts the tendency of carbon to accumulate into CO and thereby limit the abundances of the complex hydrocarbons. Thus in model calculations of Gredel et al. (1989), the inclusion of the photodissociation of CO increases the steady-state C/CO ratio from 5×10^{-5} to 6×10^{-3} and the C_3H_2/H_2 ratio from 2×10^{-12} to 2×10^{-11}.

The steady-state ratios are however considerably smaller than values found in many observations. The C/CO ratio is as large as 0.1 or 0.2 in many clouds (cf. Keene et al. 1985, Genzel et al. 1988) and the C_3H_2/H_2 abundance ratio is typically in the range $10^{-9.5}$–$10^{-8.4}$ (Cox, Walmsley and Gusten 1989). Steady-state models (cf. Graedel and Langer 1989, Herbst and Leung 1989) are fairly successful in predicting the abundances of the smaller molecules (though serious discrepancies persist between the model abundances and observations of diffuse and translucent clouds (cf. van Dishoeck and Black 1986, 1988, 1989; Viala 1986, Viala, Roueff and Abgrall 1988, Nercessian, Benayou and Viala, 1988) but fail for the complex species despite the flexibility offered by the absence of laboratory data on many of the reactions involved in the chemical models.

One explanation of the high C/CO ratio and the high abundances of complex molecules is a carbon abundance that is in excess of the oxygen abundance so that an ample supply of carbon remains after the formation of carbon monoxide is complete (cf. Graedel and Langer 1989). Mechanisms by which gaseous carbon could be more abundant than gaseous oxygen in the cloud interiors have been discussed by Blake et al. (1989). There may be difficulty in reproducing the abundances of the alcohols though cosmic ray-induced photodissociation of CO (and of O_2 and H_2O) will be a large source of oxygen atoms.

The chemistry will be modified if a substantial population of large molecules or very small grains exists in dense clouds. There is considerable spectroscopic evidence from observations of infrared sources that such systems form a major component of the diffuse interstellar medium with an abundance relative to hydrogen of the order of 10^{-7} (cf. Puget and Leger 1989). The explicit identification of the large molecules as polycyclic aromatic hydrocarbons, containing in excess of twenty carbon atoms, has been advanced.

An immediate consequence of the presence of large molecules LM in dense clouds is the neutralization of atomic ions that do not react with H_2 by charge transfer

$$C^+ + LM \rightarrow C + LM^- \tag{24}$$

and mutual neutralization

$$C^+ + LM^- \rightarrow C + LM \quad . \tag{25}$$

The chemistry is changed qualitatively as mutual neutralization replaces dissociative recombination (Lepp and Dalgarno 1988).

Models of diffuse molecular clouds (van Dishoeck and Black 1986) consistently underestimate the C^+/C ratio (Dalgarno 1988). Agreement can be obtained by postulating the presence of a component of large molecules with an abundance ratio to hydrogen of 1×10^{-7} for the cloud towards ζ Persei and of 6×10^{-7} for the cloud towards ζ Ophiuchi (Lepp et al. 1988).

Of potentially greater significance to interstellar chemistry is the contribution from chemical reactions of oxygen, carbon, nitrogen, silicon and sulphur with the large molecules. Some theoretical discussion of their chemistry has been given by Omont (1986), Duley and Williams (1986), Lepp et al. (1988) and Brown et al. (1989) and some laboratory measurements on the reactions of Si^+ with the polycyclic aromatic hydrocarbon, napthalene, have been reported (Bohme, Wlodek and Wincel 1989). Also of importance may be the interaction of cosmic rays with large molecules. Experimental studies have shown that the ions CH_3^+, $C_2H_2^+$ and $C_3H_3^+$ are major products of the ionization of polycyclic aromatic hydrocarbons by impacts of 70eV electrons (Kingston et al. 1985).

The collision frequency of large molecules with each other at an abundance ratio of 10^{-7} is high so that they will tend to grow in size and decrease in density. The chemistry will be time-dependent.

Time-dependent chemistry has been invoked to explain both the high C/CO ratio and the high abundances of hydrocarbons. Because the time scale for the accumulation of C into CO is longer than that for the formation of complex molecules, a time-dependent evolution from an initial configuration in which the carbon exists as neutral atoms produces at times of the order of 10^5 years substantial amounts of complex molecules while retaining much of the carbon in the form of C. The C_3H_2/H_2 ratio reaches peak values of about 10^{-7} (cf. Herbst and Leung 1989, Gredel et al. 1989), larger indeed than the observed ratios.

Leaving aside the question of the plausibility of such an initial condition, during the evolution of a dense molecular cloud, atoms and molecules will be lost from the gas phase by freezing on to the surfaces of grains in a time scale of less than 10^6 years. Absorption features found in the spectra of embedded infrared objects show that mantles containing H_2O, CO and other species form on grain surfaces. There should exist regions within clouds that are devoid of molecules (Williams 1985) except H_2 and the molecular ions H_3^+, H_2D^+ and HeH^+. They have not been found and there must be mechanisms for returning the grain material to the gas phase.

Desorption from interstellar grains by cosmic ray impact has been explored by Leger, Jura and Omont (1985). The explosive release of molecules in grain-grain collisions has been considered by d'Hendecourt, Allamandola and Greenberg (1985), who have presented time-dependent models of the chemical composition of the gas phase and of the grain surfaces in an interstellar cloud. The models predict large abundances of water and ammonia but carbon atoms are removed rapidly and the C/CO ratio falls steeply after a time of 10^5 years. The models do not consider complex molecules but it is probable that they follow qualitatively the longtime behaviour of neutral atomic carbon.

Grain-grain collisions may be insufficent and the chemistry may be influenced by sporadic events like star formation and by interactions with material flowing out from stars and protostars. Molecular formation in circumstellar shells has received considerable theoretical attention (cf. Nejad and Millar 1987, Glassgold et al. 1987, Mamon, Glassgold and Huggins 1988) and molecular formation in winds from protostars has been discussed by Rawlings, Williams and Canto (1988) and Glassgold, Mamon and Huggins (1989), in nova outbursts by Rawlings (1988) and in supernovae by Lepp, Dalgarno and McCray (1988, 1989), Petuchowski et al. (1989) and Latter and Black (1989).

The high velocity outflows drive shocks into the interstellar gas and compress and heat it. In the warm gas, the chemistry is driven by exothermic and endothermic reactions with H_2 of the kind

$$X + H_2 \rightleftarrows XH + H$$

and the resulting composition depends on the temperature and the H/H_2 ratio. Shocked gas is an efficient source of CH^+,

$$C^+ + H_2 \rightarrow CH^+ + H \quad ,$$

which may help to explain observations of CH^+ in diffuse clouds (Elitzur and Watson 1980, Draine 1986, Pineau des Forets et al. 1986), and of sulphur-containing molecules (Hartquist, Oppenheimer and Dalgarno 1980, Pineau des Forets, Roueff and Flower 1986) and in some environments shocks may enhance the formation of complex molecules (Mitchell 1983).

If the shock velocity exceeds 40km s^{-1}, all the molecules are destroyed. The subsequent chemistry is modified by the precursor ultraviolet radiation emitted in the cooling recombining post-shock gas (Hollenbach and McKee 1989, Neufeld and Dalgarno 1989). The CO/H_2 ratio is high and detectable populations of molecular ions may be produced as the gas passes through its warm ionized state (Neufeld and Dalgarno 1989). Complex species are not formed in any abundance in the warm gas.

The outflowing gas may itself be an important source of gas phase molecules, and its interaction with the grains is likely to remove the grain mantles and deposit further molecules into the gas phase. Observations of ammonia (Walmsley et al. 1987), water (Hinkel et al. 1987, Plaubeck and Wright 1987, Jacq et al. 1988) and methanol (Wilson et al. 1989) in the Orion-KL region all suggest that the molecules were evaporated from dust mantles heated less than 10^4 years ago. The different chemical composition in the ridge, plateau and hot core regions of the Orion molecular cloud are indications of the influence of star-forming events on the grain and gas chemistries (Blake et al. 1989).

The observations appear to demand a combination of a chemistry like that of diffuse clouds which is dominated by photoabsorption and a chemistry like that of dense clouds which is dominated by cosmic ray ionization, as would occur in a cloud containing dense clumps of material embedded in a diffuse inter-clump medium. Such a model has been postulated by Genzel et al. (1988), who attribute the high abundances of C^+ and C in M82 to ultraviolet photons penetrating into a clumpy gas.

Material will be exchanged between the diffuse and dense regions. A study of the chemical consequences of such mixing has been carried out by Chieze and Pineau des Forets (1989). Possible sources of the mixing are turbulence (Boland and de Jong 1982) and shocks (Williams and Hartquist 1984). A detailed scenario has been presented for the molecular cloud Barnard 5 which was earlier regarded as quiescent, but is now recognized as an active site (Charnley et al. 1988, 1989). According to it, ionized hydrogen and helium in stellar winds mix with molecular material ablated from the clumps.

The chemistry of molecular clouds is the result of complex interactions of stars as they evolve with the material out of which stars are formed. It should be possible to develop theoretical models with sufficient precision that the chemical composition can serve as a diagnostic probe of the history of the events that have taken place in the evolution of molecular clouds.

Acknowledgment

This work was supported by the National Science Foundation, Division of Astronomical Sciences, under Grant AST-86-17675.

References

Blake, G. A., Sutton, E. C., Masson, C. R. and Phillips, T. G. 1987 Ap. J. 315, 621.

Bohme, D.K., Wlodek, S. and Wincel, H. 1989 Ap. J. Lett in press.

Boland, W. and de Jong, T. 1982 Ap. J. 261, 110.

Brown, P. D., Duley, W.W., Jones, A. P. and Williams, D. A. 1989 M.N.R.A.S. in press.

Charnley, S. B., Dyson, J.E., Hartquist, T. W. and Williams, D.A. 1989 M.N.R.A.S. 231, 269 and 235, 1257.

Charnley, S. B., Dyson, J.E., Hartquist, T.W. and Williams, D. A. 1988 M.N.R.A.S. in press.

Chieze, J. P. and Pineau des Forets, G. 1989 Astron. Ap. in press.

Cox, P., Walmsely, C. M. and Gusten, R. 1989 Astr. Ap. 209, 382.

Dalgarno, A. 1988 Astro. Lett. Comm. 26, 153.

d'Hendecourt, L.B., Allamandola, L. J. and Greenberg, J. M. 1985 Astron. Ap. 152, 130.

Draine, B. T. 1986 Ap. J. 310, 408.

Duley, W. W. and Williams, D. A. 1986 M.N.R.A.S. 219, 859.

Elitzur, M. and Watson, W. D. 1980 Ap. J. 236, 172.

Genzel, R., Harris, A. I., Jaffe, D. T. and Stutzki, J. 1989 Ap. J. 332, 1049.

Glassgold, A., Mamon, G. A. and Huggins, P. J. 1989 Ap. J. Lett. 336, L29.

Glassgold, A. E., Mamon, G. A., Omont, A. and Lucas, R. 1987 Astron. Ap. 180, 183.

Gredel, R., Lepp, S. and Dalgarno, A. 1987 Ap. J. Lett. 323, L137.

Gredel, R., Lepp, S., Dalgarno, A. and Herbst, E. 1989 Ap. J. in press.

Hartquist, T. W., Oppenheimer, M. and Dalgarno, A. 1980 Ap. J. 236, 182.

Herbst, E. and Leung, C. M. 1989 Ap. J. Suppl. 69, 271.

Hinkel, C. Mauersberger, R., Wilson, T. L. Snyder, L. E., Menten, K. M., and Wouterlout, J. G. A. 1987 Astron. Ap. 182, 209.

Hollenbach, D. and McKee, C. F. 1989 Ap. J. 342, 275.

Irvine, W. M., Goldsmith, P. F. and Hjalmarson, A. 1987 in Interstellar Processes ed. D. J. Hollenbach and A. A. Thronson (Dordrecht: Reidel) p. 561.

Jacq, T., Jewell, P. R., Henkel, C., Walmsley, C. M. and Baudry, A. 1988 Astr. Ap. 199, L5.

Keene, J., Blake, G.A., Phillips, T.G. Huggins, P.J. and Beichman, C.A. 1985 Ap. J. 299, 967.

Kingston, R.G., Guilhaus, M., Brenton, A.G. and Beynon, J.H. 1985 Organic Mass. Spectr. 20, 405.

Langer, W. D. and Graedel, T.E. 1989 Ap. J. Suppl. 69, 241.

Leger, A., Jura, M. and Omont, A. 1985 Astr. Ap. 144, 147.

Lepp, S. and Dalgarno, A. 1988 Ap. J. 324, 553.

Lepp, S., Dalgarno, A. and McCray, R. 1988 Bull. A.A.S. 20, 671.

Lepp, S., Dalgarno, A. and McCray, R. 1989 Ap.J. in press.

Lepp, S., Dalgarno, A., van Dishoeck, E.F. and Black, J.H. 1988 Ap. J. 329, 418.

Latter, W.B. and Black, J.H. 1989 Ap. J. in press.

Mamon, G.A., Glassgold, A.E. and Huggins, P.J. 1988 Ap. J. 328, 797.

Mitchell, G. F. 1983 M.N.R.A.S. 205, 765.

Nejad, L. A. M. and Millar, T. J. 1987 Astron. Ap. 183, 279.

Nercessian, E., Benayou, J. J. and Viala, Y.P. 1988 Astr. Ap. 195, 245.

Neufeld, D.A. and Dalgarno, A. 1989 Ap. J. 340, 869.

Omont, A. 1986 Astr. Ap. 186, 184.

Petuchowski, S. J., Dwek, E., Allen, J. E. and Nuth, J.A. 1989 Ap. J. in press.

Pineau des Forets, G., Flower, D.R., Hartquist, T. W. and Dalgarno, A. 1986 M.N.R.A.S. 220, 801.

Pineau des Forets, G., Roueff, E. and Flower, D.R. 1986 M.N.R.A.S. 223, 743.

Plambeck, R.L. and Wright, M.C.H. 1987 Ap.J. 317, L101.

Prasad, S. and Tarafdar, S. P. 1983 Ap. J. 267, 603.

Puget, J.L. and Leger, A. 1989 Ann. Rev. Astr. Ap. in press.

Rawlings, J.M.C. 1988 M.N.R.A.S. 232, 507.

Rawlings, J. M. C., Williams, D. A. and Canto, J. 1988 M.N.R.A.S. 230, 695.

Sternberg, A., Dalgarno, A. and Lepp, S. 1987 Ap. J. 320, 676.

Tielens, A. and Hollenbach, D. 1985 Ap. J. 291, 722.

van Dishoeck, E. F. 1988 in Millimetre and submillimetre astronomy, Ed. W. B. Burton and R. D. Wolstencroft (Kluwer-Dordrecht) p. 117.

van Dishoeck, E. F. and Black, J. H. 1986 Ap. J. Suppl. 62, 109.

van Dishoeck, E.F. and Black, J. H. 1988 Ap. J. 334, 771.

van Dishoeck, E. F. and Black, J.H. 1989 Ap. J. 340, 273.

Viala, Y. P. 1986 Astr. Ap. Suppl. 64, 391.

Viala, Y. P., Roueff, E. and Abgrall, H. 1988 Astr. Ap. 190, 215.

Walmsley, C. M. Hermsen, W., Henkel, C., Mauersberger, R. and Wilson, T.L. 1987 Astron. Ap. 172, 311.

Williams, D. A. 1985 Quart. J.R.A.S. 26, 463.

Williams, D. A. and Hartquist, T. W. 1984 M.N.R.A.S. 210, 141.

Wilson, T.L., Johnston, K.J., Henkel, C. and Menten, K.M. 1989 Astr. Ap. 214, 321.

Discussion:

KHAN: If there is a large density contrast between the molecular cloud and the outside medium, say ρ_m/ρ_0, then the gas in the cloud experiences a shock with speed $V_0\sqrt{\rho_0/\rho_m}$, and this can be much smaller than Vo, the speed with which the cloud advances relative to the outside medium.

DALGARNO: The limiting shock velocity above which CO is dissociated is about 50 kms^{-1}.

WAMPLER: SN 1987A is incased in a rather dense, slow moving, progenitor wind. Can the CO molecules survive their encounter with this wind a few decades from now?

DALGARNO: See remark by Dr. Kahn.

CO DISTRIBUTIONS TOWARDS THE SOUTHERN DARK CLOUD OF DC 303.6+0.9

Jing-sheng Wang[1] and Robina E. Otrupcek[2]
1. Yunnan Observatory, P.O. Box 110, Kunming, 650011, P.R. China
2. Australian NRAO, P.O. Box 276, Parkes, N.S.W. 2870, Australia

ABSTRACT Dark clouds in the southern skies have been observed with CO J = 1–0 line. The mm-wave telescope of 4-m diameter was used. The beam width was $2'.7$ at 2.6 mm wavelength and beam efficiency 0.52. The authors have developed a high-resolution acousto-optical spectrograph with a velocity resolution of 0.07 kms^{-1} for this purpose. The 115.27 GHz line was detected over an area of $27' \times 21'$ including the dark cloud of DC 303.6+0.9 in Crux. Most of the positions, out of a total number of 63, show a double peak line structure. Some positions also show red and blue shifted wings. An average profile shows a higher peak at the radial velocity of -2.8 kms^{-1} with a calibrated radiation temperature of 5.5 K, a half line width of 1.5 kms^{-1}, and another peak in relative red at a velocity of -0.8 kms^{-1} with a temperature of 4.3 K, a half line widh of 1.0 kms^{-1}. Profiles distributed over this area show a bipolar diagram. It hints that a certain systematic motions may exist in this region.

1. INTRODUCTION

In the northern hemisphere the dark cloud CO line observations have already been done by many authors (e.g. Myers et al. 1983, Fukui et al. 1988, Snell et al. 1980 etc). In the southern hemisphere similar observations are still rare (Huggins et al. 1977, de Vries et al. 1984). Meanwhile the CO observation of the southern dark clouds have become more important because: (a) a number of prominent star-forming regions are situated in the southern skies, (b) an unbiases statistics for star-forming regions in our Galaxy appears significant in 1980s, and (c) some newly estabilshed powerful equipment in the southern hemisphere such as the Australian Telescope needs some preliminary survey to find the regions worth detailed investigation.

The authors commenced CO observations on the southern dark clouds in 1984 by means of a high-resolution acousto-optical spectrograph (AOS) associated with the 4-m mm-wave telescope of CSIRO Division of Radiophysics. Some preliminary results have already been published (Wang et al. 1987, Wang and Otrupcek 1987, 1988). Here we report another specimen which appears as a prominent double-peak line profile and a bipolar structure close to the direction of DC 303.6+0.9 (S 157) in Crux.

2. OBSERVATIONS

Observations were made in November and December 1986 with the 4-m mm-wave telescope at the rest frequency of 115.27 GHz which produced a half power beam width of $2'.7$ arc. The receiver provided a double side-band system temperature of 250 K. The beam efficiency was 52

A high-resolution AOS was used for these observatinos. The AOS was built by Chinese-Australian collaboration. The frequency resolution is 28 kHz. This yields a velocity resolution of 0.07 kms^{-1} at 115 GHz. It contains 1024 output channels. Each channel covers 10.5 kHz, i.e. 0.027 kms^{-1} at 115 GHz. The total bandwidth of the AOS covers a velocity range of 28 kms^{-1}. Details are described in Wang et al. (1989).

The calibration during observations was done by means of a beam-switch described in Robinson et al. (1982). The calibrated radiation temperature as defined by Kutner and Ulich (1981) was adopted. The atmosphere attenuation was precessed by regularly observing the sources of OMC-1 in Orion and the HII region of G333.1-0.43.

An area in Crux of 27′ arc (in EW) by 21′ arc (in NS) was observed at 3′ arc of grid intervals. Nine additional adjacent positions were also observed but not taken into account for the contour map. The centre = -61°40′18″, taken from Hartley et al. Southern Dark Cloud Catalogue (1986) was defined as the coordinates' origin (0,0). The four corners of the observed main aria are coded as (-7,+5), (+1,+5), (+1,-1), and (-7,-1) separately (here minus means E or N, and each step is 3′ arc). Six minute integration at each position yielded an rms noise of 1.0 K. There were some interferences during observations. These made the rms noise larger than the normal. Observations of reference position were omitted as the line profile of the dark cloud is very sharp and the total bandwidth is low. The base line was extracted from both outer sections of the observed spectrum. The centre channel of the AOS (channel No. 487.5 during these observations) was set to the radial velocity of -4.0 kms^{-1} (LSR) which was taken from Goss et al. (1980). The methods of the observation and data processing have been described in a Coalsack study by Otrupcek and Wang (1987). The data processing was carried out by a newly installed VAX-8350 computer at Yunnan Observatory, Kunming.

3. RESULTS

The CO J=1-0 line was detected in most of the 72 observed positions. This region is a small portion of a large complex of dark clouds belonging to the Southern Coalsack. However, the available observing peroid on the 4-m telescope was restrictive and we were unable to continue observations to reach the bounds of CO emission. Also we were unable to confirm the exact extent of this molecular cloud. Nevertheless, the relative results still show something interesting.

Towards the optical centre position of the dark cloud DC 303.3+0.9 some weak CO emission appears around -4.0 kms^{-1} (LSR). But at N-E of 15 arcmin from the centre position the CO line shows a prominent double peak structure. Figure 1 shows the CO J=1-0 line profile averaged from 25 positions. The 25 positions were contained within a square area. The four corners of this square were defined by coordinates codes of (- 7,+4), (-3,+4), (-3,0) and (-7,0) separately. In figure 1 the CO profile was split into two adjacent components. Comparing with its double gaussian fitting in figure 2, one can see both blue and red shifted wings overlapped on the outer slopes on the profile.

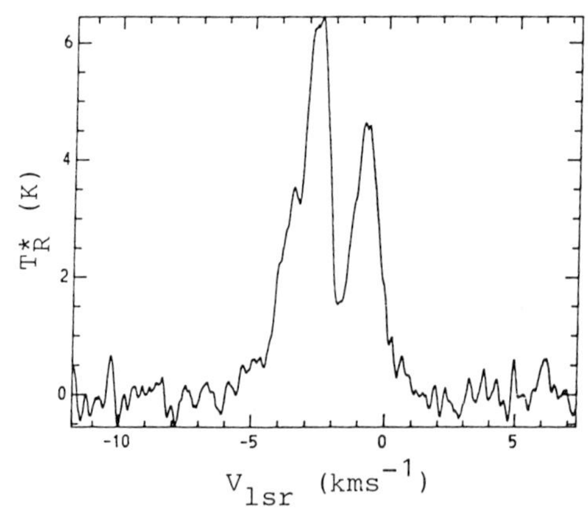

Figure 1. A double peak structure of CO J=1-0 line

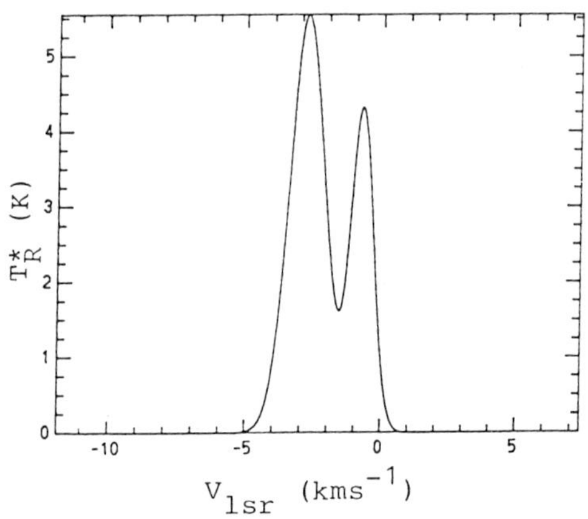

Figure 2. Double gaussian fitting of the CO profile

In figure 2 the gaussian fitted higher peask (i.e., the 'blue' one) is centred at the velocity of -2.8 kms^{-1}. The peak height of T_R^* is 5.5 K and the half width 1.5 kms^{-1}. The lower peak (the 'red' one) is centred at the velocity of -0.8 kms^{-1}. The peak height is 4.3 K and the half width 1.0 kms^{-1}.

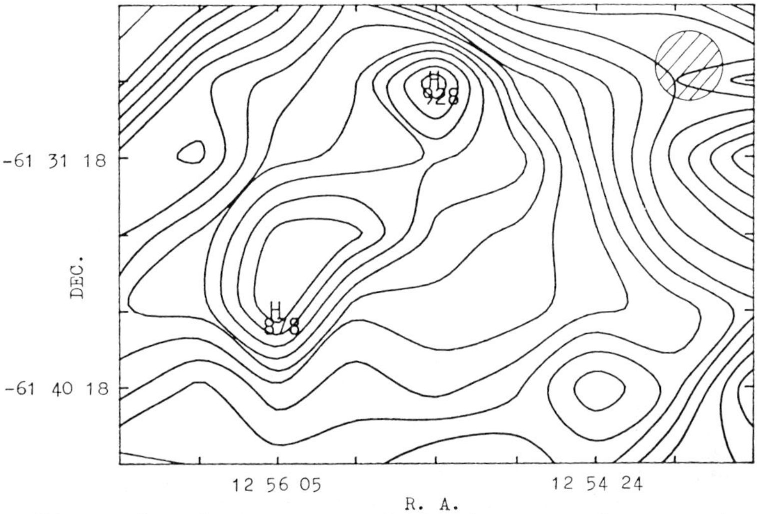

Figure 3. Contour map of the integrated temperature of the CO profiles

Figure 3 shows the contour map of the integrated temperature, $\int T_R^*\,dv$, of CO J=1-0 line profile within the 27′ × 21′ observed area. The integrating range of radial velocities (LSR) is between -5.4 kms^{-1} and +1.0 kms^{-1} covering both components. The interval between each two adjacent contours is 55 Kch. Here ch means an output channel that corresponds with 0.027 kms^{-1} in velocity. The symbol 'H' marks a local maximum at the unit of Kch. A circle shows the telescope beam (HPBW).

There appears a weak maximum at position (-1,0). It is coincident with the optical centre of DC 303.6+0.9 (0,0). But no evidence of a molecular cloud core was found.

To the north-east of the centre appears an elongated distribution along NW-SE direction. This area is consistent with the region in which profiles appear as double peaks. In order to examine the difference of distributions between the blue component and the red one, we made separate contour maps for both. The northern maximum (H 926) is coincident with both components in position (not shown in figure 4). But the other one (H 878) appears as a distinct bipolar diagram (shown in figure 4).

In figure 4 the solid contours show the blue component and the dashed lines show the red one. Assuming the boundary for the two components is at a velocity of -1.8 kms^{-1}, i.e. at the minimum between the two peaks, we take the integration ranges of velocities as (-6.4,-1.8) and (-1.8,+2.1) kms^{-1} for the blue and red ones separately. The interval in figure 4 is 36 Kch. The innermost contour level of the blue one is 648 Kch, of the red one 360 Kch. We did not reach the bounds of the cloud. The outer levels of the contour were omitted. The beam width is shown by a circle.

4. DISCUSSIONS

There are two possibilities for the explanation of the bipolar diagram:

(a) One possibility is that the two components are two independent molecular clouds situated, by chance, almost on the same radial direction. The 'blue' component has the radial velocity of

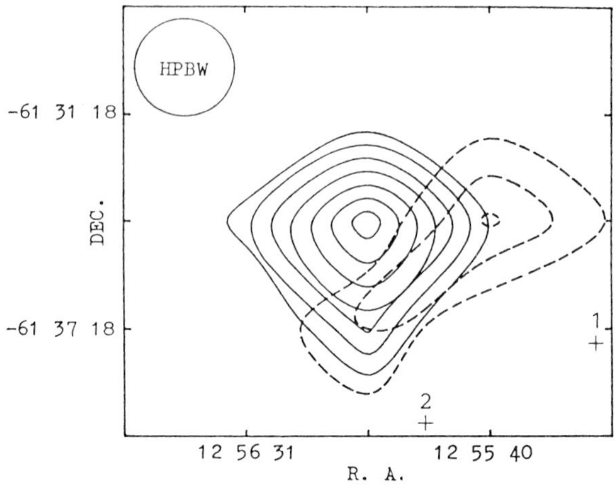

Figure 4. The bipolar diagram nearby DC 303.6+0.9

-2.8 kms^{-1} corresponding to a distance of r = 0.2 kpc estimated by menas of Oort formula. The 'red' one has the radial velocity of -0.8 kms^{-1} corresponding to a distance of 0.06 kpc. From an ESO J-plate we also see the optical extinction towards the direction of the 'blue' component is stronger than that towards the 'red' one.

(b) Another possibility is that there is a physical relation between the two components. From figures 3 and 4 one can see that the distribution of the two components is consistent with a NW-SE elongated region. Assuming the red component is at a similar distance than the blue one, i.e. about 0.2 kpc (this is perhaps acceptable considering the distance of the Coalsack, 0.18 kpc) from the Sun, the difference in velocity between the two components' is 2 kms^{-1}. The total size of this bipolar structure is about 0.5 pc. These are roughly consistent with a bipolar molecular outflow.

The positions of nearby IRAS point source are marked by crosses in figure 4. The color index of log $[F(12\mu m)/F(25\mu m)]$ for those sources are -0.27 (for No. 1 in figure 4, i.e. IRAS12554-6137) and -0.40 (for No. 2, i.e. IRAS12559-6140) respectively. We have not marked the position for IRAS source with positive color index within this region. Both IRAS sources are off the centre of the bipolar diagram. We guess source No. 2 may be more concerned. This bipolar diagram was made from separate integrations including each peak value. As the signal-to-noise ratio was not good enough to investigate the wings in detail, we can not conclude whether a bipolar molecular outflow is acutally there at the moment. Some further observations for this area are necessary.

ACKNOWLEDGEMENTS We thank Drs. B.J. Robinson and J.B. Whiteoak for help with observation arrangement, Dr. Liu Zhihuang and the VAS Group, Yunnan Observatory, for help with data processing. Our work was supported by the Exchange Agreement between Academia Sinica and CSIRO. This article was completed at IRAM Granada.

REFERENCES

de Vries, C.P., Brand, J., Israel, F.P., de Graauw, Th., Wouterloot, J.G.A., van de Stadt, H. and Habing, H.J. 1984, Astron. Astrophys. Suppl. Ser. 56, 333.

Fukui, Y., Sugitani, K., Takaba, H., Iwata, T., Mizuno, A., Ogawa, H. and Kawabata, K. 1986, Astrophys. J. Lett. 311, L85.

Goss, W.M., Manchester, R.N., Brooks, J.W., Sinclair, M.W. and Manefield, G.A. 1980, Mon. Not. Roy. Astron. Soc. 191, 533.

Hartley, M., Manchester, R.N., Smith, R.M., Tritton, S.B. and Goss, W.M. 1986, Astron. Astrophys. Suppl. Ser. 63, 27.

Huggins, P.J., Gillespie, A.R., Sollner, T.C.L.G. and Phillips, T.G. 1977, Astron. Astrophys. 54, 955.

Kutner, M.L. and Ulich, B.L. 1981, Astrophys. J. 250, 341.

Myers, P.C., Linke, R.A. and Benson, P.J. 1983, Astrophys. J. 264, 517.

Otrupcek, R.E. and Wang, J.-S. 1987, Proc. Astron. Soc. Aust. 7, 194.

Robinson, B.J., McCutcheon, W.H. and Whiteoak, J.B. 1982, Int. J. Infrared Millimetre Waves 3, 63.

Snell, R.L., Loren, R.B. and Plambeck, R.L. 1980, Astrophys. J. 239, L17.

Ulich, B.L. and Haas, R.W. 1976, Astrophys. J. 236, 192.

Wang, J.-S. and Otrupcek, R.E. 1987, Acta Astron. Sinica 28, 95.

Wang, J.-S., Robinson, B.J., Huang, G.-C. and Otrupcek, R.E. 1987, in Astrochemistry (IAU Symp. 120), Reidel, Dordrecht, 135.

Wang, J.-S. and Otrupcek, R.E. 1988, Vistas in Astron. 31, 493.

Wang, J.-S., Robinson, B.J., Huang, G.-C. and Otrupcek, R.E. 1989, Proc. Astron. Soc. Aust., in press.

Discussion:

HAYASHI: What is the linear scale of the structure which looks like a bipolar "outflow", (and how large is the velocity gradient)? If there is uncertainty about the distance, please give the number in arcminutes and $kms^{-1}(arcmin)^{-1}$.

WANG: The projected linear scale of the bipolar structure is about 9 arcmin, i.e. about 0.5 pc assuming the distance is $\sim$0.2 kpc from the sun (estimated from Oort formula). The velocity differences between the blue peak and the red one show a gradient of about 1 $kms^{-1}pc^{-1}$ from north to south.

EXTENSIVE MOLECULAR LINE SURVEY FOR DARK CLOUDS

Norio Kaifu
Nobeyama Radio Observatory, National Astronomical Observatory
Nobeyama, Minamimaki, Nagano 384-13 Japan

1. Nobeyama Spectral Line Survey

Since 1984 the molecular line search for dark clouds, mainly TMC1, for wide frequency ranges in the short cm- and mm-waverength regions has been proceeded at Nobeyama Radio Observatory using the 45-m telescope. The main collaborators are: H. Suzuki, M. Ohishi, S. Saito, Y. Yamamoto, K. Kawaguchi, and N. Kaifu.

The unbiased and wide frequency coverage molecular line survey for dark clouds is by no means basically important from the standpoint of view of the interstellar chemistry and chemical evolution of dark clouds. However such extensive line survey has not been performed yet, because of some practical difficulties. Very narrow linewidth observed in dark clouds (about 0.5 km s^{-1}) requires high frequency resolution, while the wide frequency region should be covered. The line intensity is relatively weak due to the low kinetic temperature of dark clouds. The narrow lines are very often difficult to distinguish from the confusion of weak spurious signals which are mixed from the local oscillators etc. Thus such survey is enevitably a very time consuming effort.

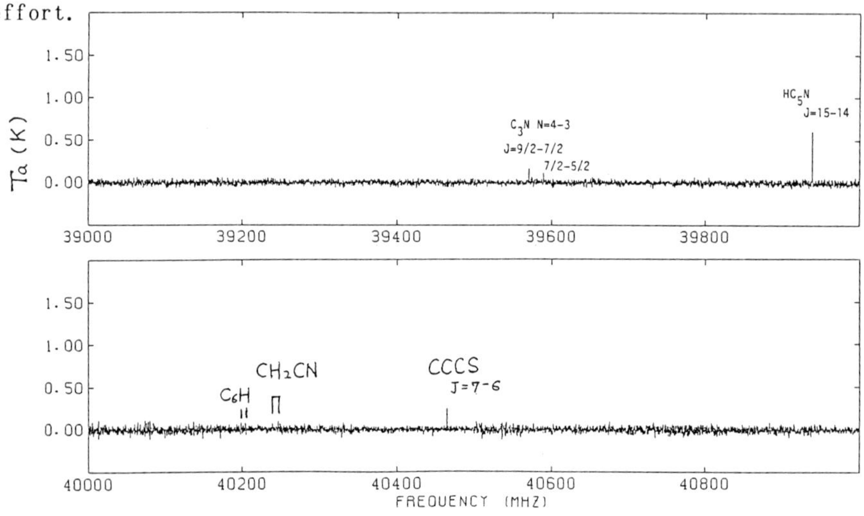

Figure 1. A sample of the 2GHz spectrum(39-41 GHz) of TMC1 taken with the 45m telescope and the wideband AOS(250KHz resolution and 16000 channel).

We have started the spectral line survey for TMC1 and some other sources by using the Nobeyama 45m telescope, wideband receivers and AOS spectrometers.

The AOS, acousto-optical radiospectrometer, has been first developed extensively at the Tokyo Astronomical Observatory (Kaifu et al. 1977) and then at Nobeyama (kaifu and Chikada 1984). The Nobeyama AOS consists of two systems, one is the wideband AOS with 16000 channel outputs, 250KHz resolution and 8 x 250MHz bandwidth. The other is the high-resolution system which provides 16000 channel, the resolution of 37KHz and the bandwidth of 8 x 40MHz. This large spectrometer system can be used simultaneously by connecting to one or two cooled receivers some of which also provide 2GHz simultaneous bandwidth. These large spectroscopic systems were essential for our survey, still the survey is by no means easy.

So far the survey has been performed in the frequency regions of 8.8-10.8GHz, 14-16GHz, 18-21GHz, 22-24GHz, 36-50GHz, and several fragmentary frequency bands in 88-110GHz. The frontend receivers used were the cooled HEMT amps for the 8-24GHz band, the cooled Schottky mixers and then a Nb SIS mixer later for the 36-50GHz band, and a cooled Schottky mixer for the 88-110GHz region. The obtained rms noise temperature of the data vary from 3mK to 30mK, depending on the receiver and observational conditions. A sample data is shown in figure 1.

The detailed results and discussion will be published elsewhere, and here we present the summary of the observed lines in table 1, and discuss some topics. As a whole we have detected 91 spectral lines of 27 molecules (including two marginal detections) and 16 lines of 14 isotopic species in TMC1. Some 25 lines are not yet identified to any known molecules.

molecule	number of lines		molecule	number of lines	
	main species	isotopic species		main species	isotopic species
CO	1	2	CH_3CN	1?	
NH_3	1		C_2H_3CN	2	
C_3N	4		$HNCO$	1	
CS	2	2	CH_3OH	1	
CCS *	4	1	CH_2CN *	11	
$CCCS$ *	3		H_2CS	2	
C_4H	10		H_2CCO	2	
C_6H *	12		CH_3COH	1?	
HCN	1	1	HCO^+	1	
HC_3N	6	4	HCS^+	1	
HC_5N	9	6	$HCCCOH$	1	
HC_7N	6		C_3H_2	2	
HC_9N	2		$c-C_3H$ *	4	
$CCCO$	1		Unidentified	25	——

Table 1. Observed molecular lines in the NRO spectral survey for TMC1
* : Molecule newly detected or identified during this survey.

During this survey we have detected several new interstellar molecules. These are: C_6H (Suzuki et al.,1986), CCS (Suzuki et al.,1984, Kaifu et al.,1987, Saito et al., 1987), CCCS (Yamamoto et al., 1987a, Kaifu et al., 1987), cyclic C_3H (Yamamoto et al., 1987b), CH_2CN (Irvine et al., 1988, Saito et al., 1989), HCCCOH (Irvine et al., 1989). Especially the detection of CCS and CCCS added a new series of carbon chain molecules containing S, which is pretty abundant in TMC1 (see figure 2). The CCS was first detected as a strong unidentified line and 4 years later it was identified with a new molecule CCS by chance. The column density of CCS in TMC1 is $9 \times 10^{13} cm^{-2}$ and can be compared with those of CS($2 \times 10^{14} cm^{-2}$) and of HCN($1.2 \times 10^{14} cm^{-2}$).

2. Carbon Chain Molecules and CCS

The detections of CCS and CCCS provided a new aspect to the formation of carbon chain molecules, which characterize the dark cloud chemistry as can be seen in table 1. In spite of relatively low cosmic abundance of S it seems that the sulphur-containing carbon chain molecules are pretty abundant. Also the series of C_nH and C_nS chains show the continuous number of carbon atoms, though the relative abundances vary with n (see figure 3). On the other hand the previously known carbon chains HC_nN are found for only odd n.

Suzuki (1987) proposed an idea to explain such observational tendencies of carbon chain molecules. The carbon chain molecules may grow first as pure carbon chains which are "bones" of observed carbon chain molecules in the partially ionized C^+ regions of dark clouds (Suzuki 1983). Then the bones are combined with H, N, O, S etc. to form stable molecules like HC_nN. C_nS could be formed simultaneously to or just after the formation of carbon bones, because the low ionization potential of S allow it to form the relatively deep S^+ region at the cloud surface, and here the sulphur chemistry should be active.

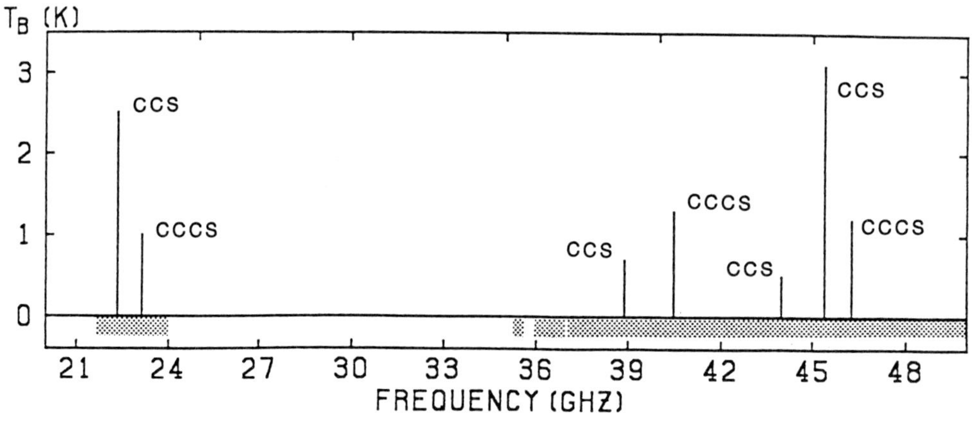

Figure 2. A schematic diagram of CCS and CCCS spectrum detected in the survey. The hatched area show the surveyed frequency range. (taken from Kaifu et al., 1987)

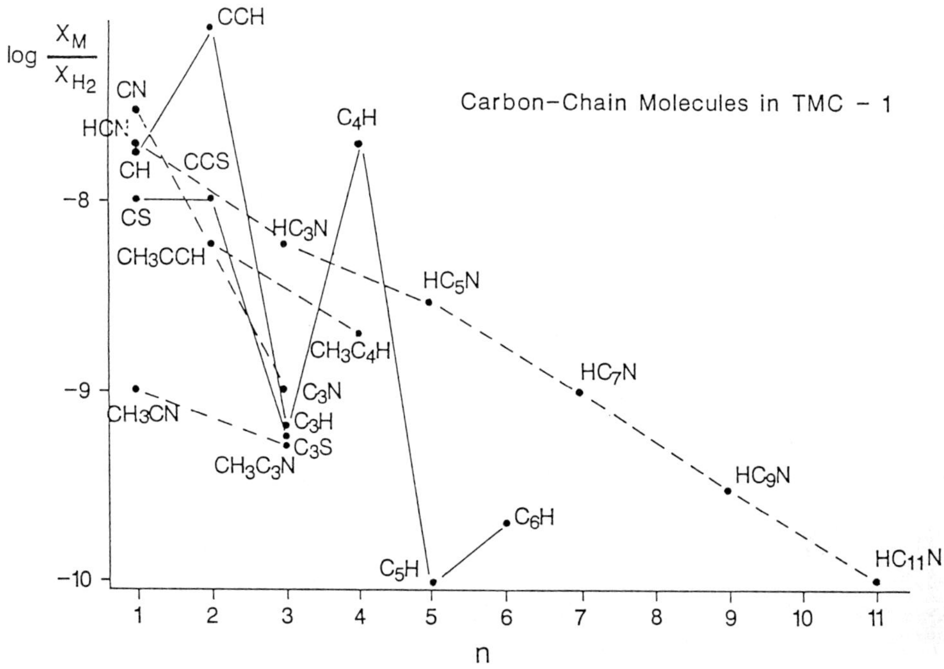

Figure 3. Abundances of carbon chain molecules in TMC1, as a function of number of carbon chain atoms n. Dashed line indicates the "odd n only" and "even n only" chain molecules. Values were taken from Irvine et al. 1987, Suzuki et al. 1986, Saito et al. 1987, Yamamoto et al. 1987a.

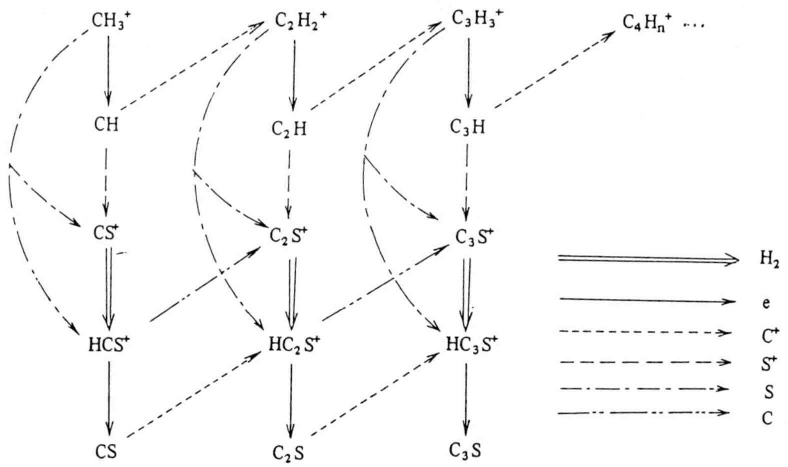

Figure 4. A schematic flow of possible C_nS formation process in the partially ionized region, taken from Suzuki (1987).

A schematic diagram of C_nS formation in the S^+ region is given by Suzuki (figure 4). The C_nS chains are not very stable and will be changed to be more stable species in denser region. Therefore the C_nS might be a good probe of the cloud evolution to trace the chemistry in the various stages of contraction of dark clouds.

We are making extensive observations to test the distribution of CCS in various dark clouds and inside some selected clouds and to compare them to those of other molecules like CS, NH_3, HC_3N etc(Yamamoto et al., 1989, Hirahara et al., 1989). Figure 5 shows a comparison of distributions of CCS and CS in TMC1, showing remarkable difference between the distributions of these two molecules. Another remarkable fact about CCS is that it cannot be found in the star forming regions where the density is high and accompanied with various shocks and heating effects. Such observational facts would be interpreted in terms of the combination of the chemical and physical evolution of the dark clouds.

Furthermore the recent detection of CH_2CN (Irvine et al.,1988, Saito et al., 1989) dramatically showed that abundant (NL of 10^{13}-$10^{14}cm^{-2}$) complex organic molecules might be detected as very weak signals because of their complex energy level structures. Such spectral lines would distribute in relatively longer wavelength regions. In NRO we plan to extend our survey to complete the frequency range from 8GHz to 50 GHz with higher sensitivity, by using wideband cooled HEMT amplifiers and SIS receivers. Also the spectral scans for 70-120GHz by using new SIS receivers will start in next winter.

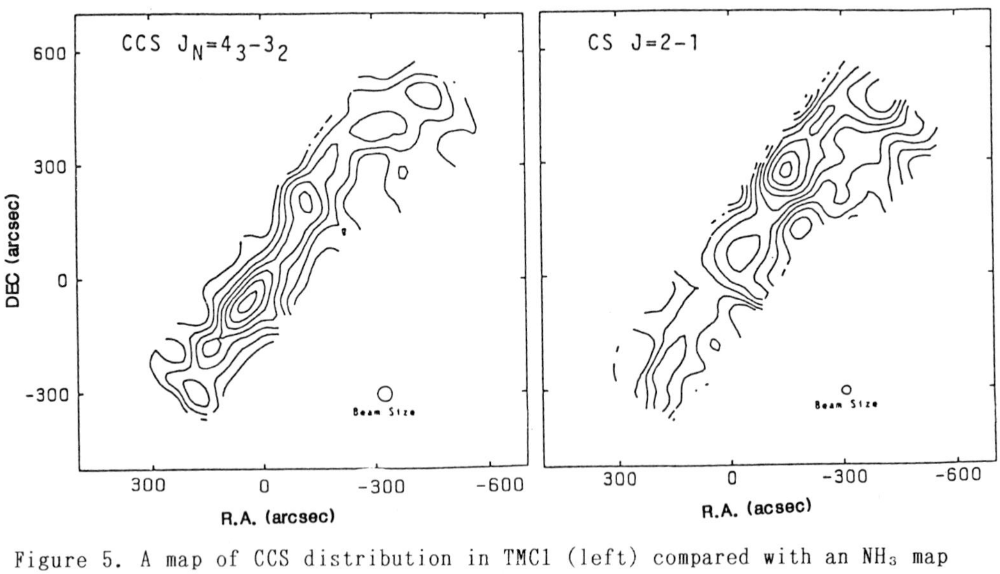

Figure 5. A map of CCS distribution in TMC1 (left) compared with an NH_3 map (right), taken with the 45m telescope (Hirahara et al., 1989).

The author offer his condolences to the late Dr. Hiroko Suzuki (1947-1987), who was a very distinguished astrophysicist, the most active staff member of Nobeyama Radio Observatory and was our best colleague and friend.
She was one of the pioneers of astrochemistry, she was at the center of the Nobeyama molecular line survey. She died on 22 November, 1987, the result of a car accident. Her great contributions to astrochemistry and to radio astronomy, and her very active life will remain in our memory forever.

REFERENCES

W.M.Irvine, F.P.Schloerb, A.Hjalmarson, and E.Herbst, in: Protostars and Planets II, (D.C.Black and M.S.Matthews, ed.), pp.579, Univ.Arizona Press, (1985).

W.M.Irvine, P.Friberg, A.Hjalmarson, S.Ishikawa, N.Kaifu, K.Kawaguchi, S.C.Madden, H.E.Matthews, M.Ohishi, S.Saito, H.Suzuki, P.Thaddeus, B.E.Turner, S.Yamamoto, and L.M.Ziurys, Astrophys J.(Letters), 334, L107, (1988).

W.M.Irvine et al., submitted to Astrophys.J., (1989)

N.Kaifu, N.Ukita, Y.Chikada and T.Miyaji, Publ.Astron.Soc.Japan, 29, 429, (1977)

N.Kaifu, and Y.Chikada, in: URSI Symposium on Millimeter and Submillimeter Astronomy, (NRO Report 63), (1984).

N.Kaifu, H.Suzuki, M.Ohishi, T.Miyaji, S.Ishikawa, T.Kasuga, M.Morimoto, and S.Saito, Astrophys.J.(Letters), 317, L111, (1987).

Y.Hirahara, Thesis, Faculty of Science, University of Tokyo, (1989).

S.Saito, K.Kawaguchi, S.Yamamoto, M.Ohishi, H.Suzuhi, and N.Kaifu, Astrophys.J.(Letters), 317, L115, (1987).

S.Saito, S.Yamamoto, W.M.Irvine, L.M.Ziurys, H.Suzuki, M.Ohishi, and N.Kaifu, Astrophys.J.(Letters), 334, L113, (1988).

H.Suzuki, Astrophys.J., 272, 579, (1983).

H.Suzuki, N.Kaifu, T.Miyaji, M.Morimoto, and M.Ohishi, Astrophys.J., 282, 197, (1984).

H.Suzuki, M.Ohishi, N.Kaifu, S.Ishikawa, T.Kasuga, K.Kawaguchi, and S.Saito, Publ.Astron.Soc.Japan, 38, 911, (1986).

H.Suzuki, S.Yamamoto, S.Saito, M.Ohishi, N.Kaifu, S.Ishikawa, and A.Murakami, a report in: IAU 3rd Asian Pacific Regional Meeting, Peking, (1987).

S.Yamamoto, S.Saito, K.Kawaguchi, N.Kaifu, H.Suzuki, and M.Ohishi, Astrophys.J.(Letters), 317, L119, (1987a).

S.Yamamoto, S.Saito, M.Ohishi, H.Suzuki, S.Ishikawa, N.Kaifu, and A.Murakami, Astrophys.J.(Letters), 322, L55, (1987b).

S.Yamamoto et al., (1989), in preparation.

MOLECULAR HYDROGEN EMISSION FROM PHOTODISSOCIATION REGIONS

TETSUO HASEGAWA
Institute of Astronomy, The University of Tokyo
2-21-1 Osawa, Mitaka, Tokyo 181, Japan

Abstract

We review new observational and theoretical developments of the understanding of the H_2 infrared emission in the last 5 years since the discovery of the fluorescent emission in NGC 2023. An excitation analysis of H_2 in a variety of Galactic sources has revealed that in many sources the excitation is expressed as a mixture of *thermal* and *fluorescent* components. This finding is in good agreement with theories of photodissociation regions, in which the population of H_2 changes its character from *pure fluorescence* to *thermal* as the density of the region increases. The ortho/para abundance ratio of the *fluorescent* H_2 is observed to lie within a limited range of $1.1 - 1.8$ which is well reproduced by depth-dependent model calculations of the ultraviolet excitation and dissociation of H_2 molecules. This may be understood as due to the independent self shielding of each of the ortho- and para-H_2, rather than the ortho/para abundance ratio of the predissociated H_2, a low formation temperature of H_2 on grains, or gas phase interchange reactions. A laser emission of molecular hydrogen discovered in the planetary nebula NGC7027 further demonstrates the nonthermal nature of the H_2 emission in photodissociation regions.

1. Introduction

Direct observation of H_2 in dense molecular clouds is very difficult and there are only two special cases in which its observation is possible : (a) thermal excitation in shocked molecular gas and (b) radiative excitation in photodissociation regions on surfaces of molecular clouds exposed to strong ultraviolet radiation.

The existence of the shocked molecular hydrogen was recognized soon after the detection of rovibrational line emission in Orion-KL (Gautier *et al.* 1976). Detailed studies of the level population of H_2 in Orion-KL have proved the collisional excitation in shocked molecular gas (e.g., Beckwith *et al.* 1983; see also Brand in this volume). Measurements in other sources have shown that the population distribution is consistent with a thermal excitation within the limits of number of observed levels and observational errors, and have led people to believe that a detection of emission from vibrationally excited molecular hydrogen is a proof of shocked molecular gas.

Detection of fluorescent H_2 emission in the reflection nebula NGC 2023 and other sources changed this situation dramatically (Gatley *et al.* 1987; Hayashi *et al.* 1985; Sellgren 1986). A detection of an H_2 emission line (e.g., $v = 1 - 0$ $S(1)$ at λ 2.12 μm) does not by itself prove an existence of a shock and any conclusion on the nature of the excitation of H_2 molecules is difficult before a detailed analysis of the population of levels with $v \geq 2$ is made.

In this paper we review recent observational and theoretical development of the understanding of the H_2 emission in the last 5 years since the discovery of the infrared fluorescence.

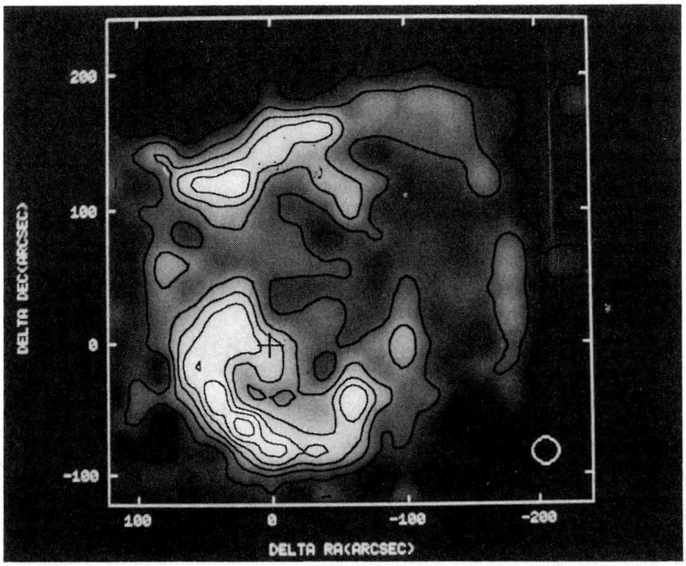

FIG. 1. — A map of the emisssion in the $v = 1-0, S(1)$ line of H_2 from the reflection nebula NGC 2023. The position of the exciting star HD 37903 is indicated by a cross. (from Gatley *et al.* 1987.)

2. Fluorescent Molecular Hydrogen

Figure 1 shows the distribution of the H_2 $v = 1 - 0$ $S(1)$ emission in NGC 2023. This reflection nebula is situated on the surface of a dark cloud facing toward us and is illuminated by a B1.5 V star. The molecular gas on the surface of the cloud is exposed to soft (i.e., non-ionizing) ultraviolet radiation from the star and a photodissociation region (Tielens and Hollenbach 1985) is formed.

In this region, H_2 molecules absorb ultraviolet photons in the Lyman and Werner bands at 912 - 1108 Å to become electronically excited. The excited H_2 molecules immediately emit ultraviolet photons to return to the ground electronic state. Ten percent of this transition leads to dissociation of H_2 molecules while the rest results in H_2 molecules in the ground electronic state with vibrational excitation, which cascade down via rovibrational transitions, i.e., fluorescence.

The detailed level population of the fluorescent molecular hydrogen first measured in NGC 2023 is shown in Figure 2 in comparison with that in Orion-KL, a prototypical shocked source . Comparison of the two panels in Figure 2 readily shows three remarkable points as follows (Hasegawa *et al.* 1987):

1. NGC 2023 shows two separate sequences of energy levels corresponding to the para (even J) and ortho (odd J) forms of H_2. As the statistical weight, g_u, applied in the figure includes the spin degeneracies, the para and ortho levels should align on a single sequence if the ortho/para abundance ratio is 3, as is indeed the case for Orion-KL shown in the lower panel. The separation between the two sequences in the $v = 1$ and $v = 2$ states corresponds to the ortho/para ratio of 2.0 and 1.4, respectively.

2. Each of the two sequences of the level population in NGC 2023 is characterized by a high vibrational excitation temperature, T_v, and a low rotational excitation tem-

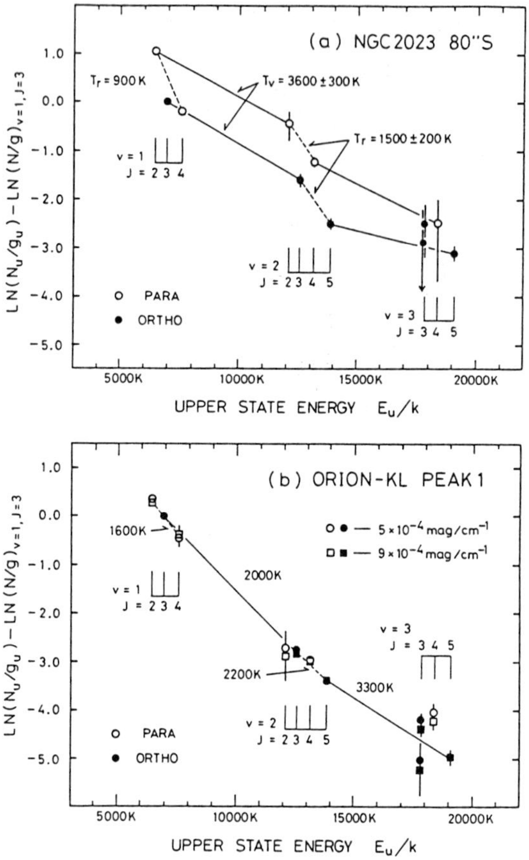

FIG. 2. — The relative level population of H_2 in NGC2023 (*a*) and Orion-KL (*b*). For Orion-KL, reddening correction has been applied based on the two estimates indicated in the upper right-hand corner. Excitation temperatures measured from the slopes of the lines are shown. (from Hasegawa *et al.* 1987.)

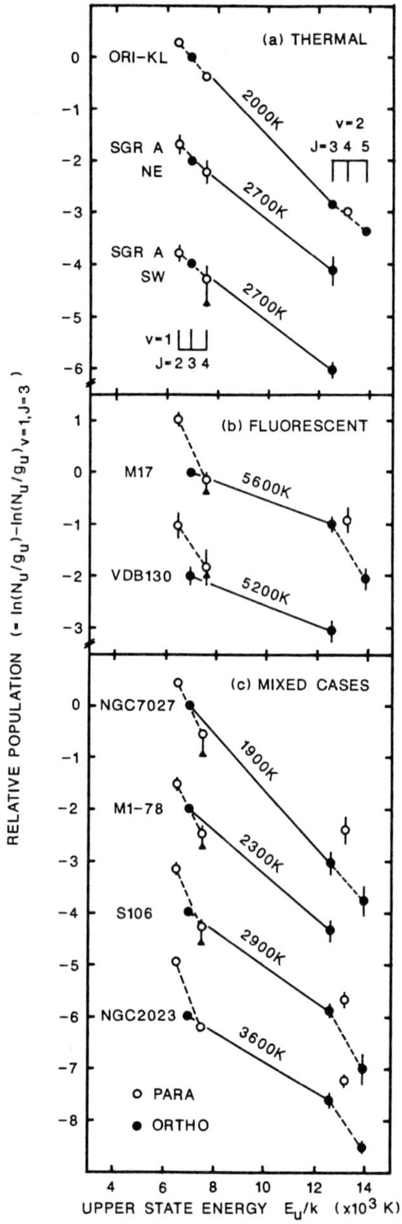

FIG. 3. — The relative level population of H_2 in a variety of Galactic objects. The sources are grouped according to the three phenomenological categories of the excitation; see Section 3 in the text. (from Tanaka *et al.* 1989.)

perature, T_r. This is in contrast to the case of Orion-KL, in which the levels at higher energy are characterized by higher excitation temperatures without any systematic difference between the vibrational and rotational temperatures.

3. The two sequences of population in NGC 2023 show similar excitation characteristics. Both the populations of para-H_2 and ortho-H_2 are consistent with the same $v = 2 - 1$ vibrational temperatures and the same rotational temperatures in the $v = 2$ state. This suggests that the ortho/para ratio measured from the separation of the two sequences may represent the ratio of the total abundance of para-H_2 and ortho-H_2.

The high vibrational temperature and the low rotational temperature found in NGC 2023 (point [1] above) are in good agreement with theoretical expectations by Black and Dalgarno (1976), Takayanagi, Sakimoto, and Onda (1987), and Black and van Dischoek (1987).

3. A Survey

A survey of fluorescent molecular hydrogen in a variety of Galactic sources were made by Tanaka *et al.* (1989). The objects included the HII regions M17 and S106, the reflection nebula vdB 130, the planetary nebula NGC 7027, the Galactic center, and the enigmatic object M1-78 which is thought to be a distant HII region. The results are shown in the population diagrams in Figure 3.

The excitation of molecular hydrogen is classified into three phenomenological categories:

Thermal Sources — Sources in this category have population distribution which align along a smooth line or curve without significant differences between the vibrational and rotational temperatures.

Fluorescent Sources — Sources in this category show population distribution in good agreement with that theoretically expected for fluorescence in low density ($n_{total} \lesssim 10^4$ cm^{-3}) photodissociation regions. The vibrational temperature is as high as $\gtrsim$5000 K while the rotational temperature is only about 1000 K. A marked departure of the ortho/para ratio from 3 is observed.

Mixed Cases — This group of sources exhibit population distribution which is in between the two extreme cases above and can be expressed as a mixture of the two excitation mechanisms. The ortho/para ratio observed in the $v = 1$ state is genarally larger than that in the $v = 2$ state. This reflects the more pronounced contribution from thermal emission in the $v = 1$ state. The prototypical fluorescent source NGC 2023 falls in this category together with the HII regions S106 and M1-78. Surprizingly, the planetary nebula NGC 7027, which has been widely accepted as a source of shocked H_2, showed a sign of fluorescent contribution; it has even a maser action in a $v = 3 - 2$ transition probably pumped by UV radiation (see Section 5).

To assess the relative importance of the thermal and fluorescent excitation in the mixed cases, Tanaka *et al.* (1989) developed a decomposition algorithm which uses the theoretical population of low density fluorescence (Takayanagi, Sakimoto, and Onda 1987; Black and van Dischoek 1987) as a template for the *purely* fluorescent component. The observed ortho/para ratio in the mixed cases becomes larger and approaches 3, the equilibrium value at high temperature, as the contribution of the thermal emission increases. After separation of the thermal contributionfrom the observed spectra, the *real* ortho/para ratio of the fluorescent H_2 can be estimated.

TABLE 1

DECOMPOSITION OF THERMAL AND FLUORESCENT EMISSION

SOURCE	THERMAL FRACTION[a] %	T_{ex}[b] K	RADIATIVE ORTHO/PARA RATIO[c] $v = 1$	$v = 2$
Orion-KL	100	2000	...	...
Sgr A NE	>90	2500	...	...
Sgr A SW	>90	2500	...	...
NGC 7027	90	1200	...	1.0 (0.3)
M1-78	80	1200	...	...
S106	65	<1400	1.4 (0.2)	1.4 (0.3)
NGC 2023	60	2000	1.1 (0.2)	1.2 (0.2)
M17	<10	...	1.8 (0.3)	1.7 (0.3)
vdB 130	<15	...	1.7 (0.4)	...

[a] Fraction of thermal component in v = 1–0 $S(1)$ line emission. Typical errors are 10%.

[b] The excitation temperature for the thermal component. Typical errors are 200 K.

[c] Numbers in parentheses are uncertainties which arise mainly from observational errors.

The results of decomposition are shown in Table 1. The ortho/para ratio of the fluorescent H_2 in the present rather diverse sample is found to lie within the limited range of 1.1 - 1.8 (we exclude NCG 7027 for now; see Section 5). The near constancy of the ortho/para ratio is rather striking, because a large range of values (including ~0) would be anticipated if the fluorescent ortho/para ratio corresponds to that in the molecular cloud before radiative excitation. The results in Table 1 may, rather, suggest that the ortho/para ratio of the fluorescent H_2 is mainly determined by elemental processes of fluorescence. Further consideration of this point wil be given in the next section.

The thermal emission separated in the decomposition described above does not necessarily originate from shocked molecular gas. Models of dense photodissociation regions have shown that when the density is as high as $\gtrsim 10^5$ cm^{-3} the gas temperature rises up to ~10^3 K and that at that high temperature the collisional excitation/deexcitation dominates the ultraviolet pumping to thermalize the level population at the gas temperature (Hollenbach 1988; Sternberg and Dalgarno 1989; Burton, Hollenbach, and Tielens 1989). The thermal component found in some of the sources may be due to collisional excitation in hot, dense photodissociation regions.

4. The Significance of the Observed Ortho/Para Ratio

The universality of the ortho/para ratio found for the fluorescent $H_2$is rather striking. Hasegawa *et al.* (1987) and Takayanagi, Sakimoto, and Onda (1987) tried to attribute the observed low ortho/para ratio in NGC 2023 to the process of H_2 formation. In the fluorescent zone, photodissociation of H_2 molecules is balanced approximately by reformation of H_2 on dust grains. If nascent H_2 molecules have a rovibrational population in a Boltzmann distribution at a *formation temperature*, T_f, which is low compared with the energy difference between the $v = 0, J = 0$ and $J = 1$ levels ($\Delta E/k = 170.5$ K), para ($J = 0$) H_2 is preferentially formed to make the ortho/para ratio considerably smaller than 3. A model incorporating the ultraviolet excitation/dissociation, the H_2 formation on grains, and the exchange reaction between H_2 and H was constructed by Takayanagi, Sakimoto, and Onda (1987) which could reproduce the observed data with $T_f = 60 - 70$ K. This model, however, did not take the dependence of the ultraviolet excitation on the depth from the cloud surface into account and used a single value for the excitation rate via absorption in each of the Lyman and Werner lines.

Black and van Dischoek (1987) performed a depth-dependent calculation for the abundance and the excitation of H_2 for a series of model photodissociation regions. They noted that (1) the exchange reaction between H_2 and H^+ is not negligible in the outermost layer of a cloud where photoionization of vibrationally excited ($v \geq 4$) H_2 supplies ample H^+, and that (2) the excitation rate is very level-specific at any particular depth to a cloud. They claimed that the low formation temperature is not required in order to reproduce the low ortho/para ratio.

Closer examination (from observer's point of view) of the models of Black and van Dischoek (1987) reveales a remarkable feature. The ortho/para ratio estimated from the intensities of the $v = 2 - 1, S(1), S(2)$, and $S(3)$ lines predicted by their calculation falls within a limited range 1.5 – 1.8 without any strong dependence on the density ($n_H = 10^2 - 10^{3.5}$ cm^{-1}) and the intensity of the ultraviolet radiation ($1 - 10^4$ times the interstellar radiation field) for their standard choice of gas temperature ($T = 100$ K), grain properties, and H_2 formation model. Its dependence on the gas temperature is also weak; at $n_T = 10^{3.5}$ cm^{-3}, the ortho/para ratio changes from 1.4 to 2.0 for a large temperature rise from 30 K to 300 K. Indeed the model calculation by Sternberg and Dalgarno (1989) which incorporates the depth dependence of the gas temperature determined by the local thermal balance gives a ratio of 1.6 for $n_T = 10^3$ cm^{-3} and the ultraviolet field 10^3 times the interstellar field, in good agreement with the ratios found in Black and van Dishoek models. This means that the interchange reactions (H_2+H and H_2+H^+) do not show a major contribution in determining the ortho/para ratio of the *fluorescent* H_2 in this density range.

The constancy of the ortho/para ratio apparent in depth-dependent models may be naturally understood if we remember that the ortho and para molecular hydrogen act as two different species with different sets of electronic transitions which absorb ultraviolet photons. The sharp transition from predominantly atomic to predominantly molecular forms of hydrogen in photodissociation regions is due to self-shielding in the ultraviolet absorption lines; the dissociation rate drops suddenly as soon as the strong absorption lines becomes optically thick (see, e.g., Figure 1 in Black and van Dischoek 1987). In other words the photodissociation stops at a depth where the ultraviolet photons in the strong absorption lines are exhausted. This situation occurs for each of ortho and para molecular hydrogen independently if overlaps of ultraviolet lines are neglected. The change in the ortho/para abundance ratio causes a slight shift of the relative position of the dissociation fronts of ortho- and para-H_2 with the ratio of the number of absorbed ultraviolet photons unchanged. As the column density of vibrationally excited ortho- or para-H_2 in a less dense ($n_H \lesssim 10^4$ cm^{-3}) photodissociation region is roughly proportional to the ultraviolet photons absorbed, the observed ortho/para ratio of the fluorescent H_2 is determined primarily by elementary processes, i.e., the oscillator strengths and wavelengths of Lymann and Werner absorption lines and the relative poulation of the lower levels of these lines which is determined by Einstein coefficients of rovibrational transitions. The universality of the observed orth/para ratio of the fluorescent component and its constancy reproduced in depth-dependent model calculations may be understood in this way.

5. Infrared Laser Emission of H_2 in NGC 7027

The nonthermal nature of the excitation of molecular hydrogen in photodissociation regions becomes most prominent in the planetary nebula NGC 7027. In 1989, Hasegawa, Tanaka, and Brand (1989) discovered an extraordinarily strong emission of the H_2 $v = 3 - 2, S(2)$ emission at λ 2.2864 μm in NGC 7027. This line was originally found in the Fourier spectrum taken by Treffers *et al.* (1976) and remained unidentified since then. The population distribution of H_2 analyzed by Hasegawa, Tanaka, and Brand exhibuts highly nonthermal nature with $v = 3, J = 4$, 5, and 6 levels and $v = 2, J = 4$ level significantly overpopulated. The $v = 3 - 2, S(2)$ transition is inverted, and the contribution of induced emission, BI_{ν}/c, is comparable to that of spontaneous emission, A, where A and B are Einstein coefficients. These are the characteristics of *Laser* emission.

Detection of the unidentified line at λ 2.2864 μm in other sources are found in the literature. Isaacman (1984) reports detection of this line in two planetary nebulae IC 5117 and NGC 6572. In addition, Thompson, Lebofsky, and Rieke (1978) reports detection of this line in the nuclear region of the Seyfert galaxy NGC 1068. Although measurements at higher spectral resolution are required to assign these reported features observed at relatively low resolution to the $v = 3 - 2, S(2)$ line of molecular hydrogen, these may suggest that the laser emission of this line is a relatively common phenomenon.

6. Conclusions

The progress of the understanding of the H_2 emission in general and especially in photodissociation regions has been huge as reviewed in this paper, and explosive increase of observational information is still going on with continuous innovation of observational capabilities in infrared astronomy. In the near future, studies of the H_2 emission in external galaxies will be made in the detail we can attain now for Galactic objects, and for the Galactic objects we will be able to visualize the detailed structure of the photodissiciation regions which we can only *observe* in theoretical models. And beynd these is a long-standing target, i.e., detection of the $v = 0 - 0, J = 0 - 2$ and $1 - 3$ lines from the bulk of a dense molecular cloud, which should open the door to an observational approach to the formation mechanism of molecular hydrogen.

Ackowledgements We wish to thank our collaborators I. Gatley, M. Tanaka, M. Hayashi, S. Hayashi, N. Kaifu, P. Brand, and R. Garden. Illuminating discussion with K. Takayanagi, K. Sakimoto, and K. Onda and inspiring comments received at the conference from A. Dalgrno and P. Solomon were very valuable in finishing this paper in its final form.

References

Beckwith, S., Evans, N, Gatley, I., Gull, G., and Russel, R. 1983, *Ap. J.*, **264**, 152.
Black, J., and Dalgarno, A. 1976, *Ap. J.*, **203**, 132.
Black, J., and van Dischoek, E. 1987, *Ap. J.*, **322**, 412.
Gatley *et al.* 1987, *Ap. J. (Letters)*, **318**, L73.

Gautier, T., Fink, U., Treffers, R., and Larson, R. 1976, *Ap. J. (Letters)*, **207**, L129.
Hasegawa, T., Gatley, I., Garden, R., Brand, P., Ohishi, M., Hayashi, M., and Kaifu, N. 1987, *Ap. J. (Letters)*, **318**, L77.
Hasegawa, T., Tanaka, M., and Brand, P., 1989, *Ap. J. (Letters)*, submitted.
Hayashi, M., Hasegawa, T., Gatley, I., Garden, R., and Kaifu, N 1985, *M. N. R. A. S.*, **215**, 31P.
Hollenbach, D. 1988, *Astr. Lett. and Comm.*, **26**, 191.
Isaacman, R. 1984, *Astr. Ap.*, **130**, 151.
Sellgren, K. 1986, *Ap. J.*, **305**, 399.
Sternberg, A. and Dalgarno, A. 1989, *Ap. J.*, **338**, 197.
Takayanagi, K., Sakimoto, K., and Onda, K. 1987, *Ap. J. (Letters)*, **318**, L81.
Tanaka, M., Hasegawa, T., Hayashi, S., Brand, P., and Gatley, I. 1989, *Ap. J.*, **336**, 207.
Thompson, R., Lebofsky, M., and Rieke, G. 1978, *Ap. J. (Letters)*, **222**, L49.
Tielens, A., and Hollenbach, D. 1985, *Ap. J.*, **291**, 722.
Treffers, R., Fink, U., Larson, H., and Gautier, T. 1976, *Ap. J.*, **209**, 793.

Cosmic Ray Induced Photodestruction of Interstellar Molecules

R. Gredel[1,2], S. Lepp[1], A. Dalgarno[1] and E. Herbst[3]
[1]Harvard-Smithsonian Center for Astrophysics; [2] European Southern Observatory; [3]Department of Physics, Duke University

ABSTRACT

Ultraviolet photons are created in the interior of dense interstellar clouds by the impact excitation of molecular hydrogen by secondary electrons generated by cosmic ray ionization. The resulting photodissociation and photoionization rates of a wide range of interstellar molecules are calculated. The effects on the equilibrium chemical composition of dense clouds are briefly discussed.

1. Introduction

The interior of a dense molecular cloud is efficiently shielded from the ultraviolet photons of the interstellar radiation field by the grains. Accordingly, the effects of the UV photons on the physical and chemical state of a dense cloud are in general neglected. However, internal sources of UV photons may be present, such as young stars or shocks from mass losing young stars. Here the discussion is focused on a diffuse source of UV photons arising from H_2 emission. In dense molecular clouds, cosmic rays with energies between 10 and 100 MeV ionize H_2 and generate secondary electrons with a mean energy around 30 eV (Cravens and Dalgarno 1978). Because the fractional ionization is generally low, the electrons lose their energy mainly by exciting, dissociating and ionizing H_2. The subsequent decay of the electronically excited states of H_2 produces UV photons within the clouds.

The idea of molecular hydrogen emission was invoked by Prasad and Tarafdar (1983) to explain the large abundance of atomic carbon which exists in several molecular clouds. In their mechanism, CO is photodissociated

$$\mathrm{CO} + \mathrm{h}\nu \rightarrow \mathrm{C} + \mathrm{O} \quad (1)$$

by the internal photons. The influence of the UV photons on the abundances of interstellar molecules has been explored by Sternberg, Dalgarno and Lepp (1987) who took into account the photons created by the excitation of the B $^1\Sigma_u^+$ and C $^1\Pi_u$ states of molecular hydrogen. A more complete description of the cosmic ray induced ultraviolet spectrum involving excitation to other singlet and triplet states of H_2 has been given by Gredel, Lepp and Dalgarno (1987) who employed it in a calculation of the photodissociation rate of CO. We use their spectrum to derive the internal photodissociation rates of a wide range of interstellar molecules for which photodissociation cross sections are available. We briefly discuss the effects on the equilibrium chemical composition of dense clouds, extending the models of Sternberg et al. (1987).

2. The ultraviolet spectrum of H_2

The collisional excitation of rotation-vibration levels in the various electronic states of H_2 by the secondary electrons lead to the emission of ultraviolet photons in the 100 - 200 nm range. The energy degradation process of the secondary electrons was calculated following the method used by Cravens, Victor and Dalgarno (1975). Excitations to the B $^1\Sigma_u^+$, B′ $^1\Sigma_u^+$, B″ $^1\Sigma_u^+$, C $^1\Pi_u$,D $^1\Pi_u$ and D′ $^1\Pi_u$ Rydberg states and to the valence E, F $^1\Sigma_g^+$ and $a^3\Sigma_g^+$ states were included together with excitations into the repulsive $b^3\Sigma_u^+$ state and into vibrational levels of the ground state. The energy loss of the injected electrons can be expressed in terms of the number of excitations, $\epsilon_i(v')$ of vibrational levels v′ of the various electronic states i. In terms of $\epsilon_i(v')$, the probability $P_i(\nu)$ for the emission of a photon of frequency ν resulting from the radiative decay out of a particular level (i v′ J′) to a lower level (j v″ J″) is given by

$$P_i(\nu) = \sum_{J_0} \frac{\epsilon_i(v')S_{J',J_0}}{\sum_{v',J'} \epsilon_i(v')S_{J',J_0}} \times \frac{A_{v'J',v''J''}}{A_{v'J'}} \times \omega_{J_0} \qquad (2)$$

where the first term is the probability of the excitation of the rotational level v′J′ in i, S_{J',J_0} is the Hönl - London factor and J_0 is the initial rotational level in the $v'' = 0$ level of the ground state of H_2.The second term is the branching ratio describing the probability that the particular transition $v'J' \rightarrow v''J''$ occurs, ν is the transition frequency, the weight $\omega_{J_0}$is the fractional population of H_2 in the initial rotational level J_0 and $A_{v'J'}$ is the total transition probability for emission from level v′J′.

The vibrational levels in the higher-lying Rydberg states predissociate. Vibrational levels $v' > 0$ in the B″ $^1\Sigma_u^+$ state and $v' > 3$ in the D′ $^1\Pi_u$ state strongly predissociate with efficiencies η near unity (Ajello et al. 1984). In the D $^1\Pi_u$ state, levels $v' = 3 - 8$ have predissociation efficiencies around 0.5 (Ajello et al. 1984). For them $(1 - \eta_i(v'))$ was included in eq. (2) as an additional factor. We calculated the corresponding fluorescence efficiencies $P_i(\nu)$ for the D $^1\Pi_u$ – X $^1\Sigma_g^+$, D′ $^1\Pi_u$ – X $^1\Sigma_g^+$ and B″ $^1\Sigma_u^+$ – X $^1\Sigma_g^+$ transitions to be 0.7, 0.4 and 0.07 respectively, in close agreement with those of Ajello et al. (1984).

The resulting spectrum was obtained for a neutral gas of H_2 by summing the probabilities for photon emission from the various electronic states i, weighted with the total excitation rate ϵ_i, according to

$$P(\nu) = \sum_i P_i(\nu) \times \epsilon_i / R_I \qquad (3)$$

where $\epsilon_i = \sum_{v'} \epsilon_i(v')$ and R_I is the total number of ionizations. The emission spectrum depends in detail on the rotational populations of the hydrogen molecules, but with the exception of CO, the resulting photodissociation and photoionization rates are little affected. We present in Figure 1 the emission spectrum for a rotational population in which the molecules are shared between the J = 0 and 1 levels in the ratio 1:3. It is shown distributed in bins of 0.1nm. The spectrum consists of numerous discrete lines superimposed on continuum emission from transitions into the vibrational continuum of the ground state and from the $a^3\Sigma_g^+ - b^3\Sigma_u^+$ transition.

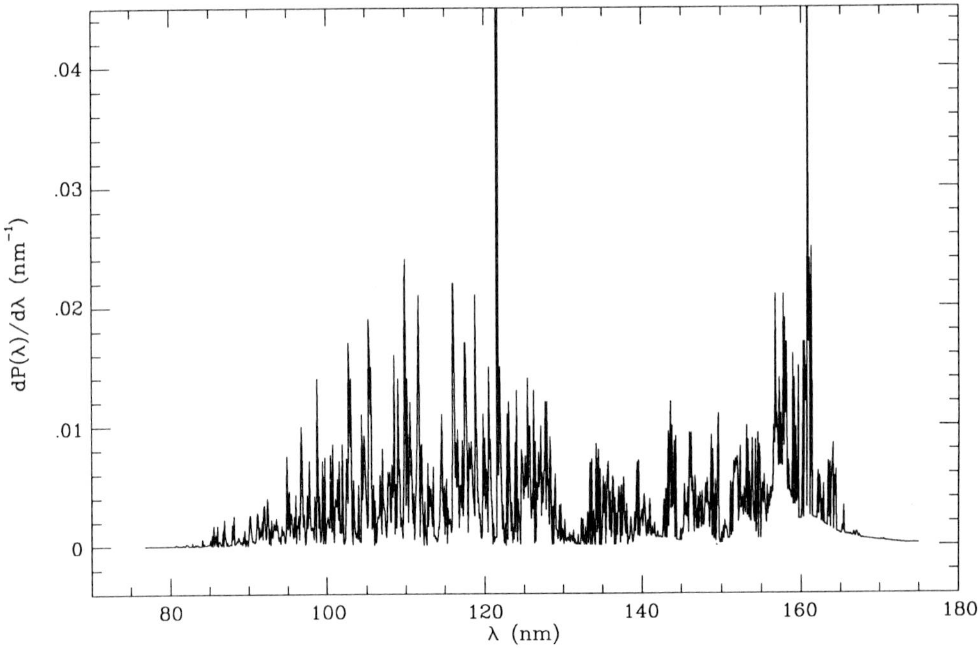

Fig. 1. The emission spectrum of H_2 between 75 nm and 175 nm shown as number of photons emitted per nm. The lines connect the points at intervals of 0.1 nm, each point representing the emission lines falling into the 0.1 nm bin and the continuum emission integrated over the bin.

3. Photodestruction of Interstellar Molecules

The photodissociation and photoionization rate R_M of a molecular species M with density n(M) corresponding to a total cosmic ray ionization rate of ζs^{-1} is given by

$$R_M = \zeta n(M) \int \frac{\sigma_M(\nu) P(\nu)}{\sigma_{tot}(\nu)} d\nu. \tag{4}$$

$P(\nu)$ is the probability for the emission of a photon at frequency ν as given in eq. (3). $\sigma_M(\nu)$ is the photodissociation or photoionization cross section and $\sigma_{tot}(\nu)$ is the total absorption cross section. It may be written in the form

$$\sigma_{tot}(\nu) = \sigma_g(1-\omega) + \sum_{M'} \kappa_{M'} \sigma_{M'}(\nu) \tag{5}$$

where $\sigma_g = 2 \times 10^{-21}\ cm^{-2}$ is the grain extinction cross section per hydrogen nucleus, ω is the grain albedo and $\kappa_M = n(M)/n_H$. In practice, for molecules in interstellar clouds, with the exception of CO (Gredel et al. 1987), $\sigma_{\rm tot}$ is dominated by $\sigma_{\rm g}(1-\omega)$ and for them we may write

$$R_M = \frac{\zeta n(M)}{(1-\omega)}\ p\ \mathrm{cm}^{-3}\ \mathrm{s}^{-1} \tag{6}$$

where the efficiency p is given by

$$p = \int \frac{\sigma_M(\nu)P(\nu)}{\sigma_g} d\nu. \tag{7}$$

The cosmic ray induced photodestruction efficiencies of 37 molecules are presented in Table 1. It includes most interstellar molecules for which at least some data on the absorption cross sections in the ultraviolet are available. For several species, there are no experimental data and theoretical values of the photoabsorption cross sections were adopted. For many molecules the data are fragmentary, especially as regards the identity of the dissociation products. The available data often refer to total absorption and in such cases we have made arbitrary estimates of the probabilities of the dissociation.

Table 1. Cosmic ray induced photodissociation and photoionization efficiencies p of interstellar molecules

Reaction			p	Reaction			p
C_2	$\rightarrow$	$C + C$	237	CH	$\rightarrow$	$C + H$	730
CH^+	$\rightarrow$	$C + H^+$	176	CN	$\rightarrow$	$C + N$	10580
OH	$\rightarrow$	$O + H$	509	O_2	$\rightarrow$	$O + O$	751
O_2	$\rightarrow$	$O_2^+ + e$	117	NO	$\rightarrow$	$N + O$	482
NO	$\rightarrow$	$NO^+ + e$	494	HCl	$\rightarrow$	$H + Cl$	610
H_2O	$\rightarrow$	$H + OH$	971	CO_2	$\rightarrow$	$CO + O$	1708
CO_2	$\rightarrow$	$CO_2^+ + e$	0	H_3^+	$\rightarrow$	$2H + H^+$	< 1
HCN	$\rightarrow$	$CN + H$	3114	HCO	$\rightarrow$	$H + CO$	421
HCO	$\rightarrow$	$HCO^+ + e$	1169	OCS	$\rightarrow$	$CO + S$	5360
OCS	$\rightarrow$	$OCS^+ + e$	1444	C_3	$\rightarrow$	$C_2 + C$	1119
H_2S	$\rightarrow$	$H_2 + S$	5154	H_2S	$\rightarrow$	$H_2S^+ + e$	1696
SO_2	$\rightarrow$	$SO + O$	1884	NH_2	$\rightarrow$	$NH + H$	80
NH_2	$\rightarrow$	$NH_2^+ + e$	649	NH_3	$\rightarrow$	$NH_2 + H$	1315
NH_3	$\rightarrow$	$NH + H_2$	541	NH_3	$\rightarrow$	$NH_3^+ + e$	576
CH_2O_2	$\rightarrow$	$CH_2O_2^+ + e$	649	CH_2O_2	$\rightarrow$	$HCO + OH$	249
CH_3N	$\rightarrow$	$HCN + H_2$	4982	CH_4	$\rightarrow$	$CH_2 + H_2$	2339
CH_4O	$\rightarrow$	$H_2CO + H_2$	3168	CH_4O	$\rightarrow$	$CH_3 + OH$	1504
CH_4O	$\rightarrow$	$CH_4O^+ + e$	1433	CH_4O	$\rightarrow$	$CH_3O^+ + H + e$	99
CH_5N	$\rightarrow$	$HCN + H_2 + 2H$	1409	CH_5N	$\rightarrow$	$CH_3N + 2H$	19
CH_5N	$\rightarrow$	$CH_3 + NH_2$	670	CH_5N	$\rightarrow$	$CN + 2H_2 + H$	367
CH_5N	$\rightarrow$	$CH_5N^+ + e$	1119	C_2H_2	$\rightarrow$	$C_2H + H$	5155
C_2H_2	$\rightarrow$	$C_2H_2^+ + e$	1309	C_2H_2O	$\rightarrow$	$CH_2 + CO$	913
C_2H_2O	$\rightarrow$	$C_2H_2O^+ + e$	1218	C_2H_3N	$\rightarrow$	$CH_3 + CN$	4756
C_2H_3N	$\rightarrow$	$C_2H_3N^+ + e$	2245	C_2H_4	$\rightarrow$	$C_2H_2 + H_2$	3700
C_2H_4	$\rightarrow$	$C_2H_4^+ + e$	778	C_2H_4O	$\rightarrow$	$CH_3 + CHO$	527
C_2H_4O	$\rightarrow$	$C_2H_4O^+ + e$	1119	CH_3OCH_3	$\rightarrow$	$H_2CO + CH_4$	1714

Table 1. continued

Reaction			p	Reaction			p
CH_3OCH_3	→	$C_2H_6O^+ + e$	1119	C_2H_5OH	→	$C_2H_5 + OH$	4307
C_2H_5OH	→	$C_2H_5OH^+ + e$	2736	C_3HN	→	$C_2H + CN$	1727
C_3H_4	→	$C_3H_3 + H$	3284	C_3H_4	→	$C_3H_4^+ + e$	5305
C_3O	→	$C_2 + CO$	6609	C_4H_2	→	$C_4H + H$	1730
C_4H_2	→	$2\ C_2H$	1730	C_4H_2	→	$C_4H_2^+ + e$	1119
H_2CO	→	$CO + H_2$	2659	HNCO	→	$NH + CO$	2361

4. The Steady State Abundances of Interstellar Molecules

We briefly explore the effects of the cosmic ray induced photons on the abundances of interstellar molecules, using a version of the chemistry of Sternberg et al. (1987), modified by the incorporation of the rate coefficients of dissociative recombination recommended by Bates (1986, 1987) and Millar, DeFrees, McLean and Herbst (1988). We present results for a uniform cloud of total hydrogen density $n_H = 10^4 cm^{-3}$, a temperature of 50K, abundances relative to H_2 for carbon of 1.46×10^{-4}, for oxygen of 3.52×10^{-4} and for nitrogen of 4.5×10^{-5}, a cosmic ray ionization rate of $\zeta = 10^{-17} s^{-1}$, a total metal abundance ratio relative to hydrogen of 1.5×10^{-8} and an albedo ω of 0.5.

Table 2. Fractional Abundances $n(M)/n_H$ at Steady State

	(a)	(b)	(c)	(d)
O	7.9(-5)	1.2(-4)	1.2(-4)	1.1(-4)
OH	3.3(-9)	1.4(-8)	1.3(-8)	1.4(-8)
H_2O	5.4(-5)	2.5(-5)	2.5(-5)	2.5(-5)
O_2	3.7(-5)	3.1(-5)	3.0(-5)	3.3(-5)
C^+	1.8(-10)	4.3(-10)	4.4(-10)	4.2(-10)
C	4.5(-9)	6.5(-9)	6.1(-7)	7.9(-7)
C_2H_2	4.2(-9)	7.2(-11)	7.5(-10)	1.9(-9)
CH_4	6.6(-8)	9.6 (-9)	9.5(-8)	1.7(-7)
C_2H	1.5(-10)	4.2(-11)	4.2(-10)	8.3(-10)
C_3H_2	2.1(-11)	8.8(-13)	9.4(-12)	1.2(-12)
NH_3	4.7(-8)	4.4(-8)	4.1(-8)	3.3(-8)
HCN	1.9(-10)	2.7(-10)	1.2(-9)	6.0(-10)

(a) no photons, no LM
(b) photons but no photodissociation of CO, no LM
(c) photons, no LM
(d) photons, LM at $n(LM)/n(H)=10^{-7}$

In the first two columns of Table 2, we show the effects on the steady-state abundances including cosmic ray induced photodissociation and photoionization, but ignoring the photodissociation of CO. The inclusion of the photon field severely diminishes the steady-state abundances of the more complex species because photodissociation is an effective destruction process at many points in the formation sequence (Sternberg et al. 1987).

The destructive effects are mitigated when the photodissociation of CO is included, as the third column of Table 2 demonstrates. The additional source of neutral carbon enhances the supply of complex hydrocarbons. Large molecules, if they exist in significant amounts in dense interstellar clouds, also enhance the abundances of the complex hydrocarbons (Lepp and Dalgarno 1988). The addition of a large molecule component with an abundance relative to hydrogen of 10^{-7} yields in our model chemistry the steady state concentrations given in column 4 of Table 2. With the inclusion of the photodissociation of CO as a source of neutral carbon, the additional source from charge transfer to large molecules and mutual neutralization to large molecular negative ions (Lepp and Dalgarno 1988) has but small consequences.

Because the time scales of the internal photodestruction processes are longer than the time scales to form molecules, the maximum abundances are not significantly affected by the presence of the internally generated photons.

References

Ajello, J.M., Shemansky, D.E., Kwok, T.L., and Yung, Y.L. 1984,*Phys. Rev.* **29**, 636.

Bates, D.R. 1986, *Astrophys. J. Letters* **306**, L45.

Bates, D.R., 1987, Recent Studies in Atomic and Molecular Processes, ed. A.E. Kingston (Plenum Press).

Cravens, T., and Dalgarno, A. 1978, *Astrophys. J.* **219**, 750.

Cravens, T., Victor, G.A., and Dalgarno, A. 1975, *Planet Spa. Sci.* **23**, 1059.

Gredel, R., Dalgarno, A., and Lepp, S. 1987, *Astrophys. J. Letters* **323**, L137.

Lepp, S., and Dalgarno, A. 1988, *Astrophys. J.* **324**, 553.

Prasad, S.S., and Tarafdar, S.P. 1983, *Astrophys. J.* **267**, 603.

Millar, T.J., DeFrees, D.J., McLean, A.D., and Herbst, E. 1988, *Astron. Astrophys.* **194**, 250.

Sternberg, A., Dalgarno, A., and Lepp, S. 1987, *Astrophys. J.* **320**, 676.

This article was processed by the author using the TEX macro package from Springer-Verlag.

THE NATURE OF SHOCKS IN MOLECULAR CLOUDS

Peter W. J. L. Brand

Department of Astronomy,University of Edinburgh

Royal Observatory Edinburgh EH9 3HJ, U.K.

Abstract Evidence is presented to suggest that the shocked molecular hydrogen emission in the brightest part of the Orion outflow is produced in a J-shock and not a C-shock; that this is true throughout the entire flow; that it may be true in many outflow sources; and that this exacerbates problems with current explanations of the very wide velocity profiles observed in molecular hydrogen emission.

1. THE DATA

This paper summarizes work done at the U.K. Infrared Telescope (supported by the Science and Engineering Research Council) in Hawaii, by a group including M.Burton, M.Bird,T.Geballe, A.Moorhouse, M.Toner, R.Wade, A.Webster, and P.Williams. Following the discovery of shocked molecular hydrogen in Orion (Gautier *et al.* 1976), difficulties faced by early hydrodynamic models (*eg* Hollenbach 1981) led to the general acceptance of magnetically-moderated C-shock models (Draine 1980, Draine & Roberge 1982, Chernoff *et al.* 1982).

The results of our measurements are shown in figure 1. These inferred column densities cover a much wider range of level energy than previously used in model fits, and clearly are at variance with published C-shock models (the dashed lines). On the other hand, the cooling zone behind a hydrodynamic shock is modelled by the continuous line, and is evidently a good fit. Two parameters are required, the overall intensity (the curve is drawn through the strongest line 1-0 S(1)) and the slope at high level energy which is determined by the pressure driving the shock.

That is the evidence that the position at which these data were obtained, Peak 1 in the Orion molecular outflow (Beckwith *et al.* 1978), is excited not by a C-shock but by a hydrodynamic (or J-) shock. The evidence that this situation is outflow-wide is

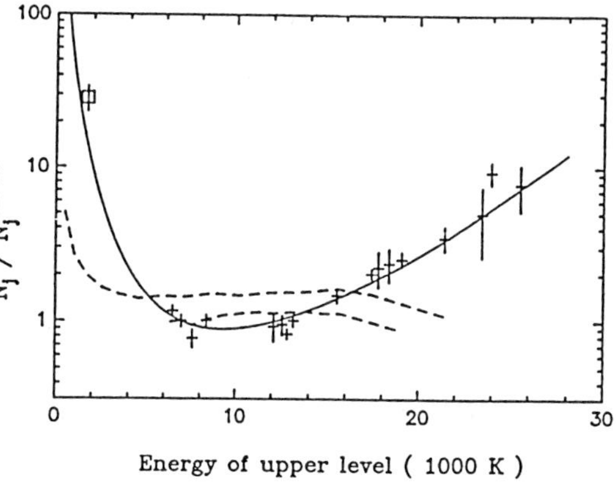

Figure 1(top). Column density derived from the observations at Peak 1 in Orion divided by those expected from 2000K gas. The dashed lines are C-shock models, the solid line is a J-shock model.

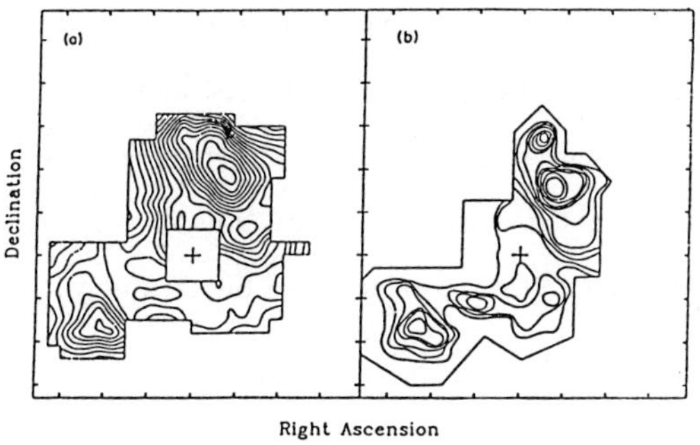

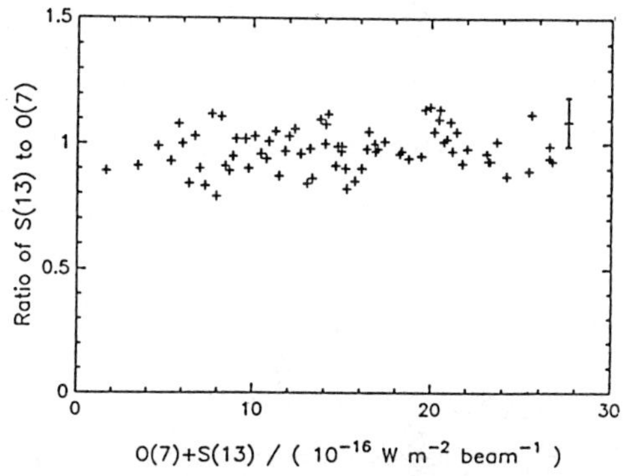

Figure 2(centre left). 3.8μm map(left) and 2.1μm map from Beckwith *et al*(right) of the Orion molecular outflow.

Figure 3(centre right). The 3.8μm line ratio from all parts of the outflow, showing no trend with intensity.

Figure 4(bottom). The same as figure 1, for data from the brightest part of the IC 443 H_2 emission

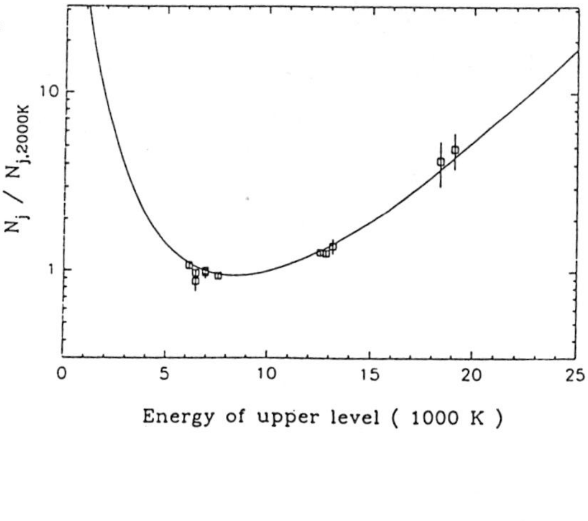

shown in figure 2, which consists of a map of intensity of the sum of two 3.8μm lines of H_2 throughout the outflow. The right-hand map is drawn from the original 2.1μm map of Beckwith *et al.* for comparison. It is clear the morphology is the same. The lines in question are the 1-0 O(7) line with an upper level energy of 8365 K and the 0-0 S(13) line with an upper level energy of 17445 K. Because the wavelengths are nearly identical, there are no problems de-reddening the data. From figure 1 it is clear that the ratio is a sensitive indicator of the shape of the right-hand part of the model curve.

Figure 3 displays this ratio from all positions in this map, ordered according to intensity. It is clear that there is no significant trend in the data, and that the curve of figure 1 applies to all parts of the outflow. That is the evidence for J-shocks throughout the outflow.

Similar data obtained from the brightest part of the H_2 emission in the supernova remnant IC 443 (Burton *et al.* 1988) is shown in figure 4. Again a J-shock-like pattern of column densities of H_2 lines is produced. Burton *et al.* (1989) show that the line ratios in several other sources are consistent with this interpretation. These are the data on which is based the assertion that J-shock excitation appears to be common in shocked molecular clouds.

2. THE THEORY

Figure 5(a) is a diagram of the cross section through a hydrodynamic J-shock, showing the cooling zone where the temperature drops at a rate determined by the local cooling function. The cooling is supposed due to collisionally excited H_2 either dissociating or de-exciting by line radiation, and to line emission from CO cooling. This can by crudely approximated by a steep power law, which is assumed in this argument. Figure 5(b) shows the same flow in temperature space, with superposed curves of the inverse cooling rate and a Boltzmann factor corresponding to a particular H_2 line. The local density of excited molecules contains these as factors, and their product is also shown in figure 5(b) as a dashed line. This peaks at a value depending only on the upper level energy of the line, and on the cooling function. It is clear from the diagram that for modest velocities (and therefore modest post-shock temperature maxima) that the integral under the dashed curve does not depend on velocity. In this approximation, and taking s to be the T-exponent in the cooling function, the column densities of the various lines with upper level energies of T_j are proportional to $(T_j)^{-s}$. This approximates closely to the solid curve in figure 1.

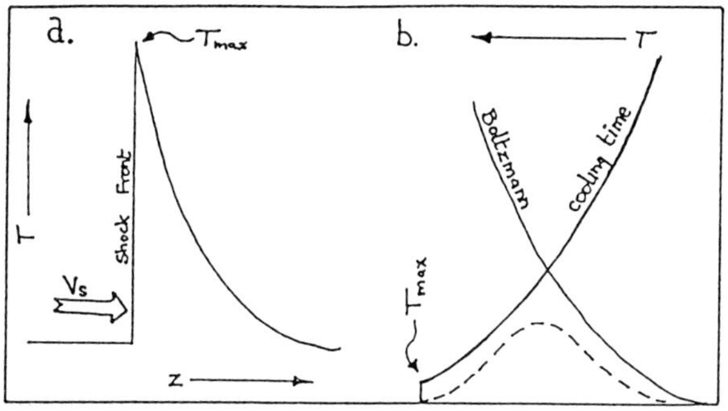

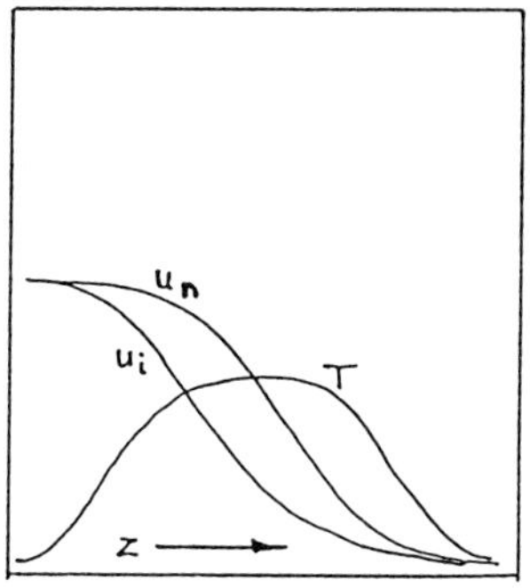

Figure 5(left). A schematic of a hydrodynamic (J-)shock. *'a'* shows the temperature profile as a function of distance through the cooling zone, while *'b'* is the same region in T-space showing the behaviour of $e^{-T_j/T}$ and of the cooling time of the shocked gas. Their product is shown as a dashed line.
Figure 6(right). shows the velocities of the ions u_i and of the neutrals u_n, and of the temperature in the frame of a C-shock.

On the other hand, a C-shock structure is quite different. The heating is by friction with the streaming ions, and is relatively slow compared with the cooling time, in contrast to the J-shock case. This means that the region where the radiation is generated reaches its highest temperature when heating balances cooling, and it behaves like a slab of gas at that maximum temperature (figure 6). Since the heating is velocity dependent, so is the maximum temperature. Thus the expected behaviour of column densities in a C-shock is that it would represent a single temperature, dependent on the shock velocity. This seems to be at variance with our observations. Because C-shock models imply a lower density in the shocked gas (there is a longer distance over which a given line is radiated), these models also show signs that the highest H_2 levels are not fully populated, and therefore the corresponding column densities fall below the Boltzmann line. On the other hand, the cooling time increases as temperature drops, and so the lowest levels find themselves in what is to all intents a cooling zone, with frictional heating no longer significant, and the pattern of column densities shows the characteristic rise at low excitation energies established also behind J-shocks. Both these phenomena are apparent in the C-shock model in figure 1.

3. INTERPRETATION

If the evidence is accepted at face value, then there are several consequences. First of all, the very good fit of the model to the data in figure 1 implies that the cooling function is correct. This means that the cooling in these shocks is dominated by H_2 dissociation above approximately 3000 K, by H_2 line cooling, and then by CO and other molecular dipole cooling below 1000 K.

Secondly, the extinction to the Orion outflow is lower than generally accepted. The best fit value to these data is $A_K = 0.8$, which implies a visual extinction of only 9 magnitudes between it and the Trapezium stars.

Thirdly, dissociational cooling depends on the square of the density while line cooling, being saturated, depends only on the first power of density. The slope of the cooling curve at high excitation energies is therefore steeper if the pressure (and therefore the density at a given temperature) behind the shock is greater. Correspondingly, the slope of the curve in figure 1 flattens. The best fit pressure at Peak 1 in Orion (or indeed throughout the Orion flow, since the 0-0 S(13) to 1-0 O(7) ratio is constant) is high, with a value of 8×10^{10} K cm^{-3}. This provides fierce constraints on the properties of the wind driving the flow.

This seems a very neat package, but there are problems. The first is that measurements of rotationally excited CO in the Orion outflow made by several groups, and summarised by McKee (1982), indicate a post-shock density an order of magnitude lower than deduced from these models. However, the region from which this radiation was collected is much greater than the region delineated by the H_2 peak, and within the vicinity there are several sources of CO heating, with very different properties. Further work is required.

The second problem appears to be a problem for any shock model, but is especially acute for J-shock models, which completely dissociate H_2 at shock speeds above 25 km s^{-1}. The width of the profile observed (Brand *et al.* 1989) implies that in the line of sight there is a significant fraction of the excited gas travelling at up to ±100 km s^{-1}, *i.e.* at over three times the dissociation speed.

However, problems or no, it is clear that the capability of high performance spectrometers on a large IR telescope to obtain accurate intensities for H_2 lines over a wide range of intensity and wavelength is providing an excellent basis for a new and detailed look at the mechanics of energy deposition in molecular clouds by young stellar objects.

4. REFERENCES

Beckwith, S., Persson, S.E., Neugebauer, G., & Becklin, E.E. 1978, *Ap. J.* **234**, L213.

Brand, P.W.J.L., Toner, M.P., Geballe,T.R., & Webster, A.S., 1989, *MNRAS* **236**, 929.

Burton, M.G., Geballe, T.R., Brand, P.W.J.L. & Webster, A.S., 1988, *MNRAS.* **231**, 617.

Burton, M.G., 1989, *MNRAS. in press.*

Chernoff, D.F., Hollenbach, D.J., & McKee, C.F., 1982 *Ap. J.* **259**, L97.

Draine, B.T., 1980, *Ap. J.*, **241** 1021.

Draine, B.T., & Roberge, W.G., 1982, *Ap. J.* **259**, L91.

Gautier, T.N. III, Fink, U., Treffers, R.R., & Larson H.P., 1976, *Ap. J.* **207**, L29.

Hollenbach, D.J., 1981, *Symposium on the Orion Nebula to honour Henry Draper,* N.Y. Acad. Sci, New York, 242.

McKee, C.F., Storey, J.V.W., Watson, D.M., & Green S., 1982, *Ap. J.* **259**, 647.

Discussion:

MÜNCH: 1) How high is Boltzmann temperature T_B derived from the intensity ratio between the two lines in the 3.6μm range?. 2) How high is T_B and the J-number of the highest excitation rotational line?.

BRAND: 1) The temperature derived from the line ratio of the H_2 lines $\emptyset - \emptyset$ S(13) and 1-$\emptyset$ O(7), is 2800K. 2) The highest angular quantum number we have observed is J=19, $\emptyset - \emptyset$ S(17). The excitation temperature is 25541K. The equivalent observed Boltzmann temperature is aproximately 3500K. These temperatures are only loosely relates to the temperature of the emitting gas, which varies from a very high temperature just behind the shock to very low temperature after a few cooling lengths. Gas at all temperatures contributes to each line, in a manner determined by the cooling rate, and the excitation temperature of the line.

FRAGMENTATION AND THE INITIAL MASS FUNCTION

Richard B. Larson
Yale Astronomy Department
Box 6666
New Haven, CT 06511, U.S.A.

1. INTRODUCTION

A central problem in the theory of star formation is to understand the spectrum of masses, or Initial Mass Function, with which stars are formed. The fundamental role of the IMF in galactic evolution has been described by Tinsley (1980), and an extensive review of evidence concerning the IMF and its possible variability has been presented by Scalo (1986). Although the IMF derived from the observations is subject to many uncertainties, two basic features seem reasonably well established. One is that the typical stellar mass, defined such that equal amounts of matter condense into stars above and below this mass, is within a factor of 3 of one solar mass. A theory of star formation should therefore be able to explain why most stars are formed with masses of order one solar mass. The second apparently universal feature is that the IMF for relatively massive stars can be approximated by a power law with a slope not greatly different from that originally proposed by Salpeter (1955). Thus we also need to understand why the IMF always has a similar power-law tail toward higher masses.

Evidence for variability of the IMF has proven slippery, but two possible trends have been suggested by a number of studies. One is that the mass spectra of open clusters may be systematically flatter at high masses than the IMF of field stars, i.e., clusters may contain a higher proportion of the most massive stars (Scalo 1986). A second possible type of variability is that some clusters or regions of star formation may contain relatively few stars with masses below one or two solar masses; this is most strongly suggested for "starburst" regions or systems. The evidence for such variability of the lower IMF, and its possible implications, have been discussed by Larson (1986) and Scalo (1987); however, since it is particularly difficult to establish differences in the lower IMF, we shall focus here on the more well established basic features noted in the previous paragraph.

Circumstantial evidence bearing on the origin of the IMF is provided by the fact that newly formed stars of different masses are typically found in different environments (Herbig 1962; Blaauw 1964; Larson 1982). Low-mass young stars are seen scattered throughout dark clouds such as the nearby Taurus clouds; the fact that they often appear in relative isolation may indicate that they form directly and not by a hierarchical fragmentation process (Herbig 1978). By contrast, massive stars form in a more strongly clustered fashion and only in much larger aggregates of gas and young stars, such as the giant molecular cloud in Orion. The most massive young stars, moreover, are often found in the dense cores of large clusters; examples are the Trapezium system in Orion and the compact groups of very massive stars seen at the centers of NGC 3603 and the 30 Doradus nebula (Moffat, Seggewiss, and Shara 1985; Baier, Ladebeck, and Weigelt 1985; Weigelt and Baier 1985). Thus the formation of condensed clusters may play an important role in the formation of the most massive stars.

The observations therefore suggest that the formation of low-mass stars is a primary process involving the fragmentation of clouds into clumps and the collapse of these clumps directly into stars. The formation of massive stars, on the other hand, may be a secondary process that occurs as a result of continuing accumulation or accretion processes in dense environments (Larson 1982). The masses of low-mass stars may then be determined by the mass scale of fragmentation, while the power-law tail of the IMF may be built up through the formation of increasingly massive stars by accretion (Larson 1986). In Sections 4 and 5, it will be suggested that protostellar interactions in a cluster of young stars also play an important role in establishing a mass spectrum, and in Section 6 it will be suggested that feedback effects associated with outflows from young stars can influence its slope.

2. FRAGMENTATION

Numerical simulations show that small density fluctuations experience little growth during the early collapse of a nearly uniform cloud, and that the cloud approaches an equilibrium disk, sheet, or filament before fragmentation into clumps becomes marked. This fact, as well as observational evidence, suggests that cloud fragmentation may primarily involve the instability and breakup of near-equilibrium structures such as sheets or filaments (Larson 1985). Detailed studies of the fragmentation of sheetlike clouds by Miyama, Narita, and Hayashi (1987a, b) suggest that sheets tend to fragment into filaments, which then become thinner and finally break into clumps. This type of evolution could account for the observed structure of some star forming clouds that, like the Taurus clouds, consist of clumps strung out along filaments (Schneider and Elmegreen 1979).

The predicted critical mass or "Jeans mass" in the Taurus clouds agrees well with the masses of the dense cores of these clouds, and also with the masses of the newly formed T Tauri stars associated with them. For a cloud temperature of 8 K, the predicted critical mass is about 1 $M_\odot$, which may be compared with a median core mass (measured within the half-maximum NH_3 contour) of 0.7 $M_\odot$, and a median T Tauri star mass of 0.6 $M_\odot$ (Larson 1985). Thus the dense NH_3 cores may indeed have formed by the fragmentation of the filamentary Taurus clouds; once formed, however, these cores apparently collapse with little, if any, further fragmentation to make Tauri stars of comparable mass.

The predicted critical mass depends mainly on the cloud temperature, and is higher in warmer clouds. For example, in the Orion cloud the temperature is typically $\sim$ 20 K, and the resulting critical mass is a few times larger than in Taurus; this is in qualitative agreement with the fact that both the masses of the molecular clumps and the masses of the known stars are systematically larger in Orion than in Taurus. The median mass of the known T Tauri stars, for example, is about 1.1 $M_\odot$ in Orion, compared with 0.6 $M_\odot$ in Taurus (Larson 1986).

It is probably also relevant to cloud fragmentation processes that, while molecular clouds are characterized on larger scales by supersonic turbulent or "non-thermal" motions, the turbulence becomes subsonic on the scale of the cloud cores discussed above (Larson 1981; Myers 1983). In addition, there is evidence that magnetic fields influence the structure and dynamics of larger regions in the Taurus clouds (Heyer et al. 1987) but are relatively unimportant for the cloud cores (Heyer 1988), possibly because the fields have largely decoupled from the cores by ambipolar diffusion (see also Myers and Goodman 1988). These properties suggest that, in addition to the primary role of self-gravity, the dissipation of turbulence and magnetic fields in molecular clouds may also be important in allowing fragmentation and star formation to occur. The development of condensed cloud cores may, in fact, proceed not in a rapid dynamical fashion but in a slow quasi-static fashion modulated by ambipolar diffusion (Shu, Adams, and Lizano 1987; Shu et al. 1988). Nevertheless, a relevant mass scale is still that set by the balance between thermal pressure and gravity, since it is on this scale that nearly spherical and centrally condensed structures capable of collapsing into stars will first form.

3. FORMATION OF LOW-MASS STARS

Once a protostellar clump begins to collapse, how does the material in it actually become condensed into a star, and how much of it goes into the star that forms? The formation and growth of a central stellar core or "embryo star" have been extensively studied in the spherical case, and the early stages of this process are probably very similar even when a modest but realistic amount of rotation is present, provided that the collapsing clump is initially centrally condensed (for recent reviews, see Shu, Adams, and Lizano 1987 and Sofia et al. 1989). When rotation is present, the infalling gas eventually begins to accumulate in a disk around the central embryo star, and a substantial fraction of the protostellar material may actually fall into this disk. Considerable evidence suggests that residual disks are in fact common around newly formed stars (e.g. Strom, Edwards, and Strom 1989). It is then of interest to know how much of the material in such a disk is eventually accreted by the central star; clearly this depends on how effectively angular momentum can be transported outward in, or removed from, the disk.

The mechanisms that might transfer angular momentum in protostellar disks have been reviewed by Larson (1989). If the mass of the disk is comparable to or larger than that of the central star, gravitational torques associated with trailing spiral density fluctuations can transfer angular momentum rapidly outward, with the result that at least half of the mass of the star-plus-disk system quickly ends up in the central star. Disk accretion may continue as a result of other effects, which may include the transport of angular momentum by wave motions such as shock waves driven into the disk by tidal interactions with neighboring stars. Another suitable wave source might be a Jupiter-like planet; if such planets form early enough, their perturbing effects may help to disperse the residual gas in protostellar disks, partly by accretion onto the central star, in a time of order 10^6 years that is comparable to the observationally inferred lifetimes of protostellar disks (Larson 1989).

If most of the gas that falls into a circumstellar disk is eventually accreted by the central star, the final stellar mass is essentially the total mass that falls into the central star and its surrounding disk. Shu et al. (1988) have argued that, since cloud cores are not sharply bounded, infall of gas from the outer parts of the core and from the surrounding cloud would continue almost indefinitely and would build up a star much more massive than a typical T Tauri star if the infall were not somehow shut off; they suggest that this is accomplished by a stellar wind (see also Shu and Terebey 1984). In this view, it is the onset of a wind associated with the beginning of deuterium burning that is responsible for determining typical stellar masses. However, Stahler (1988) points out that deuterium burning begins undramatically long before a growing embryo star reaches one solar mass, and suggests that it is probably not relevant to determining stellar masses.

The fact that the masses of the T Tauri stars observed in nearby dark clouds are similar to both the predicted "Jeans mass" and the masses of the dense cloud cores suggests, in any case, that cloud properties do play a significant role in determining stellar masses. Even though the cloud cores are not sharply bounded, there is still a central dense region that is predominantly thermally supported, surrounded by a more extended region in which turbulence and magnetic fields provide most of the support. Both turbulence and magnetic fields can inhibit continuing accretion from the region outside the central Jeans-mass core. For example, if the surrounding gas is moving supersonically with respect to the core, the accretion rate predicted by classical accretion theory is reduced by a factor that is approximately the cube of the Mach number (Bondi 1952; Hunt 1971). Magnetic stresses can also prevent the collapse of the outer parts of protostellar clouds and limit the efficiency of star formation (Mouschovias 1987). Stellar winds can further limit the growth of forming stars, either directly by blowing away cloud material (Mathieu et al. 1988) or indirectly by increasing the general level of turbulence in collapsing clouds (see Section 6). Probably a complex interplay of many effects will need to be understood before a fully quantitative theory of stellar masses is possible.

Meanwhile, it may be worth noting that the Jeans mass M_J can be expressed in terms of the sound speed c and the gas pressure P as $M_J \sim c^4/G^{3/2}P^{1/2}$. Although the thermal pressure varies strongly with location in molecular clouds, the total pressure, including turbulent and magnetic contributions, is much more nearly constant everywhere. This follows from the observation that star forming clouds always have similar surface densities, and the fact that the pressure in a self-gravitating cloud is proportional to the square of the surface density (e.g. McKee and Lin 1988). If the supporting pressure is primarily magnetic in origin, then observed cloud properties imply a field strength of the order of 30 μG in star forming clouds (Myers and Goodman 1988), but any kind of supporting pressure would be equally compatible with these observations as long as it is nearly constant everywhere. If this universal cloud pressure is combined with a sound speed of 0.2 km/s in the above expression for the Jeans mass, then the resulting mass is just under one solar mass.

4. DYNAMICS OF PROTOCLUSTERS

Unlike low-mass stars, massive stars form only in large aggregates of gas and young stars, and the most massive stars seem to form preferentially in the dense cores of massive clusters. This suggests that the dynamics and evolution of clusters of forming stars may play an important or even essential role in the formation of massive stars and in the development of the upper part of the IMF.

Numerical simulations and observations both suggest that the most massive stars form as a result of the continuing accretional growth of a few favored objects (Larson 1978, 1982). Such a process can also generate an upper IMF that is of power-law form (Zinnecker 1982). The tendency of massive stars to form in dense regions might then be a result simply of the fact that accretion proceeds fastest where the density is highest. Another effect that can occur in dense clusters of forming stars is that interactions involving protostellar disks may trigger episodes of enhanced disk accretion (Larson 1982). Interactions can also transfer material between protostellar disks, generally from smaller to larger ones, so that "the rich get richer"; for example, if a star with a disk passes through a denser and more massive disk surrounding a more massive star, its disk may be stripped away and added to the disk around the more massive star. Similar effects occur when galaxies interact: tidal interactions can trigger rapid gas inflows toward the centers of galaxies, and large galaxies can steal material from smaller ones, or even swallow them completely, thereby growing at their expense. In clusters of forming stars, as in clusters of galaxies, the sizes of disks may not be negligible compared to typical encounter distances, so that analogous phenomena may occur.

An example of an environment where interactions may be important is provided by the very dense cluster of young stars around the Trapezium in Orion, which has a density of at least 3000 solar masses per pc^3 (Herbig 1983; Herbig and Terndrup 1986). Within a period of 10^6 years, about the present age of this cluster, most of the stars will pass within 1000 AU of another star, and about 10 percent will pass within 100 AU of another star, close enough to strongly disturb a protostellar disk. Encounters would be even more important if the positions or velocities of the stars were initially correlated, for example if they had formed in subclusters (see below).

If a star passes close to or through a disk around another star, it will also experience a gravitational drag that is essentially similar to the "dynamical friction" effect of stellar dynamics. If the star is sufficiently decelerated, which can happen if the disk is massive enough or if the impact parameter of the encounter is small enough, it will be captured into a bound orbit around the other star, forming a binary system. Even when captures do not occur, energy will still be removed from the orbital motions of the stars in a young cluster by this effect, and this will cause the entire cluster to become more condensed. Since the deceleration experienced by a star is proportional to its mass, regardless of whether it interacts with gas or with other stars, the drag effect is strongest for the most massive stars, which therefore become most concentrated at the center of the cluster. The segregation of massive stars toward the center may further accelerate their relative growth, since the residual gas in the cluster will also tend to settle toward the center. Young clusters may in this way become dominated by tight central groups of very massive stars, as is indeed observed to be the case in the examples mentioned in Section 1.

If the distribution of matter in a star forming cloud is very clumpy, the stars that eventually make up a cluster may initially form in subclusters that later merge to produce a more massive and centrally condensed system, perhaps with the help of the dynamical friction effect discussed above. Again, such processes may parallel those that occur on galactic scales, since mergers can also build larger galaxies from smaller ones. Subclustering is presently observed in the Trapezium region in the form of a second tight group of massive stars that are now detected only as infrared sources in the core of the molecular cloud, but that will probably soon become visible and join the Trapezium cluster. Subclustering on various scales seems, in fact, to be a nearly universal characteristic of regions of star formation (e.g. Blaauw 1964; Larson 1982; Wilking and Lada 1985).

If the development of larger and more centrally condensed clusters is accompanied by the formation of progressively more massive stars, a power-law IMF might result if the processes involved are self-similar and have no preferred mass scale. The development of a self-similar or "fractal" hierarchy of clustering was suggested by Larson (1978) on the basis of numerical simulations of fragmentation which in some cases showed at least two levels of clustering (see, for example, Fig. 6(a) of that paper). Gravitational drag and accretion effects were also found to be important in these simulations, especially for the most massive objects, which tend to be located at the centers of groups, as expected. Although these simulations are too crude to predict an IMF with any reliability, the slope estimated by counting objects in the two largest mass bins is $x \sim 1.5$, not inconsistent with observed values (cf. Salpeter's value $x = 1.35$, where $dN/d \log m \propto m^{-x}$.)

5. FORMATION OF MASSIVE STARS

Clearly, we still need to understand in more detail how massive stars form before we can fully understand the upper IMF. The formation of massive stars is less well understood than the formation of low-mass stars, partly because the nearest examples are farther away and more heavily obscured, but it is clear that the formation of massive stars must be a more complex process. One reason is that, unlike the small quiescent cloud cores in which low-mass stars form, molecular clumps that are massive enough to form the most massive stars always contain supersonic internal turbulent motions, and they generally also contain more than one Jeans mass. Therefore they probably evolve in a less regular fashion than smaller clumps and form more than one star, consistent with the evidence that massive stars generally form in groups and clusters (Larson 1981).

Another difference is that for massive stars, radiation pressure eventually becomes important during the star formation process and prevents further gas infall from occurring, at least in the case of spherical collapse. Wolfire and Cassinelli (1987) find, in fact, that radiation pressure is even more important than was estimated by Larson and Starrfield (1971) and Kahn (1974), and prevents accretion from continuing for a star more massive than about 15 – 30 $M_\odot$. Radiation pressure can be overcome by non-spherical accretion from a dense sheet or disk of gas (Nakano 1989), but all of the matter destined to make a massive star must in this case condense into such a flattened structure before more than a small fraction of it has gone into a star. Another possible source of already condensed gas from which to make a massive star might be circumstellar disks around smaller stars, which are swept up and added to the disk around a massive star. If massive stars form in close multiple systems, actual stellar collisions and mergers might also occur, and might provide an especially effective way of building up very massive stars without hindrance from radiation pressure.

A final difference from low-mass stars is that the formation of massive stars almost certainly takes longer than the formation of low-mass stars, simply because the building materials must be assembled from a larger region that has a longer dynamical timescale. All of these differences reinforce the general conclusion that stars of larger mass form at later stages in the development of progressively more condensed aggregates of gas and stars. However, the actual processes involved remain unclear, and could include accretion from massive and dense disks, theft of additional disk material from smaller stars, and even stellar cannibalism.

A common feature of most such scenarios is that massive stars form from residual material left over from the prior formation of large numbers of less massive stars. A general implication of this is that most of the matter that condenses into stars will go into low-mass stars, and a decreasing amount will remain to make progressively more massive stars. For a power-law IMF, this implies that $x > 1$. For example, the Salpeter slope $x = 1.35$ could plausibly be produced in this way; if stars of larger mass form at later times, and if the amount of gas condensing into new stars decreases by 20 percent with each factor of 2 increase in the maximum stellar mass, then a mass spectrum with $x = 1.32$ is produced.

If clustering plays an important role in the formation of massive stars, another expectation would be that cluster mass spectra should contain a higher proportion of massive stars than the IMF of field stars. As noted earlier, there is marginal evidence for such a difference; the studies reviewed by Scalo (1986) yield $x \sim 1.2 \pm 0.5$ for open clusters, compared with $x \sim 1.7 \pm 0.5$ for field stars when the data are analyzed in the same way. There is also some evidence that the upper IMF is flatter in large clusters than in small ones (Burki 1977). Very flat mass spectra with $x \sim 0.5$ have been found by Elson, Fall, and Freeman (1989) for several Large Magellanic Cloud clusters, although in apparent conflict with this result, Mateo (1988) finds steeper spectra with $x > 2$ for a number of other Magellanic Cloud clusters. Clarification of this presently confused subject would clearly be desirable, and could yield fundamental clues regarding the origin of the IMF.

6. FEEDBACK AND THE IMF

It has often been suggested that negative feedback effects regulate the efficiency and the rate of star formation in galaxies by destroying star forming clouds or by supporting them against collapse (see, for example, Larson 1987). Self-regulation effects can also influence the IMF, especially if the more massive stars generate stronger feedback effects and if this limits the amount of matter that can condense into massive stars.

Star forming clouds may eventually be completely destroyed by ionization and strong winds from very massive stars, but even before this happens, bipolar outflows from less massive stars can stir up such clouds and increase the amount of turbulence present in them (Lada 1988). This will provide additional cloud support and may inhibit the continuing growth of forming stars. Evidence that star formation generates turbulence in molecular clouds is provided by the fact that clumps containing young stars generally have larger internal turbulent motions than clumps without stars (Fuller and Myers 1987; Benson and Myers 1989; Loren 1989). A detailed feedback cycle whereby outflows may regulate both star formation and cloud properties has been suggested by McKee and Lin (1988); they argue that outflows will tend to expand a star forming cloud and thus regulate its opacity and degree of ionization so as to keep the rate of star formation, which is presumed to be controlled by ambipolar diffusion, just sufficient to resupply the mechanical energy being dissipated in the cloud. Such a mechanism could account not only for the nearly constant surface density of star-forming clouds, but also for the typical low efficiency of star formation.

If stars of larger mass form at later stages in the development of more condensed aggregates, the amount of matter going into stars of each mass may also be regulated by feedback effects like those discussed above. If the mechanical energy deposited in the cloud per unit mass of stars formed is an increasing function of stellar mass, this could have the consequence that progressively smaller amounts of matter condense into stars of increasing mass. This might provide a more physical basis for the scheme suggested in Section 5 to produce a power-law IMF with $x > 1$, where a decreasing amount of residual matter was assumed to condense into stars of increasing mass.

To illustrate by a highly oversimplified example how feedback might control the slope of the IMF, suppose that a constant fraction (of order 10^{-3}) of the binding energy of each star is converted via outflows into mechanical energy that contributes to the support of the cloud. If the most massive star in each clump or subcluster grows in mass until the total mechanical energy generated by it is equal to the binding energy of the clump, the binding energy of the most massive star is then a constant multiple (of order 10^3) of the clump binding energy. Assuming that the radii of accreting stars are approximately independent of mass (Larson 1972; Stahler 1988) and that the binding energies of molecular clumps vary with the 1.4 power of clump mass (Larson 1981), we then predict that the maximum stellar mass increases with the 0.7 power of the clump mass. If aggregates of gas and young stars are built up in such a way that the mass of the most massive star always remains proportional to the 0.7 power of the total mass of gas and stars present, and if a nearly constant fraction of this total mass is in stars, then a power-law IMF is built up whose slope is $x = 1/0.7$, or $x = 1.4$.

Of course, this example neglects many complications, and can hardly be considered a convincing prediction of the IMF; for example, it is not fully consistent with the conclusion of Section 5 that the most massive stars form only from material that is already very condensed and presumably is little affected by feedback effects. Clearly, the full problem of understanding the IMF is exceedingly complex, and all we can really hope for at present is to gain an inkling of some of the many effects involved. In such a situation, the power of theory is limited and the need for observational input is great; therefore it is to be hoped that the increasingly detailed data now becoming available on regions of star formation can help to clarify which processes are most important in determining the IMF.

REFERENCES

Baier, G., Ladebeck, R., and Weigelt, G., 1985. *Astr. Astrophys.*, **151**, 61.

Benson, P. J., and Myers, P. C., 1989. *Astrophys. J. Suppl.*, in press.

Blaauw, A., 1964. *Ann. Rev. Astr. Astrophys.*, **2**, 213.

Bondi, H., 1952. *Mon. Not. Roy. Astr. Soc.*, **112**, 195.

Burki, G., 1977. *Astr. Astrophys.*, **57**, 135.

Elson, R. A. W., Fall, S. M., and Freeman, K. C., 1989. *Astrophys. J.*, **336**, 734.

Fuller, G. A., and Myers, P. C., 1987. In *Physical Processes in Interstellar Clouds*, eds. G. E. Morfill and M. Scholer, p. 137. Reidel, Dordrecht.

Herbig, G. H., 1962. *Advances Astr. Astrophys.*, **1**, 47.

Herbig, G. H., 1978. In *The Origin of the Solar System*, ed. S. F. Dermott, p. 219. Wiley, New York.

Herbig, G. H., 1983. *Highlights of Astronomy*, **6**, 15.

Herbig, G. H., and Terndrup, D. M., 1986. *Astrophys. J.*, **307**, 609.

Heyer, M. H., 1988. *Astrophys. J.*, **324**, 311.

Heyer, M. H., Vrba, F. J., Snell, R. L., Schloerb, F. P., Strom, S. E., Goldsmith, P. F., and Strom, K. M., 1987. *Astrophys. J.*, **321**, 855.

Hunt, R., 1971. *Mon. Not. Roy. Astr. Soc.*, **154**, 141.

Kahn, F. D., 1974. *Astr. Astrophys.*, **37**, 149.

Lada, C. J., 1988. In *Galactic and Extragalactic Star Formation*, eds. R. E. Pudritz and M. Fich, p. 5. Kluwer, Dordrecht.

Larson, R. B., 1972. *Mon. Not. Roy. Astr. Soc.*, **157**, 121.

Larson, R. B., 1978. *Mon. Not. Roy. Astr. Soc.*, **184**, 69.

Larson, R. B., 1981. *Mon. Not. Roy. Astr. Soc.*, **194**, 809.

Larson, R. B., 1982. *Mon. Not. Roy. Astr. Soc.*, **200**, 159.

Larson, R. B., 1985. *Mon. Not. Roy. Astr. Soc.*, **214**, 379.

Larson, R. B., 1986. In *Stellar Populations*, eds. C. A. Norman, A. Renzini, and M. Tosi, p. 101. Cambridge Univ. Press, Cambridge.

Larson, R. B., 1987. In *Starbursts and Galaxy Evolution*, eds. T. X. Thuan, T. Montmerle, and J. Tran Thanh Van, p. 467. Editions Frontières, Gif sur Yvette.

Larson, R. B., 1989. In *The Formation and Evolution of Planetary Systems*, eds. H. A. Weaver and L. Danly, in press. Cambridge Univ. Press, Cambridge.

Larson, R. B., and Starrfield, S., 1971. *Astr. Astrophys.*, **13**, 190.

Loren, R. B., 1989. *Astrophys. J.*, **338**, 925.

Mateo, M., 1988. *Astrophys. J.*, **331**, 261.

Mathieu, R. D., Benson, P. J., Fuller, G. A., Myers, P. C., and Schild, R. E., 1988. *Astrophys. J.*, **330**, 385.

McKee, C. F., and Lin, J.-Y., 1988. In *Origin, Structure, and Evolution of Galaxies*, ed. Fang Li Zhi, p. 47. World Scientific, Singapore.

Miyama, S. M., Narita, S., and Hayashi, C., 1987a. *Prog. Theor. Phys.*, **78**, 1051.

Miyama, S. M., Narita, S., and Hayashi, C., 1987b. *Prog. Theor. Phys.*, **78**, 1273.

Moffat, A. F. J., Seggewiss, W., and Shara, M. M., 1985. *Astrophys. J.*, **295**, 109.

Mouschovias, T. Ch., 1987. In *Physical Processes in Interstellar Clouds*, eds. G. E. Morfill and M. Scholer, p. 453. Reidel, Dordrecht.

Myers, P. C., 1983. *Astrophys. J.*, **270**, 105.

Myers, P. C., and Goodman, A. A., 1988. *Astrophys. J.*, **329**, 392.

Nakano, T., 1989. *Astrophys. J.*, **344**, in press.

Salpeter, E. E., 1955. *Astrophys. J.*, **121**, 161.

Scalo, J. M., 1986. *Fundamentals of Cosmic Physics*, **11**, 1.

Scalo, J. M., 1987. In *Starbursts and Galaxy Evolution*, eds. T. X. Thuan, T. Montmerle, and J. Tran Thanh Van, p. 445. Editions Frontières, Gif sur Yvette.

Schneider, S., and Elmegreen, B. G., 1979. *Astrophys. J. Suppl.*, **41**, 87.

Shu, F. H., and Terebey, S., 1984. In *Cool Stars, Stellar Systems, and the Sun*, eds. S. L. Baliunas and L. Hartmann, p. 78. Springer-Verlag, Berlin.

Shu, F. H., Adams, F. C., and Lizano, S., 1987. *Ann. Rev. Astr. Astrophys.*, **25**, 23.

Shu, F. H., Lizano, S., Adams, F. C., and Ruden, S. P., 1988. In *Formation and Evolution of Low Mass Stars*, eds. A. K. Dupree and M. T. V. T. Lago, p. 123. Kluwer, Dordrecht.

Sofia, S., Kawaler, S., Larson, R., and Pinsonneault, M., 1989. In *Solar Interior and Atmosphere*, eds. A. N. Cox and M. S. Matthews, in press. Univ. of Arizona Press, Tucson.

Stahler, S. W., 1988. *Astrophys. J.*, **332**, 804.

Strom, S. E., Edwards, S., and Strom, K. M., 1989. In *The Formation and Evolution of Planetary Systems*, eds. H. A. Weaver and L. Danly, in press. Cambridge Univ. Press, Cambridge.

Tinsley, B. M., 1980. *Fundamentals of Cosmic Physics*, **5**, 287.

Weigelt, G., and Baier, G., 1985. *Astr. Astrophys.*, **150**, L18.

Wilking, B. A., and Lada, C. J., 1985. In *Protostars and Planets II*, eds. D. C. Black and M. S. Matthews, p. 297. Univ. of Arizona Press, Tucson.

Wolfire, M. G., and Cassinelli, J. P., 1987. *Astrophys. J.*, **319**, 850.

Zinnecker, H. C., 1982. In *Symposium on the Orion Nebula to Honor Henry Draper*, eds. A. E. Glassgold, P. J. Huggins, and E. L. Schucking, p. 226. *Ann. New York Acad. Sci*, Vol. **395**.

Discussion:

OSTERBROCK: Is the estimate that the star density in the 30 Doradus region is 10^2 the density in the Orion Nebula cluster based on the observed numbers of most luminous stars only, or on observed numbers of less luminous stars as well?

LARSON: It is based only on the most luminous stars, with an assumption that a "standard" IMF applies. However, even for the most luminous stars alone, the density in the 30 Doradus core is much higher than in the Trapezium region.

HEILES: In a forming star cloud processes such as solar flares produce enough cosmic rays to cause enough ionization to freeze the magnetic field into the gas?

LARSON: I am not aware that this possibility has been studied carefully. Such effects might well be important close to the central star. However, cosmic rays may not penetrate far into a typical protostellar disk, which is optically highly opaque.

SOLOMON: In one of your scenarios you said that high mass stars form after the low mass stars from "leftover" gas implying high mass star formation was limited by the quantity of available gas. However molecular clouds which are gravitationally bound have 10^4, 10^5 or even 10^6 solar masses of material available to form stars. There is NO shortage of gas.

LARSON: If massive stars form in dense clusters in the cores of molecular clouds, the amount of gas in the region where a cluster forms is much less than the total mass. This gas may become strongly depleted by star formation, leaving less to make the most massive stars. However, regardless of whether gas depletion is important, I still think it is likely that the more massive stars form at later stages in the evolution of star forming regions.

KHAN: You did not mention the importance of magnetic fields in the transport of angular momentum. Is this the currently accepted view?

LARSON: It was estimated by Hayashi that throughout most of a protostellar disk, field decay by Ohmic dissipation is very rapid, and this implies that magnetic fields cannot be dynamically important. This conclusion may not apply to the outermost part of a disk, or to the part close to the central star. The outermost part (beyond 100 AU) could retain some magnetic coupling to the interstellar medium, and the innermost part could be magnetically coupled to the central star, but other mechanisms appear necessary to transport angular momentum in between.

GUIDO'S JUBILEE
IAU COLLOQUIUM N.º 120
Instituto de Astrofísica de Andalucía,
Granada (Spain)
April 17-21, 1989
Structure and Dynamics of the Interstellar Medium

STAR FORMATION

Francesco Palla
Osservatorio Astrofisico di Arcetri
L.go E. Fermi,5 – 50125 Firenze

Prologue

Judging from the poster that the Organizing Committee has selected to announce the celebration of Guido Münch Jubilee, one can easily conclude that the main characteristics of the process of star formation as emerged in recent years through the combined efforts of multiwavelengths studies of molecular clouds, were already known, here in Granada, several centuries ago to the masters who built and enriched the enigmatic palace of the Alhambra. As we can appreciate from a quick inspection of the picture, it is rather obvious to infer that stars are the byproduct of a quite complex series of phenomena, each connected to, and somewhat dependent on, the others. Also, stars do not form in isolation, but rather in clusters or associations, with a strong tendency for the largest ones, also the most massive ones, to sit in the middle of the distribution. Moreover, smaller and less massive stars outnumber their massive counterparts, apparently obeying a power–law distribution. Finally, but with the benefit of doubt, it appears that the idea that the whole process reflects an intrinsic fractal nature was also put forward at the time. With this background in mind, let us now turn to the new emerging aspects of the study of star formation.

1. Introduction

Star formation is a fundamental chapter in the study of the structure and dynamics of the interstellar medium. As circumstantial observational evidence has been accumulating over the years, we are now in the position to pinpoint the essential aspects of the complex phenomena that concur in the process of the formation of new stars out of the raw interstellar material. Observations have strongly indicated that the sites of current star formation in the Galaxy occur within the densest regions, called cores or clumps, of molecular cloud complexes. On the large scale, molecular clouds are dominated by forces (rotation, magnetic field, turbulence) that counteract gravity in the tendency to initiate dynamical collapse to form stars. At the small scale of the cold cores, eventually gravity takes over, due to the slow diffusion of the restoring forces, and star formation actually ensues. Several informative accounts on both the observational and theoretical aspects have been published recently and useful references are the reviews by Shu, Adams and Lizano (1987), Mouschovias (1987), Pudritz (1988), Walmsley (1988), and Zinnecker (1988).

In this review, rather than trying to cover all the different topics, an almost impossible task anyway, I prefer to concentrate on just one particular aspect that represents an interesting example of recent developments in the field and that has not been covered in previous summaries: the study of young stellar clusters still deeply embedded in molecular clouds. In my view, this topic is extremely instructive since *a)* it offers the possibility of addressing the problem of star formation in groups and of probing *in situ* the origin and development of the luminosity function of the newborn stars; and *b)* it provides the unique opportunity to test the predictions of the theoretical models, mainly developed for individual objects in rather idealized cases, on very large and statistically significant samples. It is important to stress that the derivation of both the luminosity and initial mass functions has been hampered by the limitation of resting on samples that were either heterogeneous, as in the case of field stars, or homogeneous but relatively old, as for open clusters, or young but rather poor in objects, such as the OB associations (cf. the discussion by Dr. Larson in this volume). In most of these cases the star formation process had long been exhausted and other dynamical phenomena might have had an influence in the shape of the observed distributions.

The recent developments brought about by the advent of near–infrared (NIR) array cameras offer the unsurpassed advantage of removing, at least in principle, some of these fundamental constraints, and the first scientific results are already circulating (cf. Gatley, DePoy and Flower 1988 for a first account). Admitedly, the whole subject of the luminosity function of embedded clusters is still in its infancy and subject to rapid, and perhaps controversial developments: therefore, some, or most, of the results and conclusions presented here must be taken *cum grano salis* and should be considerd only as tentative and preliminary. As a proof, the largest part of the references quoted in this article are still in the press, or circulate as preprints. However, the excitment for this new approach to the study of star formation makes the attempt to summarize the present status worthwhile.

2. Some observations

The primary motivation for this discussion comes from the recent publication of the luminosity function of the stellar cluster in the IR core of the ρ Ophiuchi molecular complex by Wilking, Lada and Young (1989, hereafter WLY), supplemented by a *Letter* on the same topic by Rieke, Ashok and Boyle (1989, hereafter RAB). The former study represents the last of a series of systematic studies of the embedded population of this archetype of low mass star forming region, together with the Taurus–Auriga complex (Lada and Wilking 1984, Young et al. 1986, and Myers et al. 1987). Unlike Taurus–Auriga, the ρ Ophiuchi complex is characterized by a large centrally condensed core rich of young stellar objects (YSO), and by an unusually high efficiency of star formation, that has been interpreted by Wilking and Lada (1983) as suggestive of the formation of a bound cluster.

The IR core in ρ Oph has been known to harbor a high concentration of YSOs since many years (e.g. Vrba et al. 1975; Elias 1978). A different approach to the search of embedded clusters, and to the study of the frequency of their occurence within molecular clouds, has been taken by E. Lada and collaborators (1989), who have carried out an unbiased and systematic survey for dense cores traced by the high excitation CS molecule over a large region of the Orion B molecular cloud. This has led to the discovery of several individual clumps, that when deeply imaged in the NIR have revealed the presence of four rich stellar clusters within or near the massive clumps. Similar studies, but on different objects, have been performed by other groups, such as Eiroa and Casali

(1989) who have studied the IR core of the Serpens molecular cloud, and MacCaughran et al. (1989) who have attacked the Trapezium region. All of them have proved to be extremely succesful in discovering a series of very young clusters, containing hundreds of stars each, and in allowing the derivation of a luminosity function for the first time.

Before discussing in detail the properties shown by the newly derived LFs, let us summarize the necessary steps to obtain a reliable ILF, and possibly the IMF. Briefly, three steps are needed (cf. WLY for a thorough discussion):

- to derive the spectral energy distribution (SED) of each object and obtain an estimate of the bolometric luminosity;
- to classify empirically the SEDs according to some common feature, for example their shapes;
- to transform the empirical scheme into an evolutionary sequence, to possibly ascertain the physical nature of the individual objects.

While the first two steps are basically an observational exercise, very strenuous indeed, the third one relies on theoretical models that predict the emergent SED in various evolutionary phases. In fact, it is very likely to expect that the population of the embedded clusters is represented by a mixture of the youngest protostars, still in their accretion phase, and the more mature pre–main-sequence stars. These two populations have different luminosity sources and thus different SEDs. Since theoretical models of this kind have been discussed thoroughly in the literature (e.g. Adams, Lada and Shu 1988; Myers et al. 1987; Lada 1988), they need not to be repeated here.

In analogy with the derivation of the IMF for field stars in the solar neighborhood, once the LF has been determined, the MF can be obtained by the expression:

$$dN/dm = dN/dm_\lambda \times dm_\lambda/dM_{bol} \times dM_{bol}/dm, \tag{1}$$

where, as usual, the first term on the rhs represents the distribution function at a given wavelength; the second term is the analogous of the bolometric correction for stars; and the last term implies the mass–luminosity relation, that unlike the stellar case has a temporal dependence.

The knowledge of the individual SED yields the first two terms of eq. (1). Observationally, this is not an easy task since, in order to obtain the bolometric luminosity, it is necessary to cover a very large wavelength interval. In the specific case of the ρ Ophiuchi cluster, WLY have presented results in a range from 0.4 to 100μm, even though in practice the available interval is limited to $\lambda \geq 1.25\mu$m, due to extinction problems, and to $\lambda \leq 60\mu$m, due to source confusion. Considering that some sources exhibit SEDs still raising at the longest wavelengths ($\lambda = 100\mu$m), it turns out that the bolometric correction in the worst cases would amount to a factor of two (Myers et al. 1987; Strom et al. 1988; WLY 1989). Recent measurements in the sub–mm at $\lambda = 350\mu$m, beyond the peak of the emission, have quantitatively confirmed previous estimates (Ladd et al. 1989).

Based on the theoretical background developed by Adams and Shu, Lada (1987) has suggested a classification scheme of the SEDs that would correspond to a truly evolutionary scheme. In this way, the nature of each object can be tentatively assigned and the last term of eq. (1), i.e. the mass–luminosity relation, can also be evaluated. Clearly, most of the uncertainties in the derivation of the IMF reside here. While the situation is not so bad for PMS stars, since $L_{bol} = L_{bol}(M_\star)$ and can de derived by existing evolutionary tracks if the age of the cluster is known; the case for protostars is considerably more intricate. In fact, for protostars the luminosity depends upon an additional parameter, the mass accretion rate, that in turn is related to the global properties of the ambient cloud (temperature, rotation, turbulence), which are difficult to know *a priori*.

3. The LF of embedded clusters

Luminosity functions of embedded clusters have been presented by various authors. For example, Straw et al. (1989) have made a comprehensive study of the stellar mass distribution in several centers of star formation activity associated with the molecular cloud complex NGC6334. Despite the rather high sensitivity of the survey (limited to K–mag=13.5) and the possibility of detecting faint sources, the resulting mass distribution is completely sampled only for stars with mass $\geq 4M_{\odot}$. Thus, all the conclusions are restricted only to the intermediate and high mass portion of the LF.

Most of the challenge presented by the study of embedded clusters, however, comes from the study of regions in which the low mass population can be fully sampled down to masses where the LF could start showing some interesting and unexpected features. The LF for 74 members of the ρ Oph cluster derived by WLY is shown in Figure 1. Within each luminosity interval the objects are grouped according to their SED class. Also included in the histogram are sources for which only an u.l. to the total luminosity could be derived. The solid line labelled ILF denotes the initial luminosity function corresponding to the IMF computed assuming that the embedded objects have a mass-luminosity relation appropriate to H–burning main sequence stars, an assumption hard to justify. According to the authors, the most conspicuous result is the evidence of a luminosity segregation of the objects: namely, class I objects, that in the evolutionary sequence would represent the protostar population, appear to dominate the intermediate range of luminosities ($L \geq 6L_{\odot}$); while class II objects, corresponding to PMS stars, are more numerous at low luminosities. Two other properties of the LF appear to be: a *paucity* of stars of intermediate luminosity when compared with the field stars LF, and after taking into account incomplete sampling and luminosity evolution; and a *dearth* of objects of low luminosity. The two fundamental questions raised by these results are then:

1) *Is the turnover of the LF a real feature, as in the case of field stars, or is it simply due to incompleteness?*

2) *Is the deficiency of intermediate luminosity objects something to be expected on the basis of the knowledge of PMS evolution?*

In the following, I will examine in detail these two aspects, that all future discussions on the LF of embedded clusters will have to deal with.

3a. The low luminosity tail

This is the region of the spectrum where the contribution of deep imaging at NIR wavelengths using array cameras is vital. Some authors have presented LFs that indeed indicate a turnover at some faint magnitudes. As an example, MacCaughran et al.

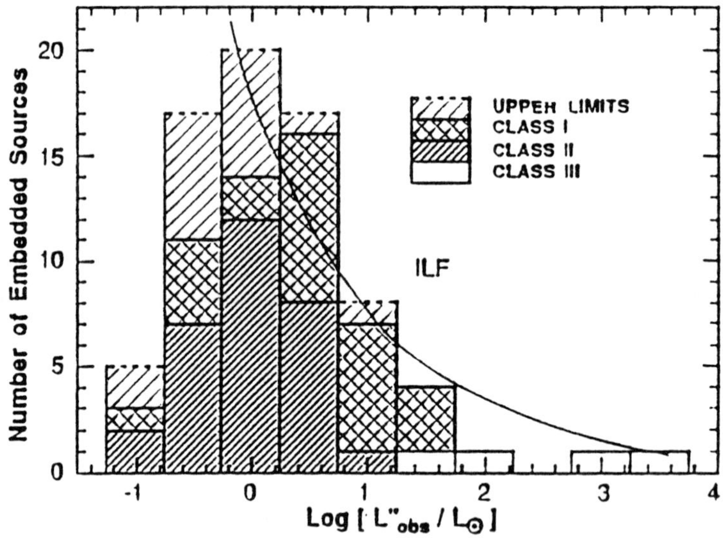

Figure 1. The luminosity function for 74 members of the ρ Ophiuchi IR core (from Wilking et al. 1989).

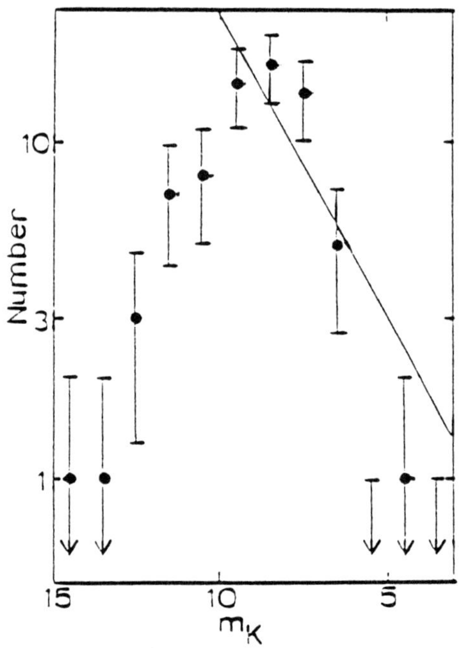

Figure 2. Distribution of K magnitudes. The straight line represents a power law fit to the source counts for $m_K < 10$ (from Rieke et al. 1989).

(1989) have surveyed the central 5'x5' of the Trapezium cluster with an almost complete sampling of the ≈ 500 members. They find a sharp cut–off at a magnitude $K \sim 13.0$, two magnitudes above the detection limit of the survey ($m_K \sim 15.5$). A similar result has been obtained by E. Lada (1989) in the case of the embedded cluster in NGC 2024.

Returning to the case raised by the ρ Oph cloud, RAB have carried out a survey of one of the regions mapped by WLY using a 68x68 pixel NIR camera. The resulting distribution function is shown in Figure 2. Despite the fact that the survey's detection limit at $m_K = 15$ is three times deeper than that of WLY , no new sources were discovered, while all the sources previously identified were indeed detected. The turnover at $m_K \sim 10$ is clearly evident, and RAB estimate that an extrapolation to magnitudes fainter than $m_K \sim 12$ would have predicted between 8 and 14 new sources , against the 2 or 3 actually detected. Of course, if this result is confirmed by further deep imaging covering a more extended region, it would represent the first independent evidence of an *intrinsic* turnover in the shape of the LF. Recently, Barsony and Burton (private communication; see also the discussion at the end of the article) claim the discovery of a large number of sources in the same field explored by WLY, but the observations apparently suffered of some problems and cannot be considered definitive as yet.

By combining the results at 2.2μm with those of WLY at longer wavelengths to derive the bolometric luminosity, RAB find that the turnover occurs at a luminosity between 0.1 and $0.01 L_\odot$, and that, therefore, the dearth in the LF refers to stars with $L \sim 10^{-2} - 10^{-3} L_\odot$, corresponding to $m_k \sim 12.5$ and $m_K \sim 14.5$ at the distance of the cloud (160 pc). The uncertainty in the exact value of the peak luminosity stems from the small statistics of the low luminosity objects, but further observations will eventually help to remove it.

At present, little can be said of the mass corresponding to the luminosity at turnover, mainly due to the uncertainties on the nature of the objects. Under the most conservative assumption that all the stars are in the PMS phase, knowing the age of the cluster (between $3 \cdot 10^6$ and 10^7 yr) and using stellar evolutionary tracks leads to an estimate of the mass in the interval $0.2 - 0.4 M_\odot$. This could be considered consistent with recent determinations of the IMF in the solar neighborhood. In this respect, it is noteworthy mentioning the great improvements in the definition of the low mass end of the IMF obtained by Leggett and Hawkins (1988) with the IR observations of M dawrfs towards the South Galactic Pole and the Hyades cluster. Their results are collected in Figure 3 and show the characteristic peak at luminosities corresponding to stars of $0.2 - 0.3 M_\odot$, and the subsequent decline down to $m \sim 0.1 M_\odot$. The reality of the rise at even fainter magnitudes, with its implications for the existence of a population of brown dwarfs, is still under debate (cf. Stobie et al. 1989, and Kroup and Tout 1989 for a different explanation of the origin of the knee in the LF); however, the turnover and the flattening appear to be an intrinsic feature of the ILF and IMF.

The indication of a similar behavior even in the case of extremely young clusters would tend to imply the existence of a preferred mass scale in the star formation process, i.e. that of the peak of the distribution. In his review, Dr. Larson argues that this scale is imprinted by the fragmentation of the large scale molecular clouds into clumps or cores of sub–stellar mass. This viewpoint differs from that advocated by Shu and collaborators (e.g. Shu, Lizano and Adams 1987) for which the mass scale is determined by processes related to the evolution of the protostar itself.

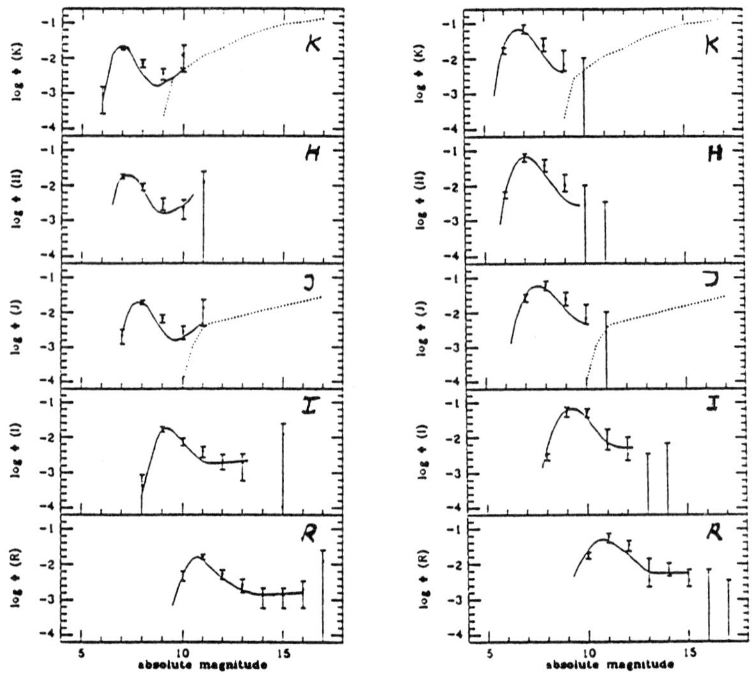

Figure 3. The Log of the luminosity function at different wavelengths for the South Galactic Pole (*left panel*) and the Hyades cluster (*right panel*) (from Leggett and Hawkins 1988).

3b. The intermediate luminosity tail

The possibility that the observed LF could show a paucity of stars of intermediate luminosity is intriguing, especially since some recent developments in the theory of protostellar and pre–main-sequence evolution of stars of mass larger than solar predict that such an effect could indeed arise (Stahler 1989; Palla and Stahler 1989). Although the observational evidence is still scant, it is worthwhile to look more in detail at the theoretical background of these results.

The basic point is that PMS stars of mass between $2M_\odot$ and $5M_\odot$, starting their contracting phase from the proper initial conditions inherited at the end of the accretion phase, are subject to a rapid *luminosity evolution* from a low to a high state. The luminosity increase takes place in a very short time compared to the typical Kelvin–Helmoltz evolutionary time, while its magnitude depends on the star mass. According to rough estimates, a $5M_\odot$ PMS star increases its surface luminosity by up to a factor of 30, going from $L \sim 15L_\odot$ at the beginning of the evolution to $L \sim 500L_\odot$ at the epoch of the flare. The most important consequence of this effect is that PMS stars of intermediate mass would first appear in the HR diagram well below the position predicted by the classical evolutionary tracks in the Hayashi phase, and only after the luminosity eruption has occured they would join the proper portion of the radiative track. This is schematically illustrated in Figure 4.

To understand how this behavior is possible one has to reconsider some basic properties of the standard theory of PMS evolution (e.g. Hayashi, Hoshi and Sugimoto 1962;

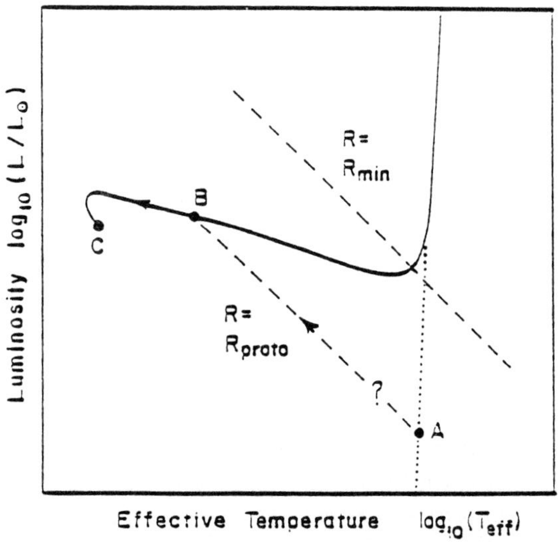

Fig 4. Schematic evolution in the HR diagram of a PMS star of $M \geq 2M_\odot$. R_{min} refers to the minimum radius at the bottom of the classic Hayashi track. R_{proto} is the initial radius of the star, acquired at the end of the protostellar phase (from Stahler 1989).

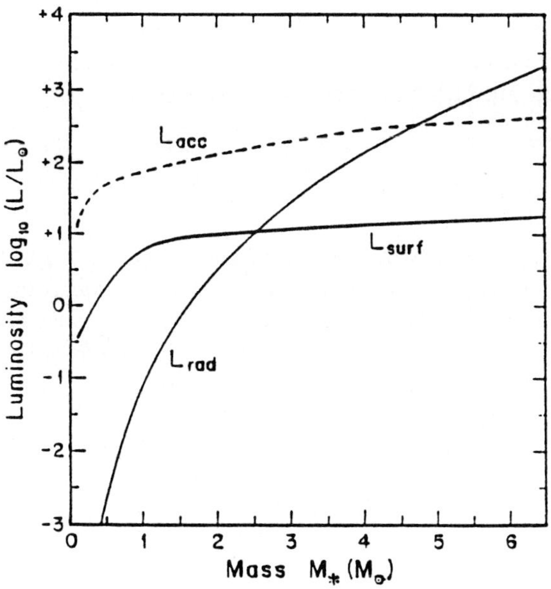

Figure 5. Luminosities of PMS stars at the beginning of their evolution, as a function of mass. L_{acc} refers to the accretion luminosity and is shown for comparison (from Stahler 1989).

Cameron 1962). According to the classical picture, PMS begin their evolution as fully convective objects due to the large initial radii, given by: $R_0 = 50R_\odot(M_\star/M_\odot)$. This relation implies that a star has a surface luminosity, L_{surf}, that greatly exceeds the luminosity that can be carried internally by radiation, L_{rad}, so that it will immediately turn convectively unstable. Stahler (1989) points out, however, that according to the results of the numerical studies of protostellar accretion (Larson 1972; Winkler and Newman 1980; Stahler, Shu and Taam 1980), radii as large as those predicted by the above relation are never reached: they typically remain an order of magnitude smaller, up to protostellar cores of $1M_\odot$. Palla and Stahler (1989) have extended previous calculations to follow the accretion phase up to cores of $5M_\odot$ and found that the radius stays fixed at $\approx 5R_\odot$, a factor of 50 less than that predicted classically. Since the surface temperature of the PMS stars in this mass range is always locked to $T_{eff} \sim 4000$K, this result implies that L_{surf} will also remain constant. On the other hand, L_{rad} is a very sensitive function of the mass, with a dependence $L_{rad} \sim M_\star^{5.5} R_\star^{-0.5}$. Therefore, as the core mass grows, L_{rad} increases dramatically and eventually overtakes L_{surf}: at this point the star will become radiatively stable, and due to the small value of the radius, will have a low luminosity when it first becomes optically visible (cf. point A of Fig. 4).

The transition where $L_{rad} \approx L_{surf}$ occurs at a core mass between $2M_\odot$ and $3M_\odot$, as can be seen in Figure 5. In order to achieve the full luminosity corresponding to L_{rad}, the cores have to wait until the internal luminosity relax to the surface due to its non-homologous contraction (classical PMS stars descend the Hayashi track homologously). The time needed for the thermal wave to reach the surface (i.e. to go from point A to point B of Fig. 4) is about the same as that needed to reach the zero–age main sequence, thus implying that there is an equal probability of finding PMS stars in the high as in the low luminosity state. Stahler (1989) discusses the possible consequences of this behavior on the mass and age estimates of young clusters, and also the predicted deficiency of sources with luminosities between 10 and $10^3 L_\odot$.

4. Conclusions and future directions

The main conclusion to be drawn from the present discussion is that the study of embedded young stellar clusters in molecular clouds, with its deep implications for the understanding of the origin and development of the IMF, has now become an integral part of the research in the field of star formation. The use of IR array cameras to map extended regions of high obscuration with high sensitivity has been instrumental in overcoming the major obstacle for the development of the field, and the advent of arrays sensitive to longer wavelengths in the near future will help in removing the uncertainties on some of the issues discussed here. On the theoretical side, these observations act as a valid test to the wealth of ideas mainly developed for different contexts, and provide an effective stimulus to better refine the numerical models on which much of the interpretation rests.

The tone of this presentation has been left intentionally optimistic, perhaps even slightly overoptimistic, and little mention has been paid to the uncertainties associated with the various steps that lead to the derivation of the ILF and IMF. Among them, serious problems are caused by the difficulty of discriminating between embedded and field stars; the clustering of a large number of sources in small fields; the fact that intracluster and foreground reddening can be very non uniform; the separation between local and global extinction, and so forth. Also, it is necessary to remove uncertainties in the estimate of the bolometric luminosities by means of very deep surveys in the far-infrared (ISO, SIRTF). In order to improve source identification and remove confusion, high resolution observations in the FIR are also required. Of utmost importance is to find reliable indicators to discriminate between *protostars* and *pre-main-sequence stars*: so far, only the SEDs have been used at this purpose, but perhaps even spectral features (at 3.1 and 9.7μm) can be used as proper and unambiguous diagnostics. Finally, the comparison between the expected flux profiles from theoretical models and the spatial emission maps obtained in the submm/mm region will prove extremely useful (Adams 1989).

Epilogue

As astronomers, we have fully appreciated the impressive images obtained with the IR cameras and presented at this Conference, that have provided us with new eyes to penetrate into the inaccessible regions where the formation of new stars takes place. As participants to the meeting to honor Guido Münch, now we can better understand the deep significance of the old local saying:

"*...no hay en la vida nada como la pena de ser ciego en Granada*".

Acknowledgements

The preparation of this review has benefited from many conversations with C. Eiroa, N. Evans, C. Giovanardi, E. Lada, S. Lizano, A. Natta, S. Stahler, and H. Zinnecker, all of whom I would like to thank. Special thanks go to Maria Massi. It is a pleasure also to thank the LOC and the SOC for their warm hospitality.

References

Adams, F.A. 1989, in *Infrared Spectroscopy in Astronomy*, XXII Eslab Symp., ESA Publ. Div., in press.

Adams, F.C., Lada, C.J., Shu, F.H. 1988, *Astrophys. J.*, **326**, 865.

Cameron, A.G.W. 1962, *Icarus*, **1**, 13.

Eiroa, C., Casali, M.M. 1989, in ESO Workshop *Low Mass Star Formation and Pre-Main Sequence Evolution*, ed. Bo Reipurth, in press.

Elias, J.H. 1978, *Astrophys. J.*, **224**, 453.

Gatley, I., DePoy, D.L., Fowler, A.M. 1988, *Science*, **242**, 1217.

Hayashi, C., Hoshi, R., Sugimoto, D. 1962, *Progr. Theor. Phys. Suppl. N.22.*

Kroupa, P., Tout, C.A. 1989, *M.N.R.A.S.*, in press.

Lada, C.J. 1988, in *Galactic and Extragalactic Star Formation*, eds. R.E. Pudritz and M. Fich., Kluwer Acad. Publ., 23.

Lada, C.J., Wilking, B.A. 1984, *Astrophys. J.*, **287**, 610.

Lada, E.A. 1989, in *Physical Processes in Fragmentation and Star Formation*, eds. R. Capuzzo–Dolcetta, A. Di Fazio and C. Chiosi, Reidel Publ. Co., in press.

Lada, E.A., Evans, N.J., DePoy, D.L., Gatley, I. 1988, *Astrophys. J.*, submitted.

Ladd, E.F., Myers, P.C., Adams, F.C., Padman, R., Davidson, J.A., Fuller, G.A. 1989, *Astrophys. J.*, in press.

Larson, R.B. 1972, *M.N.R.A.S.*, **157**, 121.

Leggett, S.K., Hawkins, M.R.S. 1988, *M.N.R.A.S.*, **234**, 1065.

MacCaughran, M.J., Aspin, C., McLean, I.S., Zinnecker, H. 1989, in ESO Workshop *Low Mass Star Formation and Pre–Main Sequence Evolution*, ed. Bo Reipurth, in press.

Mouschovias, Th. 1987, in *Physical Processes in Interstellar Clouds*, eds. G.E. Morfill and M. Scholer, Reidel Publ. Co., 453.

Myers, P.M., Fuller, G.A., Mathieu, R.D., Beichman, C.A., Benson, P.J., Schild, R.E., Emerson, J.P. 1987, *Astrophys. J.*, **319**, 340.

Palla, F., Stahler, S.W. 1989, in preparation.

Pudritz, R.E. 1988, in *Galactic and Extragalactic Star Formation*, eds. R.E. Pudritz and M. Fich, Reidel Publ. Co., 135.

Rieke, G.H., Ashok, N.M., Boyle, R.P. 1989, *Astrophys. J. Lett.*, **339**, L71.

Shu, F.H., Adams, F.C., Lizano, S. 1987, *Ann.Rev.Astr.Ap.*, **25**, 23.

Stahler, S.W. 1989, *Astrophys. J.*, in press.

Stahler, S.W., Shu, F.H., Taam, R.E. 1980, *Astrophys. J.*, **241**, 637.

Stobie, R.S., Ishida, K., Peacock, J.A. 1989, *M.N.R.A.S.*, **238**, 709.

Straw, S.M., Hyland, A.R., McGregor, P.J. 1989, *Astrophys. J. Suppl. Ser.*, **69**, 99.

Strom, K.M., Strom, S.E., Kenyon, S.J., Hartmann, L. 1988, *Astron. J.*, **95**, 534.

Vrba, F.J., Strom, K.M., Strom, S.E., Grasdalen, G.L. 1975, *Astrophys. J.*, **197**, 77.

Walmsley, C.M. 1988, in *Galactic and Extragalactic Star Formation*, eds. R.E. Pudritz and M. Fich, Reidel Publ. Co., 181.

Wilking, B.A., Lada, C.J. 1983, *Astrophys. J.*, **274**, 698.

Wilking, B.A., Lada, C.J., Young, E.T. 1989, *Astrophys. J.*, **340**, 823.

Winkler, K.H., Newman, M.J. 1980, *Astrophys. J.*, **236**, 201.

Young, E.T., Lada, C.J., Wilking, B.A. 1986, *Astrophys. J. Lett.*, **304**, L45.

Zinnecker, H. 1988, in *Evolutionary Phenomena in Galaxies*, ed. J. Beckman, Cambridge Univ. Press, 338.

Discussion:

TERLEVICH, E.: It is very dangerous to base your IMF on luminosity functions of cluster cores. It has been predicted long ago (analitically) and more recently confirmed (numerically) that the dynamical evolution of star clusters tends to deplete the core of light stars and populate the halo with them, in time scales comparable to a few (1-2) crossing times. In order to get a representative IMF therefore, you have to sample your cluster stars to regions, at least 4 or 5 core radii. In other words, you must be concerned about completeness, not only in deeth but also on width. It is very difficult to observe the haloes of open star clusters, but, as they say in the North of England: "where there is muck, there is brass".

PALLA: Although in principle your remark is correct, I think it is still premature to apply it to the ρ Oph IR cluster for which the suggestion of a forming bound cluster is only tentative. Your results clearly apply to systems where both the total number of stars and the radius are known.

BARSONY: The Rieke's result that there is a turnover at low luminosities in the initial luminosity function based on their 2μm data is wrong. We (M.Burton, A.Russell, J.Carlstrom) have unpublished data obtained with the KPNO 2μm camera which show at least 20 new sources in the same field (10'×10') as the previously published survey of Wilking and Lada. This unpublished result is entirely consistent with the expected, accepted IMF.

PALLA: Rieke's survey is admidtedly limited to a smaller region than that sampled by Wilking et al. It is clear that theirs is the first attempt to fully image the ρ Oph IR cluster: future observations, like yours, will certainly help clayifying the important issue of a possible turnover at low luminosities.

TURBULENCE IN INTERSTELLAR CLOUDS

E. Falgarone[1,2]

[1] California Institute of Technology, 405-47, Pasadena, CA91125, USA

[2] Radioastronomie, Ecole Normale Supérieure,

24 rue Lhomond, 75235 Paris Cedex 05, France

Above masses of the order of $100\, M_{\odot}$, molecular clouds have masses and sizes which scale like those of self-gravitating polytropes bounded by an external pervading pressure. It is unlikely that this scaling is due to mere observational bias. But the physics underlying this behaviour is far from being understood. In particular, the possible contribution of turbulence to both the ambient pressure and the internal pressure (whose dependence with the density would mimic a polytropic behavior) is a difficult and much debated issue. The clouds mass, size and internal velocity dispersion are such that they are observed to be in approximate virial balance between their self-gravity, the surface energy term due to the ambient pressure and their internal energy. The latter is dominated by the kinetic energy of disordered internal motions. However, there has been little evidence so far that these motions are actually turbulent rather than simply disordered. The transition to turbulence in a flow occurs when the non linear advection term in the momentum equation, $\mathbf{v}.\nabla\mathbf{v}$, considerably exceeds the viscous dissipation term, $\nu\Delta\mathbf{v}$ (where ν is the kinematic viscosity). Non linearities therefore dominate the physics of a turbulent flow and the velocities are not randomly distributed. Most of the previous attempts to determine a well-defined correlation length in the velocity field (Kleiner and Dickman 1985, *a* and *b*; Scalo 1984), which is predicted to be close to the scale at which the energy is injected, or to characterize the expected hierarchical structure (Pérault et al. 1986) have been plagued by the lack of dynamical range in the data set and the range of scales over which the correlation functions have been computed. The most plausible determination, that of Kleiner and Dickman (1987) who claim to have found a correlation length of $0.2\,pc$ in the Taurus cloud, gives a result which is so close to the angular resolution of the observations that it is doubtful.

I choose here not to review the various attempts discussed in Dickman (1985) nor to cover the topics of the sources and sinks of turbulence in molecular clouds, these topics having been reviewed quite extensively by Scalo (1987) and Falgarone and Puget (1988). Instead, I will discuss the possible observational signatures of turbulence in the interiors of clouds. One of them is the *element of surprise in the time history of the velocity field*, in other words the existence of very localized regions of enhanced ephemeral vorticity which reflects the familiar and conspicuous intermittency in space and time of the turbulent activity. I will make ample use of results of measurements on laboratory flows and in the atmosphere and compare them with observations of interstellar clouds. Molecular clouds directly interacting with energetic events like the expansion of supernovae remnants and HII regions are not considered in what follows, even though those events count as a major source of kinetic energy and momentum in interstellar turbulence.

The first part of this review provides partial answers to the question: are the disordered motions within interstellar clouds turbulent? Then, considering that there exist a few robust indications that turbulence is indeed present, the second part deals with possible important consequences of the turbulent nature of the gas motions to the physical and chemical evolution of the clouds.

1. Are the internal motions in interstellar clouds turbulent?

The maps of Figure 1 illustrate the intrinsic difficulty of defining a cloud edge without specifying the linear scale it is referred to. It displays the projected shapes at two different scales of the edges of a molecular cloud in the Taurus complex. The area selected in the large scale map of Ungerechts and Thaddeus (1987) for observations at higher angular resolution (Falgarone and Phillips 1989) has an average colomn density $N_{\mathrm{H}_2} \sim 9\,10^{20}\,cm^{-2}$ (using the galactic average conversion factor between ^{12}CO integrated brightness and H_2 column density, N_{H_2}/W_{CO} = $2.6\,10^{20}\,cm^{-2}/(K\,km\,s^{-1})^{-1}$, Bloemen et al. 1986). In the undersampled map of Figure 1b, islands of ^{12}CO (J=2–1) emission appear with a highly contrasted brightness distribution. A crude comparison of the range of H_2 column densities in these two maps can be made by assuming that the integrated brightness ratio between the ^{12}CO (J=2–1) and (J=1–0) lines ranges between 0.5 and 0.8 in the low density edges ($n_{\mathrm{H}_2} < 10^3\,cm^{-3}$). The first contour of Figure 1b thus corresponds to an H_2 column density comparable to that of the third contour of Figure 1a but is much more convoluted. The striking property of these maps is that their degree of convolutedness looks self-similar and it is tempting to use the tool of fractal geometry to try to characterize more quantitatively this kind of structure.

1.1 The fractal geometry of the cloud edges.

In his approach to fractal surfaces and curves, Lovejoy (1982) defines a fractal dimension D which relates the perimeter P of a closed curve to the area A enclosed within P by the relation $P \propto A^{D/2}$. D = 1 (resp. 2) for classical curves (resp. surfaces) like a circle (resp. sphere) and tends toward D = 2 (resp. 3) for an extremely convoluted curve (resp. surface) which would fill the entire plane (resp. space). Lovejoy measured a fractal dimension $D = 1.35 \pm 0.05$ for rain areas and clouds at scales ranging between $1\,km$ and $10^4\,km$. Table I summarizes the results obtained by the same technique on the fractal dimension of the contours of a variety of tracers of column density of interstellar clouds. The dimension derived from ^{12}CO line observations, in which the ^{12}CO (J=1–0) data set is from Ungerechts and Thaddeus (1987) and the ^{12}CO (J=2–1) and (J=3–2) from Falgarone, Phillips and Walker (1989) takes into account the effect of the various angular resolutions of the maps. N_d is the number of decades covered by the values of the perimeter P.

Table I

Tracer	Field	D	N_d	Reference
IRAS 100μm	HLCs	1.26 ± 0.04	2	Bazell and Désert (1988)
IRAS 100μm	Taurus	1.4	2	Scalo (private communication)
^{12}CO lines	Taurus	1.34 ± 0.02	4	Falgarone et al. (1989)
HI 21cm	HVCs	~ 1.6		Wakker (private communication)

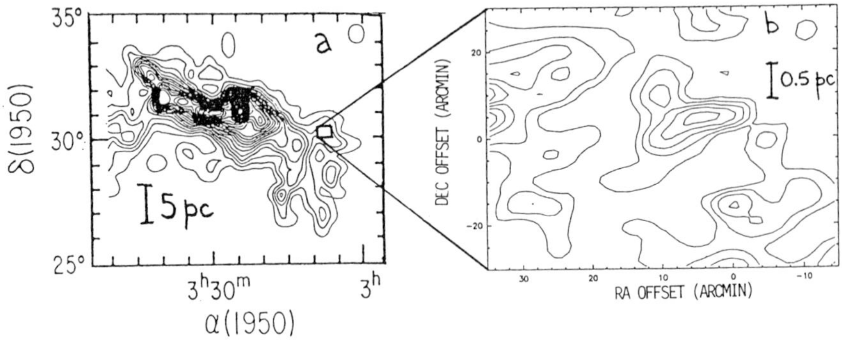

Figure 1. *(a)* Velocity-integrated intensity of ^{12}CO (J=1–0) in the Taurus complex from Ungerechts and Thaddeus (1987). First contour $0.5\,K\,km\,s^{-1}$, step $1.5\,K\,km\,s^{-1}$. *(b)* Velocity-integrated intensity of ^{12}CO (J=2–1) emission observed at the Caltech Submillimeter Observatory (HPBW=30", map sampled every 5'). First contour $2\,K\,km\,s^{-1}$, step $2\,K\,km\,s^{-1}$ (Falgarone et al. 1989).

The differences between the values of D obtained with the different tracers of the gas column density may be real and significant but it is surprising that all the tracers provide values clearly larger than 1. Indeed, the $100\mu m$ brightness emission below a level of a few $10\,MJy\,sr^{-1}$ is a tracer of the total column density of large dust grains and therefore of the total column density of H nuclei; ^{12}CO (J=2–1) traces that of the CO molecules in the $J = 2$ rotational level but may be considered as an indirect tracer of H_2 column density in low brightness clouds; the HI emission mapped by Wakker with the Westerbork interferometer is likely to be an exclusive tracer of the cold atomic component within the observed clouds. This suggests that there is no fundamental difference between the dominant physics which controls the shape of the edges of clouds, whether they are predominantly molecular or atomic.

The possible link with turbulence is the following. Sreenivasan and Meneveau (1986) and Meneveau (1989) have measured the fractal dimension of a variety of surfaces in turbulent flows. They do not measure the dimension of the projection of these surfaces onto a plane but that of the curve obtained by 2-dimensional slicing of a given interface. Their remarkable result is that, whether it is the interface between the vortex/no-vortex regions or the isoconcentration surfaces or the isodissipation surfaces, they find a similar dimension for the curve obtained by 2-dimensional slicing

$$D' = 1.36 \pm 0.05.$$

There is no mathematical argument which clearly connects the dimension D of the curve obtained by the projection on a plane of a fractal surface of dimension D_{3d} in the Euclidian space and that D' obtained by 2-dimensional slicing of the same surface by a plane, but limited experiments (Meneveau 1989) suggest that $D' \leq D$. If isotropy is assumed, the fractal dimension of the surface itself is related to that of its projection by $D_{3d} = D + 1$.

Meneveau (1989) gives a simple physical argument for such a dimension. Let $\Phi(r)$ be the flux of any quantity C which can be transported by turbulent diffusion across the surface element $S(r)$ of a fractal surface of dimension D (r is the resolution element with which the surface and therefore the flux is actually measured). The turbulent diffusivity across the element of size r is dominated by the motions at scale r and is written $\nu_{turb} \sim r\,\Delta v(r) \sim r^{4/3}$, where $\Delta v(r) \sim r^{1/3}$, *i.e.* the average velocity difference between points separated by the distance r is given by the Kolmogorov law. Then, $\Phi(r) \sim \nu_{turb}\, S(r)\, \nabla C$, where ∇C, the gradient of the quantity C over r, scales as r^{-1}. Hence, $\Phi(r) \sim S(r)\, r^{1/3}$. Since the value of $\Phi(r)$ should be independent of the value of r used, $S(r) \sim r^{-1/3}$. The definition of the fractal dimension of a surface (Mandelbrot 1982) $S(r) \sim r^{2-D_{3d}}$ then gives

$$D_{3d} = \frac{7}{3} = 2.33.$$

The question of what is the quantity of importance which would be transported by turbulence across the cloud boundaries and therefore would regulate their shape and structure at all scales will not be addressed here. But it is to be noticed that this property shared by incompressible laboratory flows of high Reynold numbers and the atmosphere is not specific to turbulence. Similar fractal dimensions are found in the physics of critical phenomena (references in Meneveau 1989).

1.2 Self-similarity of the velocity field.

In 1941, Kolmogorov predicted the self-similar behavior of the velocity field in incompressible turbulent flows by postulating the existence of a local dissipationless cascade of kinetic energy from large scales to small scales which led him to state that the average velocity difference between two points separated by a distance r should be a function only of r and the energy dissipation rate per unit mass $\epsilon_d = \nu \mid \nabla \times \mathbf{v} \mid^2$. Dimensional arguments give the scaling with r of the n-order moments of the increments of the velocity field

$$M_n = <\mid v(\mathbf{x}+\mathbf{r},t) - v(\mathbf{x},t) \mid^n> = B_n(\epsilon_d\, r)^{n/3},$$

B_n being universal constants. This prediction was later modified (Landau and Lifchitz 1959; Kolmogorov 1962) when laboratory experiments showed that the dissipation rate of specific kinetic energy was not uniform but concentrated in the limited regions of space and time where velocity gradients reach large values or diverge. A modified scaling was predicted to take into account the intermittency of ϵ_d

$$M_n \sim \bar{\epsilon}_d r^{\zeta_n},$$

where $\bar{\epsilon}_d$ is the average dissipation rate and $\zeta_n \neq n/3$. Several theoretical approaches predict the form of ζ_n (see Section 1.3).

In real observations of interstellar clouds, the half-power width of a line profile is not simply related to the second order moment of the increment distribution of the velocity field (see Scalo, 1984). It is an average over the volume sampled by the telescope beam of the velocity difference between the most probable velocity v_o (which is that of the peak temperature in optically thin lines) and all other velocities, the distance r between two such points being subject to take all possible values between a minimum value which depends on the correlation length of the velocity field and the length of the line of sight through the cloud. It is easy to show that a beam of size $d \times d$ across a medium of size l along the line of sight samples the increments of the velocity field over an average separation close to d if the gas is uniformly distributed. The separation gets closer to l when $l/d \geq 10$. This would explain why all the groups (Myers 1983; Leung, Kutner and Mead 1982; Dame et al. 1986; Solomon et al. 1987; Falgarone and Pérault 1987) which, after Larson (1981), have tried, using a variety of techniques, to relate the linewidth of molecular clouds δv to their projected size L, do find a scaling law of the type $\delta v \sim L^q$. There is no consensus however upon the slope. For Larson (1981), $q = 0.38$ while in the later investigations $q \sim 0.5$ for entities $L > 0.2\,pc$ up to $L \sim 100\,pc$. Recent sensitive observations at high angular resolution have revealed complex structures as small as $0.02\,pc$ within both opaque regions and low brightness envelopes (Pérault and Falgarone 1988). These entities, seen in the (l, b, v) distribution of ^{13}CO or $C^{18}O$ emission as closed volumes of enhanced brightness (by a factor of ~ 3 above a smooth background emission) are found to be at best weakly bound by their self-gravity. Some of them have masses $\sim 10^{-2}\,M_\odot$, two orders of magnitude below the value for gravitational binding (Falgarone and Pérault 1988). When they are included in the cloud sample, the slope q is quite close to 1/3.

This is almost an embarrassing agreement with the law of Kolmogorov since interstellar clouds have highly compressible interiors likely to be threaded by magnetic fields. Equally embarrassing is the fact that self-gravitating and non self-gravitating entities are now included in the sample. But it is to be noticed again that self-similar laws are also characteristics shared by many non-linear dissipative systems.

1.3 Intermittency of the velocity field

Decades ago, Van der Hulst (1958) and Münch (1957) had already noticed from a statistical study of the velocity of the CaII and NaI absorption lines in the interstellar medium that the distribution of these velocities was better approximated by an exponential than by a gaussian. An observational finding, which may be related to that is the detection by Shaver et al. (1982) and Anantharamaiah et al. (1984) of a long tail of HI absorption components spreading in the forbidden velocity range in the direction of galactic HII regions. This point is controversial and Kulkarni and Fich (1985) who detected this gas in emission argue that it is the in-the-plane counterpart of the galactic HI halo (Lockman 1984). A still more controversial point is the existence of non gaussian wings in the HI profiles of face-on galaxies (Lewis 1987; Dickey et al. 1989). Usually considered as tracers of HI warps due to tidal interaction with companions, it is conspicuous that in a few cases they are observed in isolated galaxies. In molecular clouds not associated with star forming activity, non gaussian wing excesses have been reported in several cases (Blitz and Stark 1986; Knapp and Bowers 1988; Magnani, Blitz and Wendel 1988). The latter show that the fast gas they detect has a column density which exceeds by 3 orders of magnitude that expected from the evaporation of the clouds in the hot interstellar medium.

In an attempt to study these excesses more systematically, Falgarone and Phillips (1989) have analyzed the line profiles of several rotational transitions of CO and variants over regions of sizes ranging between $0.02\,pc$ and $450\,pc$. Profiles come from the literature (Scoville and Young 1983; Pérault et al. 1985; Blitz and Stark 1986; Magnani et al. 1988; Falgarone and Pérault 1988), and unpublished data from the IRAM-30m telescope, the Caltech Submillimetre Observatory and other telescopes (Ungerechts; Fukui, private communications). They find that most of them have wing emission above the level expected from a gaussian line core (*e.g.* in Figure 2 and in Falgarone and Phillips 1989) and that the ratio of the velocity width of the wing, treated as a second gaussian, to that of the core is remarkably constant throughout the sample,

$$\frac{\sigma_{wing}}{\sigma_{core}} \sim 3.$$

This scaling makes it unlikely that the wing emission at all scales be due to a hot diluted component or protostellar outflows. Moreover, in two cases, Falgarone, Phillips and Walker (1989) show that the wing emission is beam diluted emission of small thermalized fragments. A tentative conclusion is that the wing emission is an intrinsic feature of the velocity field of molecular clouds at all scales, previously overlooked due to the good signal to noise ratio required to detect it. Assuming that the wing emission is a good tracer of the probability distribution of the velocity difference $v - v_o$ over the volume sampled by the beam, a possible attractive interpretation of the wing emission is that it is the signature of the intermittency of the velocity field within molecular clouds. It would reveal a higher than normal (gaussian) number of rare events, the rare events being velocity fluctuations of large amplitude about v_o.

A comparison with experimental results on duct flows at high Reynold numbers (Anselmet et al. 1984) and direct measurements in the atmosphere (Van Atta and Park 1971; Dutton and Deaven 1969) is very suggestive (Figure 2b). All the data show that the probability distribution function of the velocity increments present a significant excess at large values which is interpreted as due to the intermittency of the velocity field. They also find that the phenomenon of intermittency is as more pronounced as the scales are small but that neither the lognormal model

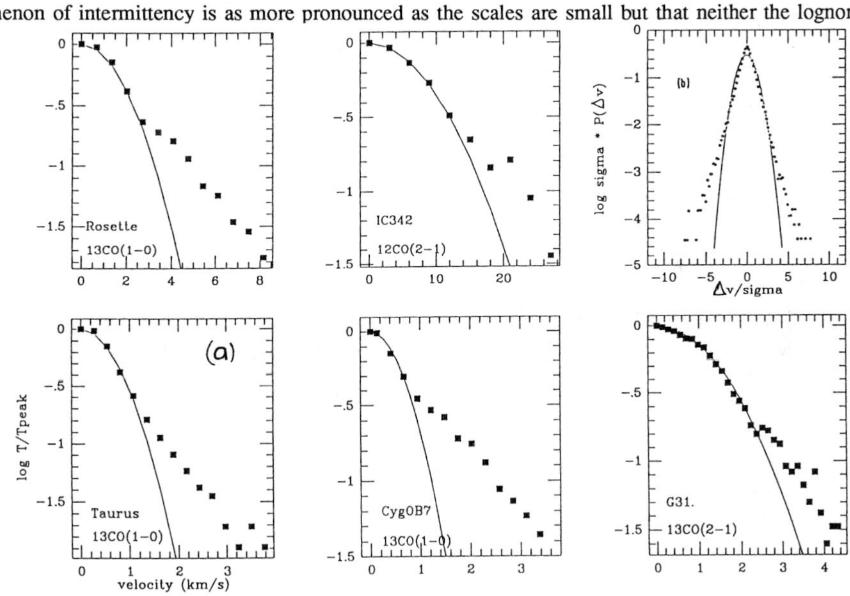

Figure 2. *(a)* Lin-log plot of a subset of normalized profiles (references in the text). The gaussian of same half power width is drawn. *(b)* Probability density of the velocity difference between two points separated by $r = 4.4cm$ from measurements in the atmospheric boundary layer (van Atta and Park, 1971).

(Kolmogorov, 1962) nor the β-model (Frisch et al. 1978) are able to reproduce the experimental data. In his multifractal approach to turbulence, Meneveau (1989) has recently been able to reproduced the values of Anselmet et al. (1984) up to the 18-order moment of the velocity probability distribution. Observations of interstellar clouds are orders of magnitude less sensitive than in situ measurements in duct flows and such a beautiful comparison with a statistical model is far beyond what can be achieved at the moment. But further sensitive observations and analysis of the lineprofiles should be able to decide upon the reality and possible ubiquity of this phenomenon. Spatial mapping of the wing emission should bring information on its nature at the different scales. If intermittency is due to non linear amplification of vorticity (Lundgren 1982), blue and red emission should be spatially correlated, at all scales. In any event, the observation of intermittency in the cloud velocity fields would provide a solid element to the proof of turbulence.

1.4 Compressibility

In 1951, Chandrasekhar and Münch derived the amplitude of the space density fluctuations in the interstellar medium from a statistical analysis of the fluctuations of the absorption of the galactic light along the plane. Using a few assumptions on the space distribution of the absorbing material, they derived from the survey data of Pannekoek and Koelbloed (1949), the root mean square amplitude of the density fluctuations $< \delta\rho^2 >^{1/2} \sim 3$ to $4 < \rho >$ ($< \rho >$ is the average density) and proposed that the interstellar medium is a continuous distribution of density fluctuations of small amplitude rather than an ensemble of discrete clouds. Within molecular clouds, regions of enhanced density far from star forming cores cannot be described entirely as islands in a sea of low brightness emission but tend to be connected in space and velocity (Falgarone and Pérault 1988; Bally et al. 1987; Swade 1989). This result seems to be valid down to scales as small as $0.02\,pc$ both within clouds areas of low average column density $\leq 10^{21}\,cm^{-2}$ and within the so-called dense cores (see Puget and Falgarone 1989).

It is interesting to note that Passot et al.(1988) find, on the basis of two dimensional simulations, that comparable density contrasts are compatible with a turbulence of low rms Mach number. The highly filamentary structures which develop in their density field are in addition very reminiscent of those observed in molecular clouds.

A provisional conclusion would then be that, on the basis of observations only, the velocity field within interstellar clouds is very likely to be turbulent with a moderate level of compressibility, somewhat smaller than would be expected from the crude but popular estimate $< \delta\rho^2 >^{1/2} \sim M^2 < \rho >$ where $M \sim 3$ is a kind of Mach number of the turbulence equal to the ratio of observed velocities to the sound speed in clouds. The gas flow within clouds would present similarities to those observed in ducts or in the atmosphere, in particular the spatial and temporal intermittency of the velocity and velocity gradient (therefore of the dissipation rate of kinetic energy) and the fractal geometry of the cloud structure. We discuss below a few consequences of the presence of turbulence within clouds.

2. The role of turbulence in the physics and chemistry of molecular clouds

2.1 The disputed issue of turbulent pressure

It is repeatedly found in the literature (*e.g.* van Dishoeck and Black 1986) that internal turbulent motions contribute locally to the gas pressure in clouds by an amount $\propto \rho\sigma_v^2$ where σ_v is a velocity dispersion of the non thermal motions usually provided by the linewidths and is therefore related to the entire volume sampled by the beam. As long as only global balances are discussed this may be correct. The additional kinetic energy in turbulent motions can be viewed as a pressure term at the scale for example of gravitationally bound entities in which self-gravity plays the role of the containing walls for the mass of gas. In unbound clouds or locally, the concept of turbulent pressure is not even intuitive. Turbulent pressure is likely to be scale dependent and a lump of matter of size $L = 2\pi/K$ will be affected differently by motions at scales $k >> K$ and $k \sim K$. The latter, which in most cases contain more energy than the former, permanently redistribute the mass contained within lumps of gas at scale K. The existence and the role of a turbulent pressure depends not only on the total power in turbulence but also on the slope of its power spectrum.

Turbulent pressure has also been invoked to play a role in the confinement of molecular clouds (Maloney 1988). If there is clearly some contribution of a surface term in the virial equilibrium of molecular clouds (Chièze 1987), the uncertainties on the parameters derived from observations and the weak sensitivity to the ambient pressure, P_0, which is predicted for the internal velocity dispersion of a self-gravitating polytrope $\sigma_v \propto P_0^{1/4} R^{1/2}$, make any value in the range $3000K\ cm^{-3} < P_0/k < 2\,10^4 K\ cm^{-3}$ consistent with the observations of molecular clouds. Such a broad range does not allow any estimate of the importance of a possible turbulent contribution to a pressure which would confine or help to confine interstellar clouds at all scales.

What observational data unambiguously show is that the large scale motions and the magnetic field are the two agents which prevent molecular clouds (which contain hundreds of Jeans masses) from collapsing in free fall (Myers and Goodman 1988a). However it is not straightforward to account theoretically for the global stability of clouds in which dense cores and stars form. The gradual loss of magnetic support in the densest and less ionized parts may explain the slow formation of dense cores (Shu et al. 1987; Myers and Goodman 1988b), but the support of the large scale and more unstable envelope has to be accounted for simultaneously. Turbulence may play a major role there.

2.2 *Gravitational instability in a turbulent medium*

The problem may be formulated as follows. In a medium of density ρ_0 and sound velocity c_s, density perturbations of wavenumber smaller than $K_J = \sqrt{4\pi\rho_0}/c_s$ grow exponentially if the only support for the gas against gravity is thermal pressure. If, instead, the equilibrium state is a solution of the equation of motion with an adequate forcing which generates stationary turbulence, one may ask how the velocity field is going to react to the density perturbation at scale K and if a gradient of turbulent pressure is going to develop to resist the gravitational forces over the unstable scales. Early attempts to solve this problem are due to Chandrasekhar (1951) and Sasao (1973) but neither of them demonstrates the existence of a turbulent pressure. Bonazzola et al.(1987) have shown in 2-dimensional numerical simulations that Jeans unstable scales can be stabilized by supersonic turbulence and that it is due to the non linear term in the equation of motion.

An analytical approach to this question is possible using renormalization group techniques (Bonazzola et al., 1989). They write an equation of motion for the velocity field built on Fourier components of spatial frequencies smaller than K. For linear density perturbations, this equation keeps the same form as the Navier-Stokes equation with pressure and viscosity terms renormalized to describe the dynamical effects of the small scales ($k > K$). The form of the turbulent viscosity they find is a rather classical result (see for example Moffat, 1981). The original result of Bonazzola et al. (1989) is that a term proportional to the density gradient appears in the equation of motion (the turbulent pressure gradient) and is given by an integral of the power in turbulent motions over spatial frequencies larger than K. The contribution of the turbulent velocity field to the pressure is given by:

$$V_K^2 = \int_K^{+\infty} dk \left[\frac{2}{3} I^{\perp}(k) + A I^{\parallel}(k)\right],$$

where $0 \leq A \leq \frac{1}{3}$ and $I^{\perp}(k)$ and $I^{\parallel}(k)$ are the power spectra of the correlations of the incompressible and compressible parts of the velocity field respectively. It should be noticed that the incompressible part of the velocity field contributes to the turbulent pressure in the same way as thermal motions of the particles to the kinetic pressure and that the same energy density in the form of a compressible field only would contribute less to the turbulent pressure.

If a linear gravitational stability analysis is carried out on a fluid controlled by this equation of motion, the following dispersion relation is derived:

$$\omega^2 - (c_s^2 + V_K^2)K^2 - 4\pi G\rho_0 = 0$$

(the viscous terms have been omitted for clarity). The stabilisation effect of the turbulence is immediately visible since the pressure gradient term is increased. Furthermore, as shown by Bonazzola et al. (1987), if the turbulent spectrum is steep enough, the usual Jeans criterion might even be reversed, the largest scales being the most stable. The turbulent pressure is only one aspect of the possible effects of supersonic turbulence on gravitational instability. A full analysis must include the generation of non linear density fluctuations by the compressible velocity field and that of their lifetime. This analysis has not been done yet.

Recently, Léorat et al.(1989) have considered the same problem and their 2D numerical simulations show a stabilisation by supersonic turbulence (see also Passot, 1987) but their interpretation of this stabilisation is different from the previous one. They claim that the compressible part of the velocity field is the only one at work in the process. A major improvement in the understanding of the problems mentioned above will occur with 3D simulations including self-gravity over a large range of spatial frequencies.

2.3 Consequences of turbulent mixing on chemical abundances

Chièze and Pineau des Forets (1989) have attempted to introduce turbulent mixing in a chemical code following an idea of Boland and de Jong (1982). They picture a cloud as a massive envelope of low column density and space density which is exchanging mass with much denser a core completely shielded from the ambient UV field. The parameter of importance is the mixing time of the matter in the core $t_{mix} = M_{core}(dM/dt)^{-1}$ where dM/dt is the mass exchange rate between the core and the envelope. Their remarkable result is that provided that $t_{mix} < 10^6 yr$ the abundances of both the neutral and ionized carbon in the dense core are enhanced by more than an order of magnitude above the value they would reach in classical models without mixing.

3. Conclusions and prospective

The self-similarity of the spatial and velocity structure of interstellar clouds, the plausibility of a Kolmogorov-type cascade of kinetic energy, the spatial intermittency of the velocity field and the existence of a turbulent pressure are all elements which, taken separately, might be the result of observational biases or attributed to any non linear process in the physics of the clouds. Taken all together, they point to the conclusion that turbulence does exist within clouds, be it purely hydrodynamic or rather hydromagnetic. The impact of such a conclusion to our understanding of the physics of the interstellar medium is yet to be determined. For example, all the processes governed by the penetration of photons in the cloud interiors would be modified by a fractal topology. This would affect considerably the heating processes, the chemistry, the radiative cooling rates and therefore the equation of state. Turbulent pressure has been shown to play a major role in the stabilisation of clouds against gravity and therefore in the star formation process.

One should not feel discouraged by the apparent intractability of turbulence but rather encouraged by the following. The recent developments of powerful instruments provide the community working on the interstellar medium with an unprecedented amount of high quality data. Also, new theoretical approaches to chaos and unpredictable behaviour of non linear systems, in conjunction with the results of powerful simulations and remarkable laboratory experiments should provide tools or guidelines to handle the complexity of the data.

References

Anantharamaiah, K. R., Radhakrishnan, V., Shaver, P. 1984, *Astr. Ap.*, **138**, 131.
Anselmet, F., Gagne, Y., Hopfinger, E. J. 1984, *J. Fluid Mech.*, **140**, 63.
Bally, J., Langer, W.D., Stark, A.A., Wilson, R.W. 1987, *Ap. J. Letters*, **312**, L43.
Bazell, D. and Désert, F. X. 1988, *Ap. J.*, **333**, 353.
Blitz L., Stark A.A. 1986, *Ap. J.(Letters)*, **300**, L89.
Bloemen, J.B.G.M. et al 1986, *Astron. Astrophys.*, **139**, 37.
Boland, W. and de Jong, T. 1982, *Ap. J.*, **261**, 110.
Bonazzola et al. 1987, *Astron. Astrophys.*, **172**, 293.
Bonazzola et al. 1989, submitted to *Physics of Fluids.*
Chandrasekhar, S. 1951, *Proc. Roy. Soc. A.*, **210**, 18 and 26.
Chandrasekhar, S. and Münch, G. 1951, *Ap. J.*, **115**, 103.
Chièze, J. P. 1987, *Astr. Ap.*, **171**, 225.
Chièze, J. P. and Pineau des Forets, G. 1989, *Astr. Ap.*, in press.
Dame, T.M., Elmegreen, B.G., Cohen, R., Thaddeus, P. 1986, *Ap. J*, **305**, 892.
Dickey, J. M., Hansen, M. M., and Helou, G., preprint.
Dickman, R. 1985, "Protostars and Planets II", eds. D. C. Black and M. S. Matthews, Univ. of Arizona Press.
Dutton, J. A. and Deaven, D. G. 1969, *Radio Science*, **4**, 1341.
Falgarone, E., Pérault, M.: 1988, *Astron. Astrophys.*, **205**, L1.
Falgarone, E., Puget, J.L.: 1986, *Astron. Astrophys.*, **162**, 235.
Falgarone, E., Puget, J.L.: 1988, *Galactic and Extragalactic Star Formation*, eds. R.E. Pudritz and M. Fich.
Falgarone, E., Pérault, M.: 1987, *Physical Processes in Interstellar Clouds*, eds. G.E. Morfill and M. Scholer.
Falgarone, E., Phillips, T. G. 1989: *Submillimetre and Submillimetre Astronomy*, ed. A. Webster, in press
Falgarone, E., Phillips, T. G., Walker C. 1989, in preparation.
Frisch, U., Sulem, P. L., Nelkin, M. 1978, *J. Fluid Mech.*, **87**, 719.
Kleiner, S. C. and Dickman, R. L. 1985, *Ap. J.*, **295**, 466 and 479.
Kleiner, S. C. and Dickman, R. L. 1987, *Ap. J.*, **312**, 837.
Knapp, G. R. and Bowers, P. F. 1988, *Ap. J.*, **332**, 299.
Kolmogorov A.N. 1941: *Dokl. Akad. Nauk.* **26**, 115.
Kolmogorov, A. N. 1962, *J. Fluid Mech.*, **13**, 82.
Kulkarni, S. R. and Fich, M. 1985, *Ap. J.*, **289**, 792.
Landau, L. D. and Lifchitz, E. M. 1959, *Fluid Mechanics*, Addison-Wesley.
Larson, R.B.: 1981, *Monthly Notices Roy. Astron. Soc.*, **194**, 809.
Léorat, J., Passot, T., Pouquet, A.: 1989, *Astron. Astrophys.*, submitted.
Leung, C. M., Kutner, M. L., and Mead, K. N. 1982, *Ap. J.*, **262**, 583.
Lewis, B. M. 1987, *Ap. J. Suppl.*, **63**, 515.
Lockman, F. J. 1984, *Ap. J.*, **283**, 90.
Lovejoy, S. 1982, *Science*, **216**, 185.

Lundgren, T. S. 1982, *Phys. Fluids*, **25**, 2193.

Magnani L., Blitz L., Wendel A. 1988, *Ap. J.(Letters)*, **331**, L127.

Maloney, P. 1988, *Ap. J.*, **334**, 761.

Mandelbrot, B. B. 1982, "The fractal geometry of nature", Freeman.

Meneveau 1989, PhD. dissertation, Yale University.

Meneguzzi, M., Frisch, U., Pouquet, A. 1981, *Phys. Rev. Letters*, **47**, 1060.

Moffat, H.K.: 1981, *J. Fluid Mech.*, **106**, 27.

Münch, G. 1957, *Ap. J.*, **125**, 42.

Myers, P.C. 1983: *Ap. J.*, **270**, 105.

Myers, P.C., Goodman, A.A. 1988a, *Ap. J. Letters*, **326**, L27.

Myers, P.C., Goodman, A.A. 1988b, *Ap. J.*, **329**, 392.

Pannekoek, A. and Koelbloed, D. 1949, *Pub. Astr. Inst. Amsterdam*, **9**.

Passot, T.: 1987, *Thèse d'Etat Université Paris VII.*

Passot, T., Pouquet, A. and Woodward, P. 1988, *Astron. Astrophys.*, **197**, 228.

Pérault M., Falgarone E., Puget J.L. 1985, *Astron. Astrophys.*, **152**, 371.

Pérault, M., Falgarone, E., Puget, J.L. 1986, *Astr. Astrophys.*, **157**, 139.

Pérault, M. and Falgarone, E.: 1988, *Molecular Clouds in the Milky Way and External Galaxies*, eds. J. Young and R. Snell.

Puget J.-L. and Falgarone, E. 1989: *Submillimetre and Submillimetre Astronomy*, ed. A. Webster, in press

Sasao, T. 1973, *Publ. Astron. Soc. Japan*, **25**, 1.

Scalo, J. M. 1984, *Ap. J.*, **277**, 556.

Scalo, J.M., Pumphrey, W.A.: 1982, *Ap. J.(Letters)*, **258**, L29.

Scalo, J.: 1987, *Interstellar Processes*, eds. D.J. Hollenbach and H.A. Thronson.

Scoville, N. and Young, J. 1983, *Ap. J.*, **265**, 148.

Solomon, P.M., Rivolo, A.R., Barrett, J., Yahil, A. 1987: *Ap. J.*, **319**, 730.

Sreenivasan, K. R. and Meneveau, C. 1986, *J. Fluid Mech.*, **173**, 357.

Shaver, P. A., Radhakrishnan, V., Anantharamaiah, K.R., Retallack, D.S., Wamsteker, W., Danks, A.C. : Astr. Astrophys., **106**, 105.

Shu, F., Adams, F.C., Lizano, S.: 1987, *Annual Review of Astron. and Astrophys.*, **25**, 23.

Swade, D. A. 1989, *Ap. J. Suppl.*, in press.

Ungerechts, H. and Thaddeus, P. 1987, *Ap. J. Suppl.*, **63**, 645.

van Atta, C. W. and Park, J. 1971, "Statistical Models and Turbulence", eds. M. Rosenblatt and C. van Atta: Springer.

van der Hulst, H. C. 1958, *Rev. Mod. Phys.*, **10**, 913.

van Dishoeck, E. F. and Black, J. H. 1986, *Ap. J. Suppl.*, **62**, 109.

Discussion:

SOLOMON: The exponent in the size-linewidth relation $\sigma_V \propto R^\beta$ is not 0.33 in either local dark clouds (Myers et al.) or in a sample of 270 galactic plane molecular clouds (Solomon et al., 1987). For both of these large samples covering a range in size from about 1 pc to 100 pc, the exponent is 0.5. Thus β does not agree with the value expected for the Kolmogorov spectrum in gravitationally bound clouds.

FALGARONE: I am aware of this point since we also found a slope 0.5 in the σ_V versus size relation for a sample of about 90 clouds or condensations observed in ^{13}CO (1-0) (Falgarone and Perault,1987). Only when we included unbound maxima of column density isolated from (l, b, V) maps of ^{13}CO and C^{18}O emission, covering the range of sizes from 0.02 pc to 0.2 pc, did we find that a slope closer to 1/3 was clearly a better fit to the ensemble of data. This change of slope may be significant and there are many reasons to expect a change in the physics of the hierarchy around the parsec scale. But I just wanted to mention that a 1/3 power law is not ruled out.

DICKEY: A Kolmogorov spectrum is ordinarily generated by a cascade of turbulent energy from large scales to smaller scales. Do you think such a thing is happening in the interestellar gas, or does some other process determine the slope of the spectrum?

FALGARONE: No, I don't think there is a cascade of energy. There are too many sources of turbulence on all different scales to tell how the energy flows.

TURBULENT AND ORDERED MOTIONS IN HII REGIONS

G. Courtès
Laboratoire d'Astronomie Spatiale, CNRS - Trav. du Siphon - 13012 Marseille
Observatoire de Marseille - 2, Place Le Verrier - 13004 Marseille (France)

Summary

The HII regions morphology have suggested, from the beginning of this century, the evidence of internal motions, very few gas radial velocities (RV) on the brightest "nebulae" had been obtained, because of obvious technical difficulties. The modern Fabry-Perot interferometers and the new Integral Field Spectrograph provide abundant RV fields at various scales in the interstellar gas. New ways in imagery, owing to these new instruments, lead to a better analyse of both, the velocities field and their precise morphology free of stellar, dusty, and non thermal continua. The richness and the quality of RV data, a better understanding of the expansion phenomena authorize again a rebirth of the turbulence observations of the ionized interstellar gas.

Key words : *HII regions - Ionized gas kinematics - turbulence - Imaging spectrography.*

Introduction

As soon as photographs of the galactic nebulae have been obtained hundred years ago, the chaotic morphology, as well as some frequent circular shapes of the gas was suggesting some evidence of various motions. However, it is at the beginning of the XXth Century that the first extended spectrographic radial velocities (RV) mapping permitted to confirm, at least for the brightest nebulae, (mainly the Orion Nebula), the reality of such a various RV.

The internal motions

Cambell and Moore (1918), using the radial velocities spectrograph of the Lick Observatory 36 inch refractor, after a laborious point by points work of 400 hours observing time, mapped 86 RV of the Orion Nebula, but earlier Fabry, Buisson and Bourget (1914) obtained, in once (45 min exposure time), owing to the first application of the Fabry-Perot "Etalon interférentiel" (F.P.) at the focus of the Foucault 80 cm telescope of Marseille Observatory (Courtès, 1986), a direct RV mapping on a field of 4' diameter. The "irregularities of speed which may amount to 1 about 10 km.s-1" had been noted. The first interpretations were oriented :

1°) **To the kinematics** of the emission clouds ; their rotation ("Circulatory movements") (Fabry and Buisson, 1911) and more seriously (by analogy with the planetary nebulae) their expansion. The 1914 observations suggested "rotary movements" that have not been confirmed by Campbell and Moore (1918). In fact, Buisson et al. had prudently noted : "rotary movements but with numerous irregularities". They noted also, in the whole body of the nebula : "areas having uniform mean RV".

2°) **To the physics** of the ionised medium. The emission lines profiles giving the temperature : $\Delta\lambda = \lambda.0.78.10^{-6} \sqrt{T/M}$; the temperature estimate corresponding to the interference limit when one increases the interference order, was 15.000 K. Buisson and al. were perfectly conscious of the "too high" evaluation of this temperature caused by the difficulty of estimation of the vanishing fringes in the complete filling of the "free spectral range" of the etalon. In fact, this measure, before the invention of the microphotometer, was wrong, especially in the evaluation of M for the 3726-29 lines. The authors missed the right interpretation, the [OII] lines, given later by Bowen (1927) and Croze and Mihul (1927), Courtès (1986).

The HII regions morphology

Before any consideration on the turbulence and ordered motions, it is important to note that the modern HII region observations permit to select three main types having their morphology related to the gas kinematics.

Type 1) The condensed classical "Strömgren spheres", very similar to the ones of the Milky Way, the LMC-SMC, M33, M31, etc. (Courtès, 1977).

Type 2) The bubbles and super bubbles, especially detected since the work of Meaburn (1978) and Meaburn (1980) as well as the high contrast Fabry-Perot pictures and the first velocities fields of 43 shells in the LMC (Georgelin et al., 1983).

Type 3) The diffuse and filamentary large scale extended emission, similar to the ones recently detected in the Galaxy (Sivan, 1974), M33 (Courtès et al., 1987) and in M31 (Boulesteix et al. 1987), (Ciardullo et al., 1988).

Type 4) Gas "ejections", small scale, the Crab. neb. "jet". Large scale nuclei activities in galaxies NGC 253, NGC 4258, etc ...

The diffuse and filamentary large scale structures are the superimposition of many very extended bubbles at various evolution and deterioration stages (for example, the differential rotation in case of M33 and M31). From the first observations in the Galaxy (Courtès, 1960), it was obvious that the filaments were most of the time, bidimensional shells, seen tangent to the line of sight.

There is between types 1,2,3 many intermediate morphologies linked to the various formation origins, but often due to simple evolution steps. The selection between 2 and 3 is often very vague (Tenorio-Taglé and Bodenheimer, 1988) in spite of a relatively clear selection between 200 pc diameter bubble and 750 to 1144 pc, super bubbles (Goudis and Meaburn, 1978) and Meaburn (1978). The stellar content, the gas density and the dust, the radio and Xray content are also very important for interpreting different initial conditions.

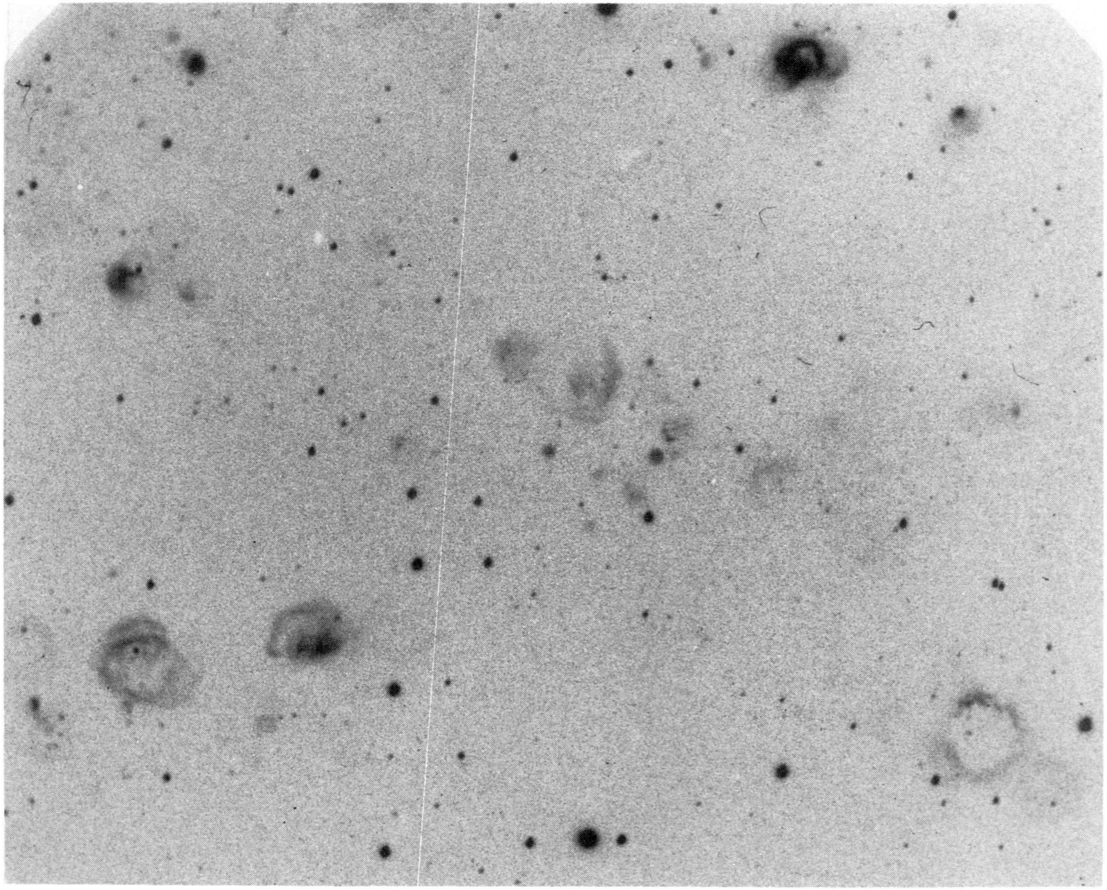

Various expansion structures in M33 (Courtès et al., 1987)
One notes Tenorio-Taglé HII regions at the top (left and right)
Secant bubbles (bottom - left). Active and "fossile" bubble (bottom - right).

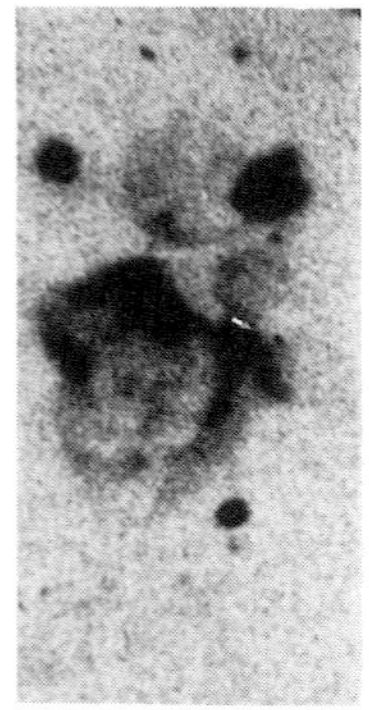

A peculiar example of large sized possible shock wave in Z 100 (top)

(Courtès et al., 1987)
(Gull and Sofia, 1979)

Picture obtained with the 6 m USSR Zelentchouk Telescope equipped of the french F/1 focal reducer.

The ionized gas motions

After numerous RV surveys of these phenomena, one could conclude that the morphologies are mainly dependant, among various kinematical phenomena, to expansions of different origins (see Plate 1) :

a) Classical symmetrical Strömgren sphere expansion, already found for type 1 (Deharveng, 1973), (Georgelin et al., 1983).

b) Interaction with stellar wind of more or less condensed star clusters in type 2 (Dyson, 1979), (Laval, 1989, this Colloquium).

c) Bow shocks generating, in composition with the velocity of the star, parabolic profiles layers (bubble p (N185) in the LMC (Georgelin et al., 1983), like in M33 (see N.N. 261-249 and Z100 in Courtès et al., 1987) ; see also Van Buren and MC.Cray (1988), Melnick, (1977), Gull and Sofia (1979).

d) Supernovae blast waves, conducting in their extreme cases to type 3.

e) "Champagne bottle effect" (Ténorio-Taglé, 1979) like in 88, 218, 632 and Z 112,115, 197 of M33 (COURTES et al., 1987) (6 case in M33),

f) Large scale galactic winds belonging to the extension of type 3.

If we want to clarify by some classification, we can, first, imagine only a morphology's coefficient of symmetry, starting from the almost perfect spherical (or ring) shape often at the outer parts of the galaxies (Courtès et al., 1987) to some degrees of segmentation and final complete dissipation in the interarm and central areas affected of the differential rotation. The relation with the diameter of the bubble in parsecs and, if possible, the nature of expansion energy would be considered (Meaburn, 1978). The best hope is in the most quantitative and accurate RV methods, given by the modern instruments.

TABLE 1

RV Expansions from Georgelin et al., 1983

Moderate expansions			Strong expansions		
N 30	35 $km.s^{-1}$		N 70	60 $km.s^{-1}$	
N 44 BC	30	Xray	N 103 B	20	
N 51 D	40	WR		75	SNR
N 75 ADE	30		N 185	70	
N 62 A	20		N 135	220	SNR
N 79	30		DEM 316		
N 84	20				
N 100	20				
N 144	25				
N 186	30	SNR			
N 204	20				
N 221	25				
DEM 89	20				

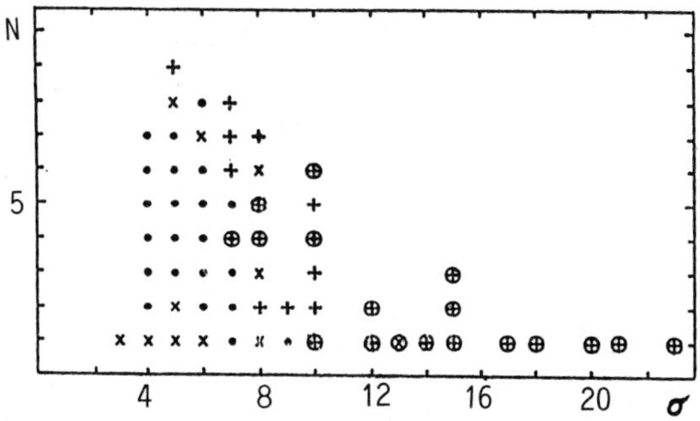

Fig. 1 : *Histogram of the velocity dispersion σ The symbols are surronded by a circle when Georgelin and al. (1983) have indicated a shock velocity.*

Interference radial velocities : The very deep and extended radial velocities mappings owing to the wide field Fabry-Perot methods (Georgelin et al., 1983), (Caulet et al., 1982) (Laval et al., 1987), show that the largest expansion RV are related to the HII regions close to a general circular shape, unique or complex.

Kinematics of the optical emission lines - Need of resolution and accuracy : interference methods have given for the first time the precise profile of the emission lines because of the fact that their resolution is independant of the brightness and can reach beyond the expected width of the line without lost of sensitivity. This is not the case in the conventional nebular spectrograph in which the line profile, for high resolution, is much broader than the geometrical width of the slit image (Courtès, 1972). Most of the recent observations have been obtained owing to the Hα and [NII] lines, some remarks have to be made on the Hα emission.

The quality of the Hα emission : The width of the Hα line is affected by :

a) The fine structure of H atomes ($\Delta\lambda$ = 0,14 A),

b) The temperature : $\Delta\lambda = \lambda\ 0.78.10^{-6}\ \sqrt{T/M}$, with, M_H = 1,008

c) The turbulence (often complicated by superposition along the line of sight of emissions completely independant of the observed structure. In this case, mean radial velocities, profiles, and other lines, [OII], [OIII], [SII], etc.. can help the discrimination). See, for example, the Münch (1958) observation of [OII] compared to [OIII] in the Orion Nebula.

d) The general internal motions at the scale of the space resolution of the instruments.

Fabry and al. (1914) noted first that the effects b) and c) can be disconnected in comparing profiles of Hα with heavier atomes like O, N or S. Münch remarked the same gain of resolution in using [OII] lines instead of Hγ in his first high resolution survey of the 3 x 3' trapezium field in the Orion Nebula with the 200" telescope coudé 4,5 A.mm-1 spectrograph (Münch, 1958). For example, the temperature broadning being function of the square root of the atomic weight, the [NII] 6584 A line is about 3.7 times narrower than Hα (Courtès et al., 1968), (Courtès, 1988). When [NII] is bright enough it constitutes the best way to detect and measure with a great accuracy the expansion splitting of the bubbles (L. Deharveng, 1973). One notes, in this case in the Orion Nebula, and with a spectral resolution of 7.10^4, the remarkable continuity of the expansion and all the considerations of Campbell and Moore as well as those of Fabry et al., and Münch (1958) or Wilson et al. (1959), found a very simple interpretation, the wide fields reveals the general expansion structure, independant of the small scale motions. Anyway, Hα is sometimes, narrow enough, to discriminate for example several components with expansion of 30 km.s^{-1} in the LMC, spectacular case of the shell situated at α_{1975} = 5 h 41 min 5 sec ; δ = - 69° 25'. The radius of the sphere is ρ = 475 pc (Caulet et al., 1982). The famous complex shell of 30 Dor Nebula shows some places of empty shells (Cox and L. Deharveng, 1983) or at least, the evidence of a larger density at the inner surface of the spheres as it was noted since the beginning of the Fabry-Perot observations (Courtès, 1960 ; Courtès et al., 1968). The LMC bubble observations of Georgelin et al. (1983) show clearly : i) a generalization of the expansion, ii) a relation between the velocity dispersion and the presence of shocks (Fig. 1).

New methods of observations

The static Fabry-Perot equipped with a focal reducer (Courtès, 1952) (Courtès, 1972) was used extensively but, an important improvement has been the Fabry-Perot Field Scanning Interferometer method designed first by B. Tully (1974) and generalized by the TAURUS instrument (Atherton et al., 1982) and by the CIGALE instrument using photon-counting detector (Boulesteix et al., 1983). A stellar wind bubble N62B was observed with this new instrument providing imagery in any choosen radial velocity (Laval et al., 1987). The authors fund a semi-spherical expanding cavity (Ve = 35 km.s^{-1}) around a 08 I star, open towards the observer. The mean diameter is 50 pc. The main advantage of CIGALE as any Fabry-Perot (Courtès, 1977) designs is to provide, in fact a very selective filter (few tenths of one Angström $\Delta\lambda = (\lambda_2 - \lambda_1)/F_{(\lambda_2 - \lambda_1)}$, the FWHM of the interference filter being $\lambda_2 - \lambda_1$ and F the finesse of the etalon) with consequently a perfect elimination of the stellar continuum. (See for example the complete elimination of the stars in Laval et al., 1987 or in Courtès (1988).

The field scanning interferometer could detect, in any field, for example 15°, by adaptation to small or large telescopes and different resolution (Caplan et al., 1985) in survey mode, all various emission features owing to their different radial velocities and their line broadning. This is a new other way of selection of the different physical natures of the HII regions.

The real geometry of the expanding shells : the high spectral resolution of the Fabry-Perot method shows that the thickness of the HII layer is very small in comparison of the mean radius of the spherical or pseudo-spherical shells. This is detected by the clear separation of the emission lines in two radial velocities components. Others arguments went for the early observations of the "rim effect" and the RV deviations in front of the "elephant truncks" (Courtès et al., 1962). A very peculiar, small scale case, the Crab neb. "jet", was discovered by van den Bergh (1970), it is, in fact, an unique cylindrical expansion case (Shull et al., 1984) and CIGALE RV mapping (Marcelin et al., 1989).

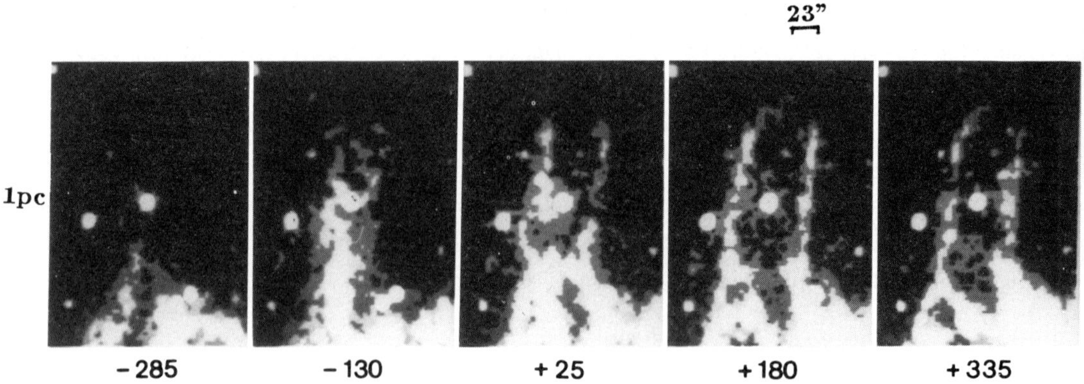

Fig. 2 : A case of cylindrical expansion : Ve = 260 km.s^{-1} [OIII], RV imagery of the Crab.neb.jet. Fabry-Perot field scanner CIGALE. (Marcelin et al., 1989)

The turbulence

The dispersion of the RV could be due to others reasons than expansions, internal group motions or ever rotations. But it was not possible to be envisaged before the turbulence concepts had been developped (Kolmogoroff, 1941, Chandrasekhar, 1949, von Karman, 1951). In the years 1950, the velocities mapping had not been developped since the 1918 last Fabry observations. Then, the first test of turbulence of the interstellar gas was made by L.H. Aller (1951), not from RV data (non existing at this time) but from the brightness fluctuations of the Hα emission on a relatively homogeneous field of filamentary and loop structures. The key idea of this method, suggested also by Pikelner (in Münch, 1958), concerns the relation between the Hα brightness $I(x)$ and its fluctuations point to point. $I' = I(x) - Io$, the density repartition of the gas supposed to be produced by the turbulence. Aller used this expression : $g(r) = \Sigma I'(x).\ I'(x+r) / \Sigma I'(x)^2$ with r taking successives values in a field of 4° wide in Cygnus. Aller found a diameter of 10 pc for the largest eddies in good agreement with the mean diameters of the loops of the ionized gas.

The turbulence theory considers a "hierarchy" of eddies, large to small with, at the end, the dissipation of energy in very small eddies in form of heat. This mechanism leads to a relation between the relative velocity of two points v and their distance l ; $v = k\ l^{1/3}$ the Kolmogoroff law. At the same period, von Weizsäcker (1951) was suggesting a direct approach from the statistics of one component of the gas velocity, the RV, in various points of the medium and their corresponding distances. He demonstrated that the observed fluctuations of velocities can be related to the Kolmogoroff theory, valid for sufficiently high Reynolds numbers and an isotropic medium. Applied with success as a first verification on the Campbell and Moore data (von Hoerner, 1951), the 86 RV of the

ORI nebula verified the v = k $l^{1/3}$ law (with possibility to reach 0,4 exponent instead of 0.33, may be due to a compressibility effect, von Weizsäcker, 1951).

The generalization of the Fabry-Perot interference method provided abundant RV mappings up to 20' from the "Trapezium". The von Hoerner results were confirmed (Fig. 3) (Courtès, 1953, 1955) and extended to several others HII regions (Courtès, 1960). The large HII region around λ ORI (12° app. diam), classical pure Strömgren sphere, was giving $l^{1/3}$ with a break of the correlation about 10 pc, close to the Aller evaluation (Fig. 4).

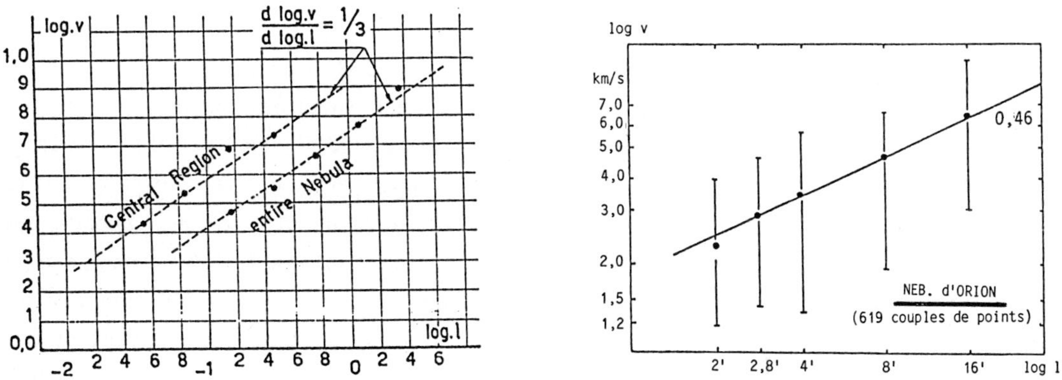

Fig. 3 : First verification of the Kolmogoroff law on the ORION nebula (from S. von Hoerner, 1951 (left) and from G. Courtès, 1953)

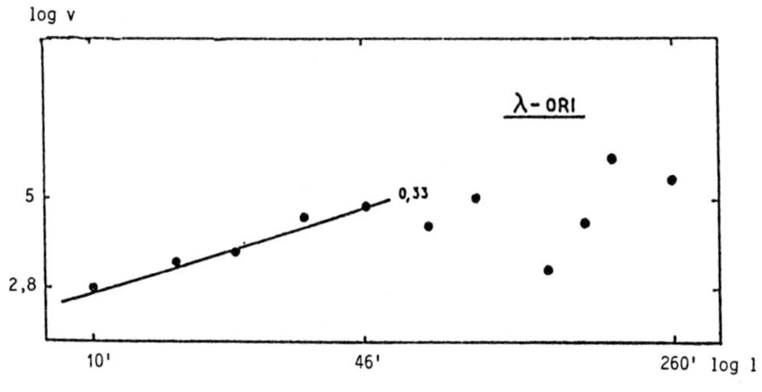

Fig. 4 : The Kolmorogoff law in λ ORI, HII regions (G. Courtès, 1953).

Former observations (Courtès, 1960) on the Horse Head NGC 434 nebula near ζ ORI having a more chaotic morphology were giving several values of the exponent n, n = 0.44, n = 0.33, n = 0.0 ± 0.1 ; n is relatively consistent with the von Hoerner results (n = 0 corresponds to the most confuse morphology of the field) (Fig. 5).

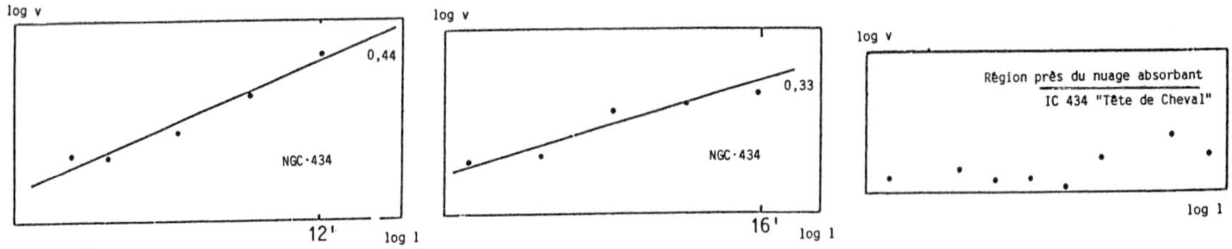

$\Delta\alpha$ = - 52 s
$\Delta\delta$ = - 25'

$\Delta\alpha$ = - 96 s
$\Delta\delta$ = + 9'

$\Delta\alpha$ = + 16 s
$\Delta\delta$ = + 25'

Fig. 5 : Differential coordinate are from ζ ORI.

Another important approach was discussed (Münch, 1958) with a rich and completely different material provided by the multislit 200" "coudé" spectrograph of Wilson et al., 1959 (600 RV on a 40x40" field). The whole survey comprised 50.000 RV. Münch clarified the turbulence problem owing to an exceptional space and spectral resolution. The splitting of the lines appears from place to place at the scale of a few arc seconds, the RV were obtained with an accuracy of 0.5 km.s^{-1}, the thermal and turbulent components were disentangled owing to the atomic weight dependance of the profile (see above). Münch found again the $l^{0,33}$ relation but remarks that the turbulent component of the line width is comparable to the variations of velocities across the gas in emission. This remark concerning the geometrical depth along the line of sight that could be of the same order than the distances of the measured points, has not any more the same meaning since the recent observations of, for example, the ORI nebula. The emitting gas is concentrated in well defined expanding layers (Courtès, 1960), (L. Deharveng, 1973). When one observes couples of RV points across the nebula, each value is the average of a relatively short depth along the line of sight and the integrated RV of each point corresponds to a relatively very small volume of the gas. The projected distance of the points is of one or two order larger and the statistics Δv versus Δl finds again a real meaning. Louise and Monnet (1970) obtened also, from about 10.000 couples of RV points, the $l^{1/3}$ Kolmogoroff law on M8. If we except the Louise and Monnet work, and Courtès et al. (1976), the turbulence has been the sleeping Lady during 27 years, because of the dissuasive effect of the outstanding Münch work. Roy and Joncas (1985) reinitiated the turbulence statistics owing to 40983 RV of S.142. Recently, Roy et al. (1986) were doing an important observation of 47 extragalactic HII regions, after similar work of Dyson (1979) and Melnick (1977). They found a D = f (W^n) with $n < 1.6$ between the linear diameter D and the velocity width W, confirming Melnick (1977). The canadian authors are suggesting the turbulence as a possible interpretation of the velocities dispersion, function of the linear size. However, L. Deharveng (1973) and Georgelin et al. (1983) have shown that the expansion of the HII regions (classics and SNR) can play (in the case of integration of the whole HII region), a considerable part of the velocities dispersion. In fact, it is necessary, to disentangle expansion, temperature (often very different from point to point and from various HII regions (Louise et Monnet, 1970), peculiar motions, especially in the case of secant HII regions or superimposition of SNR (Laval et al., 1989), before giving any interpretation of a global turbulence.

Turbulence of the ORION expanding layers : Courtès et al. (1976) were using for the first time, this well defined geometry of the expansion sheets of the ORI nebula (L. Deharveng, 1973), clearly splitted on large areas of several arc minutes, these sheets have a very small thickness in respect to their mean diameter (Fig. 6) ; they are giving, owing to the large field of the F.P., a completely different appreciation of the same kind of splitting observed at a much smaller scale by Münch. We found $l^{0.46}$ for the receding layer, may be affected by compressibility and $l^{0.06}$ for the approaching one, certainly more chaotic. The receding layer is much brighter than the approaching one, confirming the L. Deharveng (1973) model of a cavity open towards the observer. If the F.P. had been unable to select the two expanding components, we would have obtained a relation around $l^{1/3}$. One sees that the space and spectral resolution are necessary for a clear understanding of the turbulence. On the contrary, the wide aperture F.P. etalon method integrates all this broadening effects without any possibility to disentangle them. It is better, when the apparent diameter permits it, to use Fields Scanning F.P. (TAURUS or CIGALE) or the Integral Field Spectrograph I.F.S. TIGER (Courtès et al., 1987).

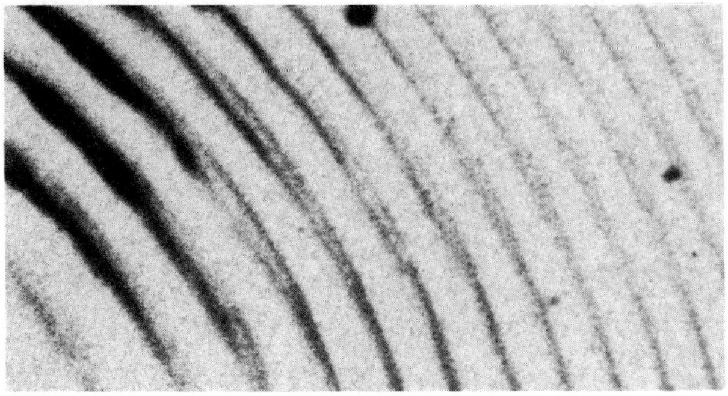

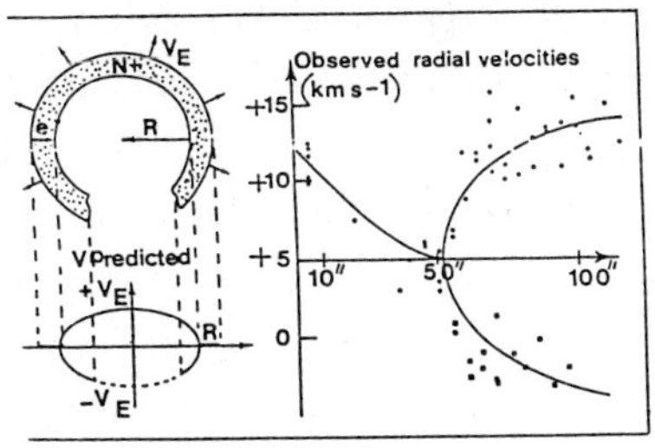

Radial velocities predicted from an expanding shell model of : Orion Nebula; radial velocities observed in region A; the ›scissa is the distance from the outline of the Nebula in a direction rpendicular to this outline

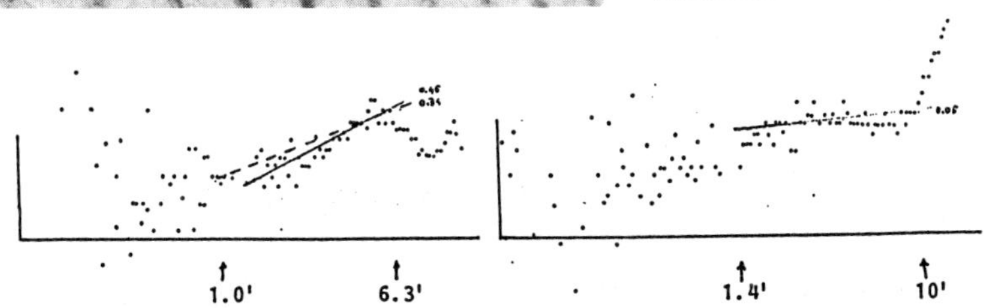

Break of linear relation log v/log I
Receding layer + Ve
(higher brightness)

Break of linear relation log v/log I
Approaching layer - Ve
(fainter brightness)

Fig. 6 : ORI NEB Statistics on 10730 RV
(G. Courtès, L. Deharveng, M. Joubert and F. Genova, 1976), (L. Deharveng, 1973)

These last months, we exploited, with Boulesteix and Lecoarer, the fantastics number of RV given by CIGALE through an automatics $\overline{Va} - \overline{Vb} = k\ (\overline{La} - \overline{Lb})^n$ computations of several hundred RV couples. It is then possible to select several areas in various parts of an HII regions, center (where expansions splitting are expected), rims (free of this effect) superimpositions of different morphological features, filaments, etc.., one obtains, in a few minutes time for each area, the RV statistics in function of the distances. We found convenient to graph in log. directly the power law. This method was applied to N 62 (Laval et al., 1987) and to Sh 305 (Y. Georgelin, 1989) (Fig. 7). The $1^{1/3}$ law appears again in some clear limited structures. Gull et al. (1974), using an echelle spectrograph and the structure function $B(r) = [v(r') - v(r'')]^2$ (Kolmogoroff, 1941), (Scalo, 1984), (O'Dell, 1986) have found extension at the $1^{1/3}$ law from 0,3 to 30 pc in NGC 7000, S252 and NGC 1499.

The NGC 434 results (Courtès, 1960) were a first attempt to see how the 1^n can change in one specific HII region. One sees that n = 0 appears often in the brightest area where several emission layers or clouds are superimposed on the line of seight, but is is not a general rule because some bright areas are often at the bright front surface of dense absorbing clouds with an effective depth, in fact very short, this is likely the case of NGC 434 ($\Delta\alpha = 96$ s ; $\Delta\delta = +\ 9'$ from ζ ORI as well as λ ORI, where the HII material is "coating" a concave cavity in the HI medium. The other case of NGC 434 with n = 0.44 is being interpreted as a compressibility effect $n = 0.33 + \varepsilon$ (von Weizsäcker, 1951). Recently Roy and al. (1986) found again $1^{1/3}$ law in M17. We agree with Roy et al. and O'Dell (1985) that the 1^n law is certainly due to the turbulence, but can not be considered as a pure verification of the Kolmogoroff theory. This test of a 1^n law has anyway the main advantage to provide a very convenient mean of appreciation of the RV repartitions and then, to understand much better the real morphology in surface and in depth.

In the case of well identified molecular clouds, this thin HII layer closely linked to the surface of this cloud could inform, not only about its own turbulence, but also at some other frequency of the molecular cloud itself, independantly of the unavoidable lenght of integration (due to the radio detection) along the line of sight in the molecular cloud.

Generalization at the galaxies scale

a) <u>The largest eddies of the turbulence :</u>

Since the first applications of the turbulence, it was suggested (von Weizsäcker, 1951) that the gas in its most extended scale, in the rotation plane of the galaxies (Courtès, 1977 ; Courtès et al., 1987), could verify some constant distorsion of the interstellar gas by the differential rotation of the galaxies. Consequently, the turbulence would create and also, destroy gas clouds in a hierarchy from large eddies to smaller

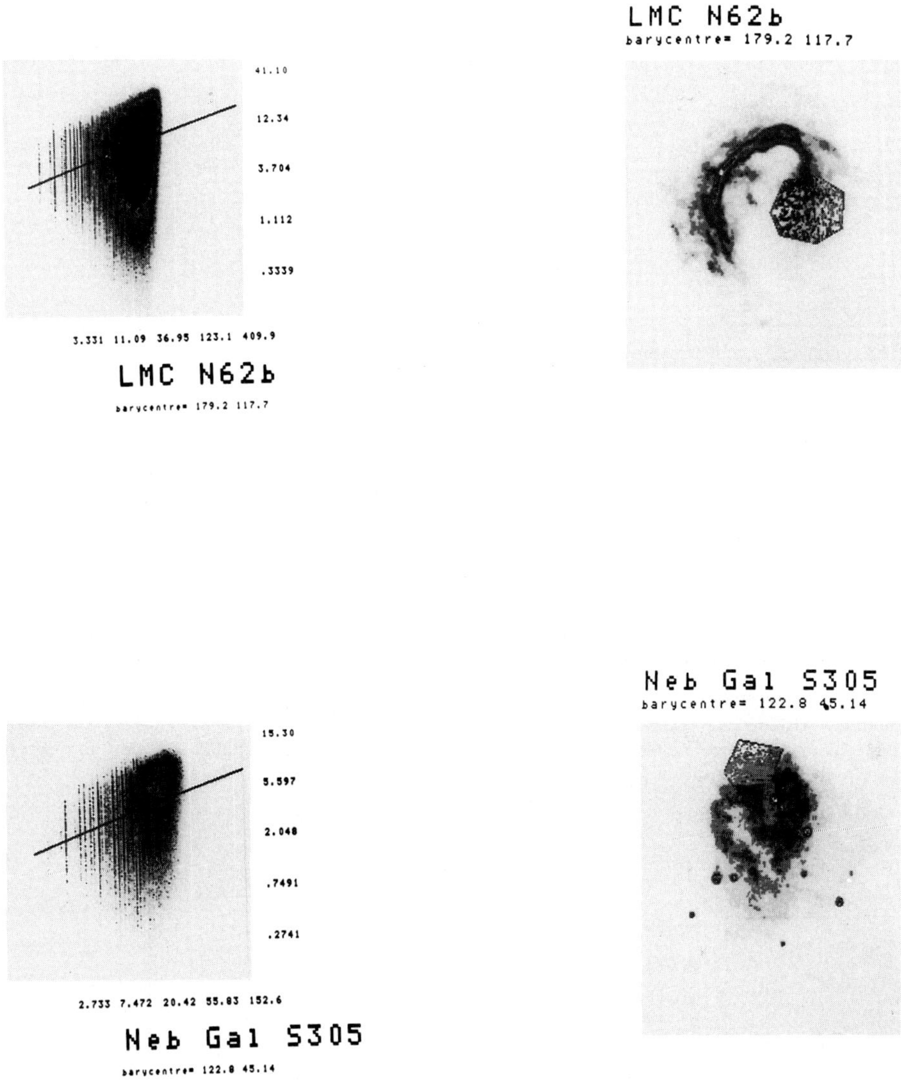

Fig. 7 :

Modern automatic evaluation of the Kolmogoroff law from CIGALE Field Scanning Fabry-Perot (see text).
The contour on the Hα picture corresponds to the RV statistics on the right. Other positions in the same nebulae are giving other slopes. The ones close to 1/3 are always in the places where the morphology suggest dissipation and eddies.

One finds $n = 0.30$ for LMC N62b (Y.M. Georgelin, 1989)
and $n = 0.31$ for Galactic S305 (A. Laval et al., 1987)

(Courtès, Boulesteix, Le Coarer, 1989)

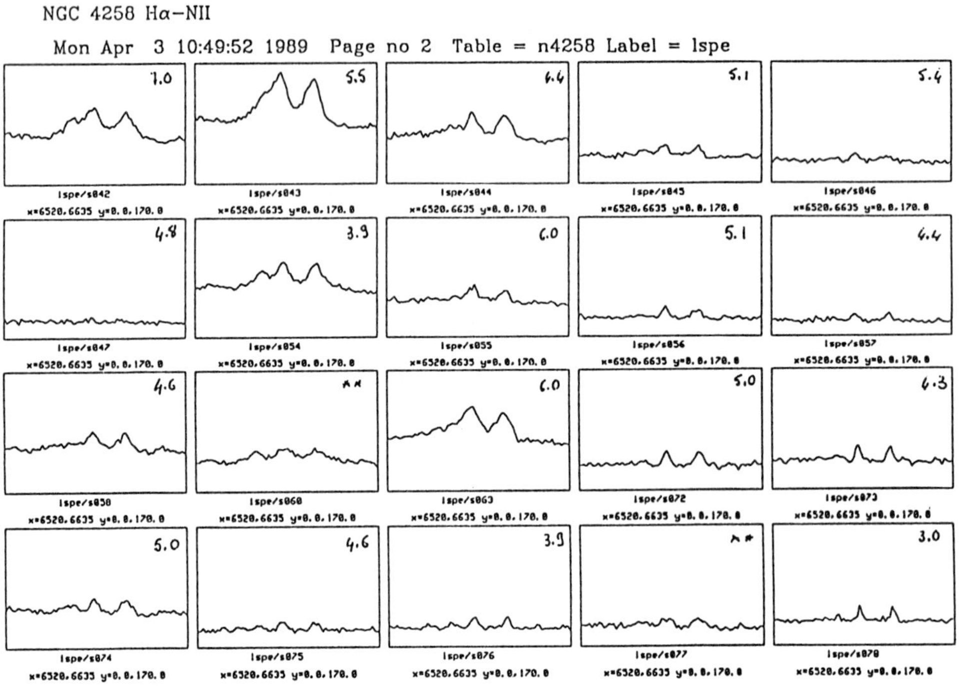

Example of simultaneous record on CCD of Hα [NII] spectra of NGC 4258 nucleus owing to the Integral Field Spectrograph TIGER. Each frame corresponds to focal image elements of 0.5 arc.sec (CFHT 3.6 m telescope, Courtès, Georgelin, Bacon and Monnet, 1989). One sees the variations of line profile and width in respect to the position from the nucleus.

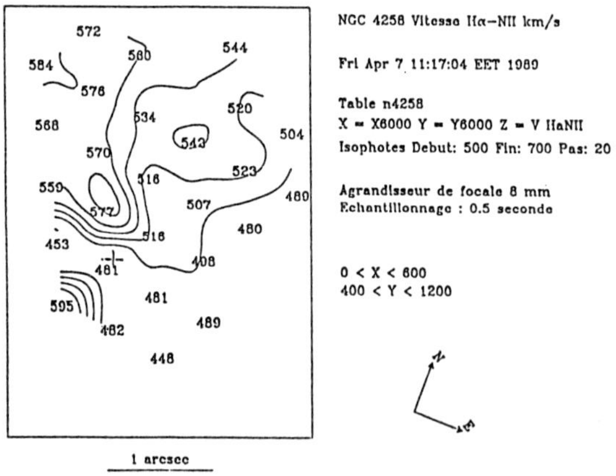

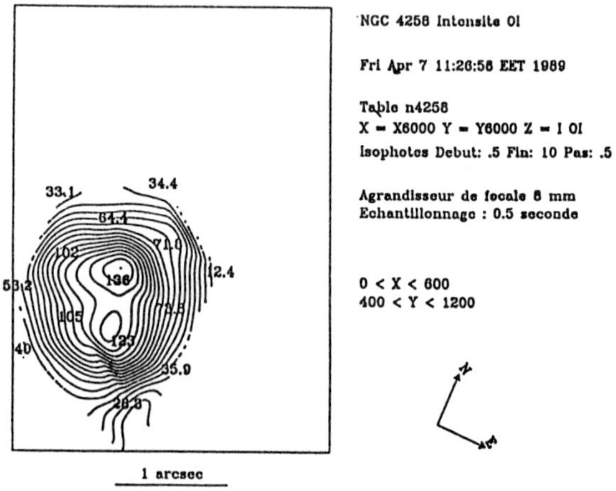

Radial velocities map (Hα, [NII])

Detection owing to OI line reconstructed imagery of a maximum 0.45 arc.sec from the nucleus on the direction of the anormalous arms. This emission is likely due to the interaction of the "jet" on the ISM.

Fig. 8 : Detection of large RV phenomena and shock excited emission at the galaxies scale.

sized eddies. We saw, from the observational point of view, how the quasi-bidimensional nature of the gas repartition is playing a fundamental role in the simplification of a relatively clear conditions for a turbulence detection (thin layer/large field, etc..). A peculiar interesting case is given by the diffuse galactic thin disk emission (340 pc according to Monnet, 1971) in some nearest galaxies like M31 and M33. The case of the central parts of M31 (Ciardullo et al., 1988) is interesting because of the lack of classical HII region, likely doted of their own expansion, compromizing the meaning of the RV statistics at the disk scale. The RV observations of Boulesteix et al. (1987) are providing an abundant material. Ciardullo et al. (1988) concluded also for a thin gas disk (80 pc), possibly more tilted than the equatorial plane of M31. We have, in spite of this difference of tilt (in fact in the good sense), a minimum of good conditions to extract from the RV CIGALE mapping the turbulence and the hope to obtain the size of the eddies along the spiral structures. The splitting of the [NII] 6584 A line found by Boulesteix et al. will reveal, may be another interpretation of the geometry and dymamic of the ionized gas in the M31 central regions.

b) The large scale motions and shocks

The galaxy nuclei and their surroundings are needing other observational approaches. The excess of continuum of their light recommands the Fabry-Perot methods, but a very efficient improvment could consist in a double pass of the same étalon owing to a corner prism. First tests give 0.8 monochromatic transmission, practically free (10^{-4}) of continuum and with a finesse F' = 1.55 F low enough to avoid a too low transmission due to the étalon flatness defacts (main risk when one looks for the same continuum rejection with a larger number of multilayers (faint emission line satellites of diffferent RV should be detected). Another way is to use an lenses array Integral Field Spectrograph (Courtès et al., 1987) in order to obtain velocities field, simultaneously for several lines and to detect shock spectrographic criteria. Recent OI detection with the CFHT 3,6 m telescope at 0,45" of the nucleus of NGC 4258 are especially encouraging for this new methodology (Fig. 8).

Conclusion

The turbulence and in general the analysis of the motions of the ionized gas can be studied now much more carefully owing to the modern interference and spectro-imagery methods. The reduction of the data owing the structure function or the simple Kolmogoroff correlation are needing more experience, but the last published results are showing that these reductions are more dependant of the instrumentation and observational conditions than of some excessive computation sophistication.

References

Aller, L.H. : 1951, Astrophys. J., 113-120

Atherton, P.D., Taylor, K, Pike, C.D., Harmer, C.F.W., Parker, N.M. and Hook, R.N. : 1982 M.N.RAS. 201, 661

Bergh, van den, S. : 1970, Astrophys. J. 160, L27

Boulesteix, J., Georgelin, Y.P., Marcelin, M., Monnet G. : 1983, SPIE Conference, Instr. Astron. V, 445, 37

Boulesteix, J., Georgelin, Y.P., Lecoarer, E., Marcelin, M., and Monnet, G. : 1987 Astron. Astrophys. 178, 91-94

Bowen I.S., 1927, PASP 39, 295

Van Buren, D. and McCray, R. : 1988, Astrophys. J. 329 L93, L96

Campbell, W.W. and MOORE : 1918, Publ. Lick Obs. 13, 96

Caulet, A., Deharveng, L., Georgelin, Y.P., Georgelin, Y.M. : 1982, Astron. Astrophys. 110, 185-197

Chandrasekhar, S. : 1949, ApJ. 110, 329

Ciardullo, R., Rubin, Vera, C., Jacoby, G.H., Ford, H.C., Ford, W.K. : 1988 Astron. Journal, Vol. 95 n° 2, 438

Courtès, G. : 1952, C.R. Acad. Sc. Paris T.234, 506 and 2424

Courtès, G. : 1953, C.R. Acad. Sc. Paris T 237, 378

Courtès, G. : 1955, IAU Symp. n° 2, Cambridge, Amsterdam North Holland Publ. Cie, 131

Courtès, G. : 1960, Ann. d'Astrophys. T23, 115

Courtès, G., Cruvellier, P. and Pottasch, S.R. : 1962, Ann. d'Astrophys. T 25, 214-217

Courtès, G., Louise, R. and Monnet, G. : 1968, Ann. d'Astrophys. T31, n° 5, 493-499

Courtès, G., Deharveng, L., Joubert, M., Genova, F. : 1976, (Private Communication)

Courtès, G. : 1972, Vistas in Astronomy T XIV, 81-161 (Invited paper) Pergamon Press, Oxford

Courtès, G. : 1977, Topics Interstellar Matter, Ed. H. van Woerden, IAU Invited Paper, 209-242, Reidel Publishing company

Courtès, G. : 1986, Journal of Optics, Vol. 17 n° 1, 29-34 (Invited paper)

Courtès, G., Petit, H., Sivan, J.P., Dodonov, S., Petit, M. : 1987, Astron. Astrophys. 174, 28-56

Courtès, G., Georgelin Y.P., Boulesteix, J., Bacon, R. and G. Monnet : 1987, IAU Sym. Santa-Cruz

Courtès, G. : 1988, The World of the Galaxies (Invited paper) Paris, de Vaucouleurs Ed.

Courtès, G., Bacon, R., Georgelin, Y.P., Monnet, G. : 1989, (in preparation)

Courtès, G., Boulesteix, J., Le Coarer, E. : 1989, (in preparation)

Cox, P. and Deharveng, L. : 1983, Astron. Astrophys. 110, 185

Croze and Mihul : 1927, C.R. Acad. Sc. Paris 185, 702

Deharveng, L. : 1973, Astron. ans Astrophys. 29, 341-346

Dyson, J.E. : 1979, Astron. Astrophys. 73, 132-136

Fabry, Ch., Buisson H. : 1911, Astrophysical Journal 33, 406

Fabry, Ch., Buisson H., Bourget H. : 1914, Astrophysical Journal Vol XV n° 3

Georgelin, Y.M., Georgelin, Y.P., Laval, A., Monnet, G., Rosado, M. : 1983 Astron. Astrophys. Suppl. 54, 459-469

Georgelin, Y.M. : 1989, (Private Communication)

Goudis, C. and Meaburn, J. : 1978, Astron. Astrophys. 68, 189

Gull, T.R., O'Dell, C.R. and Parker, R.A.R. : 1974, Icarus 21, 213

Gull, T.R. and Sofia, S. : 1979, Ap.J. 230, 782

Hoerner, S. von : 1951, Zs für Astrophys. 30, 17

Karman, T.H., von : 1949, Problems of Cormical aerodynamics IAU Symp., 129 (Central Air. Doc. Off. Dayton, Ohio)

Kolmogoroff, A.N. : 1941, Docl. Akad. Nank. SSSR, 30, 301

Laval, A., Boulesteix, J., Georgelin, Y.P., Georgelin, Y.M., and Marcelin, M. : 1987, Astron. Astrophys. 178, 199-207

Laval, A. : 1989, Colloquium n° 120 (this Colloquium)

Laval, A., Rosado, M., Boulesteix, J., Georgelin, Y.P., Marcelin, M., Monnet, G. and Le Coarer, E. : 1989, Astron. Astrophys. 208, 230-238

Louise, R. and Monnet, G. : 1970, Astron. Astrophys. 8, 486-488

Marcelin, M., Véron, Ph., Cetty, M.P., Woltjer, L., Boulesteix, J., d'Odorico, S. and Le Coarer, E. : 1989, ESO Preprint n° 642
Meaburn, J. : 1978, Astron. and Space Sci. 59, 193
Meaburn, J. : 1984, MNRAS, 211, 521
Melnick, J. : 1977, Astroph. J. 213, 15
Monnet, G. : 1971, Astron. Astrophys. 12, 379-387
Münch, G. : 1958, Rev. of Modern Phys. Vol.30 n° 3, 1035
O'Dell, C.R. : 1986, Astrophys. Journal 304, 767-770
Roy J.R. and Joncas, G. : 1985, Astrophys. J. 288, 142-147
Roy J.R., Arsenault, R. and Joncas, G. : 1986, Astrophys. J. 300, 624-638
Scalo, J.M. : 1984, Astrophys. J.277, 556
Shull, P., Carsenty, U., Scarcander, M. and Neckel, T. : 1984, Astrophys. J.285, L75
Sivan, J.P. : 1974, Astron. Astrophys. Suppl. 16, 163
Tenorio-Taglé, G. : 1979, Astron. Astrophys. 71, 59
Tenorio-Taglé, G. and Bodenheimer, P. : 1988, Ann. Rev. Astrophys. 26, 145
Tully, R.B. : 1974, Ap.J. Suppl.27, 415
Weizsäcker, C.F.von : 1951, Problems of cosmical aerodynamics IAU Symp. (Central Air Doc. Office, Dayton, Ohio, 158 ; 200)
Wilson, O.E., Münch, G., Flather, E.M., Coffeen, M.T. : 1959, Astrophys. J. Suppl. 4, 199

Discussion:

ROY: I would like to point out to the superb work of Castañeda (1988, Ap. J. Suppl.) recently published on the [OIII]λ5007 velocity field and structure functions of a large sections of the Orion Nebula. This work was done with an echelle spectrograph. How do echelle spectrographs and Fabry-Perot spectrometers compare and differ?

COURTES: The answer is simple. If you need large spectral range, spectrographs may be needed, but not absolutely true for echelle spectrograph used with long slit (needed for RV statistics along it). In this method I would recomend the Integral Field Spectrograph (IFS TIGRE) that extends to 2D the spectrographic nebular field. Anyway the obvious and unavoidable advantage of the Fabry-Perot is to have no slits. With high resolution spectrograph the line profile is larger than the geometric width of the slites image in pure monocromatic light. The loss of sensitibity is in the ratio of these two values. Low path FP with high "Finesse" can be a good compromise for these problems. The F.P field is always much larger than one or 2D Spectrographs but the scanning time is observational time consuming!

EXPANSION OF HII REGIONS IN DENSITY GRADIENTS

José Franco[1,2], Guillermo Tenorio-Tagle[2], and Peter Bodenheimer[3].

[1] Instituto de Astronomía-UNAM, México.
[2] Max-Planck Institut für Astrophysik, FRG.
[3] Lick Observatory, USA.

ABSTRACT

The main features of HII regions expanding in spherical and disk-like clouds with density gradients are reviewed. The spherical cases assume power-law density stratifications, r^{-w}, and the disk-like cases include exponential, gaussian, and sech^2 distributions. For power-law profiles, there is a critical exponent, $w_{crit} = 3/2$, above which the ionization front cannot be "trapped" and the cloud becomes fully ionized. For clouds with $w < 3/2$, the radius of the ionized region grows as $t^{4/(7-2w)}$ and drives a shock front into the ambient neutral medium. For $w = w_{crit} = 3/2$ the shock wave cannot detach from the ionization front and the two move together with a constant speed equal to about $2c_i$, where c_i is the sound speed in the ionized gas. For $w > 3/2$ the expansion corresponds to the "champagne phase", and two regimes, fast and slow, are apparent: between $3/2 < w \leq 3$, the slow regime, the inner region drives a weak shock moving with almost constant velocity through the cloud, and for $w > 3$, the fast regime, the shock becomes strong and accelerates with time.

For the case of disk-like clouds, which are assumed cylindrically symmetric, the dimensions of the initial HII regions along each azimuthal angle, θ, are described in terms of the Strömgren radius for the midplane density, R_0, and the disk scale height, H. For $y_0 = R_0 \sin(\theta)/H \leq \alpha$ (where α is a constant dependent on the assumed density distribution) the whole HII region is contained within the disk, and for $y_0 > \alpha$ a conical section of the disk becomes totally ionized. The critical azimuthal angle above which the HII region becomes unbounded is defined by $\theta_{crit} = \sin^{-1}(\alpha H/R_0)$. The expansion of initially unbounded HII regions (*i.e.* with $y_0 > \alpha$) proceeds along the z-axis and, if the disk column density remains constant during the evolution, the ionization front eventually recedes from infinity to become trapped within the expanding disk. For clouds threaded by a B-field oriented parallel to the symmetry axis, as expected in magnetically dominated clouds, this effect can be very prominent. The expanding gas overtaken by the receding ionization front maintains its linear momentum after recombination and is transformed into a high-velocity neutral outflow. In the absence of magnetic fields, the trapping has only a short duration.

I. INTRODUCTION.

For constant photon fluxes and uniform ambient densities, the evolution of HII regions has well defined formation and expansion phases (*e. g.* Kahn and Dyson, 1965; Mathews and O'Dell 1969; Spitzer 1978; Tenorio-Tagle 1982; Yorke 1986; Osterbrock 1989) During the formation phase, the stellar radiation field creates a supersonic ionization front whose velocity drops down to twice the sound speed in the ionized medium in, approximately, a recombination time. At this time, the end of the formation phase, the ionized region reaches the initial Strömgren radius (Strömgren 1939), and the pressure difference across the ionization front drives a shock wave into the ambient neutral medium. The radius of the expanding HII region, then, grows as $t^{4/7}$ afterwards. Changes in the boundary and initial conditions can result in significant departures from this simple evolution, however. A logical alternative, given that HII regions are created in molecular clouds, is to relax the assumption of a constant ambient density.

If the ionization front encounters a strong negative density gradient and overruns it, then the expansion will enter into its "champagne" phase. Numerical models of HII region evolution under these conditions have provided many details of the expected flows (Tenorio-Tagle 1982; Yorke 1986). The first models assumed discontinuities between constant density molecular and intercloud phases. Molecular clouds, however, have complex morphologies and density distributions, and they are clumpy structures composed of a variety of high-density condensations. These high-density condensations, or cloud fragments, seem to be the actual sites of star formation, and the initial shape and early evolution of the resulting HII regions depend on the corresponding fragment density distributions. Thus, further and refined numerical models have included more appropriate density gradients (*e. g.* Bodenheimer *et al.* 1983; Tenorio-Tagle *et al.* 1986). Given the complex evolution of the ionized flows, however, approximate analytical solutions are certainly useful in understanding the details of the dynamical phenomena.

Recently, the dynamical evolution in spherical clouds with power-law density distributions (Franco *et al.* 1989a, hereafter Paper I) and in disk-like clouds with a variety of stratifications (Franco *et al.* 1989b, hereafter Paper II) has been derived analytically. The results of these works, aside from reinforcing the numerical modeling, provide a physical insight to the rich variety of phenomena that can appear during the HII evolution in decreasing density gradients. These include the generation of internal shocks in clouds with power-law density stratifications and, perhaps more remarkable, the existence of receding ionization fronts in disk-like regions. Here we review the main features of the evolution reported in these papers.

II. HII REGIONS IN POWER-LAW DENSITY DISTRIBUTIONS

Following Paper I, for a spherical cloud with a molecular density distribution (including a central core, with radius r_c and constant density n_c, and an envelope with a power-law density stratification)

$$n_{H_2}(r) = \begin{cases} n_c & \text{for } r \leq r_c, \\ n_c(r/r_c)^{-w} & \text{for } r \geq r_c, \end{cases} \tag{1}$$

the size of the initial HII region is derived, as usual, by equating the total number of ionizing photons emitted by the star per unit time to the total number of recombinations per unit time within the ionized volume. For $w \neq 3/2$, this initial radius can be written as

$$R_w = g(w) R_s, \tag{2}$$

with

$$g(w) = \left[\frac{3-2w}{3} + \frac{2w}{3}\left(\frac{r_c}{R_s}\right)^3\right]^\beta \left(\frac{R_s}{r_c}\right)^{2w\beta}, \tag{3}$$

where R_s is the Strömgren radius for the density n_c. The solution for $w = 3/2$ is

$$R_{3/2} = r_c \exp\left\{\frac{1}{3}\left[\left(\frac{R_s}{r_c}\right)^3 - 1\right]\right\}. \tag{4}$$

Equation (3) defines a critical exponent for the formation phase, w_f, above which the solution for R_w does not exist

$$w_f = \frac{3}{2}\left[1 - \left(\frac{r_c}{R_s}\right)^3\right]^{-1}, \tag{5}$$

and corresponds to the maximum density gradient that is able to "trap" the ionization front (*i.e.* recombinations and new ionizations in steeper gradients are not sufficient to slow down the ionization front, which remains indefinitely as a weak R-type front). Note that for $R_s/r_c > 2$ the critical value becomes $w_f \simeq 3/2$.

After the formation phase has been completed in clouds with $w \leq w_f$, the pressure in the ionized region drives a shock into the molecular ambient medium and the HII region begins its expansion phase. For simplicity, we asssume that the shock evolution starts at $t = 0$ when R_w is achieved. Given that the expansion is subsonic with respect to the ionized gas, the density structure inside the HII region can be regarded as uniform and its average ion density at time t is

$$\rho_i(t) \simeq \mu_i \frac{(9 - 6w)^{1/2}}{3 - w}(2n_c)(R_s/R(t))^{3/2}, \tag{6}$$

where μ_i is the mass per ion, and $R(t)$ is the radius of the HII region at the time t.

For $w \leq 3/2$, the radius can be approximated by

$$R(t) \simeq R_w\left[1 + \frac{7 - 2w}{4}\left(\frac{12}{9 - 4w}\right)^{1/2}\frac{c_i t}{R_w}\right]^{4/(7-2w)}, \tag{7}$$

where c_i is the sound speed in the ionized gas. The ratio of total mass (neutral plus ionized), $M_s(t)$, to ionized mass, $M_i(t)$, contained within the expanded radius evolves as

$$\frac{M_s(t)}{M_i(t)} \simeq \left[\frac{R(t)}{R_w}\right]^{(3-2w)/2}. \tag{8}$$

Equation (8) indicates: i) for $w < 3/2$, the interphase between the ionization front and the leading shock accumulates neutral gas and its mass grows with time to exceed even the mass of ionized gas, and ii) for $w = 3/2 = w_{crit}$, the two fronts move together without allowing the formation and growth of a neutral interphase. Note that the decreasing ratio predicted by the equation for $w > 3/2$ is physically meaningless and it only indicates that the ionization front overtakes the shock front (and proceeds to ionize the whole cloud). Thus, regardless of the value of the critical exponent for the formation phase, w_f, the expansion phase is characterized by a critical exponent with a well defined value, $w_{crit} = 3/2$, which is independent of the initial conditions. Furthermore, this critical exponent $w_{crit} = 3/2$ is not affected by dust absorption (Paper I).

For $3/2 < w < w_f$, the ionization front eventually overtakes the shock and soon the whole cloud becomes ionized. At that moment the gas acquires the same temperature everywhere, and the pressure gradient simply follows the density gradient. All parts of the cloud are then set into motion, but the expanded core (now with a radius identical to the position of the overtaken shock) is the densest region and feels the strongest outwards acceleration. Thus, superimposed on the general gas expansion there is a wave driven by the fast growing core (the wave location defines the size of the expanded core), and the cloud experiences the so-called "champagne" phase. This core expansion tends to accelerate with time and two different regimes, separated by $w = 3$, are apparent: a "slow" regime with almost constant expansion velocities, and a "fast" regime with strongly accelerating shocks.

The slow regime corresponds to $3/2 < w < 3$ and the core grows approximately as

$$r(t) \simeq r_c + \left[1 + \left(\frac{3}{3-w}\right)^{1/2}\right] c_i t, \tag{9}$$

where for simplicity the initial radius of the denser part of the cloud has been set equal to r_c, the initial size of the core. For $w = 3$ the isothermal growth is approximated by

$$r(t) \simeq 3.2 r_c \left[\frac{c_i t}{r_c}\right]^{1.1}. \tag{10}$$

For $w > 3$, the fast regime, the shock acceleration increases with increasing values of the exponent and the core expansion is approximated by

$$r(t) \simeq r_c \left[1 + \left(\frac{4}{w-3}\right)^{1/2} \left(\frac{\delta + 2 - w}{2}\right) \frac{c_i t}{r_c}\right]^{2/(\delta+2-w)}, \tag{11}$$

where

$$\delta \simeq 0.55(w-3) + 2.8. \tag{12}$$

III. HII REGIONS IN DISK-LIKE CLOUDS

All stratifications likely to represent a "disk-like" cloud, or a gaseous disk, decrease faster than any power-law at large radii. Therefore, all these distributions have a critical point beyond which their density fall-off is steeper than $r^{-3/2}$, and can be completely ionized beyond that point. Following Paper II, the definition of a gaseous disk is simply an elongated cloud which can be regarded as cylindrically symmetric. The variables r, θ, and z are taken as the usual cylindrical coordinates. The gas is assumed molecular with a constant scale height, H, and the ionizing source is located at midplane. The density distribution along the z-axis is

$$n(z) = n_0 \, G(z/H), \tag{13}$$

where n_0 is the density at midplane, and $G(z/H)$ is any of the functions $e^{-z/H}$, e^{-z^2/H^2}, and $sech^2(z/H)$. Clearly, there is a critical point for each one of these cases.

The initial shape of the ionized region is defined by the equilibrium between photoionization and recombination along each solid angle. Defining the dimensionless variables $y = r \sin(\theta)/H$ and $y_0 = R_0 \sin(\theta)/H$ (where R_0 is the Strömgren radius for the midplane density), the shape is given by the integral equation

$$y_0 = \left[3 \int_0^y y^2 G^2(y)\, dy \right]^{1/3}, \tag{14}$$

and the resulting size of the initial HII region as a function of the angle θ can be written as

$$R_\theta \simeq R_0 h(\theta), \tag{15}$$

with the "shape" function

$$h(\theta) = \left(\frac{\alpha}{2y_0} \right) \ln \left[\frac{1 + (y_0/\alpha)}{1 - (y_0/\alpha)} \right], \tag{16}$$

where $\alpha = [3 \int_0^\infty y^2\, G^2(y)\, dy]^{1/3}$ and has the values 0.78, 0.88, and 0.91 for the gaussian, sech^2, and exponential cases, respectively. For small values of y_0 (*i.e.* $y_0 \leq 0.3$), the shape function reduces to

$$h(\theta) \simeq \left(1 + \frac{y_0^2}{3\alpha^2} \right). \tag{17}$$

The critical point is defined by $y_0 = \alpha$, and this corresponds to a critical angle

$$\theta_{crit} = \sin^{-1} \left(\alpha \frac{H}{R_0} \right), \tag{18}$$

which defines the conical section of the disk that is fully ionized. Note that the HII region reaches the critical point when $R_0 > \alpha H$, but it becomes completely bounded within the disk when $R_0 \leq \alpha H$.

The early stages of the expansion are controlled by the density gradient perpendicular to midplane (see Bodenheimer *et al.* 1983) but, as discussed in Paper II, the details of the general evolution are affected by lateral gas velocity components generated near the boundaries with the neutral gas. These two-dimensional effects cannot be treated with a simple analytical scheme. Note, however, that the presence of a B-field oriented in the direction of the disk symmetry axis, as expected in magnetically supported clouds, inhibits the lateral gas movements and the ionized region is forced to expand only along the z-axis (allows for a simple one-dimensional treatment of the expansion). For $B = 0$, on the other hand, two dimensional effects become important after a sound crossing time, and this restricts the discussion in the nonmagnetic case to the early stages of the expansion.

The analysis of the magnetic and nonmagnetic cases is performed only along the symmetry axis ($\theta = \pi/2$). The symbols n_0, R_0, H, and y_0 are kept as the initial values (*i.e.* at $t = 0$) of the corresponding parameters, and the discussion is restricted to initially unbounded regions ($y_0 > \alpha$).

a) Expansion with B-fields: high-velocity neutral outflows

The initial gas acceleration (defined by the density gradient) stretches the whole disk and the gas velocities can reach supersonic values. The magnetic field, oriented along the z-axis, is assumed uniform and the flow is constrained to move within the flux tubes. Thus, the evolution of the flow can be analysed in one dimension. A rough upper bound to the required field strength at midplane is $B \sim 10^{-4} n_3^{1/2} c_1$ G, where $n_3 = n_0/10^3\ \mathrm{cm}^{-3}$ and $c_1 = c_i/10\ \mathrm{km\ s}^{-1}$.

At early times, $t \leq 2H/c_i$, the scale height grows as

$$H(t) \simeq H(1 + \beta c_i^2 t^2 / H^2), \tag{19}$$

with β equal to 1, 0.76, and 0.5 for the gaussian, sech^2, and exponential stratifications, respectively. Similarly, at these early times, the midplane density decreases as

$$n_O(t) \simeq n_O(1 + \beta c_i^2 t^2 / H^2)^{-1}. \tag{20}$$

Later, for $t \geq 2H/c_i$, the late times approximation is

$$n_O(t) \simeq \frac{n_O H}{(2\beta + 0.5)c_i t}, \tag{21}$$

which is equivalent to define a late times scale height (enclosing an almost uniform density) $H(t) \sim (2\beta + 0.5)c_i t$.

The size of the HII region along midplane grows approximately as

$$R_O(t) \simeq R_O \left[\frac{n_O}{n_O(t)}\right]^{2/3}. \tag{22}$$

Note that if $R_O(t)$ becomes smaller than $H(t)$, the HII region will be trapped within the expanded disk. The growth rate for $R_O(t)$ and $H(t)$ are different, and the dimensionless variable $y_O(t) = R_O(t)/H(t)$ is a decreasing function of time. At early times it evolves as

$$y_O(t) \simeq y_O \left[1 + \beta \left(\frac{c_i t}{H}\right)^2\right]^{-1/3}, \tag{23a}$$

and at late times it is approximated by

$$y_O(t) \simeq y_O \left[\frac{H}{(2\beta + 0.5)c_i t}\right]^{1/3}. \tag{23b}$$

Thus, there exists a "trapping" time, τ_t, at which the value of $y_O(t)$ drops below unity and the expanding HII region becomes ionization bounded in the z-direction. At this time, which becomes longer with increasing values of y_O, the ionization front recedes down to a location inside $H(t)$ and the gas exterior to this point recombines.

For $y_O \leq 1.4$, the trapping occurs when the midplane density reaches the value

$$n_O(\tau_t) \simeq (\alpha / y_O)^3 \, n_O, \tag{24}$$

and the trapping time is defined by

$$\tau_t \simeq \frac{H}{c_i \beta^{1/2}} \left[\left(\frac{y_O}{\alpha}\right)^3 - 1\right]^{1/2}. \tag{25}$$

For $y_O \geq 1.4$, the trapping time is approximated by

$$\tau_t \simeq \frac{H y_O^3}{(2\beta + 0.5)c_i}, \tag{26}$$

when the corresponding midplane density is

$$n_0(\tau_t) \simeq (n_0/y_0^3). \tag{27}$$

A lower bound to the mass of the recombining zone is

$$M_{rec} \sim \mu\pi R_0^2(\tau_t) n_0 H \int_1^{\infty} G(x)dx \simeq 0.3\mu\pi R_0^2 y_0^4 n_0 H, \tag{28}$$

where μ is the mass per particle. This gas maintains its momentum after recombination and becomes a neutral high-velocity outflow. Defining $R_{17} = R_0/10^{17}$ cm, $H_{17} = H/10^{17}$ cm, $c_1 = c_i/10$ km s^{-1}, and $n_6 = n_0/10^6$ cm^{-3}, the lower limits to the mass and momentum locked in the high-velocity neutral flow are $M_{rec} \sim 0.1 R_{17}^2 n_6 H_{17} y_0^4\ M_\odot$ and $P_{rec} \sim 2R_{17}^2 n_6 H_{17} y_0^4 c_1\ M_\odot$ km s^{-1}. These approximate expressions indicate that magnetic cases with $y_0 > 1$ can generate powerful neutral outflows.

b) Expansion with B=0: short trapping stages

In this case, the expansion is affected by lateral gas motions driven by rarefaction waves. These waves reach the symmetry axis in a sound crossing time, $\tau_{2D} \simeq R_0/c_i$. Thus, the trapping occurs only at the early stages of expansion and its properties are given by equations (24) and (25). The trapping time can be written as

$$\tau_t \simeq \frac{\tau_{2D}}{y_0 \beta^{1/2}} \left[\left(\frac{y_0}{\alpha} \right)^3 - 1 \right]^{1/2}, \tag{29}$$

and given that $\tau_t \tau_{2D}$, the existence of the trapping is restricted to cases $\alpha < y_0 1$. Also, the trapping is a short lived event and the ionization front moves again outwards after a time τ_{2D}.

IV. CONCLUSIONS

The results summarized in this paper provide the main features, under a wide range of conditions, of the dynamical evolution of ionized regions in decreasing density gradients.

For spherical clouds with a power-law density distribution, r^{-w}, there is a critical exponent above which the cloud becomes completely ionized. Its value in the formation phase depends on the initial conditions, but it has a well-defined value, $w_{crit} = 3/2$, during the expansion phase. For $w \leq w_{crit}$, the radius evolves as $t^{4/(7-2w)}$. Cases with $w > w_{crit}$ follow the champagne phase, where the expansion becomes supersonic, and two well defined regimes can be separated. For $3/2 < w \leq 3$, the slow regime, the core expansion is mildly supersonic and has almost constant velocity. For $w > 3$, the fast regime, the core expansion drives a strongly accelerating shock. These results indicate, aside from the rich variety of dynamical phenomena, that optical HII regions are density bounded and that some star forming regions (those with w larger than 3) may be likely sources for cosmic ray acceleration.

For disk-like clouds, with and without magnetic fields, the evolution also presents several interesting properties. Some details depend on the particular density profile, but the main features are determined by the scale height and the initial Strömgren radius along midplane. The initial shape of the ionized region has a critical point, $y_0 = \alpha$, which defines a fully ionized conical section in the disk. For initially unbounded regions, $y_0 > \alpha$, the evolution can produce large expansion velocities. Moreover, the expanded disk can trap the ionization front and the HII region becomes bounded. If a B-field oriented parallel to the z-axis is present, the event is long lasting and produces a powerful and supersonic neutral outflow. In the absence of magnetic fields, the lifetime of the trapping stage is limited by two-dimensional effects.

All these features certainly warrant further research. Future developments and the application of these results to particular objects will be explored in forthcoming works.

ACKNOWLEDGMENTS

P. B. and J. F. acknowledge the hospitality of the Max-Planck Institut für Astrophysik. J. F. acknowledges partial financial support from CONACyT-México. G. T.-T. is supported by the Deutsche Forschungsgemeinschaft grant Mo 416.

REFERENCES

Bodenheimer, P., Yorke, H. W., Tenorio-Tagle, G., and Beltrametti, M. 1983, in *Supernova Remnants and Their X-Ray Emission, IAU Symp. 101*, eds. J. Danziger and P. Gorenstein (Dordrecht: Reidel Publ. Co.), p. 399.

Franco, J., Tenorio-Tagle, G., and Bodenheimer, P. 1989a, *Ap. J.*, in press (Paper I).

Franco, J., Tenorio-Tagle, G., and Bodenheimer, P. 1989b, *Rev. Mex. Astron. Astrofís.*, in press (Paper II).

Kahn, F. D., and Dyson, J. E. 1965, *Ann. Rev. Astron. Ap.*, **3**, 47.

Mathews, W. G., and O'Dell, C. R. 1969, *Ann. Rev. Astron. Ap.*, **7**, 67.

Osterbrock, D. E. 1989, *Astrophysics of Gaseous Nebulae and Active Galactic Nuclei* (Mill Valley: University Science Books).

Spitzer, L. 1978, *Physical Processes in the Interstellar Medium* (New York: Wiley and Sons).

Strömgren, B. 1939, *Ap. J.*, **89**, 526.

Tenorio-Tagle, G. 1982, in *Regions of Recent Star Formation*, eds. R. S. Roger and P. E. Dewdney (Dordrecht: Reidel Publ. Co.), p. 1.

Tenorio-Tagle, G., Bodenheimer, P., Lin, D. N. C., and Noriega-Crespo, A. 1986, *M. N. R. A. S.*, **221**, 635.

Yorke, H. W. 1986, *Ann. Rev. Astron. Ap.*, **24**, 49.

VARIABLE HII REGIONS

V.A. Hughes
Department of Physics, Queen's University, Kingston
Ontario, Canada K7L 3N6

Initial CO observations of the Cepheus OB cloud by Sargent (1977), identified the condensation known as Cepheus A. Other observations have shown that it has the normal empirical indications of a star forming region. In the radio continuum, using the WSRT, it consists of two thermal components (Hughes and Wouterloot 1982). The West component is associated with the Herbig-Haro object GGD-37, and will not be considered further here. The East component when observed further with the VLA, is seen to consist of two strings of about 14 compact HII regions, and is the subject of the present paper.

Observations in 1981 with the VLA in the "B" configuration, with angular resolution of 1" and 3" at wavelengths of 6 cm and 21 cm (Hughes, and Wouterloot 1984), showed that though some compact components were resolved, others were not, indicating that some have linear dimensions of less than 700 au. Further observations in 1982 using the "A" configuration with corresponding angular resolution of 0".3 and 1", (linear resolution of 200 and 700 au,) showed that some sources were not resolved at the higher resolutions. Since HII regions with diameters < 200 au were not expected to remain stable, observations were repeated in the "A" configuration in 1986 (Hughes 1988), but this time 2 cm observations were included, where the angular resolution is 0".1, (linear size of 70 au). Not only were some sources still not resolved at this shorter wavelength, but appreciable differences were clearly seen between epochs.

To eliminate artifacts which might be introduced due to procedures used for data reduction, all the data sets have been reprocessed in exactly the same way (Hughes 1988). This applies also to more recent data sets obtained in 1987 and 1988.

The chief features of the area are shown in the 20 cm and 6 cm maps of Figure 1, both from 1986 data sets. They show also the numbering of the components, including the position of the highly variable new components 8, and 9. Spectral indices are difficult to determine accurately unless the components are

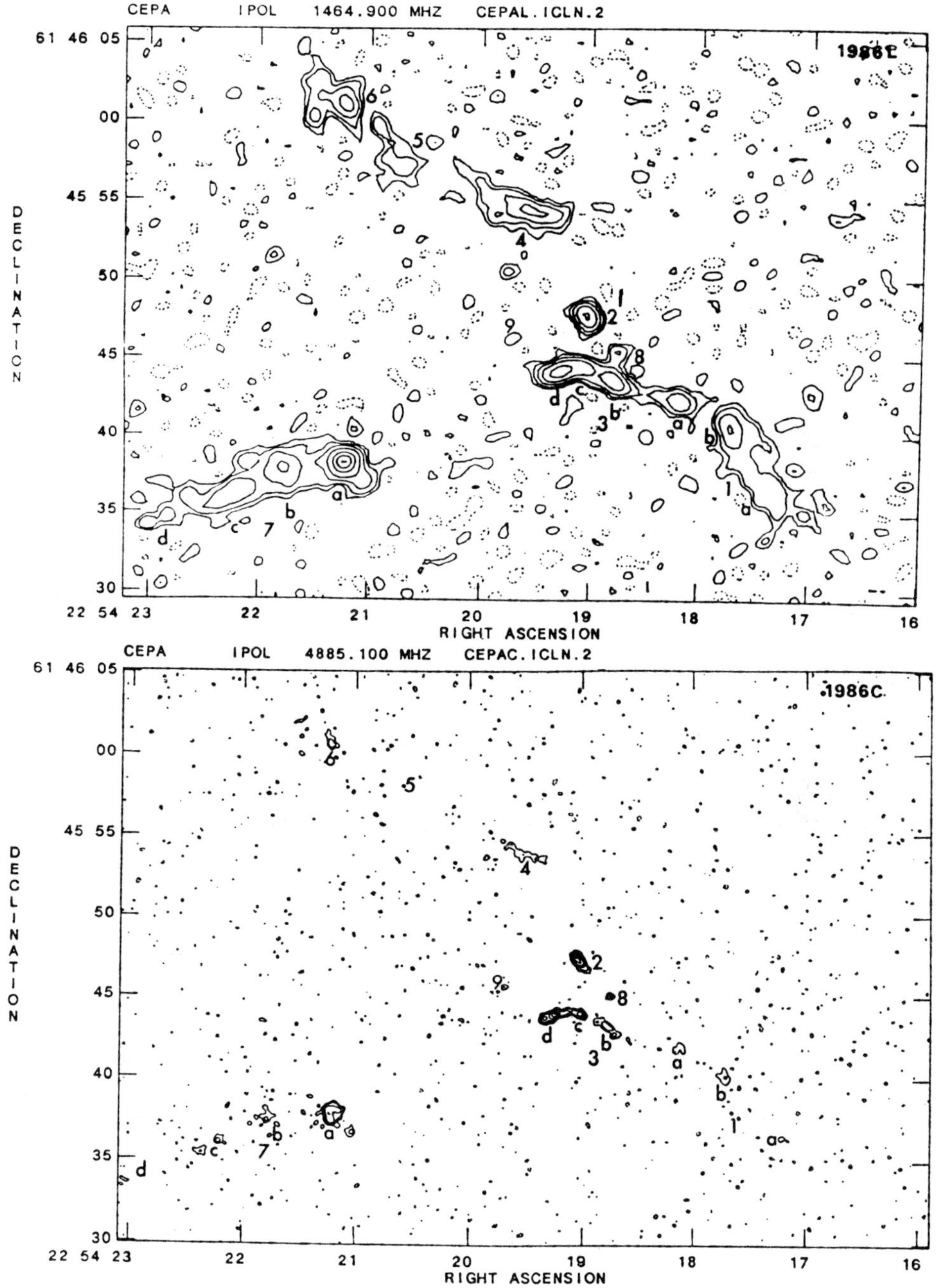

Figure 1, Upper: 20cm map of Cep A, resolution 1", from 1986 data. Contour levels are -0.1, 0.1, 0.2, 0.4, 0.8, 1.6, 2.4, and 3.2 mJy/beam.

Lower: 6 cm map with resolution 0.″1 from 1986 data. Contour levels are the same as for the upper figure.

well resolved and observed with the same angular resolution at two frequencies. The early observations at 6 cm in the "B" configuration and 20 cm in the "A" configuration showed spectral indices between 0.4 and 0.6 for components 2, 3d, and 5, and -0.1 ± 0.1 for the other outer components. The latter spectral index is indicative of an optically thin thermal (HII) region, while the former indicates the presence in the region of a part which is optically thick, and is typical of mass loss stars where the electron density varies as r^{-2}; it could equally well apply to stars which are accreting under free-fall conditions. The 2 cm observations showed, in addition, that component 2 was double, consisting of two unresolved sources separated by 70 au, component 3d had two similar sources separated by 200 au, while component 3c was a single unresolved source. Of possible significance is that the orientation of the sources of component 2 is in the direction of Component 8, as is also that of Component 3d.

Associated with the compact components are H_20 and OH masers (Cohen et al 1984). Component 2 has a number of OH masers outside the HII region, as have components 3c and 3d, while component 2, 3a and 3d have H_2O masers at the projected edge of the HII regions.

The most remarkable aspect of the region is the variability of the components. This was suspected in the early observations, but is very pronounced with components 8 and 9. Component 8 was present in 1981, and 1986 with a peak flux density of a few mJy, but below the detection level of 0.1 mJy in 1982, 1987 and 1988. Component 9 was seen as a barely significant source in 1986, was not evident in 1987, but is clearly present in 1988. The spectral indices of both appear to be 0.6. Maps of the region at 6 cm containing these sources for epochs 1986, and 1988, are shown in Figure 2. The variation in flux density of those components which remain well separated from their neighbours is shown in Figure 3. Some characteristics are that the two sources of component 2 appear to vary independantly of each other; component 3 has decreased in time to a level below the noise level at 6 cm, but not at 20 cm, apparently due to an increase in size; component 7 has the appearance of a shell-type object.

The initial interpretation was based on the assumption that the individual HII regions were internally excited, so that, provided they are optically thin, the radio flux density enables the excitation parameter to be determined, and thus the spectral type of the assumed exciting star. This led to an interpretation of the region as consisting of two lines of about 14 B3 stars. However, some reservation was felt about this, since the HII regions must be very young and 14 stars would have to all turn on inside a period of about 1,000 years, and all be in the same stage of evolution.

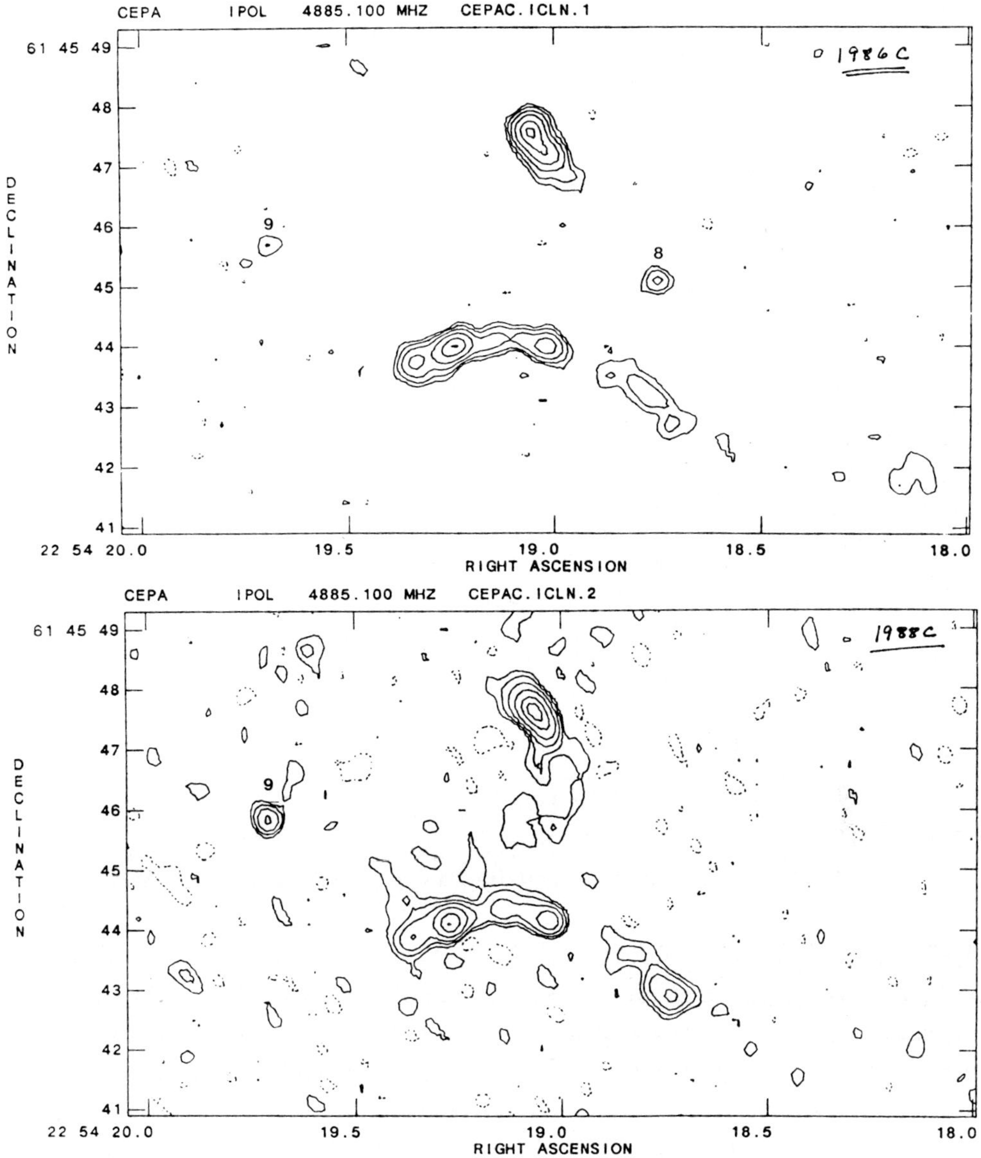

Figure 2. Maps of the central region of Cep A from the 1986 and 1988 data bases; wavelength 6 cm, resolution 0".3. Contour levels are -0.1, 0.1, 0.2, 0.4, 0.8, 1.6, 2.4, and 3.2 mJy/beam. The varibility of components 8 and 9 is clearly seen. The higher noise level in the 1988C map is due to a shorter integration time.

Additional evidence for the presence of stars depended on IR observations, which showed the total 100 μm flux to be that expected from the assumed 14 B3 stars (Evans, et al. 1981). On the other hand, observations at 2 μm by Lenzen

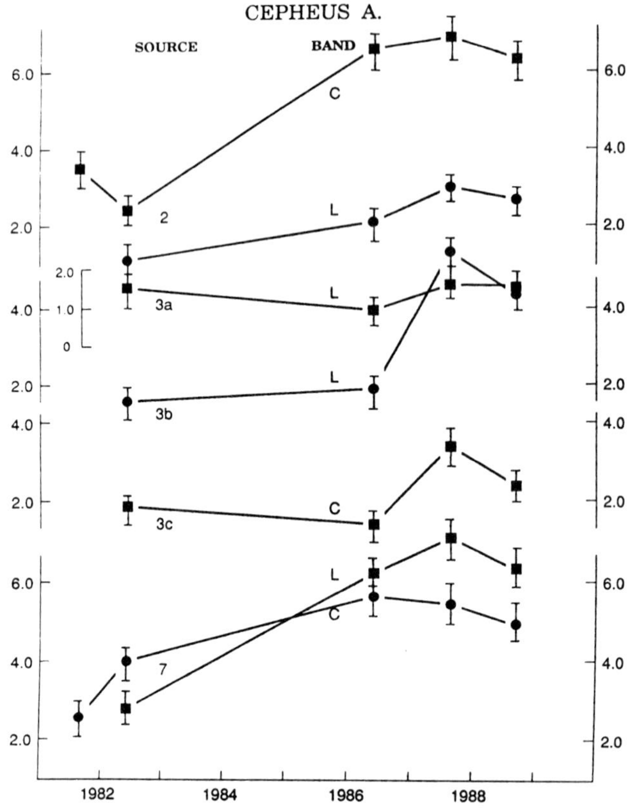

Figure 3. Plot of variation in flux density, in mJy/beam, against time.

(1988) failed to detect emission from any of the sources, except component 3a. Also, scattered polarized IR radiation indicated a source somewhere in the region of component 2. An analysis of IRAS Point Source Catalog by Hughes and MacLeod (1988) indicated extinction of 3-5 mag at 12 μm, suggesting > 100 mag in the visible, and consistent with similar estimates by Lenzen, Hodapp, and Solf (1984).

As would be expected, densities in the region are high, being estimated at $10^3 - 10^4$ by Sargent (1977), but more recent observations with Moriarty-Schieven (unpublished) of the CS (7-6) transition gives values > 10^6 cm^{-3}, and possibly as high as 10^7 cm^{-3}. The fact that an HII region can recombine in < 1 year, also requires densities > 10^6 cm^{-3} (Hughes 1988).

One explanation of how stars might produce variable HII regions is suggested from the work by Yorke and Krügel, who simulated a model for stellar accretion. Essentially, if a star of > 5 $M_{\odot}$ forms, then the core could come onto the main sequence while the outer parts are still accreting. A shock is produced with high IR luminosity, sufficient to slow accretion, which in turn reduces the IR luminosity. Oscillations are produced with periods of between 5 and 1,000 years, depending on the surrounding densities, so that it would appear possible to obtain periods of a few years. The more diffuse HII regions would correspond to the ending of the instability phase and develpment of a stable HII region. In addition, shell sources could be produced as for Source 7 (a), since recombination rates in the denser core would be greater than in the outer parts. Chief objections to the model are that isotropic accretion is unlikely, and in any case, there is no explanation of why stars should form in lines.

An alternative model is one where only a few stars are present, possibly Sources 2, 3c and 3d, which are the source of collimated jets. It is possible that these jets can be focussed along their length, as described in the paper in these proceedings by Cantó. If this is so, then the model will have to explain why two compact (< 70 au) HII regions are in the line of the jets, and form inside a few tens of au from the source. Alternatively, the compact objects could be stars, in which case the jets would have to emanate along the line of the two stars.

REFERENCES

Cohen, R.J., Rowland, P.R., and Blair, M.M. 1984, Mon. Not. Roy. Astr. Soc., **210**, 425.

Evans, N.J. et al. 1981, Ap.J., **244**, 115.

Hughes, V.A. 1985, Ap.J., **298**, 830.

Hughes, V.A. 1988, Ap.J., **333**, 788.

Hughes, V.A., and MacLeod, G.C. 1988, Astron. J., **97**, 786.

Hughes, V.A., and Wouterloot, G.G.A. 1982, Astron. Ap., **106**, 171.

Hughes, V.A., and Wouterloot, J.G.A. 1984, Astron. Ap., **276**, 204.

Lenzen, R. 1988, Astron. Ap. 1988, **190**, 269.

Lenzen, R., Hodapp, K-W., and Solf, J. 1984, Astron. Ap., **137**, 202.

Sargent, A.I. 1977, Ap.J., **318**, 736.

Yorke, H.W., and Krügel, E. 1977, Astron. Ap., **54**, 183.

Discussion:

YORKE: In the paper Yorke, Krugel (1977) oscillatory behaviour was found on two time scales, one at about 5 years and another at about 30000 years. Did you look for indications of periodicity in your observations of variability?

HUGHES: Yes, but over the period of 8 years, source 8 is the only one with a clear periodicity. It appears to be a relaxation oscillator, being present in 1981, not in 1982, present in 1986 but not in 1987 or 1988. As far as periodicity, your analysis was dependent on local density, and our density of $\sim 10^6 cm^{-3}$ is far higher than any you considered - one might expect a higher periodicity than 5 years.

SCHWARZ: How sure are you about the reality of these variations? Can these not be due to self-cal in a poor S/N regime?

HUGHES: I am very confident that the major variations that were discussed are real. Each set of observations contained more than 100000 visibilities, there were all reduced using the same self-cal procedures; in no case was the self-cal procedure overdone. As far as S/N ratio is concerned, peak fluxes are 2-3 mJy, but the 3σ noise level was < 0.1 mJy. This did not provide a poor S/N regime.

SARGENT: My first question echoes that of U. Schwarz in that I am also concerned about the effects of "clean" and "self-cal" on the maps. I would also like to know the range of percentage variation in the source fluxes. Are there differences at different wavelengths?

HUGHES: My answer echoes a former answer, in that I see no difficulties in applying the clean procedure, and "self-cal" was not overdone. In some cases, variations in individual sources were difficult to measure due to not being completely resolved; in others, sources may be over-resolved at 6 cm but not at 20 cm, as for Source 3. For those sources for which confident flux densities were determined, the figure in the text shows the variations. Source 8 and the new source (9) were completely resolved, and their flux densities are indicated in the paper.

TERLEVICH, R.: Have you looked for proper motions, and if not detected, can you give an upper limit to the tangential velocity in your objects?

HUGHES: Yes, but the present upper limit is 400 kms^{-1}. We hope in time to lower this value, but one limit is the accurate position of calibrations sources.

HARI OM VATS: How sure are you that variability time scales less than 1 year are not present in HII regions?.

HUGHES: The highly variable sources could have variability of less than 1 year, but at present the minimun interval between observations is about 1.25 years. We have observations taken in 1989 Jan, not yet reduced, wich should show if there is variability over a period of 0.5 years.

THE SHARPLESS 187 GAS COMPLEX : A STUDY OF THE MOLECULAR, ATOMIC, IONIZED AND DUST COMPONENTS

Gilles Joncas[1], Daniel Durand[2], C. Kömpe[3] and R.S. Roger[4]
1 Dépt. de Physique, Université Laval, Québec, Canada, G1K 7P4
2 Dominion Astrophysical Observatory, Victoria, B.C., Canada, V8X 4M6
3 Herzberg Institute of Astrophysics, Ottawa, Ont., Canada, K1A 0R6
4 Dominion Radio Astrophysical Observatory, Penticton, B.C., Canada, V2A 6K3

1. Introduction

Star forming regions are one of the active components of the interstellar medium and as such play an important role in the galactic "ecosystem". When massive young stars are borned they have a strong impact on their environment through their radiation flux and stellar wind. We are then facing complicated interplays between gas in different states (ionized, atomic and molecular), dust particles and the young stars. The understanding of these interplays can only be done using multifrequency observations. Such an endaveour is already in progress (see Joncas et al. 1988 and Kömpe et al. 1989). We will describe here the young star forming region S187 (Sharpless 1959). This gas complex is nearby (≈1 kpc) thus permitting high spatial resolution with medium size instruments. It contains a faint optically visible HII region ionized by an unidentified B0 or B0.5 star. The associated molecular cloud, discovered by Blair et al. (1975), contains a molecular outflow (Bally and Lada 1983) to which an H_2O maser is associated (Henkel et al. 1986).

2. Observations

The following is a summary, the instruments and data handling will be described in more details elsewhere (Joncas et al. 1989b).

Ionized gas. Three Hα interferograms were obtained with the Université Laval Fabry-Pérot camera. The instrument was installed on the Observatoire Astronomique du mont Mégantic 1.6 m telescope. The detector was the observatory's RCA CCD chip. The scanning Fabry-Pérot interferometer (FP) has a resolution (FWHM) of 0.024 nm. Each interferogram results from the summation of five 2000 sec exposures.

Ionized and atomic gas. The radio centimetric observations were secured with the aperture synthesis radio telescope of the Dominion Radio Astrophysical Observatory (DRAO). It is designed to make HI spectral line observations simultaneously with the observation of 21 cm and 74 cm continuum emission. The inner portion of the uv planes were filled using single dish observations. The resolution (FWHM) was 1'0 EW x 1'.1 NS x 1.32 km s^{-1} at 21 cm. The 74 cm observations were discussed in Joncas et al. (1989a).

Molecular gas. The molecular cloud was observed with the Observatoire de Bordeaux 2.5 m millimeter antenna in the ^{13}CO J=1-0 line. The HPBW of the telescope is 4'.4 and the spectrometer channel width we used was 0.27 km s^{-1}. The spectra have a noise level (1σ) of ≈0.2 K. We surveyed ≈900 square arcminutes but the molecular cloud is not entirely mapped yet.

Dust particles. We used the survey data obtained by the Infra-Red Astronomical Satellite (IRAS). In order to gain higher spatial resolution and sensitivity, the survey scans were reprocessed with a technique called two-dimensional survey coaddition available at the Infrared Processing and Analysis Center (IPAC). The final product are maps with pixel size of 15", 15", 30" and 1' at 12, 25, 60 and 100 μm.

3. Results

3.1 The HII region

Because of the limited size and faintness of the nebula and the large amount of stars which contaminated the FP rings, only a small number of velocity points were confidently measured. From a total of 27 Hα radial velocity points, we obtained a mean LSR velocity of -18.6 km s^{-1} (1σ=4.5 km s^{-1}). Unfortunately there are to few data points to check for the presence of velocity gradients across the nebula.

Figure 1 is a superimposition of our 21 cm continuum emission map from a 1° square field containing S187 on a reproduction of an E plate of the POSS. At the 3σ level (the first contour) the nebula is 7' in diameter and roughly circular. The dented contour at the southern edge comes from the subtraction of a strong extragalactic radio source. The lack of agreement between the optical and radio pictures shows that the HII region suffers from a large amount of absorption. Actually, from Fig. 1, we see this absorption to be present across a good portion of the field. Note also that some diffuse optical emission has no radio counterpart. The measured radio flux is 1.15 ± 0.12 Jy. Assuming the nebula to be spherical (see Fig. 1) we derived from this flux and from an electron temperature of 6200 K (Rossano 1978), an r.m.s. density of 47 cm^{-3}, a mass (H + He) of 8.2 $M_{\odot}$ and an excitation parameter of 13.1 cm^{-2} pc (B0-B0.5 star, Panagia 1973).

3.2 The atomic gas

Although the DRAO synthesis telescope has a 2° field of view at 21 cm, we will concentrate here on the HI line emission in the S187 area. Inspection of the 128 HI maps produced from the observations reveals the presence of HI emission probably related with our object. Figure 2 is a map of integrated HI emission in a 1° square field around S187 with overlaid continuum contours of S187. Apparent is an HI feature taking the form of a shell partially surrounding the nebula over the

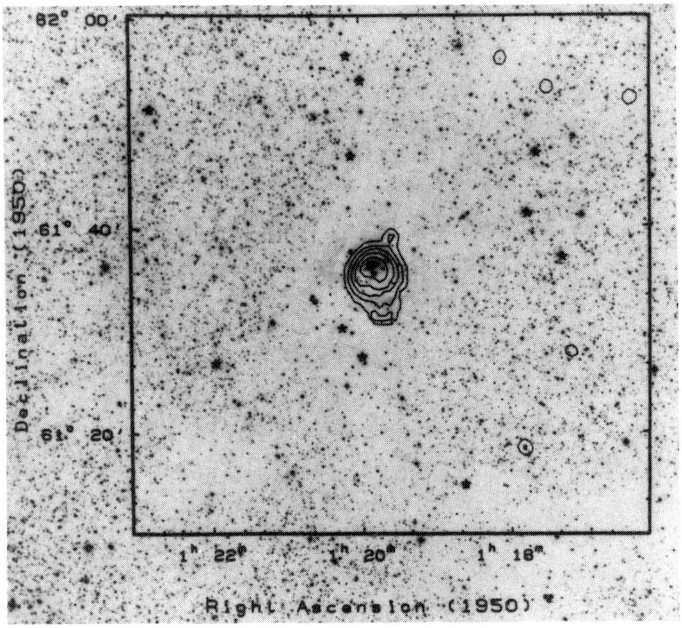

Fig. 1. Superimposition of a reproduction of an E plate of the POSS with a DRAO 21 cm continuum map of the S187 area.

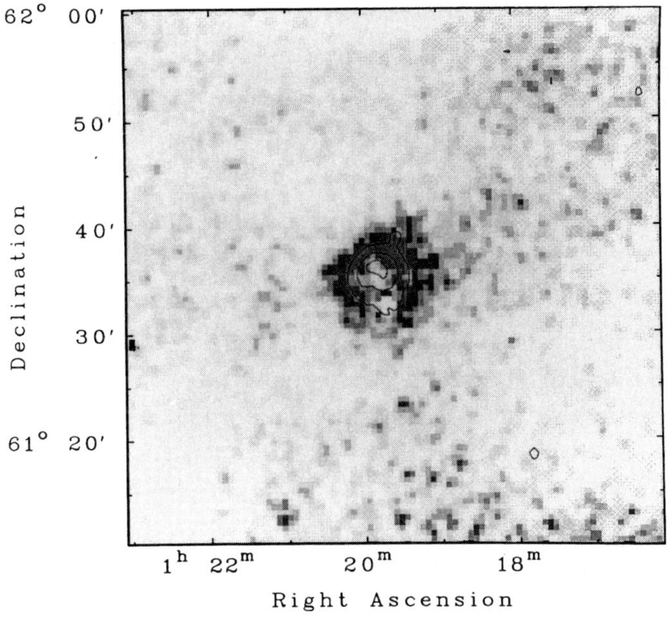

Fig. 2. Map of HI emission in a 1° x 1° field around S187 integrated over the velocity range -10.1 to -22.4 km s^{-1}. Overlaid is the radio continuum contours of S187.

integrated velocity interval (-10.1 to -22.4 km s^{-1}), its diameter is ≈12.5'. Starting at -10.1 km s^{-1} the HI first follows the northwestern contour of the nebula. As the velocity becomes more negative the HI emission migrates to follow the NE and SW contours. At Vlsr ≤ - 18.3 km s^{-1}, the HI is now found at the SE edge and overlaps part of the nebula. The shell is ≈0.7 pc thick on its eastern half and ≈1.6 pc thick on its western half (FWHM). The total mass of the shell is ≈$70M_\odot$ (assuming optically thin conditions). The velocity range of the ionized gas (-10.5 to -27.0 km s^{-1}) is roughly within the limits of the shell's velocity field.

3.3 The molecular cloud

Figure 3 shows the integrated intensity map of our ^{13}CO observations. The LSR velocity range is -7 to -18 km s^{-1}. The molecular cloud is ≈12 pc x ≈6 pc in extent and is elongated along a NW-SE axis. Recall however that the cloud may be somewhat bigger and differ in shape since we have not finished its mapping yet. The summation of all our spectra reveals the presence of 4 velocity components located at -14.5, -13.0, -11.5 and -8.5 km s^{-1}. No velocity gradient is apparent across the cloud at a spatial resolution of 4.4'. The origin of the different velocity components is still conjectured. We calculated the mass of the cloud to be ≈$4600M_\odot$.

Eventhough they have different spatial resolution, we compared the velocity field of the HI shell with the overlapping ^{13}CO spectra. For the area covered by the nebula, the HI gas is blueshifted with respect to the molecular gas. It is impossible to say if the HI is in front or behind the HII region. In the area covered by the shell, the atomic gas has the same velocity or is blueshifted with respect to molecular gas. It thus appears that the HI feature surrounding S187 is expanding in part into or/and out of the molecular cloud. We believe the atomic gas to originate from the dissociation of the molecular gas by the UV photons emanating from the exciting star. Let us recall here the mean velocity of the ionized gas, -18.6 km s^{-1}. The ionized gas also seems to be expanding away from the molecular cloud towards the observer.

4.4 Dust particles

The 2-D coadd technique produces maps with spatial resolution varying with wavelength. For the sake of comparaison we show in Fig. 4 IR continuum maps at 12, 25, 60 and 100 μm having the same pixel size, 1'. The first contour is at the 3σ level. All four maps have the same peak position and it corresponds to the position of the 21 cm radio continuum peak. The IR emission has the same orientation as the ^{13}CO emission and is as extended. We derived the following fluxes 360, 680, 6700, 10000 Jy for the emitting dust. From these the total IR luminosity (8 to 2200μm) was derived to be 2.3 x 10^4 $L_\odot$. Assuming that the IR radiation comes from the absorption of every Ly α photons, a B0 star would supply all the necessary radiation.

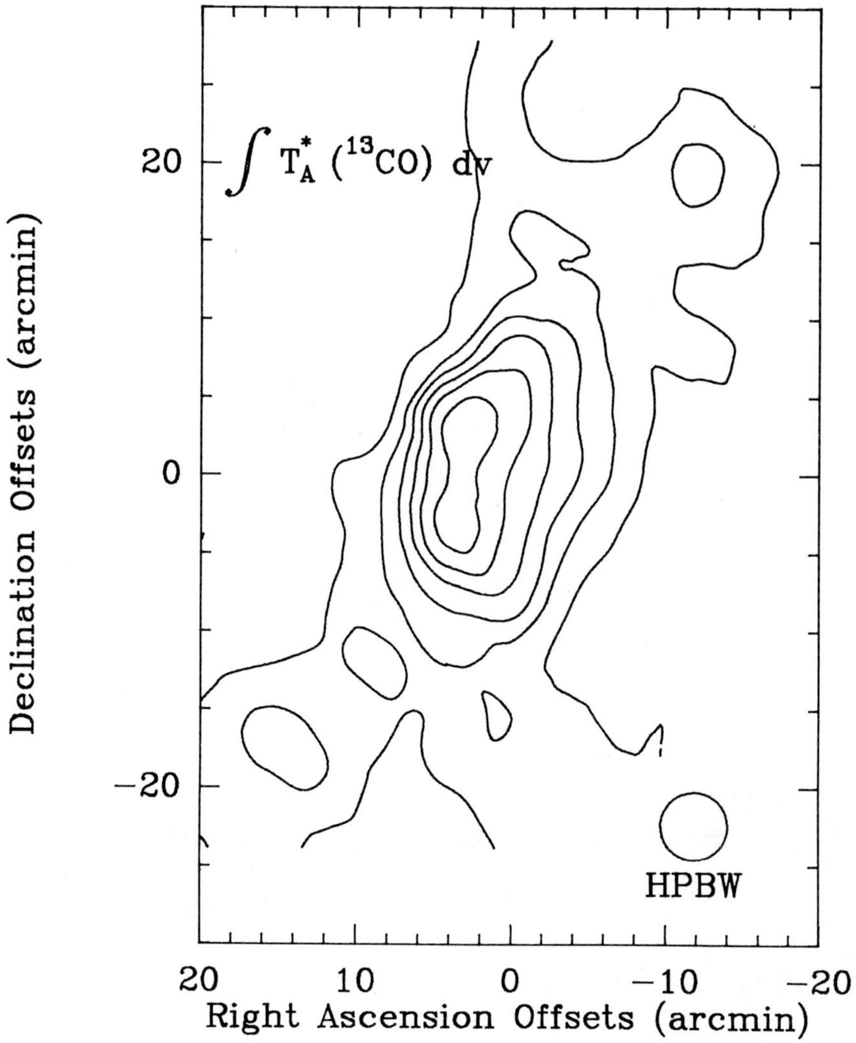

Fig. 3. Integrated intensity map of the ^{13}CO (J=1-0) emission line of the S187 molecular cloud. The angular resolution is 4'.4 and the reference position is : α_{1950} = $1^h19^m48^s$, δ_{1950} = $61^\circ 35'$.

Following the usual technique (Evans 1980), we calculated the mean dust temperature to be $\langle T_{12/25} \rangle$ = 191 K (1σ=22 K) and $\langle T_{60/100} \rangle$ = 33 K (1σ=5 K). $T_{12/25}$ and $T_{60/100}$ dust temperature maps (not shown here) present temperature gradients of opposite behaviour. This is consistent with the presence of two grain populations, one with thermal emission and the other with non-thermal emission. The total amount of dust emitting at 100 μm was derived to be 1.6 $M_\odot$.

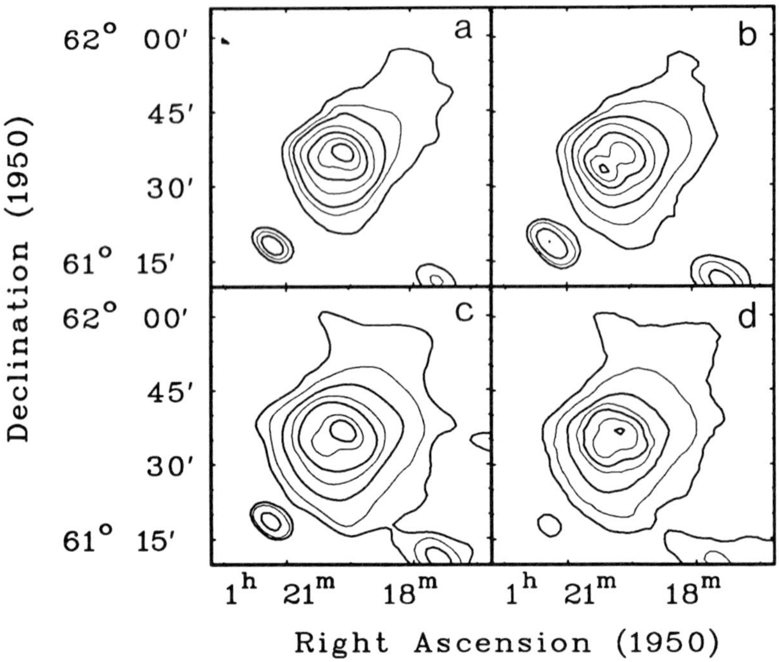

Fig. 4. The IR continuum maps of S187 (IRAS co-added data) (a) 12μm, (b) 25μm (c) 60μm, (d) 100μm.

5. Discussion

What we have shown here is an overview of part of our S187 data bank. We have not fully explored it yet, we still have monochromatic Hα CCD frames, BVR stellar photometry of the area and the IRAS point source catalog to look into. We also have to integrate the multifrequency data together. This will be delt with in a forthcoming paper. However it appears for now that the most massive star in the area is a B0 star, that S187 may be partly a reflection nebula, that the HI gas is closely linked to the molecular gas and that the exciting star seems to be heating dust across the entire surface of the molecular cloud facing the observer.

References

Bally, J., Lada, C.J.: 1983, Astrophys. J. 265, 824

Blair, G.N., Peters, W.L., Vanden Bout, P.A.: 1975, Astrophys. J. 200, L161

Evans, N.J., II: 1980, "Interstellar molecules", IAU Symposium 87, ed. B. Andrew, Reidel, Dordrecht, 1

Henkel, C., Haschich, A.D., Gusten, R.: 1986, Astron. Astrophys. 165, 197

Joncas, G., Kömpe, C., de la Noë, J.: 1988, Astrophys. J. 332, 1030

Joncas, G., Roger, R.S., Dewdney, P.E.: 1989a, Astron. Astrophys. in press

Joncas, G., Dewdney, P., Durand, D., Kömpe, C., Roger, R.S.: 1989b, in preparation

Kömpe, C., Joncas, G., Beaudry, A., Wouterloot, J.G.A.: 1989, Astron. Astrophys. in press

Panagia, N.: 1973, Astron. J. 78, 929

Rossano, G.S.: 1978, Astron. J. 83, 1214

Sharpless, S.: 1959, Astrophys. J. Suppl., 4, 257

Turbulent Mixing in Wind-Blown HII Regions

D. Breitschwerdt[†] and F.D. Kahn[‡]
[†]*Max-Planck-Institut für Kernphysik, D-6900 Heidelberg, Postfach 10 39 80, FRG*
[‡]*Department of Astronomy, The University, Manchester M13 9PL,England*

Abstract

Turbulent mixing between an ionization bounded HII region and a hot shocked stellar wind (HSSW), which keeps it under pressure, is examined. Recently we have shown that acoustic disturbances can grow there to finite amplitude in a time scale which is comparable to the sound crossing time in the HII layer. The resulting turbulence will then stretch fluid elements and the frozen-in magnetic field. A condition under which turbulence can decay down to the viscous scale, where mixing is very efficient, is derived. For a uniform and plane parallel magnetic field B_0 and a constant density ρ_0 of the ambient medium, we find that efficient mixing takes place near the polar regions. Subsequently the rate of mass addition to the hot bubble is calculated and it is shown that catastrophic cooling is likely to occur. In the case of NGC 6334(A) it seems that this has just happened and we predict an upper limit for B_0 of 4×10^{-5} gauss there. This model may also explain the existence of highly ionized species (e.g. OVI), soft X-rays and high velocity flows of the order of 100 km/s.

1. Introduction

According to well-known theoretical models (e.g. Dyson and deVries, 1972; Castor et al., 1975; Dyson, 1989) OB stars and Planetary Nebulae should generally have HII regions which are squeezed into thin shells held under pressure by hot shocked stellar winds (HSSWs). However, quite often observations show rather extended HII regions, such as the Rosette Nebula (Dickel 1974) or NGC 6334(A) (Rodriguez et al. 1988). In addition, the classical model fails to explain high velocity flows in excess of 100 km/s (López et al. 1989), the existence of highly ionized species, such as OVI, NV and also soft X-rays. These are indicative of temperatures of the order of 10^5 - 10^6 K, which are neither present in the hot bubble ($> 10^7$ K) nor in the HII region ($\approx 10^4$ K).

In a recent paper (Breitschwerdt and Kahn, 1988) we have shown that acoustic disturbances travelling parallel to the HII/HSSW interface can grow to finite amplitude provided that the radiation is not exactly incident at right angles and that the HII region is ionization bounded. Hence the classical model is unstable and the growth rate of waves is fairly large, viz. inversely proportional to the sound crossing time in the HII layer. We will show in the following that the resulting turbulence can induce local mixing and cooling, thus altering the dynamics of wind-blown HII regions drastically.

2. Model for Turbulent Mixing with Magnetic Field

We assume that the wind expands into an ambient medium with a magnetic field which is uniform and plane parallel on a scale of tens of parsecs. It should also be sufficiently weak, so that similarity solutions for the hot bubble are still applicable.

At largest scale l, which corresponds to the thickness of the HII layer, the magnetic energy density shall be less than the turbulent energy density. All the speeds u are less than the isothermal speed of sound c_i in the ionized gas, so that the turbulent flow is incompressible and the characteristic speed at length scale l' follows a Kolmogorov law

$$u' = u \left(\frac{l'}{l}\right)^{1/3} . \tag{1}$$

The effect of turbulence is to stretch fluid elements and the frozen-in magnetic field which, viewed on smaller scales, leads to an increase in surface area and field strength. However, there will also be a reduction of magnetic energy density due to magnetic reconnection.

Let us divide the length scales into octaves:

$$\frac{l}{l'} = 2^n , \tag{2}$$

and let B_n^2 be the mean squared field at the nth octave. It is then easy to see that in the absence of reconnection B_n^2 scales like

$$B_n^2 = B_i^2 \left(\frac{l}{l'}\right)^{1/3} = B_i^2 e^{\eta n}, \tag{3}$$

where $\eta = (1/3)\log 2$.

Therefore the change in B_n^2 from octave n to $n+1$ in the presence of magnetic reconnection is

$$\frac{dB_n^2}{dn} = \eta B_n^2 - C\left(\frac{dB_n^2}{dn}\right)\frac{B_n}{l'\sqrt{4\pi\rho_i}}\frac{l}{u} . \tag{4}$$

ρ_i is the mass density in the HII region and C is a proportionality constant usually taken to be 0.1 (Petschek 1964). In this paper we prefer the value 0.05 since the lines of force are not necessarily antiparallel here.

For convenience we write $B_n = fB_i$ and $\sqrt{4\pi\rho_i u^2}/B_i = M_A$, the Alfvén Mach number. In a forthcoming paper (Kahn and Breitschwerdt, 1989) we solve equation (4) in terms of f. Here we just briefly discuss the two asymptotic limits for small and for large n.

In the first case the solution without reconnection is recovered

$$f = e^{\eta n/2} , \tag{5}$$

and stretching is the dominant process. For large n, i.e. small length scales, reconnection is more important and we find

$$f = \left(\frac{7M_A}{C}\right)^{6/7} e^{-5\eta n/2} . \tag{6}$$

This leads to the conclusion that there is a critical length scale l'_c, where the magnetic energy density has a maximum, which is given by

$$\frac{l}{l'_c} \equiv \left(\frac{7M_A}{C}\right)^{6/7} . \tag{7}$$

Thus on smaller scales the magnetic field strength becomes progressively weaker and turbulence can cascade down until viscous dissipation takes place.

In order that this process works, two important restricions have to be imposed. First, the accumulated rate of dissipated magnetic energy down to scale l'_c due to reconnection must be less than the rate of dissipated mechanical energy. It is shown elsewhere that this can be expressed as $M_A \gg 1.8\,C^{-1/6} \approx 3$. Second, the magnetic energy density must always be less than the kinetic energy density on *any* scale. This criterion yields $M_A \gg (7/C)^{3/4} \approx 41$, which is the more compelling.

This will limit the total surface area of the HII/HSSW interface, where efficient mixing can take place, simply because the magnetic field has to be weak enough to satisfy the above conditions.

3. Mixing Rate and Catastrophic Cooling

Using our assumptions of the preceding section we describe the large scale expansion of the bubble by similarity solutions (e.g. Dyson 1981):

$$R_b = 0.76 \left(\frac{L_w}{\rho_0}\right)^{1/5} t^{3/5}\,, \tag{8a}$$

where R_b is the radius of the bubble and L_w is the wind mechanical luminosity and

$$\rho_i = 0.78\rho_0 \left(\frac{\dot{R}_b}{c_i}\right)^2 . \tag{8b}$$

Here and in the following all dimensional quantities are in CGS units. For a plane parallel magnetic field geometry, the field strength in the compressed HII region, B_i, varies with polar angle like

$$B_i = 1.15 B_0 \left(\frac{\dot{R}_b^2}{c_i^2}\right) \sin\theta\,. \tag{9}$$

This follows from conservation of magnetic flux across the shell and the thicknes l of the ionized part (which has been used in equation (9)) can be derived from ionization balance.

A typical growth rate for unstable waves in the ionized layer is $\sigma = 0.1c_i/l$ and therefore the rate at which energy is fed into the largest eddies is $0.1c_iu^2/l$, which when equated to the rate of dissipation, yields $u = 0.13c_i$. This value has to be used when the above condition $M_A > 41$ is applied. It implies that there is a critical field strength B_i or by virtue of equation (9) a critical angle θ_c, increasing with time, so that for $\theta < \theta_c$ free mixing occurs at a rate $0.13\rho_ic_i$ per unit area. In our model with its simple field configuration this happens near the polar caps of the HII/HSSW interface and the rate of mass addition to the bubble is

$$\dot{\mathcal{M}} = 2.7 \times 10^{-5} L_w^{2/5} \rho_0^{8/5} c_i^3 B_0^{-2} t^{6/5}\,. \tag{10}$$

The flow of cooler (10^4 K) gas per unit time overwhelms the HSSW gas generated by the terminal stellar wind shock by a large factor and therefore induces local radiative cooling in an otherwise always adiabatic bubble. A catastrophic loss of pressure near the polar caps seems unavoidable and therefore large scale motions should occur. It is interesting to note that the magnetic field geometry automatically produces bipolar outflow structures once the shell breaks up; these have been observed in many cases.

If the mass of the bubble is obtained in this fashion, an averaged density and pressure and thus the time scale for radiative cooling, t_c, can be worked out. Cooling of the bubble may become catastrophic in the sense that its expansion is severly disturbed and it may lead to a collapse in the extreme case. This happens when $t \leq t_c$ or

$$t \geq t_{crit} \equiv 2.73 \times 10^9 \left(\frac{\rho_0}{\rho_{-24}}\right)^{-23/16} \left(\frac{L_w}{L_{36}}\right)^{11/32} \left(\frac{B_0}{B_{-6}}\right)^{25/16} \text{ yr}, \tag{11}$$

where $c_i = 10$ km/s has been used and ρ_{-24}, L_{36} and B_{-6} are 10^{-24}g/cm^3, 10^{36} erg/s and 10^{-6} gauss, respectively.

To illustrate this effect, we apply these results to NGC 6334(A), which has been observed by Rodriguez et al. (1988). They infer $L_w = 10^{36}$ erg/s, $\rho_0 = 4 \times 10^{-19}\text{g/cm}^3$, $R_b = 3.7 \times 10^{17}$ cm, $l = 0.1 R_b$ and a Ly c photon rate $S_* = 4 \times 10^{48}\text{s}^{-1}$, which yields an age of about 2.2×10^{11} s. It can be shown easily that the HII region is ionization bounded and therefore our model holds. The bubble on the radio map appears slightly distorted and is embedded in a bipolar flow. Therefore our tentative conclusion is that catastrophic cooling has just occured but it has not yet destroyed the structure. Alternatively, the loss of pressure increases the thickness of the HII layer and therefore the growth time of the instability as well. As a consequence the mixing rate drops and therefore cooling becomes less efficient. In any case it allows us to set an *upper* limit to the magnetic field strength in the surrounding molecular cloud:

$$B_0 \leq 4 \times 10^{-5} \text{ gauss}. \tag{12}$$

It would be highly desirable to obtain measurements both of the field strength and its orientation for NGC 6334(A) or the ambient media of other early type stars.

4. Conclusions

We suggest that turbulent mixing around early type stars should be a common phenomenon. In the view of the obstruction of conductive heat flux by even weak magnetic fields it is a possible way of exchanging mass and energy between the hot bubble and the HII region. It has been shown that the mixing rate itself depends mainly on B_0 and ρ_0 and only weakly on stellar parameters such as the Ly c photon rate or the mechanical luminosity of the wind. The model presented here can explain naturally the existence of OVI and NV lines and soft X-rays which should arise in cooling regions and also high velocity flows, which we are currently investigating.

References

Breitschwerdt, D. and Kahn, F.D., 1988, *Mon. Not. R. astr. Soc.*, **235**, 1011.

Castor, J., McCray, R., and Weaver, R., 1975, *Astrophys. J.*, **200**, L107.

Dickel, R.H., 1974, *Astr. Astrophys.*, **31**, 11.

Dyson, J.E., 1981, *Investigating the Universe*, ed. Kahn, F.D., D. Reidel Publ. Comp., Dordrecht, Holland, p.125.

Dyson, J.E., 1989, *these proceedings.*

Dyson, J.E. and deVries, J., 1972, *Astr. Astrophys.*, **20**, 223.

Kahn, F.D. and Breitschwerdt, D., 1989, *Mon. Not. R. astr. Soc.*, (submitted).

López, J.A., Falcón, L.H., Ruiz, M.T. and Roth, M., 1987, *Planetary Nebulae*, IAU Symposium No. 131, eds. Torres-Peimbert, S., Kluwer Acad. Publ. Comp., Dordrecht, p.179.

Petschek, H.E., 1964, *NASA SP-50*, U.S. Govt. Printing Office, p.425.

Rodriguez, L.F., Cantó, J. and Moran, J.M., 1988, *Astrophys. J.*, **333**, 801.

Discussion:

DYSON: Do your results imply that energy driven bubbles are rather rare?

BREITSCWERDT: At present I cannot give a clearcut answer. The mixing process which I have discussed drives the bubble into the momentum driven phase for some time. However the following increase in thickness of the HII region reduces the instability rate and hence turbulence. At the same time hot gas is fed into the bubble continuosly by the stellar wind and thus raise the pressure there and possibly lead to an energy driven flow again. We are going to investigate whether there exists a self-limiting process or whether the bubble will undergo an oscilatory behaviour switching from momentum to energy driven flow.

JONCAS: In an HII region undergoing a Champagne phase, would the flow of ionized gas eliminate any effect the stellar wind bubble would have? Is it time dependent?

BREITSCWERDT: Certainly not. On the contrary, if there is a density jump at some interface like the edge of a molecular cloud, the formation of a secondary bubble will occur which will also be driven by the stellar wind.

ROY: Could heat conduction play any role in the energy transfer? Why?

BREITSCWERDT: Although heat conduction has been advocated quite often in the past as a mechanism for energy transfer I would like to stress that there are some severe problems. First, any magnetic field with a component perpendicular to the temperature gradient will reduce the mean free path of electrons considerabily. Second, the normal Spitzer conductivity is certainly not apropiate in the case of a very steep temperature gradient where the scale height is less than the mean free path, like at the HII/ bubble interface.

ACCRETION FLOWS IN HIGH-MASS STAR FORMATION

Eric Keto
Lawrence Livermore National Laboratory
Institute of Geophysics and Planetary Physics

ABSTRACT

We compare observed and simulated images of the accretion flows associated with high mass star formation in the regions G10.6–0.4 and DR21. We describe, as a result of the comparison, the temperature, density, and velocity fields. Our results indicate that the G10.6–0.4 cloud core is strongly condensed and has approximately equal velocities in rotation and infall at its current evolutionary state. The rapid collapse and lack of rotational support suggests that significant angular momentum transfer is occuring over scales at least as large as those observed (0.5 pc). A milligauss magnetic field would have sufficient energy to supply the required braking torque of 10^{47} ergs. The DR21 core shows approximately spherically symmetric radial accretion with no detectable rotation. Unlike the G10.6–0.4 core, the DR21 core does not contain an embedded HII region. Thus this core may represent a molecular cloud condensation undergoing gravitational collapse and accretion just prior to the formation of massive stars.

RADIATIVE TRANSFER

For comparison of any model with observations we need to compute a three dimensional image of the line brightness where two of the three dimensions are the x and y map positions, and the third is the line-of-sight velocity. We adopt the following simple scheme. From the model H_2 density, temperature, and fractional abundance of our molecular tracer we compute the source function and opacity assuming LTE. The local line profile is specified by the thermal and turbulent broadening and the model line-of-sight velocity. The line brightness at a particular frequency is determined by integrating the radiative transfer equation along the line of sight. The resulting image is then convolved with the beam pattern and autocorrelator response appropriate to the observing apparatus.

G10.6–0.4

Spectral line observations of the dense gas immediately around the G10.6–0.4 HII region suggest infall and rotation consistent with gravitational collapse in the potential well of the gas and stars. The flow represents a continuation beyond the onset of nuclear burning of the original accretion responsible for star-formation. The rotational and infall velocities increase with decreasing radius suggesting spin-up and accelerating infall. Comparison of the $NH_3(1,1)$ to $NH_3(3,3)$ line ratios indicates that the molecular cloud core is heated by dust reradiation of the UV continuum from the newly formed stars and that the density is centrally condensed (Ho and Haschick 1986, Keto *et al.* 1987, 1988).

Our observations of G10.6–0.4 include four data sets. We observed the $NH_3(1,1)$ and (3,3) rotational states using a compact configuration (D) of the VLA to achieve 10″ resolution, and an extended configuration (B) for 0.3″ resolution. In figures 1a and 2a, we plot the line brightness as a function of velocity and position in the equatorial plane of the rotating gas from two of these data sets. In figures 1b and 2b, we show plots generated from our model. Table 1 lists the equations and parameters which constitute the model. The model has been arbitrarily chosen on the basis of mathematical simplicity guided by the analysis of the observations. The model is purely diagnostic in the sense that it is designed to match the observations rather than describe the physics of the accretion flow.

Table 1. Model Parameters for G10.6–0.4

HII radius[a]	r_{HII}	0.025		
electron temperature	T_e	8000 K		
continuum opacity	κ_{HII}	3.25×10^{-18} cm^{-1}		
stellar mass	M_{st}	235 $M_\odot$		
systemic velocity	V_{LSR}	–3.5 km s^{-1}		
cloud radius	r_c	1.0		
inclination angle	i	zero		
ellipticity	ϵ	1.0		
temperature	$T = T_0(r_T/r)^{\alpha_T}$	$T_0 = 40$ K	$r_T = 0.35$	$\alpha_T = 0.5$
density	$n = n_0(r_n/r)^{\alpha_n}$	$n_0 = 2 \times 10^4$	$r_n = 0.35$	$\alpha_n = 2.5$
azimuthal velocity[b]	$v_{az} = v_{0az}(r_{0az}/r)^{\alpha_{0az}} - v_{1az}(r/r_{1az})^{\alpha_{1az}}$			
		$v_{0az} = 4.0$	$r_{0az} = 0.025$	$\alpha_{0az} = 0.1$
		$v_{1az} = 0.25$	$r_{1az} = 0.025$	$\alpha_{1az} = 0.8$
turbulent velocity	$v_t = v_{0t}(r/r_{v_t})^{\alpha_{v_t}}$	$v_{0t} = 0.25$	$r_{v_t} = 0.025$	$\alpha_{v_t} = 0.8$
radial velocity	$v_r^2 = GM(r)/r - v_{az}^2 - v_t^2$			
$[NH_3/H_2]$	f_{NH_3}	1.4×10^{-6}		

[a] All radii are in pc.
[b] All velocities are in km s^{-1}.

Figure 3a shows a vector plot of the modeled velocities in the equatorial plane and Fig. 3b, the three velocity components with radius. The gas spirals into the center quite rapidly, completing less than one orbit over the two decades in radial distances accessible by the observations. The failure to achieve rotational support

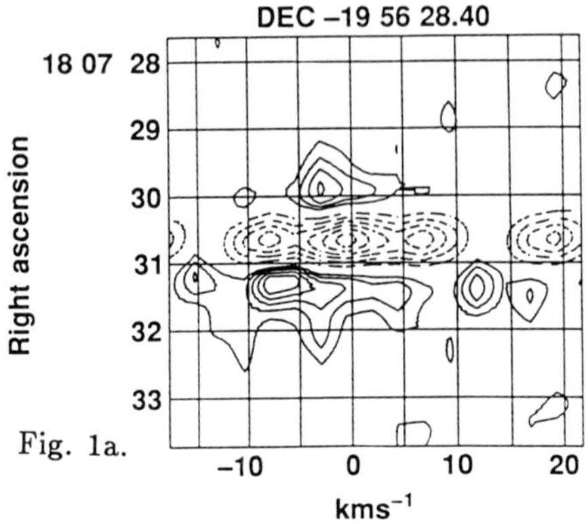

Fig. 1a.

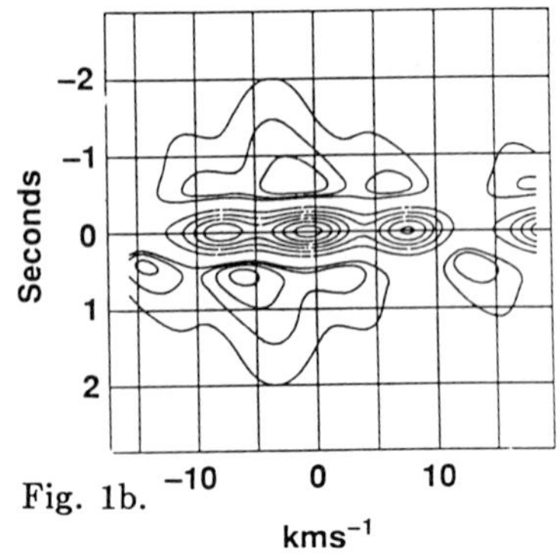

Fig. 1b.

Fig. 1a. Position-velocity plot of the $NH_3(1,1)$ line brightness in the equatorial plane of G10.6–0.4. The resolution is 10″ and 2.4 km s^{-1}. The contour interval is 50 mJy/beam in emission and 200 mJy/beam in absorption (dashed contours). The multiple components are due to the hyperfine structure of NH_3. Fig. 1b. Model position-velocity plot. The contour interval is 8.5 in degrees K in emission and 21 K in absorption.

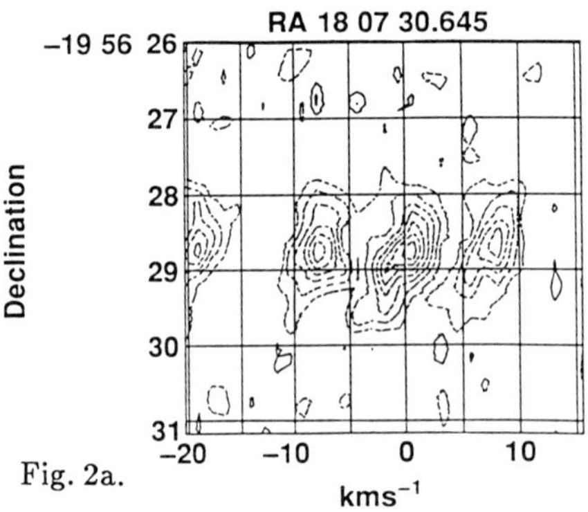

Fig. 2a.

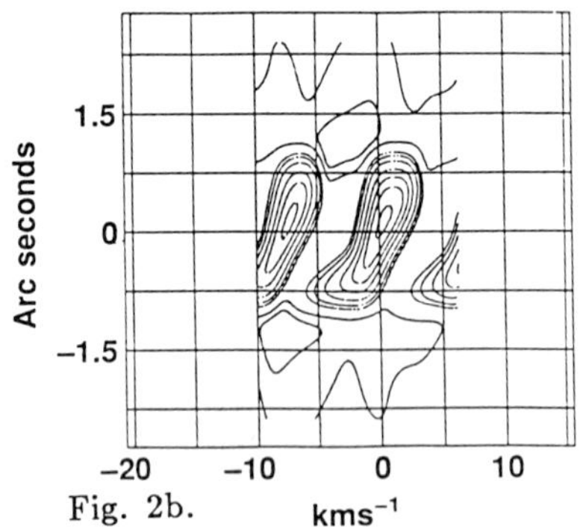

Fig. 2b.

Fig. 2a. Position-velocity diagram of the $NH_3(1,1)$ line at 0.4″ and 1.2 km s^{-1} resolution. The contour interval is –30 mJy/beam. Fig. 2b. Model position-velocity diagram. The contours are –625, –500, –375, –250, –125, –62, –31, 12.5, 25 K. Only a portion of Fig. 2a has been modeled.

suggests either collapse from a state of negligibly low angular momentum, or more likely, strong angular momentum braking. The time rate of change of the angular momentum may be estimated as $dL/dt = mv_r(v_{az1}r_1 - v_{az2}r_2)/(r_1 - r_2)$. This estimate strictly applies only assuming a steady state velocity field which we know cannot be correct. Nevertheless, the required torque is about 5×10^{46} ergs from 0.35 to 0.1 pc and 10^{47} ergs from 0.1 to 0.025 pc. The average field strength, $\langle B^2/8\pi \rangle$, required to equal this torque is 0.5 mG within 0.35 pc and a factor of ten higher within 0.1 pc. The required field strengths are within the range measured by Zeeman splitting of OH masers around similar HII regions and are also consistent with the increase expected from flux freezing in the condensed core.

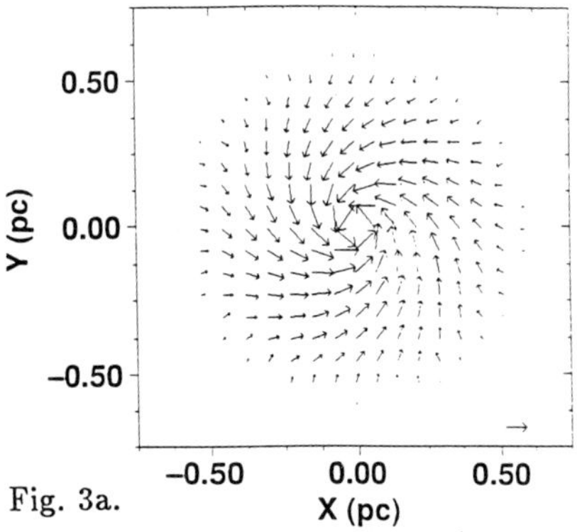

Fig. 3a.

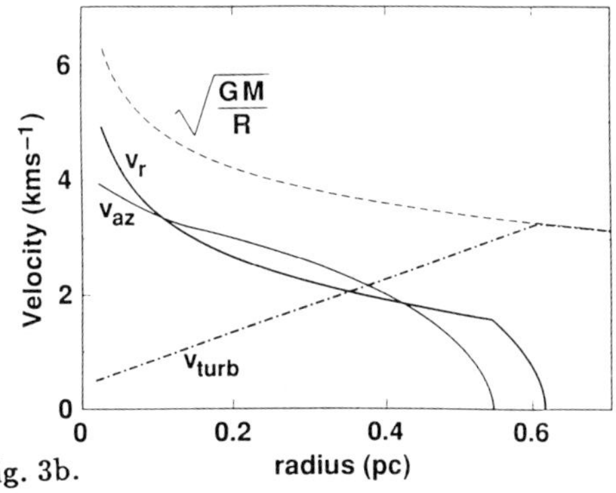

Fig. 3b.

Fig. 3a The velocity directions and magnitudes in the equatorial plane of the G10.6–0.4. The arrow lower right is 1.3 km s^{-1}. *3b.* The three velocity components used in the model and the gravitational energy in km s^{-1}.

DR21

Continuum observations show that massive star forming regions typically contain clusters of compact and ultracompact HII regions of a range of sizes and ages. Thus there should be dense cores in the molecular cloud surrounding these newly formed HII regions which are at an earlier evolutionary stage. This stage would be characterized by an accretion flow similar to that observed around G10.6–0.4 but existing prior to the formation of ZAMS stars and the ionization of the immediate molecular gas.

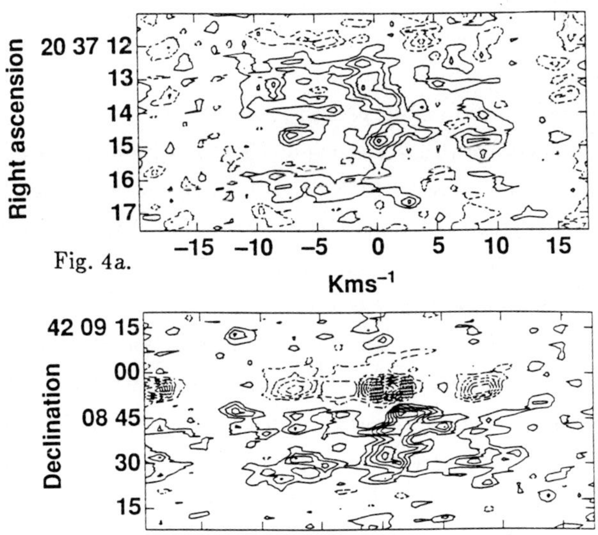

Fig. 4a.

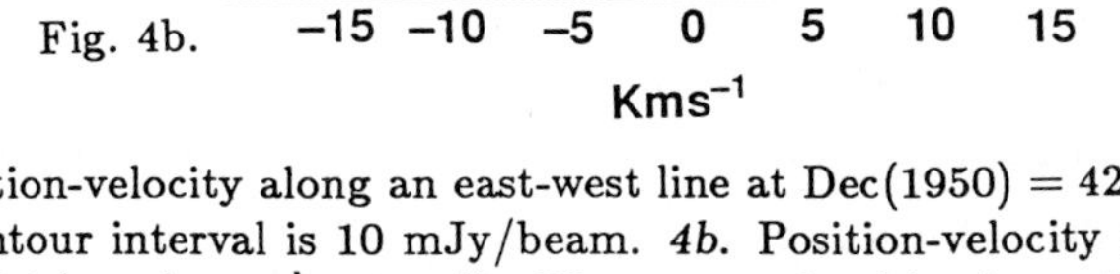

Fig. 4b.

Fig. 4a. Position-velocity along an east-west line at Dec(1950) = 42°08′37″ near DR21. The contour interval is 10 mJy/beam. *4b.* Position-velocity along a north-south line at RA(1950) = $20^h37^m13\overset{s}{.}7$ The contour level in absorption is 20 mJy/beam.

Observations in the (1,1) and (3,3) rotational lines of NH_3 suggest that a dense concentration of gas to the south of the DR21 cluster of HII regions is such an isolated molecular cloud core undergoing gravitational collapse in the earliest stages of star formation.

In Figures 4a and 4b we show velocity-position diagrams made along east-west and north-south lines across the molecular core. The prominent curved structures seen in Fig. 4a in emission and in 4b, if the northern portion of the core seen in absorption is taken into account, are indicative of projection effects associated with radial flow. The fact that the projection effects are seen in perpendicular orientations implies that the radial flow is approximately spherically symmetric. The position of the most redshifted velocity defines the collapse center, RA(1950) = $20^h37^m13\overset{s}{.}7$, Dec(1950) = $42°08'37''$.

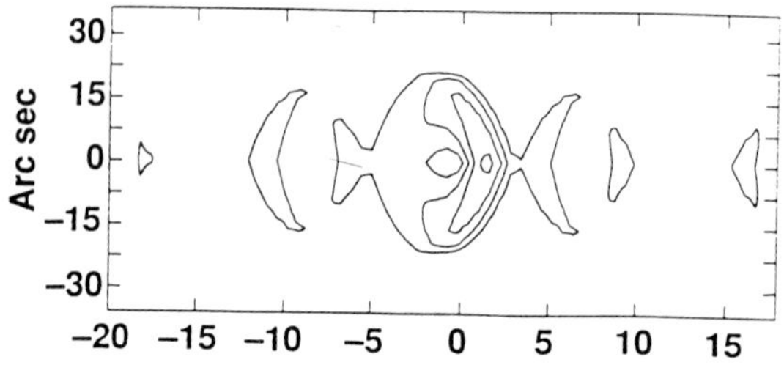

Fig. 5.

Fig. 5. Model velocity-position diagram to be compared with Figures 4a and 4b.

Figure 5 shows a model velocity-position plot generated assuming LTE radiative transfer and a spherical core with constant radial velocity of 2.5 km s^{-1}, $n(H_2)$ density of 10^5 cm^{-3}, kinetic temperature of 40 K, and linewidth of 1.0 km s^{-1}. The model includes a foreground cloud with constant line of sight velocity of 2.5 km s^{-1}, $n(H_2)$ density of 10^4 cm^{-3}, temperature of 10 K, and linewidth of 2.0 km s^{-1}. In the absence of this cold foreground cloud, one should expect to see both the front and back hemispheres of the spherically symmetric radial flow. This would manifest itself on the velocity-position plots as a mirror image of the curved structure around V_{LSR} = +2.5 km s^{-1}, but at a V_{LSR} of about −2.5 km s^{-1}. We suggest that the missing emission at −2.5 km s^{-1} is absorbed by the large scale, cold, foreground cloud. This cloud is not apparent in our data in emission because structure larger than 75″, or twice the southern core diameter, will be resolved out. However, we do detect a component at this V_{LSR} in absorption. In addition, there is considerable evidence from single dish and lower angular resolution observations for a large scale molecular cloud detected in NH_3(1,1) and (2,2), OH, HCN, and H_2CO at −2 to −3 km s^{-1}(Dickel *et al.* 1978; Matsakis *et al.* 1981; Guilloteau *et al.* 1983; Johnston *et al.* 1984; Dickel, Ho, and Wright 1985). Since the modeled flow is spherically symmetric, Fig. 5 should apply equally well to both orientations in Fig. 4 provided that the fact that the northern portion of the core is seen in absorption in Fig. 4b is taken into account.

From these observations it is not possible to determine whether we are observing the redshifted front or back hemisphere of the radial flow. However, the observed radial velocity of 2.5 $km\ s^{-1}$ at 0.2 pc is consistent with infall in the self-gravity of a 290 $M_{\odot}$ core. The mass of the core estimated independently from the column density of NH_3 is 270 $M_{\odot}$ to an accuracy of a factor of two.

Work performed under the auspices of U.S. Department of Energy at Lawrence Livermore National Laboratory under contract number W-7405-ENG-48.

REFERENCES

Dickel, H. R., Ho, P. T. P., and Wright, M. C. H., 1985, *Ap. J.*, **290**, 256.

Guilloteau, S., Wilson, T. L., Martin, R. N., Batrla, W., and Pauls, T. A., 1983 *Astr. Ap.*, **124**, 322.

Ho, P. T. P., and Haschick, A. D. 1986, *Ap. J.*, **304**, 501.

Johnston, K.J., Henkel, C., and Wilson, T.L., 1984 *Ap. J. (Letters)*, **285**, L85.

Keto, E. R., Ho, P. T. P., and Haschick, A. D., 1987a, *Ap. J.*, **318**, 712.

Keto, E. R., Ho, P. T. P., and Haschick, A. D., 1988, *Ap. J.*, **324**, 920.

Matsakis, D. N., Hjalmarson, A., Palmer, P., Cheung, A. C., and Townes, C. H., 1981, *Ap. J. (Letters)*, **250**, L89.

AN OPTICAL STUDY OF THE STAR FORMATION REGION NGC 7129

A.I. Gomez de Castro and C. Eiroa

Observatorio Astronómico Nacional, Spain

NGC 7129 is a region where many signs of active star formation have been observed. Three water vapor masers, five Herbig-Haro objects, two FIR sources, two molecular outflows and several PMS stars have been detected. We have obtained very deep CCD images of the GGD nebulosities found in the region. The observations have been carried out with the 3.5 m. telescope at the Calar Alto Observatory (Almeria, Spain). The main conclusions inferred from the data are: (1) The Herbig-Haro nature of GGD 32, GGD 34, GGD 35, HH 103 and HH 105 is confirmed, however GGD 33 has shown to be a PMS star, GGD 33a, with a cometary nebulosity, GGD 33b; (2) GGD 32 and HH 103 are embedded in an extended structure (at least as large as 0.465x1.36 pc), where another Herbig-Haro object and an arc connecting GGD 32 and HH 103 have been detected. The whole ensemble seems to be linked to a third molecular outflow which is not well resolved in the available radiomaps; (3) The arc connecting GGD 32 and HH 103 contains low excitation material and probably traces a low density cavity within the molecular cloud; (4) GGD 34 is constituted by three condensations aligned in the East-West direction, the western one is a bow-shock where the cooling region is resolved and has a size of around 2000 AU; (5) the PMS star associated to GGD 35 has been identified. The mass outflow is highly collimated close to the star, but at a distance of 0.072 pc from the star it is deflected by an external agent which deviates it in the direction of the molecular outflow associated to LkHo 234.

* A.I. Gomez de Castro is supported by Posdoc fellowship of the M.E.C.

STAR FORMATION IN THE NGC 2071 MOLECULAR CLOUD

N.J. Evans

Departtment of Astronomy, University of Texas

The NGC 2071 molecular cloud has been studied with a broad array of techniques, including a large scale study of CS emission, high resolution scans in the far-infrared, NH_3 studies with the VLA, and near-infrarred imaging. The far-infrared emission constrains the density distribution to fall off approximately as r^{-1}. The NH_3 data strongly supports the presence of a disk oriented perpendicular to the molecular outflow, while the CS emission indicates the presence of dense gas in the region. The results will be combined into a coherent picture of this region of current star formation and molecular outflow.

A THEORY ON THE SLOPE OF THE IMF

H. Zinnecker

Max-Planck-Institut für Extraterr. Physik, Garching

Molecular clouds are clumpy, and the mass spectrum (dN/dM) of clumps scales with the clump mass M to the -1.5 power, as determined from CO observations of molecular clouds (Blitz 1988, Stutzki et al. 1989). The basic idea is to translate the mass spectrum of clumps into a mass spectrum of stars (that are assumed to form from these clumps) by virtue of a clump-star (i.e. initial-final) mass relation.

If the final, stellar mass m scales with the p-th power of the initial clump mass M, one finds that the stellar mass spectrum (dN/dm) is a power law with index x=(p+0.5)/p; for example, for p=0.4 one obtains x=2.25, very close to the index of the Salpeter IMF.

The physical justification for the power p in the final-initial mass relation comes from the theory of ambipolar diffusion (for low-mass stars) and from the theory of radiation pressure acting on dust grains (for high mass atars): magnetic fields or radiation pressure prevent complete accretion of the clump mass onto a protostellar core and induce a clump-mass dependent inefficiency factor in the accretion process. Fragmentation of the clumps into smaller masses is ignored in this simplified picture, but could be incorporated.

There are several predictions following from this concept, among them that triggered star formation occurring in starburst galaxies should lead to an IMF directly reflecting the clump mass spectrum, since the magnetic and radiative forces against protostellar accretion are overwhelmed by pressure-induced or radiatively-induced implosion, causing very efficient accretion (p=1) and thus a rather flat IMF.

S266: A DISTANT HII REGION IN THE GALAXY

A. Manchado, C. Esteban, J.M. Vilchez

Instituto de Astrofísica de Canarias, La Laguna, Tenerife, Spain

Long slit high and low resolution spectroscopy of S266 was conducted in order to investigate its precise nature. The H_α high resolution profile shows two different components; the narrow one extends all over the nebula while a very broad ($\Delta V_{FWHM} = 700\ km\ s^{-1}$) indicates that a strong wind is associated with the central star. From the narrow component we have worked out radial velocity ($V_{LRS} = 23\ km\ s^{-1}$) which yields to a kinematic distance of a 9 kpc. This value implies a galactocentric distance of 19 kpc which makes this region one of the most distant in the Galaxy.

The location of the observed lines ratios in a diagnostic diagram indicates that the object is an HII region class nebula. From the existing photometric data we found that the exciting star appears to be a late B supergiant.

SILICON BEARING MOLECULES IN MOLECULAR CLOUDS

J. Martín-Pintado, J. Gómez-González, R. Bachiller, P. Planesas, V. Bujarrabal

Centro Astronómico de Yebes (IGN), Apdo. 148, E-19080 Guadalajara, Spain

We have used the IRAM 30-m telescope to make high angular resolution observations of the silicon-bearing molecule, SiO and SiS in region of massive star formation. We have mapped the J=2-1 and the J=5-4 lines of SiO with angular resolutions of 26 and 12″ respectively. For all the sources mapped, the SiO emission is more extended that the beam. The extreme cases are found toward NGC7538 and W49N where the SiO emission has sizes of 1x0.5 pc and 2.8x1 pc respectively. This result is in contrast with the size of the SiO region in Orion-IRc2 which is of the order 0.03 pc. The maxima of the SiO emission are found at the position of the H_2O masers. This indicates, as previously thought, that SiO emission probes the high temperature regions and/or the shocked gas and dust surrounding young stars. In the DR21(OH) region we have found the unique SiO source which is not associated to H_2O maser emission. This SiO condensation is 40″ south of DR21(OH) and it was first detected as a source of 350μm emission. The SiO emission is elongated in the east-west direction. From an analysis of the FIR continuum emission this object appears to be cool ($T_K < 30$) and it has been proposed that this source represents a pre-stellar condensation which may be evolving toward the star formation stage. The detection of SiO, howevwer, casts some doubts about the evolutionary stage of this source. Futher high angular resolution observations are necessary to establish the nature of this source.

SiS emission have been detected in the J=5-4 and J=6-5 lines toward the strongest radio continuum sources in Sgr B2, Sgr B2(OHM) and Sgr B2(OHN). The SiS lines are observed in emission. This is in contrast with the J=2-1 line of SiO which is observed in absorption against the continuum sources. Since the physical conditions required to excite the observed transitions of SiO and SiS are very similar, the SiO and SiS emissions arise from different regions. Absorption lines toward Sgr B2 seem to arise in a hot low density envelope. If the SiO absorption lines arise in this envelope, the SiO/SiS ratio in the diffuse envelope is $\gtrsim$60, somewhat larger than the cosmic O/S ratio.

DENSE CLUMPS IN NGC 2024 - PROTOSTELLAR CONDENSATIONS ?

A. Schulz, R. Zylka, R. Güsten

Max-Planck-Institut für Radioastronomie, Bonn

It has been suggested that in NGC 2024 star forming processes are still going on, and dense cores found by 1.3 mm continuum observations are identified as possible protostellar condensations. To trace the molecular gas of high density, we observed the CS (J=7-6) line using the 30 m IRAM MRT for the first time at submillimetre frequencies (spat. resol. 8.5″). These and complementary observations of lower J CS/$C^{34}S$ transitions as well as the temperature probing NH_3(1,1) and (2,2) non-metastable lines enable us to determine temperature and density of 3 selected cores, their kinematics and their masses. These parameters are compared with what is expected for protostellar condensations.

RADIO MAPS OF THE REGIONS RCW 57 AND W 49

N.S. Sabalisck[1], Z. Abrahan[1] and C.E. Tateyama[2]

[1]Instituto Astronomico e Geofisico, Sao Paulo, Brasil

[2]Instituto de Pesquisas Espaciais, Sao Jose dos Campos, Brasil

We present maps of the regions RCW57 and W59 in the frequency of 22 GHz with 4.′6 angular resolution. The thermal radio sources RCW57 II are well resolved in the map. RCW57 I presents, at this frequency a diameter twice as large as that reported at 5 GHz, obtained with similar resolution, but coincides with the size of the source at the wavelenght of 1 mm. In the map of the RCW57 region we detected also the supernova remnant G291.0-0.1, of plerionic type. We find a break in the spectrum of this source between 8 and 22 GHz, with the spectral index changing from -0.29 to -0.77. This behaviour is also found in other remnats of the same type.

The region W49 includes the giant HII region W49A and the supernova remnant W49B. The structure of the map is similar to the structure found at lower frequencies with similar angular resolution. For the HII region we find an integrated flux density much larger than that of the embeded compact sources, indicating that most of the emission originates in the extended, optically thin component. The integrated flux density of the supernova remnant W49N lies well in the extrapolated spectrum of the source, with a spectral index of -0.45 between 408 MHz and 22 GHz.

IONIZATION OF THE GALACTIC CENTER ARCHED FILAMENTS

R. Rubin[1], M. Morris[1], E.F. Erickson[2], S. Colgan[2] and J. Simpson[3]

[1]UCLA

[2]NASA Ames Research Center

[3]UC Berkeley

The remarkable filament system seen in radio observations in the vicinity of the galactic center includes two thin filaments which arch away from the galactic plane (E.G. Yusef-Zadem et al 1984). The brightest part of each of these thermal structures is located at GO.10+0.02 and GO.07+0.04. Morris and Yusef-Zadem (1989) reason that photoionization by OB stars is unlikely on geometrical and morphological grounds. They suggest a magnetohydrodynamic mechanism to account for the radio emission and ionization. Erickson et al. (1968) were able to explain most of their observations of the far infrared (FIR) fine structure line emission from these locations in terms of a photoionization model.

Here we expand on the ionization mechanism of Morris and Yusef-Zadem, which is based on a retrograde high velocity (~100km/s). Cloud encountering the strong ambient poloidal magnetic field. Any charged particles go into helical motion with velocities of ~0 - 200 km/s relative to the neutral material in the cloud. Collisional ionization produces more ions by the critical ionization velocity phenomenon (E.G. Fornisano et al. 1982). The average energy of proton collisions is ~52 eV.

Through plasma waves, there is a collisionless transfer of part of the kinetic energy of the cloud, involving the magnetic field, to the electrons. At the densities prevalent, the distribution of newly produced ions is isotropic along the gyro orbits and the enrgy transfer efficiency is expected to be near the lower limit (Galeev et al., 1986), yielding an average velocity of the free electrons of a few eV. Assuming a Maxwellian distribution, definite predictions can be made of the collisional ionization equilibrium. At the densities involved, this should be independent of density and only depend on electron temperature, T_e.

Measurements of the electron density, N_e, from [OIII] 52/88 μm at peaks of the radio emission indicate a value $\approx 300\ \mathrm{cm}^{-3}$ (Erickson et al., 1988). Under this scenario, it is reasonable to assume that this N_e applies for all species that coexist in the same space. The intensities of various IR and radio lines will then depend on T_e and, of course, an assumed elemental abundance set. Predictions of a set of these line intensities have been made. By comparing with observed FIR line flux ratios (Erickson et al, 1988) it is apparent that only a narrow range in $T_e \lesssim 40000K$ satisfies most of the observations.

However, the collisionally ionized volume is not sufficient by itself to match all the data: (1) The predicted OI (63μm) flux is much to weak; (2) The $T_e \sim 5300K$ derived from the radio is much less than the value of $T_e \sim 40000k$ required by ionization equilibrium; and (3) in the collisionally ionized volume nearly all He would be He^+. Hence the expected radio recombination line ratio $< He^+/H^+ >$ would be the He/H abundance ratio ~ 0.1 and not the upper limit of $<$ 0.03 observed by Pauls and Mezger (1980). These observations suggest that a "secondary" HII region/PDR (Photodissociation Region) has formed in the underlying neutral/molecular cloud, which is photoionized/dissociated by the diffuse photons produced in the volume of collisional ionization.

Using the same elemental abundances and N_e, photoionization models similar to those of Erickson et al (1988) are also found to fit the FIR and radio data equally well and to be narrowly confined to a radiation field characterized by a Kurucz atmosphere $T_{eff} \lesssim 35000K$, $\log g = 4$ sequence.

THE CHAMAELEON DARK CLOUDS COMPLEX: PRELIMINARY ANALYSIS OF THE COLOUR EXCESSES E(b-y) TOWARDS THE SELECTED AREA 203

G.A.P. Franco

Copenhagen University Observatory, Copenhagen, Denmark and Departamento de Física, UFMG, Belo Horizonte, Brasil

The Chamaeleon dark clouds form a large complex of interstellar obscuring material situated at $\approx$ 15° below the galactic plane. Although it is accepted as being one of the closest low-mass star formation region to the Sun, its distance has been debated issues. The proposed distance is in general dependent on the value assumed for the ratio of total-to-selective extinction, which in the Chamaeleon clouds has proved controversial, leading to distances estimates ranging from 115 to 215 pc.

Selected Area 203 ($l = 300^\circ.0, b = 13^\circ.1$) lies approximately in the geometric center of the Chamaeleon cloud complex, in a relatively less obscured area, when compared with the surrounding dark region. As part of an investigation of the interstellar dust distribution towards the Chamaeleon dark clouds complex, all stars earlier than G0 and brighter than $m_{pg} \approx 10^{\rm m}.4$ in the Potsdam Spektral Durchmusterung of SA 203 were selected for observation. About 200 selected stars were observed in four-colour uvby and H_β photometry.

A preliminary analysis of the distribution of the colour excesses, E(b-y), as a function of the distance is presented. The obtained E(b-y) versus distance diagram clearly suggests the presence of a sheet-like structure at a distance of $\approx$137 pc, indicating that at least part of the Chamaeleon complex is located nearer than 140 pc.

MECHANICAL ENERGY SOURCES

INTERSTELLAR WIND-BLOWN BUBBLES

J. E. Dyson
Department of Astronomy, University of Manchester
England

Summary

'Classical' stellar wind-driven bubbles are either energy or momentum driven, and evolve in smooth media containing only radial density gradients. Real bubbles are produced from winds from moving stars and which blow into non-homogeneous media. The resulting mixing of clump material can produce a variety of thermal, dynamical and chemical effects. In this review we discuss some of the modifications to classical bubbles which ensue, using observational examples where appropriate.

1. Introduction

Continuous mass loss from stars is ubiquitous throughout the H-R diagram. The mass loss rates, $\dot{M}_*$, range from the insignificant $10^{-14}\ M_\odot\ \mathrm{yr}^{-1}$ for the Sun to perhaps as much as $10^{-3}\ M_\odot\ \mathrm{yr}^{-1}$ in some red giants and supergiants. $\dot{M}_*$ varies from one star to another and depends on the evolutionary stage of a particular star. Evolutionary tracks in the H-R diagram can be strongly influenced by mass loss. Mass loss may even determine the initial main sequence mass of low mass stars. The velocity, V_*, of the outflowing material can range from the 20 km s^{-1} or so characteristic of red supergiants to perhaps as much as 3000 km s^{-1} in the winds from some planetary nebulae nuclei. This variation reflects differing escape velocities and differing ejection mechanisms. As will be discussed later (Section 2), the wind velocity is the single most important parameter of the wind for determining the resulting dynamical interaction between the wind and the circumstellar environment—at least as long as the environment is smoothly distributed.

The physical conditions in winds also vary. Winds may be fully ionized (e.g. from early type stars) or may be largely neutral (e.g. from some pre-main sequence objects). It is doubtful if such a thing as a perfectly smooth wind occurs. Blue supergiant winds appear to be highly structured. Similarly, the structures of some planetary nebulae and Wolf-Rayet nebulae clearly indicate that the slow winds ejected prior to the fast wind stage were extremely clumpy.

The mechanisms driving mass loss are varied and sometimes highly problematic. Radiation pressure driven winds (utilising ion resonance line or dust absorption) are well known. Alfvén wave or acoustic wave driven winds may occur. The winds may sometimes not originate on the actual stellar surface. For example, winds from accretion discs have been suggested to account for the exceptionally high wind momenta apparently associated with some molecular cloud flows. It is arguable that the most fundamental and important questions relating to stellar mass loss concern the wind driving mechanisms and the wind morphology. In connection with this latter point, the collimation and confinement of winds into jets and bipolar outflows are as yet not satisfactorily understood.

This review is concerned with topics involving a somewhat easier aspect, in the sense that the increasing body of exceptional observational data makes it possible to really discriminate between different models. This is the interaction of winds with the circumstellar gaseous environment. We first briefly review the basic assumptions and results of 'classical' bubble evolution and then relax various of these basic assumptions. Specific observational examples are given which illustrate the various effects.

2. Classical Wind-Blown Bubbles

These bubbles are produced when isotropic stellar winds blow into smooth circumstellar media containing only radial density variations. The basic two-shock flow pattern was first described by Pikelner (1968). The resulting dynamics of the swept-up circumstellar matter are determined by the thermal behaviour of the shocked wind and the form of the radial density distribution (Dyson 1984).

'Energy' driven flows occur when the shocked wind loses energy only by adiabatic expansion. The outer shock, which accelerates the circumstellar matter, has a radius R_s and velocity $V_s(\equiv \dot{R}_s)$ given by

$$R_s = \phi^{1/(5+\beta)} \left(\frac{2\dot{E}_*}{\rho_*}\right)^{1/(5+\beta)} t^{3/(5+\beta)} \tag{1}$$

$$V_s = \left(\frac{3}{5+\beta}\right) \frac{R_s}{t} \tag{2}$$

$$\phi = \frac{(3+\beta)(5+\beta)^3}{12\pi(11+\beta)(7+2\beta)} . \tag{3}$$

In deriving these equations, the circumstellar gas has a density distribution $\rho(r) = \rho_* r^\beta$ and the wind mechanical luminosity, $\dot{E}_* \, (\equiv \frac{1}{2}\dot{M}_* V_*)$, is taken as constant.

'Momentum' driven flows occur when the shocked wind cools in a time much less than the dynamical timescale. The accelerated interstellar gas is now driven by wind momentum rather than by shocked wind pressure. The outer shock radius and velocity are now given by

$$R_s = \chi^{1/(4+\beta)} \left(\frac{\dot{M}_* V_*}{\rho_*}\right)^{1/(4+\beta)} t^{2/(4+\beta)} \tag{4}$$

$$V_s = \left(\frac{2}{4+\beta}\right) \frac{R_s}{t} \tag{5}$$

$$\chi = \frac{(3+\beta)(4+\beta)}{8\pi} . \tag{6}$$

The wind momentum output rate is here taken as constant.

In both cases, the shell accelerates if $\beta < -2$ and, in principle, Rayleigh-Taylor instability should lead to its break up. We return to this point in Section 3.

Energy driven flows are initiated if the cooling time in the shocked wind is greater than the dynamical timescale at the time when the wind has swept up its own mass of ambient gas. If the converse is true, a momentum driven flow is initiated. Energy driven or momentum driven flows are produced according to whether the wind speed exceeds or is less than a critical value V_{CR} respectively. For the uniform density case, $V_{CR} \approx 100(n_0 \dot{M}_6)^{1/9}$ km s^{-1} (Dyson 1984), where n_0 is the ambient number density and $\dot{M}_6 \equiv \dot{M}_*/10^{-6} M_\odot$ yr^{-1}. This criterion implies that interactions between fast winds (e.g. from O stars) and their environment are initially energy driven. The velocities of winds from stars found in molecular clouds tend to lie in the interestingly intermediate range 100–500 km s^{-1} approximately.

Flows may evolve from one type to another for two reasons. Firstly, the wind velocity may be time dependent. Breitschwerdt and Kahn (in preparation) have discussed planetary nebula dynamics allowing for the increasing wind speed between the post-AGB phase and the constant luminosity phase of the nucleus' evolution. Such evolutionary effects may be important elsewhere. For example, Dyson (1984) suggested that high luminosity stars power up to their main sequence values of $\dot{M}_*$ and V_* through a higher $\dot{M}_*$ and lower V_* phase.

Secondly, the circumstellar density distribution can induce flow changes. Dyson (1984) has shown that the ratio of the radiative cooling time in the wind to the dynamical timescale varies as $t^{-\beta/(4+\beta)}$ for an initially momentum driven flow and as $t^{-(1+2\beta)/(5+\beta)}$ for an initially energy driven flow. Hence momentum driven flows are maintained only if $\beta > 0$ otherwise they become energy driven. Energy driven flows are maintained only if $\beta \leq -\frac{1}{2}$. It is interesting to note (Dyson 1984) that for $\beta = 0$, the time at which an energy driven flow becomes momentum driven scales as V_*^{21}!

In principle, discrimination between the two types of flow can be made by observationally determining the two efficiency factors ϵ and μ for the conversion of wind kinetic energy and momentum into swept-up gas kinetic energy and momentum respectively. In an energy driven flow, $\epsilon = 3(5+\beta)/(11+\beta)(7+2\beta)$ and $\mu = \epsilon(V_*/V_s)$. In a momentum driven flow, $\epsilon = V_s/V_*$ and $\mu = 1$. An estimate of $\epsilon \ll 0.1$ is usually regarded as an indication that the flow is momentum driven. However a number of important reservations regarding derived efficiencies are necessary.

Shells driven by winds from stars with strong radiation fields may be only partially ionized. Both the masses of the neutral and ionized components must be known to derive the efficiency factors. For example, if an efficiency ϵ^i is derived using only the ionized gas mass, $\epsilon^i/\epsilon = (t_c/t)^{(5+4\beta)/(5+\beta)}$ where t_c is the time when the flow becomes fully ionized. For $\beta < -5/4$, $\epsilon^i/\epsilon < 1$ for $t < t_c$ and $\epsilon^i/\epsilon = 1$ for $t \geq t_c$. If $\beta > -5/4$, $\epsilon^i < 1$ for $t > t_c$ and $\epsilon^i/\epsilon = 1$ for $t \leq t_c$. Here the flow is assumed energy driven since such stars have high speed winds.

Dyson and Smith (1986) and VanBuren (1986) have given observational examples where these effects are of importance.

Cappa de Nicolau and Niemela (1984) estimated the neutral mass of the Wolf-Rayet nebula θ Mus from an apparent hole in the 21cm H emission as $M_n \approx 10^4\, M_\odot$, giving an estimated $\epsilon^n \lesssim 0.01$, implying momentum driving. Dyson and Smith (1986) reanalysed the flow and obtained an ionized mass $M_i \approx 4.4\, M_n$ giving $\epsilon \approx 0.05$. The interpretation of the dynamics of this particular nebula is complicated by the presence of an O type companion to the W-R star, but Dyson and Smith (1986) concluded that energy driving was marginally possible.

VanBuren (1986) noted that the ionized mass of the nebula around the O8 III star λ Ori is $M_i \approx 2.10^3\, M_\odot$ leading to $\epsilon^i \approx 0.007$. Momentum driving is implied. However, again analysis of the flow dynamics leads to a neutral gas mass $M_n \approx 13\, M_i$ and hence an $\epsilon \approx 0.1$, consistent, within observational and theoretical uncertainty, with energy driving.

Analysis of molecular flow dynamics can, in principle, give important data on the wind characteristics of the stars driving the flows. The expressions for the efficiencies clearly show that the type of flow must be known and further, as pointed out by Dyson (1984), the wind momentum and mechanical luminosity can be independently derived from the same set of data only if an additional parameter, the wind velocity V_*, is simultaneously known. A related problem concerns the use of measurements of the luminosity of the extended IR emission seen around some molecular flows to discriminate between the two types of flow. In an energy driven flow, the radiated energy can only derive from the cooling swept-up gas. The luminosity is constant in time and is $L_E = (3(3+\beta)/(5+\beta))\epsilon\dot{E}_*$ which ranges from $27/77\ \dot{E}_*\,(\beta = 0)$ to $1/3\dot{E}_*\,(\beta = 2)$. In a momentum driven flow, the luminosity is dominated by the energy radiated from the shocked wind, $L_M \approx \dot{E}_*$. It is clear that a very accurate independent determination of $\dot{E}_*$ is needed to discriminate between the two types of flow on this basis.

In the remainder of this review, we examine departures from the above description with illustrative examples where appropriate.

3. Shell Break-Up and the Circumstellar Density Distribution

As noted previously, radial density distributions with $\beta < -2$ lead to immediate shell acceleration and consequent Rayleigh-Taylor instability. More interesting examples occur when the acceleration occurs later on in the bubble evolution.

The general principles are easily established. For energy driven flows, the internal bubble pressure P_i is spatially uniform because of the high sound speed there. (This is the basis of the variant of the Kompaneets (1960) explosion theory first applied to stellar wind bubble evolution by Dyson (1977a.) For simplicity we assume a plane stratified ambient density distribution, $\rho = \rho_0 f(z)$ where $f(0) = 1$. The shell velocity in the direction $\theta\,(\equiv \cos^{-1}(z/R_s))$ satisfies $V_s^2(\theta) \sim P_i/f$. Since $P_i \sim t/R_s^3$, $V_s^2 \sim t/R_s^3 f$ or $V_s^3 \sim 1/R_s^2 f$. In the plane $z = 0$, $V_s^3 R_s^2 \sim$ const., i.e. the standard uniform density equation. If now f has an exponential (or related) form, $f = e^{-z/\Delta}$, where Δ is a scale height, $V_s^3 \sim e^{R_s/\Delta}/R_s^2$ in the direction $\theta = 0$ (R_s along the z axis). The shell velocity decreases to a minimum at $R_s \approx \Delta$ and acceleration then occurs with $\dot{V}_s$ increasing. Similar behaviour obtains in all directions for a radial density distribution $\rho(r) = \rho_0 e^{-r/\Delta}$. (This phenomenon is often referred to as 'blow out'.)

The examples cited below are for energy driven flows, but very similar behaviour occurs for momentum driven flows. For the plane stratified distribution, $V_s(\theta) \sim t/\mu(\theta)$ where $\mu(\theta)$ is the mass swept up per unit solid angle in the direction θ. For small R_s or near $z = 0$, $\mu \sim R_s^3$ and $V_s^2 R_s^2 \sim$ const., the uniform density equation. At large R_s along the z axis, $\mu \to$ const. and hence $V_s \sim t$. Again the transition from deceleration to acceleration occurs at $R_s \approx \Delta$. An important distinction between the two types of flow is that now $\dot{V}_s \to$ constant.

We now discuss three examples. Dyson (1981) noted that winds from groups of O stars could produce large scale bubbles in the galactic plane which could blow-out because of the finite scale height. Since acceleration commences at $R_s \approx \Delta$, the associated time and shell velocity are respectively $t_a \approx (\Delta^5\rho_0/\dot{E}_*)^{1/3}$, $V_a \approx 0.6\ (\dot{E}_*/\Delta^2\rho_0)^{1/3}$, where $\dot{E}_*$ is the integrated mechanical luminosity of the stellar group. As a simple example, if we require a group of stars to blow out in the solar neighbourhood ($\Delta \approx$ 100pc, $\rho_0 \approx 10^{-24}$ gm cm^{-3}) in a time $\sim 5\,10^{13}$ sec, the necessary integrated mechanical luminosity is $\sim 3\,10^{36}$ erg, i.e. some tens of O stars. The associated shell velocity near blow-out is $V_a \approx 20$ km s^{-1}. More realistic calculations include the effects of the inevitable accompanying supernovae (MacLow and McCray 1988).

Optical line splitting observed across the south lobe of the bipolar nebula S106 (Solf and Carsenty 1982) is clearly indicated of an expanding shell structure. Dyson (1983) modelled the nebula in terms of the interaction between a fast stellar wind and a surrounding stratified disc. The ragged end of the south lobe is attributed to the operation of Rayleigh-Taylor instability as blow out occurs. The gas dynamical model and the optical emission combine to require a stellar mechanical luminosity $\dot{E}_* \approx 2\,10^{35}$ erg s^{-1} and a Lyman continuum photon output $S_* \approx 4\,10^{47}$ s^{-1}, consistent with observational data on O9.5V stars. The observed infra-red luminosity implies a spectral type O9-B0V (Eiroa et al. 1979). S106 is an interesting example where gas dynamics can be applied to stellar spectral typing.

Molecular hydrogen features with velocities exceeding 250 km s^{-1} are observed towards the proto-planetary nebula CRL618 (Burton and Geballe 1986). Other high velocity features include the H_2O maser spots with velocities greater than about 74 km s^{-1} (line-of-sight) seen towards W51 MAIN (Genzel et al. 1981). Serious problems exist with simple shock acceleration of molecule bearing gas to such velocities since molecules are dissociated at shock velocities greater than about 50 km s^{-1} even with magnetically moderated shocks. Ram pressure acceleration of non-gravitationally bound clumps (e.g. Norman and Silk 1979) is a widely invoked but mythical process (e.g. Hartquist and Dyson 1987).

Hartquist and Dyson (1987) suggested that the coherent acceleration of a shell to high velocities is a more dynamically plausible mechanism. They argued that density distributions with scale heights $\sim 10^{17}$ cm should exist around some stars and that molecular material would be directly incorporated into the expanding shell once the shell velocity dropped below about 50 km s^{-1}. This molecular material would then be reaccelerated as the shell reaccelerates down the density gradient.

They argued that Rayleigh-Taylor instability is in itself not a sufficient condition for the shell fragmentation needed to produce discrete features. They conjectured that the shells would try to reseal themselves by the ablation of material off the fragments. This conjecture has received observational support from high resolution echelle spectroscopy of filaments in the Vela supernova remnant (Meaburn, Hartquist and Dyson 1988) where small scale velocity blisters are clearly seen. Since self-gravity of these shells is never important, it is likely that only genuine thermal instability (as well as Rayleigh-Taylor instability) can result in shell fragmentation (Hartquist and Dyson 1987). Thermal instability behind the shock sets in at velocities $\geq 100-130$ km s^{-1} (e.g. Innes, Giddings and Falle 1986), consistent with the maser data. (The model for S106 discussed above also satisfies this criterion.)

4. Distortion of Bubbles by Stellar Motion

Energy driven bubbles eventually distort into high asymmetric structures as a result of stellar motion (Weaver et al. 1977). The stellar wind acts as a snowplough with a high brightness bow wave of gas in the direction of motion. Initially momentum driven shells distort similarly. The subsequent flow pattern has been described by Dyson (1977b). The shocked wind does not radiate as it flows through the bow-shock pattern if the wind speed $V_* \gtrsim 100\,(V_0/20 \text{ km s}^{-1})^{4/11}\, n_0^{1/11}\, \dot{M}_6$ km s^{-1}, where V_0 is the stellar velocity with respect to the interstellar gas. Under most circumstances of interest, the shocked stellar wind occupies a thick hot region. The distance of the stand-off shock follows from momentum balance with allowance made for the finite thickness of the hot region and is $d_0 \approx 2(V_*/100 \text{ km s}^{-1})^{1/2} \dot{M}_6^{1/2} (V_0/20 \text{ km s}^{-1})^{-1} n_0^{-1/2}$pc (cf. Dyson 1977b). Thus arc-like structures with dimensions of a few pc should be produced. VanBuren and McCray (1988) have shown that such structures can be found in the IRAS data.

Dyson and Ghanbari (1989) have modelled the Wolf-Rayet nebula NGC 3199 as just such an interstellar snowplough. The observationally determined mass of NGC 3199 ($\sim 200 M_\odot$) clearly shows that it is of interstellar and not stellar origin. Dyson and Ghanbari (1989) show that the snowplough was set up in the O phase preceding the W-R phase, and obtain an excellent fit to the nebular morphology by assuming the ambient gas is of uniform density. From the nebular dimensions and density and the kinematic data of Whitehead, Meaburn and Goudis (1988), they derive a stellar velocity $V_0 \approx 60$ km s^{-1} and a wind mechanical luminosity $\dot{E}_* \approx 4\,10^{37}$ erg s^{-1}. This latter value is in excellent agreement with derived data on other WN5 stars (Barlow, Smith and Willis 1981). The ambient gas density $n_0 \approx 10$ cm $^{-3}$, and the nebula contains about 110 $M_\odot$ of ionized gas and about twice this amount of neutral gas. The difference between the derived and measured ionized mass is probably due to clumping. Although this model suggests that the central star of NGC 3199 is a runaway, no compact companion has yet been observed. It is worth noting here that estimates of ϵ for this nebula imply that it is momentum driven, but ϵ has no relevance to this steady flow pattern (Dyson and Smith 1986).

5. Effects of Non-Homogeneities on Bubble Structure

Hartquist and Dyson (1988) have reviewed the effects of mass pickup from clumps on many kinds of flow of astrophysical interest. If the general flow relative to a clump is subsonic in the clump frame, material is ablated into the flow by the Bernouilli effect. If the flow is supersonic, expansion of the clump into the wake behind a bow shock results in mixing. Mass addition affects flow dynamics, thermodynamics and chemistry.

A very simple example of the modification of flow dynamics is given by Hartquist et al. (1986—henceforth HI). Momentum driven bubbles moving into clumpy media and where the ablated mass dominates the shell mass have radii which evolve as $t^{1/4}$ instead of the usual $t^{1/2}$ (provided that the mass input rate per unit volume is constant).

The dynamics of the clumpy Wolf-Rayet nebula RCW 58 exemplifies the effects of mixing on flow dynamics and thermodynamics (H1; Smith, Pettini, Dyson and Hartquist 1984—henceforth S1—and 1988—henceforth S2). The low ($\sim 5 M_\odot$) mass, clumpy structure and observed He, N enrichment imply that RCW 58 is composed of stellar ejecta from the red supergiant phase preceding the W-R phase (Maeder 1984). The conversion efficiency is $\epsilon \sim 0.001$ (S1) and the likely cause is the enhancement of radiative losses by the mixing of cool clump gas with hot shocked stellar wind. IUE absorption data show a linear velocity/ionization potential correlation which cannot be contained by standard bubble theory (S1). Analysis of the flow dynamics behind the wind shock show a bubble structure far removed from the classical model (H1). Moving outwards from this shock, there are four distinct zones (H1):-

1. A post-shock zone where mass loading from the clumps occurs and there are no radiative losses. The temperature of the gas changes from $5\,10^7$K to $2.5\,10^7$K and the Mach number (in the shock frame) increases to about unity.

2. Continuing outwards there is next a zone, again with mass loading and no radiative losses, but where the Mach number stays around unity in some statistical sense. The temperature drops from $2.5\,10^7$K to about 10^5K. The mixing has by now

added about 40 times as much material to the flow as came from the wind. (This figure is required to explain the velocity/I.P. correlation.)

3. At a temperature of 10^5K, cooling by collisional excitation and ionization of H from the clumps becomes dominant. A zone too thin to allow mass loading ensues. The region is isobaric and the temperature falls from 10^5K to the 10^4K characteristic of material radiatively ionized by the stellar radiation field. The UV absorption lines are formed here with the correct velocity variation.
4. Finally, there is a shell of radiatively ionized gas expanding with the same velocity as the lowest UV absorption line velocity (about 100 km s^{-1} with respect to the star).

Recent optical echelle spectroscopy (S2) strongly supports this model. Zone (4), never previously observed, is clearly identifiable. Features relevant to the mixing process are also apparent. For example, good mixing at the clump/mass loaded wind interfaces is indicated by the similarity between the [OIII] and [NII] velocity shifts and line widths over the clearly identifiable clumps. Very interestingly, the line widths from both clumps and shell are about 30 km s^{-1}. These line widths may be subsonic for the clumps since the effective sound speed in the mixing zone could be appreciably higher than 10 km s^{-1}. The line widths are supersonic for the shell and S2 speculate that turbulence in the shocked wind might drive supersonic turbulence in the shell.

Molecular cloud lifetimes are greater than the depletion time of molecules onto grains (cf. Dalgarno, these Proceedings). The correlation between molecular abundances and line widths implies a connection between the abundances and internal cloud dynamics (Williams and Hartquist 1984). To explain the high abundance of C^o in dense molecular clouds, they argued that the dynamical cycling of material between clumps and inter-clump gas prevents equilibrium chemistry from being attained. This dynamical cycling occurs as stellar winds dissipate clumps and produce a shell which then fragments (Norman and Silk 1980). Charnley et al. (1988a) noted that the stellar winds would ablate H_2 and mantle bearing grains from cold clumps and mix them into the mass loaded wind. Bow shocks in the wind around these clumps may produce H^+ and He^+, and mixed-in grains also can lose molecules by sputtering as they pass through these shocks. A complex mixture of grains, molecules, ions and atoms passes eventually through a weak shock where the wind reaches pressure equilibrium with the surrounding cloud. Chemical evolution of the shocked gas occurs, clumps are formed, molecules deplete onto grains and the clumps are then affected by the next generation of stellar winds. Charnley et al. (1988b) followed the chemistry through several evolutionary cycles and have had some success in producing abundances compatible with those observed. More recent work (Charnley et al. 1989) focusses on the chemistry in the clump-stellar wind interface.

6. Conclusions and Acknowledgements

'Classical' stellar wind bubbles have had considerable success in explaining many of the general features of wind-circumstellar interactions. However, the increasing availability of a wide range of high quality observational data has shown that real objects show significant variations from these rather simple models. Of particular importance are the various effects of mixing as the winds blow into non-uniform media. The study of these mixing interfaces, both theoretical and observational, is probably the single most important area of work in bubble dynamics for the immediate future.

I am grateful to the Organizing Committee for their invitation to present this paper in honour of Guido's Jubilee. I have enjoyed endless hours of discussion with my collaborators and thank them for this and all their hard work. This paper was largely written whilst a guest of the Academia Sinica, Beijing, China, to whom I am very grateful.

References

Barlow, M. J., Smith, L. J. and Willis, A. J., 1981. Mon. Not. R. astr. Soc., **196**, 101.

Burton, M. G. and Geballe, T. R., 1986. Mon. Not. R. astr. Soc., **223**, 13P.

Cappa de Nicolau, C. and Niemela, V. S., 1984. Astron. J., **89**, 1398.

Charnley, S. B., Dyson, J. E., Hartquist, T. W. and Williams, D. A., 1988a. Mon. Not. R. astr. Soc., **231**, 269.

Charnley, S. B., Dyson, J. E., Hartquist, T. W. and Williams, D. A., 1988b. Mon. Not. R. astr. Soc., **235**, 1257.

Charnley, S. B., Dyson, J. E., Hartquist, T. W. and Williams, D. A., 1989. Mon. Not. R. astr. Soc. (in press).

Dyson, J. E., 1977a. Astron. Astrophys., **59**, 161.

Dyson, J. E., 1977b. Astrophys. Space Sci., **51**, 197.

Dyson, J. E., 1981. In *Investigating the Universe*, ed. F. D. Kahn, p.125. D. Reidel Publ. Co., Dordrecht, Holland.

Dyson, J. E., 1983. Astron.. Astrophys., **124**, 77.

Dyson, J. E., 1984. Astrophys. Space Sci., **106**, 181.

Dyson, J. E. and Ghanbari, J., 1989. Astron. Astrophys. (in press).

Dyson, J. E. and Smith, L.J., 1986. In *Cosmical Gas Dynamics*, ed. F. D. Kahn, p.173. VNU Science Press, Utrecht, Holland.

Eiroa, C., Elsässer, H. and Lahulla, J. F., 1979. Astron. Astrophys., **74**, 89.

Genzel, R., Downes, D., Schneps, M. H., Reid, M. J., Moran, J. M., Kogan, L. R., Kostenko, V. I., Matveyenko, L. I. and Ronnaug, B., 1981. Astrophys. J., **247**, 1039.

Hartquist, T. W. and Dyson, J. E., 1987. Mon. Not. R. astr. Soc., **228**, 957.

Hartquist, T. W. and Dyson, J. E., 1988. Astrophys. Space Sci., **144**, 615.

Hartquist, T. W., Dyson, J. E., Pettini, M. and Smith, L. J., 1986. Mon. Not. R. astr. Soc., **221**, 715.

Innes, D. E., Giddings, J. R. and Falle, S. A. E. G., 1986. In *Cosmical Gas Dynamics*, ed. F. D. Kahn, p.143. VNU Science Press, Utrecht, Holland.

Kompaneets, A. S., 1960. Soviet Phys. Doklady, **5**, 46.

MacLow, M.-M. and McCray, R., 1988. Astrophys. J., **324**, 776.

Maeder, A., 1984. In *Observational Tests of the Stellar Evolution Theory*, IAU Symposium 105, eds. A. Maeder and A. Renzini, p.299. D. Reidel Publ. Co., Dordrecht, Holland.

Meaburn, J., Hartquist, T. W. and Dyson, J. E., 1988. Mon. Not. R. astr. Soc., **230**, 243.

Norman, C. A. and Silk, J., 1979. Astrophys. J., **228**, 197.

Norman, C. A. and Silk, J., 1980. Astrophys. J., **238**, 138.

Pikelner, S. B., 1968. Astrophys. Letters, **2**, 97.

Smith, L. J., Pettini, M., Dyson, J. E. and Hartquist, T. W., 1984. Mon. Not. R. astr. Soc., **211**, 679.
Smith, L. J., Pettini, M., Dyson, J. E. and Hartquist, T. W., 1988. Mon. Not. R. astr. Soc., **234**, 625.
Solf, J. and Carsenty, U., 1982. Astron. Astrophys., **113**, 342.
VanBuren, D., 1986. Astrophys. J., **306**, 538.
VanBuren, D. and McCray, R., 1988. Astrophys. J., **239**, L93.
Weaver, R., McCray, R., Castor, J., Shapiro, P. and Moore, R., 1977. Astrophys. J., **218**, 377.
Whitehead, M. J., Meaburn, J. and Goudis, C. D., 1988. Astron. Astrophys., **196**, 261.
Williams, D. A. and Hartquist, T. W., 1984. Mon. Not. R. astr. Soc., **210**, 141.

Discussion:

CASTAÑEDA: How sensitive are the conclusions of your model depending of clumps size distribution, density and number density?

DYSON: Well, we need at least one clump!. To be serious, I believe the gross features are probably rather insensitive to such details. However detailed local features certainly are.

TENORIO-TAGLE: Would you comment on how thermal conductivity affects the clumps of matter immersed in the hot wind region.

DYSON: I dont believe in thermal conductivity. I am very convinced that hydrodynamic mixing is much more important.

PECKER: In your cumply approach to the bubble problem, how does evolve the ambiguity between energy-driven winds and momentum-driven winds? Presumably, the question almost loses sense!

DYSON: My guess is that generaly mixing in the cold gas enhances radiative cooling this leading to momentum driven flows even for wind interactions which one might expect were energy driven (e.g., RCW58). However it does depend on how many clumps one has, their distribution etc. and one could certainly imagine situations were although clumps are present, no more than, say 50% of the energy was radiated away and the flow was essentially energy driven.

STELLAR WINDS IN A-TYPE SUPERGIANTS

A. Talavera [1]* A.I. Gomez de Castro [2]

[1]ESA IUE Observatory, VILSPA P.O.Box 54065, 28080 Madrid, Spain
[2]Observatorio Astronomico Nacional. Alfonso XII, 3, 28014 Madrid, Spain

1. INTRODUCTION

The contribution of A supergiant stars to the return of mass and energy to the interstellar medium is not very important. Abbott (1982) analised a sample of early stars and concluded that B and A supergiants provided less than 8 % of the mass input to the ISM by stellar winds. However, A-type supergiants are important in the framework of the stellar winds-mass loss phenomenology since they are located at the boundary between hot and cool stars, where radiative acceleration may not be sufficiently efficient to drive the wind.

We have studied the ultraviolet spectrum of all the A type supergiants observed with IUE (Talavera and Gomez de Castro 1987, 1988). This work allowed us to classify them in two groups. The *Group I* contains the less luminous A supergiants; these stars show indications of mass outflow only in the Mg II resonance lines where an absorption component of variable blueshift and intensity has been detected. The stars included in the *Group II* show strong evidence of wind and mass loss in the UV resonance lines of Mg II, C II, Si II and Fe II, and the terminal velocity of the wind derived from the Mg II lines is about 250 km/s; the most luminous A supergiants belong to this group.

There are also two important indicators of mass loss in the visible: a) emission : Hα, Hβ, He I 5876Å, and b) dependence of the radial velocity on the excitation potencial of the lines and/or regular progression of the radial velocity along the Balmer series (Hutchings 1970, Wolf 1983).

The only systematic study of Hα in A supergiants was made by Rosendhal (1970). He found an evolution of the emission in the line with the absolute magnitude of the stars. He found emission in Hα for magnitudes brighter than Mv=-6.8.

In 1988 we started an observing programme whose goal is the detailed analysis of the envelope of the A type supergiants. As a first step we have obtained high resolution profiles of lines formed at different depths in the atmosphere of the stars. These data will allow us to model the wind and the structure of the outer layers of these stars.

We present here the first results of this programme.

---

*Affiliated to the Astrophysics Division, Space Science Department

2. OBSERVATIONS

We have selected our stars from the compilation of Humphreys (1978). Since most of them belong to OB associations, we have a good estimate of the distance and therefore of the absolute magnitude. Part of the stars we studied with IUE are included in this first sample which is listed in Table 1.

The observations have been done with the Isaac Newton Telescope (INT) of the Observatorio del Roque de los Muchachos in the Canary Islands and with the 2.2 m telescope of the Calar Alto Observatory in September and October 1988. At the INT we used the Intermediate Dispersion Spectrograph (IDS) equipped with a GEC CCD and in Calar Alto we used the Coude Spectrograph with an RCA CCD.

The spectral intervals we have selected correspond to the following lines: Ca II K line (3933Å), Mg II (4481Å), Hβ (4861Å), Na I and He I (5890Å) and Hα (6563Å). We have observed all these lines in most of the stars of the sample with a resolving power around 20000.

Table 1. Hα Observations of A-type Supergiants

STAR	Sp.T.	Mbol	Observations INT	Observations CA	Profile	IUE
BD +60 51	A2 Ib	-5.01	27,29/Sep		AB	
HD 3283	A4 III	-5.07	27,29/Sep		AB	
HD 236995	A0 Ia	-5.08	27,28/Sep		AB	
HD 46300	A0 Ib	-5.18	26,28/Sep	20/Oct	AB	Group I
BD +60 2542	A2 Ib	-5.31	27,29/Sep		AB	
HD 4717	A0 Ib	-5.58	27,29/Sep		AB	
BD +61 153	A0 Ib	-5.58	27,29/Sep		AB	
HD 209900	A0 Ib-II	-5.68	27,29/Sep	23/Oct	AB	
HD 211971	A2 Ib	-5.71	27,29/Sep	23/Oct	EM	
HD 5776	A0 Ib	-5.88	27,29/Sep		AB	
HD 2928	A0 Iab	-5.88	27,29/Sep		AB	
HD 14535	A2 Ia	-6.51	27,29/Sep		PCyg	
HD 13744	A0 Iab	-6.78	26,28/Sep	23/Oct	EM	
HD 16778	A2 Ia	-6.81	27,29/Sep		EM	
HD 207260	A2 Iab	-6.91	27,29/Sep	23/Oct	EM	Group II
HD 20041	A0 Ia	-6.98	27,29/Sep	20/Oct	EM	
HD 13476	A3 Iab	-7.04	26,28/Sep	23/Oct	EM	
HD 3940	A1 Ia	-7.10	27,29/Sep	23/Oct	EM	
HD 212593	B9 Iab	-7.22	27,29/Sep	23/Oct	AB	
HD 14433	A1 Ia	-7.30	26,28/Sep	23/Oct	EM	Group II
HD 15316	A3 Iab	-7.44	27,29/Sep		EM	
HD 223960	A0 Ia	-7.48	27,29/Sep	23/Oct	Pec	Group II
HD 21389	A0 Iab	-7.48	26,28/Sep	20/Oct	Pec	Group II
HD 213470	A3 Ia...	-7.64	27,29/Sep	23/Oct	EM	
HD 14489	A2 Ia	-7.81	26,28/Sep		EM	Group II
HD 17378	A5 Ia	-7.89	26,28/Sep	20/Oct	EM	Group II
HD 210221	A3 Ib	-7.94	27,29/Sep	23/Oct	EM	
HD 12953	A1 Ia	-8.20	26,28/Sep	20/Oct	PCyg	Group II
HD 223385	A3 Ia	-8.44	27,29/Sep	23/Oct	PCyg	Group II
HD 197345	A2 Ia	-8.61	27,29/Sep		EM	Group II

Mbol and Spec. Type from Humphreys (1978); INT=Isaac Newton Telescope; CA=2.2m Calar Alto; all the observing dates are 1988.

3. PRELIMINARY RESULTS: Hα

3.1 Data reduction

The spectra have been extracted from the CCD frames after dark noise correction and flatfielding. The one dimensional spectra so obtained have been rebinned into a wavelength scale by using a polinomial fit to the positions provided by the thorium-argon lines of the comparison lamp.

The resolution measured in the comparison spectra is 0.25 Å(11 km/s at Hα) and 0.40 Å(18 km/s) in Calar Alto and the INT respectively. Both set of data have a signal to noise ratio ranging between 100 and 200.

All the spectra have been divided by a linear continuum placed by eye.

The data have been processed using the IHAP and MIDAS systems.

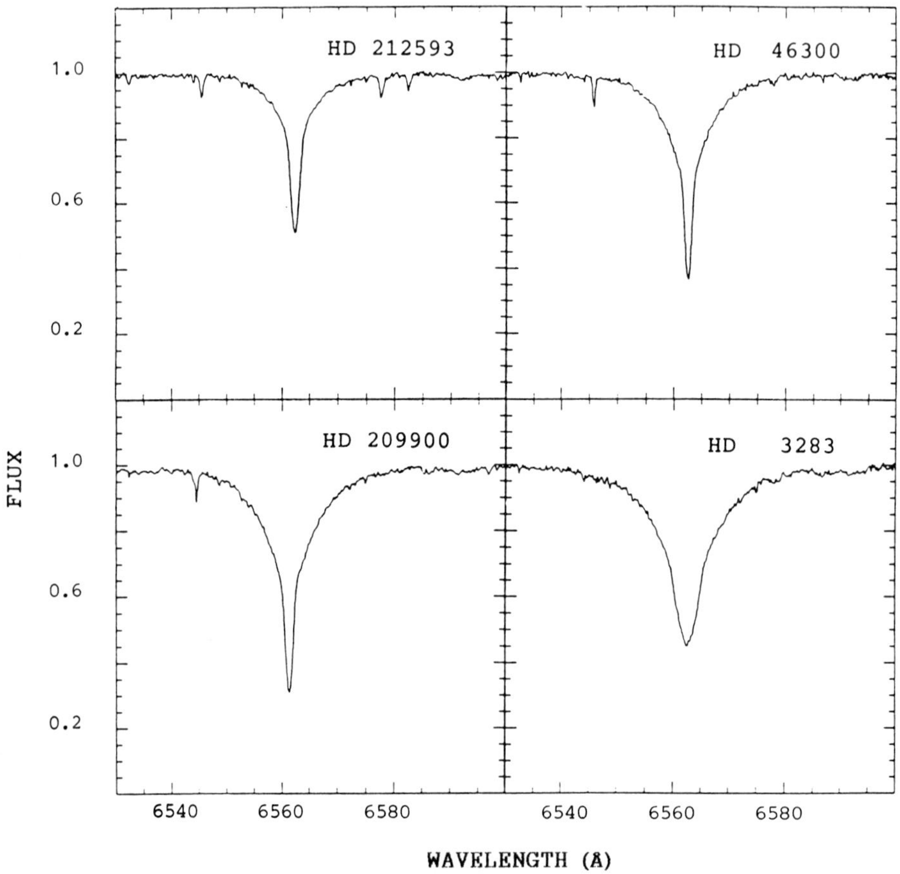

Fig.1. Hα profiles in some of the less luminous A supergiants.

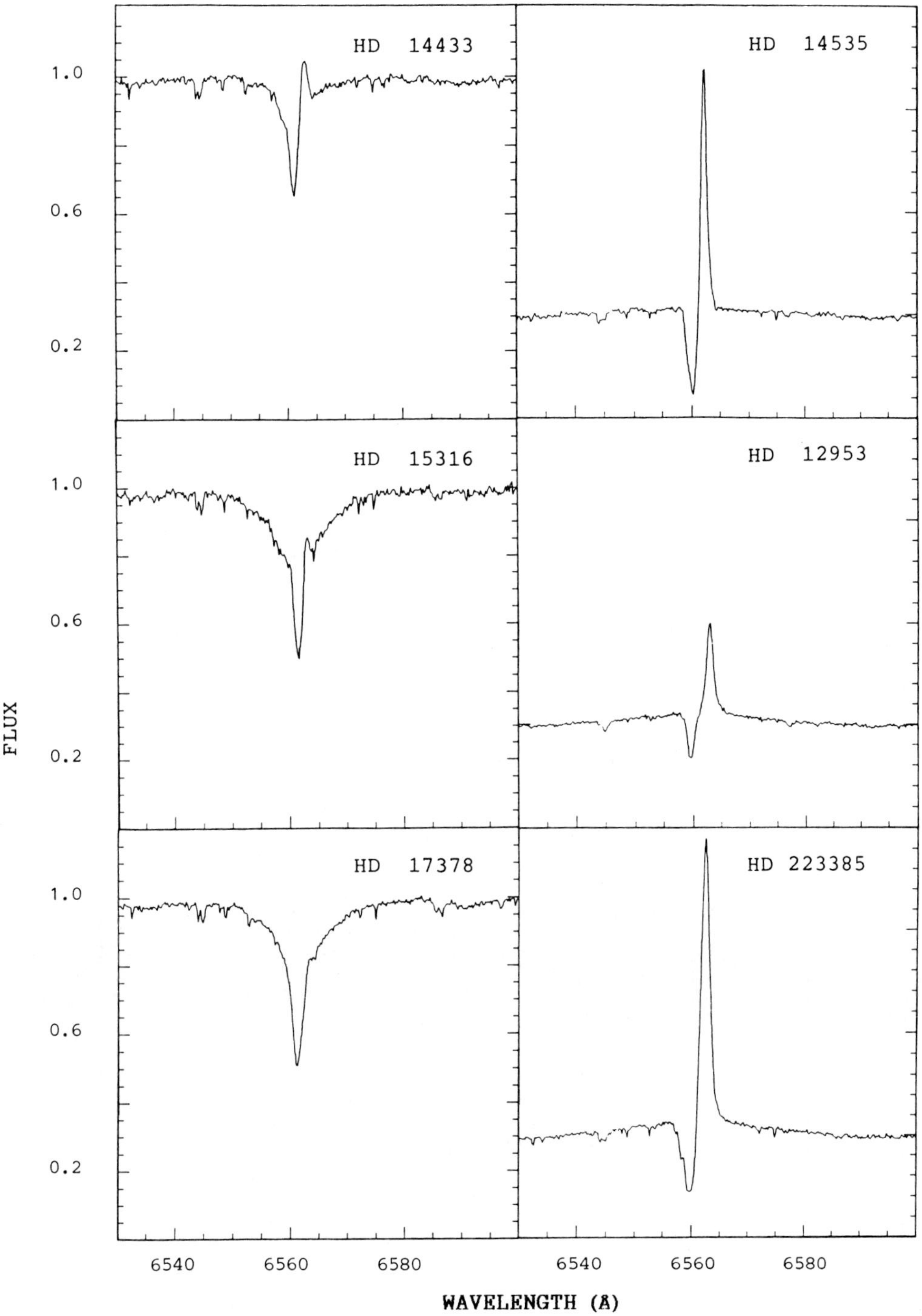

Fig.2. Evolution of the Hα line in luminous A supergiants.

3.2 Observed profiles

The Hα profiles in A-type supergiants present very different shapes going from pure symmetric absorption to emission. The observed profile type is given for each star in Table 1. From that table it can be seen that the profiles evolve according to the absolute magnitude of the stars as it was noticed by Rosendhal (1970). This trend is ilustrated in Figs. 1 and 2. The less luminous stars show symmetric absorption profiles; then when luminosity increases an asymmetry starts developing in the core of the absorption which becomes a P Cyg type III profile for the brightest stars. In two "peculiar" stars the observed profile is in pure emission.

The less luminous stars do not show variability in the profiles, which have been observed for all the stars at least twice, at different epochs (see Table 1). However we have observed variations in some of the stars showing asymmetric-emission profiles. Some of these variations are very strong. They will be studied in future work.

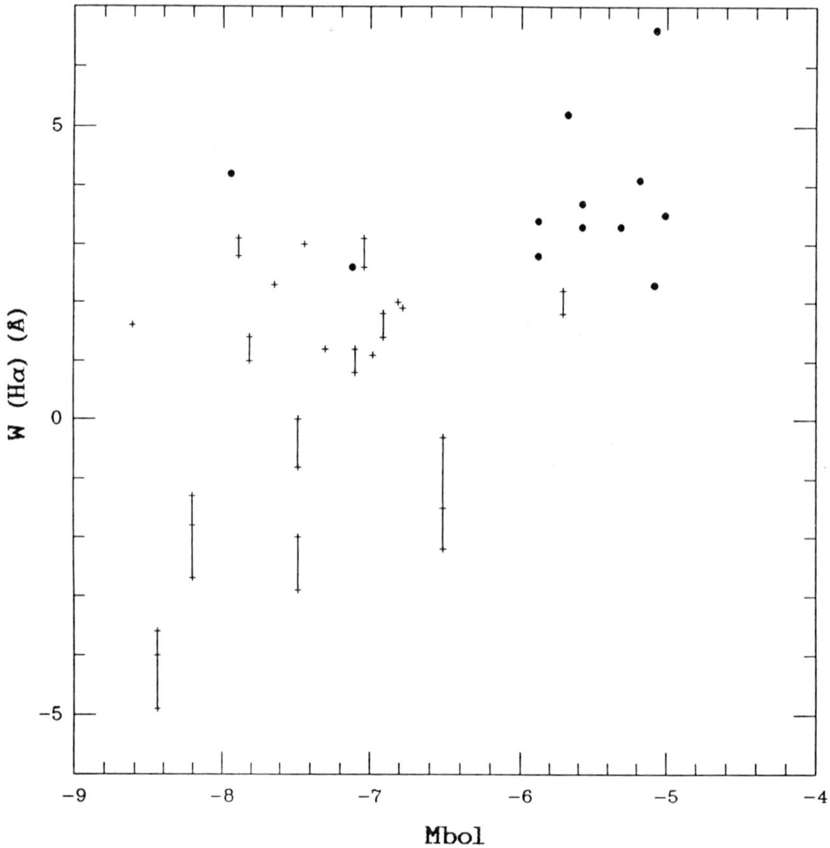

Fig.3. Equivalent width of Hα versus bolometric magnitude. Symmetric absorption lines are represented with dots and lines showing emission with crosses. Vertical lines indicate observed variation.

3.3 Equivalent widths

We have measured the equivalent width of the observed Hα lines. The results are illustrated in Fig. 3, where we have represented W(Hα) against Mbol. The reported values correspond to absorption below the continuum minus emission above the continuum. In that figure we have indicated the variations observed in the equivalent width by vertical lines.

Although we cannot say there is a clear correlation W(Hα) - Mbol, it can be seen from Fig. 3 that there is a trend of increasing equivalent width with increasing magnitude. In other words the absorption in Hα is cancelled by emission in the core of the line when we go to more luminous stars. We have measured also the emission peak intensity for the stars showing that feature and we have obtained also the same trend.

From our sample of stars we can say that emission in the core of Hα is observed for stars brighter than Mv=-5.5.

3.4 Comparison with the U. V.

The sample of stars we have observed is larger than the one on which we based our studies with IUE. The first result of our new observations in Hα is the confirmaton of the division between Groups I and II established from IUE data. The less luminous A supergiants which showed weak signs of stellar wind in the U.V., those included in group I, have Hα in absorption, while the stars of group II, the most luminous ones, present emission signs in the Hα profiles.

4. CONCLUSIONS

We have presented some very preliminary results of a large programme intended to study the winds of A-type supergiants. Our data confirm and extend to a larger sample the results pointed out by Rosendhal (1970) and provide further evidence for the conclusions derived from our work in the U.V.

The project is in progress and we are continuing the analysis of the lines mentioned in Section 2. In the second part of our project we shall try the modelling of the envelope-wind complex in A-type supergiants.

REFERENCES

Abbott, D.C. 1982, *Astrophys.J.*, **263**, 723

Humphreys, R.M. 1978, *Astrophys.J.Suppl.*, **38**, 309

Hutchings, J.B. 1970, *Mon.Not.Roy.Astr.Soc.*, **147**, 161

Rosendhal, J.D. 1970, *Astrophys.J.*, **186**, 909

Talavera, A., and Gomez de Castro, A.I.1987, *Astron.Astrophys.*, **181**, 300

Talavera, A., and Gomez de Castro, A.I.1988, in *Proceedings of The 10th European Regional Astronomy Meeting of the IAU*, ed. P. Harmanec, **Vol. 5**, p.207

Wolf, S.C. 1983, in *The A-type stars: problems and perspectives*, NASA SP-463, p.118

Supernova Remnants

S.A.E.G.Falle,
Department of Applied Mathematical Studies,
The University,
Leeds LS2 9JT, U.K.

Abstract.

In this review I will concentrate on older remnants, by which I mean those in which radiative cooling is important somewhere and the swept up mass is sufficiently large for the details of the initial explosion not to matter. For such remnants it is the optical emission which is crucial since it allows us to deduce a great deal about the physical state of the emitting gas provided we are careful about how we interprete it. Without discussing any particular remnant in detail, I will consider how large and small scale density variations in the ambient medium affect the appearance and energetics of such remnants.

Introduction.

For spherical remnants expanding in a uniform environment it is possible to divide the evolution into three phases, free expansion, Sedov-Taylor and radiative (Woltjer 1972). Such a simple picture ignores the fact that the interstellar medium is inhomogeneous on many scales , some of which correspond to the sizes of supernova remnants and that Type II supernovae can significantly modify the interstellar medium in their neighbourhood. It is nevertheless a useful way of classifying remnants even though there are many that do not fall clearly into any one category.

For the purposes of this article, I will call a remnant old if radiative cooling is important somewhere, but will not insist that a significant part of the explosion energy has been radiated away. Observationally this means that a remnant is old if it has a filamentary structure whose emission is characteristic of radiative shocks. Examples are the Cygnus Loop, Shan 147 and Vela. As far as theory is concerned we have to look at the effects of radiative cooling on the dynamics and appearance of the remnant.

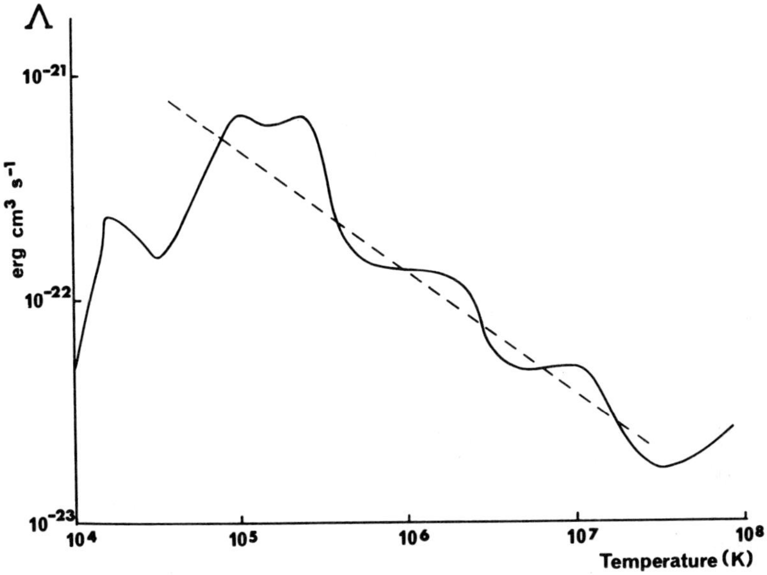

Figure 1. The radiative cooling rate per unit volume for an optically thin plasma. The straight line is the approximation (1) (Kahn 1976).

The radiative cooling rate for an optically thin gas in the relevant temperature range is shown in figure 1. Although this is not the most recent calculation, it has a maximum at about 10^5 K , which is what is important as far as the dynamics is concerned. Note that it does not include the effect of dust cooling which might well dominate above 10^5 K, depending upon the dust to gas ratio (Dwek 1987).

One of the nice things about this cooling law is that in the range 5×10^4 K < T < 5×10^7 K it is very well approximated by $T^{-1/2}$ power law,

$$\Lambda = A\rho^2 (p/\rho)^{-1/2} \quad (A = 3.9\times10^{32} \text{ c.g.s}). \qquad (1)$$

Kahn (1976) showed that this assumption makes it possible to calculate the effect of radiative cooling on the overall energetics independently of the details of the dynamics. This can not only be applied to spherical remants, but also to those in a plane stratified medium (Falle ,Garlick and Pidsley

1984). Unfortunately this cooling curve also has some nasty features which, as we shall see later, makes the transition to the radiative phase horribly complicated.

Spherical Remnant.

Suppose that a supernova explosion has energy E_0, ejects mass M_e and occurs in a uniform medium with density ρ_0. Then there will be a Sedov-Taylor phase provided

$$\left(\frac{E_0}{10^{51}}\right)^{-0.74}\left(\frac{M_e}{M_\odot}\right)^{5/6}\left(\frac{\rho_0}{10^{-24}}\right)^{0.2} < 4 \ . \qquad (2)$$

This is simply the condition that the remnant enters the Sedov phase before radiative cooling becomes important. It is based on Gull's (1973) calculations, which show that it looks like a Sedov solution once it has swept up about 50 M_e, combined with Cox's (1972) estimate of when radiative cooling becomes important.

One would prefer a Sedov phase to exist, because then all the details of the original explosion can be ignored and only the energy E_0 matters. Condition (2) is satisfied for all plausible values of E_0, M_e and ρ_0, but unfortunately it ignores the fact that a Type II supernova can modify its surroundings, either because of its ionizing radiation (Shull, Dyson, Kahn & West 1985), or its stellar wind (Charles, Kahn & McKee 1985). There is a good deal of observational evidence that this occurs (Braun 1987).

I am going to ignore these complications and assume that the original state of the ambient medium is more important than the details of the explosion. Then in a uniform medium radiative cooling becomes important when the post shock temperature is

$$T = T_{sg} = 1.2\text{x}10^{6}\left(\frac{E_0}{10^{51}}\right)^{0.1}\left(\frac{\rho_0}{10^{-24}}\right)^{0.2} \text{K} \qquad (3)$$

Cox (1972).

If we now ignore magnetic fields and assume that all shocks are strong, then as long as $T \geq 5\text{x}10^4$ K everywhere, things only depend on E_0, ρ_0 and A. From these we can form a characteristic mass, length and time given by

$$m_c = \frac{(2.02E_0)^{6/7}}{\rho_0^{2/7}A^{3/7}} = 7.3\text{x}10^{36}\left(\frac{E_0}{10^{51}}\right)^{6/7}\left(\frac{\rho_0}{10^{-24}}\right)^{-2/7} \text{gm},$$

$$l_c = \frac{(2.02E_0)^{2/7}}{\rho_0^{3/7}A^{1/7}} = 1.9\text{x}10^{20}\left(\frac{E_0}{10^{51}}\right)^{2/7}\left(\frac{\rho_0}{10^{-24}}\right)^{-3/7} \text{cm}, \qquad (4)$$

$$t_c = \frac{(2.02E_0)^{3/14}}{\rho_0^{4/7}A^{5/14}} = 1.2\text{x}10^{13}\left(\frac{E_0}{10^{51}}\right)^{3/14}\left(\frac{\rho_0}{10^{-24}}\right)^{-4/7} \text{s}.$$

We expect radiative cooling to become important when the swept up mass is about m_c and the radius and age will then be aproximately l_c and t_c respectively. Notice that these numbers are about right for the Cygnus Loop and IC443.

Radiative Instabilities.

We can write the cooling rate shown in figure 1 in the form

$$\Lambda = A\rho^2\Phi(p/\rho c_*^2) , \qquad (5)$$

where

$$c_*^2 = \frac{kT_*}{\mu m_h} , \qquad (6)$$

and T_* is a reference temperature. T_* can be chosen to be the temperature at the maximum of Φ ($T_* = 10^5$ K). If we then set $\Phi(T_*) = 1$, we get $A = 2\text{x}10^{26}$ c.g.s.

For a spherical remnant the flow is now governed by the parameters E_o, ρ_o, A and c_* and from these we can form a dimensionless parameter

$$\alpha = \frac{T_{sg}}{T_*} = 10\left(\frac{E_0}{10^{51}}\right)^{0.11}\left(\frac{\rho_0}{10^{-24}}\right)^{0.22}. \qquad (7)$$

Here T_{sg} is the temperature defined by equation (3). α only affects the evolution of the remnant if there is radiatively cooling gas at temperatures below T_*.

Let us now look at the stability of gas whose cooling rate is given by (5). Suppose that $\Phi(T) \propto T^s$. Then if cooling occurs at constant pressure, the cooling time increases with increasing temperature if $s < 2$, while for constant density this is true for $s < 1$. This suggests that there is instability if $s < 2$ for constant pressure cooling and $s < 1$ for constant density.

Now the pressure will remain roughly constant if the cooling time $t_{cool} \gg t_{dyn}$ where t_{dyn} is some dynamical timescale. Conversely the density will remain constant if $t_{cool} \ll t_{dyn}$. Suppose that a region of initial size ℓ begins to cool and that the resulting compression is one dimensional. Then

$$\ell(t) \propto \frac{1}{\rho(t)} .$$

The relevant dynamical time is obviously the sound crossing time, so

$$t_{dyn} = \frac{\ell}{c} \propto \frac{1}{\rho T^{1/2}} .$$

On the other hand we have for the cooling time

$$t_{cool} \propto \frac{p}{\rho^2 T^s} \propto \frac{1}{\rho T^{s-1}} .$$

Hence

$$\frac{t_{dyn}}{t_{cool}} \propto T^{s-3/2} , \tag{8}$$

and so increases as the gas cools if $s < 3/2$. If t_{dyn} ever becomes much smaller than t_{cool}, then we expect a large pressure imbalance to occur which will lead to the formation of shocks. A necessary condition for this is $s < 3/2$. For the interstellar cooling law this condition is satisfied for $T > T_*$ and so we expect this kind of instability for spherical remnants if $\alpha > 1$. From (7) we can see that this should happen for almost all such remnants.

Various authors have looked at radiative instabilities. Both Avedisova (1974) and McCray, Stein & Kafatos (1975) carried out a linearised stability analysis with the post shock pressure held fixed. They found that density fluctuations grow if $s < 3$ for perturbations with wavelength much greater than the cooling length, while $s < 2$ is required if the wavelength is much shorter than the cooling length. However, these are isobaric instabilities and do not lead to the formation of additional shocks.

In my numerical calculations of thin shell formation in spherical remnants (Falle 1975, 1981), I found that cooling led to the formation of multiple shocks which caused large variations in the speed of the primary shock. Langer, Chanmugam and Shaviv (1981,1982) found a similar effect in their calculations of radiative accretion onto white dwarfs.

These results have stimulated a lot of interest in such instabilities. Chevalier & Imamura (1982) used a linearised stability analysis to show that the shock speed will not be constant if s < 0.8, even if it is driven by a constant speed piston. To some extent this is confirmed by numerical calculations (Imamura, Wolff & Durisen 1984). Recently Bertschinger (1986) has extended this analysis to two dimensions and shown that in that case instability occurs if s < 1.

It has become common practice to deduce the velocity of radiative shocks by comparing the observed optical and UV line ratios with those predicted by steady shock models with various shock speeds (e.g. Raymond et.al. 1980). Unfortunately the above considerations suggest that radiative shocks will not be steady if the shock speed is high enough for cooling to occur in the unstable region of the cooling curve.

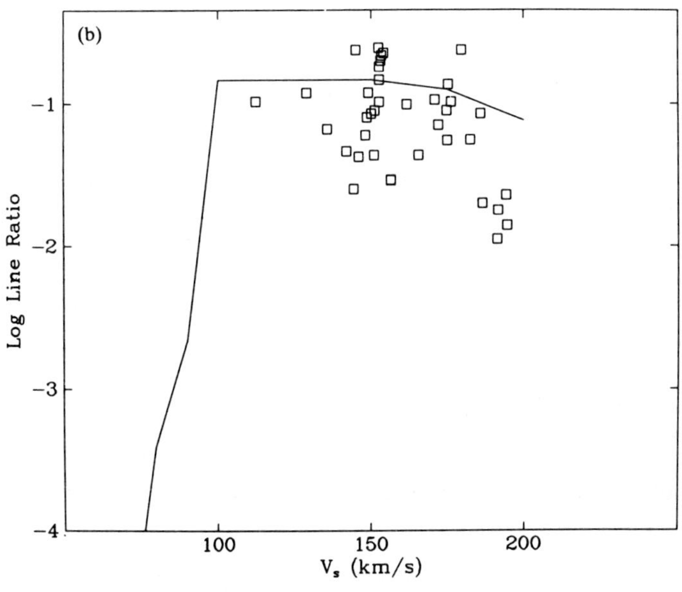

Figure 2. Instantaneous [OIII]5008/[OII]3728 line ratio for an unsteady shock whose mean speed is 175 km s^{-1}. The solid line is the ratio for a steady shock (Innes, Giddings & Falle 1987).

Recently Innes, Giddings & Falle 1987 have shown that, if the detailed atomic physics is included, then radiative shocks will be unsteady if their speed is greater than 130 km s^{-1}. The line ratios then do not correlate with the primary shock speed, nor even with the mean fluid speed, but vary dramatically on the cooling timescale. This effect can be seen in figure 2 which shows the instantaneous [OIII]5008/[OII]3728 line ratio plotted against the instantaneous primary shock speed for a shock driven by a constant speed piston such that the mean shock speed is 175 km s^{-1}. The variations in shock speed were induced by a single sinusoidal density perturbation upstream. The perturbation had an amplitude of 50% of the upstream density and a wavelength 1.4 times the thickness of the cooling region.

Small Scale Inhomogeneities.

The appearance of remnants like the Cygnus Loop suggests that the blast wave is interacting with irregularities with quite small scales. Indeed McKee & Cowie (1975) have argued that in the Cygnus Loop we only see optical filaments when shocks propagate into small clouds.

The interaction of a plane shock with density inhomogeneities has been looked at by many authors (e.g. Sgro 1975; Chevalier & Theys 1975; Woodward 1976; Nittmann, Falle & Gaskell 1982; Hamilton 1985; Heathcote & Brand 1983). Although we have a rough idea of what happens, at least in the adiabatic case, there are a number of important details which are not clear.

Rather than looking at deformable clouds, I will consider what happens when a shock hits a rigid object. Of course supernova remnants do not encounter rigid objects, at least not of significant size, but from pressure balance we have

$$V_c = V_e \left(\frac{\rho_e}{\rho_c} \right)^{1/2}, \qquad (9)$$

where V is a shock velocity and the subscripts e and c refer to the exterior and cloud respectively. So if the cloud is much denser than its surroundings, it deforms slowly compared to the timescale of the exterior flow. The cloud therefore behaves like a rigid body, at least as far as the transient stage of the exterior flow is concerned.

Because of this it makes sense to treat clouds which are much denser than the ambient medium as rigid bodies. This is more efficient from a computational point of view since it avoids the large disparity in timescales between the interior and exterior flows. Another advantage is that there is a wealth of experimental data upon which we can draw.

An obvious case to look at is that of a plane shock hitting a rigid sphere. Figures 3 and 4 show the results obtained with a second order Eulerian

Godunov scheme on a high resolution spherical polar grid. The advantages of such a grid are that not only can the sphere be represented exactly, but also that the highest resolution is near the surface of the sphere where most of the action takes place. There is, however, some degradation of the solution due to the non-uniformity of the grid.

Figure 3 shows the sequence of events. Initially there is a regular reflection at the surface which evolves into a Mach reflection at $\theta \simeq 135^0$ (here $\theta = 0$ corresponds to the direction of motion of the incident shock). The reflected shock associated with this eventually becomes a stationary bow shock in this case since the flow behind the incident shock is supersonic in the frame of the obstacle.

Further round the sphere the incident shock is diffracted into a funnel shape before reflecting off the symmetry axis. Because of the axial symmetry this is a Mach reflection from the very beginning. The Mach disc which moves down the axis initially has zero size, but it grows enough to become significant before the triple point associated with it disappears and it merges with the incident shock.

The solution at the latest time is shown in greater detail in figure 4. This should be compared with the experimental results obtained by Bryson & Gross (1961) for shock interaction with a rigid sphere in air. Most of the features of the experiment are well reproduced by the simulation, the exception being the vortex formed when the slip line produced by the first Mach reflection rolls up. It would, however, be too much to expect to get this sort of thing right at this resolution.

We must now consider how much of this is relevant to supernova remnants. Clearly the flow is the essentially the same as the one that arises when a non-radiative shock encounters a very dense spherical cloud and so our results can be used to interpret observations of such shocks (Raymond, Davis, Gull & Parker 1980). The first thing to note is that large portions of both the reflected and diffracted shocks are oblique. This means that the usual practice of interpreting the observations in terms of normal shocks can be very misleading. Fortunately, for non-radiative shocks it is possible to calculate the flow as we have done here and then use the appropriate physics to deduce the spectrum and velocity dispersion. In this way we ought to be able to get a good idea of the nature of some of the filaments in, for example, the Cygnus Loop.

If the shocks are radiative, then life becomes much more difficult. Raga & Böhm (1987) have computed the flow past a hemisphere using a McCormack

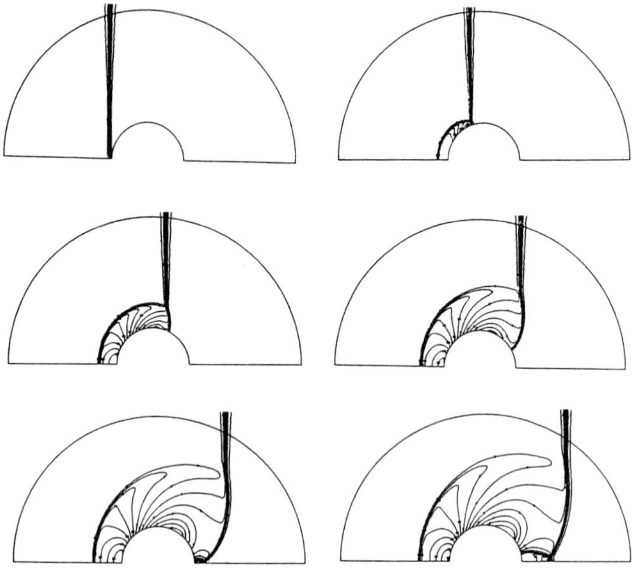

Figure 3. Pressure contours for a plane shock with an initial Mach number of 2.81 interacting with a rigid sphere (γ = 5/3). Calculated on spherical polar grid with 100 cells in the radial direction and 180 in polar angle.

scheme. Since they had a priori knowledge of where the radiative shock would be, they were able to resolve the cooling region by an ingenious choice of grid so that their results are almost certainly reliable. Clearly this kind of thing is much harder to do for unsteady problems and for the moment it is probably better to try and guess how cooling modifies the flow. We have little chance of doing this if the cooling length is of the same order as the size of the obstacle or if the shock speeds are in the range in which the radiative instability is important, but we can do something if the cooling length is very short and the shock is stable. The shocks are then isothermal and to some extent one can get an idea of the trends by looking at lower values of the ratio of specific heats γ. Reducing γ increases the shock compression so that the bow shock is closer to the surface of the sphere. Its shape is therefore similar to that of the obstacle and, since this is true generally, we should be able to deduce something about the shape of a dense cloud if we can identify the bow shock. Another consequence of the greater compression is that the transition to Mach reflection is less likely. For a sphere a Mach shock will always occur both on the surface and on the axis downstream, but as the

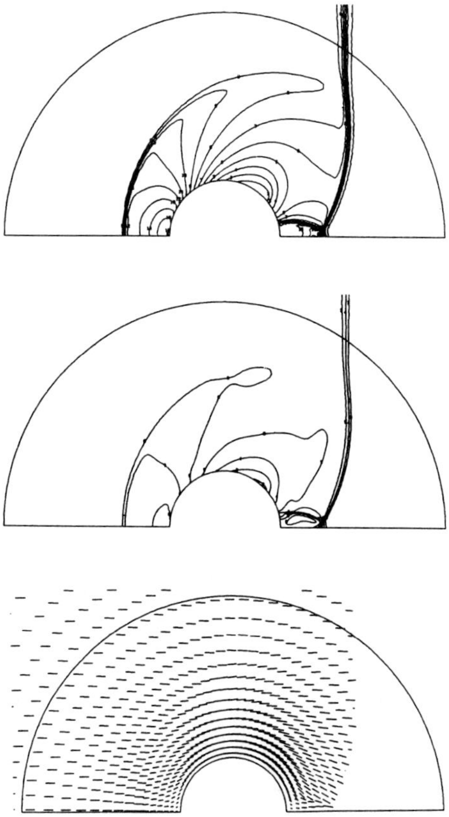

Pressure Contours

Density Contours

Velocity Vectors

Figure 4. Details of the flow at the latest time calculated.

shock compression tends to infinity the point at which Mach reflection starts approaches $\theta = 90^0$ and the size of the Mach disc on axis tends to zero.

Conclusions.

I have discussed some of the effects that radiative cooling and density inhomogeneities can have on the evolution of supernova remnants. I have indicated that radiative instabilities must exist in radiative remnants and that these make it very difficult to interpret the spectra of radiative shocks. They may also be responsible for at least some of the complex structure seen in old remnants.

Interactions with small scale inhomogeneities are more difficult to deal with, but we can use laboratory experiments, numerical simulations and perhaps Whitham's area rule (Whitham 1974) to deduce how clouds of various sizes and densities affect the appearance of remnants. We clearly need a reliable quantitative theory of such interactions in order to make the proper use of the detailed observations that are now possible.

References.

Avedisova, V.S. 1974. Sov. Astron., 18, 283.

Bertschinger, E. 1986. Astrophys. J., 304, 154.

Braun, R. 1987. IAU Colloquium 101, Penticton, Canada, p.363.

Bryson, A.E. & Gross, R.W.F. 1961. J. Fluid Mech., 10, 1

Charles, P.A, Kahn, S.M. & McKee, C.F. 1985. Astrophys. J., 295, 456.

Chevalier, R.A. & Imamura, J.N. 1982. Astrophys. J., 261, 543.

Chevalier, R.A. & Theys, J.C. 1975. Astrophys. J., 195, 53.

Cox, D.P. 1972. Astrophys. J., 178, 159.

Dwek, E. 1987. IAU Colloquium 101, Penticton, Canada, p.363.

Falle, S.A.E.G. 1975. Mon. Not. R. astr. Soc., 172, 55.

" 1981. Mon. Not. R. astr. Soc., 195, 1011.

Falle, S.A.E.G., Garlick, A.R. & Pidsley, P.H. 1984. Mon. Not. R. astr. Soc., 208, 925.

Gull, S.F. 1973. Mon. Not. R. astr. Soc., 161, 47.

Hamilton, A.J.S. 1985. Astrophys. J., 291, 523.

Heathcote, S.R. & Brand, P.W.J.L. 1983. Mon. Not. R. astr. Soc., 203, 67.

Imamura, J.N., Wolff, M.T. & Durisen, R.H. 1984. Astrophys. J., 276, 667.

Innes, D.E., Giddings, J.R. & Falle, S.A.E.G. 1987. Mon. Not. R. astr. Soc., 226, 67.

Kahn, F.D. 1976. Astr. & Astrophys., 50, 145.

Langer, S.H., Chanmugam, G. & Shaviv, G. 1981. Astrophys. J., 245, L23.

" 1982. Astrophys. J., 258, 289.

McCray, R., Stein, R.F. & Kafatos, M. 1975. Astrophys. J., 196, 565.

McKee, C.F. & Cowie, L.L. 1975. Astrophys. J., 195, 715

Nittmann, J., Falle, S.A.E.G. & Gaskell, P.H. 1982. Mon. Not. R. astr. Soc., 201, 833.

Raga, A.C. & Böhm, K.H. 1987. Astrophys. J., 323, 193.

Raymond, J.C., Black, J.H., Dupree, A.K. & Hartmann, L. 1980. Astrophys. J., 238, 881.

Raymond, J.C., Davis, M., Gull, T.R. & Parker, R.A.R. 1980. Astrophys. J., 238, L21.

Sgro, A.G. 1975. Astrophys. J., 197, 621.

Shull, P., Dyson, J.E., Kahn, F.D. & West, K.A. 1985. Mon. Not. R. astr. Soc., 212, 799.

Whitham, G.B. 1974. Linear and Nonlinear Waves, Wiley Interscience, chapter 8.

Woltjer, L. 1972. Ann. Rev. Astron. Astrophys., 10, 129.

Woodward, P.R. 1976. Astrophys. J., 207, 484

Discussion:

ZINNECKER: Given these impressive numerical calculations, can you tell us under which conditions gaseous clumps can be imploded after being hit by a supernova remnant shock, i.e. for which parameters of the clumps (e.g. size, density) can star formation be triggered?

FALLE: I can't tell you off the top of my head, but basically the denser the cloud the smaller it can be and still be induced to collapse. This is not only because of the dependence of the Jeans' mass on density, but because a dense cloud can more easily survive until it reaches the low velocity region in the interior of the remnant.

PALOUS: Even if the SNR is non spherical for any reason, entering the Sedov phase of evolution, it will be more spherical after, since the parts moving with a higher velocity will be more decelerated from the ambient medium than the other parts. This was pointed out in a paper by Bisnovatyg-Kogan and Biliunikov (1982).

FALLE: You are quite right. In the Sedov-Taylor phase the remnant has lost all memory of the details of the explosion, including any asymetry. One can therefore conclude that if an old remnant is not spherically symmetric then this must be due to the surrounding medium.

DUST CONDENSATION IN THE EJECTA OF SN 1987A

L.B. Lucy, I.J. Danziger, C. Gouiffes, and P. Bouchet
European Southern Observatory, Karl-Schwarzschild-Str. 2
D-8046 Garching bei München, Federal Republic of Germany

ABSTRACT

An asymmetry of optical emission lines that appeared in Sept. 1988 is interpreted as evidence of dust condensation within the metal-rich ejecta of SN 1987A. A quantitative analysis of this spectroscopic effect is given and shown to be compatible with the photometric record. Moreover, observational and theoretical estimates of the bolometric light curve come into agreement when the far-IR excess is interpreted as thermal emission by grains in the ejecta. A grain population comprising small silicate grains with an admixture of graphite or amorphous carbon particles is suggested by the data. The relevance of this discovery to suggestions that supernovae are major sources of interstellar dust is briefly discussed.

1. INTRODUCTION

Various indirect arguments have been adduced that supernovae are major sources of interstellar dust. The initial suggestion (Cernuschi et al. 1967) that SN ejecta are promising sites for grain condensation followed from the recognition that their cooling trajectories bring about appropriate thermodynamic conditions. Subsequent consideration of the potential condensable mass per SN and of the frequency of SN led Hoyle and Wickramasinghe (1970) to note that this source alone could in ~ 10^9 yr produce the inferred mass of dust in the Galaxy. Further support for this formation site comes from the argument (Dwek and Scalo 1980; Clayton 1982) that the large depletions of refractory elements in the interstellar gas require that, when injected into interstellar space by SN explosions, these elements are already locked up in grains. The isotopic anomalies in meteorites also point to grain condensation occurring before SN ejecta are mixed and diluted with isotopically normal interstellar gas (Clayton 1982).

The prospect that SN 1987A should allow these arguments to be tested has been widely recognized, with attention focussing on the photometric consequences of dust formation. In particular, Dwek (1988) and Kozasa et al. (1989) have stressed the possibility noted already by Hoyle and Wickramasinghe (1970) of a total black-out in the UV and optical, with the SN then becoming a purely IR object. But in fact no evidence has been presented of any extinction at short wavelengths, and the observed brightening in the IR has been interpreted as a thermal echo from pre-existing circumstellar dust (Roche et al. 1989; Felten and Dwek 1989).

In contrast, the starting point of this investigation was our recognition of a spectroscopic signature of dust within the expanding ejecta, namely the line profile asymmetry caused by the greater attenuation of radiation received from receding matter. The analogous effect in novae was first discussed by McLaughlin (1935) and is now known to appear simultaneously with the more familiar photometric signatures of dust formation (Smith et al. 1979). Chevalier (1981) has previously invoked this nova analogy when explaining an Hα asymmetry in SN 1979c as due to dust in the expanding envelope.

A brief report on this work appeared in IAUC No. 4746. The present account has been up-dated to the time of writing - late May.

2. OBSERVED PROFILES

Fig. 1 shows the change in the [OI]λ6300,6363Å profile between 5 Aug '88 and 3 Mar '89. The earlier spectrum shows the [OI] components to be essentially unshifted with respect to the LMC, whereas in the later spectrum both components are blueshifted by $\simeq$ 600 km s^{-1}.

Because this effect occurs without perturbing the secular trend in the intensity ratio of the [OI] components, it cannot plausibly be attributed to self-absorption; consequently, the blue shift must be imposed by an effect distinct from local line formation. An appealing possibility is the onset of significant attenuation within the expanding ejecta due to dust condensation. Of course, other absorption mechanisms could produce the same asymmetry; but, under the prevailing conditions of low temperature - as indicated by the prior formation of several molecular species (CO, SiO, CS) - dust is the obvious choice.

3. THEORETICAL LINE PROFILES

We take the metal-rich ejecta to be a spherically symmetric nebula (Fig. 2) expanding radially with velocity law $V = r/t$, where t is the time since core collapse. Surfaces of constant line-of-sight velocity are then simply planes $z = -v$, where the unit of length is the sphere's instantaneous radius $R_E = V_E t$ and the unit of velocity is V_E.

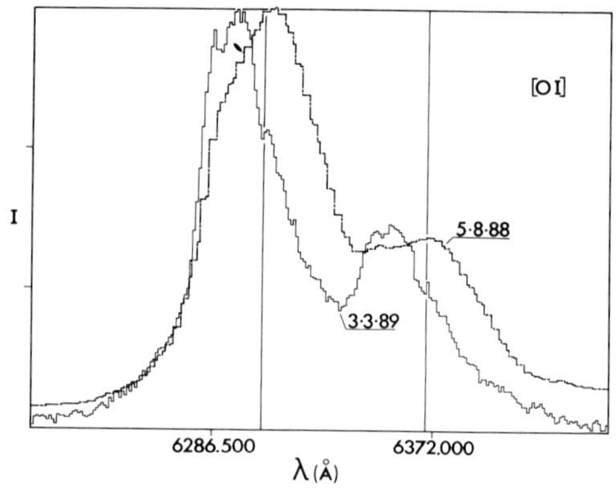

Fig. 1: Emission profile of [OI]λ6300,6363 Å on 5 Aug '88 and 3 Mar '89. Expected wavelengths at velocity of LMC are indicated. The profiles are scaled to same peak intensity for violet component.

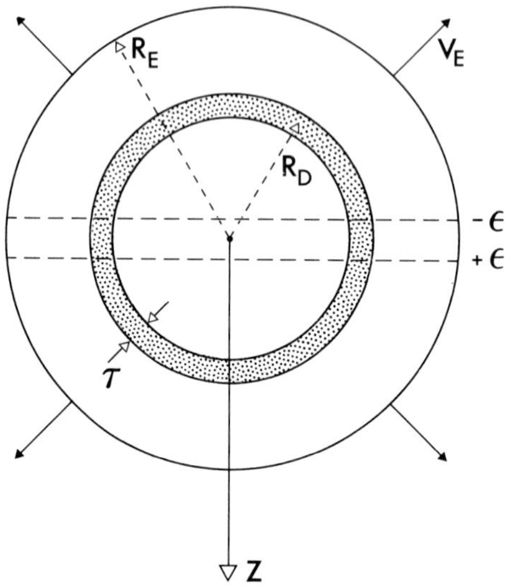

Fig. 2: Model for expanding ejecta with dust condensation in thin shell (Model I). The observer is at $z = +\infty$.

Model I. On the assumption that the spectroscopic detection of dust closely followed its initial formation, we first suppose that dust is still restricted to a thin spherical shell at the radius (unknown) where condensation was first allowed. In addition, we suppose that the dust grains are still small ($2\pi a \ll \lambda$) so that dust scattering may be neglected - i.e., the albedo $\omega \ll 1$. Now, from Fig. 2, we see that, if the dust shell is geometrically infinitesimal, the numbers of line photons received from planes $z = \pm \varepsilon$ become identical as $\varepsilon \to 0$. Accordingly, with $\omega = 0$, the line profile $I(v)$ is such that

$$\frac{dI}{dv} = 0 \quad \text{at } v = 0 \ , \tag{1}$$

a result that is evidently independent of τ and holds whatever the radial distribution of line emissivity. But this prediction of zero slope at $v = 0$ when the dust is in a thin shell is contradicted by the observed profiles (Fig. 1). Further evidence of the poor fit of this model is provided by Fig. 3, which shows line profiles for various dust-shell radii in the case of constant line emissivity.

<u>Model II</u>. The obvious implication of this failure is that dust was already widely distributed when detected spectroscopically. Accordingly, retaining the assumption $\omega = 0$, we now suppose that the dust is uniformly distributed throughout the ejecta ($r < R_E$). In this case, radiation received from a point on the surface $z = +v$ is attenuated relative to that from the corresponding point on $z = -v$ by the factor $\exp(-2\tau v)$, where $\tau = k\rho R_E$. The line profile is therefore such that

$$I(v) = I(-v)\exp(-2\tau v) \ , \tag{2}$$

from which it follows that

$$\frac{d\ln I}{dv} = -\tau \quad \text{at } v = 0 \ . \tag{3}$$

If we now make the further assumption that the line emissivity is independent of r, the blue side ($v < 0$) of the line profile is given by

$$\tau^2 I(v) = 1 - v\tau - (1+\tau)\exp\{-(1+v)\tau\} \ . \tag{4}$$

From this, one readily finds that maximum intensity occurs at velocity

$$v = -1 + \tau^{-1} \ln(1+\tau) \ . \tag{5}$$

Examples of line profiles derived from equations (2) and (4) are given in Fig. 4.

<u>Model III</u>. Given that dust may have formed well before its spectroscopic detection, the possibility of large grains with non-negligible albedo must be considered. Accordingly, in Fig. 5 we present Monte Carlo calculations of line profiles for uniform distributions of line emissivity and dust, with the dust grains now being isotropic scatterers with $\omega = 0.6$, an albedo typical of local Galactic dust in the visual (Witt 1988). These profiles reveal a potentially useful diagnostic for dust scattering in the form of an extended red wing comprising photons that escape after several scatterings. These photons are redshifted because they have done work on the expanding sphere of dust.

4. ANALYSIS OF OBSERVED LINE PROFILES

The above simple models can be used to quantify the amounts and other properties of the dust required by our explanation of the line profile asymmetry. If this

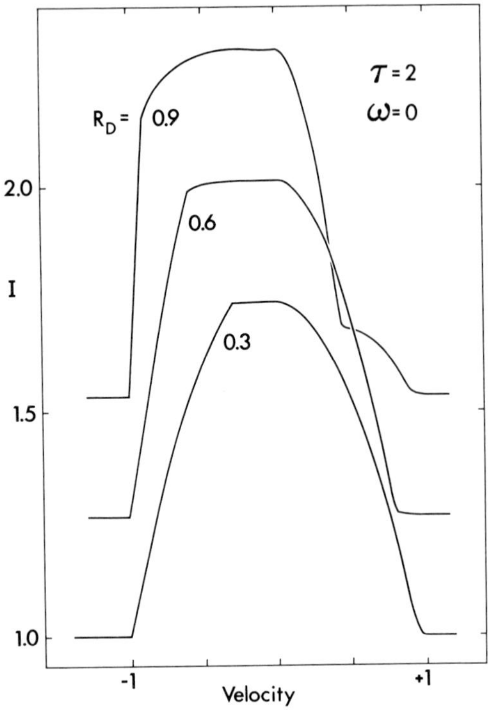

Fig. 3: Theoretical line profiles for Model I. Thin shells of zero-albedo dust with τ = 2 at various radii R_D. Line emissivity is constant.

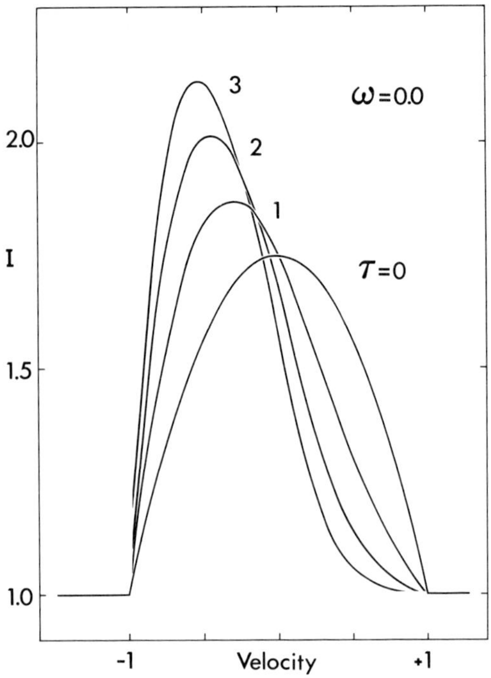

Fig. 4: Theoretical line profiles for Model II. Zero-albedo dust uniformly distributed in the ejecta with various values of $\tau = k\rho R_E$. Line emissivity is constant.

spectroscopically derived data proves compatible with the photometric record, the dust condensation hypothesis will be greatly strengthened, and a more sophisticated modelling of the line profiles will then be justified.

Albedo. The diagnostic of dust scattering exhibited in Fig. 5 is not convincingly seen when line profiles before and after dust condensation are compared. However, because of the difficulty of defining a continuum, the predicted effect for ω = 0.6 is roughly the limit of what could in fact be detected. Accordingly, we conclude that $\omega \lesssim 0.6$ in the interval λλ4600-9800Å. If we now model the dust with a single particle species and size, this albedo limit implies a grain radius a $\lesssim$ 0.05 μ for grains of "astronomical silicate" (Draine and Lee 1984). But for grains of iron (Wickramasinghe 1973) or graphite (Draine and Lee 1984), the albedo limit is not sufficiently restrictive to limit particle sizes. Because of the non-detection of this albedo effect, we subsequently take ω = 0.

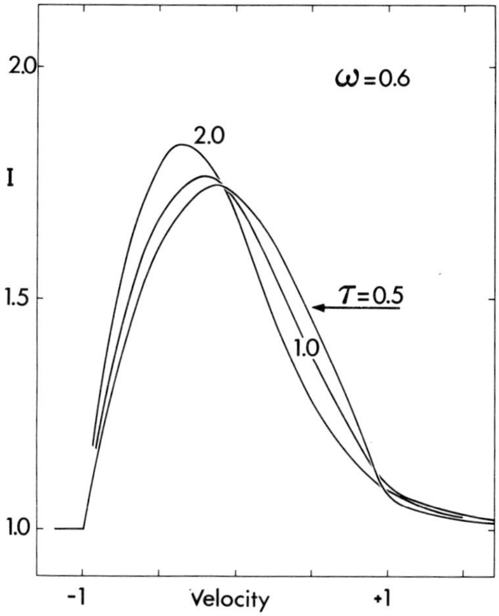

Fig. 5: Theoretical line profiles for Model III. Dust with albedo = 0.6 is uniformly distributed through the ejecta with indicated values of $\tau = k\rho R_E$.

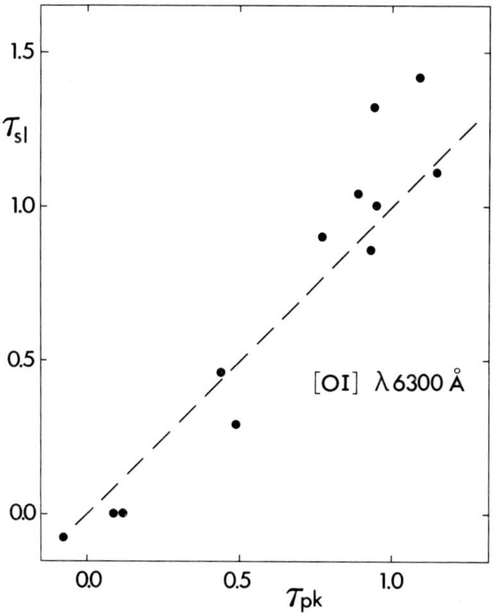

Fig. 6: Comparison of independent estimates of $\tau = k\rho R_E$ for Model II from observed [OI]λ6300 Å profile.

Expansion Velocity. With $\omega = 0$ and uniform distributions of dust and line emissivity (Model II), line profiles $I(\lambda)$ depend on two parameters V_E and τ. We now determine V_E by requiring that the parabolic profile predicted when $\tau = 0$ has the same half width at half intensity as does the violet side of the [OI]λ6300Å line before dust condensation. From measurements on three spectra, we thus obtain $V_E = 1870$ km s^{-1}, a value used throughout the remainder of this investigation.

Distribution of Dust. In Sect. 3, we noted the superiority of line profiles computed under the assumption of a uniform distribution of dust (Model II) as against the thin-shell model (Model I). This can be tested quantitatively by comparing τ_{sl} and τ_{pk}, independent estimates of $\tau = k\rho R_E$ that can be derived from an observed profile by application of equations (3) and (5), respectively. This comparison is shown in Fig. 6 for the [OI]λ6300Å line. (Note that here and below negative values of τ have been accepted.)

Because Fig. 6 reveals that $\tau_{sl} \approx \tau_{pk}$ and because the thin-shell model predicts $\tau_{sl} = 0$, we conclude that, when spectroscopically detected via blue shifts, dust was already widely distributed in the ejecta. Nevertheless, since simultaneous condensation throughout the ejecta is implausible, a more refined analysis of line profiles might well yield earlier evidence of dust within a restricted range of

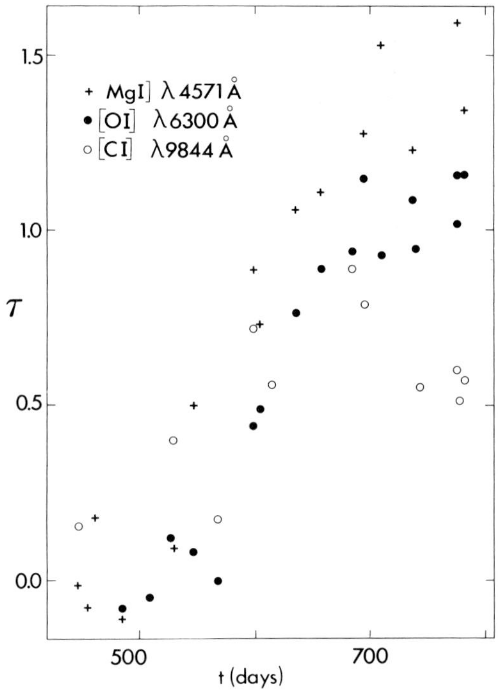

Fig. 7: Optical depth τ_λ as function of time t derived from blue shifts of indicated lines using Eq.(5).

radii. In view of the successful outcome of this test, Model II is used throughout the remainder of this paper.

Evolution of τ_λ. Model II is now used to derive the dependence of $\tau = k\rho R_E$ on both time and wavelength from the varying blue shifts of different lines. The lines measured are: MgI]λ4571Å, [OI]λ6300Å, and [CI]λ9844Å. For the unblended Mg and O lines, τ_λ is given directly by Eq. (5). For the C line, which is a triplet with overlapping components, the corresponding τ_λ is obtained by assuming the individual components' profiles are given by Eqs. (2) and (4) with intensities proportional to their respective Einstein A coefficients. The resulting values of τ_λ are plotted against t in Fig. 7. Unfortunately, the spectra from which these data derive are not of uniform S/N. As is evident from the scatter, the OI measures are the most, and the CI the least, reliable. The four CI measures after t = 740d are, however, of greater weight than the earlier CI data.

Inspection of the OI data suggests that the dust optical depth in the ejecta has been a smooth and monotonically increasing function of time, being negligible prior to t ≃ 530d, increasing rapidly after t ≃ 580d, and then only slowly after t ≃ 670d.

Fig. 6 also reveals clear evidence of selective extinction, especially when low weight is assigned to the CI measures prior to t = 740d. The observed increase in τ_λ

with decreasing λ requires that extinction in the optical is not dominated by large grains. Using the same sources as above, we find that these data imply $a \lesssim 0.5\mu$ for grains of astronomical silicate, and $a \lesssim 0.1\mu$ for either graphite or iron. Note that the silicate limit is less restrictive than that derived from the albedo diagnostic.

For later testing of compatibility with photometric data, it is useful to derive the optical depth $\tau_V(t)$ at $\lambda = 5500Å$, the effective wavelength of the V band. This has been done by fitting a smooth curve to the OI data and then correcting to 5500Å by assuming that $\tau_\lambda \propto \lambda^{-1}$, a formula roughly consistent with the MgI data. The resulting values are given in Table I, together with the implied mean extinction $A_V = -2.5 \log p$, where $p(\tau_V)$ is the escape probability. Under the assumptions of Model II, Osterbrock's (1974, p.242) formula for p applies.

Table 1

Optical depth and mean extinction at $\lambda = 0.55\mu$

t(dys)	τ_V	A_V(mag.)	t(dys)	τ_V	A_V(mag.)
500	0.00	0.00	650	0.97	0.68
525	0.02	0.02	675	1.08	0.74
550	0.08	0.06	700	1.15	0.78
575	0.21	0.17	725	1.19	0.80
600	0.54	0.40	750	1.22	0.82
625	0.81	0.58	775	1.25	0.84

Condensation Efficiency. On the assumption that the grains are all of the same size and composition, the dust mass required by Model II to produce optical depth τ_V is

$$M_d = \frac{16}{9} \pi R^2 \frac{\rho a}{Q_V} \tau_V , \qquad (6)$$

where ρ is the density of the grains and $Q_V(a)$ is their extinction efficiency at $\lambda = 0.55\mu$. For small grains, $Q \propto a$ and so M_d is then independent of a. Note also that when grain condensation stops τ_V decreases as t^{-2}.

Because the spectroscopic data on blue shifts does not decisively favour a particular grain composition, we apply Eq. (6) to several plausible candidates. At t = 775d, the end of the reported observing period, the still increasing grain masses are: 3.2×10^{-4}, 6.7×10^{-5}, 1.6×10^{-5}, and 4.4×10^{-6} solar masses for small grains of astronomical silicate, iron, graphite, and amorphous carbon, respectively. The requisite grain properties were obtained from the sources cited earlier, except for amorphous carbon where the Rowan-Robinson and Harris (1983) fit to the experimental data of Koike et al. (1980) was used.

Using the abundances predicted for the ejecta by Hashimoto et al. (1989), we can compute the maximum mass possible for grains of a given composition and thus convert the above grain masses into condensation efficiencies. The resulting efficiencies up to t = 775d are 6.2×10^{-4}, 8.1×10^{-4}, 1.4×10^{-4} and 3.9×10^{-5} for grains of astronomical silicate, iron, graphite and amorphous carbon, respectively. The result for astronomical silicate derives from a maximum mass of 0.51 $M_{\odot}$ obtained by assuming it to be Mg_2SiO_4. At first sight, these low efficiencies suggest that far more condensation will occur in the coming months and that the predicted black out might yet occur. Another possibility, however, is that the condensation efficiency is already much higher but is not evident because of the clumpiness of the dust distribution.

5. ADDITIONAL SPECTROSCOPIC EVIDENCE

Dust absorption may also explain the weakening of the [CoII]λ10.5μ line below its predicted strength. At t = 280 and 400d, this line was observed (IAUC No. 4575) at the strengths expected from the decay of an initial 0.075 $M_{\odot}$ of ^{56}Ni. But at t = 638d, its strength was only 0.37 of that predicted - this calculation assumes an ionization ratio CoII:CoI of 1:2, which is supported by observations of FeII and FeI lines. If this deficiency is indeed due to dust absorption, then τ_{λ} = 1.8 at 10.5μ at an epoch when τ_{λ} = 0.9 at 0.55μ - see Table I. On the assumption of a single grain species, this gives $Q(10.5\mu)/Q(0.55\mu) = 2.0$. For small grains of astronomical silicate, this ratio is 1.85. But for graphite and amorphous carbon this ratio is 0.011 and 0.052, respectively; and graphical data given by Wickramasinghe (1973) indicates that this ratio is $\lesssim$ 0.01 for Fe. This appears to be a decisive argument in favour of silicate grains, but its crucial dependence on the ionization ratio of cobalt must be noted.

Depletion. In Sect. 4, we found the condensation efficiency to be low but remarked that clumpiness allows higher values. If the condensation efficiency were in fact to approach unity for some grain species, then the depletion of the least abundant contributing element might become evident spectroscopically. At first sight, an observed weakening after day 550 of the [SiI]λ1.64μ line relative to the continuum suggests depletion, and this would strongly support silicate as the dominant grain species. However, this line is temperature sensitive, and so the observed effect may simply be due to declining excitation in the cooling ejecta. Clearly, independent and accurate temperature estimates are needed before depletion can be claimed.

6. PHOTOMETRIC EVIDENCE - OPTICAL

We now test our interpretation of the line profile asymmetry by examining its compatibility with the photometric record.

Extinction. In their analysis of the superb Geneva photometry of SN 1987A, Burki et al. (1989) do not invoke extinction when interpreting the behaviour of the V light curve up to t = 700d. Moreover, Roche et al. (1989), in favouring time-delayed emission from pre-existing circumstellar grains - a thermal echo - when explaining their IR observations on day 580, exclude prompt emission from newly created grains within the ejecta in part because the concomitant optical extinction is not seen. Nevertheless, Fig. 8 shows that, when corrected for the slowly increasing mean visual extinction (Table I) derived from our analysis of the McLaughlin effect, the resulting dust-free V light curve is not implausible: it shows the expected (e.g. Pinto et al. 1988) monotonically increasing departure from the earlier exponential decay in consequence of the ejecta's increasing transparency to γ-rays.

Note that the corrected light curve shows no evidence of energy input from the pulsar (cf. Burki et al. 1989). On our interpretation, the slowing decline of the observed V light curve after $\simeq$ day 650 is due to a slowing in the rate of dust condensation.

Fluctuations. Burki et al. (1989) have called attention to variability on a time scale shorter than roughly 10-20 days of the U and B light curves after day 450 and note that the V curve is more stable. This short time scale implicates the ejecta, and the coincidence of its onset with the brightening at 10μ (IAUC No. 4645; Roche et al. 1989) suggests dust. A possible interpretation is as follows: The continuum is emitted by discrete clumps, some of which after day 450 are occulted by dust clouds. These clouds are relatively opaque in the U band and relatively transparent in the V band. Light curve spikes occur when a clump emerges from occultation - as when the sun suddenly shines through a gap in the clouds. This interpretation is qualiatively consistent with our suggestion (Sect. 4) that dust condensation might have started well before it became evident via the McLaughlin effect.

7. PHOTOMETRIC EVIDENCE - INFRARED

The AAT Experiment. As noted earlier, Roche et al. (1989) attribute the 10μ excess to a thermal echo. The principal reason for this preference was their discovery that SN 1987A had a measurable angular size at 8-10μ on day 580. Specifically, they measured a FWHM of 2".2 for the supernova as against 1".6 for comparison stars. But they also found the IR profiles to be centred on the supernova to within 0".25.

Accordingly, while convincingly establishing a contribution from distant dust, their experiment does not exclude models that assign a substantial fraction of the 10μ emission to dust within the ejecta.

IR Emission from the Ejecta. In computing IR emission from dust within the ejecta, we again adopt Model II from Sect. 3. Thus we suppose the dust to be uniformly distributed in (0,R) and to have $\omega = 0$. In addition, we assume the emissivity of the gas to be constant for $r < R$ and zero for $r > R$. With this model, we calculate as a function of t the maximum amount of dust of a given type permitted by the photometric record. The steps in the calculation are the following:

A) Specify t and choose grain model - e.g. graphite with $a = 0.01\mu$.
B) Guess $\tau_V(t)$, the optical depth $k\rho R$ at 0.55μ.
C) From τ_V and the known function $Q_\lambda(a)$, compute τ_λ and hence the escape probabilities $p_\lambda = p(\tau_\lambda)$ at the effective wavelengths of the photometric bands.
D) Derive the generated luminosity density $\Lambda_\lambda^d+\Lambda_\lambda^g$ due to emission by dust and gas from the observed luminosity density L_λ via the equation

$$L_\lambda = p_\lambda(\Lambda_\lambda^d+\Lambda_\lambda^g) \ . \tag{7}$$

E) Assume an isothermal stratification for the dust. The dust's temperature may then be derived iteratively from the condition of its global thermal equilibrium,

$$\int (1-p_\lambda)(\Lambda_\lambda^d+\Lambda_\lambda^g)d\lambda = \int \Lambda_\lambda^d \, d\lambda \ . \tag{8}$$

F) With Λ_λ^d now determined, Eq. (7) is solved for Λ_λ^g.
G) If $\Lambda_\lambda^g > 0$ at all λ, then τ_V is increased and steps C)-F) are repeated.
H) A converged model has $\Lambda_\lambda^g = 0$ at some λ.

The photometric record used for these calculations comprises broad band light curves, smoothed and read off at 25 day intervals. The raw data at short wavelengths - $UBVR_cI_c$ - is from the South African observers (Whitelock et al. 1989, and references therein), while at longer wavelengths - $JHKLMN_1N_2N_3Q_0$ - the data is from ESO La Silla. The quantities $L_\lambda(t)$ were derived from the photometric magnitudes using LMC distance modulus = 18.5 and after correcting for foreground extinction assuming $E(B-V) = 0.15$.

The maximum permitted τ_V's as a function of time are plotted in Fig. 9 for the grain species considered earlier (Sect. 4), all for $a = 0.01\mu$, as well as for grey particles ($Q_\lambda \equiv 1$). Also shown is $\tau_V(t)$ derived spectroscopically (Table I). These results show that the $\tau_V(t)$ functions for small silicate grains and for grey

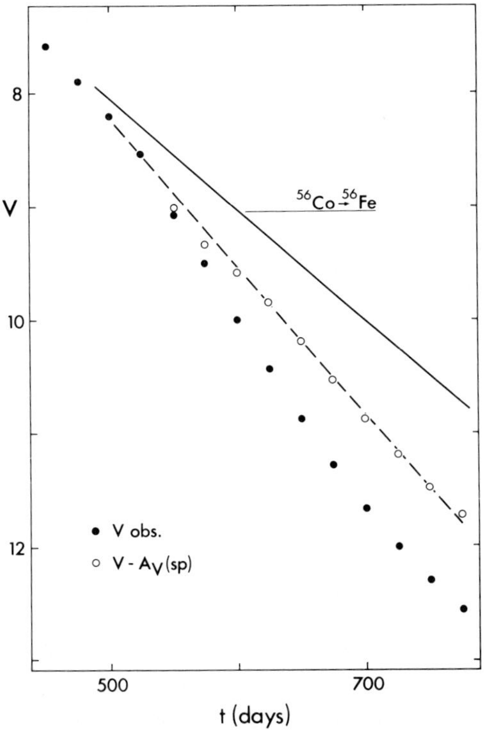

Fig. 8: Visual light curve corrected for extinction by internal dust. The spectroscopically derived correction A_V(Sp) is from Table I. The observed points are a smoothed version of South African data (Sect. 7).

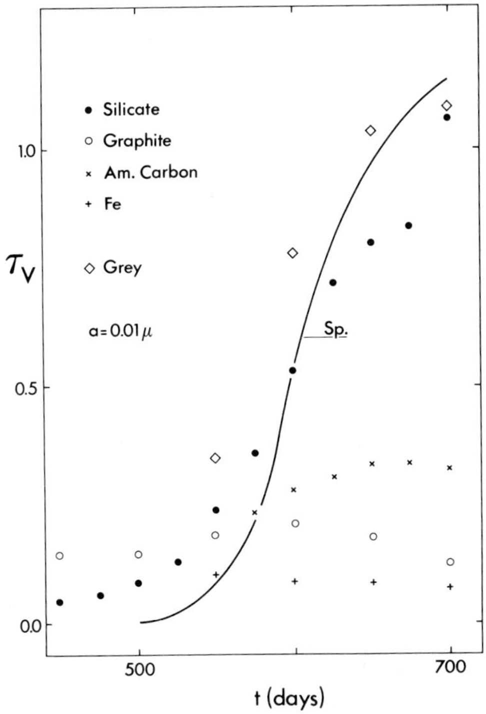

Fig. 9: Maximum optical depth τ_V at 0.55μ for different grain species, all for a = 0.01μ. Also plotted is the spectroscopically-derived τ_V from Table I.

particles track the spectroscopic function rather well. But this is not true for small grains of graphite, amorphous carbon, or iron; and so small grains of these substances can only be minor constituents of the ejecta's dust population. Of course, large grains of any substance approximate grey absorbers and so would be admitted as major constituents by this photometric argument. But the spectroscopic evidence of selective extinction in the optical (Sect. 4; Fig. 7) forbids this.

The differences in maximum τ_V's between the various grain species (and sizes) are due to differences in emission efficiencies in the far-IR. This is illustrated in Fig. 10 where emission by silicate and graphite grains with a = 0.01 μ is compared with the photometric data on day 625. The silicate grains, being efficient radiators, have low equilibrium temperature T_d = 356K and are limited to τ_V = 0.72 by the flux in the N_2 band (λ_{eff} = 9.69μ). For graphite grains, on the other hand, low emission efficiency requires T_d = 676K and they are limited to τ_V = 0.20 by the flux in the L band (λ_{eff} = 3.78μ). Fig. 10 also shows the effect on the silicate emission of dust in the ejecta being clumped with filling factor $\alpha = 0.1$ and with the individual blobs having $\tau_b = k\rho R_b = 3$. Further details of such calculations will be given elsewhere.

Fig. 9 perhaps suggests that compatibility of the spectroscopic and photometric data can be achieved simply by assuming all the grains to be small silicate particles. But this would not allow 10μ emission from distant dust, as required by the AAT experiment. Fortunately, a mixture satisfying this further constraint can be constructed: On day 625, spectroscopy (Table I) requires $\tau_V = 0.81$, which is fulfilled if we assign $\tau_V = 0.55$ to silicates and 0.26 to amorphous carbon. This choice allows ≃ 24% of the 10μ emission to be attributed to circumstellar dust, an amount probably consistent with the AAT experiment. If graphite is preferred as the minor constituent, the silicate contribution must increase to $\tau_V = 0.61$, and only 15% of the 10μ emission is then circumstellar. These simple exercises show that the dust condensation hypothesis can satisfy the observational constraints and that it does so with an astrophysically plausible mix of grain types and sizes.

The above results are somewhat in conflict with work of Moseley et al. (1989). These authors in analyzing data from days 623 and 639 conclude that emission from graphite particles with $a = 0.1\mu$ and $T_d \sim 300\text{-}400K$ dominates the IR continuum from 18-35μ and suggest that this is the newly-formed dust that we inferred (IAUC No. 4746) from line profile asymmetries. However, for graphite grains in the ejecta, our calculations give $T_d = 693K$ on day 625 for $a = 0.1\mu$ and thus negligible emission for $\lambda > 10\mu$ - cf. Fig. 10. If their radii are increased to $a = 1\mu$, T_d drops to 406K and then the far IR can indeed be accounted for. But then the near perfect greyness of such particles at short wavelengths is incompatible with the inferred selective extinction in the optical (Sect. 4; Fig. 7).

8. PHOTOMETRIC EVIDENCE - BOLOMETRIC

Pinto et al. (1988) have stressed that, because of the simplicity of the physics of the transfer and degradation of γ-rays in the SN's envelope, the prediction of the luminosity generated by the thermalization of γ-rays emitted by ^{56}Co and ^{57}Co is reliable, given their initial radial distribution. They therefore emphasized the diagnostic possibilities of such calculations for detecting departures from a model in which the "bolometric" light curve (excluding X- and γ-rays) is powered solely by promptly thermalized input from unstable isotopes.

In the present context, this diagnostic is of interest for testing whether the far-IR excess is due predominantly to newly-formed dust in the ejecta or to pre-existing circumstellar dust. In the former case, the IR excess represents prompt

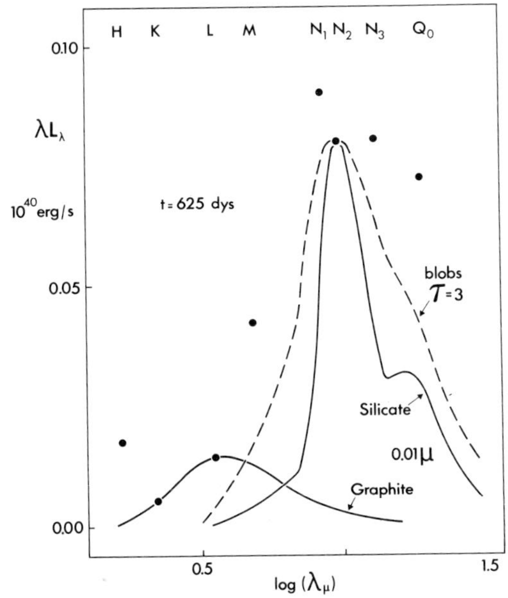

Fig. 10: Grain emission on day 625. Maximum emission by silicate and graphite grains with a = 0.01μ is shown together with photometric data. Effect of clumping on silicate emission is also shown.

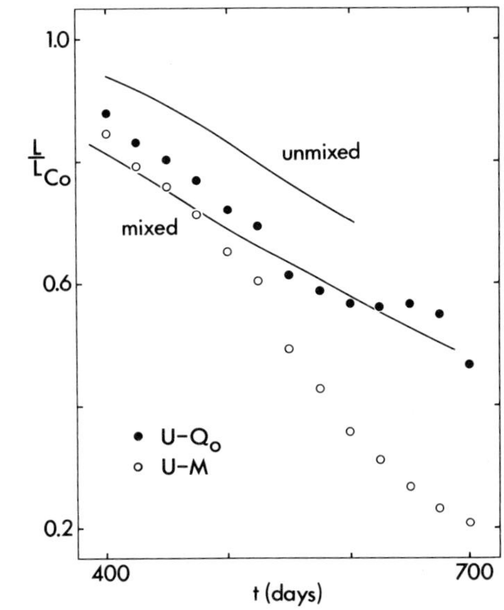

Fig. 11: Bolometric emission as fraction of energy released by ^{56}Co. Effect on observational estimates including ($U-Q_0$) and excluding (U-M) far-IR is shown. Theoretical calculations from Pinto et al. (1988).

emission and so must be included in estimating the bolometric luminosity. But in the latter case, the IR excess is a thermal echo of radiation emitted at ≃ t - 400 days and so must be excluded.

These two alternative observational estimates of the bolometric luminosity are compared with the Pinto et al. (1988) calculations in Fig. 11. The 14 broad-band light curves identified in Sect. 7 are again used and the IR excess whose origin is in question is defined as the flux in the N_1-Q_0 bands. (The quadrature formula used here and for the integrals in Eq.[8] has end-point weights derived by assuming black body emission at T = 2000K for $\lambda < 0.4\mu$ and emission from grains with $Q \propto \lambda^{-2}$ for $\lambda > 20\mu$.) Fig. 11 shows that assigning the IR excess to the ejecta gives a bolometric light curve that agrees well with the preferred model ("mixed") of Pinto et al. (1988). Accordingly, this test provides strong support for the dust condensation hypothesis. Interestingly, it also supports our earlier suggestion (Sect. 7) that only a modest percentage ($\lesssim$ 25%) of the IR excess derives from circumstellar dust.

That L(U-M) had begun to fall below model predictions was first pointed out by Catchpole et al. (1988), and they noted the possibility of significant flux at other wavelengths. More recently, Moseley et al. (1989) have used their far-IR observations to discuss the SN's energy budget on day 613 and find this to be balanced if the IR excess is attributed to dust in the ejecta.

Some structure is evident in the $L(U-Q_0)$ bolometric light curve in Fig. 11. If not due to quadrature errors, this could be due to departures from simultaneity in dust condensation in a given spherical shell. This would result in a failure of the instantaneous balancing of optical extinction and IR excess. The thermal echo is of course a further complication.

9. CONCLUSION

The aim of this investigation has been to collect and analyze evidence relating to the possible condensation of dust in the metal-rich ejecta of SN 1987A. In view of the number of independent supporting arguments as well as their mutual consistency, both quantitative and chronological, we conclude that dust has indeed formed in the ejecta.

In thus establishing that dust can in fact form in the ejecta of Type II supernovae, we have added direct observational support to the indirect arguments summarized in Sect. 1 that identify supernovae as major sources of interstellar dust. Moreover, our crude initial attempt to determine the mix of grain species is not inconsistent with this hypothesis. However, on the negative side, our estimates of the condensation efficiencies fall short by factors $\gtrsim 10^2$ of those demanded by this hypothesis, which therefore requires that the bulk of the grain formation lies in the future. Moreover, there is the question of the dust's survival when a reverse shock slows the ejecta's expansion. Perhaps a new phase of dust condensation follows this event or perhaps the surviving grain fragments reaccrete refractory elements immediately after the shock's passage (Seab 1987). If this phase of grain formation or growth goes to completion before mixing occurs across the contact discontinuity, then the observational constraints summarized in Sect. 1 are met.

ACKNOWLEDGEMENTS

For information and comments, we thank R.A. Chevalier, B.T. Draine, A.F.M. Moorwood, H.J. Völk, and E.J. Wampler.

REFERENCES

Burki, G., Cramer, N., Burnet, M., Rufener, F., Pernier, B., Richard, C.: 1989, Astron. Astrophys. **213**, L26

Catchpole, R.M., Whitelock, P.A., Menzies, J.W., Feast, M.W., Marang, F., Sekiguchi, K., van Wyk, F., Roberts, G., Balona, L.A., Egan, J.M., Carter, B.S., Laney, C.D., Laing, J.D., Spencer Jones, J.H., Glass, I.S., Winkler, H., Fairall, A.P., Lloyd Evans, T.H.H., Cropper, M.S., Shenton, M., Hill, P.W., Payne, P., Jones, K.N., Wargau, W., Mason, K.O., Jeffery, C.S., Hellier, C., Parker, Q.A., Chini, R., James, P.A., Doyle, J.G., Butler, C.J., Bromage, G.: 1989, Monthly Notices Roy. Astron. Soc. **237**, 55P

Cernuschi, F., Marsicano, F., Codina, S.: 1967, Ann. d'Astrophys. **30**, 1039

Chevalier, R.A.: 1981, Astrophys. J. **251**, 259

Clayton, D.D.: 1982, Q.J. Roy. Astron. Soc. **23**, 174

Draine, B.T., Lee, H.M.: 1984, Astrophys. J. **285**, 89

Dwek, E., Scalo, J.M.: 1980, Astrophys. J. **239**, 193

Dwek, E.: 1988, Astrophys. J. **329**, 814

Felten, J.E., Dwek, E.: 1989, Nature **339**, 123

Hashimoto, M., Nomoto, K., Shigeyama, T.: 1989, Astron. Astrophys. **210**, L5

Hoyle, F., Wickramasinghe, N.C.: 1970, Nature **226**, 62

Koike, C., Hasegawa, H., Manabe, E.: 1980, Astrophys. Sp. Sc. **67**, 495

Kozasa, T., Hasegawa, H., Nomoto, K.: 1989, Astrophys. J. (in press)

McLaughlin, D.B.: 1935, Pub. Am. Astron. Soc. **8**, 145

Moseley, S.H., Dwek, E., Glaccum, W., Graham, J.R., Loewenstein, R.F., Silverberg, R.F.: 1989, Nature (in press)

Osterbrock, D.E.: 1974, Astrophysics of Gaseous Nebulae (W.H. Freeman; San Francisco)

Pinto, P.A., Woosley, S.E., Ensman, L.M.: 1988, Astrophys. J. **331**, L101

Roche, P.F., Aitken, D.K., Smith, C.H., James, S.D.: 1989, Nature **337**, 533

Rowan-Robinson, M., Harris, S.: 1983, Monthly Notices Roy. Astron. Soc. **202**, 797

Seab, C.G.: 1987, Interstellar Processes, eds. D.J. Hollenbach and H.A. Thronson, Jr. (Reidel), p.491

Smith, S.E., Noah, P.V., Cottrell, M.J.: 1979, Pub. Astron. Soc. Pac. **91**, 775

Whitelock, P.A., Catchpole, R.M., Menzies, J.W., Feast, M.W., Woosley, S.E., Allen, D.A., van Wyk, F., Marang, F., Laney, C.D., Winkler, H., Sekiguchi, K., Balona, L.A., Carter, B.S., Spencer Jones, J.H., Laing, J.D., Lloyd Evans, T., Fairall, A.P., Buckley, D.A.H., Glass, I.S., Penston, M.V., da Costa, L.N., Bell, S.A., Hellier, C., Shara, M., Moffat, A.F.J.: 1989, Monthly Notices Roy. Astron. Soc. (in press)

Wickramasinghe, N.C.: 1973, Light Scattering Functions for Small Particles, (Adam Hilger; London)

Witt, A.N.: 1988, Dust in the Universe, eds. M.E. Bailey and D.A. Williams (Cambridge University Press), p.1

Discussion:

CESARSKY, C.: The minimum rate of grain condensation corresponds to the case of uniform distribution - no clumpling. What is this rate for SN1987a.

LUCY: In the printed text, the masses and condensation efficiencies refer to the uniform distribution model and are therefore the minimum value you request.

BECKER: Fesen and I have recently taken spectra of SN1980k eight years after maximum. The emission lines show a strong assymetry, consistent with the formation of dust in the ejecta.

LUCY: I noted with interest you recent account of these spectra in BAAS. I believe that you are indeed observing the same phenomenon that we have described for SN1987a.

OBSERVATIONS OF THE PROGENITOR WIND OF SN 1987A

E.J. Wampler[1], A. Richichi[2], D. Baade[1]

1. European Southern Observatory, Garching
2. Arcetri Observatory, Firenze

As part of a general ESO program to monitor the spectrum of SN 1987A, we are collecting intermediate high resolution ($R = \lambda/\Delta\lambda \approx 2\times10^4$) spectra of the supernova. By the end of the year 1987 it was possible to see narrow emission features due to [OIII] and hydrogen in spectra taken with the ESO Cassegrain echelle spectrograph (CASPEC) (Wampler and Richichi 1988, 1989). At that time, the velocities of these lines were about 286 km/sec and their widths were about 14 km/sec (FWHM). It is almost certain that these lines arise from a cloud of highly ionized gas that was first detected in ultraviolet spectra obtained by IUE in May 1987 (Fransson et al. 1989). This cloud is presumably a circumstellar shell that surrounded the progenitor of SN 1987A and was ionized by the initial UV flash from the supernova explosion. In the ideal case of an expanding spherical shell that is large enough for light travel times to be important, one would expect the intensity of an emission line to increase and then decrease again as the light from the front of the shell fades. At the same time the velocity of the line would increase as the integrated light became dominated by light from the back (receding) hemisphere. The IUE light curve (Fransson et al. 1989) for the ultraviolet [NV] line was consistent with the photometric behavior expected from a nebula about 400 light days in diameter. At the distance of the LMC, one light year is equal to a span of one arc-second, and the photometric behavior of [NV] is consistent with the extent of the nebula measured by Wampler and Richichi (1989).

This paper gives the results of observations extending from July 18 1988 to Feb 27 1989. The July data are from a single echelle spectrogram obtained by Dr. George Wallerstein using the CTIO echelle spectrograph attached to their 4-meter telescope. The other spectra are from observations using CASPEC on the ESO 3.6-meter telescope at La Silla. These consist of a single spectrum obtained Aug 28 1988 and merged spectra obtained from short observing runs centered on Oct 1, 1988, Jan 1, 1989 and Feb 25, 1989. Because the spectra, and in particular the narrow lines, show little change over a period of a few days it is acceptable to group the spectra in this way.

The spectra were calibrated by comparing the CASPEC data with flux calibrated spectra obtained at much lower resolution with the ESO Boller and Chivens spectrograph attached to the ESO 1.5-meter telescope. This ratio of the CASPEC spectrum to a calibrated low resolution spectrum taken at about the same date was used to construct a smooth response function to rectify the CASPEC data. Fig. 1 shows a comparison of a rectified CASPEC spectrum overplotted on a low resolution calibrated

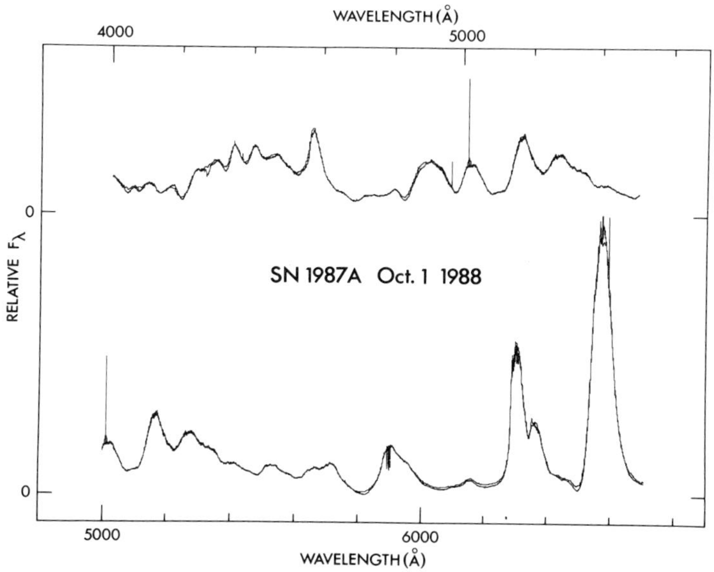

Fig. 1. A comparison between the low resolution flux calibrated spectrum and the CASPEC spectrum after it was corrected for the relative instrumental response function.

spectrum. Because the calculated response function is a very low frequency function the detailed agreement of the two spectra is an indication of the photometric accuracy of the two spectrographs. This procedure results in a relative calibration of the echelle data. An absolute calibration was then obtained by convolving the rectified echelle spectra with either a B or V filter function and comparing the derived "flux" with the appropriate B or V magnitude (Burki et al. 1989) for the date in question. Thus, although the calibration is not independent of these other data, it is sufficiently accurate for our purposes.

By the end of 1988 the Supernova had faded enough that many narrow nebular lines were visible. Fig. 2 shows the Feb 27, 1989 spectrum with some of the narrow lines identified. Table 1 lists the measured line intensities normalized to Hβ = 100. The absolute intensity of Hβ was $jH\beta = 2.8\times10^{-14}$ ergs s^{-1} cm^{-2} at the end of Dec 1987 and increased by about a factor of 3 by Feb 1989 (see Table 2 below). There is, in addition, a change in the temperature of the nebula and the level of ionization. An inspection of Table 1 shows that the [OIII]λλ4959,5007 lines have been decreasing in intensity relative to Hβ over the interval covered by these data. Thus, at the present time, the [OIII] lines have been nearly constant in strength. As noted above, [NV] peaked in intensity at about day 400. Therefore, the high ionization lines are weakening while the low ionization lines are increasing in strength.

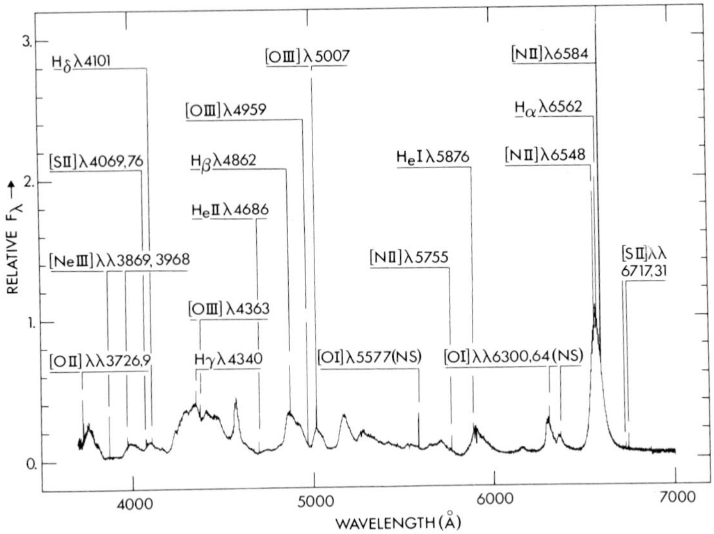

Fig. 2. The spectrum of SN 1987A in Feb 1989. Narrow nebular lines are indicated.

Table 1: Relative Narrow Line Ratios

Ion	λ	18 Jul'88	Aug'88	1 Oct'88	1 Jan'89	27 Feb'89
[OII]	3726	...	...	...	...	54
[OII]	3729	...	...	...	...	23
[NeIII]	3869	...	...	...	...	56
[NeIII]	3968	...	...	...	...	13
Hε	3970	...	...	...	...	5
[SII]	4069	...	...	21	14	11
[SII]	4076	...	...	13:	6:	4
Hδ	4101	...	...	36	10.8:	14
Hγ	4340	...	42	48	33	21
[OIII]	4363	52	54:	58	33	18
HeII	4686	...	38	27	19	24
Hβ	4862	100	100	100	100	100
[OIII]	4959	350	384	280	213	221
[OIII]	5007	...	1137	910	690	604
[NII]	5755	...	...	28	32	23
HeI	5876	...	...	24	25	15
[OI]	6300	...	...	...	...	8:
[OI]	6364	...	...	...	...	2.5:
[NII]	6548	...	...	243	263	220
Hα	6562	...	...	393	363	324
[NII]	6584	...	...	857	812	718
[SII]	6717	...	...	...	17	12
[SII]	6731	...	...	...	32	29

Using the graphs published by Cantó et al. (1980) we find that the nebular density given by the [OII]λ3727 doublet is about 1×10^4 cm^{-3} while the [SII]λ6724 doublet gives a density of 4×10^4 cm^{-3}. This is in good agreement with the density determined from the ultraviolet CIII]λ1909 doublet (Fransson et al. 1989). Table 2 lists the temperature determinations for the nebula. The temperature is obtained from the equations:

$$\frac{j\lambda 6548 + j\lambda 6583}{j\lambda 5755} = \frac{6.91 \exp\left[(2.50\times 10^4)/T\right]}{1 + 2.5\times 10^{-3}(N_e/T^{1/2})}$$

for the [NII] lines and

$$\frac{j\lambda 4959 + j\lambda 5007}{j\lambda 4363} = \frac{7.73 \exp\left[(3.29\times 10^4)/T\right]}{1 + 4.5\times 10^{-4}(N_e/T^{1/2})}$$

for the [OIII] lines (Osterbrock 1989). The density was taken to be 3×10^4 cm^{-3} for all epochs; this might be a slight overestimate and could result in a slight underestimate of the electron temperature. The [OIII] temperature has decreased by about a factor of 2 since Dec 1987 (Wampler and Richichi, 1989). The [NII] temperature is significantly lower than that found from the [OIII] lines.

Table 2: Hβ Flux and Temperature Determinations

Date	jHβ (ergs s^{-1})	$\frac{j5007+j4959}{j4363}$	T[OIII]	$\frac{j6584+j6548}{j5755}$	T[NII]
Jul 18 1988	4.6×10^{-14}	27	2.5×10^4	...	....
Aug 28 1988	6.4×10^{-14}	28	2.4×10^4	...	....
Oct 1 1988	7.2×10^{-14}	21	3.0×10^4	39	1.11×10^4
Jan 1 1989	7.4×10^{-14}	27	2.5×10^4	34	1.18×10^4
Feb 27 1989	1.0×10^{-13}	39	1.9×10^4	41	1.09×10^4

It is worth noting that the Balmer decrement is slightly steeper than the value predicted for case B radiative recombination. The observed decrement could be reconciled with the expected case B value if one allows reddening corresponding to E(B-V) = 0.25 mag. This value is within the errors of the estimated reddening of the supernova (Fransson et al. 1989; Kirshner and Gilmozzi 1989; Wampler et al. 1987). There is, in addition, the possibility that there is dust mixed with the line emission region. In the λλ6700-6800 region the continuum level of the Supernova is very low (see fig. 2). Here the spectrum appears slightly wider and displaced to the west compared to the spectrum of the Supernova in spectral regions dominated by bright emission lines coming from the expanding supernova envelope. As the narrow emission lines from the circumstellar shell are also displaced slightly to the west these observations are compatible with dust mixed with the gas in the circumstellar shell. In the violet CaII absorption region, λλ3800-3900, where the Supernova spectrum is also faint, the continuum is narrow and undisplaced. This suggests that while the dust is reflecting the supernova light in the red, in the violet the early onset of CaII absorption in the Supernova spectrum extinguished the violet light that otherwise could be reflected from the dust in the λλ3800-3900 region.

The velocities of the lines have been changing. The velocity of the [OIII] lines is now about 290 km/sec, compared to 286 km/sec in Dec 1987. We have sufficient data for the two epochs that we believe this difference is significant.

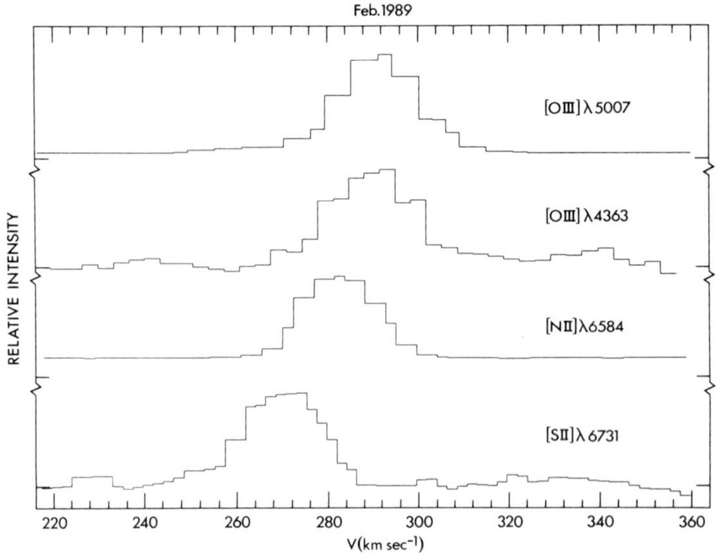

Fig. 3. The velocity profiles for lines of differing excitation.

The low ionization lines are at a lower velocity, about 260 km/sec. Fig. 3 shows the velocity structure found for several lines in the Feb. 1989 spectrum. If the lowest velocities represent the velocity of the near side gas while the high velocities represent the back (receding) gas then the expansion velocity of the nebula is about 15 km/sec, somewhat larger than the expansion velocity estimated earlier (Wampler and Richichi 1989). These velocity differences, together with the differences in the gas temperatures determined from the [NII] or [OIII] lines, show that lines from different ionization states are coming from physically separate regions. Thus it is not possible to determine the relative abundances of the ions from these data.

Finally, the structure of the narrow emission line region is revealed in a series of spectra taken with CASPEC configured with a 15×15 arc second aperture. In this "slitless" mode the nebula is seen as a small "C" shaped nebulosity about 1 arcsec in diameter and with a differential expansion of about 5 km/sec across it. A similar profile is seen in [OIII], [NeIII], [NII] and [OII] lines. As the velocity gradient across the nebula in each line is smaller than the spread in velocity from ion to ion each ion must show only a portion of the true extent of the nebula.

We are indebted to the ESO SN 1987A archive that is maintained by J. Danziger and R. Fosbury for the low resolution, flux calibrated spectra of SN 1987A that were used to rectify our high resolution echelle spectra. We also thank the Geneva observing team and J. Danziger for their unpublished B, V magnitudes which we used to determine the absolute calibration of our spectra. Finally, we thank J. Melnick for the Aug 1988 SN spectrum and G. Wallerstein for the July 1988 one.

References

Burki, G., Cramer, N., Burnet, M., Rufener, F., Pernier, B., and Richard, C. 1989, Astron. Astrophys., **213**, L26.

Cantó, J., Elliott, K.H., Meaburn, J., and Theokas, A.C. 1980, MNRAS, **193**, 911.

Fransson, C., Cassatella, A., Gilmozzi, R., Kirshner, R.P., Panagia, N., Sonneborn, G., and Wamsteker, W. 1989, Ap.J., **336**, 429.

Kirshner, R.P., and Gilmozzi, R. 1989, "Observing SN 1987A with IUE", preprint.

Osterbrock, D.E. 1989, "Astrophysics of Gaseous Nebulae and Active Galactic Nuclei" [University Science Books: Mill Valley, Calif.].

Wampler, E.J., Truran, J.W., Lucy, L.B., Höflich, P., and Hillebrandt, W. 1987, Astron. Astrophys., **182**, L51.

Wampler, E.J., and Richichi, A. 1988, The Messenger, No. **52**, 14.

Wampler, E.J., and Richichi, A. 1989, Astron. Astrophys., in press.

Discussion:

TRIMBLE: What is this stuff?! That is, is it the same gas seen by IUE with a velocity even smaller than the IUE limit of 30 km sec^{-1}?

WAMPLER: Presumably yes, since it is enriched in nitrogen. The real velocity, of course, has a projection factor in it.

PARESCE: Can you tell from your data how the C structure you showed is actually oriented on the sky? Can the opening of the C be oriented towards the observer, for example?

WAMPLER: The answer to the first part is that the blue shifted side of the nebula is roughly North and the open part of the "C" is facing the East. The answer to the second part is, "not completely, but it may be partly open to the observer".

THE DYNAMICAL EVOLUTION OF A CLUMPY MEDIUM

H.W. Yorke, R. Kunze, R. Spurzem
Institut für Astronomie und Astrophysik
der Universität Würzburg, Am Hubland,
D-8700 Würzburg, Federal Republic Germany

Abstract: We consider the evolution of a multicomponent medium: stars and gas clouds embedded in an ambient gaseous medium. The stars and gas clouds are treated stellar dynamically, using a velocity moment expansion of the Boltzmann equation. The basic equations and interaction terms are discussed. Examples of 2-D numerical calculations are presented (see also Kunze et al., this meeting).

1.0 Introduction

The evolution of interstellar gas is often studied by numerical solution of the gas dynamic equations. Due to memory and CPU time, the discretization of these gas dynamical equations is restricted to $\lesssim 10^5$ grid cells in an explicit calculation. This means that for a 2-D calculation the grid size is $\Delta r \gtrsim 10^{-2.5} L$ where L is the overall length scale of the problem. When considering the nucleus of a galaxy, for instance, $L \sim 1$ kpc and $\Delta r \gtrsim 3\,\mathrm{pc}$ (for implicit calculations it is possible to use varying grid scales over several orders of magnitude). Considering a volume of the order $(\Delta r)^3$, the number of particles is sufficiently large to be treated statistically and the velocity distributions generally sufficiently Maxwellian — except in the case of shocks — that the gas dynamical equations apply.

If, however, the interstellar gas is clumpy on scales of the order of a few times the grid size or smaller, then the discretized gas dynamical equations are no longer valid. The numerical resolution of gas dynamical codes is insufficient to follow the evolution of the interstellar gas on both large and small length scales. It is important to realize that this may significantly affect the overall evolution. In the following we describe a numerical procedure for treating such a clumpy medium.

2.0 The gas model for stellar systems

Consider the Boltzmann equation (see e.g. Binney and Tremaine 1987)

$$\frac{\partial f}{\partial t} + (\mathbf{u} \cdot \nabla_{\mathbf{r}}) f + (\mathbf{a} \cdot \nabla_{\mathbf{u}}) f = \left(\frac{\delta f}{\delta t}\right)_c \quad ,$$

where $f = f(\mathbf{r}, \mathbf{u}, t)$ is the one-particle distribution function (phase space density of particles), $\mathbf{a} = d\mathbf{u}/dt$ is the acceleration, and $(\delta f / \delta t)_c$ is the collision term, describing the sudden jump of particles from one part of phase space to another. We define velocity moments of the distribution function by $\langle lnm \rangle = \int u_x^l u_y^n u_z^m f du_x du_y du_z$. Multiplying the Boltzmann equation by the velocity components to various powers and integration over velocity space yields for e.g. Cartesian coordinates the moment equations

$$\frac{\partial}{\partial t}\langle 000\rangle + \frac{\partial}{\partial x}\langle 100\rangle + \frac{\partial}{\partial y}\langle 010\rangle + \frac{\partial}{\partial z}\langle 001\rangle = \cdots$$

$$\frac{\partial}{\partial t}\langle 100\rangle + \frac{\partial}{\partial x}\langle 200\rangle + \frac{\partial}{\partial y}\langle 110\rangle + \frac{\partial}{\partial z}\langle 101\rangle + \cdots$$

$$\vdots$$

By defining $\rho_* = \langle 000\rangle$, $\rho_* v_x = \langle 100\rangle$, etc. the first equation in this series resembles the equation of continuity

$$\frac{\partial \rho_*}{\partial t} + \nabla \cdot (\rho_* \mathrm{v}) = \sum_i \left(\frac{\delta \rho_*}{\delta t}\right)_i$$

In order to be computer-solvable, this sequence of moment equations of ascending order must be truncated at some level and complemented by a closure relation, which relates the moments of the maximum order to moments of lower order. This set of equations resembles those of gas dynamics, except for the effects of anisotropy, possible collisional terms and the particular choice of the closure relation. Such a system of moment equations has been successfully applied to interesting stellar dynamical problems (Bettwieser and Spurzem 1986) and compares well with other methods of calculating stellar dynamics (cf. e.g. Cohn 1985, Heggie 1985, Bettwieser and Sugimoto 1985).

3. The physical description of a multi-component medium

We interprete a "clumpy" medium as a system of "clouds" moving in a system of point masses (stars) and an ambient intercloud medium (ICM). Considering the equations of the ICM component our treatment is equivalent to the work of Cowie et al. (1981); their method is extended here by the assumption that a separate Boltzmann equation applies for every component. In addition to the usual gas dynamical equations for the ICM this yields a set of moment equations up to second order (see section 2 and also Scalo and Struck-Marcell 1984). Moreover, our code includes moment equations for the stellar component (cf. Langbein et al. 1989), but for the present investigation it suffices to consider the stellar component as stationary. "Phase transitions" and interactions among the components (drag forces on clouds moving relative to the ICM) as depicted in Fig. 1 are considered similarly as in Cowie et al. (1981). The decrease of cloud kinetic energy by cloud-cloud collisions is also taken into account.

We calculate the evolution of supernova remnants expanding into a medium in which a fraction of the total mass is in form of clouds. Depending on its temperature, the ICM can depict either the hot ($T \sim 10^6$ K) or warm ($T \sim 10^4$ K) component of the ISM. Cooling of the gaseous component is considered according to the cooling function of Raymond et al. (1976), modified in our 2-D calculations by the assumption that any attempt to cool below 10^4 K only results in an increased rate of cloud condensation (process 3 in Fig. 1). This process is only important in the dense shell of the remnant. For simplicity we only consider clouds of a single mass ($m_2 = 10\,\mathrm{M}_\odot$) and a single radius ($a = 1.7$ pc). We shall give a more complete description of our numerical model elsewhere.

4.0 The numerical model

4.1 Basic assumptions

The gas dynamic equations which describe the ICM simultaneously with the moment equations for the cloud system are solved with an explicit 2-D second-order scheme

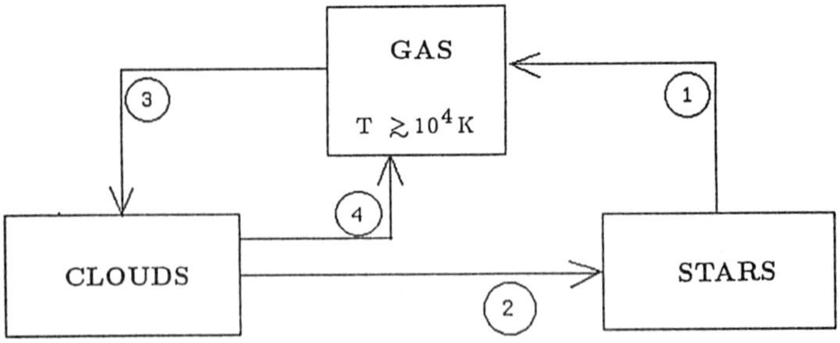

Fig. 1. *The three components considered in the present investigation and some of their associated phase transitions and interactions: 1) mass loss from aging and dying stars; 2) star formation; 3) condensation of clouds; 4) destruction of clouds by champagne flows, evaporation (heat conduction at cloud/gas interface).*

(see Różyczka, 1985, for more details). The calculations are performed on a cylindrical grid with 100×50 uniformly distributed cells with a resolution of 5 pc in both Z- and R-directions. Axial symmetry with respect to the Z-axis and mirror symmetry with respect to the Z=0 plane are assumed.

Magnetic fields and heat conduction in the ICM gas are presently neglected. Cooling of the intercloud gas is included although we do not allow the gas to cool below 10^4 K. The intercloud gas is assumed to be fully ionized while the gas inside the clouds is completely neutral. Gravity is neglected except for *case 2* where a gravitational potential was chosen in order to hold the stratified medium in hydrostatic equilibrium.

We model the simultaneous explosions of 10 supernovae (an "OB association") by the sudden release of thermal energy and density within a small volume. Per supernova $10\,M_\odot$ of mass loss and 10^{51} erg of kinetic energy are liberated. Mass and energy are smeared out over the initial volume of the remnant to give the equivalent amounts of mass and energy density. In *case 2* a stratified medium is chosen to account for the nonuniformity of the galactic disk. The spatial gas and cloud density are distributed like a Gaussian with scale heights of 200 pc and 140 pc, respectively.

4.2 Results

case 1. 10 supernovae with a total energy of 10^{52} erg explode into a medium with an ICM density of $\rho_1 = 1.03\,10^{-25}\,\mathrm{g\,cm^{-3}}$ and a smeared out cloud density of $\rho_2 = 1.02\,10^{-24}\,\mathrm{g\,cm^{-3}}$ in a considered volume of 500×250 pc.

We compare one calculation without any interaction processes and phase transitions (#1) with another including these effects (#2). The results of calculation #1 show qualitative agreement with other 2-D calculations (e.g. Bodenheimer et al., 1984; Tenorio-Tagle et al., 1987). In calculation #2 the evolution was qualitatively similar to #1. The ICM density interior to the remnant, however, grew up to a value about four times larger and the temperature was lowered by the same factor compared to calculation #1. When the material of the expanding shell has cooled down, its temperature and density again resembled more closely the situation yielded with calculation #1. In Fig. 2 the density profiles of the ICM and the clouds together with their velocity fields are displayed for the last model of #2. A comparison of the evolution of the remnants' radii (Fig.3) shows that the remnant in calculation #2 evolved to a larger radius.

Evaporative mass loss of the clouds is the most effective process during early phases of the remnant's evolution in calculation #2 caused by the very high temperature of the hot interior. The mean cloud density ρ_2 falls off there by about one order of magnitude.

Drag between clouds and ICM and the acceleration of clouds by the pressure gradients across the shock front add approximately a same amount of momentum to the clouds, which achieved a maximum bulk velocity of about $10\,\mathrm{km\,s^{-1}}$ outwards near the remnant's outer boundary.

The smallest effect has been seen by collisions between clouds. During the evolution the velocity dispersion of the clouds changed by only a few percent.

A complementary 1-D implicit calculation has been performed in spherical symmetry for comparison. It is capable for taking into account clouds of different mass simultaneously. Preliminary results show the same basic features of the remnant's evolution as in the 2-D calculations. A central cloud-free zone is established by evaporation and the acceleration of clouds outwards to $\sim 10\,\mathrm{km\,s^{-1}}$. The more massive clouds are less affected by this processes than the lighter ones.

case 2. With the same free parameters as in *case 1* we now let 10 supernovae explode at $z = 0$ into a stratified medium. Again we compare the evolution of the remnant with and without interaction terms between the cloud and ICM components. In both calculations the remnant became clearly aspherical after an evolution of $\sim 10^6$ yr. The shell cooled down to $\approx 10^4$ K even in the parts of the remnant perpendicular to the plane. Here, cloud condensation is enhanced.

5.0 Discussion and conclusions

In the presence of a "clumpy" medium consisting of "H I clouds" and a warm ($\sim 10^4$ K) intercloud medium the evolution of a supernova remnant seems not to be greatly affected. Substantial cloud evaporation takes place only in the early evolutionary phases of the evolution. A relatively large amount of matter is added to the ICM. Due to the evaporation of clouds the remnant does not become as strongly rarefied as in

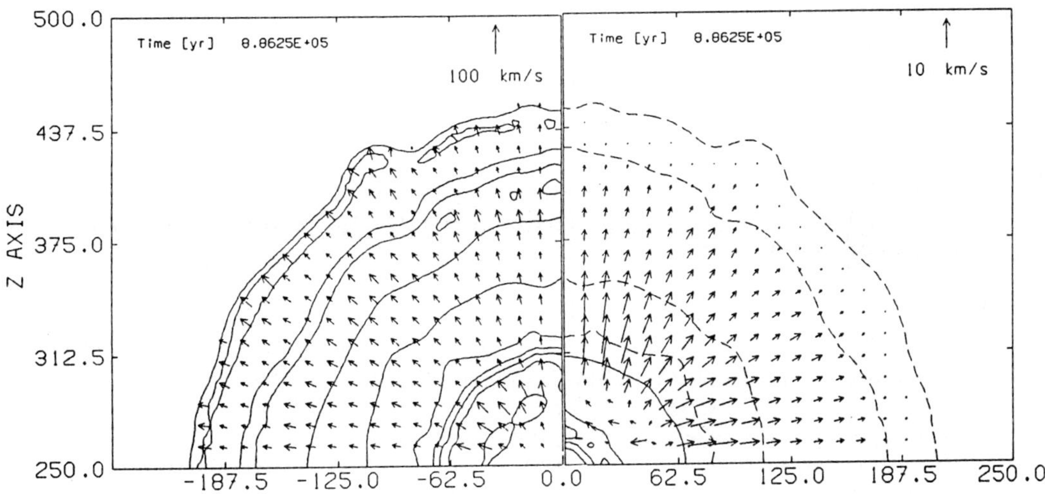

Fig. 2. *The supernova remnant of the last model from calculation #2 of* **case 1**. *Lines of constant ICM density are plotted in a logarithmic scale of* $\Delta\log\rho_1 = 0.2$. *For some contours the value of* $\log\rho_1$ *(in* $\mathrm{g\,cm^{-3}}$*) is given explicitly. In the second part of the figure isodensities of the cloud component are displayed. Here it is* $\Delta\log\rho_2 = 0.05$. *For comparison one ICM density contour is added (dashed line).*

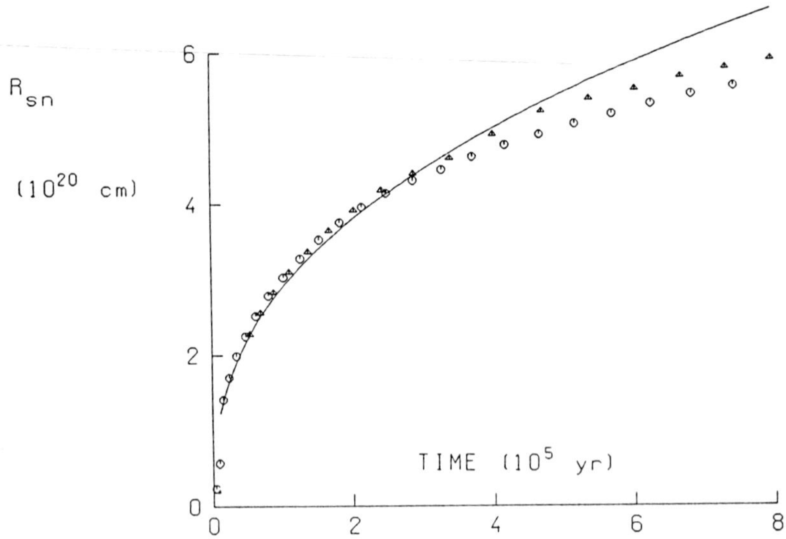

Fig. 3. *Evolution of a supernova remnant's radius for* **case 1**. *Circles refer to calculation #1, triangles to #2. The straight line is the Sedov solution.*

the case without clouds; density contrasts interior to the remnant are smeared out. Higher ICM densities inside the remnant lead to lower gas temperatures. Since multiple supernova explosions are often taken into account as a heating source of the galactic halo gas (see eg. Tomisaka and Ikeuchi, 1986; Mac Low et al. 1989), this may affect the heating rate of the galactic halo gas. Further investigation is necessary in order to clarify this point.

The clouds near the outer shell of the remnant are accelerated up to velocities of about $10\,\mathrm{km\,s^{-1}}$ outwards. Consecutive supernova explosions may lead to a further acceleration of the clouds and thus enhance their overall velocity dispersion.

Acknowledgements We acknowledge support by the DFG under grants Yo 5/5–1 and Yo 5/5–2. The calculations have been performed on a VAX 8810 of the "Rechenzentrum der Universität Würzburg".

References

Bettwieser, E. and Sugimoto, D.: 1985, Monthly Notices Roy. Astron. Soc. **212**, 189

Bettwieser, E. and Spurzem, R.: 1986, Astron. Astrophys. **161**, 102

Binney, J. and Tremaine, S.: 1987, *Galactic Dynamics*, Princton Series in Astrophysics

Bodenheimer, P., Yorke, H.W. and Tenorio-Tagle, G.: 1984, Astron. Astrophys. **138**, 215

Cohn, H.: 1985, in *"Dynamics of star clusters", IAU Symp. 113*, eds. J. Goodman, P. Hut, Dordrecht, Reidel, p. 161

Cowie, L.L. and McKee, C.F.: 1977, Astrophys. J. **211**, 135

Cowie, L.L., McKee, C.F. and Ostriker, J.P.: 1981, Astrophys. J. **247**, 908

Heggie, D.C.: 1985, in *"Dynamics of star clusters", IAU Symp. 113*, eds. J. Goodman, P. Hut, Dordrecht, Reidel, p. 139

Kunze, R., Yorke, H.W., Spurzem, R.: 1989, this meeting

Langbein, T., Spurzem, R., Fricke, K.J. and Yorke, H.W.: 1989, submitted to Astron. Astrophys.

Mac Low, M.-M., McCray, R. and Norman, M.L.: 1989, Astrophys. J. **337**, 141

Raymond, J.C., Cox, D.P. and Smith, B.W.: 1976, Astrophys. J. **204**, 290

Różyczka, M.: 1985, Astron. Astrophys. **143**, 59

Scalo, J.M. and Struck-Marcell, C.: 1984, Astrophys. J. **276**, 78

Tenorio-Tagle, G., Bodenheimer, P. and Różyczka, M.: 1987, Astron. Astrophys. **182**, 120

Tomisaka, K. and Ikeuchi, S.: 1986, Publ. Astron. Soc. Japan **38**, 697

Discussion:

VÖLK: You have plotted the radii of the SNRs vs. time in a linear form. Is there a Sedov phase reconcilable in the figure and if so, at what times does it occur with and without clumps?

YORKE: We find the classical phases of SNR evolution in sequences calculated with and without clumpiness. For the particular set of parameters we choose for our test calculations of an explosion in a homogeneous (or homogeneously clumpy) medium there was little modification. The main differences was an abscense of very low density gas in the SNR interior. Evaporating clumps provided a source of gas here. Also, in a radius versus time diagram, the clumpy calculations got a little further.

PECKER: How do you parametrize the clumpiness? How is its characteristic size defined?

YORKE: We have treated "clumpiness" by considering the evolution of a two component medium. There is a "cloud" component which we treat stellar dynamically and a "gaseous" component which is treated by solving the gas dynamical equations. The interaction terms, of course, implicity contain our cloud model. For this we have to specify typical cloud sizes and internal cloud densities. For one series of calculations we specified the mass of the clouds and their internal temperature. The sizes and internal densities were calculated by assuming presure equilibrium with the gaseous component.

DYSON: I find it hard to understand how putting in clumps and enhancing radiative cooling makes a supernova travel further.

YORKE: By injecting material (evaporating clumps) into the SNR at an early evolutionary phase, when the remmant is still adiabatic, we have enhanced the momentum of the expanding material at the cost of having the radiative phase begin somewhat earlier. The net result for the particular set of parameters we used was that the calculation with cumps got a little further after the snow-plough phase began. Unfortunatelly, I cannot say whether this effect will continue to occur when we begin varying the parameters or whether this is a singular result. We need to do a parameter study.

INFRARED ENVIRONMENT OF THE BE STAR 6 CEPHEI: INTERACTION OF STELLAR AND INTERSTELLAR WINDS

M. Kun

Konkoly Observatory, Hungary

Several arclike structures are seen on the high resolution IRAS images (processes via GEISHA system of Groningen) centred on the Be star 6 Cephei, among others an extended bow shock which indicates the rapd motion of the star and the interstellar matter relative to each other. 6 Cephei is embedded in a giant infrared-emitting shell which is thougth to be a three million year old supernova remnant, therefore in this case the presence of the bow shock around the star shows the interaction of a supernova blast wave with the stellar wind of a Be star. In this paper an attempt is made to reveal the history of the interactions of the star with its surroundings during different periods of its life which are recorded on the infrared maps. Some aspects of the formation and evolution of 6 Cephei can be inferred also from its infrared environment.

CARBON MONOXIDE EMISSION FROM YOUNG PLANETARY NEBULAE

R. Bachiller, V. Bujarrabal, J. Martín-Pintado, P. Planesas and J. Gómez-González

Centro Astronómico de Yebes (I.G.N.), Apartado 148,
E-19080 Guadalajara (Spain)

We have mapped the CO J=2-1 and J=1-0 emission with high angular resolution (about 10" and 20", respectively) from the young planetary nebulae NGC2346, M2-9 and NGC6720 (the Ring nebula in Lyra). The observations were carried out by using the IRAM 30-m dish at Pico Veleta (near Granada, Spain).
In NGC2346 the dense gas appears to constitute the walls of a cavity or tube with clumpy structure. The more prominent clumps are concentrated on a kind of expanding ring which surrounds the central star. The kinematical age of this ring is about 1600 years. Emission in lines of ^{13}CO, HCN, HNC and HCO^+ has been detected toward one of these clumps. In M2-9 the molecular emission is concentrated in a region smaller than 10" around the central star. In the Ring nebula, the emission extends over a region of 100"x80", indicating that the molecular mass is about 0.1 solar masses, i.e. much higher than values estimated before. Our observations demonstrate that the CO is not distributed in a close spheroidal shell, but in a cylindrical or barrel-like structure. Its youngness (age = 3000 yr) and the moderate luminosity of its stellar progenitor during the AGB phase, explain why CO remains abundant in the Ring nebula.
More detailed reports on our work can be found in the papers by Bachiller et al. (1988, Astron. Astrophys., 196, L5 ; 1989, Astron. Astrophys, 210, 366 ; 1989, Astron. Astrophys., in press ; 1989, Astron. Astrophys., submitted)

INTERACTION BETWEEN A STELLAR WIND AND THE IONIZED GAS IN N120 (LMC)

A. Laval[1], M. Rosado[2], J. Boulesteix[1], Y.P. Georgelin[1], M. Marcelin[1], A. Greve[3], J. Larsen[4], A. Viale[1]

[1]Observatoire de Marseille, France

[2]Universidad Nacional Autónoma de México, México

[3]IRAM, Granada, Spain

[4]Sterrewacht Leiden, Holand

The western part of the nebula N120, in the LMC, was observed in H_α with the equipment CIGALE at ESO. This equipment consists mainly of a scanning Perot-Fabry interferometer and a two-dimensional photon counting system. The spectral step used here is 0.41 Åover free spectral range of 8.24 Å. The total field of view of $7' \times 7'$ is covered by 256 x 256 pixels (each pixel being there $2.6'' \times 2.6''$).

Inside the intense condensation named N120A which has a diameter of 40″ from deeply exposed H_α photographs, violent gaseous motions are detected. The average velocity found over a 32″ circle centered on the knot is 233 km/s, which is close to the value 235 km/s of the diffuse medium in this field. Inside a 8″ circle surrounding the inner star BI141, the line splittings reach values between 65 and 95 km/s and the H_α fluxes are lower than in the surroundings. Beyond 8^{primel} from the star the fluxes and the velocities remain constant at 600-800 events/pixel and 235-240 km/s though the line FWHM may be as high as 110 km/s. (The fluxes are just being turned into absolute values using published absolute measures for N120).

No radio emission has been detected at this position, which rules out a origin from an explosion of supernova which ought to be recent to explain the small size of the moving zone. The presence of the inner star suggests an interaction of the gaseous medium with a stellar wind.

Previous photometric observations of the star lead to a spectral type B1-2 Ib which is rather late to explain the observed ionized atomic species. Several observations are necessary now to complete our knowledge of this star; a visible spectrum has already been obtained (it is under reduction) and UV spectrum will be soon available (IUE shift on May 5th 1989). Then the eventual mechanical power due to the star will be compared with the energy injected into the interstellar medium.

DISCS, OUTFLOWS, JETS AND HH OBJECTS

DISKS AND OUTFLOWS

Luis F. Rodríguez
Instituto de Astronomía, UNAM
Apdo. Postal 70-264
04510 México, D. F., MEXICO

ABSTRACT

This paper reviews recent developments in the study of bipolar outflows and their possible collimation by disklike structures. The exciting stars of most bipolar outflows are embedded in dense cores, self-gravitating structures with masses in the range of 1 to 10^3 $M_\odot$. In many cases, these dense cores are flattened, with its major axis aligned perpendicular to the axis of the outflow. These flattened structures, the interstellar toroids, appear to play a role in the large scale (0.1 pc) collimation of the bipolar outflows. However, a considerable number of observations point to the presence of collimation on much smaller scales. In particular, the results of Sargent and collaborators have revealed the existence of circumstellar disks (with radii of ~1000 AU) that are gravitationally bound to their stars. I also review recent models of accretion disks around T Tauri stars have been successful in explaining the general features of the continuum spectra of T Tauri stars from the ultraviolet to the infrared. Some radio results that have not been considered by these models are discussed.

A discussion of radio continuum observations of Herbig-Haro objects and their exciting sources is also given. In particular, it is emphasized that the Herbig-Haro objects associated with massive stars are more luminous in the radio and exhibit larger line widths in the optical than the classic HH objects (that is, those associated with low-luminosity, young stars).

I. INTRODUCTION

Perhaps the most remarkable property of the molecular outflows is that they frequently exhibit a bipolar geometry. A natural explanation for this characteristic is that the young stars undergoing the outflow phase are surrounded by disks, toroids, or in general some kind of flattened gaseous structures that could provide the observed collimation. The first part of this paper reviews some recent observational and theoretical contributions to our understanding of the disk-outflow connection. Considerable advances have been made, and it seems likely that in the near future observers and theoreticians will even get to be talking of the same thing.

In the second part of this paper I review a field that has experimented rapid growth in the last couple of years: the observations of radio continuum emission from Herbig-Haro objects and their exciting sources. After the detection of HH1 and HH2 and their exciting source VLA 1 (Pravdo *et al.* 1985), searches for radio continuum emission toward other classic HH objects (by classic HH objects I mean those excited by a low-mass, low-luminosity young star) proved unsuccessful (Lane 1989; Curiel *et al.* 1989). However, it has been known for many years (Gull *et al.* 1973; Münch 1977; Gyulbudaghian, Glushkov, and Denisyuk 1978; Cantó *et al.* 1980; Meaburn and White 1982; and Strom *et al.* 1986) that the HH objects can also be found in association with more massive stars. These HH objects associated with massive stars have turned out to be intrinsically much brighter than their smaller brothers and they, as well as their exciting sources, can be detected and studied in the radio continuum.

II. DENSE GAS AROUND STARS THAT POWER OUTFLOWS

I will discuss the topic of the presence of gaseous structures around young stars that are powering bipolar outflows starting from what appears to be more firmly established and moving to progressively more speculative realms. In general, it is known that the exciting stars of outflows are, in most cases, embedded in *dense cores* (Torrelles *et al.* 1983; 1986; Myers *et al.* 1988; Anglada *et al.* 1989). These are objects that typically have dimensions of a few tenths of a parsec, masses from 1 to 10^3 solar masses, and molecular hydrogen densities exceeding a few times 10^3 cm^{-3} (Myers 1985).

The dense cores are better studied in the line emission of molecules like ammonia (NH_3) and carbon monosulphide (CS), which are tracers of dense gas. The association of stars with dense cores has two interesting implications. First, the dense cores are usually as or more massive than the embedded star. It then seems reasonable to conclude that the star condensed from gas that was originally in the core. Second, the stars usually appear projected very close ($\leq$0.1 pc) to the peak of the core. Since drift velocities of $\sim$ 1 km s^{-1} between the star and the core are not unreasonable, one reaches the remarkable conclusion that the embedded star is younger that 10^5 years.

In many of the sources studied the dense core exhibits a flattened geometry, with its major axis aligned perpendicular to the outflow axis. Torrelles *et al.* (1983) refer to these flattened cores as *interstellar toroids*, interstellar because they have large, interstellar dimensions, and toroids because they are usually fairly thick structures and the term disk, favored by many authors, seems more adequate for thinner structures. A recent example of an interstellar, flattened structure aligned perpendicular to the axis of a bipolar outflow (Yamashita *et al.* 1989) is shown in Figure 1, and a review of other similar cases is given by Rodríguez (1988). Under the interstellar toroid hypothesis the bipolar collimation is a result of large scale ($\sim$0.1 pc) structures whose density gradients channel the expansion of the gas. Despite a number of problems that are discussed by Rodríguez (1988), it remains an observational fact that most well-studied bipolar outflows show evidence for the presence of an interstellar toroid.

The interstellar toroids are relatively large, self-gravitating structures that appear to be in approximate virial equilibrium. This implies that the ram pressure of the wind should be similar to the gas pressure in the toroid. This can be found to be a reasonable conclusion. Assume for wind parameters a mass loss rate of $3 \times 10^{-7}\ M_\odot\ \mathrm{yr}^{-1}$ with a velocity of 200 km s^{-1}. At a distance of 0.1 pc the wind will have a ram pressure of $3 \times 10^{-10}\ g\ cm^{-1}\ s^{-2}$. This ram pressure is similar to that expected for the gas pressure in the toroid, with typical molecular hydrogen densities

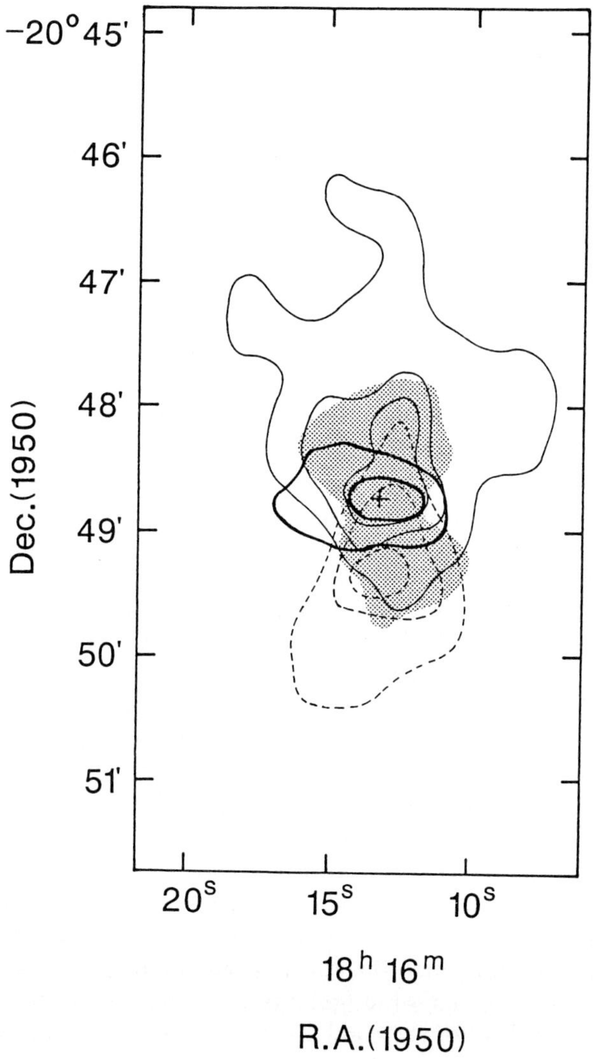

Figure 1. A schematic diagram of the Nobeyama map of the blueshifted (solid line) and redshifted (dashed line) CO J=1-0 bipolar emission associated with the source GGD27 IRS (marked with a cross in the figure). In this figure it is also shown, with thick solid contours, the CS J=2-1 emission around GGD27 IRS. Note the nearly perpendicular alignment between the axis of the bipolar outflow (traced by the CO) and the major axis of the dense gas (traced by the CS). The shaded area outlines the extent of the infrared reflection nebula in the region. Data from Yamashita *et al.* (1989).

of $\sim 10^4$ cm^{-3} and turbulent velocities of ~1 km s^{-1}. In some cases, however, one finds evidence for small, expansive motions (of order of one km s^{-1}) induced by the stellar winds that are powering the bipolar outflows (Torrelles *et al.* 1987; Marcaide *et al.* 1988).

III. EVIDENCE FOR SMALLER STRUCTURES COLLIMATING THE OUTFLOWS

During the last years it has become evident that, in addition to any possible large scale collimating mechanisms, there is collimation taking place on scales much smaller than 0.1 pc. Rodríguez (1988) has reviewed the *direct* (images of structures obtained with arc second resolution that resemble circumstellar disks or that can be interpreted as being created with the help of a circumstellar disk) and *indirect* (usually observations of lower angular resolution that can be interpreted as requiring a disk) evidences for circumstellar disks. Among the direct evidences one can list the optical and radio jets that appear to emanate from some young stars, the infrared halos detected with speckle observations or with maximum entropy reconstructions, and the disklike structures detected with millimeter interferometry.

The best example of a circumstellar disk around a young star is that found around HL Tau by Sargent and Beckwith (1987). The HL Tau disk was detected in the J=1-0 transition of ^{13}CO and has a mass of $\sim$0.1 $M_{\odot}$ and a major axis of about 4000 AU (that is, a radius of about 0.01 pc, a tenth of that of the typical interstellar toroids). What makes this structure unique is that its kinematics are consistent with Keplerian rotation around a 0.5 $M_{\odot}$ central mass. In contrast with the self-gravitating character of the interstellar toroids, the HL Tau disk appears to be gravitationally bound to its star. Then, the adjective *circumstellar* seems well justified for this object. The high velocity CO outflow found in association with HL Tau (Calvet, Cantó, and Rodríguez 1983) can be interpreted as a bipolar outflow with its axis perpendicular to the major axis of the circumstellar disk (Sato *et al.* 1989).

Another possible circumstellar disk has been mapped around L1551 IRS 5 by Sargent *et al.* (1988). This structure is shown in Figure 2. The observations are consistent with the gas being gravitationally bound to the star and its overall properties are fairly similar to those of the HL Tau system.

With the existence of collimation on a small scale one may wonder if larger scale collimation is even needed. An extreme view can be summarized as follows. Assume that the dense core from which the star formed was originally spherical. A star forms at its center and a bipolar wind (collimated by an unspecified mechanism at a scale much smaller than that of the core) starts blowing. This bipolar wind accelerates gas in the core, producing a bipolar outflow. A bipolar cavity will then be cleared, and the remaining dense gas will resemble an interstellar toroid. In this scheme the interstellar toroids are not the cause of the bipolar outflows, but one of the results. Three considerations argue against this point of view. First, for this scheme to work one would require to clear with the bipolar outflow an amount of mass comparable with that present now in the core. Even when a detailed study of this topic has not been undertaken, in general the mass in the high-velocity gas is about one to two orders of magnitude less than the mass in the interstellar toroid. Second, even before star formation takes place, Benson and Myers (1989) have found that dense cores exhibit flattening. This result implies that the future star will find an anisotropic gaseous environment and that its wind will escape more easily in

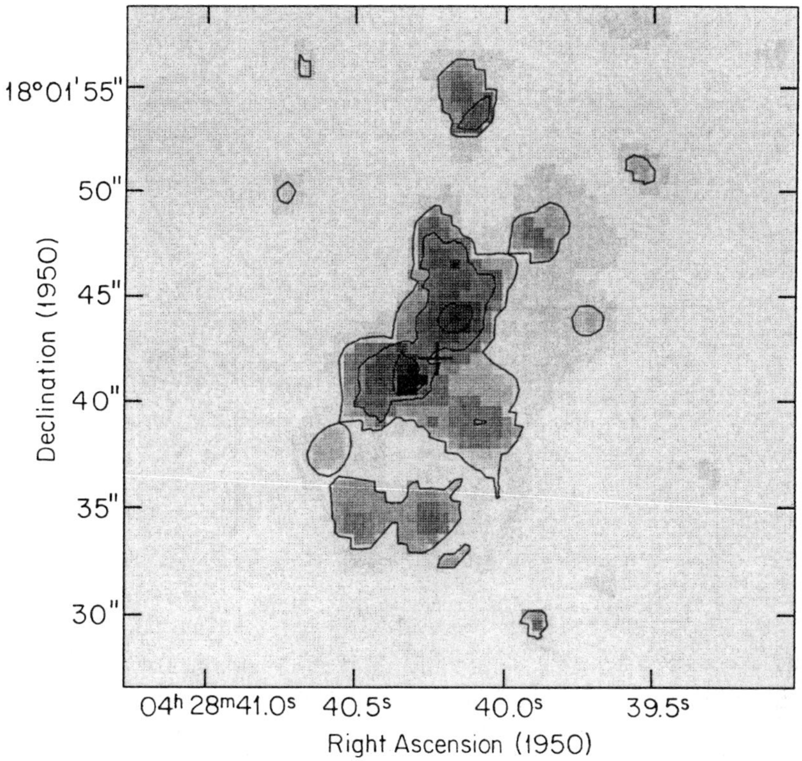

Figure 2. Gray-scale, aperture synthesis map of the integrated $C^{18}O$ emission from L1551 IRS 5 at 3.5 arc sec resolution taken from Sargent *et al.* (1988). The cross marks the position of the 2 cm compact continuum sources.

the poles of the configuration. Finally, the numerical models of Boss (1987) for the collapse of rotating, axisymmetric, isothermal clouds show that these objects form a protostellar core surrounded by a flattened structure with a central evacuated region, that is, an interstellar toroid.

IV. THEORETICAL WORK ON CIRCUMSTELLAR DISKS

During the last few years there has been also considerable advancement in the theoretical modeling of circumstellar disks. I will present now a brief summary of the main results of this work, following the papers of Adams and Shu (1986) and Bertout, Basri, and Bouvier (1988). A thin accretion disk in Keplerian rotation around a star is assumed. If the disk's luminosity comes mainly from reprocessing of stellar photons the disk is considered to be *passive*. An optically thick, but physically thin passive disk will have a luminosity of about 1/4 of that of the star. If viscous dissipation in the accretion disk results in significant heating and production of intrinsic luminosity, the disk is considered to be *active*. The luminosity of an active disk can equal and even exceed that of the star. In both passive and active disks the temperature of the disk depends with distance to the star as $T_D \propto r^{-3/4}$. This temperature dependence

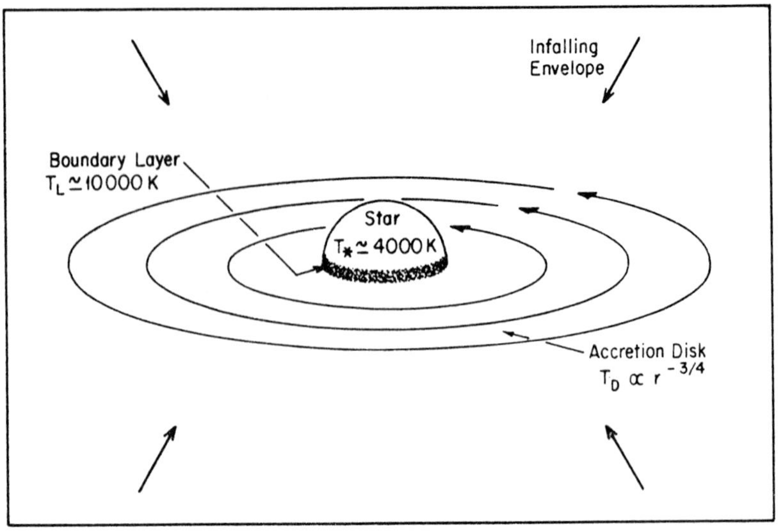

Figure 3. Schematic diagram of the accretion disk model for T Tauri stars.

and the assumption that the disk is optically thick produce a near- and middle-infrared spectrum with a flux density depending on frequency as $S_\nu \propto \nu^{1/3}$. Even when some sources do show this spectral index, the most extreme T Tauri stars (that is, those with larger infrared excesses, ultraviolet excesses, variability, and line emission) have a flat near- and middle-infrared spectrum. To alleviate this problem, it has been proposed that the outer parts of the disk must radiate comparatively more that in the orthodox model (Kenyon and Hartmann 1987; Adams, Lada, and Shu 1988).

One additional ingredient in these disk models is the presence of a *boundary layer*. The gas being accreted will dissipate one half of its gravitational energy by viscous processes in the disk. The other half will be dissipated as the rapidly rotating (a few hundred km s^{-1}) gas in the inner part of the disk is abruptly incorporated to the slowly rotating (a few tens of km s^{-1}) equator of the star. The characteristics of the boundary layer where this transformation of kinetic energy into heat and radiation takes place are poorly known, but it is believed that temperatures as high as 10^4 K could be obtained. This relatively hot region could account for the ultraviolet excesses known to exist in some T Tau stars. Finally, an infalling envelope surrounds the star and disk. A schematic depiction of the components of the accretion disk model for T Tauri stars is shown in Figure 3.

Bertout, Basri, and Bouvier (1988) have discussed that the spectral shape of the continuum of T Tau stars, from the ultraviolet to the infrared, can be explained in a general sense if we assume that the stars are chromospherically active and that they have an accretion disk. Reasonable accretion rates of $\sim 10^{-7}$ $M_\odot$ yr^{-1} are required. This remarkable achievement leads one to try to use this model as a paradigm that could explain other observations. As a radio observer, I would like to emphasize some radio results that are not being considered by the present accretion disk models and that could perhaps be explained with additional theoretical work.

IV a. *The Size of the Disk*

The theoretical modeling available refers in most cases to disks with a radius of about 100 AU. It is interesting to point out the arguments that are usually given by the theoreticians to justify their adoption of a 100 AU radius for the disk.

1) A dimension of 100 AU is, roughly, the size of our planetary system.

2) The lack of redshifted forbidden line emission observed in many T Tau stars suggests that an absorbing screen hides the receding part of the wind from our view (Edwards *et al.* 1987). From the emission measure of the lines it is estimated that the minimum size of the screen must be on the order of 100 AU.

3) At large scales, the infalling gas in the envelope moves toward the disk in a more or less spherically symmetric fashion. However, because of conservation of angular momentum, rotation becomes progressively more important in the kinematics of the infalling gas. The radius at which rotation dominates the motion of gas can be considered as a rotational barrier to infall and this is estimated again to happen at around 100 AU from the star. For the gas to continue falling toward the star an accreting disk must exist up to at least that radii.

4) Finally, the continuum spectrum of the active T Tau stars shows a bend at around 3×10^{13} Hz (10 μm) that can be modeled if the disk has an outer radius of about 100 AU (Adams, Lada, and Shu 1988).

Even when the 100 AU radius may actually turn out to be the typical value for accretion disks, it is important to note that the observational results of Sargent and collaborators indicate the presence of structures about an order of magnitude larger. Even when the object is at a different evolutionary stage, the famous dust disk around β Pictoris has a radius of about 400 AU (Smith and Terrile 1985), again significantly larger than 100 AU. I consider very important to understand the relation between the inner 100 AU, that have been modeled in great detail, and the larger structures found observationally.

IV b. *The Collimation of the Outflows*

While the interstellar size structures have been modeled in considerable detail to explain how they could collimate the observed bipolar outflows (see, for example, the papers of Barral and Cantó 1981; Königl 1982; and Pudritz and Norman 1983), very little theoretical work has been made with circumstellar accretion disks in what refers to their role as collimators of bipolar outflows and jets. Shu, Adams, and Lizano (1987) note that, gradually, the gas infalling from the envelope will fall preferentially onto the disk rather than onto the star. The stellar wind will then break out along the rotational axis of the system (the channels of weakest resistance), creating a bipolar outflow. A detailed modeling of this scenario would be most valuable.

IV c. *The Radio Continuum Excess*

The models for the continuum emission from accretion disks can be extrapolated to the radio wavelengths, where they predict a negligible flux. It is known, however, that T Tau stars are frequently associated with radio continuum emission at levels that are orders of magnitude larger than expected from dust emission in the disk.

Furthermore, high angular resolution Very Large Array observations of some objects indicate that the radio emission is coming from within 100 AU from the star (Bieging and Cohen 1985; Rodríguez *et al.* 1986; Rodríguez *et al.* 1989), the same volume that is supposed to contain the accretion disk. This radio continuum excess probably is free-free emission arising from shock-ionized gas produced as the stellar wind interacts with the infalling gas. But again, successful modeling has not been achieved.

V. RADIO CONTINUUM FROM HERBIG-HARO OBJECTS AND THEIR EXCITING SOURCES

The Herbig-Haro objects are optical nebulosities excited by a shock wave produced by the collimated wind of a young star. Since they are formed by ionized gas it is expected that they will emit free-free radiation in the radio wavelenghts at some level. Herbig-Haro 1 and 2 were detected with the VLA by Pravdo *et al.* (1985) at the mJy level. These authors also detected the central exciting source of the system, VLA 1, a heavily obscured pre-main-sequence object with a luminosity of $\sim 50\ L_\odot$ (Harvey *et al.* 1986). The detailed VLA study of the HH1-2 region by Rodríguez *et al.* (1989) has produced a significant advance in our understanding of the HH phenomenon. When observed with sub arc sec angular resolution, VLA 1 appears elongated and its major axis is aligned within a few degrees with the HH1-2 axis. The dependences with frequency of the flux density and of the angular size of the major axis of VLA 1 can be explained in terms of a confined, thermal jet of bipolar geometry. The radio emission from HH1 and 2 is optically-thin free-free radiation and it has been possible to confirm their proper motions, first observed in the optical by Herbig and Jones (1981), comparing VLA data taken only two years apart.

Searches for radio continuum emission from other classic HH objects proved unsuccessful (Lane 1989; Curiel *et al.* 1989) most probably because HH1 and 2 are the more luminous objects of the class. However, as I discussed in the Introduction, the HH phenomenon is not restricted to low luminosity stars and one could expect the HH objects associated with more luminous stars to be intrinsically more luminous and thus easier to detect in the radio. This expectation has proved to be correct and during this year there have been reports of radio continuum emission from several HH objects associated with massive stars. Along these lines of research, HH80-81 (Rodríguez and Reipurth 1989), Cep A West (=the HW object; Hughes 1989), M42/HH3-4 (Yusef-Zadeh 1989), and M16/HH1 (Curiel *et al.* 1990) have been detected. Figure 4 shows a VLA image of the Orion region that shows, to the southeast of the Orion bar, radio continuum emission from M42/HH3-4 (Yusef-Zadeh 1989).

One problem in the study of these HH objects is that in some cases their nature as *bona fide* HH objects is uncertain. In particular, in some of these objects the radiative contribution to the excitation of the HH object could be important. It is perhaps because of this problem that the study of suspected HH objects associated with massive stars did not receive much attention. However, the HH nature is well established at least in the cases of HH80-81 (Reipurth and Graham 1988; Heathcote and Reipurth 1989) and Cep A West (Hartigan, Raymond, and Hartmann 1987). The objects HH80-81 are intrinsically 10 times more luminous in the radio than HH1-2,

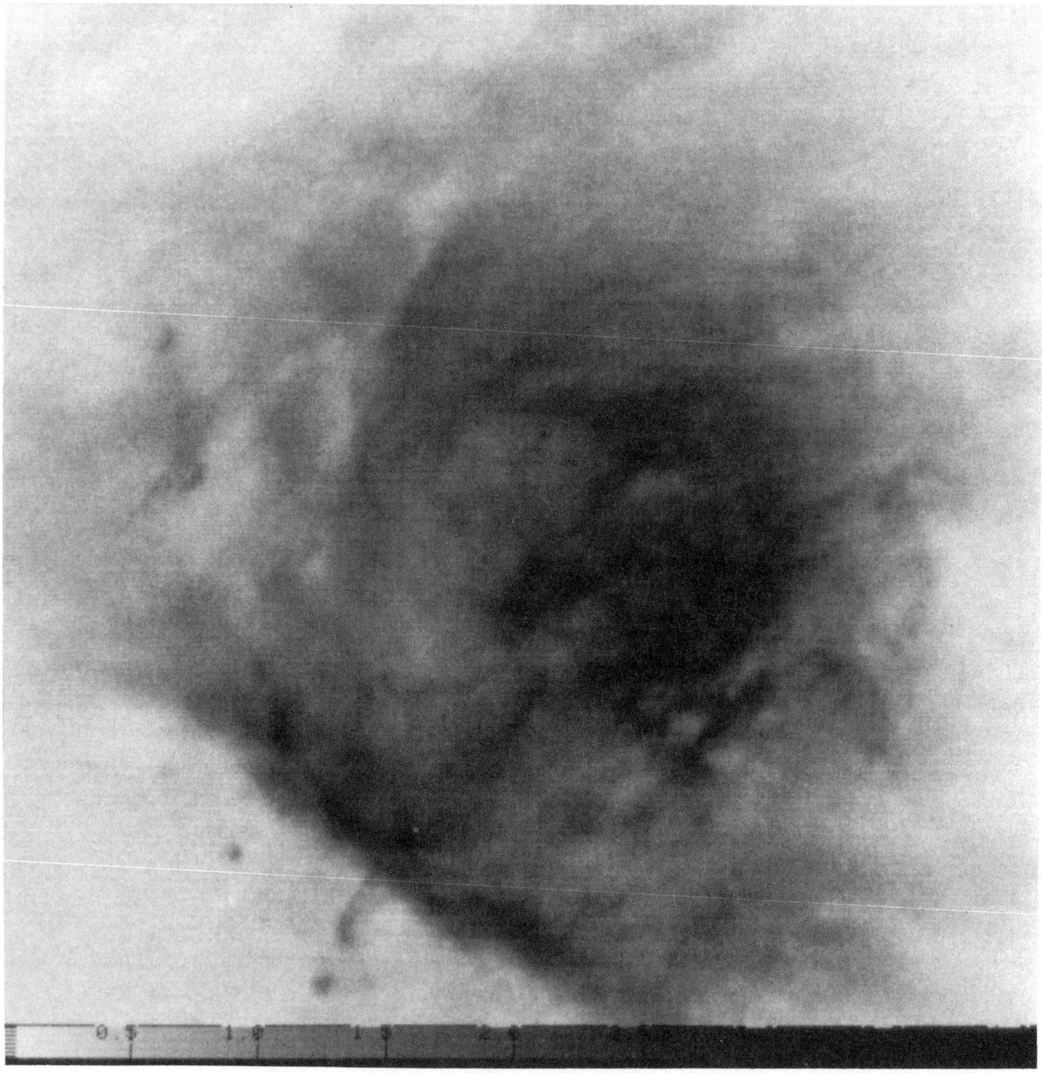

Figure 4. VLA image of the Orion region made at 6 cm. The Orion nebula occupies most of the figure. M42/HH3-4 (Walsh 1982) are the faint nebulosities to the southeast of the Orion bar (bottom left corner of the image). Data from Yusef-Zadeh (1989).

while Cep A West is intrinsically 30 times more luminous in the radio than HH1-2. Since the radio emission is expected to be correlated, to first approximation, with

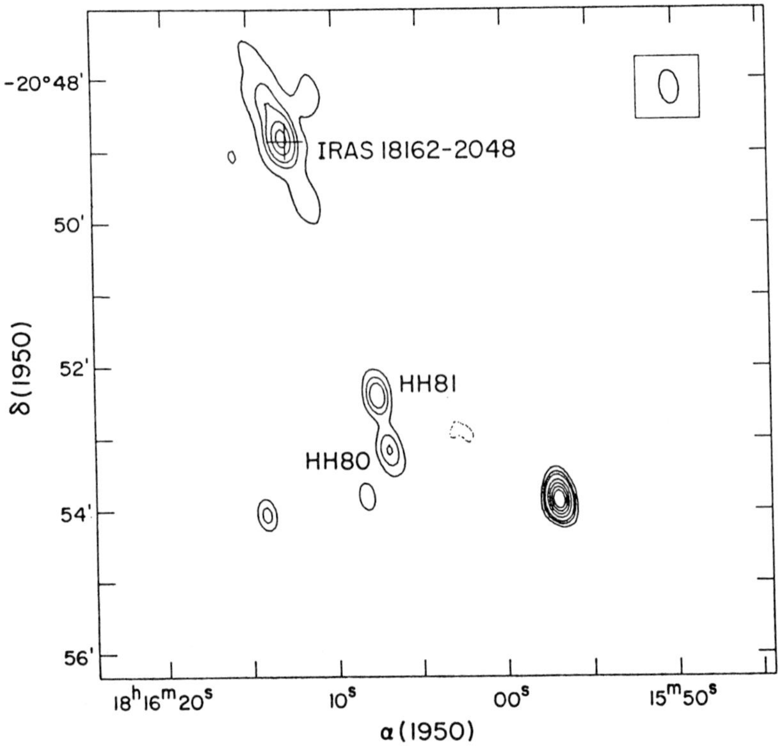

Figure 5. VLA map of the 6-cm emission from the HH80-81 region. The positions of these two HH objects, as well as that of IRAS 18162-2048 (=GGD27 IRS), the suspected exciting source of the region, are indicated in the figure. The large cross indicates the IRAS position. The half power contour of the beam is shown in the top right corner. Data from Rodríguez and Reipurth (1989). Note that additional data on this source are shown in Figure 1.

the optical emission (Pravdo *et al.* 1985), it is probable that similar ratios hold in the optical (after correction for extinction). An additional characteristic of these HH objects is that they possess optical line emission with widths much broader than those of classic HH objects. While classic HH objects have line widths of 100-200 km s^{-1}, HH80-81 have line widths of 700 km s^{-1} (Heathcote and Reipurth 1989) and Cep A West has line widths of 500 km s^{-1} (Hartigan, Raymond and Hartmann 1987). This result is quite important because it may imply that the winds powering these HH objects have higher velocities than those powering classic HH objects, as expected since the former are associated with more massive stars than the latter.

The HH objects 80 and 81 have been studied in some detail recently (Reipurth and Graham 1988; Rodríguez and Reipurth 1989). These HH objects are being excited by a 10^4 $L_\odot$ infrared source, GGD27 IRS (=IRAS 18162-2048), the same object presented in Figure 1. This exciting source is also detectable in the radio continuum (Figure 5), and exhibits elongated morphology with its major axis pointing approximately to HH80-81. The projected separation between IRAS 18162-2048 and HH80-81 is ~2 pc, while that between VLA 1 and HH1 or HH2 is ~0.2 pc. The HH80-81 system may be, in some sense, a scaled-up version of HH1-2. Of course, HH1 and 2 are located symmetrically with respect to VLA 1 and to consolidate the analogy one would expect to find HH objects to the northeast of IRAS 18162-2048.

VI. CONCLUSIONS

I reviewed two topics in the area of outflows from young stars. The observational evidence for a disk-outflow connection (in the sense that the disks help to collimate the bipolar outflows) appears to be considerable, in particular for structures in the 1000 AU to 0.1 pc size range. It is possible that circumstellar accretion disks of even smaller dimensions could be collimating the outflows at the 100 AU and smaller scales, but there is not yet a quantitative model available. Given the relative success of the accretion disk models in explaining the continuum of T Tauri stars from the ultraviolet to the infrared, I have pointed out several radio results that have not been considered by these models.

The second and last part of this review discusses recent radio continuum observations of Herbig-Haro objects and their exciting sources. In particular, the field of radio observations of HH objects associated with massive stars has grown rapidly in the last couple of years. These HH objects are intrinsically much more luminous in the radio and possess larger line widths in the optical than the classic HH objects.

ACKNOWLEDGEMENTS

I gratefully acknowledge support from a Guggenheim Fellowship. I am also thankful to F. C. Adams, J. Cantó, and P. Hartigan for their valuable comments. F. Yusef-Zadeh kindly provided the VLA image shown in Figure 4.

REFERENCES

Adams, F. C. and Shu, F. H. 1986, *Ap. J.*, **308**, 836.
Adams, F. C., Lada, C. J., and Shu, F. H. 1988, *Ap. J.*, **326**, 865.
Anglada, G., Rodríguez, L. F., Torrelles, J. M., Estalella, R., Ho, P. T. P., Cantó, J., López, R., and Verdes-Montenegro, L. 1989, *Ap. J.*, **341**, 208.
Barral, J. F. and Cantó, J. 1981, *Rev. Mexicana Astron. Astrof.*, **5**, 101.
Benson, P. J. and Myers, P. C. 1989, *Ap. J. Suppl.*, **000**, 000.
Bertout, C. Basri, G., and Bouvier, J. 1988, *Ap. J.*, **330**, 350.
Bieging, J. H. and Cohen, M. 1985, *Ap. J. (Letters)*, **289**, L5.
Boss, A. P. 1987, *Ap. J.*, **316**, 721.
Calvet, N., Cantó, J., and Rodríguez, L. F. 1983, **268**, 739.
Cantó, J., Goudis, C., Johnson, P. G., and Meaburn, J. 1980, *Astr. Ap.*, **85**, 128.
Curiel, S., Rodríguez, L. F., Cantó, J., and Torrelles, J. M. 1989, *Rev. Mexicana Astron. Astrof.*, **17**, 000.
Curiel, S. *et al.* 1990, in preparation.
Edwards, S., Cabrit, S. Strom, S. E., Heyer, I., Strom, K. M., and Henderson, E. 1987, *Ap. J.*, **321**, 473.

Gull, T. R., Goad, L., Chiu, H. Y., Maran, S. P., and Hobbs, R. W. 1973, *Publ. A. S. P.*, **85**, 526.
Gyulbudaghian, A. L., Glushkov, Yu. I., and Denisyuk, E. K. 1978 *Ap. J. (Letters)*, **224**, L137.
Hartigan, P., Raymond, J. and Hartmann, L. 1987, *Ap. J.*, **316**, 323.
Heathcote S. and Reipurth, B. 1989, in preparation.
Herbig, G. H. and Jones B. F. 1981, *A. J.*, **86**, 1232.
Hughes, V. A. 1989, *A. J.*, **97**, 1114.
Kenyon, S. J. and Hartmann, L. 1987, *Ap. J.*, **323**, 714.
Königl, A. 1982, *Ap. J.*, **261**, 115.
Lane, A. P. 1989, in preparation.
Marcaide, J. M., Torrelles, J., Güsten, R., Menten, K., Ho, P. T. P., Moran, J. M., and Rodríguez, L. F. 1988, *Astr. Ap.*, **197**, 235.
Meaburn, J. and White, N. J. 1982, *M. N. R. A. S.*, **199**, 121.
Münch, G. 1977, *Ap. J. (Letters)*, **212**, L77.
Myers, P. C. 1985, in *Protostars and Planets II* , ed. D. C. Black and M. S. Matthews (University of Arizona), p. 81.
Myers, P. C., Heyer, M., Snell, R. L., and Goldsmith, P. F. 1988, *Ap. J.*, **324**, 907.
Pravdo, S. H., Rodríguez, L. F., Curiel, S., Cantó, J., Torrelles, J. M., Becker, R. H., and Sellgren, K. M. 1985, *Ap. J. (Letters)*, **293**, L35.
Pudritz, R. E. and Norman, C. A. 1983, *Ap. J.*, **274**, 677.
Reipurth, B. and Graham, J. A. 1988, *Astr. Ap.*, **202**, 219.
Rodríguez, L. F., Cantó, J., Torrelles, J. M., and Ho, P. T. P. 1986, *Ap. J. (Letters)*, **301**, L25.
Rodríguez, L. F. 1988, in *Galactic and Extragalactic Star Formation* , ed. R. E. Pudritz and M. Fich (Kluwer Academic Publishers), p. 97
Rodríguez, L. F., Ho, P. T. P., Torrelles, J. M., Curiel, S. and Cantó, J. 1989, submitted to *Ap. J.*
Rodríguez, L. F. and Reipurth, B. 1989, *Rev. Mexicana Astron. Astrof.*, **17**, 59.
Sargent, A. I. and Beckwith S. 1987, *Ap. J.*, **323**, 294.
Sargent, A. I., Beckwith S., Keene, J., and Masson, C. 1988, *Ap. J.*, **333**, 936.
Sato, S. *et al.*. 1989, preprint.
Shu, F. H., Adams, F. C., and Lizano, S. 1987, *Ann. Rev. Astr. Ap.*, **25**, 23.
Smith, B. A. and Terrile, R. J. 1985, *Science*, **226**, 1421.
Strom, K. M., Strom, S. E., Wolff, S. C., Morgan, J., and Wenz, M. 1986, *Ap. J. Suppl.*, **62**, 39.
Torrelles, J. M., Rodríguez, L. F., Cantó, J., Carral, P., Marcaide, J., Moran, J. M., and Ho, P. T. P. 1983, *Ap. J.*, **274**, 214.
Torrelles, J. M., Ho, P. T. P., Moran, J. M., Rodríguez, L. F., and Cantó, J. 1986, *Ap. J.*, **307**, 787.
Torrelles, J. M., Ho, P. T. P., Rodríguez, L. F., Cantó, J., and Moran, J. M. 1987, *Ap. J.*, **321**, 884.
Walsh, J. R. 1982, *M. N. R. A. S.*, **201**, 561.
Yamashita, T., Suzuki, H., Kaifu, N., Tamura, M., Mountain, C. M., and Moore, T. J. T. 1989, submitted to *Ap. J.*
Yusef-Zadeh, F. 1989, in preparation.

Discussion:

DYSON: Is it clear whether your "super" HH objects are shock or radiatively excited?.

RODRIGUEZ: Not in all, and this problem has certainly impeded the advancement in the understanding of this interesting variety of HH objects. However, in the case of HH80-81, the studies of Reipurth and collaborators strongly support a shock-excited nature.

MÜNCH: In relation to the comment made by J. Dyson I would say that some of the condensations seen near θ^2 Ori, south of the Bar, are not quite alike the HH-objects seen in low temperature regions. One of them has a central star, of G-K type, a roundish appearence and with an extension in directions opposite to that of θ^2 Ori. It looks as if the UV-radiation of θ^2 has substantially altered the surrounding medium against which the objects is expanding at low velocity. The object near the BN and Low-Kleinman nebula, on the other hand, is a true HH-object, with obvious signatures of shock (240 kms^{-1}) in its spectrum. The question I would like to ask to Luis Felipe is whether in the molecular disks any indications of anomalous isotopic composition have been found. In our planetary system isotopic anomalies have been found and are explained by hard particle radiation (soft cosmic rays) that may have been emitted by the Sun in its T Tau phase.

RODRIGUEZ: We are only starting to study these disks. The isotopic anomalies may be present only very close to the star, involving relatively little gas. I belive that in the future, with increased sensivity, your question could be addressed.

HUGHES: In addition to the HH objects, for which a model is that of an expanding shock, there appears to be HH-objects produced by bullets. For example, GGD-37 appears to contain bullets with speeds of up to 500 kms^{-1}. Do you have any comment on this, and are there 2 types of HH objects?

RODRIGUEZ: You are correct in pointing out GGD37 as an HH object powered by a massive star. As is discussed in the review by Cantó, the HH phenomenon can appear at least in four different scenarios.

NATTA: You said that probably the momentum in the wind is much lower than the momentum in the CO outflow. I think that most of the wind is in the form of neutral gas, and that indeed the two may be very close.

RODRIGUEZ: I agree with you that the winds could be largely neutral. In the case of HH7-11 the results of Lizano et al. (1988) support this view.

THE STRUCTURE OF DENSE CLOUD CORES

Alwyn Wootten
National Radio Astronomy Observatory (NRAO[1])
Edgemont Rd., Charlottesville, Va 22903, USA

Things are seldom what they seem.
Skimmed milk masquerades as cream.
Little Buttercup, *H. M. S. Pinafore*

Open slit spectra of planetary nebulae, in which images of the object are recorded in the light of several spectral lines on a single plate, have long proven a useful diagnostic of nebular properties and morphology. Fortunately, the reasonably simple structure of most planetaries greatly aids interpretation of the images. The dust-enshrouded mass-losing asymptotic giant branch stars from which planetaries evolve have now also been imaged at millimeter wavelengths. These high-resolution images have demonstrated the role of photochemistry in molding the composition of circumstellar shells. This powerful techinique is less well-developed as a tool for analyzing the structure of localized density concentrations in molecular clouds, the cores in which stars form. Even pre-astral cores, in which stars have not yet formed, may have an extended and intricate geometry which renders mapping tedious and masks their true structure. Their basic pre-astral structure may be complexly contorted by the character and extent of star formation within them. How, then, does our perception of the structure of a core depend upon the line in whose light it is imaged? Which lines optimally determine physical structure? How should chemical differences, perceived by comparisons of images in different lines, be used to determine the physical characteristics of a core?

In this paper I summarize results of surveys of the cores of the ρ Ophiuchi molecular complex in an attempt to provide preliminary answers to these questions. In this nearby (160 pc) cloud, 78 young stars have been catalogued in the central region alone (Wilking, Lada and Young 1989). Loren (1989) has described 89 dense clumps throughout the complex. Within these clumps, Loren, Wootten and Wilking (1989) have described 12 very dense cores, some of which are currently active sites of star formation. A few appear to be undisturbed pre-astral cores whose structure, hopefully, could reveal how they have formed.

As star formation unleashes an important new energy source in a pre-astral core, the core warms. Conventional gas thermometers fail to monitor the early phases of this change. Lower CO

[1] The NRAO is operated by Associated Universities, Incorporated, under cooperative agreement with the National Science Foundation

transitions fail to penetrate a dense core owing to the optical depth of the envelope. Ammonia fails to supply sufficient accuracy for our needs; its abundance also apparently varies dramatically in some denser cores. A potentially much more sensitive temperature monitor can be established through comparison of emission from optically thin deuterium-bearing molecules to other optically thin isotopomers. The $[DCO^+]/[H^{13}CO^+]$ ratio R varies especially strongly over the 15-30K temperature range which is of most interest (Wootten. Loren and Snell 1982). To establish this ratio as an effective probe, Butner *et al.* (1988) compared maps of DCO^+, $H^{13}CO^+$ and warm dust (IRAS 12μm map). They found pockets of depressed R values near known stars, even those currently emitting little energy (*e.g.*E32, at 0.5 $L_\odot$). Using this tool, they were able to pinpoint the location of the pre-astral core B2 (Figure 1). This tool has now been applied to other cores (Loren, Wootten and Wilking 1989) which have now been established as pre- or post-astral. Unfortunately, excitation effects can bias the interpretation of individual maps and several-transition images may be crucial to proper understanding of core structure.

In L1689N, a cold pre-astral core lies adjacent to the warm core harboring the young star IRAS16293-2422 (Wootten and Loren 1986). DCO^+ maps (Wootten, Butner, Loren, Kaifu, Yamashita and Hayashi, in preparation) show similar emission at both locations, confounding expectations. R , measured from the J=1-0 lines, varies insignificantly, increasing from 1.7 at the IR source to 2 at the cold core. In contrast, the warm core vanishes in images in the J=3-2 DCO^+ line (Wootten and Loren 1986). An image in the J=3-2 HCO^+ line barely shows the cold pre-astral core, so dominated is it by powerful emission from the warm core. Here, clearly, excitation plays a major role in the determination of the structures of these cores. However, when the effects of excitation are properly accounted, it is clear that **the deuterium enhancement in HCO^+ provides an excellent identifier of cold dense pre-astral cores where stars have not yet formed.**

Ammonia maps also reveal interesting aspects of chemical differentiation within L1689N. Single-antenna maps of the (1,1) line (Wootten and Loren 1986) and the (2,2) line (unpublished data) show a cloud completely dominated by emission from the cold core. Measures of the cloud mass, using a standard ammonia analysis (Ho and Townes 1983) fix the cold core mass at 6 $M_\odot$, while the warm star-forming core contains less than 0.1 $M_\odot$! Not surprisingly, perhaps, high resolution maps (Mundy, Wootten and Wilking, 1989) of other mass tracers require much more material in the post-astral core. Using various estimators, masses range from >0.22 $M_\odot$(from $C^{18}O$ flux measurements) to 1.6 $M_\odot$(suggested by the dynamics of disk rotation which includes a central object) to 0.9 – 6 $M_\odot$(from dust emission measurements). High resolution maps of the (1,1) lines of NH_3 reveal a hole in emission near the central source, where $C^{18}O$ and dust emission peaks. Nowhere does the temperature derived from high resolution (2,2) ammonia observations

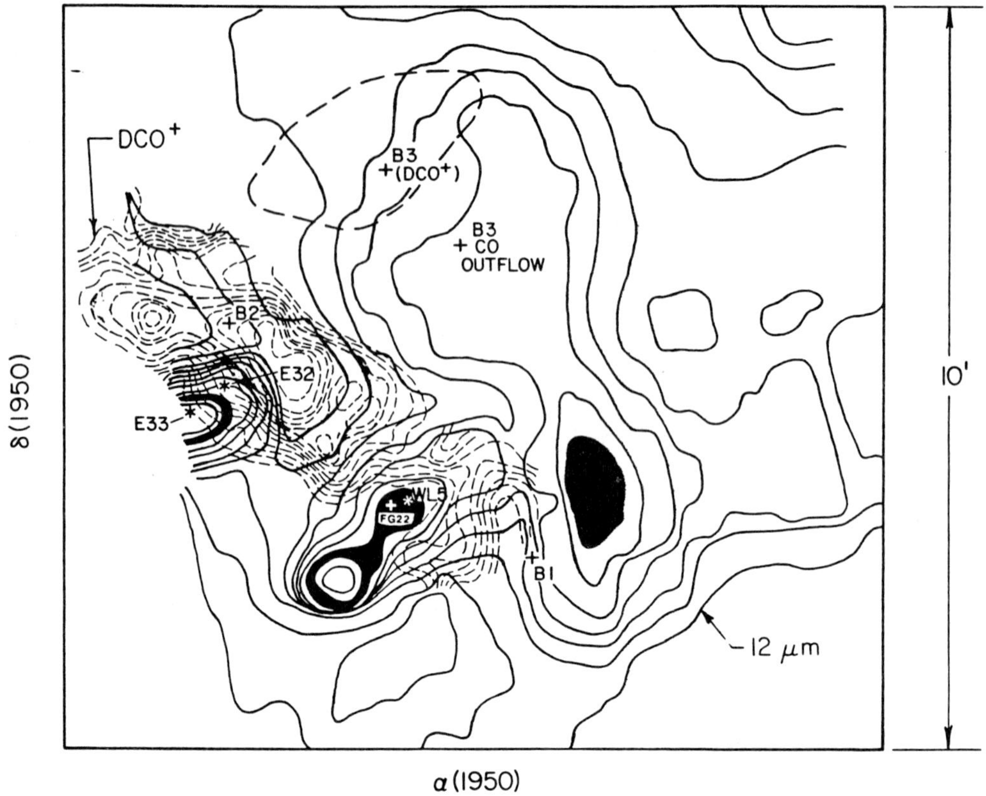

Fig. 1. Map of the region around the pre-astral core ρ Oph B2 (Butner *et al.* 1988). Solid contours, taken from IRAS 12 μm maps, trace warm dust. Highlighted contours show local peaks near point sources Elias 32 and 33, WL5, and extended source YLW10A (right of B1). At E32 and E33, $H^{13}CO^+$ J=1-0 emission peaks, demonstrating a gas column density enhancement occurs there, as well as a dust temperature peak. Light dashed contours trace DCO^+ J=1-0 intensity, obtained with 15" resolution. Note that these contours avoid E32 and E33. R then locates the unheated material and defines core B2. Near the most luminous source in the region, WL5 and its thermal radio source FG22, the column density of all measured molecules declines, illustrating the evacuation of molecules from its vicinity. Dense core B1, and modest density core B3, defined by lower resolution DCO^+ observations, are marked, as is a region of low-velocity CO flow of uncertain origin near B3.

exceed 15K, reinforcing our conclusion that ammonia emission is lacking from the central disk. Lastly, ammonia only weakly reveals the strong velocity field of the disk, a striking characteristic of the $C^{18}O$ maps. The most plausible reason for the disappearance of ammonia emission is that the ammonia molecules have accreted onto grains and core warming has not sufficiently progressed to release them back to the interstellar gas. One can only conclude that **ammonia offers an exceptionally poor probe of the densest regions of this cloud core.**

Next, consider the emission from sulfur-bearing molecules in these cores. Sulfur chemistry is initiated by endothermic reactions. Hence it could offer a foil to the cold-sensitive deuterium enrichment probe discussed earlier. A reasonable working hypothesis holds that maps of sulfur compounds will pinpoint regions of active star-forming activity. Wootten and Loren (1988, and in preparation) measured several low-lying transitions of SO_2 and SO in the Oph A, Oph B1 and B2, and IR16293 cores. $C^{34}S$ emission was measured as a control against which to contrast possible changes in relative abundance. Within the B1 and IR16293 cores, and at the peak of 1.3mm dust emission in Oph A, the $[SO_2]/[C^{34}S]$ ratio was stable, showing neither enhancement nor depression. However, near the young star GSS30, lying northwest of the Oph A core, strong emission from SO_2 and SO was found, in a region where emission from no other molecules was known to peak. The line brightness temperature in the unresolved region lies near 8K in the 3_{13}-2_{02} line, similar to the excitation temperature measured between this and the 3_{22}-2_{11} line, indicating the line could be optically thick. These observations require at least an order of magnitude enhancement of the SO_2 abundance at GSS30 over other positions in the cloud. It is difficult to attribute this to a shock, although high-velocity wings are known in CO (Loren et al. 1980) and HCO^+ (Loren and Wootten 1982) at a position to the southwest. The strong SO_2 line shows a profile substantially narrower (0.8 km s^{-1}) than the $C^{34}S$ line (1.6 km s^{-1}) at GSS30! Although our expectation that SO_2 would pinpoint active star-forming regions appears justified, a detailed explanation for the character of the emission remains perplexing.

Lastly, we consider an oft-used density probe, the 2cm line of formaldehyde. Usually seen in absorption against the cosmic background radiation, in fairly dense regions ($n(H_2)$ $\sim 10^6$ cm^{-3}) the line appears in emission. Only a few such regions are known–in OMC1 and OMC2, DR21(OH) and in the Oph B core. A recent survey (Wootten, Butner, Mangum and Loren in preparation) has revealed two more emission regions in the ρ Oph complex. One is associated with a dense clump of gas which contains an embedded radio continuum source, L1689S. This clump also emits strongly in the 1.3mm formaldehyde lines. An excitation analysis confirms that this clump is quite dense. As in the other regions where it has seen, here the 2cm line has proven an effective locator of a dense clump, probably active in star formation. The second region, however, is more enigmatic. Lying considerably east of both the cold and warm cores in L1689N, there is only weak emission in the millimeter formaldehyde lines at its position. An excitation analysis fails to account for its formaldehyde spectrum. Ammonia emission is also quite weak, again suggesting high densities are absent. It would seem quite difficult to argue on any line of evidence except that of the 2cm emission that a dense clump was located here. As with the other probes of core structure, the 2cm line also appears to have vexing exceptions to its general usefulness as a density probe.

We conclude that indeed the perception of the structure of dense cores depends critically upon the transition in whose light it is imaged. Although a number of probes seem useful for discriminating cores harboring active star formation from those which are quiet, in each situation a complexity enters the interpretation. Furthermore, each core illustrates a different stage of star formation. Only with suitably detailed maps of well-chosen molecules will we be able to piece together the whole time sequence to describe the changes in cloud structure which herald core formation and presage the creation of a new star.

REFERENCES

Butner, H. M., Wootten, A., Loren, R. B., Kaifu, N., Suzuki, H., Yamashita, T. and Hayashi, S. 1988 in *Molecular Clouds in the Milky Way and External Galaxies,* ed. R. Dickman, R. Snell and J. Young, Springer-Verlag: Berlin, p. 32.

Ho, P. and Townes, C. 1983 *Ann. Rev. Astr. Ap.*, **21**,, p. 239.

Loren, R. 1989 *Ap. J.*, **338**,, 902.

Loren, R. and Wootten, A. 1982, ESA SP-192 *Galactic and Extragalactic Infrared Spectroscopy,* p. 93.

Loren, R. and Wootten, A. 1986 *Ap. J.*, **306**, 142.

Loren, R., Wootten, A. and Wilking, B. 1989, submitted.

Loren, R., Wootten, A., Sandqvist, A., and Bernes, C. 1980, *Ap. J. (Letters)*, **240**, L65.

Mundy, L., Wootten, A. and Wilking, B. 1989, in preparation.

Wilking, B., Lada, C. and Young, E. 1989 *Ap. J.*, **340**,, 823.

Wootten, A. and Loren, R. 1987 *Ap. J.*, **317**, 220.

Wootten, A. and Loren, R. 1989 in *The Physics and Chemistry of Interstellar Molecular Clouds,* ed. T. Armstrong and G. Winnewisser, in press.

Wootten, A., Loren, R. B., and Snell, R. L. 1982, *Ap. J.*, **255**, 160.

MOLECULAR DISKS AROUND YOUNG STARS

Anneila I. Sargent,[1] and Steven V. W. Beckwith,[2]
[1] California Institute of Technology, Pasadena, CA 91125
[2] Cornell University, Ithaca, NY 14853

I. Introduction

There is now a substantial body of evidence that disk-like structures of gas and dust surround many protostars and young stellar objects (*e.g.* Rodriguez, these proceedings). We are currently studying a number of these disks to establish whether they have properties compatible with those attributed to the proto-solar nebula, in the early stages of evolution of our solar system. The frequency with which such proto-planetary disks are associated with young stars is also being investigated.

With the 30-meter IRAM telescope, and in collaboration with R. Chini and R. Güsten of the Max-Planck-Institut für Radioastronomie, Bonn, we have carried out a survey of T Tauri stars in the Taurus-Auriga cloud complex to ascertain the fraction that show 1.2 mm continuum emission. As discussed below, the detection of millimeter-wave emission from the environs of these stars implies the presence of gaseous circumstellar disks. Aperture synthesis observations of selected objects from this sample are under way at Owens Valley Radio Observatory (OVRO), using the Millimeter Wave interferometer; from the resulting maps the morphology and kinematic properties of individual disks can be determined.

II. The Disk around HL Tauri

Interferometric molecular line observations not only provide high spatial resolution information about the morphology and mass of the gas around young stars, but also permit studies of the velocity field. Because such measurements are insensitive to structures greater than about 30″ in size, the circumstellar environments of objects which are embedded in much larger clouds can be investigated.

Aperture synthesis observations of HL Tauri, for example, in the J = 1→0 transitions of CO and ^{13}CO, have demonstrated the presence of an elongated structure around this young T Tauri star (Beckwith *et al.* 1986; Sargent and Beckwith 1987). About 0.1 $M_\odot$ of gas appears to be moving in Keplerian orbits around an approximately 1 $M_\odot$ star. Even in the ^{13}CO J = 1→0 line the emission appears to be optically thick and can be detected as far as 2000 AU from the star. Taken together, these properties suggest that here the very early stages of evolution of a planetary system like our own are being viewed.

III. 1.2 mm Continuum Survey

Since aperture synthesis mapping of systems such as HL Tauri is a time-consuming project, the selection of optimum candidates for interferometric observations is critical. Continuum emission from the disks under consideration is expected to be optically thin at millimeter wavelengths. Thus, observations at 1.2 mm should provide a *direct* measure of the amount of circumstellar material associated with stars. Other indicators of circumstellar disks, such as infrared spectral energy distributions (Adams, Lada and Shu 1987, 1988; Kenyon and Hartmann 1987; Bertout, Basri and Bouvier 1988) or infrared excesses (Strom *et al.* 1989; Cohen, Emerson and Beichman 1989), preferentially reflect disk temperatures. The magnitude of the 1.2 mm flux is thus a better discriminant of candidates for the relatively time-consuming, interferometric observations.

Our 1.2 mm continuum survey of the Taurus-Auriga cloud encompassed 87 young stars (Beckwith *et al.* 1989). The $\approx 10''$ resolution of the 30 meter IRAM telescope at 1.2 mm corresponds to a spatial scale of 1400 AU at the Taurus distance, 140 pc, and discrimates against source confusion. The sample included both classical T Tauri (TT) and weak-line T Tauri stars (WTT). WTT's are defined as those having Hα equivalent widths of less than 10Å (*c.f.* Strom *et al.* 1989). Of the TT stars, 53% showed 1.2 mm emission, while 29% of the WTT's were detected. Similar fractions of TT's are suggested by searches which rely on detections of infrared emission much in excess of that expected from the stellar photospheres (Strom *et al.* 1989), or on models of the infrared spectra (Cohen, Emerson, and Beichman 1989). For two stars, HL Tau and DG Tau, continuum emission at 1.4 mm has been measured independently at OVRO (Woody *et al.* 1989). Within the uncertainties, the fluxes agree very well with those measured at IRAM, 880 mJy and 450 mJy for HL Tau and DG Tau respectively. It is noteworthy that the proportion of detected WTT's is not very different from the fraction of TT's seen, refuting the hypothesis that Hα weakness implies a lack of circumstellar material (Walter 1987).

It is assumed that these 1.2 mm fluxes are due mainly to thermal emission from particles around the stars. A comparison of our results with radio fluxes measured at the VLA by Bieging, Cohen, amd Schwartz (1984) indicates little contribution from free-free emission. The 6 cm flux of DG Tau, for example, one of the strongest sources at cm wavelengths, suggests a 1.2 mm flux of at most 36 mJy, only 8% of the observed value. For thermal emission, the expected visual extinction, A_V, to each star can be calculated, assuming various particle distributions. In general, the 1.2 mm fluxes imply high values of A_V. However, the measured A_Vs are rarely greater than 3^m, indicating that the emitting particles must be distributed very non-uniformly, most probably in flattened disks.

IV. Interferometric Observations of DG Tau

As discussed above, the 1.2 mm fluxes provide a good diagnostic for circumstellar disks suitable for interferometer mapping. With a flux of 440 mJy, DG Tau is, like HL Tau, one of the strong 1.2 mm sources. Also, [SII] observations of DG Tau (Mundt 1988) show a blue-shifted jet emanating from the star in the direction PA 226°, almost parallel to the magnetic field, as might be expected if the circumstellar mass distribution were anisotropic (see Hodapp 1984). Aperture synthesis mapping of DG Tau in the J = 1→0 ^{13}CO line at 110 GHz was carried out at OVRO in 1988-89. Continuum observations at 2.7 mm were acquired simultaneously and show a 71 mJy source, unresolved in the $6'' \times 5''$ (800 × 700 AU) beam. Following the arguments of Beckwith *et al.* (1986), with excitation temperature 50 K and dust emissivity directly dependent on frequency (see also Woody *et al.* 1989), this continuum measurement leads to a mass of approximately 0.05 $M_\odot$ for the circumstellar material.

A map of the ^{13}CO emission integrated over the velocity range 6.4 to 7.6 km s^{-1} is presented in Figure 1. No emission above a 2σ level of 300 mJy/beam was detected outside this velocity range. An elongated feature of radius 2000 AU is centered approximately on the star. The elongation is along PA

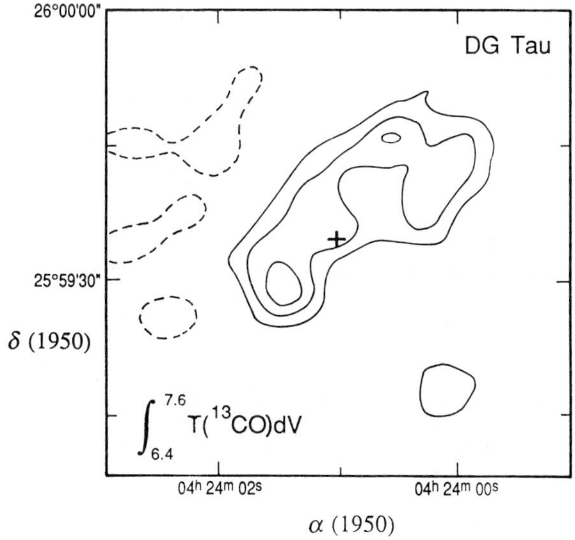

Figure 1. A map of the integrated ^{13}CO emission from DG Tau. Contours begin at 300 mJy/beam, the 3σ level, and are separated by 200 mJy. A cross marks the location of the 2.7 mm continuum source, which coincides with the stellar position.

130°, perpendicular to the optical jet detected by Mundt (1988). The total integrated flux, 6.2 Jy km s^{-1}, indicates a mass around 0.02 $M_\odot$ (*c.f.* Sargent and Beckwith 1987).

Maps of the ^{13}CO emission at 0.26 km s^{-1} resolution demonstrate that the position of the peak changes with velocity; above 7.0 km s^{-1} emission tends to be confined to the the south-east half of the elongated structure of Figure 1, while below 7 km s^{-1} emission is predominantly from the north-west region. For gravitational binding of this circumstellar gas, a stellar mass of 1.4 $M_\odot$ is required. Although uncertain, the mass of DG Tau inferred from its position on the HR diagram is very close to this value, 1.8 $M_\odot$, suggesting that the gas is bound to the star. Although the evidence for Keplerian rotation is not as compelling as in the case of HL Tau, the two systems are remarkably similar and it is likely that DG Tau is also surrounded by a proto-planetary disk.

V. Discussion

It seems reasonable to expect that the detectable mass in a circumstellar disk will diminish with time, the circumstellar environment becoming more diffuse and possibly accumulating into larger structures like planets. While estimates of the disk masses may be derived from continuum and line interferometry, as described above, the requisite observations exist for only a few sources. However, it has recently been shown that the mass of a circumstellar disk may be determined relatively accurately from only a few observed parameters (Beckwith *et al.* 1989). Briefly, the infrared spectral energy distribution shortward of about 100 μm, where emission is optically thick, specifies the temperature distribution in the disk; knowing this, and the strength of the millimeter flux (where emission is predominantly optically thin), the disk mass may be calculated.

This method has been applied to all of our 1.2 mm sources for which infrared fluxes are available. We find disk masses ranging between 0.001 $M_\odot$ and about 0.5 $M_\odot$. Average disk temperatures range from a few $\times$ 10 K to a few $\times$ 100 K. The sensitivity was such that most systems with masses comparable to that of the proto-solar nebula, 0.01 $M_\odot$, should have been deteceded. Calculations of the spectral energy distribution of HL Tau fit well with the observations for a disk mass of 0.09 $M_\odot$. Likewise, DG Tau requires a disk mass of 0.03 $M_\odot$. Both values are well within the ranges suggested by the interferometer measurements.

Comparisons of the disk masses and temperatures with the ages of the stars (based on their positions with respect to isochrones in temperature-luminosity plots) indicate no correlation between mass and age. Over a time-scale of 10^5 to 10^7 years there is no evidence of the mass in the disks decreasing. There is a very slight tendency for the disks to become cooler with age. Strom *et al.* 1989 have suggested on the basis of IRAS flux measurements that disk masses do show some signs of evolution over this time span; their results, deriving from the optically thick infrared regime may reflect decreasing temperatures rather than decreasing mass.

In summary, we have found that approximately 50% of a sample of solar-type, pre-main sequence stars have circumstellar disks. There is little evidence of mass evolution for stars with ages between 10^5 and 10^7 years, although older disks could be cooler. Aperture synthesis mapping of two of these stars has confirmed the presence of disk-like structures around them. The masses, radii and velocity patterns observed are consistent with those of the proto-solar nebula, suggesting that planetary systems may be forming around some of these objects.

References

Adams, F. C., Lada, C. J., and Shu, F. H. 1987, *Ap. J.*, **312**, 788.

Adams, F. C., Lada, C. J., and Shu, F. H. 1988, *Ap. J.*, **326**, 865.

Beckwith, S. Sargent, A. I., Chini, R. and Gusten, R. 1989, *Ap. J.*, submitted.

Beckwith, S., Sargent, A. I., Scoville, N. Z., Masson, C. R., Zuckerman, B., and Phillips, T. G. 1986, *Ap. J.*, **309**, 755.

Bertout, C., Basri, G., and Bouvier, J. 1988 *Ap. J.*, **330**, 350.

Cohen, M., Emerson, J. P., and Beichman, C. A. 1989, *Ap. J.*, in press.

Hodapp, K.-W. 1984, *Astr. Ap.*, 141, 255.

Kenyon, S. J., and Hartmann, L. 1987, *Ap. J.*, **323,** 714.

Mundt, R. 1988, in NATO Advanced Study Institute *"Formation and Evolution of Low Mass Stars"*, ed. A. K. Dupree and M. T. V. T. Lago.

Sargent, A. I., and Beckwith, S. 1987, *Ap. J.*, **323**, 294.

Strom, K. M., Strom, S. E. Edwards, S., Cabrit, S., and Skrutskie, M. F. 1989, *A. J.*, **97**, 1451.

Walter, F. M. 1987, *Pub. A. S. P.*, **99**, 31.

Woody, D. P., Scott, S. L., Scoville, N. Z., Mundy, L. G., Sargent, A. I., Padin, S., Tinney, C. G., and Wilson, C. D. 1989, *Ap. J. (Letters)*, **337**, L41.

Discussion:

ELSÄSSER: Is there any correlation between the orientation of the disks you resolved, projected on the sky?

SARGENT: I presume you are referring to their orientations not just relative to one another, but also to the local magnetic field. In fact, the three disks we have examined in the Taurus cloud, HL TAU, L1551 IRS5 and DG TAU are all similary oriented, close to PA 135, which is perpendicular to the local magnetic field direction.

LARSON: I would guess that the inner parts of protostellar disks evolve faster than the outer parts, and that your data probe mainly the outer parts. What radii do your results typically refer to?

SARGENT: It is true that at present we are investigating large radii. The disks I have shown have radii of order 1000 AU. The smallest disk we have detected in molecular gas, L1551 IRS5, has a radius around 500 AU. But of course the continuum sources, which are mostly unresolved in 6" beams, show that the bulk of the material is confined to regions of radii $\sim$ 500 AU. With higher resolution, especially at 1mm, we should be able to probe a little deeper.

SCHULZ: The "usual" dust emission β is about 2. Could you comment on your value of 1?

SARGENT: For most of our sources we can fit the 1.3mm and 2.6mm points in the spectral energy distributions only by adopting $\beta = 1$. A study of the 1.4mm emission from a variety of galactic sources carried out at Owens Valley Radio Observatory (Woody et al.,1989) also suggested $\beta = 1$ when comparison with the 2.6mm fluxes was made. Although $\beta = 2$ is generally expected for both crystalline and amorphous materials (Tielens and Allamandola,1987), $\beta = 1$ can result for layared materials where phonons are limited in two dimensions. Values of 1 and 1.5 have been found for amorphous carbon and layered silicates (Koike, Hasegawa and Manabe, 1980; Day, 1976).

CAMERON: Do you have a feeling for how far out into the far-IR/ submillimetre wavelength regime you have to go before the continuum emission from the disks you have observed becomes optically thin?

SARGENT: Our data on HL TAU and DG TAU show that they are still optically thick at 100 μm and thin by 1mm . We have no observations at intermediate wavelengths, although we expect to acquire these at the Caltech Submillimeter Observatory next season.

ZINNECKER: Your result that about 50% of the T Tauri stars are detected at 1mm (indicative of cool dusty circumstellar disks) leads me to ask about the other 50% that you did not detect: could those be binary stars (whose incidence is of the order of 50%)? Did you detect continuum emission from T Tauri binary stars (e.g., T Tau)?

SARGENT: Yes, we did detect 1.3mm continuum emission from T Tauri. Weintraub et al. (1989) have in fact argued for a circumbinary disk on the basis of OVRO measures. I doubt if the presence of a companion would prevent there being a disk. Recent speckle measures of Z CMa also appear to show a close companion and a disk (Koresko, Beckwith and Sargent 1989). What I find quite interesting is that Becklin and Zuckerman (1988), in their studies of Vega-like objects, also find a 50% detection rate for systems with circumstellar structure.

On the Nature of the Beta Pictoris Circumstellar Nebula

*Francesco Paresce** and *Pawel Artymowicz*

Space Telescope Science Institute
3700 San Martin Drive
Baltimore, Maryland 21218 U.S.A.

Summary

We summarize the observed properties and deduced physical characteristics of the Beta Pictoris circumstellar nebula obtained from a detailed analysis of existing IR and optical data. On the basis of these results, we reject the hypothesis that the observed feature is a bipolar nebula surrounding an evolved star. We claim the nebula is a flattened disk of orbiting particles making up a planetary system in its clearing out phase as small grains collide, erode, and are swept out by radiation pressure.

Introduction

The nearby A5V star Beta Pictoris (HR2020) has attracted some attention lately due to the discovery of a spatially resolved IR excess (Gillett, 1986) and a thin extended bipolar reflection nebulosity in the near IR and optical (Smith and Terrile, 1984, 1987; Paresce and Burrows, 1987). There are several possible *a priori* explanations for the observations with the most plausible being either a true bipolar nebula (BPN) as observed around evolved stars (Morris, 1981; Herbig, 1989) or a flattened disk of orbiting material in a proto or planetary system, possibly somewhat like our own. This latter possibility, of course, is, by far, the most exciting and, if confirmed, could represent an unparalleled opportunity to observe another solar system in considerable detail due to its relative proximity to the Sun. It is, consequently, quite important to be able to place this nebulosity in its proper astronomical context. In this brief report, we investigate this issue on the basis of the known facts about the system and the logical deductions that can be made therefrom.

* Affiliated with the Astrophysics Division of the Space Science Department of ESA.

Physical Characteristics of the Beta Pic Nebulosity

The basic question, then, is: what is the elongated feature at p.a. 30° shown in figure 1? This image was taken by Paresce and Burrows with their coronograph at the ESO 2.2m telescope in the Ic band centered at 7900Å. The spacing of the isophote contours in this figure corresponds to an 0.5 magnitude surface brightness difference normalized to an intensity of 15.9 mag arcsec^{-2} in this band at 6" from the center of the star located at the intersection of the N, E axes and hidden by the 2" thick occulting wedge. This feature can be observed out to at least 1100 AU from the star with a roughly constant resolved FWHM thickness of 30 - 50 AU. The observed broad band fluxes from B to I_c are consistent with scattering of Beta Pic starlight by neutrally reflecting particles. This wavelength independent albedo and a recently discovered decrease in U band reflected light over the gray case is overwhelming evidence for large (a > 1 micron) particles as the dominant scatterers. This is not surprising since smaller particles would certainly not survive long in the Beta Pic environment before being blown out by radiation pressure (Artymowicz, 1988). The nebulosity has been reliably observed down to ≈ 5" (85 AU) from the central star, a limit set, presently, by the seeing dominated profile of the V = 3.87 central object.

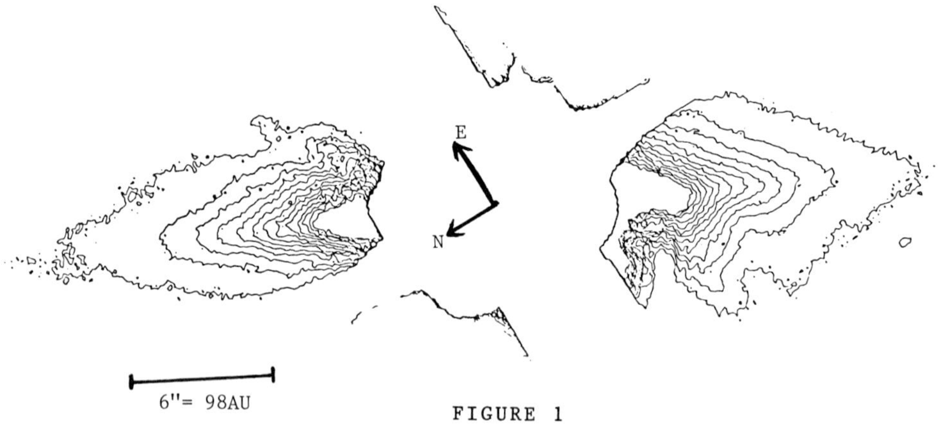

FIGURE 1

The Beta Pic system exhibits the largest excess far IR flux of any main sequence star observed by IRAS *i.e.*, 1.59, 10.1, 18.8 and 11.2 Jy at 12, 25, 60 and 100 microns, respectively, after subtraction of a suitable photospheric component. The deconvolution of spatial IRAS scans at 60 microns yield a < 14" FWHM size in the in-scan (0° p.a.) and a nominal 22" ± 6" FWHM size in the cross-scan (70° p.a.) direction (Gillett, 1986) although a subsequent analysis (Gillett *et al.*, 1988) revised the error bar to justify only a marginal spatial IR resolution. Attempts to resolve Beta Pic in the 1 - 10 micron near IR range have been inconclusive, so far.

Beta Pictoris has been known for some time as a shell star with strong narrow absorption lines of Ca II and Na I intrinsic to the star's immediate environment because about one to two orders of magnitude stronger than plausibly attributable to the intervening ISM (Hobbs *et al.*, 1985; Hobbs, 1986). Temporally varying UV lines of Al, Mg and Fe are also characteristic of this system (Bruhweiler and Kondo, 1987; Lagrange-Henri *et al.*, 1989 and references therein). These observations imply a surprisingly small amount of neutral gas in the circumstellar environment ($N_{HI} \simeq 10^{20}$ atoms cm^{-2}) with a strong concentration in the very near environment ($r \simeq 1$ AU) of the star's surface. Nebular emission in the Hα line, for example, is undetectable. The most interesting aspect of these results concerns the appearance of sporadic temporal variations in the red wing of some UV absorption lines corresponding to up to $\simeq 400$ kms^{-1} infall velocities. These data are presently best explained as being due to km-sized bodies, possibly comets, breaking up and evaporating as they collide with the star and temporarily populating a narrow inner region with inward spiralling gaseous debris. Since this is a well documented phenomenon in our own system (see Morrison and Owen, 1988 p. 116 for example) such an explanation does not seem as far-fetched as it may sound. The implied mass flow rate is small ($\simeq 10^{16}$ g yr^{-1}) and, thus, does not affect the stability or lifetime of a possible outer disk.

There is certainly enough observational data available on this object to make it worthwhile to attempt to construct a comprehensive model of the nebulosity. This should be possible even though the IR and optical data refer to somewhat different regions with a small but significant overlap (Norman and Paresce, 1989). Past use of oversimplified *ad hoc* models addressing only a subset of the data base has resulted in apparent contradictions and inconsistencies. We believe that a better approach is to develop self-consistent, physically realistic models of the Beta Pic nebulosity that satisfy both the optical and the IR results preferably with a single population of scattering and emitting grains. We have described in detail how a family of just such models can be found using non parametric maximum entropy techniques (Artymowicz, Burrows, and Paresce, 1989). Although a specific model depends mainly on the precise IR grain emissivity law which has to be assumed to be of the standard form $\varepsilon \sim \lambda^{-1}$, a very reasonable range of input parameters yields very similar final results. These are: a clear dust-free zone around the star of radius 10 - 20 AU with a peak dust density located at 20 - 40 AU at a grain temperature of $\simeq 95$ K (165K at the edge of the clear zone), a normal optical thickness at peak of $\simeq 10^{-2}$, an equivalent scattering area of $10^{30} cm^2$, and a minimum grain albedo of 0.5. An absolute minimum total mass of grains is 10^{26}g if all the mass is in 1 micron diameter particles, the minimum size allowed by the observations. A more realistic order of magnitude estimate of the nebular mass can be made by using the solar system meteoroid $a^{-3.5}\,da$ distribution and a particle size upper limit of 0.1 km. In this case, one obtains 10^{30} g, a value remarkably similar to the total mass of solids in our solar system. We emphasize here that all these results are basically independent of the 3 dimensional geometry assumed for the Beta Pic nebula.

The True Nature of the Beta Pic Nebulosity

At this point, we can make some progress in sorting out the crucial question of whether the scattering and emitting material is distributed as a flattened edge-on disk reminiscent of an orbiting planetary system or, as Herbig (1989) has recently suggested, as a narrow jet-like bipolar structure similar to those encountered in classical BPNs. In the absence of reliable information on the velocity field of the observable Beta Pic emitters, the only parameter which could settle the issue definitively, the best way to proceed is to compare the known properties of the Beta Pic nebula with the average properties of BPNs. We have used data from Morris, 1981 and Calvet and Cohen, 1978 and references therein for this comparison. The relevant data are listed in Table 1. It is quite obvious from even a cursory examination of this table that, notwithstanding the associated uncertainties, it would be difficult indeed to classify the Beta Pictoris system as a BPN. Moreover, Beta Pic falls squarely on the MS for Population I stars line on the HR diagram of 13 BPNs published by Calvet & Cohen, 1978 confirming the liminosity class V classification of this object given by Hoffleit and Jaschek, 1982 and its clean distinction from the BPN class. Actually, a more detailed examination of the typical BPN morphology usually consisting of wide ragged structures many with extended horns shows that Beta Pic would be out of place even in this respect and is morphologically much more similar to the jets emanating from young stellar objects which this object is certainly not.

Table 1. A Comparison of Physical Properties: Beta Pic vs. BPNs

Property	Beta Pictoris	Bipolar Nebulae
Size (cm)	$5 \cdot 10^{15}$	$5 \cdot 10^{17}$
Mass (g)	10^{26} - 10^{-30}	10^{31} - 10^{32}
Gas to Dust Mass Ratio	< 1	$\simeq 10^2$
Grain Temperature	< 250K	$\geq 10^3$K
Expansion Velocity (km s^{-1})	?	10 - 100
Lifetime (yr)	$\simeq 10^8$ - 10^9	10^3 - 10^4
Molecules	none	OH, CO, H_2O, PAH, etc.
Mass loss Rate of Central Star ($M_\odot$ yr^{-1})	$\simeq 0$	10^{-6} - 10^{-4}
Intrinsic Balmer A_V(mag)	$\simeq 0$	2 - 6
v sin i	139	?

Another way to look at this problem is to ask whether a quasi-linear dusty feature could actually be theoretically supported in the Beta Pic system as we know it if one arbitrarily stretches the boundaries of the BPN category to encompass it.

There is, presently, enough circumstantial evidence of a dynamical nature to make this hypothesis extremely unattractive. Here, we briefly enumerate some of the most obvious difficulties. First, the optical coronographic images show a structure with roughly constant cross section out to $\approx$ 1000 AU. Only an unobserved and very dense ambient medium would confine a jet of outflowing grains to a rapidly narrowing solid angle. Second, since the star has an appreciable rotation velocity of v sin i = 139 km s^{-1}, the jets must originate from the poles of the star and the equatorial region in our line of sight would have to contain a massive, extremely optically thick disk. This possibility is completely ruled out by the relative transparency of the Beta Pic line of sight. Third, the absence of grains hotter than about 200K in the nebula imply that the grains must form on dynamical timescales in a cold environment rather than being ejected from the hot stellar envelope. We cannot think of any plausible mechanism that could accomplish this.

Perhaps most decisive is the fact that in order to support a dynamical jet we need to invoke a stellar mass loss rate for which there is absolutely no supporting evidence. A rough minimum estimate of $\dot{M}$ is 10^{26}g per 10^3 yr, the dynamical time-scale, or 10^{-10} $M_\odot$ yr^{-1}. Since this refers only to the directly observable solid grains of ~1 micron size, the total implied $\dot{M}$ is more likely to be in the range 10^{-6} - 10^{-4} $M_\odot$ yr^{-1}, values that certainly would not escape detection. We conclude on the basis of these considerations and the data shown in Table 1 that the observed Beta Pictoris nebulosity is almost certainly not a standard bipolar reflection nebula. As long as a simpler and more plausible explanation for this structure exists, we see no further benefit to be derived from pursuing this hypothesis any further.

The Evolutionary State of the Beta Pictoris Disk

Having convinced ourselves that the Beta Pic nebulosity is most likely the visible and IR manifestation of an orbiting swarm of medium to large sized particles confined to a disk of 30 - 50 AU thickness viewed nearly edge-on, it is fair to ask in what evolutionary state the system finds itself presently, and whether larger planets or planetesimals exist in the occulted inner part of the disk. Although we cannot directly observe this part, all the existing data point to an unambiguous value of the normal optical thickness $\tau \approx 10^{-2}$ at the peak of the grain density profile around 30 AU. Together with the observed disk thickness, this value implies typical grain collision frequencies of 10^{-4} years at speeds of order of $\approx$ 1 km s^{-1}. For these conditions, grain growth by mutual coalescence or snowballing cannot occur and the dominant effect is erosion and destruction of the larger grains. The low gas densities ensure that gas accretion is unimportant anywhere in the disk. We conclude that it is very unlikely that the system is, at present, building large bodies from the dust but is in a clearing-out phase as the larger bodies are ground down to smaller ones until they can be blown out by radiation pressure.

There are several independent indications that the Beta Pic disk contains a whole range of particle sizes, up to cometary or even larger. The first direct evidence is provided by the sporadic infall events of km sized bodies in which a large amount of redshifted gas is seen near the star as discussed earlier. Second, the importance of gas drag on long timescales requires the disk to be heated dynamically to its observed scale height by the gravity of the embedded bodies. The mass of a typical perturber capable of stirring the system to the observed thickness (between 100 AU and 200 AU from the star) is very roughly equal to one lunar mass. Third, the efficient erosion of grains requires a large reservoir of mass to be present in the disk for it to be a relatively long lived phenomenon. We can estimate that during a plausible $\simeq 2 \cdot 10^8$ yr lifetime, the system might have lost a total mass of order 10^{29}g or $\simeq 10^{-5}$ of the star's mass and only an order of magnitude less than the expected total mass of a possible planetary system formed in the primordial Beta Pic nebula. Thus, this high erosion rate may imply that in the next few 10^8 yr the Beta Pic disk will evolve toward a less conspicuous phase cleared of much of the present solid circumstellar material, possibly leaving behind only the largest surviving bodies.

References

Artymowicz, P., 1988, *Ap. J. Lett*, **335**, L79

Artymowicz, P., *et al.*, 1989, *Ap. J.*, **337**, 494

Bruhweiler, F., and Kondo, Y., 1987 in *Physics of Be Stars*, eds. A. Slettebak and T. Snow (Cambridge U. Press; Cambridge) p. 526

Calvet, N., and Cohen, M., 1978, *M.N.R.A.S.*, **182**, 687

Gillett, F. C., 1986, in *Light on Dark Matter*, Ap. Sp. Sci. Lib., **124**, ed. F. Israel (Dordrecht: Reidel), p. 61

Gillett, F. C., *et al*, 1987, *B.A.A.S.*, **19**, 829

Herbig, G., 1989, in *The Formation and Evolution of Planetary Systems*, eds. H. A. Weaver and L. Danly (Cambridge U. Press; Cambridge), p. 296

Hobbs, L. M., *et al.*, 1985, *Ap. J. Lett.*, **239**, L29

Hobbs, L. M., 1986, *Ap. J.*, **308**, 854

Hoffleit, D., and Jaschek, C., 1982, *The Bright Star Catalogue*, (Yale U. Obs.; New Haven)

Lagrange-Henri, A. M., *et al.*, 1988, *Astr. Ap.*, **190**, 275

Morris, M., 1981, *Ap. J.*, **249**, 572

Morrison, D., and Owen, T., 1988, *The Planetary System*, (Addison-Wesley; New York), p. 116

Norman, C., and Paresce, F., 1989, in *The Formation and Evolution of Planetary Systems*, eds. H. A. Weaver and L. Danly, (Cambridge U. Press; Cambridge), p. 151

Paresce, F., and Burrows, C., 1987, *Ap. J. Lett.*, **319**, L23

Smith, B. A., and Terrile, R. J., 1984, *Science*, **226**, 421

________________, 1987, *B.A.A.S.*, **19**, 829

NEAR-INFRARED IMAGES OF THE SERPENS MOLECULAR CLOUD CORE

C. Eiroa(1) and M. Casali(2)

(1) Observatorio Astronómico-IGN, c/ Alfonso XII, 3
28014 Madrid, Spain

(2) Joint Astronomy Centre, Hilo, Hawaii, USA

ABSTRACT:

Near-infrared images of the Serpens molecular cloud core have been carried out at UKIRT (Mauna Kea Observatory) using the infrared array camera, IRCAM. A large-scale diffuse nebulosity extending over the central part of the core is observed. Over 100 K-sources are detected in the 30 arc min^2 cloud core. Some of them are PMS objects which were previously unknown. For the first time, a near-infrared counterpart of the far-infrared source Serpens FIRS1 has been detected. The 2.2 μm source appears as a point like object at the apex of an extended knotty, jet-like nebulosity oriented towards the northwest. In addition, a group of 11 stellar objects is seen in the position of the IR/radio source SVS4. These objects are embedded in a very faint nebulosity and form one of the densest clustering of young stars found in dark clouds.

1. Introduction

Studies of young stellar associations or clusters still embedded in their parent molecular clouds are in many respects of fundamental interest for star formation theories. In the optical, these kinds of studies have been limited to the brightests members of the stellar aggregates, for the extinction produced by the interstellar material of the cloud is very high. A usual way to avoid this inconvenience has been the use of infrared techniques. In the infrared the extinction is considerably lower than in the optical, allowing us to penetrate deeper in the dark cloud environments. In addition, many of the stellar objects embedded in the clouds are still in pre-main sequence evolutionary stages or are surrounded by dense circumstellar dusty envelopes; in both cases it is expected that they mainly radiate in the infrared.

The young stellar population embedded in several dark clouds, e. g. Taurus, Chamaeleon, Ophiucus, has been succesfully studied in the infrared up to typical limiting magnitudes K $\sim$ 11-12 mag. The recent advent of infrared arrays enables us to extend these studies to considably higher sensitivity limits and, consequently, to penetrate deeper in the cloud and detect objects of much lower luminosity. In this way, parameters like luminosity function, star formation efficiency, etc., can be estimated in a more reliable manner.

The Serpens dark cloud is a well known star-forming region. Observations at near-infrared wavelengths have been carried out by Strom et al. (1976), Churchwell and Koornneef (1986) and Gómez de Castro et al. (1988). The latter authors found that at least 12 young objects are embedded in the 3.3 arc min radius dense cloud core and suggested the likely existence of other undetected members of the Serpens young stellar aggregate. In this work we briefly describe the results of a near-infrared study of the Serpens core using the infrared camera IRCAM at UKIRT. A complete description of our results will be given in a forthcoming paper (Casali and Eiroa, 1989).

2. Observations

JHKL' infrared images of the Serpens core were carried out in April 1988 using the infrared camera, IRCAM, of the United Kingdom Infrared Telescope (Mauna Kea Observatory). The camera is equipped with a 58x62 detector array. 30 frames of $\sim$1'x1' at each wavelength were obtained, thus covering the whole Serpens core. The scale used was 1.2"/pixel and the limiting magnitude of the observations is K $\sim$17 mag. Details will be given in Casali and Eiroa (1989).

3. Results

New and unexpected aspects of the Serpens core are revealed by the infrared images. These are connected with the diffuse medium of the cloud and with the young population of stars.

A large-scale, diffuse nebulosity embedded in the core is detected at K. It extends from the north of the Serpens bipolar reflection nebula to the south in the direction of the source SVS4. No hint at all of this nebulosity is seen in the very deep 0.9 μm images of Gómez de Castro et al. (1988). The nebulosity coincides quite well with the south-east region of the NH_3 and far-infrared emission (e.g. Zhang et al., 1988).

130-140 stellar objects are detected in the observed area. Only around 30 objects are known from previous optical and infrared works. Many of the new stars are likely background stars, but in some cases the infrared images indicate that they are Serpens embedded objects. Three new cometary-like nebulae have been detected. One of them is located towards the north-west of the Serpens reflection nebula and the two others towards the south of this latter object. Interesting is that the new cometaries are approximately oriented in the SE-NW direction, reinforcing the alignment of young objects within the Serpens core noticed by Gómez de Castro et al. (1988).

Perhaps the two most interesting objects discovered by the infrared images are related with the sources Serpens FIRS1 and SVS4. They are the only sources of the cloud where radio continuum emission has been detected (Rodriguez et al., 1980), and in both cases the luminosity deduced from the radio observations is much larger than the total infrared luminosity (Snell and Bally, 1986).

An extended jet-like nebular object with a point source located at its south-east apex is detected at the position of FIRS1. This is the first time that an object is detected at this position at wavelengths shorter than the far-infrared range. The nebulosity is oriented in the SE-NW direction, as other Serpens young objects are. The source is only observed at 2.2 μm; it is probably a reflection nebula embedded in the core, the illuminating star being the point source at the south-east end. We cannot excluded, however, the possibility that shock excited molecular hydrogen is contributing to the K nebulosity.

The frames taken at the position of the source SVS4 reveal a group of 11 objects surrounded by a very faint nebulosity. The IR colours of these stars show an IR excess. Thus, we suggest that they are members of the Serpens young star population. Taking into account this result, the number of PMS objects in the Serpens core is multiplied by a factor of two and, consequently, the star formation efficiency and the stellar mass density are increased considerably with respect to the estimate by Gómez de Castro et al. (1988). In addition, the SVS4 stars are grouped in a region of projected size ~40" (~ 0.05 pc after the distance estimate of Chavarría et al., 1988), and therefore they represent one of the densest clustering of young stars found in dark clouds.

C. E. is supported in part by grant DGICYT PB87-0167

References

Casali, M., Eiroa, C.: 1989, in preparation
Chavarría-K, C., de Lara, E., Finkenzeller, U., Mendoza, E., Ocegueda, J.: 1988, Astron. Astrophys. 197, 151
Churchwell, E., Koornneef, J.: 1986, Astrophys. J. 300, 729
Gómez de Castro, A. I., Eiroa, C., Lenzen, R.: 1988 201, 299
Rodriguez, L. F., Moran, J. M., Ho, P. T. P., Gottlieb, E. W.: 1980, Astrophys. J. 235, 845
Snell, R. L., Bally, J.: 1986, Astrophys. J. 303, 683
Strom, S. E., Vrba, F. J., Strom, K., M.: 1976, Astron. J. 81, 638
Zhang, C. Y., Laureijs, R. J., Clark, F. O.: 1988, Astron. Astrophys. 196, 236

MOLECULAR OUTFLOWS

Ronald L. Snell
Five College Radio Astronomy Observatory and the Department of Physics and Astronomy, University of Massachusetts, Amherst

I. Introduction

The development of millimeter and submillimeter spectroscopy has allowed astronomers to probe the cold, dense component of the interstellar medium. This medium, primarily composed of gas in molecular form, is gravitationally bound into relatively distinct clouds that are sites of star formation within our Galaxy. The most ubiquitous tracers of these molecular clouds are the rotational transitions of carbon monoxide. Observations of CO emission have been commonly used to estimate the size, temperature, mass, and density of molecular clouds; in addition, the spectral line profiles can be used to study the internal dynamics of these clouds. Although the sound speed within molecular clouds is only about 0.2 km s^{-1}, observed CO line widths are more typically 1 to 5 km s^{-1}. Thus, the internal dynamics of molecular clouds are characterized by supersonic gas motions whose nature is poorly understood.

In addition to the generally supersonic gas motions present in molecular clouds, there are localized regions of much faster hypersonic gas motions associated with newly forming stars. These motions are understood to represent expanding gas flows from young stellar objects. An example of hypersonic molecular flow is illustrated in Figure 1a, in which a series of CO spectra obtained along a line from northwest to southeast through the infrared source associated with HH7–11 are displayed (Snell and Edwards 1981). The line widths away from the position of the infrared source (3N–3W and 2S–2E) are dominated by the supersonic gas motions in the cloud, but towards the infrared source the total velocity extent of the emission is much greater, arising from gas outflowing from the young star embedded within the molecular cloud.

One of the most interesting aspects of molecular outflows is their bipolar nature (Snell, Loren, and Plambeck 1980). This can be seen in the maps of the redshifted and blueshifted emissions presented in Figure 1b for the HH7–11 region; the redshifted emission extends to the northwest while the blueshifted emission extends to the southeast. Presumably, the winds from the young star at the center of HH7–11 are collimated into two oppositely directed jets that accelerate the surrounding molecular material. This produces the bipolar signature seen in the high velocity CO emission. Further evidence for highly collimated stellar winds is also provided by detection of optical jets (see Mundt 1988; Strom *et al.* 1986).

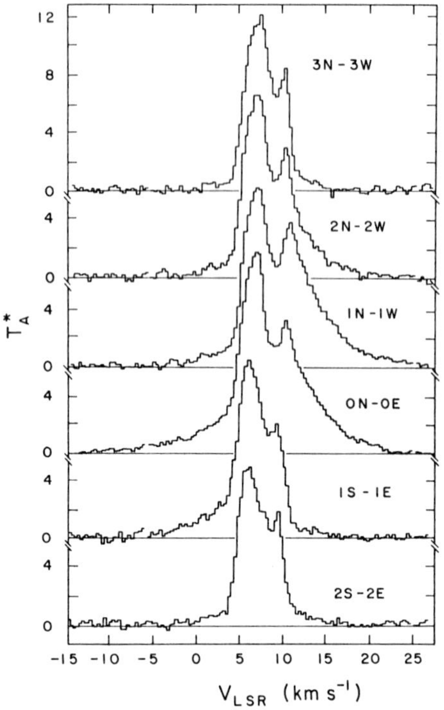

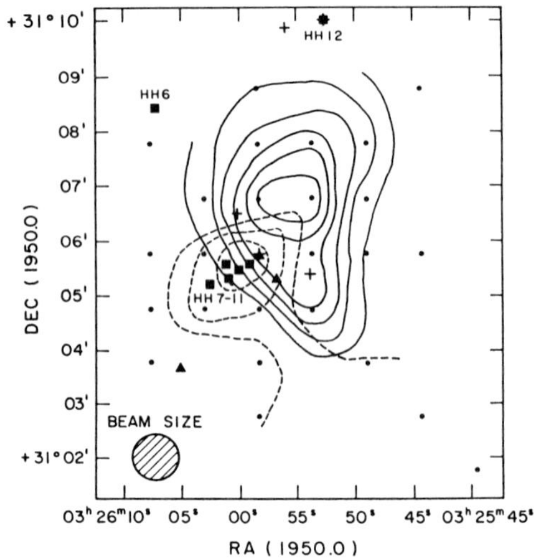

Figure 1a (left). – CO J=2–1 spectra obtained at six locations along a southeast to northwest axis through the infrared source associated with HH7–11. The offset position of each spectrum in arcminutes north, south, east, and west of the IR sources are indicated.
Figure 1b (right). – A map of the integrated intensity of the high velocity redshifted (solid contours) and blueshifted (dashed contours) CO emissions in the HH7–11 region. Both figures are from Snell and Edwards (1981).

A vast body of data has been collected over the past decade which addresses the physical properties and nature of molecular outflows. Only a brief summary of this literature will be presented, since it is thoroughly covered in the recent reviews by Lada (1985) and Snell (1987). Instead, most of this review will be devoted to a summary of the more recent developments in this field.

II. Summary of Molecular Outflow Properties

The properties of molecular outflows have been summarized by Lada (1985) and Snell (1987), and I refer readers to these reviews for a more detailed description of the properties of molecular outflows. Briefly, molecular outflows have hypersonic gas motions with velocities of 10 to 100 km s^{-1} and are associated with young stellar objects (YSOs). The mass of the hypersonic molecular gas (0.1 to 100 $M_\odot$) contains kinetic energies of 10^{43} to 10^{47} ergs. The outflow phenomenon is short lived with dynamical timescales $\lesssim$ few x 10^5 years; thus, massive stellar winds (with mass loss rates of 10^{-7} to 10^{-3} $M_\odot$ yr^{-1} for wind velocities of 200 km s^{-1}) are required. Approximately 75% of all known outflows show bipolar morphologies, suggesting that the winds from young stars in general are collimated. Outflows are associated with both high and low luminosity young stars (1 to 10^5 $L_\odot$) and the energies and momenta in the outflows are correlated with the radiant luminosity. Typical properties of an outflow are summarized in Table 1. Finally, a large number of outflows have been detected, suggesting that the outflows represent a common evolutionary phase of many young stars.

TABLE 1

Typical Molecular Outflow Properties

Mass	3 $M_\odot$
Velocity extent	25 km s^{-1}
Size	0.5 pc
Energy	10^{45} ergs
Momentum	20 $M_\odot$ km s^{-1}
Dynamical timescale	5x10^4 years
Force	10^{-3} $M_\odot$ km s^{-1} yr^{-1}
Mass loss rate	5x10^{-6} $M_\odot$ yr^{-1}

The data on molecular outflows summarized in the reviews of Lada (1985) and Snell (1987) are extensive, but still a number of fundamental questions concerning the nature of these outflows remain to be answered. Some of these are:

- What is the evolutionary status of the YSOs responsible for molecular outflows ?
- What is the frequency of outflow events and what type (mass) stars produce outflows ?
- What is the structure and kinematics of the outflowing molecular gas ?
- What impact do outflows have on stellar evolution and on the evolution of the surrounding molecular cloud ?
- What is the nature of the stellar wind that drives the molecular outflows ?
- How are these stellar winds generated and collimated ?

A summary of recent studies that have attempted to answer many of these questions will be presented in the following sections. Three topics will be emphasized: 1.) outflow surveys, 2.) the structure of molecular outflows, and 3.) the nature of the stellar winds.

III. Outflow Surveys

Past searches for outflows were neither systematic nor statistially complete, owing primarily to the lack of a good census of YSOs in molecular clouds. While past surveys either relied on optical catalogs, or incomplete infrared surveys to identify YSOs, the data available from the Infrared Astronomical Satellite (IRAS) have now allowed for a more complete census. In addition, totally unbiased searches for outflows have also been undertaken in a number of molecular clouds; these allow for the detection of outflows associated with YSOs below the IRAS detection threshold or in confused regions of the sky in the far–infrared. Together, such surveys have provided a more definitive determination of the frequency of energetic outflows and the evolutionary status of the young stars responsible for them.

a.) *Outflows Associated with Luminous FIR Sources*

A flux limited sample of IRAS sources has been studied by Snell *et al.* (1988) and Snell Dickman, and Huang (1989). Their sample consisted of all 100 μm sources with flux densities > 500 Jy found in the range $0^h < \alpha < 12^h$ and $\delta > 0°$. Based on both their far–infrared colors and their close association with strong CO emission, all members of this sample are believed to be relatively luminous (10^4 $L_\odot$) YSOs still deeply embedded in the clouds from which they formed. Among the 51 sources in the sample, 12 were already known to have associated molecular outflows (Lada 1985). The 39 remaining sources were surveyed and 11 new outflows detected. Thus, the detection rate of outflows associated with luminous YSOs is nearly 50 %, similar to detection rate of outflow candidates by Casoli *et al.* (1986).

The high incidence of outflows in these surveys can be used to set an upper limit on the lifetime of objects as bright far–infrared sources. If these sources lived considerably longer than the duration of the outflow phase, then one would expect to have a very low rate of outflow detection and based on the 50% detection rate, the maximum lifetime of a FIR source is roughly twice the outflow lifetime. The average age of the 11 outflows studied is estimated to be $2x10^5$ years; therefore, the material surrounding these young O and B stars must be dispersed within about $4x10^5$ years. The cloud dispersal and outflow phenomena must be concurrent and begin shortly after a young star begins to generate any substantial luminosity.

b.) *Far–infrared Sources with $L < 100\ L_\odot$*

A number of searches have also been undertaken to identify outflows associated with low luminosity far–infrared sources. A flux limited sample of IRAS sources in the Taurus Dark Clouds was surveyed for outflows by Heyer et al. (1987). Their detection rate of 3/30 (outflows detected/sources observed) was relatively low, but most of the sources observed were optically visible stars. In fact, their detection rate was similar to an earlier outflow survey of an optical sample of T Tauri stars (Edwards and Snell 1982). Since low luminosity, and presumably low mass, stars have long pre–main sequence lifetimes and because the outflow phenomenon is short lived, it is not surprising that the detection rate is very low.

Studies of outflows associated with IRAS sources within NH_3 cores have been completed by Myers *et al.* (1988) and Terebey, Vogel, and Myers (1989). These sources are thought to be the youngest of the YSOs because they are still associated with the dense gas out of which they were formed. For IRAS sources embedded within NH_3 cores they found an outflow detection rate of 16/30. Five additional sources showed broad line widths, but were not mapped to confirm their outflow nature; thus, as many as 21/30 IRAS sources in cores could be associated with outflows. On the other hand, for IRAS sources outside of cores the detection rate was 0/8, and for cores without stars the detection rate was 0/4. Thus, outflow activity is apparently associated with only the youngest infrared sources. Based again on the outflows' short duration, the outflow phenomenon must begin early in the evolution of young, low mass stars, probably within the first 10^5 years. Myers *et al.* (1988) also found that outflows have sufficient energy, duration, and frequency to be the main agent of core dispersal in molecular clouds. Thus, after the outflows have removed the surrounding dense and visually opaque material, the stars still have long pre–main sequence lifetimes as optically visible T Tauri stars.

c.) *Unbiased Surveys*

It is necessary to have a complete census of outflows within a molecular cloud in order to assess their impact on cloud evolution. Since the IRAS Point Source Catalog may be incomplete and since outflow activity may begin before appreciable radiant energy is generated, totally unbiased surveys may be needed to accurately obtain a census of outflows. The first survey of this kind was made by Margulis and Lada (1986) in the Mon OB1 molecular cloud. They surveyed roughly 1.5 square degrees encompassing a cloud mass of 3×10^4 $M_\odot$. Additional studies by Margulis, Lada, and Snell (1988) and Margulis, Lada, and Young (1989) provided detailed maps of the outflows and attempted to identify the YSOs responsible. Nine molecular outflows were detected, associated with the youngest and most luminous IRAS sources within the cloud. The energy and momentum injected into the molecular cloud are also apparently sufficient to maintain the turbulent support of the clouds against gravity, even if the turbulent energy is dissipated on a free–fall timescale.

One of the most energetic outflows identified in the Mon OB1 survey, that labeled G, does not have any obvious optical or IRAS emission associated with it and, thus, must have a relatively low bolometric luminosity. This emphasizes that unbiased surveys may be essential in order to obtain a complete inventory of outflows within molecular clouds.

A more ambitious outflow search by Fukui and collaborators (Fukui *et al.* 1986; Fukui 1989) is in progress on the Nogoya 4 m telescope. They have completed an unbiased survey of four clouds (Orion N, Orion S, Mon R2, and S287) and have begun similar surveys in a number of other clouds. They have detected 50 new outflows, 41 of which lie within 1 kpc of the sun. In the clouds that have been completely sampled, outflows are found associated with the coldest (based on 12 and 25 μm flux densities) and most luminous YSOs in these clouds. In the Orion S cloud a total of 12 outflows have been identified and these may well contribute substantially to the turbulent support of this cloud.

d.) *Summary*

The number of known outflows is increasing steadily as more molecular clouds in the solar vicinity are surveyed. The number of outflows known to lie within 1 kpc of the sun is > 100; their average lifetime is approximately 3×10^4 years. Thus, the formation rate of outflow progenitor stars is $> 10^{-3}$ yr^{-1} kpc^{-2}. Comparing this rate of formation to that of field stars in the solar neighborhood (Miller and Scalo 1979), implies that all stars greater than 1 $M_\odot$ must produce *detectable* outflows in their lifetime. This not only indicates that both high and low mass stars produce molecular outflows, but that basically *all* stars must pass through an evolutionary phase in which they produce energetic winds. This phase for both high and low mass stars must occur within about 10^5 years after a source begins generating any substantial luminosity. In addition, outflows may be the primary agent to disperse the dense gas surrounding these stars. Finally, the momentum carried by the dispersed gas may play an important role in maintaining the turbulent support of clouds; this turbulence may be the origin of the supersonic gas motions in clouds.

IV. Structure of Molecular Outflows

The structure and kinematics of outflowing molecular gas provide important clues to its energetics and origin. A summary of recent observations of three molecular outflows, L1551, B335, and NGC2071 is presented below.

L1551 is a nearby molecular cloud with a well collimated outflow of large angular size (30'). Recent studies by Moriarty–Schieven *et al.* (1987), Uchida *et al.* (1987), Rainey *et al.* (1987), Moriarty–Schieven and Snell (1988), and Fridlund *et al.* (1989) have provided high resolution images of the CO emission. One of the most important conclusions reached by these investigators, primarily from the limb–brightened appearence of the molecular emission, is that the high velocity molecular emission arises from an expanding shell of material. The high velocity blueshifted and redshifted emissions are shown in Figure 2. The blueshifted lobe shows clearly that the largest column density of high velocity gas is found at the periphery of the

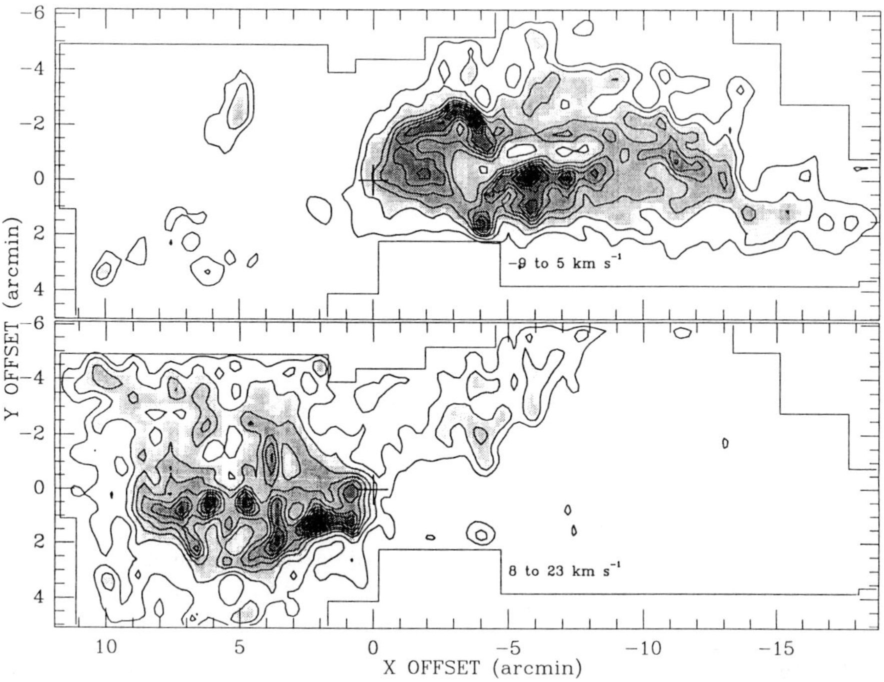

Figure 2. – A map of the high velocity blueshifted (upper) and redshifted (lower) CO emissions in the L1551 outflow obtained by Moriarty–Schieven and Snell (1988). The x–axis is parallel to the outflow axis (at a position angle of 45°, see Figure 2) and the origin of the x–y coordinate system is the position of L1551 IRS–5 (maked by a cross).

outflow. Further evidence for a shell structure is provided by the detection of optical reflection nebulosity coincident with the regions of high column density in the blueshifted lobe. Presumably, light from IRS–5 is able to propagate across the nearly evacuated cavity created by the stellar wind and reflect off the walls of swept–up molecular gas. Expansion of this shell perpendicular to the outflow axis can be seen in the spatial–velocity diagram shown in Figure 3. The diagram shows the ambient cloud emission at 6.5 km s^{-1} as well as the high velocity blueshifted emission; the circular pattern is indicative of expansion with a velocity of about 3 km s^{-1}. The total mass in the molecular outflow is estimated to be 3.5 $M_{\odot}$. Both the large mass and the morphology argue strongly for the high velocity molecular gas to be primarily accelerated ambient cloud material.

Extended far–infrared emission has also been detected from the L1551 outflow (Clark and Laureijs 1986; Edwards *et al.* 1986). The total luminosity is estimated to be 7 $L_{\odot}$, 35 times larger than the mechanical luminosity in the molecular outflow, but still significantly less than the bolometric luminosity of IRS–5. The extended infrared emission is unlikely to be due to

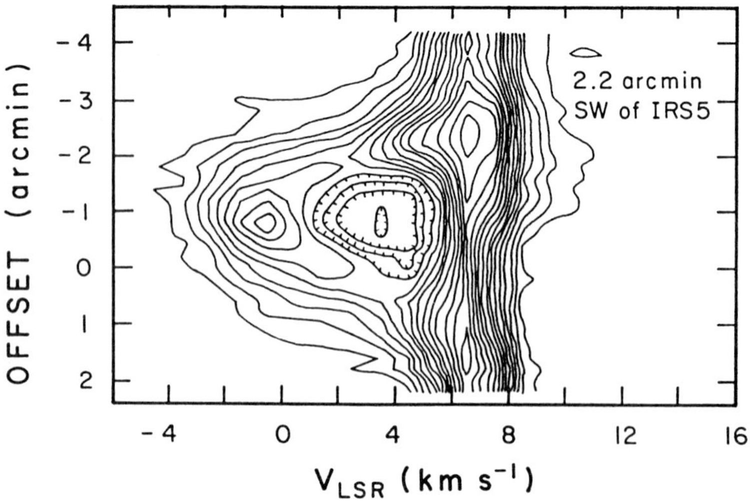

Figure 3. – A spatial–velocity diagram of the CO emission in the blueshifted lobe of the L1551 outflow (Moriarty–Schieven *et al.* 1987). The spatial axis is oriented perpendicular to the outflow axis and located 2.2' (X offset = –2.2) from IRS–5.

dust heated by stellar photons and probably arises from the stellar wind from IRS–5. If most of the stellar wind energy is radiated away at the wind shock located at the boundary with the swept–up molecular shell, this could account for the observed far–infrared luminosity.

The molecular outflow in B335 is also relatively nearby (250 pc). Several recent papers (Cabrit, Goldsmith, and Snell 1988; Hirano *et al.* 1988; Moriarty–Schieven and Snell 1989) have presented high resolution images of the high velocity CO emission. The data show that, like L1551, the high velocity emission appears to be confined to a shell. One of the most peculiar aspects of the B335 outflow is the multiple patterns of superimposed blueshifted and redshifted emissions. These patterns have been attributed to the presence of two independent outflows, but the new data demonstrate that B335 is in fact a single, highly collimated outflow, oriented nearly in the plane of the sky. The superimposed blueshifted and redshifted emissions arise from the front and back sides of the expanding shell. The angle of inclination out of the plane of the sky is estimated to be between 8 and 10°. Thus, the B335 outflow is very similar to that in L1551, except viewed at a smaller inclination angle.

NGC2071 lies at a distance of 390 pc and like L1551 and B335, its outflow shows a very collimated structure. Moriarty–Schieven, Snell, and Hughes (1989) have recently obtained a high resolution map of CO emission from the J=2–1 transition. The outflow as a shell morphology similar to L1551, but is much more inclined out of the plane of the sky. The similarity to the L1551 outflow is quite striking and many of the observed differences between the two outflows can be attributed to the larger inclination angle at which the NGC2071 outflow is viewed.

Generally, high velocity molecular emission appears to arise primarily from an expanding shell driven by a stellar wind. Many of the observed differences in molecular outflows can be attributed to the inclination angle at which they are viewed. Small inclination angle outflows (such as B335 and L1551) have low velocity emission only, are highly collimated, have little overlap of their outflow lobes, and have a readily detectable expansion component. Intermediate inclination angle outflows (such as NGC2071 and HH7–11) have moderate velocity emission, are bipolar but not well collimated, have lobes that overlap, and have their observed kinematics dominated by their axial motion. Large inclination angle outflows (possibly AFGL490) have high velocity emission, are poorly collimated, have total overlap of their lobes, and have their expansion directed perpendicular to line of sight.

V. The Stellar Wind

The basic morphology of molecular outflows is illustrated in Figure 5. The most plausible origin of the high velocity molecular gas is ambient cloud material accelerated by a much faster

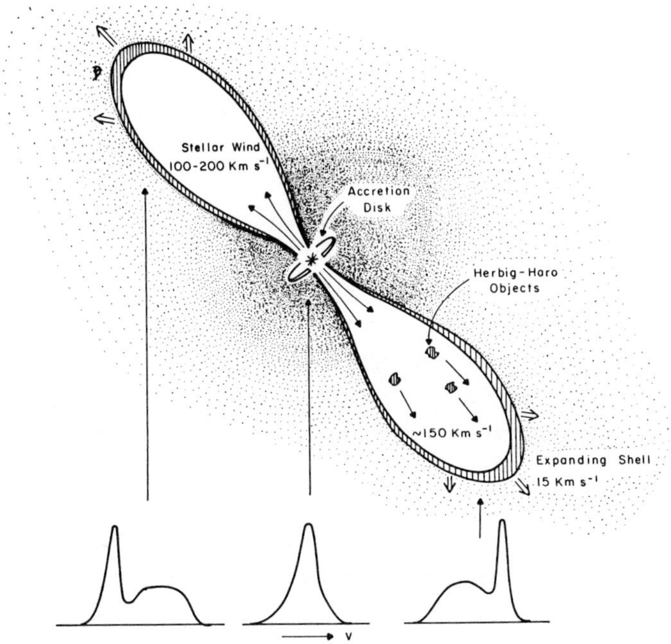

Figure 5 – A schematic diagram showing the morphology of the stellar wind driven molecular shell and the location of the shock–excited Herbig–Haro objects.

stellar wind. What is the nature of this stellar wind ? Rodriguez and Canto (1983), Snell *et al.* (1985), and Snell and Bally (1986) have searched for the radio continuum emission that would be produced if the stellar winds were ionized. In the cases where free–free emission was detected the mass loss rate of ionized gas was not sufficient to drive the molecular outflows. Levreault (1985) and Strom *et al.* (1986) found similar results in their studies of optical jets. Thus, although collimated jets are seen at radio and optical wavelengths, the ionized gas responsible for the emission must represent a small fraction of the total mass present in the stellar wind.

Recently, extremely high–velocity atomic hydrogen emission has been detected from HH7–11 by Lizano *et al.* (1988). The emission implies wind velocities as large as 170 km s^{-1} and indicates a mass loss rate of $3x10^{-6}$ $M_{\odot}$ yr^{-1}, sufficient to drive the associated molecular outflow. Thus, in this one outflow the stellar wind is clearly neutral and atomic. In L1551, the HI observations made by Lizano *et al.* are more ambiguous but there is also evidence for high velocity HI emission that suggests the possibility for a neutral stellar wind with a line of sight velocity of 50 km s^{-1} (190 km s^{-1} after applying an inclination correction). The mass loss rate in this stellar wind may also be sufficient to drive the more massive molecular outflow and account for the excess far–infrared emission seen in this source.

An equally remarkable discovery made by Lizano *et al.* was the detection of extremely high velocity *molecular* emission in the HH7–11 outflow. The CO emission has a velocity extent of ± 160 km s^{-1}, considerably larger than the velocities seen in Figure 1b. Lizano *et al.* suggested that the CO emission arises from molecular gas entrained in a neutral stellar wind. Alternatively, chemical modeling of neutral winds (Glassgold, Mamon, and Huggins 1988) suggests that it is possible to form an appreciable amount of CO from the atomic gas. The abundance of CO relative to H in the stellar wind is strongly dependent on the mass loss rate of the driving source, but for large mass loss rates ($> 10^{-5}$ $M_{\odot}$ yr^{-1}) the abundance of CO approaches that found in the ambient molecular material.

Observations to detect extremely high velocity CO emission from stellar winds have been carried out by Koo (1989) and by Margulis and Snell (1989). They detected weak, but extremely high velocity, emission in a number of sources known to have molecular outflows. An example is shown in Figure 5 for NGC2071 where emission is detected over a velocity range $>$ 110 km s^{-1}. In fact, the high velocity emission appears to be readily separated into contributions from the molecular shell and the stellar wind. The break in the power law shape of the line wings, indicated in Figure 5, may indicate the division of the gas into these two components; emission at temperatures below about 0.1 K may arise primarily from molecular gas in a stellar wind. Extremely high velocity emission is detected only in those outflows with mass loss rates in excess of 10^{-6} $M_{\odot}$ yr^{-1}. Thus, these observations suggest that significant stellar winds exist and that these winds are primarily neutral and atomic. Such winds may be sufficient to drive even the more massive known molecular outflows.

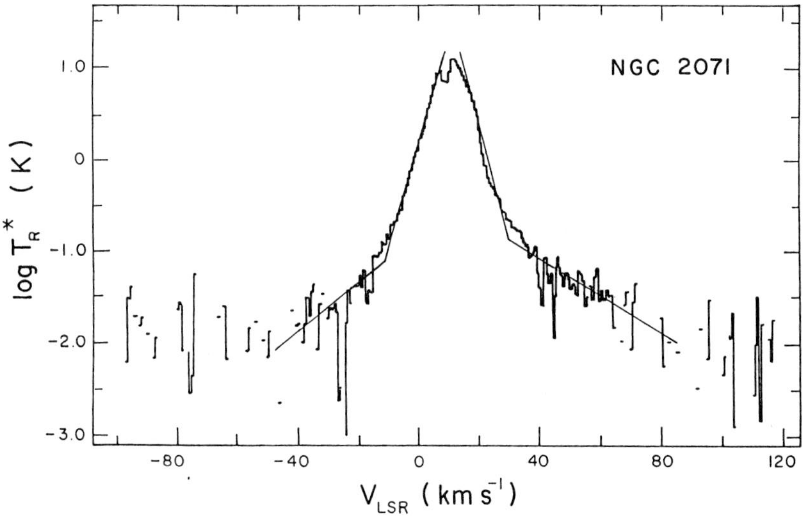

Figure 5. – Logarithm of the spectrum of NGC2071 from Margulis and Snell (1989). Data points in which the antenna temperature dropped below the baseline are not plotted.

VI. Conclusions

Outflows are evidently a common evolutionary phase for stars of all masses. The outflow phenomenon occurs within the first 10^5 years of a YSO's life and is likely responsible for dispersing the dense gas surrounding it. The bulk of the high velocity molecular emission in an outflow arises from ambient cloud material swept up into an expanding shell by a faster, and more energetic stellar wind. This wind is largely neutral, atomic, and highly collimated into opposed jets. Finally, the energy and momentum carried away by the wind may have important consequences for both the star formation process and for the turbulent support of molecular clouds against gravity.

References

Cabrit, S., Goldsmith, P.F., and Snell, R.L. 1988, *Ap. J.*, **334**, 196.
Casoli, F., Dupraz, C., Gerin, M., Combes, F., and Boulanger, F. 1986, *Astron. Astrophy.*, **169**, 281.
Clark, F.O., and Laureijs, R.J. 1986, *Astron. Astrophys.*, **154**, L26.
Edwards, S., and Snell, R.L. 1982, *Ap. J.*, **261**, 151.
Edwards, S., Strom, S.E., Snell, R.L., Jarrett, T.H., Beichman, C.A., and Strom, K.M. 1986, *Ap. J. (Letters)*, **307**, L65.
Fridlund, C.V.M., Sandqvist, Aa., Nordh, H.L., and Olofsson, G. 1989, *Astron. Astrophys.*, **213**, 310.
Fukui, Y. 1989, private communication.

Fukui, Y., Sugitani, H., Takaba, H., Iwata, T., Mizuno, A., Ogawa, H., and Kawabata, K 1986, *Ap. J. (Letters)*, **311**, L85.
Glassgold, A.E., Mamon, G.A., and Huggins, P.J. 1989, *Ap. J. (Letters)*, **336**, L29.
Heyer, M.H., Snell, R.L., Goldsmith, P.F., and Myers, P.C. 1987, *Ap. J.*, **321**, 370.
Hirano, N., Kameya, O., Nakayama, M., and Takakubo, K. 1988, *Ap. J. (Letters)*, **327**, L69.
Koo, B.–C. 1989, *Ap. J.*, **337**, 318.
Lada, C.J. 1985, *Ann. Rev. Astron. Astrophys.*, **23**, 267.
Levreault, R.M. 1985, Ph.D. thesis, University of Texas.
Lizano, S., Heiles, C., Rodriguez, L.F., Koo, B.–C., Shu, F.H., Hasegawa, T., Hayashi, S., and Mirabel, I.F. 1988, *Ap. J.*, **328**, 763.
Margulis, M., and Lada, C.J. 1986, *Ap. J. (Letters)*, **309**, L87.
Margulis, M., Lada, C.J., and Snell, R.L. 1988, *Ap. J.*, **333**, 316.
Margulis, M., Lada, C.J., and Young, E.T. 1989, preprint.
Margulis, M., and Snell, R.L. 1989, *Ap. J.*, in press.
Miller, G.E., and Scalo, J.M. 1979, *Ap. J. Suppl.*, **41**, 513.
Moriarty–Schieven, G.H., Snell, R.L., Strom, S.E., Schloerb, F.P., Strom, K.M., and Grasdalen, G.L. 1987, *Ap. J.*, **319**, 742.
Moriarty–Schieven, G.H., and Snell, R.L. 1988, *Ap. J.*, **332**, 364.
Moriarty–Schieven, G.H., and Snell, R.L. 1989, *Ap. J.*, **338**, 952.
Moriarty–Schieven, G.H., Snell, R.L., and Hughes, V. 1989, *Ap. J.*, in press.
Mundt, R. 1988, in NATO–ASI on *Formation and Evolution of Low Mass Stars*, eds. A. Dupree and M.T.V. Lago (Dordrecht: Reidel).
Myers, P.C., Heyer, M., Snell, R.L., and Goldsmith, P.F. 1988, *Ap. J.*, **324**, 907.
Rainey, R., White, G.J., Richaredson, K.J., Griffen, M.J., Cronin, N.J., Monterio, T.S., and Hilton, J. 1987, *Astro. Astrophys.*, **179**, 237.
Rodriguez, L.F., and Canto, J. 1983, *Rev. Mexicana Astr. Ap.*, **8**, 163.
Snell, R.L. 1987, in *IAU Symposium 115, Star Forming Regions*, eds. M. Peimbert and J. Jugaku (Dordrecht: Reidel) p. 213.
Snell, R.L., and Edwards, S. 1981, *Ap. J.*, **251**, 103.
Snell, R.L., Loren, R.B., and Plambeck, R.L. 1980, *Ap. J. (Letters)*, **239**, L17.
Snell, R L., Bally, J., Strom, S.E., and Strom, K.M. 1985, *Ap. J.*, **290**, 587.
Snell, R.L., and Bally, J. 1986, *Ap. J.*, **303**, 683.
Snell, R.L., Huang, Y.–L., Dickman, R.L., and Claussen, M.J. 1988, *Ap. J.*, **325**, 853.
Snell, R.L., Dickman, R.L., and Huang, Y.–L. 1989, in preparation.
Strom, K.M., Strom, S.E., Wolff, S.C., Morgan, J., and Wenz, M. 1986, *Ap. J. Suppl.*, **62**, 39.
Terebey, S., Vogel, S.N., and Myers, P.C. 1989, *Ap. J.*, **340**, 472.
Uchida, Y., Kaifu, N., Shibata, K., Hayashi, S.S., and Hasagawa, T. 1987, in *IAU Symposium 115, Star Forming Regions*, eds. M. Peimbert and J. Juku (Dordrecht: Reidel), p. 287.

Discussion:

FRANCO: I have a question and a comment. There is a similar analysis recently made by Russ Levreault, but he reached the opposite conclusion (that outflows are not suitable to support the cloud). Can you clarify the reasons for this apparent contradiction? The comment, on the other hand, is that the results presented by G. Tenorio-Tagle and myself show that both photoionization and photodissociation fronts are capable of creating high-velocity neutral outflows. This mechanism can provide the momentum required to drive the molecular outflows.

SNELL: How important molecular outflows are in supporting clouds is certainly an open question. They must provide some support, the question is if they are the dominant source of support. The best way to address this question is through unbiased surveys of molecular clouds for outflows to provide a complete census. Russ Levreault has not done such a survey and thus does not have a coupled census of outflows in the clouds he studied. Complete surveys of clouds coupled with accurate measurements of their energy and momentum are necessary to answer this question.

SOLOMON: The outflow lifetime is critical to the question of cloud support. How accuratly is the lifetime determined and what are the assumptions?

SNELL: The outflow lifetimes are very poorly known, since they require a detailed knowledge of the structure of kinematics of the molecular gas as well as the inclination angle of the outflow. Lifetimes are usually computed from the projected size of the outflow divided by the radial velocity (often a mass weighted radial velocity is usual). It is important to find more accurate means to derive lifetimes.

LARSON: You mentioned the possible role of outflows in generating turbulence that helps to support molecular clouds. Is there evidence that clouds containing outflows have been stirred up or otherwise influenced by the outflows in regions that are not obviously part of an outflow?

SNELL: I believe there is such evidence, but it is difficult to prove it is due to molecular outflows. In the Taurus dark clouds there are several old outflows that can still be connected to their stellar source; these outflows are decelerating and have relatively large masses and low velocities. The spectra of these old outflows have, instead of high velocity wings, two components; one componet due to the ambient gas and a second one due to the swept-up massive shell. Much of the multiple velocity structure seen in Taurus and other clouds may well be due to these decelerating shells.

Submillimetre Mapping and Photometry of Bipolar Flows - Evidence for Compact Disks

Göran Sandell
Joint Astronomy Centre
665 Komohana Street, Hilo, HI 96720, USA

ABSTRACT: The mm/sub-mm dust emission from bipolar outflow sources is discussed. In nearby outflows the stars driving the flows appear point like and unresolved, i.e. apparent angular sizes $\lesssim$ 7" (i.e. 700AU at 100pc). The spectra are surprisingly flat, suggesting that the dust emission is affected by temperature and density gradients or that the dust surrounding these stars differs from that of the general interstellar medium. If the spectra are characterized with a conventional dust emissivity index, our sample gives an average value of 1.1, rather than the figure of 2, which is usually claimed at these wavelengths for dust in molecular clouds. The total dust and gas masses associated with these stars are a few tenths of a solar mass for nearby low luminosity sources and the gas densities are $>10^7$ cm^{-3}, if the gas is in a shell, and the densities could be much higher if the gas densities fall off as a function of radius, or if the mass is distributed in a disk. These masses correspond to extremely high extinctions, which are not observed, and we therefore argue that the mass distribution must be anisotropic, most likely as an inclined disk surrounding the star. Hence, stars driving outflows are surrounded by a dense disk, possibly an accretion disks and/or collimating disk. If these stars are compared to the more evolved T Tauri stars, we find that the difference in disk mass could account for the mass loss observed in these stars. Because of the high gas densities in the disk, most commonly used molecular lines will be optically thick, and molecular observations should be done in rare isotopes and high density tracers in order to yield reliable results of gas densities and gas dynamics.

The unresolved continuum emission is often found to be surrounded by weaker extended emission, which is typically orthogonal to the outflow direction. There may also be extended emission in the the direction of the outflow, either due to hot dust in the beginning of the flow (jet) or dust compressed by the flow.

INTRODUCTION

During the last two years we have been studying bipolar outflow sources and PMS-stars in the mm/sub-mm continuum using the 15m James Clerk Maxwell Telescope (JCMT) on Mauna Kea, Hawaii. The results from these observations are now starting to appear: SSV13/HH7-11: Sandell et al., 1989a; NGC2071IR, LkHα234 and G35.2: Dent et al., 1989; B335: Chandler et al., 1989, although results on the spectacular Rho Oph outflow 16293-2422 and on high luminosity outflows are still in the process of being written up (Sandell et al., 1989b,c). Results on visible PMS stars (T Tauri, FU Ori and Herbig Ae/Be stars) which are more evolved than the deeply embedded bipolar outflow sources, but for which mass loss still occur, are reported on by Weintraub et al. (1989) and Sandell (1989a).

This talk is a first attempt to summarize the general results emerging from these studies and the same time we present some new results from recent observations. The mm/sub-mm emission provides an extremely powerful tool to study the properties of the dust surrounding bipolar outflows. With the current generation of new sub-mm telescopes we can reach much higher spatial resolution than KAO and IRAS, and even though the sub-mm sky at Mauna Kea is at best semitransparent, the large collection area of the JCMT, we can also go fainter and detect sources not seen by IRAS. By mapping we can study the outflow morphology and from multi-wavelength observations we can derive temperatures, densities, masses and the properties of the dust grains in the immediate vicinity of the exciting stars.

RESULTS AND DISCUSSION.

Data obtained from detailed high spatial resolution mm/sub-mm mapping are still rather limited, but one can already see some clear trends. In all nearby low-luminosity sources that have been observed, the star driving the outflow appears point like and unresolved, although there is normally also some extended emission present (Fig. 1). The initial sample was chosen with the aims of searching for disks around the outflows using the following criteria: 1) the bipolar outflow should lie in or close to the plane of the sky, 2) the source should be nearby, 3) there should be some evidence for the presence of a disk, either from molecular line observations or from near-IR imaging and polarization and 4) the star driving the flow should be of low luminosity. The first three criteria were made in order to maximize the detection rate of disks. If the outflow is in the plane of the sky, the disk will be viewed edge on, thus giving the maximum optical depth along the line of sight. It will enhance the likelihood of seeing the disk and perhaps resolving it. Restricting the observations to low luminosity sources makes the heating of the dust from the star much more localized, and the continuum emission will therefore be a more direct measure of the distribution of dust column density.

Although the initial results were quite promising - we found disk like structures in every source we observed (Sandell et al., 1988; Dent et al., 1989), we now believe that these extended orthogonal structures surrounding the outflows have nothing to do with the collimating disks that we were looking for. The fact that these extended structures appear to be perpendicular to the outflow direction suggests that they are remnants from the initial star formation phase, but we cannot see that they would have any influence on the current outflow phase.

What is significant, however, is the unresolved point-like source coinciding with the star driving the flow. Mm/sub-mm photometry of the central sources in the wavelength regime 2mm - 800μm, supplemented with 3mm aperture synthesis continuum data, (when available), suggest that the dust emissivity is close to 1 (Table 1). This is very similar to the results obtained by Weintraub et al. (1989) for T Tauri and FU Orionis stars and to Owens Valley aperture synthesis data on embedded PMS objects (Woody et al., 1989). It should be noted that the dust emissivity (β-index) derived by us is computed using standard assumptions, i.e. assuming single temperature (50 K, except for L723 and B335, for which we adopt T_d = 25K), homogeneous, optically thin dust emission, where the wavelength dependence of the dust emissivity can be approximated as a power law. As pointed out by Weintraub et al. (1989) effects like geometry, temperature or density gradients, can lower the apparent β-index, while if the grains are large compared to the wavelength, the power law assumption is no longer valid, and the spectrum starts to mimick a black body spectrum (β = 0). Model calculations by Weintraub and Sandell (1989) show that temperature and density gradients alone can explain the hitherto observed mm/sub-mm spectra, although the low apparent β-indexes observed in our sample may also be real (the grain population of dust in the extremely dense surroundings of a young star is expected to differ from dust grains in molecular and dark clouds). However, whatever the cause is, it is rather fortunate, because the less steeply falling spectrum makes the source easier to detect. It should be noted that so far we have a detection rate of a 100%, i.e. stars with outflows are easy to detect in the sub-mm continuum.

Table 1 summarizes the results of our mm/sub-mm obsevations on JCMT. It includes all low luminosity sources that have been observed by us, but gives only a few examples of higher luminosity outflow sources. Some more data can be found in Sandell (1989a). The masses quoted in Table 1 are total masses obtained by assuming a gas to dust mass ratio of 100. The masses are derived using the formula given in Woody et al. (1989) using the assumptions given above and by taking an average of all observed bands. We feel that his is justified, because β cannot be less than 1 (see e.g. Emerson, 1988). We will need to run model calculations in order to obtain more reliable estimates. The densities (n) are calculated by assuming that all the mass is in a spherically homogeneous dust shell, and the visual extinction (A_V) is the extinction that such a dust shell would produce along the line of sight to the star. Both these parameters are lower limits, because the dust will not survive close to the star, and the linear size of the dust shell is in almost all cases an observed upper limit. Even when the star appears resolved, the size may be overestimated due to contamination from fainter extended emission. The star may also appear to be resolved because it is a binary system, which seems to be the case for 16293-2422 (Wootten, 1989). If the mass was distributed in a disk, both the density and the extinction may be severely underestimated. Even a very modest homogenous disk, with an aspect ratio of 3:1, would already increase the density by a factor of two.

The mean value of β from the sample in Table 1 is 1.1, if we exclude the data on HH1/HH2, which may be in error (single observation, only 1.1mm and 800μm). These values are less extreme than those on T Tauri stars (Weintraub et al., 1989), but nevertheless significantly different from 2. In the case of SSV13, the 1.1mm and 800μm mapping by Sandell et al. (1989a) indicates that the β-index of the extended emission surrounding SSV13 is close to 2. This suggest that if the outflow sources are associated with extended emission, we may not be able to derive the true properties of the disk emission using single dish data.

Another striking fact in Table 1 is the high visual extinctions in the line of sight to the stars, which in most cases exceed observed extinctions by more than an order of magnitude, c.f. SSV13: 250^m vs. 15^m; L1551IRS5: 340^m vs. 19^m. If we could put more stringent size constraints on the sub-mm emission, the discrepancies would probably be even higher. We therefore argue that the sub-mm emission cannot originate from a shell, it must be an anistropic distribution, presumably some kind of a disk. This would already be expected from the strong molecular outflows that are seen, because these are driven by the star. At this stage we cannot really predict what the disks look like. We only know their approximate mass and we have a limit to their overall size ($\lesssim$ 2000 AU) for nearby stars. What is crucially needed is to learn about their detailed morphology (thin or thick disk) and dynamics. These are data which must come from mm-aperture synthesis

Table 1: Physical properties of outflow sources observed in continuum on JCMT with the common user bolometer UKT14 (for details about UKT14; see Duncan et al., 1989). Luminosities and distances are taken from Mozurkewich et al. (1986), or occasionally from other sources.

Outflow	Size ["]	d [pc]	L [$L_\odot$]	β-index	M_{tot} [$M_\odot$]	n [cm^{-3}]	A_V [mag]	Wavelength[a]
Low luminosity sources:								
SSV13[b]	≤4x4	350	58	0.7±0.2	0.25±0.05	$2x10^7$	250	3.06/2.73/1.4
HH6VLA	<7	350	18	1.4±0.3	0.28±0.02	$5x10^6$	90	1.1/.79
L1551IRS5	6x8	160	38	1.2±0.3	0.11±0.01	$5x10^7$	340	1.1/.79
16293-2422	11x<5	160	27	1.2±0.1	0.64±0.04	$1x10^8$	910	2/1.3/1.1/.79
L723	<7	300	3	1.0±0.2	0.24±0.02	$7x10^6$	110	1.3/1.1/.79
B335	<6x7	250	1.7	1.1±0.2	0.26±0.05	$1x10^7$	190	1.3/1.1/.79
HH1/HH2VLA	<10[c]	460	50	1.8±0.4	0.47±0.13	$2x10^6$	60	1.1/.79
PVCep	<7	500	80	1.4±0.4	0.32±0.07	$2x10^6$	50	1.1/.79/.45/.35
Intermediate luminosity sources:								
NGC7129FIR	8x10	1000	230	1.2±0.3	4.2 ± 0.1	$2x10^6$	110	1.1/.79
LkHα234	<7x9	1000	10^3	1.5±0.5	3.6 ± 0.3	$2x10^6$	120	1.1/.79
High luminosity sources:								
M8E	15x17	1500	$2x10^4$	0.8±0.3	31 ± 6	$6x10^5$	110	2.0/1.1/.79
NGC7538B[e]	19x21	2800	$2x10^5$	1.1±0.2	215 ± 5	$3x10^5$	140	2.0/1.1/.79

[a] This column gives the center wavelength for the filters used to derive β-indexes and masses.
[b] Mass and β-index derived from data in Grossman et al. (1987) and Woody et al. (1989), see Sandell et al. (1989a)
[c] Source size uncertain
[d] β = 1.2 if 450 and 350μm data are omitted.
[e] Fluxes corrected for free-free emission

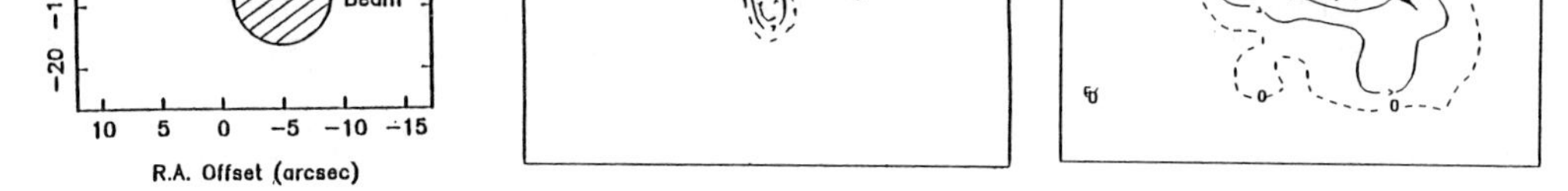

Figure 1. a. 450μm map of B335 (from Chandler et al., 1989). **b.** 800μm map of L1551IRS5. c. 450μm map of L1551IRS5. The two 450μm maps have a resolution of 8.5", while the resolution of the 800μm map is 13.5". Note that in both cases one can see emission perpendicular to the outflow (marked by arrow), as well as in the flow direction.

telescopes. Sargent (1989) presented data on outflow sources, which suggest that the gas appears to move in Keplerian orbits around the stars, but the data base is very limited. She also noted that one has to observe these disks in ^{13}CO or $C^{18}O$, because the gas is optically thick. This is very clear from the gas densities derived in Table 1. Since the gas densities could well be in excess of 10^{10} cm^{-3} while the gas is quite cold, the molecules could be frozen onto grains, and the molecular abundances severely depleted. The ideal molecular tracer should have low excitation transitions and very high dipole moment, in order to trace adequately cold, very high density gas.

The deduced masses of the disks in the low luminosity sample are very similar. This may partly be due to the way we have computed the masses, because we chose to neglect differences in the observed β-index, but to a large extent the results are probably real. The sample is quite homogeneous, and most of the outflows are of about the same age. If these masses are compared to the disk masses seen in T Tauri stars (Weintraub et al., 1989), which are presumably much more evolved, but similar objects, one can note that the T Tauri disk are about a few x .01 $M_\odot$, i.e. about 10 times smaller than the embedded stars discussed here. If one crudely assumes that T Tauri stars are ~ 10^6 -10^7 years older, this would correspond to a mass loss rate of 10^{-7} - 10^{-8} $M_\odot$ $years^{-1}$, if all of the disk mass went into the flow. These mass loss rates appear plausible for T Tauri stars, but are definitely too low for the deeply embedded outflows. However, it is also likely that part of the disk mass will fall onto the star and serve to increase its mass.

As mentioned earlier, bipolar outflow stars are also associated with fainter extended emission. Some of this emission is perpendicular to the flow, but there is also emission in the flow direction. This is illustrated in Figure 1, where we show high resolution 450μm images of B335 and L1551IRS5. In both cases there is evidence for emission perpendicular to the flow, which is more clearly seen in the larger 800μm images (see e.g. Chandler et al., 1989 for an 800μm map of B335). The L1551 800μm map is of poor S/N but appears to confirm the presence of more extended N-S emission. Other outflows that reveal clear extended emission are SSV13, 16293-2422, and LkHα234 (Sandell et al., 1989a,b; Dent et al., 1989). Some of the emission is clearly associated with the large "interstellar disks" seen in ammonia and carbon monosulfide, while for example B335, L1551 and SSV13 also show emission in the outflow direction. This could be due to dust compressed by the outflow along the cavity walls , like in SSV13, or it could be hot dust associated with the beginning of the flow. In order to determine the properties of this relatively faint dust emission one needs deep, well calibrated maps at three wavelengths, which have not yet, as yet, been obtained.

Acknowledgements: The James Clerk Maxwell Telescope is operated on a joint basis between the United Kingdom Science and Engineering Research Council (SERC), the Netherlands Organisation for the Advancement of Pure Research (ZWO), the Canadian National Research Council (NRC), and the University of Hawaii (UH). I want to thank not only all my collaborators in this project: W.D. Duncan, S. Hayashi, C. Aspin, P.A.R. Ade, M.J. Griffin, W.K. Gear, C.J. Chandler, W.R.F. Dent, and E.I. Robson, but also the whole staff at JCMT.

References

Chandler, C.J., Gear, W.K., Sandell, G. Hayashi, S., Duncan, W.D., Griffin, M.J., Hazell, A.S.: 1989, MNRAS, submitted

Dent, W.R.F., Sandell, G., Duncan, W.D., Robson, E.I.: 1989, MNRAS, in press

Duncan, W.D., Robson, E.I., Ade, P.A.R., Griffin, M.J., Sandell, G.: 1989, MNRAS, submitted

Emerson, J.: 1988, in "Formation and Evolutin of Low Mass Stars", Proc. of the NATO Adv. Study Inst., Eds. A.K. Dupree and M.T.V.T. Lago

Grossman, E.N., Masson, C.R., Sargent, A.I., Scoville, N.Z., Scott, S., Woody, D.P.: 1987, Astrophys. J. **320**, 356

Mozurkewich, D., Schwartz, P.R., Smith, H.A.: 1986, Astrophys. J. **311**, 371

Sandell, G., Duncan, W.D., Dent, W.R.F., Robson, I., Gear, W.K.: 1988, in "Molecular Clouds in the Milky Way and External Galaxies" Eds. R.L. Dickman, R.L. Snell, and J. Young (Springer Verlag: Berlin)

Sandell, G.: 1989b, in the "Proceedings of the International Symposium on Submillimetre and Millimetre Astronomy, Kona, Hawaii, Oct 3–6, 1988" Ed. A. Webster (Springer Verlag: Berlin)

Sandell, G., Aspin, C., Duncan, W.D., Robson, E.I., Dent, W.R.F.: 1989a, Astron. Astrophys. in press

Sandell, G., Hayashi, S., Chandler, C.J., Gear, W.K.: 1989b, in preparation

Sandell, G., et al.: 1989c, in preparation

Sargent, A.I.: 1989, this volume

Weintraub, D.A., Sandell, G., Duncan, W.D.: 1989, Astrophys. J. **340**, L69

Weintraub, D.A., Sandell, G.: 1989, in preparation

Woody, D.P., Scott, S.L., Scoville, N.Z., Mundy, L.G., Sargent, A.I., Padin, S., Tinney, C.G., Wilson, C.D.: 1989, Astrophys. J. **337**, L41

Wooten, A.: 1989, Astrophys. J. **337**, 858

Discussion:

CAMERON: In your near-IR camera image of the flow from SSV13 there appeared to be emission apparently extending away from HH7. Do you believe that this emission is associated with SSV13 in which case it would imply that the HH7 knot is not the working surface of the optical outflows?

SANDELL: It is not clear whether the faint emission seen in front of HH7 in the IRCAM image by Graden, Burton and Russell relates to SSV13. It may be the case, but there is also another outflow further south.

SOLOMON: Is it possible that the grain emissivity at $350\mu m$, 1mm or $\lambda = 2mm$ is simply not known and that the column density (or density) you deduce from these observations is substantially in error? Do we really know the very long wavelengh emissivity of the dust? The grains could be large or elongated.

SANDELL: There are uncertainties, but not large enough to explain these huge discrepancies. The dust emissivity law can be measured, and for young stars associated with outflows or surrounded by dust disks we measure $\beta \sim 1$, regardless of luminosity. Density and temperature gradients may act to lower the measured β-index. If the grains are large (c.f. DG Tau; Weintraub et al, Ap. J. Letters, in press), the definition of the β index breaks down and we essentially just see a black body.

ZINNECKER: Is there a correlation between the derived dust masses and the luminosities of the sources?

SANDELL: Yes, there is, but there is considerable scatter. This is probably due to age effects, i.e., the youngest stars have larger dust masses than more evolved ones, and luminous stars also have more massive disks.

AFGL 2591 and Monoceros R2: Cavities in the Molecular Cloud

José M. Torrelles*, Paul T. P. Ho†, Luis F. Rodríguez‡, and Jorge Cantó‡
* *Instituto de Astrofísica de Andalucía, CSIC*
† *Harvard-Smithsonian Center for Astrophysics*
‡ *Instituto de Astronomía, UNAM*

We observed with the VLA(D) the (J,K)=(1,1) and (2,2) inversion transition lines of ammonia toward two regions with molecular outflow: AFGL 2591 and Monoceros R2.

We detected a clumpy medium in AFGL 2591. Two of the detected clumps may be part of a shell-like structure surrounding the near IR bubble previously detected by Forrest and Shure (1986), and Lenzen (1987) (Figure 1, Torrelles *et al.* 1989a). By comparing the (2,2) and (1,1) lines we find that heating is more important for the eastern side of the shell-like structure, with temperatures of $\sim$ 200 K near to AFGL 2591 (Figure 2, Torrelles *et al.* 1989a). A significant enhanced emission in the NH_3 line wings up to $\pm$ 3 km s^{-1} is observed near the peak temperature, indicating a local perturbation of the molecular gas. Our observations are consistent with the interpretation given by Forrest and Shure (1986) and Lenzen (1987) in terms that the near IR bubble is reflected light of a cavity, traced by the ammonia shell-like structure, and created in the molecular cloud by a central outflow probably related to AFGL 2591. The location of the radiocontinuum sources (1), (2), and (3) (Campbell 1984) on the edges of the neutral shell-like structure (see Figure 2) suggests that these could be bright rims externally excited by AFGL 2591 in a similar way as in Cepheus A (Torrelles *et al.* 1986).

The ammonia emission in Mon R2 shows a clumpy structure with multiple condensations. These condensations are mainly distributed in a remarkable arclike structure (Figure 3, Torrelles *et al.* 1989) located $\sim$ 40″ SW of the "blyster" type HII region (Massi, Felli, and Simon 1985). Within this arc we find a velocity gradient with the highest velocities *away* from the stellar activity center. By comparing the (2,2) and (1,1) lines, we also find a temperature gradient, with the highest temperatures *toward* the stellar activity center (see Figure 3). These properties imply an interaction of the stellar activity center with the high-density molecular gas. We favour two possible explanations for the VLA ammonia arclike structure:

(1) It may be part of an inner molecular wall of a toroid's cavity created by the pressure of the stellar wind driving the molecular outflow. (A molecular toroid was suggested to be in Mon R2 to collimate the bipolar outflow in the SE-NW direction, Torrelles *et al.* 1983.) This model predicts the presence of another arclike structure located $\sim$ 40″ NE of the HII region. However we could not detect this counterpart structure in the present experiment due to primary beam response, 0.1 at this position.

(2) It may be produced by the expansion of the "blyster" type HII region in the *champagne* phase. In this case, a more complete molecular shell structure bordering the HII region may be present, but undetected in the present experiment by the primary beam response.

We think that VLA ammonia observations with a new phase center are required to differentiate between models (1) and (2).

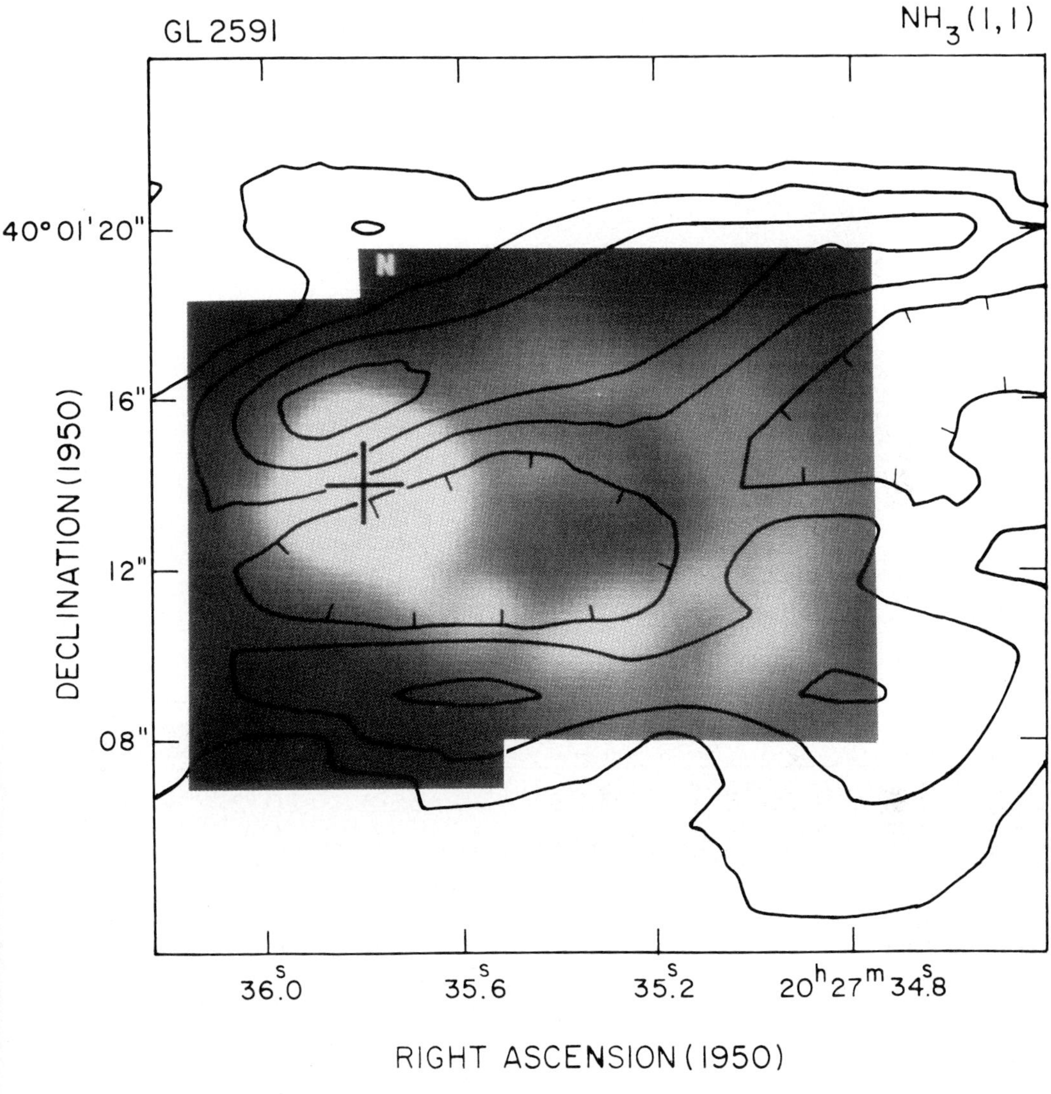

Figure 1. Contour map of the integrated NH_3 (1,1) emission over the -6.3 → -4.5 km s^{-1} range (synthesized beam= 6″) superposed on the K image of Forrest and Shure (1986). Contour levels are = -2, 2, 3, 4, 5, 6 × (30 mJy/beam km s^{-1}). Cross denotes the position of the AFGL 2591 source. (Figure adopted from Torrelles *et al.* 1989a.)

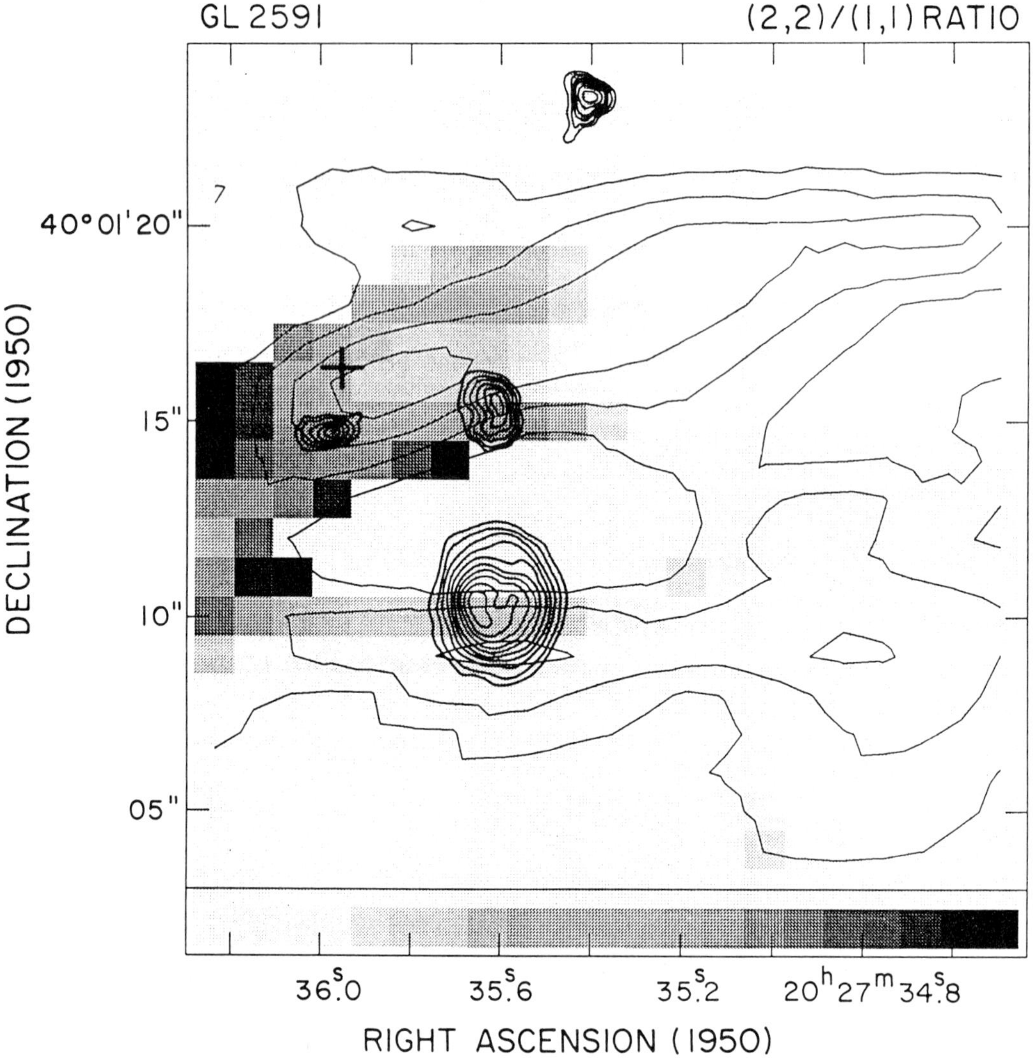

Figure 2. $S_\nu(2,2;m)/S_\nu(1,1;m)$ ratio (grey scale) superposed to the $NH_3(1,1)$ emission (Figure 1). The grey scale is linear and ranges from 0.6 to 3 for the flux ratio from left to right of the grey bar (down of figure). Radio continuum sources (*3, 1, 2, 4,* in decreasing Right Ascension) from Campbell (1984) are also superposed. Cross denotes the H_2O maser position (Wynn-Williams *et al.* 1977). (Figure adopted from Torrelles *et al.* 1989a.)

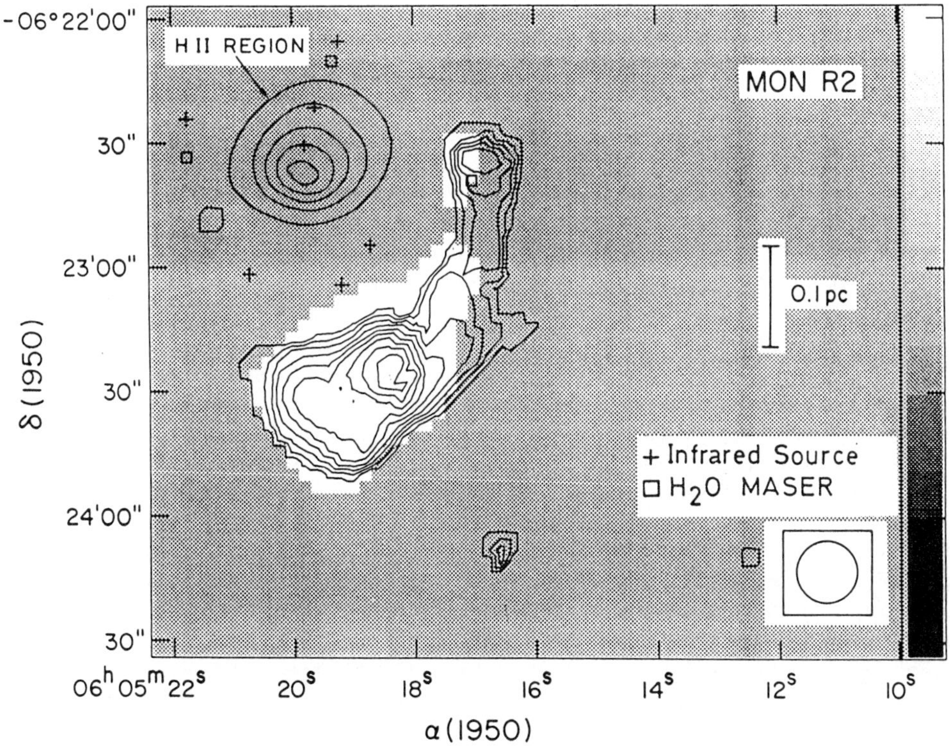

Figure 3. Contour map and grey scale image of the integrated intensity of the (1,1) and (2,2) ammonia main component line, respectively, over the $V_{LSR} = 9.0 \rightarrow 12.6$ km s^{-1} range (synthesized beam = 15″). Contour levels are = 1, 2, 3, 4, 5, 6, 7, 8, 9 × (43 mJy/beam km s^{-1}). The grey scale is linear and ranges from -43 to 43 mJy/beam km s^{-1} from down to up of the grey bar (right of figure). Continuum map (λ=1.3 cm) of the "blyster" type HII region is also superposed. Note that the (2,2) arclike structure is shifted to the NE of the (1,1) one, indicating an important heating at that edge. (Figure adopted from Torrelles *et al.* 1989b.)

References

Campbell, B. 1984, *Ap.J.*, **287**, 334.

Forrest, W. J., and Shure, M. A. 1986, *Ap. J. (Letters)*, **311**, L81.

Lenzen, R. 1987, *Astr. Ap.*, **173**, 124.

Massi, M., Felli, M., and Simon, M. 1985, *Astr. Ap.*, **152**, 387.

Torrelles, J. M., Ho, P. T. P., Rodríguez, L. F., and Cantó, J. 1989a, *Ap. J.*, in press (July 15).

Torrelles, J. M., Ho, P. T. P., Rodríguez, L. F., and Cantó, J. 1989b, *Ap. J.*, in press (November 15).

Torrelles, J. M., Ho, P. T. P., Rodríguez, L. F., and Cantó, J. 1986, *Ap. J.*, **305**, 721.

Torrelles, J. M., Rodríguez, L. F., Cantó, J. Carral, P., Marcaide, J., Moran, J., and Ho, P. T. P. 1983, *Ap. J.*, **274**, 214.

Wynn-Williams, C. G., Becklin, E. E., Forster, J. R., Matthews, K., Neugebauer, G., Welch, W. J., and Wright, M. C. H. 1977, *Ap. J. (Letters)*, **211**, L89.

APERTURE SYNTHESIS OBSERVATIONS OF CS, NH_3, AND CONTINUUM IN THE BIPOLAR FLOW SOURCE NGC2071-IRS

R. KAWABE, Y. KITAMURA, M. ISHIGURO, T. HASEGAWA(+), Y. CHIKADA, and S.K. OKUMURA

NOBEYAMA RADIO OBSERVATORY, NATIONAL ASTRONOMICAL OBSERVATORY.
(+) INSTITUTE OF ASTRONOMY, UNIVERSITY OF TOKYO.

ABSTRACT. We have made aperture synthesis observations of CS(J=1-0, 2-1) and NH_3(1,1) lines and 49, 98, and 110 GHz continuum in NGC2071-IRS with the Nobeyama Millimeter Array. We have obtained maps of these lines and continuum maps with 2".7-20" resolution. We have found that dense molecular gas has a disk structure with a radial scale ranging 0.01 pc - 0.1 pc and has a ring-like structure with expanding motion at the central 5000 AU region. We also have found that there exists double dust continuum sources which are separated by 2500 AU in projection and are apparently located at the inner edges of the ring. Our observational results suggest that the disk of molecular gas has a central hole formed by wind and UV radiation from a central young stellar object, the central part is expanding, and that dust continuum emission comes from tangential parts of the shock compressed ring ($r \sim 1300$ AU, $M(H_2) \sim$ 21-34 Mo, and $n(H_2) \sim 10^9$) at the most inner side of the disk structure. The other possible model of the dust continuum sources is a binary system of self-luminous young stellar objects.

I. Introduction

NGC2071-IRS is a young stellar object (or objects) with $L_{IR} \sim 900$ Lo (Harvey et al. 1979) and a typical example of CO bipolar flow sources (Bally 1982, Snell et al. 1984). Elongated structure of dense molecular gas suggesting the existence of a large scale gaseous disk was found by single dish observations (Bally 1982, Takano et al. 1984, 1986). VLA 5 GHz observations (Snell and Bally 1986) revealed the exsistence of three compact HII regions, each of which is associated with each of 10 μm sources (Persson et al. 1981) and each of H_2O and OH masers (Sandel et al. 1984). From these coincidences, it was suggested that each of these sources may be a young stellar object (Snell and Bally 1986). Excess of 3 mm continuum emission was found (Scoville et al. 1986). The emission is associated with two of three HII regions, but not spatialy resolved. It has been unclear whether the 3 mm emission comes from dust or a HII region.

Investigating the structure of dense molecular gas and dust cloud very close to young stellar objects helps us to understand star formation in the molecular gas disk and the evolution of dense molecular gas. However, there has been a large gap of spatial resolutions between the VLA sub-arcsecond structure of compact HII regions and the 40"-2' structure of dense molecular gas. We have made 2."7-20" resolution observations of NGC2071-IRS with the Nobeyama Millimeter Array in NH_3 and CS lines and 49-110 GHz continuum. The observations revealed new structures of dense molecular gas and dust. The results and interpretations are presented.

II. Observations

We have made aperture synthesis observations of CS(J=1-0, 2-1) and NH_3(1,1) emission from NGC2071-IRS with the Nobeyama Millimeter Array (Ishiguro et al.

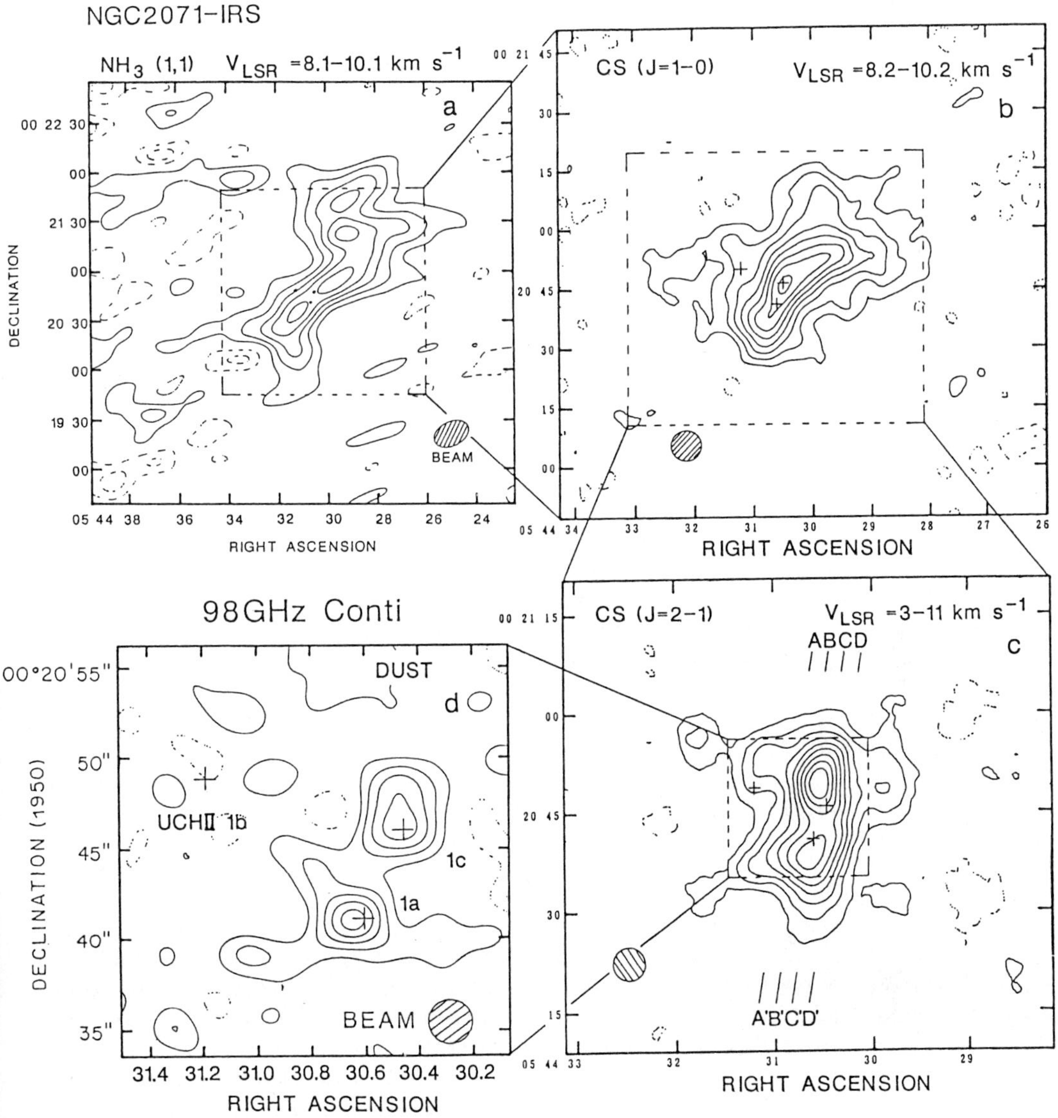

Figure 1. Montage of obtained maps of NH3(1,1), CS(1-0), CS(2-1), and 98 GHz continuum in NGC2071-IRS. Velocity ranges of the former three maps are V_{LSR}=8.1-10.1, 8.2-10.2, 3-11 km s^{-1}, respectively. Contour intervals are 88 mJy/beam, 144 mJy/beam, 165 mJy/beam, and 10.1 mJy/beam for the maps of NH3(1,1), CS(1-0), CS(2-1), and 98 GHz continuum, respectively. Three crosses (or dots) indicate ultracompact HII regions (1a-1c) in the VLA 5GHz map (Snell and Bally 1986), each of which corresponds to each of 10 um sources, IRS-1, 2, and 3 (Persson et al. 1981).

1984, 1989) during Apr. 1987 to May 1989. We used HEMT receivers for 23.7 GHz (Kasuga et al. 1986) and dual frequency SIS receivers for 40/100 GHz (Kawabe et al. 1989). Their system noise temperatures (SSB) at the zenith were around 200 K (23.7GHz), 300 K (49GHz and 98GHz), and 500 K (110GHz). We used an FFT spectro-correlator (FX) with 1024 channels per baseline (Chikada et al. 1987). The bandwidths were 80 MHz for NH_3 and CS(1-0), and 320 MHz for CS(2-1) and 110 GHz continuum. Velocity resolutions were 1.0 km s^{-1} for NH_3 and CS(2-1), and 0.5 km s^{-1} for CS(1-0). Line free channels were used for mapping of continuum emission at 49, 98, and 110 GHz. The number of baselines and synthesized beam were 20, 60, 40, and 10, and 16"x23", 8", 5"(2.7" for 98 GHz continuum), and 7" for NH_3, CS(1-0), CS(2-1), and 110 GHz continuum, respectively.

III. Results

(i) Quiescent Component; Disk of Molecular Gas

Figure 1a and 1b show maps of intensities of NH_3 and CS(1-0) quiescent components integrated over narrow velocity ranges of V_{lsr} = 8.1-10.1, and 8.2-10.2 km s^{-1}, respectively. NH_3 emission was not detected outside this velocity range in our observations. The maps of NH_3 and CS(1-0) indicate flat, almost edge-on disk structures with sizes of 90"x<20" and 35"x10"(0.085 pc x 0.025 pc), respectively. The recent VLA result of NH_3(2,2) (Zhou et al. 1989) is very consistent with our NH_3 result. The disk structures are just perpendicular to the bipolar outflow (Snell et al. 1986, Scoville et al 1986) as already noted by Takano et al. (1984, 1986). Two of three compact HII regions associated with IRS-1 and IRS-3 are located at the center of the disk.

(ii) Wing Component; Expanding Ring inside the Disk

Figure 1c shows a map of the CS(2-1) intensity of an intermediate wing component (Kitamura et.al. 1989) integrated over a velocity range, V_{lsr} = 3-8 km s^{-1}. Wing component at higher velocity in CS(2-1) with an extent of 20 km s^{-1} has the same bipolarity as the CO high velocity flow (Bally 1982; Kitamura et al. 1989). The map indicate a more compact ridge than the NH3 and CS(1-0) disk, with a size of 20" x 10" (0.05pc x 0.025). The two peaks on the ridge are located outside two compact HII regions (i.e., VLA 5GHz sources, "1a" and "1c", associated with IRS-1 and IRS-3, respectively; Snell and Bally 1986). Figure 2 shows four position-velocity maps along four cuts (shown in Fig.1c) parallel to the ridge. A main feature in the maps is an oval pattern corresponding to a large non-circular motion (~4-5 km s^{-1}), which is traced with a dotted line in Fig 1c. This pattern and ridge structure suggest the existence of expanding molecular ring inside the disk. The dynamical center of the expansion is between two compact HII regions (see Fig. 2). Its expanding velocity and dynamical time scale are about 5 km s^{-1}, and $3x10^{3}$ years, respectively. The time scale is shorter than the dynamical age of the high velocity CO outflow, $1.6x10^4$ years (Snell et al. 1986) by a factor of five. Such an expanding motion of dense molecular gas in the inner part of the disk was detected so far in Orion-KL (Plambeck et al. 1982) and GL490 (Kawabe et al. 1984).

(iii) 3 mm continuum; double dust continuum sources

Obtained total flux densities of contionuum emission at 49, 98, and 110 GHz are plotted in Figure 3 with other data (see Dent et al. 1989). The cm to mm spectrum of the continuum emission from NGC2071-IRS indicates that the 98 and 110 GHz emission almost comes from dust. The flux density at 49 GHz is explained well as a combination of free-free emission and dust emission.

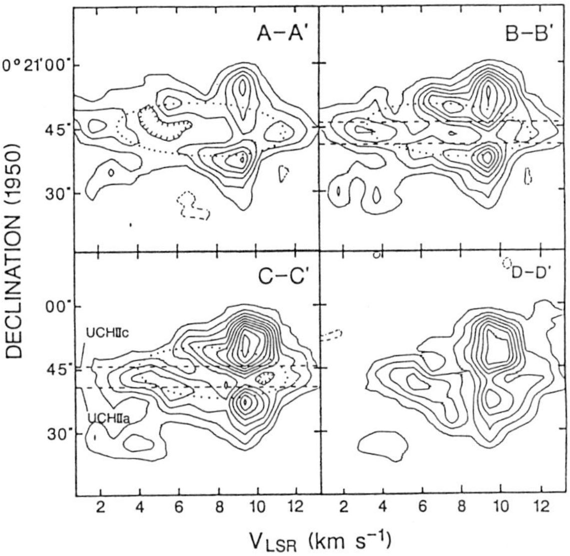

Figure 2. Four position-velocity diagrams in CS(2-1) along four cuts shown in Fig.1. Dotted line indicates a oval pattern of an expanding ring. Dashed lines indicate positions of two ultracompact HII regions.

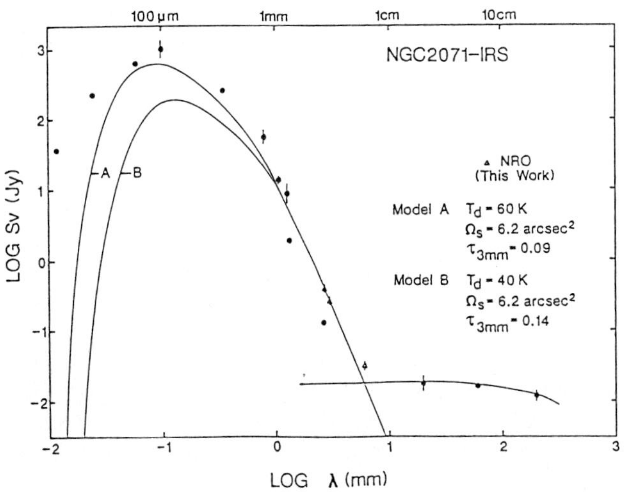

Figure 3. Spectrum of contimuum emission in NGC2071-IRS. Our data at 110, 98, and 49 GHz (△) are plotted with other data (see Dent et al. 1989).

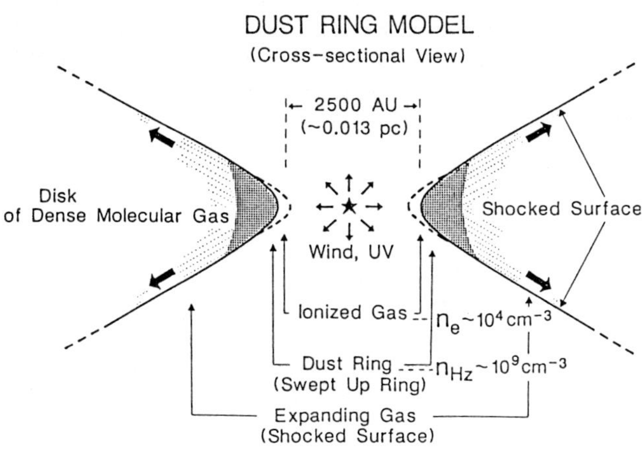

Figure 4. Schematic diagram of a dust ring model (cross-sectional view).

Figure 1d shows a map of 98 GHz continuum emission. The emission is resolved to double point-like sources, each position of which coincides with each of the two compact HII regions (associated with IRS-1 and IRS-3). The two peaks at 98 GHz fall between the two CS(2-1) peaks. The CS(2-1) peaks and the 98 GHz point-like sources are aligned with the disk structure of the NH_3 and CS(1-0) emission.

We have fitted an uniform temperature model of the dust source to the spectrum in order to estimate a dust temperature, T_d, and a dust opacity at 3 mm, τ_{3mm}. We assumed a dust emissivity index of two and the source size in total, 6.2 $arcsec^2$, which is roughly estimated from the 98 GHz map (Fig. 1d). Fitting was done for fitted lines not to exceed over the lower envelope of single dish data at $\lambda < 1.3$ mm, because the data sample extended emission missed in our interferometric data. The dust temperature is estimated to be lower than ~60 K. Total mass and number density of H_2 molecule of the two sources are derived to be ~21-34 Mo and $\sim 10^9$ cm^{-3}, respectively, assuming T_d = 40-60 K (lower boundary, 40 K, is taken based on Takano 1986), an empirical relation between an H_2 column density and a dust opacity (Hildebrand 1983), and assuming that an extent of the dust sources along the line of sight is about 1200 AU.

IV. Models of Double Dust Continuum Sources

We have found two dust continuum sources with high H_2 density and large H_2 mass at the center of the disk and "expanding ring" of dense molecular gas ($n(H_2) \sim 10^4$-10^6 cm^{-3}). There are two possible model to explain the structure; one is a dust ring model and the other is a binary model of self-luminous young stellar objects.

(i) A Dust Ring Model

The positional relations between the double dust sources and the two CS(2-1) peaks suggest that the dust emission is strong at the tangential parts of the dust ring most inside of expanding ring and disk. A schematic diagram of this model is shown in Figure 4. The formation of the dust ring is explained as follows. The strong wind and UV radiation from a central newly formed star interact with high density gas in the disk region to form shock compressed ring in the equatorial plane, and accelerate the gas. The gas sweeping effect of shock raises the H2 mass of the ring according to the expansion of the ring. The expanding ring in the disk is an evidence of the interaction between the wind and the disk. The positional coincidence among HII regions, OH and H_2O masers, and dust continuum sources does not conflict with this model. Two HII regions and faint emission between them (Snell and Bally 1986) can be interpreted as an ionized inner edge of the dust ring. The masers would be excited at the inner edge. This ring model can produce the observed structures. A similar model to explain two 5 GHz continuum sources was also proposed for L1551-IRS5 (Rodrigues et al. 1987). The main argument against this model is that there is no observational evidence of single central driving source between IRS-1 and IRS-3.

(ii) A Binary Model

There is a possibility that each of the double dust continuum sources corresponds to a self-luminous young stellar object which is formed through fragmentation in the singel molecular gas disk. Simulations of isothermal rotationg clouds suggested that a flattened disk fragments and each of fragments has roughly equal mass, and that in total the flagments contain ~30%-40% of the cloud mass (Miyama et al. 1984).
The molecular gas disk in NGL2071 is very flat (see Fig. 1). Each H_2 mass of binary dust clouds is almost equal and is estimated to be 10-15 Mo, and the dust clouds contains ~40%-80% of the total H_2 mass of the 30" region, because the 30" scale disk

of molecular gas has a mass of 5-30 Mo (Takano et al. 1984). These properties of the disk and the double dust sources are roughly consistent with the results of the simulations. On the other hand, from positional coincidences of H_2O and OH masers, ultra compact HII regions , it was suggested that each of these sources may be a young stellar object (Snell and Bally 1986). These support the possibility of a binary. However, it is unclear which source drives the high velocity outflow and the "expanding ring", because each of two sources are shifted from the dynamical center of the "expanding ring" and the origin of the CO outflow (Scoville et al. 1986).

References

Bally, J. 1982, Ap.J., 261, 558

Chikada, Y. et al., 1987, Proc. of the IEEE, 75, No.9, 1203

Dent, W.R.F. et al. 1989, preprint.

Harvey, P.M., Campbell, M.F., Hoffmann, W. F., Thronson, Jr., H. A., and Gatley, I. 1979, Ap.J., 229, 990.

Hildebrand, R.H. 1983, Quant.J.R.A.S., 24, 1267.

Ishiguro, M. et al., 1984, in the proceedings of the International Synposium on Millimter and Submillimeter Wave Radio Astronomy, Granada, 11.

Ishiguro, M. et al., 1989, in preparation.

Kasuga, T., Kawabe, R., Ishiguro, Yamada, K., Kurihara, H., Niori, H., and Hirachi, Y. 1987, Rev.Sci.Instrum., 58, 379.

Kawabe, R. et al. 1984, Ap.J.(Letters), L73, 1984.

Kawabe, R. et al. 1989, in preparation.

Kitamura, Y., Kawabe, R., Yamashita, T., and Hayashi, M., 1989, preprint.

Miyama, S.M., Hayashi, C., and Narita, S. 1984, Ap.J., 279, 621.

Persson, R.L., Geball, T.R., Simon, T.L., Lonsdale, C.J., and Baas, F. 1981, Ap.J.(Letters), 251, L85.

Plambeck, R.L., Wright, M.C.H., Bieging, J.H., Baud, B., Ho, P.T.P., and Vogel, S.N. 1982, Ap.J.(Letters), 259, 617.

Rodriguez, L.F., Canto, J., Torrelles, J.M., and Ho, P.T.P. 1986, Ap.J.(Letters), 301, L25.

Sandel, Y., Nyman, L.A., Haschick,A., and Winnberg, A. 1984, in Lecture Notes in Physics (Nearby Molecular Clouds), 237, 234.

Scoville, N.Z., Sargent, A.I., Sanders, D.B., Claussen, M.J., Masson, C.R., Lo, K.Y., and Phillips, T.G. 1986, Ap.J., 303, 416.

Snell, R. and Bally, J., 1986, Ap.J., 303, 683.

Snell, R., Scoville, N.Z., Sanders, D.B., and Erickson, N.R. 1984, Ap.J., 284, 176.

Takano, T., Fukui, Y., Ogawa, H., Takaba, H., Kawabe, R., Fujimoto, Y., Sugitani, K., and Fujimoto, M. 1984, Ap.J.(Letters), 282, L69.

Takano, T., Stutzki, J., Fukui, Y., and Winnewissar, G. 1986, Astr.Ap., 167,333.

Takano, T. 1986, Ap.J., 303, 349.

Wootten, A., Loren, R.B., Sandquist, A., Friberg, P., and Hjalmarson, A. 1984, Ap.J., 279, 633.

Zhou, S., Evans II, N.J., and Mundy, L.C. 1989, preprint.

Discussion:

SANDELL (Comment): I think you should review your interpretion of only one source carefully. The continuum appears to peak on the two free-free and IR sources. The discrepancy in spatial coincidence in CS may just be due to optical depth effects- the densities are extremely high. Polarization imaging or UKIRT with the infrared camera suggest that there is more than one source.

A SWEPT-UP MOLECULAR BUBBLE IN L1551

SAEKO S. HAYASHI
Joint Astronomy Centre, 665 Komohana Street
Hilo, Hawaii 96720, U.S.A.

MASAHIKO HAYASHI
Department of Astronomy, University of Tokyo
Bunkyo, Tokyo 181, Japan

NORIO KAIFU
Nobeyama Radio Observatory,
Nobeyama, Nagano 384–13, Japan

1. Introduction

The environment of the young stellar object IRS5 in L1551 dark cloud is a representative of the "protostellar disk and outflow" systems found in star forming regions. The bipolar molecular outflow there was discovered as the first of its kind a decade ago (Snell, Loren, and Plambeck 1980). Its location in the sky, that is, its proximity (160 pc), isolation, and its almost edge-on inclination have favored the observation in great detail. IRS5 is thought to have the spectral type G - K (Mundt *et al.* 1985) similar to the Sun, with its dominant activity in the stellar wind, and not in the UV radiation as in massive protostars. The blueshifted and redshifted outflow lobes are clearly resolved into a pair of shell structures. The successive studies, mostly in CO lines, have led to a model of the outflow in which the molecular material is accelerated at the edge of a cavity evacuated by the protostellar wind (*e.g.* Uchida *et al.* 1987, Rainey *et al.* 1987, Moriarty-Schieven and Snell 1988).

Meanwhile, another set of molecular lines were used to identify the central protostellar core, the presumed agent for energizing and collimating the outflow. The intensity of CS J=1-0 and J=2-1 transitions peaks at two bright spots, to the north and south of IRS5. The origin of the outflow spatially coincides with the CS depression and the axis of the outflow is orthogonal to the elongation of the CS emission (*e.g.* Kaifu *et al.* 1984). The J=2-1 transition of CS peaks closer to the center than the distribution of the J=1-0 emission, a clear evidence of the density gradiant increasing toward the protostar. The map of NH_3 shows an elongated structure along the outflow in addition to an elongation similar to that present in the CS distribution, indicating the vigorous interaction between the outflow and the ambient dense disk (*e.g.* Menten and Walmsley 1985). FIR continuum emission from the dust core also shows an elongation both along and perpendicular to the molecular outflow (Cohen *et al.* 1985).

2. Description of the current approach

A hole in the molecular emission was found in the south-west of IRS5, via recent sensitive ^{13}CO observations (Bally and Hayashi 1989; also noticeable in Moriarty-Schieven and Snell 1988). In order to locate this ^{13}CO structure with respect to other features, especially the outflow and the disk, we have carried out a higher spatial resolution observation with Nobeyama 45-m radio telescope. An area of $\sim 15' \times 10'$ along and across the outflow axis (position angle of $\sim 45^o$) has been covered in ^{13}CO J=1-0 emission line at 110.201 GHz.

The emission from the rare isotope CO is moderately thin ($\tau = 0.5 - 1$) even in the line velocity at $V_{\rm LSR} = 6.5\ km\ s^{-1}$. Because of this opacity, the ^{13}CO distribution shows enhanced features against the quiescent cloud, in contrast to the distribution of the low velocity ^{12}CO emission. The current data presents the following characteristics :

- A tuning-fork like structure to the south-west of IRS5 – reminiscent of the blueshifted ^{12}CO emission.
- A broken symmetry of the structure in the south-west and the north-east – unlike in the ^{12}CO bipolar distribution.
- Spatial relation with the optical/IR compact sources – indicating interaction with other young stellar objects.

The most conspicuous feature, the tuning-fork like ridge in ^{13}CO emission, is limited to the narrow velocity range. Its dimension is most clearly measured in the intensity distribution integrated between $V_{\rm LSR} = 5$ and 8 $km\ s^{-1}$. The length of both north and southern "legs" is approximately $4'$, separated by a similar distance and with a thickness $20 - 40''$. The ridge of this enhanced ^{13}CO emission consists of clumps of less than $1'$ scale size. This characteristic tuning-fork shape can be traced in the individual velocity maps at $V_{\rm LSR} = 5-6$ and $6-7$ $km\ s^{-1}$, namely in the line core and in the slightly blueshifted velocity. The ^{13}CO intensity, integrated over a wide range of blueshifted velocity, does show a visible mixing with the outflow; a series of high velocity ^{12}CO emission. The northern "leg" is brighter than the southern one, and has some (anti-)correlation with the intricated optical nebulosity HH102. The prominent ^{13}CO ridge is situated to the north of HH102, and there is another faint ridge of ^{13}CO between the two streaks of this Herbig-Haro object.

As a natural expectation, one may look for a counterpart of this southern arch. At first glance, there is no significant arch nor shell toward the north and east of IRS5, where the redshifted lobe of ^{12}CO is situated. However, one can pick up a depression of ^{13}CO to the east, as a bay like structure opening to the east. This depression zone does not make a perfect pair with the south-west tuning-fork; it is detached from IRS5, its scale is smaller and its axis is offset from the extention of the south-east arch, and not aligned with the ^{12}CO redshifted lobe either. The location of ^{13}CO with respect to the redshifted lobe of ^{12}CO ridge falls into the ^{13}CO bay, and the northern ridge of ^{12}CO also encounters the lower ^{13}CO emission. The area between HL & XZ Tau and the bay is strong in ^{13}CO emission.

A spur, a short and bright ridge of ^{13}CO of $\sim 2'$ extends form IRS5 to the north-east. It is aligned to the radio continuum jet (*e.g.* Bieging *et al.* 1984), and located opposite to the optical jet present in the south-west of IRS5 (*e.g.* Mundt and Fried 1983). The south of this spur delineates the edge of the newly identified IR source L1551 NE (Emerson *et al.* 1984). Also there is some correlation between ^{12}CO and L1551 NE; the curved ridge of blueshifted ^{12}CO and a peak of redshifted ^{12}CO are seen in the east of L1551 NE. The spatial coincidence of L1551 NE and those molecular features strongly suggests the activity form this infrared source interacting with the ambient cloud. The ^{13}CO distribution exhibits a correlation (or anticorrelation, rather) with other optical/IR sources as well. HL and XZ Tau, and HH30, which are to the north of the redshifted ^{12}CO, and the compact knot HH29 in the south, are accompanied by small clumps of ^{13}CO. HL & XZ Tau themselves are in the depression.

3. Implication

The current data have evoked two arguments on the outflow structure. So far, the model based mainly on ^{12}CO outflow data is qualitatively sufficient to reproduce the morphology of the outflow phenomena, although it has severe deficit in the momentum supply. The present ^{13}CO measurement requires a significant amount of energy and momentum deposit into the molecular material, which further exaggerates the quantitative difficulty of the outflow models.

The most striking result is that the south-west arch of ^{13}CO is located immediately outside the blueshifted ^{12}CO emission arch, as shown in figure 1. The distance between the two legs of

^{12}CO tuning-fork feature is 2.5′ to 3′ (0.11 – 0.13 pc) while the ^{13}CO is 4′ (0.18pc) further out. Even the low velocity ^{12}CO shell is surrounded by the arch of ^{13}CO. The clear offset between these two archs implies that ^{13}CO is not merely a limb brightening of the ^{12}CO outflow shell. This ^{13}CO structure is most likely a shell, not just filaments, sitting outside the ^{12}CO "shell", namely, the ^{13}CO emission arises from the region surrounding the outflow in a 3 dimensional sense. In order to discriminate between the ^{12}CO and ^{13}CO molecular archs, they are here called ^{12}CO shell and ^{13}CO bubble, respectively. The name bubble is meant to show preference to the hypothetical expanding structure which surrounds the stellar wind cavity. This is reminiscent of the bubble created by the stellar wind from a massive star, phenomena which has been throughly studies before protostars (*e.g.* Weaver *et al.* 1977). We suppose that this bubble is ambient material swept-up by the stellar wind (*e.g.* Moriarty-Schiven and Snell 1988), rather than the material directly blown out from the protostellar disk. The emission in the higher velocity range could arise from the clumps entrained in the outflow, which are abrases from the cavity wall.

The expansion timescale in the case of L1551 IRS5 is estimated to be $\sim 10^5$ years, if one assumes an outward velocity of the order of 1 $km\ s^{-1}$, and a distance of 0.09 pc. The mass of the ^{13}CO bubble is estimated to be $3 - 7M_\odot$, derived from the column density of ^{13}CO with the adopted excitation temperature ranging between 10 K for the dark cloud and 25 K for the outflow. In order to sweep up the ambient material into this bubble, the energy budget for the outflow mechanism is significantly increased, not to mention the difficulty with the momentum deficit. The mass of ^{13}CO bubble is comparable to that of the entire outflow, 3.5 $M_\odot$ (Moriarty-Schieven and Snell 1988), and to that of the protostellar disk of IRS5 (Kaifu *et al.* 1984).

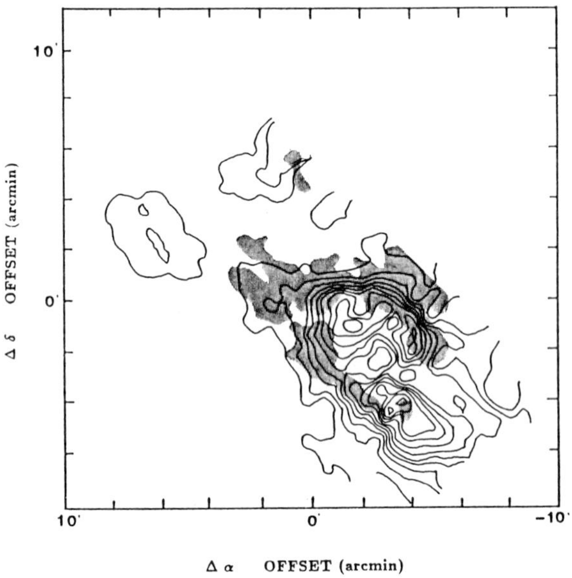

Figure 1.
Molecular bubble surrounding the outflow from L1551 IRS5

shaded area : ^{13}CO J=1-0 (total intensity above 5 K km/s)

contours : ^{12}CO J=1-0 (intensity for -5 to 5.5 km/s min. 1.5, step 1.5 K km/s)

The second issue is on the bipolarity, or the symmetry of the distribution with respect to the driving source L1551 IRS5. The ^{12}CO structure has been described as a pair of shells, similar in shape to each other. The blueshifted shell is in the south-west of IRS5, and the redshifted shell is to the north-east of it, respectively. The ^{13}CO bubble could be still bipolar in nature, but not in appearance. The bubble structure is seen in the south-west of IRS5, while there is no counterpart in the north-east. This asymmetry in ^{13}CO could be due to the large scale asymmetry of the dark cloud structure as well as to the influence of the activity from other young stellar objects.

The bipolar nature of the ^{13}CO bubble is shown in relation to the ^{12}CO distribution. There is a remarkable coincidence between the ^{12}CO peaks in the ^{13}CO bay. The blueshifted ^{12}CO emission peaks at the north-east bay, while the redshifted ^{12}CO delineates the northern edge of it. The ^{13}CO looks like a basin of the ^{12}CO water fall. However, the presence of ^{13}CO ridge in the middle of the ^{12}CO redshifted "lobe" brought some conflict when interpretating that lobe as a "shell". Perhaps the outflow to the redshifted velocity range goes more eastward, particularly at the distance of the bay area. This is in accordance with the ^{13}CO spur, as well as optical and radio continuum jets (e.g. Mundt and Fried 1983, Bieging *et al.* 1984). In addition to this shift, other protostars (HL & XZ Tau, L1551 NE) seem to disturb this area.

Summary

The observations toward L1551 dark cloud exhibited a vivid image of the star formation site where the placental cloud is dissipating. Mapping in ^{13}CO emission resolved a bubble to the south-west of L1551 IRS5, which surrounds the ^{12}CO shell blueshifted outflow. We interpret this feature as material swept up by the protostellar wind. There are some other indications of the interaction between the ambient cloud and other yound stellar objects to the north-east of IRS5. The discovery of the massive swept-up bubble doubled (at least) the estimate of the energy and momentum budget of the driving mechanism.

The authors are grateful to the support at Nobeyama Radio Observatory of the National Astronomical Observatory of Japan, which is a common user facility for the international community, as well as to the staff of the Joint Astronomy Centre which runs United Kingdom Infrared Telescope and James Clerk Maxwell Telescope atop of Mauna Kea on behalf of Science and Engineering Research Council, Nederlandse Organisatie Voor Wetenschappelijk Onderzoek, and National Research Council of Canada.

References

Bally, J. and Hayashi, M. 1989, *in preparation.*
Bieging, J.H., Cohen, M., and Schwartz, P.R. 1984, *Ap. J.*,k 282, 699.
Cohen, M., Harvey, P.M., and Schwartz, R.D. 1985, *Ap. J.*, 296, 633.
Emerson, J.P. *et al.* 1984, *Ap. J. Lett.*, 278, L49.
Kaifu, N. *et al.* 1984, *Astr. Ap.*, 134, 7.
Lada, C.J. 1985, *Ann. Rev. Astron. Astrophys.*, 23, 267.
Menten, K.M., and Walmsley, C.M. 1985, *Astr. Ap.*, 146, 369.
Moriarty-Schieven, G., and Snell, R.L. 1988, *Ap. J.*, 332, 364.
Mundt, R., and Fried, J.W. 1983, *Ap. J. Lett.*, 274, L83.
Mundt, R. *et al.* 1985, *Ap. J. Lett.*, 297, L41.
Rainey, R. *et al.* 1987, *Astr. Ap.*, 179, 347.
Snell, R.L., Loren, R.B., and Plambeck, R.L. 1980, *Ap. J. Lett.*, 239, L17.
Uchida, Y. *et al.* 1987, *P.A.S.J.*, 30, 907.
Weaver, R. *et al.* 1977, *Ap. J.*, 218, 377.

Discussion

PALOUS: The snow-plough model can be applied in the case of the strong shock, when the thickness of the shock is much less than the radius of the cavity. In our case $0.03/0.1 = 0.3$, which means that probably the weak shock approach would be more appropriate.
HAYASHI: I agree with the above comment. In particular, the overlap (co-existence) between ^{12}CO (outflow) and ^{13}CO (bubble) implies that the shock will not be totally dissociative. Also I admit the name "snow-plough" may be misleading for this case, since the energetics and time scales in star forming regions are different from supernova explosions the phenomena for which this word was first used.

ON THE FORMATION AND PROPAGATION OF INTERSTELLAR JETS

Guillermo Tenorio-Tagle
Max-Planck-Institut für Physik und Astrophysik
D-8046 Garching bei München, F. R. G.

Summary. The flow that results from stationary supersonically converging conical streams is studied by means of two dimensional numerical calculations. Convergency leads to a "colimator" conical shock and this to the formation of an underexpanded hydrodynamical jet. The properties of the conical shock and of the collimated stream are summarized for a variety of initial conditions. The resultant jets are followed to large distances ($\geq 7.5\times10^{17}$ cm) away from the injection point, and are shown to develop a complicated structure that eventually approaches a stationary state. Stationary features such as crossing shocks are shown to evolve from working surface shocks, left behind the head of the jet to ensure self-collimation of the beam matter.

Transient, as opposed to stationary, supersonically converging conical streams are also shown to generate collimator shocks, and under special conditions, long-lasting hydrodynamical jets and bipolar flows. Such flows are shown to emanate from compact disc funnels, if they are suddenly inmersed into a thermal bath. These calculations suggest that in many cases, the outflow phenomena (*e.g.* HH objects, jets and bipolar molecular flows) may simply be signaling the disruption of the compact discs, out of which exciting stars form.

I. Hydrodynamical jets powered by stellar winds. The solution to stationary supersonically converging conical flows has recently been worked out both analytically (Cantó *et al.* 1988) and numerically (Tenorio-Tagle *et al.* 1988). The converging streams have been assumed to result from an originally isotropic stellar wind redirected at an oblique "inner" shock which, in steady state and in the presence of a density gradient (see Cantó 1980 and Cantó and Rodriguez 1980), acquires an ovoidal "bishops hat" configuration. The wind gas upon crossing the oblique inner shock, acquires a pressure P_0 , the original pressure of the ambient gas (as required by the steady state solution), and is redirected to move towards the axis of symmetry, generating in this way a stationary supersonically converging conical flow. The boundary and initial conditions for the numerical calculations have assumed a constant velocity (U_0), uniform density (n_0) stream, in pressure equilibrium with the surrounding medium and supersonically converging towards the symmetry axis with an incidence angle Θ (see figures 1 and 2). Once the density of the incoming flow (n_0) is defined, the width of the stream can be derived from the assumption that all ejected wind matter ($\dot{M}_{wnd}$), is recollected by the inner shock. The stream has been continuously replenished at the bottom of the computational grid to model the stationary converging flow.

The stationary supersonic flow inmediately develops a steady density enhancement caused by convergency alone, and a sudden further compression is provided by a conical shock that forms near the symmetry axis. Both features can be identified in figure 1 by noticing first the well spaced isodensity contour lines, drawn across the incoming stream, and

Table 1

Θ	β	U_{jet} km s^{-1}	A
15	3	96.4	1.6
30	1.8	84.9	2.7
45	1.5	68.3	3.3
60	1.8	49.3	3.5
75	4.0	25.0	3.6

Aperture angle β

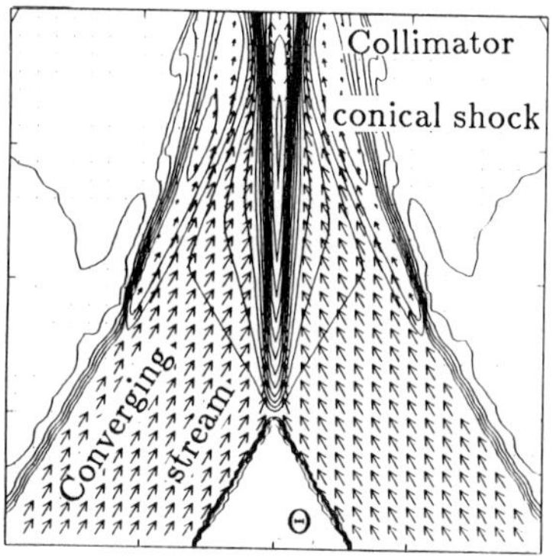

Fig.2.

the surface (the conical shock) across which the density contours crowd and the velocity field is sharply redirected. The two features become almost inmediately stationary, regardless of the assumed incidence angle Θ (see Tenorio-Tagle *et al.* 1988). Note that the conical shock is also an oblique structure to the incoming stream and thus across it, the perpendicular component of the incoming velocity becomes thermalized while the parallel component is transmitted, bending the stream towards the shock, and defining the velocity of injection (U_{jet}) of the resultant flow. Strong radiative cooling (an isobaric process) leads also to a large compression behind the conical shock and in all calculated cases to a temperature of about 10^4K. Such a temperature sets the value of the adiabatic speed of sound (c_{jet}) of the compressed matter before it is injected into the surrounding gas, and thus defines the Mach number M_{jet} (= U_{jet} / c_{jet}) at the injection point. It is in fact M_{jet}, the one that accounts for the morphological sequence as a function of angle Θ (see figure 1), *i.e.* whether a thin elongated skinny fish or a fat french frog will result depends only on M_{jet}, as when the compressed gas exits the conical shock it inmediately expands into the surrounding gas. Its velocity of expansion is in all cases c_{jet} (about 10 km s^{-1}), and thus the larger the M_{jet} the more elongated structures that will result.

The analytic and numerical results have shown that: i) the solution to a supersonically converging stationary conical flow is the formation of a stationary "collimator" conical shock, with its apex placed at the point of first contact. ii) In all cases where strong radiative cooling takes place (those with an incoming velocity $U_0 \cos\Theta < 200$ km s^{-1}, which lead to large values of the interstellar cooling rate), the collimator conical shock presents a small aperture angle β (< 5°, see figure 2 and table 1). The solution to quasi-adiabatic cases can also be found in the paper that describes the numerical work (Tenorio-Tagle *et al.* 1988). iii) The velocity of the shocked gas U_{jet} is directed along the symmetry axis and thus given the small values of the aperture angle β, the relation $U_{jet} = U_0 \cos\Theta$, is a good approximation in all cases. iv) The incoming stream is compressed by a large factor A (= $\log(n_{jet}/n_0)$) which in all radiative cases amounts to more than two or three orders of magnitude (see Table 1). v) Radiative cooling behind the shock leads, in all cases, to a temperature $T_{jet} \sim 10^4$K. This, together with the large compression factor results into a pressure $P_{jet} \gg P_0$, and thus given (iii) the well directed velocity, it implies that the stationary collimator conical shock leads to the formation of an underexpanded ($P_{jet} \gg P_0$) hydrodynamical jet. Finally, vi) the radiative flow behind the collimator shock may be identified with the so called "optical jet structures", seen in the vicinity of some young stellar objects.

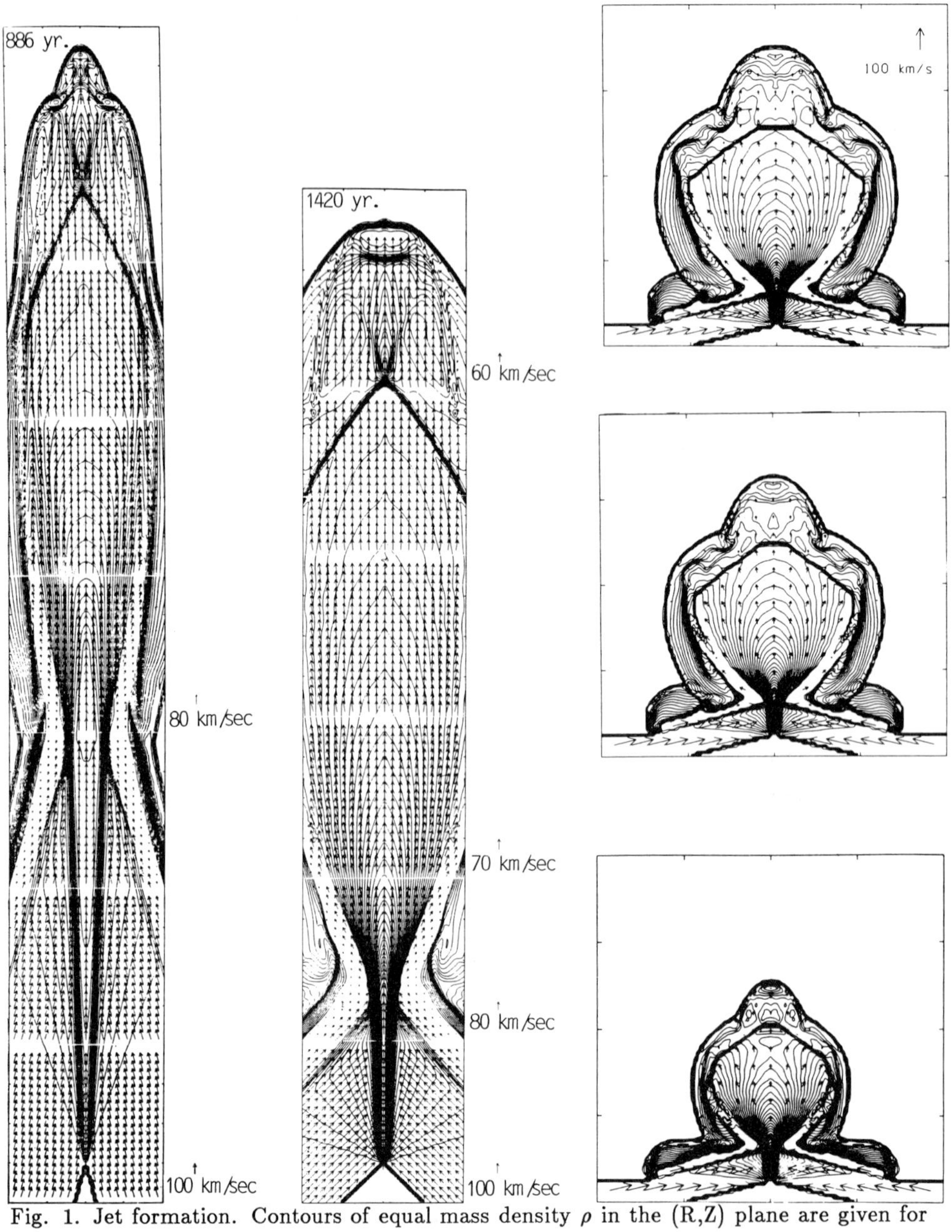

Fig. 1. Jet formation. Contours of equal mass density ρ in the (R,Z) plane are given for cases with $n_0 = 100\ cm^{-3}$, $U_0 = 100\ km\ s^{-1}$, and an incidence angle $\Theta = 15°$, $45°$ and $75°$, respectively. The contour interval $\Delta \log \rho = 0.2$ in the first two frames and 0.1 in the last case. The evolutionary times in yr are 886, 1420, and 1360, 2050, and 2320 for the $\Theta = 75°$ sequence. Superimposed is the velocity field to be compared with the standard velocity vector in km s^{-1} given at the right hand side of the plots. Changes in the absolute magnitude of this vector are also indicated. The distance between consecutive tick marks is 10^{16} cm in the first two frames and 4×10^{16} cm in the $\Theta = 75°$ panels.

II. The propagation and structure of interstellar jets.

Figure 3 displays a fully developed hydrodynamical jet, produced by the mechanism described in section I ($n_0 = 100\ cm^{-3}$, $U_0 = 100\ km\ s^{-1}$, $\Theta = 30°$, initially in pressure equilibrium with an outside medium where $n = 10^3\ cm^{-3}$and $T = 10\ K$), after some 7000yr of evolution. Perhaps, it is the time dependent jet with the largest length to width (at injection point) ratio, so far calculated in an astrophysical context. A full description of its density, temperature and pressure structure as it propagates through interstellar space is to appear in a forthcoming paper (Tenorio-Tagle and Różyczka 1989). Here only a brief summary of the results is given. Before that however, one should make the point that the structure left by the jet as it propagates away from the injection point is independent of the formation mechanism. *i.e.* once an overpressured, preferentially well directed supersonic beam forms, by whatever means, and it is injected into the surrounding gas, it will propagate while digging a channel and causing a structure similar to the one shown in figure 3. How does such a structure develop? and/or how is collimation at all distances from the injection point achieved? are some of the main issues presently in this field.

As mentioned in section I, as soon as the collimated matter is injected into the surrounding gas, it begins to expand. Expansion is due to the overpressure ($P_{jet} \gg P_0$), which causes the divergency of the beam at a speed c_{jet} ($\sim 10 km\ s^{-1}$). As matter moves away from the injection point its density drops as $1/R^2$, and it cools down both by radiation and by its own expansion. Thus, the diverging beam eventually reaches pressure values which are even lower than P_0. However, its supersonic motion leads to the establishment of another oblique shock (an "incident" shock) across which the beam pressure is restored. Across any oblique shock, only the component of the gas velocity perpendicular to the shock becomes thermalized and enhances the pressure. The other component of the velocity (the one parallel to the shock), is almost fully transmitted and thus the gas is effectively deflected towards the shock. The first of these properties defines the shape of the incident shock, which extends across to overtake the whole expanded beam, while giving it the same pressure (P_{shock}). Clearly the expanded beam matter closer to the injection point retains a larger value of the original pressure and would meet a section of the incident shock which is almost parallel to its own motion, and thus hardly modifies its pressure. However, rarer and

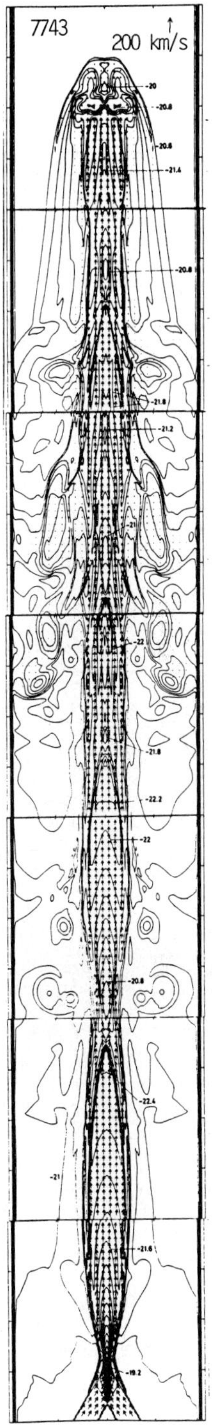

Fig. 3. The structure of interstellar jets. The same as figure 1 for a stream converging with an incidence angle $\Theta = 30$, after 7700 yr of evolution. The distance between consecutive tick marks is 10^{16} cm. Some of the contours are labelled with the value log ρ ($g\ cm^{-3}$).

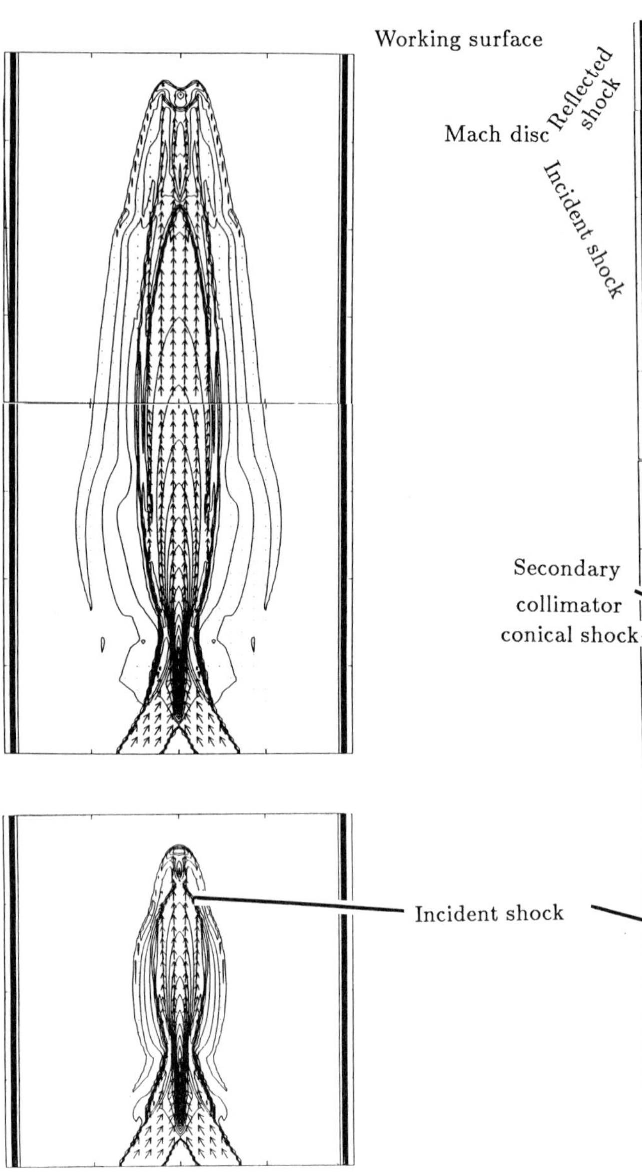

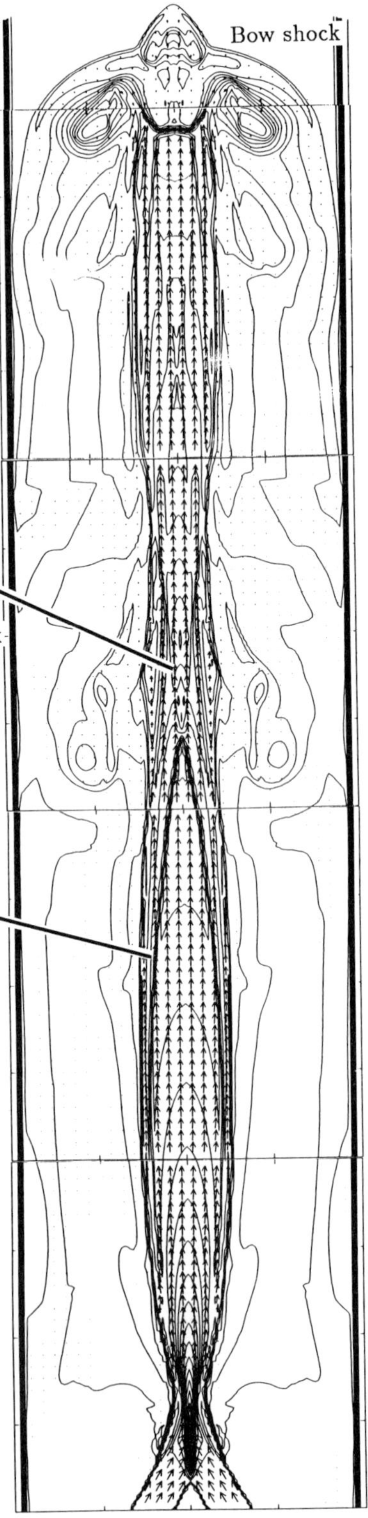

Fig. 4. The structure of interstellar jets. The same jet shown in figure 3 after 456yr, 1270yr, and 4355yr of evolution. The distance between consecutive tick marks is 10^{16} cm.

further away beam matter, will cross a more frontal section of the incident shock in order to acquire the same pressure value (P_{shock}). Meanwhile, the shocked beam is redirected to slide behind the incident shock. As long as P_{shock} is larger than P_0, the incident shock will continuously adjust, mainly by moving further away from the injection point, until the steady state condition, $P_{shock} = P_0$, is achieved. At that moment, the incident shock becomes a standing feature in the flow, and the shocked matter behind it evolves into a stationary supersonically converging conical stream. The solution to such a flow is given in section I. Clearly, a secondary collimator conical shock will form. This will replenish the beam and will re-inject it into the surrounding gas, mimicking our initial condition. Therefore, the flow pattern is to repeat itself, many times, as the head of the jet continues to penetrate the interstellar gas.

Figure 4 shows the same jet shown in figure 3, at earlier times. The two features the incident shock and secondary collimator conical shock (referred to usually as the reflected shock) are clearly indicated. Notice however that at such former evolutionary times, both structures continuously move with the flow. Furthermore, both features are originally sitting at the head of the jet, and in fact form part of the so called working surface. *i.e.* a collection of incident, reflected and Mach disk shocks, which while decelerating the beam matter, provide it with the large overpressure that drives a leading bow shock. As soon as the reflected shock begins to replenish the beam, even before its stationary state is reached, a new set of working surface shocks appears to lift the pressure of the beam matter and push further the leading bow shock. Such an evolution therefore also causes drastic temporal changes at the head of the jet which in fact become periodic as the jet grows. *i.e.* they repeat themselves every time the head of the jet crosses a location to be eventually occupied by a stationary set of incident and reflected shocks. Figure 4c displays for example a well developed set of working surface shocks with a wide Mach disk joining at triple points the new set of incident and reflected shocks. The Mach disk presents its maximum extent when the working surface lies half-way between the last collimation point and the future stationary location of the incident shock. It then decreases in size, to become unresolved by the numerics, as the incident shock approaches its new stationary state. Such changes also affect the shape, and thus the strength, of the leading bow shock. Therefore, the large variety of strengths and possible shapes acquired by the shocks at the head of a hydrodynamical jet could certainly explain the variety of degrees of excitation observed in IIII objects, while accounting for their observed proper motions. From the study of the jet structure one can conclude that:

i) In an underexpanded jet, self-collimation is achieved by means of oblique incident shocks which drive the beam matter to compose converging conical streams which then lead to the appearance of reflected (secondary collimator conical) shocks that replenish the beam, provide it with an overpressure, and re-direct it to move along the symmetry axis.

ii) Every pair of incident and reflected shocks form a so called "crossing" shock within the jet structure. The time dependent solution has shown us that crossing shocks are evolved working surface shocks that approached a stationary state to provide the required self collimation.

iii) The separation between stationary crossing shocks depends only on the value of the initial ambient pressure P_0. When the pressure behind an incident shock $P_{shock} = P_0$, the crossing shock becomes stationary and thus, the larger the initial pressure P_0 the smaller the separation.

iv) Several crossing shocks may be found along a jet structure. However, from these only the most outer ones would be moving, adjusting the flow to become stationary. Thus, their relation to the emitting knots seen in optical jet structures near young stellar objects may be an over-interpretation of the theory. Note for example that the measured velocity of observed neighbouring emitting knots is vastly different, and both accelerating as well as decelerating velocity patterns have been detected (Reipurth, 1989).

III. Jets and outflows driven by a thermal bath.

No doubt nature provides several other ways of generating supersonically converging conical streams, and thus a variety of events capable of producing hydrodynamical jets. This section of the paper deals with a mechanism, and the range of parameters required, to cause a jet from the conical supersonic convergence of matter originally belonging to a compact disc (CD). The project, presently being carried out in collaboration with Drs. J. Franco and P. Bodenheimer, assumes a radiative flux produced by a recently formed star, or at its accretion shock, that strongly modifies the temperature structure of the CD. Upon perturbation of the disc, large pressure gradients result and these while overwhelming the forces that originally keep the disc together, lead to a strong acceleration of the irradiated matter. For the case of massive stars, Bally and Scoville (1982) and Pudritz (1985) have suggested that supersonic outflows should result upon photo-ionization of the disc matter out of which the stars formed. Photo-ionization leads to a temperature (T_{gas}) of about 10^4K, and this to strong pressure gradients, both at the border of the HII region (at the ionization front) and within the ionized volume, between all neighbouring parcels of gas presenting strong density gradients The pressure gradients lead then to expansion, both into the remaining neutral section of the disc, causing a strong compression of the neutral matter, and into the lower density sections of the HII region volume. The latter (as shown by Tenorio-Tagle, 1979) leads to supersonic motions directed along the initial density gradients, and cause the rapid disruption of the CD (see also the contribution of Dr. Franco). Thus, a crucial issue in such a model is the true density stratification of a CD. A self consistent approach should account for rotation, the central gravitational force, the presence of a B field and for the thermal pressure of the CD. Such a detailed structure is beyond the present initial approach to the problem. Instead, adhoc density distributions, similar to those shown in figure 5 have been assumed. Such density stratifications present two regimes: sections of the disc, far from the central source, have a plane stratification with a constant scale height (H). Closer to the center, the distribution may still be described with a plane stratification but with a decreasingly smaller scale height, causing a well defined funnel with a decreasingly smaller aperture angle α , as one approaches the axis of symmetry.

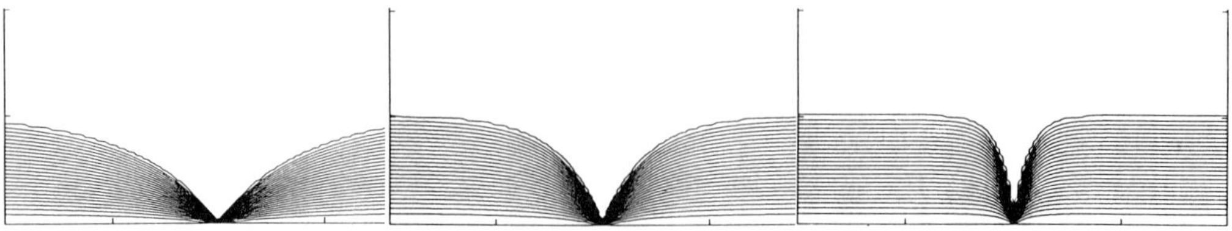

Fig. 5. Adhoc CD density distributions

Note that such an stratification may not be identical to the actual true density distribution in CDs, however our approximations are likely to cover the range of parameters representing the true aspect of CDs. Given such a density distribution, upon photo-ionization the pressure gradients within the ionized volume would lead to two well directed outflows: one of them, perpendicular to the disc plane, would emanate from the ionized section of the disc with an initial constant H. A second faster flow, would develop from the disc funnel where the density gradient is even more abrupt. This latter flow will clearly be a supersonic conical stream, converging towards the symmetry axis with an incidence angle Θ , pre-set by the local aperture of the CD ($\Theta = 90 - \alpha$). Figure 6 shows two dimensional calculations of these flows. In both cases the original density at the disc plane is $10^6 cm^{-3}$while outside the interstellar matter density is 1 cm^{-3}. The ionizing flux ($F_\star$) = 10^{49} photons s^{-1} , and the ionized section of the discs is indicated in the figures by a dashed line. Expansion leads to large velocities and thus within a few hundred years the irradiated disc is disrupted. The largest velocities occur at the skin of the disc, where the density gradient is largest however, the whole outward flow becomes supersonic (u $\gg$ 10km s^{-1}) and soon it reaches large dimensions (see figure 6).

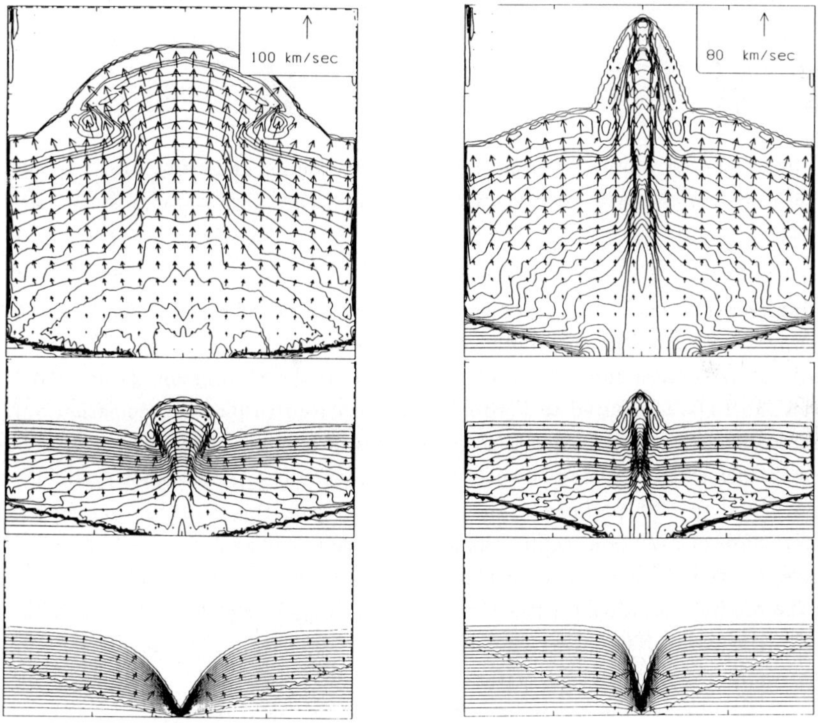

Fig. 6. CD disruption. Similar to figure 1, The original CD density stratification and evolution of two cases are shown at times 147yr, 522yr, 1089yr, and 150yr, 814yr, and 1453yr, respectively. Distance between consecutive tick marks is 7.5 $\times 10^{16}$ cm.

All models for the formation of either jets, HH objects and/or molecular outflows when confronted with the observations present energy problems. This is also the case when applying the above results to the outflows seen in the vicinity of low mass stellar objects. In such a case, surely both the IAU and some of the respectable members of the establishment, will not allow the usage of the expensive ionizing radiation ($F_\star$) assumed in the above calculations, and therefore several steps have been taken to alleviate the energy problem. First of all, the photon flux has been reduced by six orders of magnitude. Note that an O6 star produces 10^{49} photons s^{-1} , while a B0.5 has an output of 10^{46} ionizing photons s^{-1} (see Osterbrock 1989). In the following calculations $F_\star$ has been set equal to 10^{43} photons s^{-1} .

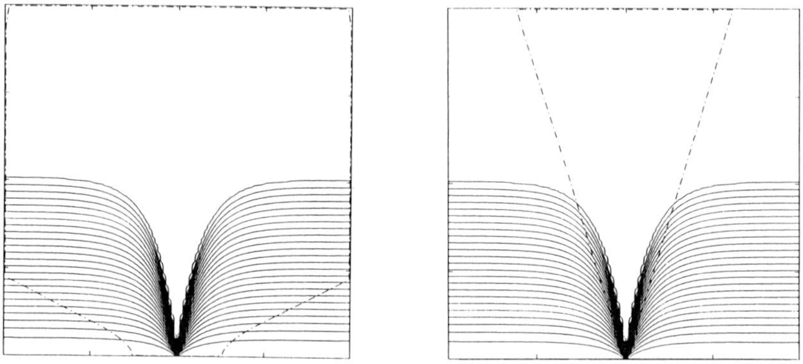

Fig. 7. The irradiated volume. The initial HII region volume caused by an $F_\star = 10^{49}$ and 10^{43} photons s^{-1} are shown. Distance between consecutive tick marks is 3.75 $\times 10^{16}$ cm.

Such a flux causes a much smaller ionized initial volume as indicated in figure 7. Secondly, the temperature caused by the photon flux, has been set to $T_{gas} = 2000$K, instead of the 10^4K resultant from photo-ionization. This assumes that the average energy per photon is smaller than the 13.6eV required to ionize H, and may lead to photo-dissociation only. How many of these photons are produced at the surface of a young stellar object or at its accreation shock, say when it undergoes a Θ Orionis phenomena, is not known and thus it has been set equal to 10^{43} photons s^{-1} . A further step in this direction has been to simulate a short flash of radiation, rather than the constant photon flux condition used in the case of massive stars. This has been achieved by means of a boundary condition which sharply cuts the photon flux after a pre-set critical time (t_{crit}). *e.g.* $F_\star = F_\star$ (time/t_{crit})$^{-5}$. Values of t_{crit} equal to 100yr, 20yr, and 5yr have been used without leading to noticible changes in the resultant flows. The calculation shown in figure 8 resulted from an $F_\star = 10^{43}$ photons s^{-1} , which provided the irradiated matter with a temperature $T_{gas} = 2000$K, and was sharply reduced after a $t_{crit} = 5$yr. The irradiation of the disc leads to similar flows to those obtained for the case of massive stars (compare figures 8 and 6). The irradiated funnel responds inmediately to the pressure gradient and collapses supersonically towards the symmetry axis. The self-collision starts at the "surface" of the star causing the formation of a collimator shock. This however, given the finite amount of converging matter, cannot be supported at such location and thus it effectively moves upwards along the symmetry axis, with a phase velocity $U = U_0 \ / \cos \Theta$, as matter originally further away approaches it with a velocity U_0 . The collimator shock presents under these circumstances a narrow cylindrical shape the length of which is determined by the local duration of the transient converging flow, and the phase velocity U. The cylindrical shock is also an oblique shock to the incoming funnel matter. Note that given the continuous change in the local value of α the shocked matter would retain larger values of the incoming velocity, the further away from the stellar surface that it encounters the cylindrical shock. Concecuently, the velocity of the collapsing material originally deep inside the funnel, where the aperture angle α (= α_0) is small, will be almost fully termalized leading, upon radiative cooling, to a dense rod of shocked matter slowly moving along the symmetry axis. It is in fact not until matter with an original angle $\alpha \geq 45°$ reaches the cylindrical shock that a fast well directed stream ($U_{jet} = U_0 \ \cos \Theta$) begins to be injected into the undisturbed surrounding gas, pouring out of the tip of the cylindrical collimator shock. The shock is still moving upwards with its phase velocity U. However by then, U has become smaller than the transmitted velocity U_{jet} , allowing for the injection of the well directed supersonic beam and thus for the formation of a hydrodynamical jet (see figure 8). Not all density distributions lead to such feature (see figure 6). If the original aperture angle α_0 , at the deepest part of the funnel is large (say $\geq 45°$), then the converging flow has time to fan away even before it reaches the cylindrical shock, and thus inhibiting the formation af a jet. Cases with a small α_0 (see figures 6 and 8) lead to a better collimation.

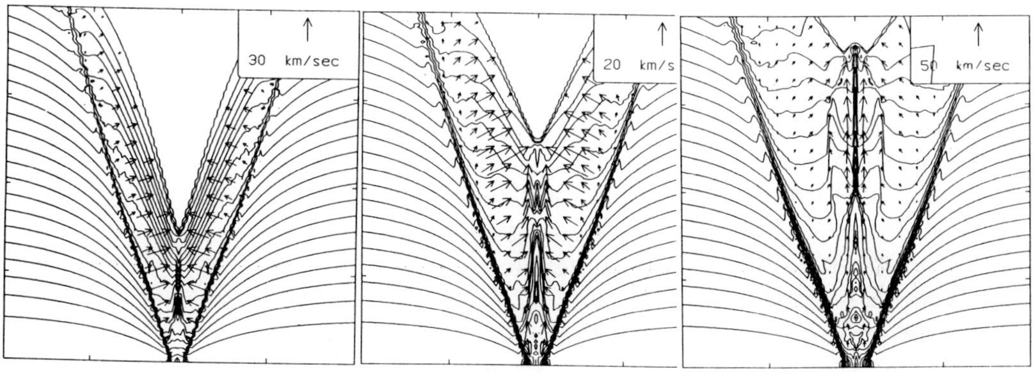

Fig. 8. The supersonic collapse of the CD funnel. A time secuence obtained at t = 94yr, 140yr, and 200yr, showing the rapid development and upward motion of the cylindrical collimator shock. The last frame also shows the collimated flow begining to move ahead of the converging conical stream. Distance between consecutive tick marks is 1.5×10^{16} cm.

The fact that all calculated flows look alike under all circunstances considered *i.e.* under the assumption of photo-ionization or photo-dissociation, and even in the cases where a short flash of radiation was assumed, result from the lengthly recombination and cooling times both of which are much larger than the dynamical time promoted by the pressure gradients. Thus, even if the densest parts of the irradiated disc cool fairly quickly, by then the flow pattern would be fully developed. *i.e.* matter may cool and/or recombine however, as the momentum gained is not radiated away, the cooling stream will continue to move supersonically along the track defined by the pressure gradients. Therefore, from the explored parameter space which include the CD density distribution, the value of $F_\star$, its duration, and the temperature acquired by the irradiated gas (T_{gas}), it seems likely that bipolar flows, jets and their resultant HH objects could in many cases be simply signaling the disruption of the CD, remnant of the stellar formation process. The disruptive radiation may arise from either the star or from its accretion shock, but all irradiated sections of the CD would be disrupted leading to well directed outflows. The event could in principle be repeated many times in the same source, if the remains of the disc have time to readjusts before another massive accretion event takes place.

The model for jet formation here described differs from all other models that suggest the funnel matter to be mechanically pushed aside until a nozzle forms and a rarer interior gas streams through. Instead, these calculations have shown the unavoidable supersonic collapse of the disc funnel, which in some favorable cases leads to the formation of a hydrodynamical jet, upon the sudden inmersion of the disc matter into a thermal bath.

Acknowledgments.

I am gratefull to Drs. M. Różyczka , J. Franco and P. Bodenheimer for allowing me to show some of our results prior to publication.

References

Bally, J. , Scoville, N. Z. 1982.: Astrophys. J.255,497.

Cantó, J.: 1980, *Astron. Astrophys.* **86**, 327

Cantó, J., Rodriguez, L.F.: 1980, Astrophys. J 239, 982.

Cantó, J., Tenorio-Tagle, G., Różyczka, M.: 1988, *Astron. Astrophys.* **192**, 287

Osterbrock, D. E. 1989. Astrophysics of Gaseous Nebulae and Active Galactic Nuclei. University Science Books. Mill Valley, California. p 22.

Pudritz, R. E. 1985.: Astrophys. J. 293,216.

Reipurth, B.: 1989, *Astron. Astrophys.* (in press)

Tenorio-Tagle, G. 1979, Astron. Astrophys. 71, 59.

Tenorio-Tagle, G., Cantó, J., Różyczka, M.: 1988, *Astron. Astrophys.* **202**, 256

Tenorio-Tagle, G., Różyczka, M.: 1989, *Astron. Astrophys.* (in press)

Discussion:

ELSÄSSER. What one observes in several cases, for instance in L1551, is a rather broad flow pattern besides a strongly collimated jet within this. Can this phenomenon be understood within the frame of your model?

TENORIO-TAGLE. First of all allow me to say that in observational astronomy everything that appears luminous and straight is called a jet. A true hydrodynamical jet must present the structure discussed in section II, otherwise it is not a jet. I do see great resemblance between the flows detected in several sources and the computed models shown in section III (see figures 6-8). A well developed flow emanating from the disrupted CD contains an elongated, straight and denser rod of matter, formed during the collapse of the deepest part of the CD funnel. This, given its larger density cools faster and thus it should appear more luminous than the overall flow, in agreement with your observations.

DYSON. What happens if you include stellar winds?

TENORIO-TAGLE. As shown in figures 6-8, the thermal bath leads to a supersonic motion and thus within a few hundred years (see figure 8) the flow or thermal wind, which in some cases may be interpreted as a stellar wind!, is well established. Therefore I will expect that by the time your winds become important, if at all, the whole density distribution will already be largely modified by the thermal disruption of the CD. This is particularly true for all models based on the steady state wind configuration which requires a much longer time to be reached.

PECKER. You obtain splendid collimation! But you start from a strictly symmetrical cylindrical geometry. Would the actual initial condition be less strictly regular, let us say with fluctuations in density distribution, would you keep the collimating effect, or on the contrary, would this collimating effect be entirelly destroyed?

TENORIO-TAGLE. All models here described for the formation of hydro jets do required a supersonic conical stream converging towards the symmetry axis. I suppose, slight deviations in the symmetry of this flow could still yield a jet.

MELNICK. When you look at detailed narrow band optical pictures of jets you can get the impresion that they have helical structures that, at least to me, suggest the presence of magnetic fields. Can you comment on how would magnetic fields modify your model?

TENORIO-TAGLE. Perhaps one should wait for a definite confirmation before jumping into B field wild speculation. In the mean time I shall try to cure my present phobia agains magnetic fields.

THE COLLIMATION OF NONADIABATIC WINDS FROM YOUNG STARS

A. C. Raga

Canadian Institute for Theoretical Astrophysics
60 St. George Street, Toronto, Ontario, Canada

J. Cantó

Instituto de Astronomía, U.N.A.M.
México 20, D. F., México

1. INTRODUCTION

Outflows from young stars in many cases show evidence for collimation. For example, bipolar molecular outflows are sometimes quite well collimated (see, e. g., Snell 1982), and stellar jets and Herbig-Haro objects show an even higher degree of collimation (see, e. g., Mundt 1987).

Different models have been proposed to explain this quite remarkable characteristic of stellar outflows. The better explored theoretical scenario consists of a source surrounded by a stratified "thick disk" environment. The stellar wind, which is assumed to be initially isotropic, is collimated by the interaction between the wind and the environment.

Two different regimes for this flow have been explored in the past (see, e. g., the review of Dyson 1987) :

1 - Cantó (1980) showed that in the limit of very high radiative losses (*i. e.* , cooling distances much smaller than other characteristic distances of the flow) the above configuration leads to the formation of two elongated cavities (see figure 1). These cavities are surrounded by a narrow layer of cold gas, which flows towards the apex of the cavities,

2 - Königl (1982) studied the adiabatic limit, in which two oppositely directed de Laval nozzles are formed (see figure 2).

Raga and Cantó (1989) have recently extended the work of Königl (1982) to the case of nonnegligible radiative energy losses. The results from these calculations describe in an approximate way the regime of stellar wind collimation with moderate radiative energy losses, intermediate between the assumptions of the models of Cantó (1980) and Königl (1982).

Figure 1 :

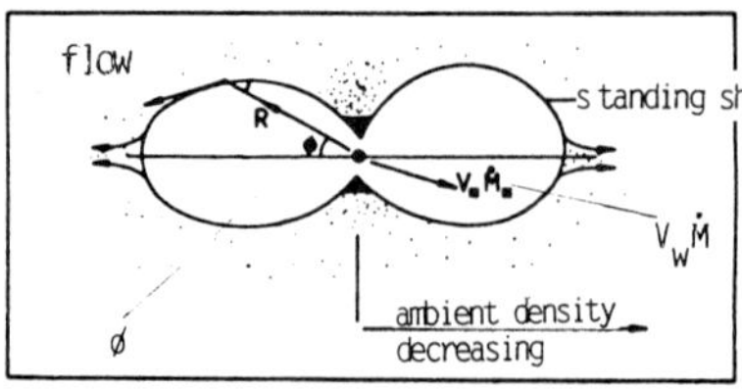

Figure 2 :

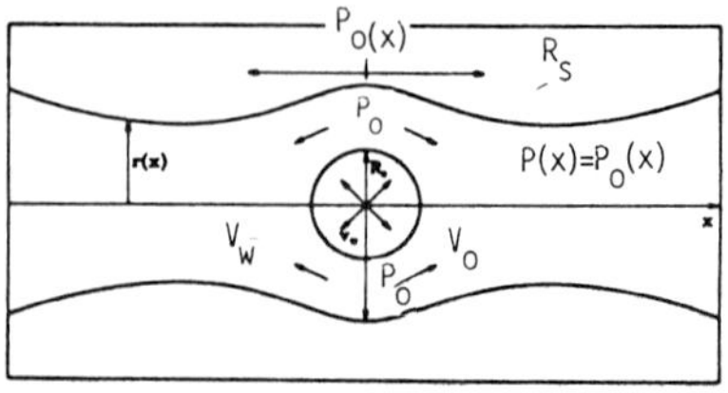

2. ADIABATIC VS. NONADIABATIC COLLIMATION

In order to decide which regime is relevant, we have to evaluate the magnitude of the cooling distance of the postshock gas relative to other characteristic distances of the flow. We can do this by studying a very simplified description of the flow.

In a de Laval nozzle flow, the stellar wind goes through a shock at a radius R_S (see figure 2). The position of this shock and the postshock characteristics are determined by the balance between the ram pressure ($\rho_w v_w^2$) of the wind and the environmental pressure at the equatorial plane (P_0) :

$$P_0 = \rho_{e,0}\,\Delta v_0^2 = \rho_w\, v_w^2\,, \tag{1}$$

where $\rho_{e,0}$ and Δv_0 are the environmental density and turbulent velocity (respectively) at $x = 0$ (see figure 2).

We define a dimensionless parameter $\kappa = d_{cool}/H$, where d_{cool} is the postshock cooling distance of the stellar wind, and H is the environmental scale height. The limit $\kappa \to \infty$ corresponds to the conditions required for the formation of an adiabatic de Laval nozzle. For $\kappa \to 0$ the flow should resemble the highly nonadiabatic cavity model of Cantó (1980).

Following Cantó (1980), for the environment we assume an "isothermal disk" pressure stratification, so that :

$$H = \frac{P_0^{1/2}}{(8\pi G)^{1/2}\rho_{e,0}} \approx 1.7\times 10^{16}\text{cm}\left(\frac{\Delta v_0}{\text{km s}^{-1}}\right)\left(\frac{n_0}{10^7\text{cm}}\right)^{-1/2}, \tag{2}$$

where n_0 is the number density of the environment at $x = 0$ (see figure 2). For the case of stellar environments, this implies possible scale heights in the range of $10^2 - 10^4$ astronomical units.

Using equations (1) and (2) and the cooling function of Raga and Cantó (1989), we obtain :

$$\kappa = \frac{d_{cool}}{H} \approx 8.3\times 10^{-3}\left(\frac{\Delta v_0}{\text{km s}^{-1}}\right)^{-4}\left(\frac{H}{10^{16}\text{cm}}\right)\left(\frac{v_w}{100\,\text{km s}^{-1}}\right)^{6.26}, \tag{3}$$

for wind velocities in the range $v_w \sim 80 - 800$ km s^{-1}.

From this equation we see that κ is extremely sensitive on the wind velocity. For example, for an environment with $H = 10^{16}$ cm and $\Delta v_0 = 1$ km s^{-1}, a $v_w = 100$ km s^{-1} wind corresponds to $\kappa \approx 10^{-2}$, while a somewhat faster $v_w = 300$ km s^{-1} wind already implies $\kappa \approx 10$! This result illustrates the fact that nozzles with a very large range in κ values are possible for stellar outflows, ranging from almost adiabatic to highly nonadiabatic flows.

Figure 3 :

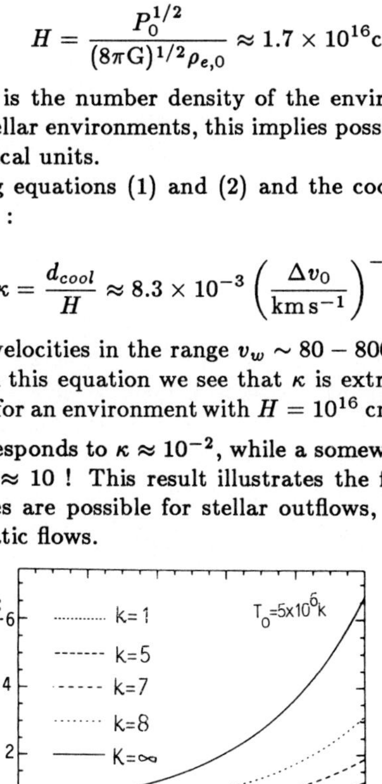

Figure 4 :

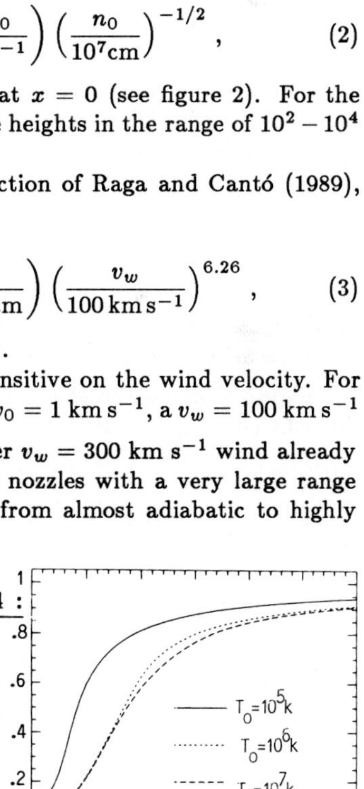

3. GENERAL PROPERTIES OF NONADIABATIC NOZZLES

As we have described in §1., both the adiabatic (*i. e.* , $\kappa \to \infty$) and the highly nonadiabatic (*i. e.* , $\kappa \to 0$) regimes have been studied in the past. The recent work of Raga and Cantó (1989) extends these calculations to the finite κ regime.

However, the models of Raga and Cantó (1989) are extremely simplified, and based on assumptions almost completely identical to the ones of Königl (1982), the only difference being the introduction of a radiative energy loss term in the equations. Because of this, while the Raga and Cantó models do reproduce the adiabatic results for the $\kappa \to \infty$ limit, they do not converge to the Cantó (1980) cavity solution in the $\kappa \to 0$ limit. This is a direct result of the fact that the approximations of the Raga and Cantó models break down for $\kappa < 1$.

In the following, we will limit our discussion to outflows with $\kappa > 1$, so that results based on the calculations of Raga and Cantó (1989) are applicable.

3.a Morphology of the nozzle

We find that the nozzle flow has two distinct regimes.

1 - "Quasi-adiabatic" nozzles : flows with $\kappa = d_{cool}/H > 6$ show general characteristics which are qualitatively not very different from the adiabatic (*i. e.* , $\kappa \to \infty$) nozzle (see Figure 3). These nozzles result in the formation of an overpressured jet of relatively low Mach number ($\sim 2 - 5$).

2 - "Collapsed" nozzles : flows with $\kappa < 6$ show a remarkably different behaviour. These highly nonadiabatic nozzles look similar to the adiabatic nozzle close to the star, but at larger distances, the nozzle radius "collapses" towards the axis. This collapse (which is caused by the strong temperature drop due to the radiative cooling) halts when the gas is cool enough to recombine and the cooling rate decreases drastically (see figure 3). These nozzles produce a very highly collimated, narrow, high Mach number ($\sim 10 - 100$) overpressured jet.

3.b Efficiency and luminosity

Figure 4 shows the efficiency of a pair of nozzles as a function of $\kappa = d_{cool}/H$. The efficiency ϵ_{KE} is defined as the ratio between the kinetic energy flux of the jets (produced by a pair of nozzles) and the kinetic luminosity of the wind :

$$L_w = \frac{\dot{M}\,v_w^2}{2} \approx 8.1 \times 10^{-3}\,\mathrm{L_\odot} \left(\frac{\dot{M}}{10^{-8}\mathrm{M_\odot yr^{-1}}} \right) \left(\frac{v_w}{100\,\mathrm{km\,s^{-1}}} \right)^2 , \tag{4}$$

where $\dot{M}$ is the mass loss rate of the outflow. Curves are given for three values of the post-stellar wind shock temperature $T_0 = 10^5$, 10^6 and 10^7 K, corresponding to wind velocities $v_w \approx 82$, 260 and 820 km s^{-1}, respectively.

Figure 5 shows the radiative luminosity ϵ_L (measured in units of the wind kinetic luminosity L_w) of a nozzle pair as a function of κ. From figures 4 and 5 it is clear that in nozzles with large κ most of the kinetic energy of the stellar wind is converted into kinetic energy of the collimated outflow. On the other hand, for low κ values, most of the kinetic luminosity of the wind is lost by emission of radiation.

Figure 5 .

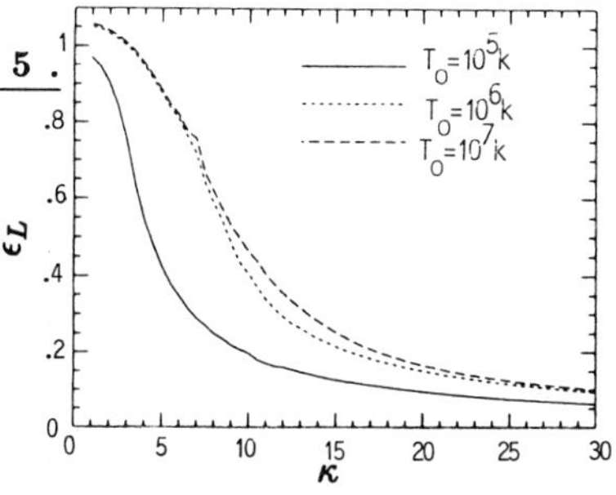

Figure 6 :

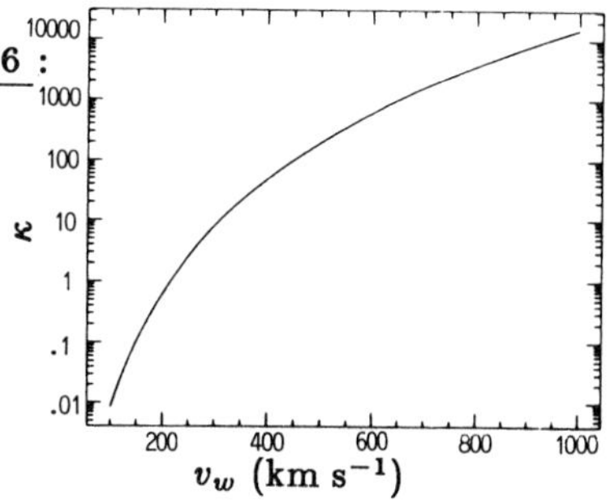

3.c Evolution of a nozzle flow

It is well known that while young stars (e. g., T Tauri stars) have very massive outflows, the mass loss of more evolved stars (e. g., main sequence stars) is much less substantial. Also, it seems that protostars (in earlier evolutionary phases than T Tauri stars) have even more massive outflows. Because of this, it appears to be reasonable to assume that through the lifetime of a collimated outflow, the stellar wind feeding the outflow has a monotonically decreasing $\dot{M}$. Provided that the change in mass loss rate is slow compared to the timescale for setting up a stationary de Laval nozzle flow, we can qualitatively describe the evolution of a bipolar outflow with our stationary models.

A decreasing mass loss rate $\dot{M}$ can be the result of changes in the density or in the velocity of the wind (or both at the same time). Let us discuss separately the effects of varying the density and the velocity of the wind.

3.c.1 *Wind with decreasing density*

If $\dot{M}$ decreases with time because of a decreasing wind density, the principal effect on the nozzle flow will be a gradual decrease in the radius R_S of the stellar wind shock (see figure 2 and equation 3). The κ value of the nozzle, however, will remain unchanged (see equation 3), implying that the general properties of the nozzles remain qualitatively unchanged. The radiative luminosity of the nozzles is proportional to the kinetic luminosity of the wind (see equation 4 and figure 5).

3.c.2 *Wind with decreasing velocity*

If the wind velocity v_w decreases with time, a much more interesting result is obtained. Because of the high dependence of κ on v_w (see equation 3), such an evolution can produce all the range from almost adiabatic nozzles (in an early, high v_w phase) to highly nonadiabatic nozzles (in a later, lower v_w epoch). This is illustrated in figure 6, in which we show κ as a function of v_w. These results have been obtained for a nozzle with an initial velocity of 1000 km s^{-1} and initial mass loss rate $\dot{M} = 10^{-7} M_\odot yr^{-1}$ (the environment is assumed to have $H = 10^{16}$ cm and $\Delta v_0 = 1$ km s^{-1}), but can be scaled to other values using equations 2-4.

In figure 6 we see the drastic change of κ during the evolution of the wind. In particular, we see that for $v_w \sim 280$ km s^{-1} we have $\kappa \sim 6$, the point at which the transition between "quasi-adiabatic" and "collapsed" nozzles occurs (see §3.a and figure 3). In other words, while the wind has $v_w > 280\,\mathrm{km\,s^{-1}}$ the nozzles are "quasi-adiabatic", resulting in the production of relatively low Mach number, not very highly collimated outflows. On the other hand, as soon as the wind velocity drops to $v_w < 280\,\mathrm{km\,s^{-1}}$, the nozzles enter the "collapsed" regime, so that very highly collimated, high Mach number jets are produced. This result agrees very well with the wind velocity predicted by Dyson (1987) for the transition between an adiabatic and a nonadiabatic stellar outlfow.

Figures 7 and 8 show the radiative luminosity and average temperature (weighted with the emissivity of the gas) of the nozzles as a function of the wind velocity. We see that for large wind velocities the luminosity of the nozzles is low (a result of the high efficiency of the adiabatic collimation, see figure 4). For low wind velocities, the radiative luminosity is also low (due to the low kinetic luminosity of the wind, see equation 4 and figure 4). A sharp peak in the luminosity is obtained for $v_w \sim 280\,\mathrm{km\,s^{-1}}$, the velocity at which the transition between the "quasi-adiabatic" and "collapsed" nozzles occurs. The radiation field emitted at the luminosity maximum corresponds to the spectrum of a plasma with a temperature $T \sim 5 \times 10^5$ K (see figure 8).

Figure 7 :

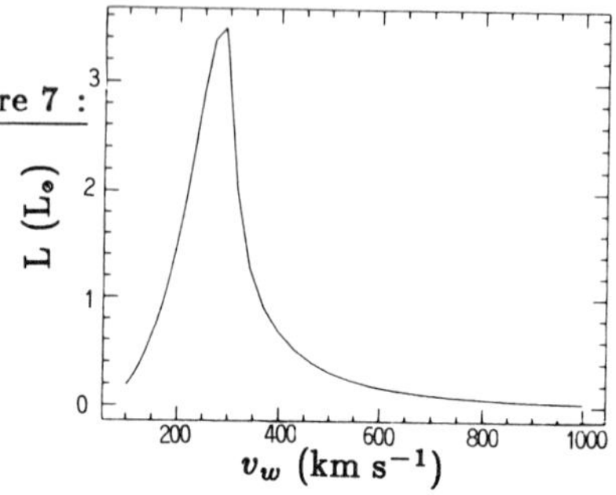

Figure 8 :

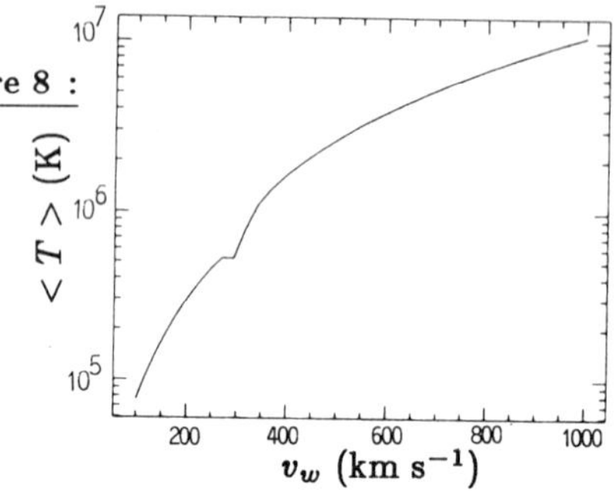

4. CONCLUSIONS

Depending on the importance of the radiative cooling, the collimation of a stellar wind by the interaction with a stratified environment produces qualtiatively different outflows. Three distinct regimes are observed :

- for $d_{cool} > 6 \times H$, "quasi-adiabatic" nozzles are formed, producing a relatively low Mach number outflow,
- for $1 < d_{cool} < 6 \times H$, "collapsed" nozzles are formed, producing very highly collimated, narrow, high Mach number jets,
- for $d_{cool} << H$, the outflow results in the formation of elongated cavities surrounded by a layer of cold gas (as described by Cantó 1980).

The regime of cooling distances lower but comparable to the environmental scale height has still not been explored theoretically.

We find that the nozzle flows that are most likely to be directly observed are the ones with $\kappa \sim 6$ (marginally collapsed nozzles), because they show the highest radiative luminosities. Also, these nozzles are the ones that produce the fastest highly collimated jets (nozzles with $\kappa > 6$ can produce higher velocity outflows, but with the poorer collimation and lower Mach number characteristic of the quasi-adiabatic regime).

An outflow source with a wind velocity that decreases with time could initially produce "quasi-adiabatic" nozzles, and later "collapsed" nozzles. The initial phase would produce a high velocity outflow, but with low radiative luminosity and relatively poor collimation (this phase could be associated with the formation of a bipolar molecular outflow). The final phase would produce a lower velocity outflow, but with a much higher degree of collimation (this phase could be associated with the formation of stellar jets or Herbig-Haro objects).

The transition between these two phases would occur (for the typical environmental parameters chosen above) at a velocity of a few hundred kilometers per second. This might provide an interesting explanation of why very high velocity stellar jets are never observed.

We would like to thank Alberto López for allowing us to use the diagram shown in figure 1. A. Raga would also like to thank Sue Terebey and Ron Snell for pleasant discussions which were extremely helpful for this work. This research was supported by the Connaught Fund of the University of Toronto and the Natural Sciences and Engineering Research Council of Canada.

REFERENCES

Cantó, J. (1980) *Astron. Astroph.* **86**, 327

Dyson, J. E. 1987. In *Circumstellar Matter*, IAU Symposium No. 122, edited by I. Appenzeller and C. Jordan (Reidel, Dordrecht), p. 159.

Kõnigl, A. 1982, *Ap. J.* **261**, 115.

Mundt, R. 1987. In *Circumstellar Matter*, IAU Symposium No. 122, edited by I. Appenzeller and C. Jordan (Reidel, Dordrecht), p. 147.

Raga, A. C., and Cantó, J. 1989, *Ap. J.* (September issue, in press).

Snell, R. 1983, *Rev. Mexicana Astron. Astrof.* **7**, 79.

Discussion:

DULTZIN: Extragalactic people worry very much about stability problems of accretion disks in active galactic nuclei. Do you ever worry (about such problems)?

RAGA: I have really not thought about disk stability problems.

PALLA: Given the importance of radiative cooling in your model, shoudn't non-equilibrium effects in the recombination process be taken into account? Namely, the gas can cool to temperatures well below 10^4K, if the assumption of ionization equilibrium is relaxed.

RAGA: Definitely, non-equilibrium ionization effects would be important. The result of including these effects would indeed be to lower the temperature at which the radiative energy losses became negligible, so that the "collapsed" nozzle solutions would produce narrower, higher Mach number jets.

HERBIG-HARO OBJECTS

Karl-Heinz Böhm
Astronomy Department, University of Washington
Seattle, Washington 98195

Abstract

After a brief introductory discussion of the statistics of known Herbig-Haro (HH) objects we present a survey of recent spectroscopic results in the ultraviolet, the optical and the near infrared range (the latter mostly in connection with H_2 observations). We emphasize the importance of the use of spatially resolved line profiles (position-velocity diagrams) in the optical range for the purpose of testing hydrodynamic models of HH objects. Such observations have now been supplemented by the measurement of spatially dependent intensity ratios for a large number of optical lines ~200 in HH 1) and of some ultraviolet lines (including fluorescent H_2 lines) which are very useful for diagnostic purposes. The relevance and importance of spectroscopy and imaging in the infrared H_2 lines is discussed.

In the second part of the paper we review the present status of the interpretation of spatially resolved spectra and of monochromatic images by hydrodynamic models. We emphasize the successes as well as the shortcomings of the bow shock interpretation of HH spectra and point out that there are a few cases (*e.g.*, HH 43) in which a "shocked" cloudlet model is more appropriate than the model of the working surface (plus bow shock) of a jet. We discuss the intriguing [Fe II] problem.

I. Introduction

In recent years our understanding of Herbig-Haro objects has increased rapidly. In the middle to late 70s it became completely clear that optical HH spectra are spectra of shock waves (Schwartz 1975; Böhm, Siegmund and Schwartz 1976; Dopita 1978a, b; Raymond 1979) and that they occur under a wider range of environmental conditions than was originally thought (Münch 1977). Guido Münch has contributed to both results in an important way. His discovery of an HH object in the Orion nebula was of special importance and was the first of a series of such discoveries in H II regions (see *e.g.*, Axon a. Taylor 1984). In the early to mid eighties the idea that HH objects are tracers of highly collimated bipolar jet-like flows from young stars or even protostars became more or less generally accepted. (See *e.g.*. Mundt and Fried 1983; Strom, Strom and Stocke 1983; Mundt *et al.* 1984; Mundt 1985; Strom *et al.* 1986.) It also became clear that often (but not always) the brightest HH objects occur near th end of jet and might be identified with the "working surface" (either the bow shock or the jet-shock, see *e.g.*, Blondin, Königl. Fryxell 1989).

We feel that one of the most important discoveries which led to the general acceptance of these ideas were the proper motion studies by Herbig and Jones (see *e.g.*, 1981, 1983) which showed convincingly that the velocity vectors in the jet flows point indeed radially away from the central star.

At present we have therefore a more or less accepted scenario which gives a very qualitative explanation of Herbig-Haro objects, their relation to bipolar jets and to the circumstellar matter of young stars in general. There remain a large number of unanswered questions, of which we mention three:

1. By which physical mechanisms are the highly collimated jets generated?
2. What is the detailed radiation-hydrodynamics of the HH objects. How can it explain the important observational facts including the spatial variation of line intensities in the UV, optical range and the infrared, the radial velocity field and the radial velocity dispersion as well as the different proper motions of the individual clumps of a single HH object. There is certainly some discrepancy between the optimism which comes from the fact that we do understand Herbig-Haro objects in a very crude way and the difficulties which one encounters if one wants to explain spatially resolved spectra even in an approximate way.
3. Is the standard scenario with HH objects being due to shocks at the working surface of the jet or to internal jet shocks really applicable to all HH objects? Since the detection of an HH spectrum really indicates only the presence of a shock we would not necessarily expect that all HH objects must be explained by the standard scenario. Recent detailed studies of some HH objects more or less force us to accept a different explanation in a few cases (*e.g.*, in HH 43 and HH 24, see Schwartz, Dopita and Cohen 1985; Böhm and Solf 1989; Solf 1987).

In this review I shall concentrate on the observations relevant to points 2 and 3 and their interpretation. Certainly point 1 is the most fundamental one. It is, however, treated in Tenorio-Tagle's (1989) and in Cantó's (1989) paper (see also Dyson 1987). The coverage of our review will be more restricted than that of Mundt (1987) who also included a discussion of jets and their central stars.

II. The Number of Known Herbig-Haro Objects

This paper will be mostly concerned with the study of physical processes and the interpretation of spectra applied to a few selected interesting HH objects.

In order to remind ourselves that HH objects are much more common than indicated by the few objects mentioned below it seems appropriate to state at least very briefly how many HH objects are known. This is, of course, not a very precisely defined question since it is a matter of definition whether the individual condensations in a small compact HH complex is counted as a single object or whether the whole complex should be counted as one HH object.

Usually HH objects are assumed to occur only in regions of recent star formation and we follow this assumption. More general definitions have also occasionally been used (*e.g.*, Cohen 1987).

As is well known Herbig (1974) published the original and extremely useful "standard" catalog of HH objects. It contains 77 objects if we consider every individual condensation listed (*e.g.*, HH 24A, B, C, D) as a separate object. In the meantime many new HH objects have been discovered through the efforts of Graham, Hartigan, Meaburn, Mundt, Reipurth, Schwartz, the Stroms and many others. There is now a new, very useful catalog of HH objects by von Hippel, Burnell and Williams (1988) which lists 184 HH objects (again counting individual condensations as separate objects). The authors have restricted themselves to objects for which their identification as HH objects is not in doubt.

One might have some doubt about considering individual condensations as individual objects. If they really owe their existence to hydrodynamic instabilities in a bow shock (as considered by Raga and Böhm 1987 and Raga *et al.* 1988) then they are, of course, short-lived transient features only (in agreement with some observations (see Herbig 1969)). But this is really a minor point and a matter of opinion.

III. Spectroscopy and its Interpretation

Optical: In view of the fact that spectroscopic studies of HH objects have been carried out already in the 1950s (Herbig 1951; Böhm 1956; Osterbrock 1958) it would seem surprising that it should be possible to get still important new insights from optical spectroscopy. But this is definitely the case. The progress has been based mostly on a) high spatial resolution spectroscopy (Hartmann and Raymond 1984; Böhm and Solf 1985; Schwartz, Dopita and Cohen 1985; Solf, Böhm and Raga 1986; Solf 1987; Solf and Böhm 1987), b) high spectral resolution obtained by using echelle or coude spectrographs and c) highly improved spectrophotometry using CCDs and other modern detectors. Typically spectral resolutions of up to ~0.3 Å (corresponding to ~15 $\mathrm{km\,s^{-1}}$ velocity at $\mathrm{H\alpha}$) and spatial resolutions of $1''$ (seconds of arc) have been achieved for a number of brighter emission lines in the moderately bright HH objects. The amount of information contained in these spatially resolved line profiles ("position-velocity diagrams") is considerable and can be used for rather sensitive tests of theoretical models. The position-velocity diagrams as well as spatially integrated line profiles have played an especially large role in the test of bow shock model of HH objects. (See *e.g.*, Choe, Böhm and Solf 1985; Raga and Böhm 1985, 1986, 1987; Hartigan, Raymond and Hartmann 1987; Raga *et al.* 1988; Solf, Böhm and Raga 1986.)

Spatially resolved high resolution spectroscopy (with 15-20 $\mathrm{km\,s^{-1}}$ resolution) is available 12 of the brighter HH objects (HH 1, HH 2, HH 3, HH 7, HH 8, HH 10, HH 11, HH 12, HH 24, HH 32, HH 34, HH 43, HH 47). Of these HH 1, HH 2, HH 24, HH 32 and HH 34 have been studied in specially great detail and it is well known that specially for HH 1, HH 32, and HH 34 relatively detailed bow shock models have been developed which reproduce position-velocity diagrams fairly well whereas the position-velocity diagrams of HH 7-HH 11 so far have not permitted any simple hydrodynamic explanation.

More recently spatially resolved echellette spectra of high spatial but moderate (1-2 Å) spectral resolution have supplied us with a wealth of new information (Solf, Böhm and Raga 1988; Böhm and Solf 1989). They permit the simultaneous coverage of the wavelength range $3720 \lesssim \lambda \lesssim 10830$ Å and have lead to the discovery of many new lines and the measurement of the spatial variation of intensities and radial velocities of these lines. The measurement of the spatial distribution permits us to decide immediately whether a line is formed in the HH object or in its environment. Newly discovered lines include extremely low ionization like the NaD lines as well as quite high ionization lines like the A IV lines. Since the instrument is very sensitive the spatial variation of quite faint lines can be studied. This includes the "auroral" lines of [O I], [C I], [N II], [O II], [S II], [O III] permitting us to determine the spatial variation of the electron temperature in the line forming regions of these ions from the spatial variation of the ratio of "auroral" to "nebular" lines. A comparison of the electron temperatures for the line-forming regions for different ionization stages gives us a new and very direct proof of the presence of shock waves. It varies (*e.g.*, in HH 1) monotonically from $T_e \sim 1.0 \times 10^4$ K for the [C I] regions to $\sim 4.5 \times 10^4$ K in the [O III] regions. The information about the spatial variation of up to 200 lines in a single object permits very sensitive tests of the hydrodynamics of an HH object.

Let me arbitrarily select three interesting problems which have resulted from recent spectroscopy studies.

1. If we interpret the spectra in terms of bow shock models (which is certainly appropriate in many cases, see above) we find that there are a few objects for which the jet model (or a bullet model) is not acceptable. HH 43 is the best example it as three condensations A, B and C. The radial velocity is relatively large and negative (50 to -60 km s^{-1}) in most places. Only "in front" of each condensation does the velocity go to values of ~ -10 to -20 km s^{-1}. This can be only understood if the flow (stellar wind) is brought to rest in front of the three obstacles HH 43A, B and C which is exactly the situation which is considered in the "shocked cloudlet model" (Schwartz 1978).

2. Studying the spatial variation of a larger number of line ratios in bow shock-like HH objects we find *e.g.*, in HH 1 excellent to good agreement in a fairly large number of line ratios including [O II] 3727/Hα, [O III] 5007/Hα, ([S II] 4069/76)/Hα, ([S III] 6716/31)/Hα, [N II] 6583/Hα, [S II] 9532/Hα, ([N I] 5198/5200)/Hα. There are other ratios (*e.g.*, [O I] 6300/63/Hα) for which the agreement is qualitative only. For a very few lines there is fairly strong disagreement between prediction and observation. These include the [A III] lines though the fairly similar [O III] and [S III] lines show agreement between observation and theory.

3. The sensitivity of the instrument permitted us to study a large number of [Fe II] lines in several HH objects. At present there is not yet enough information about collision strength available in order to carry out a convincing abundance analysis. Nevertheless, the following quite intriguing result can be easily obtained. The average intensity ratio of the [Fe II] lines to, say, Hβ is drastically different for different objects It varies about a factor 10 in the objects observed so far. However, contrary to our naive expectation the [Fe II]/Hβ line ratio is not correlated with the excitation of the HH object (*i.e.*, low excitation objects do not on the average show [Fe I]/Hβ ratios different from those of high excitation objects). This indicates that these average line ratios are neither a direct consequence of the presently observed excitation or ionization nor they are a consequence of the present grain destruction which should be related by the presently observed shock strength. Surprisingly the [Fe II]/Hβ ratio seems to be constant for objects which belong to the same HH complex. For instance, the three objects HH 43A, B and C which lie very close together in space (within $\sim 30''$) all show the same unusually low [Fe II]/Hβ ratio although HH 43A is a high excitation object (comparable to HH 1) and HH 43B and C are definitely low excitation objects

Ultraviolet: The earlier I.U.E. observations of HH objects have been discussed in preceding review papers, *e.g.*, Schwartz 1983c, 1985; Böhm 1983. One of the basic "old" I.U.E. results was the dramatic difference between "high excitation objects" (represented by HH 1, HH 2, HH 32, see Ortolani and d'Odorico 1980; Böhm, Böhm-Vitense and Brugel 1981; Brugel, Shull and Seab 1982; Böhm-Vitense *et al.* 1982; Böhm and Böhm-Vitense 1984) and "low excitation objects" (represented by HH 43 and HH 47, see Schwartz 1983a, b). High excitation objects show C IV 1550 and C III] 1909 as the strongest lines in the IUE short wavelength range (1250 Å $\lesssim \lambda \lesssim$ 1950 Å) whereas low excitation HH objects show almost only fluorescent H_2 lines (probably pumped by the very strong Lα emission of the shock. The fact that no intermediate objects are seen is impressive but the HH object sample accessible with IUE is still too small to draw convincing conclusions in this respect.

During the last few years very long (2-shift) exposures of I.U.E. spectra of HH objects have been obtained and (together with archival data) been used mostly in connection with two problems:

1. Attempts to determine the spatial distribution of continuum and line emission also in the ultraviolet and a comparison of the results to analogous optical data (see above).
2. A more detailed study of the wavelength dependence of the ultraviolet continuum and a test of the hypothesis that the ultraviolet continuum as well as the optical continuum is due to collisionally enhanced two-photon emission. (Dopita, Binette and Schwartz 1982; Brugel, Shull and Seab 1982.)

With regard to point 1, special precautions are necessary. The point-spread function of I.U.E. and the variation of the sensitivity (see Clarke and Moos 1981) have, of course, to be taken into account. The results of 1 agree to a large extent at least qualitatively with expectations (Böhm *et al.* 1987; Lee *et al.* 1988). Emission regions of C IV 1550 and C III] 1909 are typically small and comparable to [O III] emission regions. Surprisingly the emission regions of Mg II 2800 and C II] 2326 are also relatively small. The continuum emission region in the short wavelength range (1250 Å $\lesssim \lambda \lesssim$ 1950 Å) of IUE is (spatially) considerably more extended than the emission regions of IUE long wavelength and of the optical range (which agree with each other). We consider this as an additional evidence that in the SW region of IUE a second continuum emission process is important in addition to the collisionally enhanced two-photon emission which essentially determines the continuum in the optical and long-wavelength IUE ranges.

The measurements described under 2 showed that the observed wavelength dependence in the SW range of IUE is not compatible with a pure two-photon continuum (Figure 1). Our tentative conclusion was that a superposition of a collisionally enhanced two-photon continuum and a fluorescent H_2 continuum (see Dalgarno, Herzberg and Stephens 1970) would fit the observations much better.

We are now also studying the spatial emission distribution of the fluorescent H_2 lines in the low excitation objects HH 43 and HH 47. Interestingly, we find that, in both HH 43 and HH 47, the lines are formed in considerably smaller regions than the continuum.

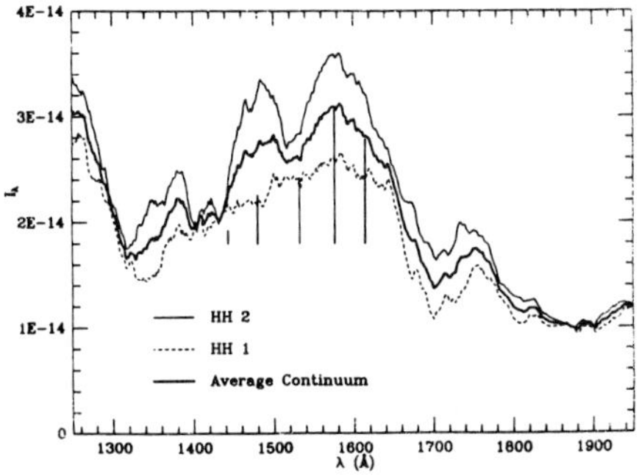

Figure 1. The observed continuum of HH1 and HH2 in the short wavelength range of IUE. The length of the vertical lines in the range 1440 Å $\lesssim \lambda \lesssim$ 1660 Å give a crude description of the expected wavelength dependence of the fluorescent H_2 continuum (from Böhm *et al.* 1987).

Near Infrared H_2 Emission. The emission of HH objects in the infrared quadrupole H_2 lines is of great interest and so far has supplied us with interesting but only partially understood information. Schwartz, Cohen and Williams (1987) have detected H_2 IR emission in many HH objects most of which are low excitation objects.

Imaging of HH objects in the light of H_2 quadrupole lines is of great interest. Results by Zealey *et al.* (1986) showed already the very close association between H_2 and optically emitting regions in HH 32. High resolution imaging of HH 43 in the 0-1 S(1) line of H_2 (2.12 μ) by Schwartz *et al.* (1988) shows a very strong correlation between the H_2 and the optical emission of HH 43.

Of very great interest in this context is the H_2 emission of the chain of HH objects HH 7-HH 11 which was observed first in the H_2 lines by Zealey, Williams and Sandell (1984) and Lightfoot and Glencross (1986). Recently Zinnecker *et al.* (1989) have studied the line profiles of the 2.12 μ line in these objects and have found suprising similarities with optical lines in the following sense. The optical line profiles are known (with a velocity resolution of 15 – 20 $\mathrm{km\,s^{-1}}$) from coude spectra (Solf and Böhm 1987). They show somewhat complex line profiles which differ from object to object. The H_2 line profile looks (at least qualitatively) similar to the optical profiles for each individual object. However, quantitatively the lines are consistently somewhat narrower and less blue-shifted than the optical lines. The authors suggest entrainment of molecular gas into the jet forming HH 7-HH 11 as the probable (qualitative) explanation of observed line profiles. The explanation is complicated by the fact that HCO^+ line profiles observed by Rudolph and Welch (1988) show clumps of high density matter ($N \sim 2 \times 10^5$ $\mathrm{cm^{-3}}$ or larger) just "downstream" of HH 10 and HH 8. Rudolph and Welch (1988) argue that this favors the "shocked-cloudlet" model (Schwartz 1978). However, the study of optical position-velocity diagrams (Solf and Böhm 1987) and line profiles (Hartigan, Raymond and Hartmann 1987) indicates that HH 7-HH 11 can be described neither by the shocked cloudlet nor by the bullet model. It is my definite impression that at least in the case of HH 7-HH 11 more theoretical insight is needed before we can have models which permit a synthesis of optical, IR (*i.e.*, H_2) and radio (*e.g.*, HCO^+) observations.

It would be useful to make a comparison of the spatial distribution of the IR H_2 emission and the fluorescent H_2 emission in the UV. We are carrying out such a study for HH 43 using the IR data by Schwartz *et al.* (1988) and new two-shift I.U.E. spectra which we have obtained recently.

IV. The Interpretation of HH Spectra in Terms of Hydrodynamic Models: Success and Unsolved Problems

In the preceding chapters we have discussed ultraviolet, optical and infrared spectroscopy of HH objects and its interpretation, using some illustrative examples.

How successful are hydrodynamical models in general in explaining HH spectra? Before we can answer this we have to state more precisely what we want to explain. Our aims are ambitious. We want to predict

a. the total line fluxes integrated over the whole object (see *e.g.*, Hartmann and Raymond 1984)

b. high resolution line profiles (Hartigan, Raymond and Hartmann 1987)

c. position-velocity-diagrams (see *e.g.*, Choe, Böhm and Solf 1985; Böhm and Solf 1985; Raga and Böhm 1985, 1986; Solf, Böhm and Raga 1986)

d. the spatial variation of the intensity ratios of many lines (Solf and Böhm 1988; Böhm and Solf 1989)

e. the proper motion of "clumps" (condensations) of HH objects (see *e.g.*, Herbig and Jones 1981, 1983; Schwartz, Jones and Sirk 1984; Jones and Walker 1985) In practice position-velocity diagrams have been determined for only a small number of lines because otherwise the required observing time becomes very large. It is much easier to study the spatial variation of the total intensity of a given line (rather than of the line profile).

In a number of objects either a part or all of the above listed observations have been explained successfully by bow shock models. This is especially true for HH 1, HH 32, HH 34, HH 43, HH 47 and (to a lesser extent) HH 2 (see *e.g.*, Choe, Böhm and Solf 1985; Raga and Böhm 1986; Reipurth *et al.* 1986; Solf, Böhm and Raga 1986; Hartigan, Raymond and Hartmann 1987; Raymond, Hartigan and Hartmann 1988; Raga *et al.* 1988). If we restrict ourselves to the interpretation of (spatially integrated) line profiles and to line flux ratios. the bow shock models have been tested and shown to be applicable to many more HH objects (Hartigan, Raymond and Hartmann 1987). For the interpretation of flux ratios, integrated line profiles, and position-velocity diagrams simplified stationary bow shock models have usually and successfully been used. In these the geometrical shape of the bow shock is determined in advanced and then every small piece of the bow shock is approximated by a plane (usually oblique) shock for whose recombination region the line emission can be studied in detail. This approach was first used by Hartmann and Raymond (1984) for the study of flux ratios and by Choe. Böhm and Solf (1985) and by Raga (1985) and Raga and Böhm (1985) for the study of high resolution position velocity diagrams.

This simplified approach cannot be used for the study of the formation of individual condensations (clumps) of a HH object and of their proper motions. Raga and Böhm (1987) therefore have carried out numerical simulation of 2-dimensional time-dependent bow shocks.

While adiabatic flows approach (after some time) the known stationary solution the non-adiabatic (radiative) flows show at high Mach number thermal instabilities somewhat similar to those discussed by Falle (1981) and Innes (1985) for the one dimensional case. Raga *et al.* (1988) found that calculations of this type for a bow shock of 185 $\mathrm{km\,s^{-1}}$ stagnation velocity predict a pattern of condensations, a distribution of proper motions (of the individual condensations), monochromatic Hα and [O III] images and position-velocity diagrams of Hα which are all in (at least qualitative) agreement with the observations (Figure 2). This is probably the most detailed and critical test of the bow shock theory for an HH object so far. One has, of course, to keep in mind that the computations are restricted to two space dimensions (axial symmetry) while the real hydrodynamics of the object is three dimensional. Other very critical tests of the bow shock theory have been done for HH 32 (Solf, Böhm and Raga 1986; Hartigan, Mundt and Stocke 1986). In this case, the rather complex position velocity diagrams could be explained. Also the surprising "double layer" effect (the fact that the "low" velocity component of the line has its spatial maximum always $\sim 0.''5 - 0.''7$ farther away from the central star than the spatial maximum of the high velocity component) has been explained in terms of bow shock models (Raga, Böhm and Solf 1986).

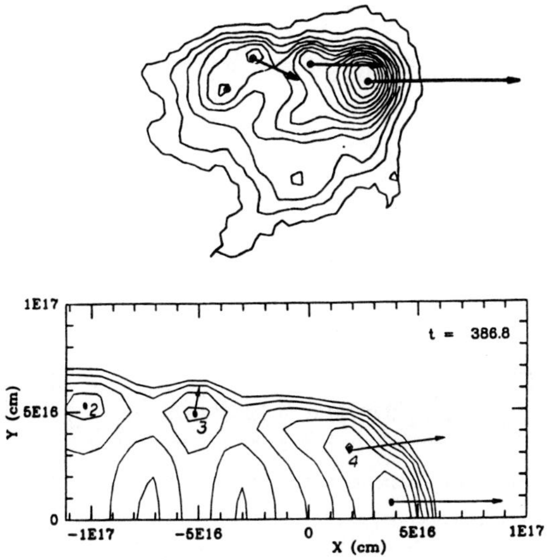

Figure 2. Comparison of observational data for HH1 (upper part) and theoretical predictions (lower part for a time-dependent non-adiabatic bow shock model (V_s = 185 $\mathrm{km\,s^{-1}}$). The contour lines show the spatial intensity distribution (indicating the condensation structure of the HH object), the arrows show the proper motions of the individual condensations. (The proper motions have been transformed to a system in which the condensation with the smallest motion has a velocity of zero.) The theoretical model is axisymmetric (only half of it is shown). Because of this and because of the time dependence the theory and the observations are not fully comparable. It is, however, impressive that the theory reproduces the observations qualitatively well. Based on data from Raga *et al.* (1988).

A more sophisticated approach to the interpretation of the observations and specifically of monochromatic imaging has recently been used by Raga (1988) and in the work of Blondin, Königl and Fryxell (1989), in which the hydrodynamics of the heads of radiative jets including the outer bow shock has been studied. In principle, this is, of course, a more convincing approach than the above mentioned studies of pure bow shocks. It shows that the observed HH emission may be due either to the bow shock or the jet shock or both. At present such calculations still have to make drastic approximations about the details of the formation of the emitted spectrum (as is the case also for the time dependent two dimensional simulations of bow shocks). Consequently the calculations are not yet well suited for a prediction of flux ratios or of position velocity diagrams in many different lines. For these one would at present still have to go back to the simplified stationary bow shock models described above.

There are objects which do not seem to be connected with working surfaces of jets. Let us quote a few examples. If the condensations in jets are internal shocks (see Binette, Raga and Cantó 1989) the clumps in the HH34 jet are a good example (Reipurth *et al.* 1986; Bührke, Mundt and Ray 1988; Raga and Mateo 1988). They often show very low excitation with [S II]/Hα ratios up to 10. Then there are objects (with HH 43A, B and C being the best examples) which definitely seem to be shocked cloudlets (Schwartz, Dopita and Cohen 1985; Böhm and Solf 1989). The strongest evidence for this comes from the mapping of the radial velocity field (see above). In this case bow shock models are still applicable, but the usual jet models are not.

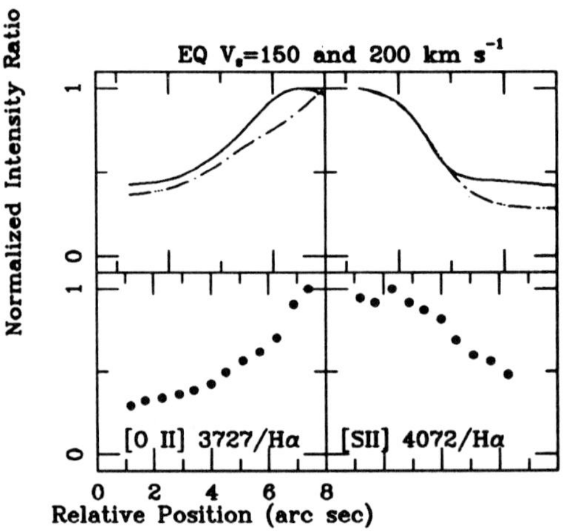

Figure 3. Comparison of the observed (lower part of figure) and predicted spatially dependent intensity ratios [O II] 3727/Hα, [S II] (4068 + 4076)/Hα. The observations refer to HH 1, the theoretical predictions are made for bow shock models with V_s = 150 km s^{-1} (solid line) and V_s = 200 km s^{-1} (broken line). The bow shock axis lies in the plane of the sky. A comparison of theory and observations for many more line ratios will be presented and discussed in detail in a forthcoming paper by Noriega-Crespo, Böhm and Raga (1989).

Finally, there are objects like the HH 7-HH 11 chain of objects which (at least in our own opinion) so far have not been explained in a satisfactory way. It has been suggested that they are shocked cloudlets (Rudolph and Welch 1988) or a jet with entrainment of molecular matter (Zinnecker *et al.* 1989). Blondin, Königl and Fryxell (1989) have interpreted the HH 7 part as the working surface of a "light" jet in which the ratio of the cooling distance to the jet radius is small. All these suggestions explain certain aspects of the observations well but we feel that the explanation *e.g.*, of the detailed velocity field, the very abrupt transition in the velocity field from HH 11 to HH 10, the unusual position-velocity diagram of HH 11 have not yet been explained (see Solf and Böhm 1987).

Even in the cases in which most of the observations are compatible with bow shock models (*e.g.*, HH 1, HH 32, HH 34, HH 43 etc.) some intriguing problems remain. Recently many new data have been obtained for these and other objects (for instance, in the case of HH 1 the spatial emission distribution can now be studied in about 200 emission lines, see Solf, Böhm and Raga 1988). Taking HH 1 as an example we might ask: Do we continue to get good agreement with bow shock model predictions if we test many emission lines? We are presently trying to answer this question by using an approach analogous to that of Raga and Böhm (1986) but calculating spatially dependent line intensities (instead of profiles, see Figure 3) and doing this for many lines (Noriega-Crespo, Böhm and Raga 1989). The results which we have obtained so far are not as simple as we had expected and are in fact quite intriguing. In order to make the test as sensitive as possible we have studied the spatial dependence of the intensity ratio of the lines to Hα. The curves describing the spatial variation of this ratio differ drastically from emission line to emission line and it is improbable to get even the qualitatively correct behavior by accident. We find very good agreement between theoretical predictions and observations of the spatial intensity distribution of Hα and the line intensity ratios of Hα of to the [O III], [S III], [O II], [S II], and [N II] lines. It is especially impressive that the completely different variations of the [S II] 4068/76 Hα and [S II] 6716/31/Hα are both very well explained. Qualitative agreement of varying degrees between observation and theory is found for the ratios involving [O I], [N I] and (marginally) [C I]. It is interesting to note that there is qualitative disagreement in two cases namely [Ca II] 7291/Hα and [A III] 7136/Hα. Because of the very low ionization energies of both Ca I and Ca II (6.1 and 11.9 eV) and the relatively high transition probability of the 7291 line (1.3 s^{-1}) this line tests probably different parts (namely low T, high g) of the shock recombination regions than all the other lines mentioned above. Therefore, it is at least possible that shock models with greater accuracy in the relatively cool parts of the recombination regions may lead to better agreement. The discrepancy for the [A III] 7136/Hα ratio is considerably enigmatic because the ratios of the [O III] 5007 and [S III] 9532 lines to Hα can be explained rather well. In principle observational material for the spatial emission distribution of $\sim$ 200 lines is available. The program which is used for the theoretical predictions (basically due to Alex Raga) permits the use of any (axially symmetric) bow shock like shape. We intend to find out whether some (not too unreasonable) modification of the bow shock shape would lead to even better agreement with observations. The result of such an attempt would (in combination with further hydrodynamic studies) be very helpful for finding reliable hydrodynamic models of HH objects.

I thank Alberto Noriega-Crespo and Alejandro Raga for very helpful discussion. A. Noriega-Crespo has kindly drawn Figure 3.

A part of the research reported here has been supported by NSF grant AST 87-17867.

References

Axon, D. J. and Taylor, K. 1984, *M.N.R.A.S.*, **207**, 241.

Binette, L., Raga, A. C. and Cantó, J. 1989, in preparation.

Blondin, J. M., Königl, A. and Fryxell,, B. A. 1989, *Ap. J. (Lett.)*, **337**, L37.

Böhm, K. H. 1956, *Ap. J.*, **123**, 379.

Böhm, K. H. 1983, *Rev. Mex. Astron. Astrof.*, **7**, 55.

Böhm, K. H. and Böhm-Vitense, E. 1984, *Ap. J.*, **277**, 216.

Böhm, K. H., Böhm-Vitense, E. and Brugel, E. W. 1981, *Ap. J. (Lett.)*, **245**, L113.

Böhm, K. H., Bührke, T., Raga, A. C., Brugel, E. W., Witt, A. N. and Mundt, R. 1987, *Ap. J.*, **316**, 349.

Böhm, K. H., Siegmund, W. A., and Schwartz, R. D. 1976, *Ap. J.*, **203**, 399.

Böhm, K. H. and Solf, J. 1985, *Ap. J.*, **294**, 533.

Böhm, K. H. and Solf, J. 1989, *Ap. J.* (submitted).

Böhm-Vitense, E., Böhm, K. H., Cardelli, J. A. and Nemec, J. M. 1982, *Ap. J.*, **262**, 224.

Brugel, E. W., Shull, J. M. and Seab, C. G. 1982, *Ap. J. (Lett.)*, **262**, L35.

Bührke, T., Mundt, R. and Ray, T. P. 1988, *Astr. Ap.*, **200**, 99.

Cantó, J. 1989, these Proceedings.

Choe, S.-U., Böhm, K. H. and Solf, J. 1985, *Ap. J.*, **288**, 338.

Clarke, J. T. and Moos, H. W. 1981, in *The Universe at Ultraviolet Wavelengths*, ed. R. D. Chapman, NASA CP No. 2171, p. 787.

Cohen, M. 1986, *Irish A. J.*, **17**, 238.

Dalgarno, A., Herzberg, G. and Stephens, T. L. 1970, *Ap. J. (Lett.)*, **162**, L49.

Dopita, M. A. 1978a, *Ap. J. Suppl.*, **37**, 117.

Dopita, M. A. 1978b, *Astr. Ap.*, **63**, 237.

Dopita, M. A., Binette, L. and Schwartz, R. D. 1982, *Ap. J.*, **261**, 183.

Dyson,, J. E. 1987, in Proceedings I.A.U. Symp. 122, Circumstellar Matter, ed. I.. Appenzeller and C. Jordan (Dordrecht: Reidel), p. 159.

Falle, S. A. E. G. 1981, *M.N.R.A.S.*, **195**, 1011.

Hartigan, P., Mundt, R. and Stocke, J. 1986, *A. J.*, **91**, 1357.

Hartigan, P., Raymond, J. and Hartmann, L. 1987, *Ap. J.*, **276**, 560.

Hartmann, L. and Raymond, J. 1984, *Ap. J.*, **276**, 560.

Herbig, G. H. 1951, *Ap. J.*, **113**, 697.

Herbig, G. H. 1969, in *Nonperiodic Phenomena in Variable Stars*, ed. L. Detre (Dordrecht: Reidel), p. 75.

Herbig, G. H. 1974, Lick Obs. Bull. No. 658.

Herbig, G. H. and Jones, B. F. 1981, *A. J.*, **86**, 1232.

Herbig, G. H. and Jones, B. F. 1983, *A. J.*, **88**, 1040.

Innes, D. E. 1985, Ph.D. Thesis, University College, London.

Jones, B. F. and Walker, M. F. 1985, *A. J.*, **90**, 1320.

Lee, M. G., Böhm, K. H., Temple, S. D., Raga, A. C., Mateo, M. L., Brugel, E. W., and Mundt, R. 1988, *A. J.*, **96**, 1690.

Lightfoot, J. F. and Glencross, W. M. 1986, *M.N.R.A.S.*. **221**, 993.

Münch, G. 1977, *Ap. J. (Lett.)*, **212**, L77.

Mundt, R. 1986, *Can. J. Phys.*, **64**, 407.

Mundt, R. 1987, in Proc. I.A.U. Symp. 122, Circumstellar Matter, ed. I. Appenzeller and C. Jordan (Dordrecht: Reidel), p. 147.

Mundt, R., Bührke, T., Fried, J. W., Neckel, T., Sarcander, M. and Stocke, J. 1984, *Astr. Ap.*, **140**, 17.

Mundt, R. and Fried, J. W. 1983, *Ap. J. (Lett.)*, **274**, L83.

Noriega-Crespo, A., Böhm, K. H., and Raga, A. C. 1989, *A. J.* (submitted).

Ortolani, S. and d'Odorico, S. 1980, *Astr. Ap.*, **83**, L8.
Osterbrock, D. W. 1958, *P.A.S.P.*, **70**, 399.
Raga, A. C. 1985, Ph.D. Thesis, University of Washington.
Raga, A. C. 1988, *Ap. J.*, **335**, 820.
Raga, A. C. and Böhm, K. H. 1985, *Ap. J. Suppl.*, **58**, 201.
Raga, A. C. and Böhm, K. H. 1986, *Ap. J.*, **308**, 829.
Raga, A. C. and Böhm, K. H. 1987, *Ap. J.*, **323**, 193.
Raga, A. C., Böhm, K. H. and Solf, J. 1986, *A. J.*, **92**, 119.
Raga, A. C. and Mateo, M. 1988, *A. J.*, **95**, 543.
Raga, A. C., Mateo, M., Böhm, K. H. and Solf, J. 1988, *A. J.*, **95**, 1783.
Raymond, J. C. 1979, *Ap. J. Suppl.*, **39**, 1.
Raymond, J. C., Hartigan, P. and Hartmann, L. 1988, *Ap. J.*, **326**, 323.
Reipurth, B., Bally, J., Graham, J. A., Lane, A. and Zealey, W. J. 1986, *Astr. Ap.*, **164**, 51.
Rudolph, A. and Welch, W. J. 1988, *Ap. J. (Lett.)*, **326**, L31.
Schwartz, R. D. 1975, *Ap. J.*, **195**, 631.
Schwartz, R. D. 1978, *Ap. J.*, **223**, 884.
Schwartz, R. D. 1983a, *Ap. J. (Lett.)*, **268**, L87.
Schwartz, R. D. 1983b, *Ann. Rev. Astr. Ap.*, **21**, 209.
Schwartz, R. D. 1983c, *Rev. Mex. Astron. Astrof.*, **7**, 27.
Schwartz, R. D. 1985, in *Protostars and Planets II*, ed. D. C. Black and M. S. Mathews (Tucson: Univ. of Arizona Press), p. 405.
Schwartz, R. D., Cohen, M. and Williams, P. M. 1987, *Ap. J.*, **322**, 403.
Schwartz, R. D., Dopita, M. A. and Cohen, M. 1985, *A. J.*, **90**, 1820.
Schwartz, R. D., Jones, B. F. and Sirk, M. 1984, *A. J.*, **89**, 1735.
Schwartz, R. D., Williams, P. M., Cohen, M. and Jennings, D. G. 1988, *Ap. J. (Lett.)*, **334**, L99.
Solf, J. 1987, *Astr. Ap.*, **184**, 322.
Solf, J. and Böhm, K. H. 1987, *A. J.*, **93**, 1172.
Solf, J., Böhm, K. H. and Raga, A. C. 1986, *Ap. J.*, **305**, 795.
Solf, J., Böhm, K. H. and Raga, A. C. 1988, *Ap. J.*, **334**, 229.
Strom, K. M., Strom, S. E. and Stocke, 1983, *Ap. J. (Lett.)*, **271**, L23.
Strom, K. M., Strom, S. E., Wolff, S. E., Morgan, J. and Wenz, M. 1986, *Ap. J. Suppl.*, **62**, 39.
Tenorio-Tagle, G., These Proceedings.
von Hippel, T., Burnell, S. J. and Williams, P. M. 1988, *Astr. Ap. Suppl.*, **74**, 431.
Zealey, W. J., Williams, P. M. and Sandell, G. 1984, *Astr. Ap.*, **140**, L131.
Zealey, W. J., Williams, P. M., Taylor, K. N. R., Storey, J. W. V. and Sandell, G. 1986, *Astr. Ap.*, **158**, L9.
Zinnecker, H., Mundt, R., Geballe, T. R., and Zealey, W. J. 1989, *Ap. J.* (in press).

Discussion:

OSTERBROCK: The excellent agreement between the observed and predicted spectra is highly convincing that the general model is correct. Concerning the [FeII]/Hβ ratio problem, can you say whether the ratios of the individual [FeII] lines to one another are the same in various objects, or if they differ?.

BÖHM: The [FeII] line ratios agree qualitatively in different objects (in the sense that e.g., [FeII]λ7155 is always the strongest [FeII] line) but not quantitatively.

MÜNCH: Because, to my knowledge, no calculation of collisional excitation cross sections for the metastable levels of C^0 has ever been made, may I ask what kind of estimates have you used to determine T_e from the [CI] lines?

BÖHM: I am not sure whether your statement applies also to the long-lying metastable levels which are the upper levels of the [CI] λ9849, λ9823 and λ8727 lines. In any case Aller in his 1984 book on gaseous nebulae quotes collision strengths for these lines. In the moment I do not remember on which original work these results are based but I shall check this.

PECKER: From the analysis of line intensity variations, what can you say about the variations of N_e is the variations of T_e? which effect dominates?

BÖHM: In the "auroral" to "nebular" line ratios the T_e variation is of course the most important one. The spatial dependence of many line ratios is also strongly influenced by variations in the ionization equilibrium and by the question whether the lines are formed in the low density or the high density limit or in an intermediate range.

A CO Search for Molecular Gas in High Mass Post-Main-Sequence Nebulae

J.P. Phillips and A. Mampaso
Instituto de Astrofisica de Canarias
La Laguna, Tenerife, Spain

N. Ukita
Nobeyama Radio Observatory
Tokyo Astronomical Observatory
University of Tokyo, Japan

P.G. Williams
Physics Department, Queen Mary College,
London E1 4NS, England

High mass post-main-sequence nebulae are characterised by a set of unusual, and in certain cases extreme physical characteristics, including large outflow velocities (cf. Phillips and Mampaso, 1988a), extremely compact high emission measure cores (Phillips and Mampaso, 1988b), a high incidence of bipolar morphology (Peimbert and Torres Peimbert, 1982), and evidence for anomalously high levels of shock excited H_2 S(1) emission towards both the source cores (Phillips et al 1983, 1985) and nebular peripheries (Zuckerman and Gatley, 1988). The large central star masses also predispose these sources to rapid evolution within the H-R plane (perhaps one or two orders of magnitude more rapid than for typical PN (Schonberner 1981, 1983), and the acquisition of stellar temperatures $T_* > 10^5$ K, giving rise to correspondingly high levels of nebular excitation.

Finally, we may note that the high progenitor mass range $2 \leq M/M_\odot \leq 10$ not only results in anomalous abundance ratios C/O, He/H, and N/O (Becker and Iben, 1980; Peimbert and Torres Peimbert, 1982), a primary identifier for such nebulae, but also leads to relatively high shell masses M_s approaching perhaps $\sim 9\ M_\odot$. The characteristics of these shells are rather ill-defined, although it seems clear that given the rapid evolution of the central stars, and the high masses sloughed off during the OH/IR mass-loss phase, then only a relatively small proportion can as yet have been ionised (cf. Phillips and Mampaso, 1989). Similarly, if the progenitor temperatures at time of shell ejection were less than $\sim$ 2500 K, then much of the shell is likely to have been molecular (Glassgold and Huggins, 1983) - a position which for most sources is unlikely to be radically altered through subsequent photo- or shock-dissociation.

That such is the case for at least certain nebulae appears to be confirmed by a high incidence of H_2 S(1) emission (cf. table 1), and several detections in the lower order rotational transitions of CO (e.g. Huggins and Healy, 1986; Knapp and Morris, 1985)

- although the general incidence of such emission is rather poorly established. In the following, therefore, we have sought to survey a broad sample of such sources in the J=1-0, J=2-1, and J=3-2 transitions of CO, using both the Kitt Peak NRAO 12 meter antenna (November 1987, March 1989), and the 45 meter antenna of the Nobeyama Radio Observatory (April 1989). The results of this survey are summarised in table 1.

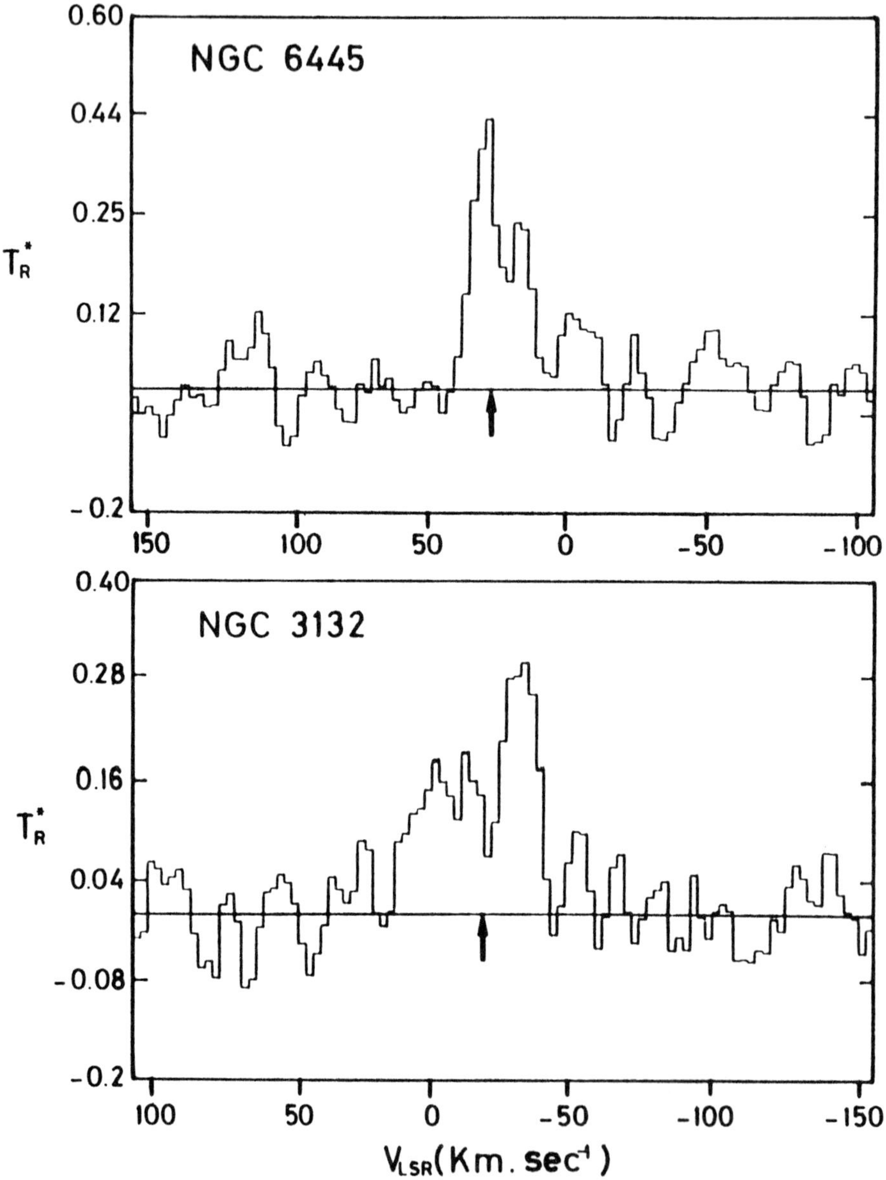

Figure 1 Two new detections of CO J=2-1 emission in planetary nebulae, where baselines are fixed at $T_R^* = 0$ K, and arrows indicate the central V_{LSR} of the HII regions. The results have been smoothed with a three-point Hanning function, and were in both cases acquired using the 12 metre NRAO antenna at Kitt Peak.

It is clear, from this, that our present efforts have greatly enhanced the overall numbers of such sources measured in these transitions, although the crop of further detections appears rather meagre (we find new sources of emission in NGC 3132 and NGC 6445; see fig 1) - implying an overall detection rate of $\sim$ 28 %. This, in part, may of course arise from the rather poor limiting temperature derived for certain low latitude sources (cf. NGC 2452, NGC 6629), for which noise temperatures became prohibitively large. It is clear that several of these results would benefit from repetition at more favourable observing locations.

Nevertheless, and allowing that as many as half these lower latitude sources are eventually detected, it is clear that the overall detection limit (to $\sim$ 0.1 K) is unlikely to exceed $\sim$ 36 % .

Such a figure is of course considerably greater than for surveys of the generality of PN - although it is perhaps still surprisingly low when considering likely shell masses. It is also of interest to notice the strong correlation between detected CO, the incidence of H_2 S(1) emission, and (less comprehensively) the presence of 21 cm HI in either emission or absorption.

Superficially, therefore, one might conclude that yes, the high masses of such sources are indeed reflected in a high incidence of molecular emission - but that the majority of high mass shells contain little if any molecular gas.

Whether this is true in reality is less easy to discern. In particular, the absence of H_2 S(1) emission in non-CO sources may be a partial artefact of incomplete surveys in this transition; and, although it is surely suggestive, the apparent correlation noted above may therefore be somewhat misleading. Similarly, and given that some such relationship is genuine, then this may reflect no more than an approximate correlation between H_2 S(1) and CO emission terms; a presumption that gains some support from the similar CO and H_2 S(1) emission distributions in certain sources (cf. Phillips and Mampaso, 1989) , and evidence for related CO and H_2 shock emission environs in nebulae such as NGC 2346 (cf. Phillips and Mampaso, 1989). Allied to similar levels of source detectability in the two transitions, then it is clear that a correlation such as observed in table 1 would be by no means surprising.

Finally, we note that reduced levels of CO (and H_2) emission may arise not only from a reduced incidence of molecular gas, but also perhaps from variations in emission zone characteristics. Where CO emission derives primarily from a post-shock cooling zone, for instance, then high velocity gradients would lead to optically thin emission - a characteristic of at least certain of these nebulae (Phillips and Mampaso, 1989). Thus, many non-detections recorded here may simply possess rather smaller values of X(CO)/dv/dr, and correspondingly reduced CO antenna temperatures.

In short, it is clear that whilst a large fraction of such nebulae may appear to possess little molecular gas - the evidence is rather persuasive - further observations would be useful in order to assess whether this is artefact of our present, somewhat limited detection capabilities.

Table 1

CO Survey of High Mass Nebulae

Non-Detections

Source	$2\sigma^1_{rms}$	J→J-1	Notes
NGC 650	0.04	1-0	2
NGC 2440	0.07	1-0	2,3
NGC 2452	0.40	3-2	2
NGC 2818	0.14	1-0	2
NGC 6532	0.20	2-1	2
NGC 6629	0.36	2-1	2
NGC 6741	0.17	2-1	2
NGC 6751	0.22	2-1	2
NGC 6778	0.05	1-0	2
NGC 6853	0.11	2-1	2
NGC 6894	0.32	3-2	2
NGC 7008	0.10	2-1	2
HB 5	?	2-1	2,6
Hu 1-2	0.12	2-1	2,3
M 2-55	0.06	1-0	2
M 3-3	0.12	2-1	2
Me 2-2	0.09	1-0	2
VV 47	0.06	1-0	2

Detections

Source	T^*_R	J→J-1	Notes
NGC 2346	0.4	2-1	2,3,4
NGC 3132	0.3	2-1	2,4
NGC 6302	0.4	2-1	2,3,4,5
NGC 6445	0.45	2-1	2
NGC 7027	3.2	3-2	2,3,4,7
CRL 618	1.5	3-2	2,3,4,7
M 2-9	0.1	2-1	3,4

Notes: 1) Results correspond to $2\sigma_{rms}$ upper limits; 2) Source measured in this survey; 3) Source measured previously; 4) Associated with H_2 S(1) emission; 5) HI 21cm emission measured in absorption/emission; 6) Source may be strongly interacting with enveloping region - emission is present at V_{LSR} of the HII region, but association with source is unclear; 7) Results based on recent J=3-2 CO observations by the authors at Kitt Peak.

In either case, it is clear that most detected PN fall within our present subset of high mass PN, and a goodly fraction of such sources reveal a penchant for strong CO emission; although the characteristics of such emission zones remain woefully ill-determined, and further observations (in other transitions) would prove of considerable interest and utility.

References

Becker, S.A., and Iben, I., 1980. Astrophys. J. **237**, 111

Glassgold, A.E., and Huggins, P.J., 1983. Mon. Not. Roy. Astron. Soc. **203**, 517

Huggins, P.J., and Healy, A.P., 1986. Astrophys. J. **305**, L29

Knapp, G.R., and Morris, M., 1985. Astrophys. J. **292**, 640

Phillips, J.P., and Mampaso, A., 1988b. Astron. Astrophys. **190**, 237

Phillips, J.P., and Mampaso, A., 1988a. in IAU Symposium No. 131, ed S. Torres Peimbert, D. Reidel Publishing Co., Dordrecht, Holland

Phillips, J.P., Reay, N.K., and White, G.J., 1983. Mon. Not. Roy. Astron. Soc. **203**, 977

Phillips, J.P., White, G.J., and Harten, R., 1985. Astron. Astrophys. **145**, 118

Schonberner, D., 1981. Astron. Astrophys. **103**, 19

Schonberner, D., 1983. Astron. Astrophys. **272**, 708

Zuckerman, B., and Gatley, I., 1988. Astrophys. J. **324**, 501

A REMARKABLE BIPOLAR FLOW IN THE CENTER OF THE Rho OPHIUCHI CLOUD

Ph. Andre[1], J. Martin-Pintado[2], D. Despois[3] and T. Montmerle[4]

[1]Institut d'Radio-Astronomie Millimetrique, Granada, Spain

[2]Centro Astronómico de Yebes, Spain

[3]Observatoire de Bordeaux, France

[4]C.E.A. Saclay, France

Using the IRAM 30-m telescope in August and December 1988, we have discovered the first molecular outflow in the central part (L1688) of the nearby ρ Ophiuchi dark cloud. This outflow, found in the $J = 2-1$ line of 12CO near the cloud core A, is an extreme case, weak (outflow mass-loss rate $\approx 5 \times 10^{-8}\ M_{\odot} yr^{-1}$) and highly collimated (lenght to width ratio > 14), which explains why it has escaped previous detections with smaller telescopes. The high-velocity molecular gas is hot and optically thin, making the $J = 2-1$ line of 12CO $\approx$ 3-4 times stronger than the $J = 1-0$ line. Unexpectedly, this outflow does not appears to be driven by any of the embedded near-IR sources known in this region previous deep VLA surveys of the cloud (André, Montmerle, and Feigelson, 1987; Stine *et al.*, 1988; André *et al.*, in prep.). The outflow exciting source is thus probably a very low-luminosity (($L < 0.1 L_{\odot}$) young stellar object. Using the 30-m equipped with the MPIfIR bolometer, we have very recently found (March 1989) that this object is the strongest continuum point source of L1688 at 1.3 mm. By analogy with L1551-IRS5 and HL Tau, the radio properties of this source suggest that it possesses a weak, possibly collimated, ionized wind and a relatively massive, cold circumstellar disk ($M_{disk} \approx 0.1 M_{\odot}$).

Outflow activity does not appear to be widespread within the highest density regions and/or around luminous near-IR sources and seems a rare phenomenon in the core of the ρ Oph cloud. The fact that only one outflow has been discovered so far in this region, rich in embedded IR sources, is at variance which the current ideas on low-mass star formation (e.g., Lada 1988 and references therein)

References

- André, Ph., Montmerle, T., and Feigelson, E.D.: 1987, A.J., 93, 1182.
- Lada, C.J.: 1988, in *Formation and Evolution of Low Mass Stars*, eds. A.K. Dupree and M.T.V.T. Lago, Kluwer Academic Publishers, p. 93.
- Stine, P.C., Feigelson, E.D., André, Ph., and Montmerle, T.: 1988, A.J., 96, 1394.

MASSIVE DUST DISKS SURROUNDING HERBIG Ae/Be STARS

G. Sandell

Joint Astronomy Centre, Hilo, Hawaii, USA

I present preliminary results from a small mm/submm continuum survey, mostly photometry (1.1mm, 800μm, 450μm and 350μm), but also some mapping at 800μm of Herbig Ae/Be stars and a few peculiar stars. These observations were commenced on the 15m JCMT submillimetre telescope on Mauna Kea, Hawaii.

The sample included the following stars: HR 5999, AB Aur, R Mon, LkH_α 198, LKH_α 234, BD+40°4124, HD 200775, MCW 1080, R CrA, TY CrA, PV Cep, V 645 Cyg, and MWC 349. All stars were detected except HR 5999. In the R Corona Australis region it is difficult to separate the stars from the extended strong emission of the surrounding dust cloud.

If I restrict myself to stars, which have been observed at more than two wavelenghts, and for which the emission is sufficiently compact, I find that the apparent β-index is $\sim$1, i.e. less extreme than for T Tauri stars (Weintraub et al., 1989, Ap.J. Lett (in press)) where $\beta < 1$. The total masses (gas and dust), deduced form the observed continuum emission, range from a few 0.1 $M_\odot$ to 8 $M_\odot$ for the most extreme case (LkH_α 234). Although the emission is unresolved, I argue that the emission must originate from asymmetric disk like structures surrounding the stars, because if the matter was in a shell, the stars would not be visible. Only one star, LkH_α 198, appears resolved and disk like, while LkH_α 234, which itself is unresolved, is surrounded by a large disk perpendicular to the outflow. MCW 349, appears to be associated with very little dust, most of the emission we see is probably due to free-free emission.

NEW YOUNG OBJECTS FROM THE IRAS POINT SOURCE CATALOGUE

A. Manchado[1], P. García Lario[1], K. C. Sahu[2], S.R. Pottasch[2]

[1]Instituto de Astrofísica de Canarias, La Laguna, Tenerife, Spain

[2]Kapteyn Astronomical Institute, Groningen, The Netherlands

Three new objects have been found, when carrying out a survey of IRAS sources with colours like planetary nebulae, via low resolution spectroscopy. IRAS 14592-6311 and IRAS 07173-1733 are bright cometary nebulae, showing metal line emission very similar to V1331 Cyg (LkH_α 120), which is the prototype of a certain number of T Tauri stars. They present strong P-Cygni profiles and some other peculiar characteristics. IRAS 14592-6311 is associated with a molecular cloud at a distance of 2.9 kpc, while IRAS 07173-1733 is located at the edge of the dark cloud KHAV 201. In the case of IRAS 05506+2414, it presents a bipolar structure with Herbig-Haro emission characteristics in one of the lobes, produces by shocked gas, and is located near Orion.

IUE OBSERVATIONS OF HERBIG-HARO OBJECTS 7, 11 & 29

M. Cameron[1] and R. Liseau[2]

[1]Max-Plank Institut für Physik und Astrophysik, Garching

[2] Instituto de Fisica dello Spazio Interplanetario, Frascati

UV spectra of H-H 7 & 11 (L1450, NGC1333) and H-H 29 (L1551) have been observed with both the IUE long- and short- wavelength cameras. These data were obtained using long ($\leq$ 12 hours) exposure shifts over a period of three years. As expected, the S/N quality of the raw data is relatively poor and a special IUE extraction procedure has been developed within the MIDAS environment to deal with this problem. The data on H-H 7 & 11 confirm the very low excitation state of these objects, although the absence of Lyman band H_2 line emission is somewhat enigmatic. The continuum emission from HH 29 reveals considerable structure which is inconsistent with a pure 2-photon model for this source. The interpretation of these results are discussed in the light of conflicting optical data.

H_2 2.12 μm SPECTROSCOPY AND IMAGING OF HH OBJECTS

H. Zinnecker, R. Mundt, A. Moneti, T.R. Geballe, W.J. Zealey

Max-Planck-Institut für Estraterr. Physik, Garching

We have obtained high spectral resolution observations of a number of Herbig-Haro (HH) objects in the H_2 v=1-0 S(1) line at 2.12μm. Objects observed included HH1/2, HH7-11, HH19, HH32A, HH40, and HH43, all associated with jet-like features or collimated optical outflows. Here we present velocity-resolved 2.12μm spectroscopy for HH40 (an HH-objects moving close to the line of sight) an for HH43B (an HH-object moving close to the plane of the sky). The full set of observations including interpretation is given in Zinnecker et al. (1989). We also present high spatial resolution H_2 2.12μm images of HH40 and HH43. The 2.12μm H_2 line profiles were obtained with the UKIRT infrared Fabry-Perot system (effective resolution 30-35 km/s) using a diaphragm of diameter 11" for both HH40 and HH43. The H_2 images were obtained with the IR-array imager at the CTIO 1.5m telescope through a narrowband filter centred on the v=1-0 S(1) line. The IR-camera used had 58x62 pixels and a resolution of 0.9"/pixel. The exposure time was 2x120sec and 5x60sec for HH40 and HH43, respectively. The images are sky subtracted, but not flat fielded, and should be viewed as preliminary test images.

NEW OVRO RESULTS SHOW DISKS ARE NOT NECESSARY FOR FOCUSSING BIPOLAR OUTFLOWS

M. Barsony

University of California, Berkeley

In order to better understand the outflow phenomenon in young, high mass stars, the three sources S87, LkH$_\alpha$101, and S106 were chosen for closer study. High spatial resolution (5″ millimeter line ^{13}CO (J=1→0) and CS (J=2→1) maps of these sources were obtained with the Owens Valley Radio Observatory's (OVRO) millimeter array at 0.15 and 3.0 km s^{-1} resolutions, respectively.

The OVRO (Owens Valley Radio Observatory) maps were combined with data from the 14 m FCRAO (Five College Radio Astronomy Observatory) millimeter wave radio telescope, the VLA, IRAS, and the Palomar 5 m and 1.5 m telescopes. A synthesis of the data reveals that although all three pre-main sequence objects are the sources of powerful, ionized, stellar winds, only one, S87/IRS1, LkH$_\alpha$101, and S106 IR are $1.8 \times 10^{-5}, 1.7 \times 10^{-6}, and 1.1 \times 10^{-5}$ $M_\odot yr^{-1}$, with corresponding wind velocities of 160, 350, and 220 km/s. In all cases, the wind velocities are lower, and the mass loss rates higher, than for main sequence stars of the same spectral types. Radiation pressure in inadequate to drive these winds, which can be anisotropic in their velocity fields.

The existence of massive, large scale (r≈ 10^{16} cm) disks, required by numerous proposed molecular outflow models, can now be ruled out. This conclusion is based on the results of high (5″) spatial resolutions millimeter line observations of S106. This source, which has heretofore been the best disk candidate in the context of high-mass star formation, turns out not to have a large, molecular gas disk. Instead, the kinematic and spatial structure of the molecular material in S106 is indicative of swept-up, turbulent, remnant cloud gas. Only one of the many proposed molecular outflow models is consistent with the new observations (Königl, 1982).

THE SPECTRUM OF A PARTIALLY IONIZED JET

L. Binette[1], A. Raga[1], J. Cantó[2]

[1]Institute for Theoretical Astrophysics, Canada

[2]Universidad Nacional Autónoma de México, México

Observed stellar jets have the following characteristics : a high degree of collimation, a structure of bright, quasi-periodic knots, a very low excitation spectrum (red [S II] to Hα line ratios of 1–10). Assuming that the latter are indicative of the structure of a supersonic jet, we have made a detailed prediction of the emission line spectrum within each radiating *knot* using the multi-purpose code *MAPPINGS* which includes all the atomic processes important in low velocity shocks. The structure (density, temperature, ionization) of the stationary, supersonic, nonadiabatic stellar jet was computed separately using a hydrodynamical code developed by A. Raga. An initially overpressured supersonic jet tries to adjust its pressure to the environmental pressure. In this process of pressure adjustment, a series of expansion fans and incident/reflected shock pairs are formed along the jet. We associate the emission from the recombination region behind these *crossing shocks* with the *knots* observed in stellar jets. We find that the calculated emission spectrum is of remarkably low excitation.

SODIUM IONIZATION IN T-TAURI STARS

A. Natta[1] and C. Giovanardi[2]

Osservatorio Astrofisico di Arcetri, Italy

Centro per l'Astronomia Infrarossa, CNR, Italy

We present the results of a study of the sodium ionization and excitation in the winds of low-luminosity, pre-main-sequence stars. Line profiles for the NaI doublet at 5990,5986 Åare discussed and compared with the observations for those T Tauri stars with P-Cygni profiles. We find that the observed shape of the lines put significative constraints on the rate of mass-loss ($M \geq 3 \times 10^{-8} M_{\odot}/yr$).

By comparing the properties of the NaI lines with the observed luminosity of infrared hydrogen recombination lines, such as Br_7, it is possible to determine at the same time the rate of mass-loss and the temperature of the inner regions of the wind, where both lines are formed. The results for four stars show that the wind is in general cool ($T_{gas} \lesssim 7000K$), and that the momentum in the wind is comparable to the momentum measured from CO lines.

INVERTING THE POSITION-VELOCITY DIAGRAMS OF MOLECULAR DISCS

J.S. Richer and R. Padman

Mullard Radio Astronomy Observatory Cambridge, England

The molecular discs around Young Stellar Objects can be detected through high spatial and spectral resolution observations of their position-velocity diagrams. By mapping a high lying transition of a density-tracing molecule (e.g. HCN, HCO^+, NH_3), the lines should be optically thin and we can expect to see the signature of rotation of the protostellar accretion disc. We present single-dish observations of the S106 system which show evidence for the presence of such a rotating disc. We also discuss a simple disc model which predicts the form of the resulting l-v diagram. With high quality data, we should be able to do the inverse problem that of recovering the disc properties from the l-v diagram. We discuss approaches to this inverse problem both by a Singular Value Decomposition (least-squares) method, and by a Maximum Entropy Method.

IDENTIFICATION OF OUTFLOW EXCITING SOURCES THROUGH AMMONIA OBSERVATIONS

G. Anglada[1,2], L.F. Rodriguez[3,4], J. M. Torrelles[5], R. Estalella[1,2], P.T.P. Ho[3], J. Cantó[6], R. López[1,2], and L. Verdes-Montenegro[5]

[1]Department de Física de l' Atmósfera, Universitat de Barcelona

[2]Grup d'Astrofísica, Societat Catalana de Física, IEC

[3]Harvard-Smithsonian Center for Astrophysics

[4]On sabbatical leave from Instituto de Astronomía, UNAM

[5] Instituto de Astrofísica de Andalucía, CSIC

[6]Instituto de Astronomía, UNAM

Using the 37-m telescope of the Haystack Observatory, we observed the (J,K)=(1,1) ammonia transition towards the suspected exciting sources of twelve regions with molecular or optical outflows: L1448, L1455, L1524 (Haro 6-10), RNO 43, HH 34, HH 38,43, Haro 4-255 FIR, NGC 2264 (HH 14-4,6), L43, R CrA, HH 32a (AS353A), and V1331 Cyg. We detected and mapped ammonia emission in nine of these regions.

In five of mapped regions, the spatial coincidence of the high-density gas with the objects previously proposed as exciting sources gives support to those identifications. For L1524 (Haro 6-10) and HH 38,43 we propose a new location for their exciting sources. We detected a radio continuum source and an unusually strong H_2O maser coinciding with the maximum of the ammonia emission in L1448. This region appears to be the site of very recent star formation.

THE MOLECULAR ENVOLOPE OF MIRA

P. Planesas, R. Bachiller, J. Martín-Pintado, V. Bujarrabal

Centro Astronómico de Yebes, OAN-IGN, Spain

We have mapped the CO $J = 2 \rightarrow 1$ and $J = 1 \rightarrow 0$ emission of the circumstellar envelope of Mira. Emission in both transitions extends to a distance of $\sim 4 \times 10^{16}$cm from the star. In the inner 2×10^{16}cm the lines show the presence of three velocity components. The main component has the intermediate velocity and extends over the whole envelope. From this component we have estimated a mass loss rate of $3 \times 10^{-7}\ M_{\odot} yr^{-1}$ and a total molecular mass of $1.2 \times 10^{-3}\ M_{\odot}$. We have shown that, from our data, the only reliable explanation for the lowest and highest velocity components is that they are due to an outflow located within the envelope.

NARROWBAND PHOTOMETRY OF PHOTOMETRICALLY PECULIAR OBJECTS

Eugenio E. Mendoza V.
Instituto de Astronomia, UNAM

This work is based upon $\alpha(16)\Lambda(9)$-photometry for 2 Planets, 11 Wolf-Rayet stars and 7 Planetary Nebulae. The results show anomalous $\alpha(16)$ *and* $\Lambda(9)$-indices for these objects. Thus, they are photometrically peculiar in this system. The main results are:

1) Callisto, Jupiter IV, shows $\alpha(16)$ and $\Lambda(9)$-indices which can be considered as excellent representatives of solar type stars (G2 V).

2) Uranus and Neptune have anomalous $\Lambda(9)$-index, because of a spectral feature in absorption around λ 7805 Å, most likely due to a carbon compound. Thus, they are off the main sequence in the $\alpha(16)\Lambda(9)$-array.

3) The Hα-line is possibly stronger in Uranus (marginal) and Neptune than in the Sun.

4) Wolf-Rayet stars have anomalous $\alpha(16)$-index, because of an extremely wide He II-line (λ 6560 Å), lack of hydrogen and the presence of spectral emission features that fall in the continuum used to determine this index.

5) Most WC stars have anomalous $\Lambda(9)$-index, because the presence of spectral emission features in the short wavelength continuum that defines this index.

6) Some W-R stars show variations in the strength of He II-line (λ 6560 Å).

7) Most WC stars are separated from WN stars in the $\alpha(16)\Lambda(9)$-array.

8) Planetary Nebulae have anomalous $\alpha(16)$-index, because the continuum around the hydrogen line is probably contaminated by [N II]-lines at λ 6548 Å and λ 6583 Å.

9) Planetary Nebulae have anomalous $\Lambda(9)$-index, because the short wavelength continuum is heavily contaminated by the [Ar III]-line at λ 7751 Å.

10) Planetary Nebulae lie far apart from all kind of stellar objects in the $\alpha(16)\Lambda(9)$-diagram.

This work was partially supported by a grant from CONACYT, No. P228CCOX880202.

THE ORION NEBULA

THE STRUCTURE OF MOLECULAR CLOUDS FROM LARGE SCALE SURVEYS OF CO AND CS

John Bally
AT&T Bell Laboratories
HOH-L245, Holmdel, NJ 07733

ABSTRACT

The molecular gas lying in the interior of the Orion superbubble consists of sheets, filaments and bubbles in which the dense active star forming cloud cores are embedded. Many regions have a wind-swept or cometary appearance suggesting strong interaction with the Orion OB association. External heating dominates the cloud energy budget, resulting in cold (~10K) cloud cores surrounded by hot (>20K) envelopes. Non-ionizing UV radiation emitted by late B and A stars may be the source of the kinetic energy needed to form some of the large scale cavities and contributes to the supersonic line widths seen in many clouds.

INTRODUCTION

The formation, evolution, star formation history, and disruption of molecular clouds can be studied by mapping their spatial structure and kinematics using millimeter-wave spectral lines. During the past 5 years, 70% of the observing time on the 7-meter diameter millimeter wave telescope at Crawford Hill, which has a 100" beam at 110 GHz, has been dedicated to the large scale mapping of molecular clouds. This effort has involved the entire Bell Labs group consisting of R. W. Wilson, A. A. Stark, W. D. Langer, and myself. Of the 350,000 spectra, about 85% are in the J=1-0 ^{13}CO line, 10% in the J=2-1 CS line, and 5% in the J=1-0 ^{12}CO line. In the Orion region alone, about 180,000 spectra have been obtained. Since each spectrum consists of over 60 independent velocity channels, the display and analysis of this data presents a major challenge. One approach is to utilize a three color display in which the line intensity in three adjacent narrow velocity ranges are simultaneously displayed by the red, green, and blue guns of an RGB monitor. This method renders small scale fluctuations in the velocity field as true color and presents the line intensity as the brightness of the display.

THE LARGE SCALE STRUCTURE OF THE ORION REGION

Orion contains one of the nearest sites of active massive star formation, located at a distance of about 500 pc. Over the last 10^7 years, an OB association with at least several dozen O stars and many more B and later type stars has formed (see Goudis 1982 for a review of early work). A large 10 by 25 degree diameter bubble surrounding this region is visible in optical emission lines (Sivan 1974). Barnard's Loop is located at the eastern edge of the bubble and a large "U"-shaped emission nebula in Eridanus marks its western edge. Faint emission connecting these features may represent the outer boundary of a "superbubble" blown in the interstellar medium by the combined effect of ionizing radiation, stellar winds, and supernova explosions over the past 10^7 years (Cowie, Songaila, and York 1979). Most of the interior may be filled with hot (10^5 to 10^6 K) gas. The oblong shape of the bubble is probably produced by the decreasing pressure gradient toward high galactic latitudes. At its high latitude end (in Eridanus) the bubble may have "burst", pumping hot gas into the galactic corona, making Orion the nearest example of a "galactic fountain" (Bregman 1980).

Two giant molecular clouds reside in the interior of the Orion bubble. The Orion A cloud in the south contains the HII regions M42 (Orion Nebula) and NGC1977. The Orion B

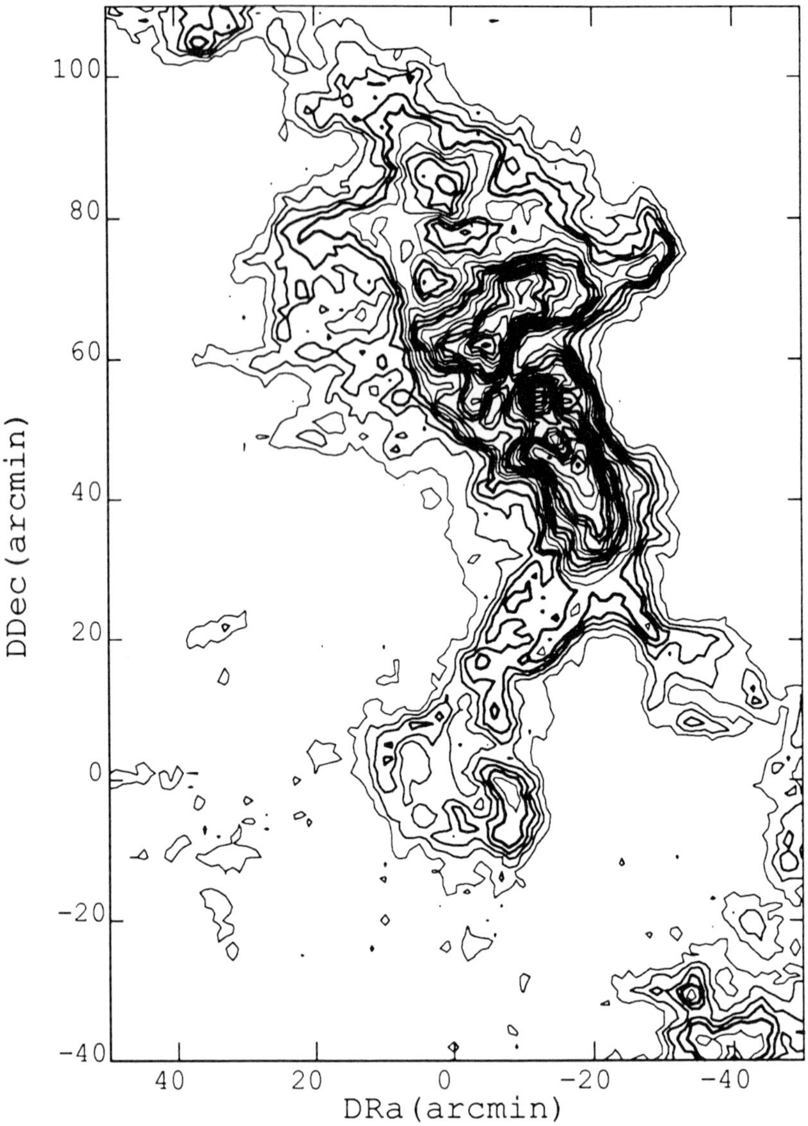

Figure 1: A ^{13}CO contour map showing the emission integrated from V_{LSR} = 9 km s^{-1} to V_{LSR} = 11 km s^{-1}. The contour interval is 1 Kkm s^{-1} and ranges from 1 to 20 Kkm s^{-1}. All co-ordinates are arc-minute offsets from $\alpha = 05^h\ 45^m\ 0.1^s$ and $\delta = -0^\circ\ 47'\ 00''$. Note the bubbles near the bottom of the figure and near the right edge. The reflection nebulae NGC2071 and NGC2068 are located in the main concentration above the center of the figure.

cloud in the north contains two separate star forming centers; NGC2023 and NGC2024 (the radio source Orion B) and a separate region of lower mass star formation associated with the reflection nebulae NGC2068 and NGC2071 to the north. In ^{13}CO emission, the cloud is highly fragmented into clumps, filaments, sheets, and partial shell structures, down to a size scale of 0.2 pc. Figure 1 shows a map of ^{13}CO emission in the region containing NGC2068 and NGC2071. South of the reflection nebulae, several cavities with bright walls and relatively empty interiors are seen. North of NGC2071, there are a series of arches of molecular gas which may be fragments of shells or filaments. Several streamers of gas extend from the eastern and western ends of the dense ridge of gas containing NGC2071 and trail off in a direction pointing away from the center of the Ori OB association. Such "wind-swept" structure can also be seen in the cometary globules and clouds west of the main Orion molecular clouds.

The Orion A cloud has a systematic velocity gradient along its length. The mean gas velocity varies from V_{LSR} = 4 km s^{-1} in the south to V_{LSR} = 12 km s^{-1} in the north. There is a smaller amplitude chaotic component to the velocity field superimposed on the large scale motion. The velocity gradient may indicate rotation, or can be interpreted as sheer. It is possible that the northern end of the cloud, which lies close to the older OB subgroups, and contains some of the youngest ones, has been accelerated by interaction with ionizing radiation and the resulting rocket effect. The positive velocity shift in the north requires that the cloud lies mostly on the far side of the association, which is consistent with the visibility of Barnard's Loop in front of the southern portion of Orion A cloud and with the position of the Orion Nebula in front of the molecular cloud.

Many filaments lace the Orion A cloud; some of the largest ones are several degrees long and are aligned with the major axis of the cloud, including the integral-sign shaped high density ridge containing the OMC1 cloud core lying behind the Orion Nebula. Some smaller filaments lie nearly parallel to the cloud minor axis, are less than 100" wide but over 1° degree long. Like in Orion B, close inspection of the images reveal many cavities and partial shells, suggesting local disruption of the cloud by embedded stellar objects. One prominent 20' radius shell and partial cavity surrounds NGC1999 and the HH 1-2 cloud core lying roughly 1° south of the Orion nebula. Smaller cavities and shells are superimposed on this structure at different velocities.

In addition to the over $10^5 M_\odot$ of gas in the Orion A and B clouds, there are many smaller clouds lying to the west of the main cloud concentration at even lower galactic latitudes. Many of these clouds as well as the entire Orion A cloud have cometary shapes whose axes of symmetry point towards the center of the Orion OB association. The entire northern portion of the Orion A cloud appears compressed relative its southern part. CS observations can be used to infer that the mean gas density in the "integral" shaped filament is much higher than in the southern portion of the cloud. The cloud shapes and velocity gradients suggest that the large scale structure of the cloud has been determined to a large extent by interaction with the OB association. Ionizing radiation can dissociate, heat, and ionize the cloud forming both high and low density HII regions. The ablated gas may be further heated by the shocks produced by fast stellar winds and supernova explosions and will be eventually incorporated into hot, high pressure, but low density interior medium of the Orion superbubble, driving its expansion. In return, the high pressure of the hot medium, and ablation by ionizing radiation can drive shock waves into the colder molecular cloud, resulting in compression and acceleration. Lower mass regions may be accelerated more, resulting in the wind swept cometary appearance of some of the clouds.

Compression of the northern portion of Orion A and the western portion of Orion B may have triggered massive star formation occurring at these sites. If so, the shocks may not be always plane- parallel as suggested by Elmegreen and Lada (1977). If a density enhancement retards the shock velocity relative to the surrounding medium, a conical pinch may compress the medium from all sides, producing a filament pointing toward the point of origin of the shock source. This may be the way the compressed filament at the northern end of the Orion A cloud formed.

EXCITATION OF CO IN THE ORION A MOLECULAR CLOUD

Alain Castets and Gilles Duvert at the University of Grenoble have mapped many square degrees in the Orion A molecular cloud in the J=2-1 transitions of ^{12}CO and ^{13}CO using the 2.5-meter POM-2 antenna located on the Plateau de Bure. Combined with the Crawford Hill J=1-0 observations, this data can be used to analyze the excitation conditions of the CO molecule in an entire molecular cloud. As described in Castets et al. (1989), there are problems with the model frequently used to analyze CO data. Usually, the ^{12}CO line is assumed to measure the excitation temperature along the line-of-sight, along which it is assumed to be constant. Since the ^{13}CO line is always fainter, it is assumed to be optically thin and used to determine the column density of the gas, under the assumption that the ^{12}CO and ^{13}CO excitation temperatures are the same and uniform along the line-of-sight. Castets et al. (1989) show that this assumption is wrong over much of the Orion A cloud.

The J=2-1 and J=1-0 ^{13}CO lines are observed to have a nearly identical peak intensity and line shape along 90% of the lines-of-sight through the Orion A cloud. Since the J=2-1 CS line can be detected $T_A > 0.1$ K along most directions, the gas density is likely to be sufficient to thermalize ^{13}CO. Castets et al. (1989) conclude that the ^{13}CO lines are optically thick and abandon the assumption that the excitation temperatures of ^{13}CO and ^{12}CO are equal. The observed J=1-0 $^{12}CO/^{13}CO$ line ratio is explained by requiring the ^{12}CO emitting layer to have a higher excitation temperature than the ^{13}CO layer. This can occur if clouds are mostly heated from the outside. The penetration depth of the ultraviolet radiation responsible for heating cloud surfaces by means of photoelectrons emitted from grains is similar to the depth into the cloud where ^{12}CO saturates (τ=1) for a typical 1 to 4 km s^{-1} line width and a gas temperature of order 20 to 30 K. The column density lying in front of the τ=1 surface for the J = 1 - 0 transition of CO, (the penetration depth) is given by

$$N_{\tau=1}(H_2) = \frac{2.4\times10^{14} T_{rot} \Delta v(\mathrm{km\ s^{-1}})}{X_{CO}[1-\exp(-h\nu/kT_{ex})]}$$

Where X_{CO} is the relative abundance of the CO species relative to hydrogen, T_{rot} is the excitation temperature, and Δv is the line width. For Δv = 2km s^{-1}, T = 20K, and $X_{CO} = 5\times10^{-5}$ (appropriate for ^{12}CO), this gives $N_{\tau=1}(H_2) \approx 4\times10^{20}$cm^{-2}. For T = 10 K and $X_{CO} = 1\times10^{-6}$ (for ^{13}CO) with the same line-width, $N_{\tau=1}(H_2) \approx 9\times10^{21}$cm^{-2} Thus the ^{12}CO lines probe the externally heated cloud surface while the ^{13}CO lines probe the shielded cloud interior.

The ^{13}CO lines appear to be optically thin the immediate vicinity of luminous embedded sources, such as irc2 in the Orion A cloud core. Here, the gas is hot and the CO population is spread over many levels in the rotation ladder. The radius of influence of an embedded heating source of luminosity L_{100} (measured in units of 100 $L_\odot$) is given by $R(T<10K) = 0.4\ T_{10}^{-5/2}\ L_{100}^{1/2}$ (parsecs) where T is in units of 10 K. In a typical GMC, there are only several dozen sources with luminosities above 100 $L_\odot$, so only a small fraction of the mass and volume of a GMC can be internally heated. The rest of the interior of molecular clouds are probably heated by cosmic rays, dissipation of turbulence or shock waves, or possibly by ambipolar diffusion and remain cold.

CAN NON-IONIZING UV RADIATION PRODUCE BUBBLES AND TURBULENCE?

The superthermal line widths of spectral lines in molecular clouds has been a long standing mystery. In order to maintain the supersonic turbulence for more than a dissipation time scale τ_{dis}, which is the order of 10 free fall-times (10^6 to 10^7 years), some mechanism must supply kinetic energy at a rate $M_{cloud}\Delta V_{turb}^2/\tau_{dis}$ where ΔV_{turb} is the line width and τ_{dis} is the dissipation time scale. External agents are inefficient as a source of energy input since the large density and temperature difference between the molecular cloud and the surrounding medium provides an impedance mismatch. Most of the energy of supernova explosions, stellar winds from main sequence stars, or radiation is thought to couple into the intercloud medium or to accelerate the clouds as a whole. A source of energy which resides inside the cloud is more efficient.

With the discovery of bipolar outflows and the recognition that virtually all young stellar objects undergo a period of energetic mass loss, it has become evident that young stars are a source of the energy to support the turbulence (Bally, 1982). If clouds had smooth internal structure, flows would be capable of efficiently injecting kinetic energy into a cloud. However, the observed fragmented structure implies that the energy may be more efficiently coupled to the intercloud medium than to the dense CO emitting gas. The early calculations on support of clouds by outflows neglected the inefficiency of the coupling of the flow energy to a cloud which is highly fragmented on a scale smaller than the characteristic size of the flow. In this case, the flow is expected to blow out of the dense cloud, and its energy will not couple directly into the dense gas.

Although it is possible that some bubbles and partial shells are fossil outflows, their large size and lack of bipolarity suggest that some other mechanism is at work. Main sequence stars can be found near the centers of some bubbles and partial shells. The partial rings surrounding NGC1999 are nearly concentric with the bright reflection nebulosity, suggesting that the stars producing the scattered light may also be responsible for the large scale structure in the gas. Although winds produced during both the pre-main- sequence and main-sequence life of a star may help inflate a bubble, it is intriguing to to investigate the possible role of non-ionizing radiation.

Soft UV radiation is responsible for heating the surface layers of clouds. In regions near B or A type stars, the radiation field can be orders of magnitude stronger than in average locations in interstellar space. Tielens and Hollenbach (1985) and Hollenbach (1988) have shown that the temperature of the boundary layer of a cloud can be raised to hundreds of degrees by a combination of photoelectric and other heating processes. As discussed above, there is evidence for an increase in the average temperature of the CO emitting layer relative to the shielded cloud interior. Heated gas at the cloud surface can stream away from the cloud at a velocity of a few km/sec, resulting in a reaction force on the cloud, analogous to the rocket effect produced by ionizing radiation. The effect of the non-ionizing component is expected to be similar in nature but milder in amplitude. A column density N of gas can be accelerated to a velocity V by a source of soft UV of luminosity L_* in a time τ given by $\tau = 2\pi r^2 \mu m_H N v^2 / X_{UV} \varepsilon L_*$ which for $N = 10^{21} cm^{-2}$, $r = 10$ pc, $L_* = 100\ L_\odot$ and a UV fraction of 10%, and an overall UV heating efficiency of 0.1% implies that a velocity of 1 km s^{-1} can be reached in 10^5 years. Soft UV might produce some shells and bubbles, and together with outflows, and externally generated shocks, may contribute kinetic energy required to support the supersonic cloud line widths.

REFERENCES

Bally, J 1982 *Ann.N.Y.Acad.Sci.*,**395**,191.

Bally, J., Langer, W.D, Wilson, R.W. and Stark, A.A. 1987 *Ap.J.(Letters)*, **312**,L45.

Bregman, J. 1980 *Ap.J.*,**236**,577.

Castets, A., Duvert, G., Bally, J., Wilson, R.W., Langer, W.D., and Dutrey, A. *Astron.Astrophys.*,(in preparation).

Cowie, L.L., Songaila, A., and York,D.G. 1979 *Ap.J.*,**230**,469.

Elmegreen, B.G. and Lada, C.J. 1977 *Ap.J.*,**214**,725.

Goudis, C. 1982 *The Orion Complex: A Case Study of Interstellar Matter.* (Dordrecht: D. Reidel).

Sivan, J.P. 1974 *Astron.Astophys.(Suppl.)*,**16**,163.

Tielens, A.G.G.M. and Hollenbach, D.J. 1985 *Ap.J.*,**291**,722.

Hollenbach, D.J. 1988 *I.A.U. Symposium #135, Interstellar Dust*, ed. A.G.G.M. Tielens and L. Allamandola.

Discussion:

FRANCO: Photodissociating radiation (and perhaps even radiation pressure on dust grains) may add some energy to the case of the Pleiades.

BALLY: Yes I agree. The main point I want to make is that the TOTAL heating rate (sum of photodissociation, photoelectric, etc.) may have important dynamical consequences for the structure and kinematics of clouds. Hot gas in the boundary layer will expand away form the cloud surface, and the resulting "rocket" effect can accelerate the cloud. This may produce bubbles such as seen in Orion, surrounding the Pleiades. In enviroments such as Orion, where there are many later than B3 stars, this mechanism may dominate input from outflows in providing cloud support.

MÜNCH: The photodissociation of H_2 should occur in a narrow range fairly near the "surface" of a cloud ($\tau_{opt} \sim 1$). In order to maintain the cloud heated we badly need cosmic rays.

BALLY: The UV penetration depth is roughly $\tau_{UV} \sim 1/3$ to $1/5\tau_{opt}$. The dissociation of CO occurs within this region. Since τ_{opt} corresponds to a hydrogen column density of 10^{21}cm^{-2}, the ^{12}CO line saturates well within the region where UV heating plays a dominant role (via photoelectric heating and H_2 dissociation heating). Visual and near IR photons, however, can still penetrate to $\tau_{opt} \sim$ few with sufficient efficiency to dominate cosmic ray heating, and this "long wavelength" heating will be substantial in environments where the ambient radiation field is stronger than the average value. The interior of the cloud can be heated to 5 to 10K required to explain the observed intensity of ^{13}CO by either cosmic-rays or the dissipation of supersonic turbulence which dominates the millimeter wavelenght line-widths. The proposed rocket effect produced by the ablation of the hot surface layers of the cloud may sustain this turbulence.

The Radio Continuum Morphology of the Orion Nebula

Marcello Felli
Osservatorio di Arcetri, Largo E. Fermi 5, 50125, Florence, Italy

Ed Churchwell
University of Wisconsin, 475 N. Charter St., Madison, WI 53706, U.S.A.

Douglas O. S. Wood
Center for Astrophysics, 60 Garden St., Cambridge, Mass 02138, U.S.A.

The Orion Nebula is the bench mark galactic HII region in which major advances have been made in our understanding of the ionization structure, energy balance, plasma dynamics, star formation and interactions of ionized plasmas with dense molecular clouds. Most of these studies require accurate comparison with radio continuum images.

The basic properties of the radio emission from the Orion Nebula, such as the overall shape and flux density, have been accurately determined with single dish observations for a long time. The highest resolution single dish map has been obtained with the 100 m Bonn radiotelescope at 23 GHz, with a resolution of 43" (Wilson and Pauls, 1984). This map corresponds very well with Hα emission. The emission from M42 and M43 are clearly separated and some structure is found in M42, in particular: a bright bar (I-front) close to $\theta^2 Ori$ and a central core about 3' in diameter. The total flux density in M42 is 400 Jy and the mean Emission Measure (EM) is $\sim 5\ 10^6$ pc cm^{-6}.

Several high resolution interferometric radio maps of the HII region have been published (Martin and Gull, 1976; Righini-Cohen, Simon and Felli, 1981; Garay et al, 1982; Johnston et al, 1983; Garay, Moran and Reid, 1987; Churchwell, Felli, Wood and Massi, 1987; Barvainis and Wootten, 1987; Ohashi et al, 1989). These maps are incomplete because an interferometer acts as a spatial frequency filter and selects only a portion of the spatial frequency spectrum contained in the true source brightness distribution. This effect translates into filtering out the low EM structures as the interferometer baseline increases. Consequently, interferometers with longer baselines are sensitive only to compact, high EM components. This effect is, in many ways, similar to the relationship between sensitivity to low EM and exposure time in optical photographs. In this case, the longer the exposure time, the lower the EM that can be detected, but at the same time the high emission regions are usually saturated.

Another consequence of this effect is that radio maps made with different resolutions will contain different total flux densities. For extended sources like the Orion Nebula, the higher the resolution, the lower the total flux that will be detected.

In this paper we present a comparative study of the radio structure of the nebula over scale sizes from 0.1" to 45". The data were obtained with the VLA in its four standard configurations, from the most extended (A) to the most compact (D), at two widely spaced wavelenghts. At 20 cm we imaged a field of about 30' with resolutions ranging from 45" to 1", while at 2 cm we were able to observe only the inner 3' central core with resolutions from 4" to 0.1". Here we explore the surface brightness distribution of the nebula from single dish resolutions to the highest presently achievable.

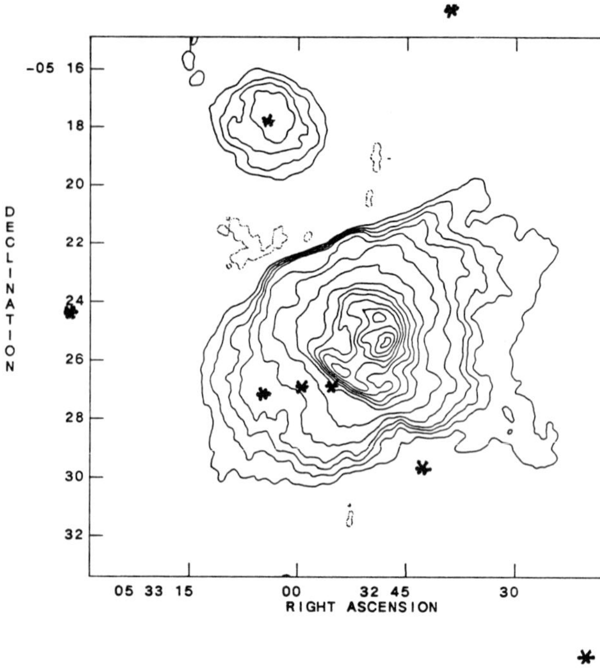

Figure 1a

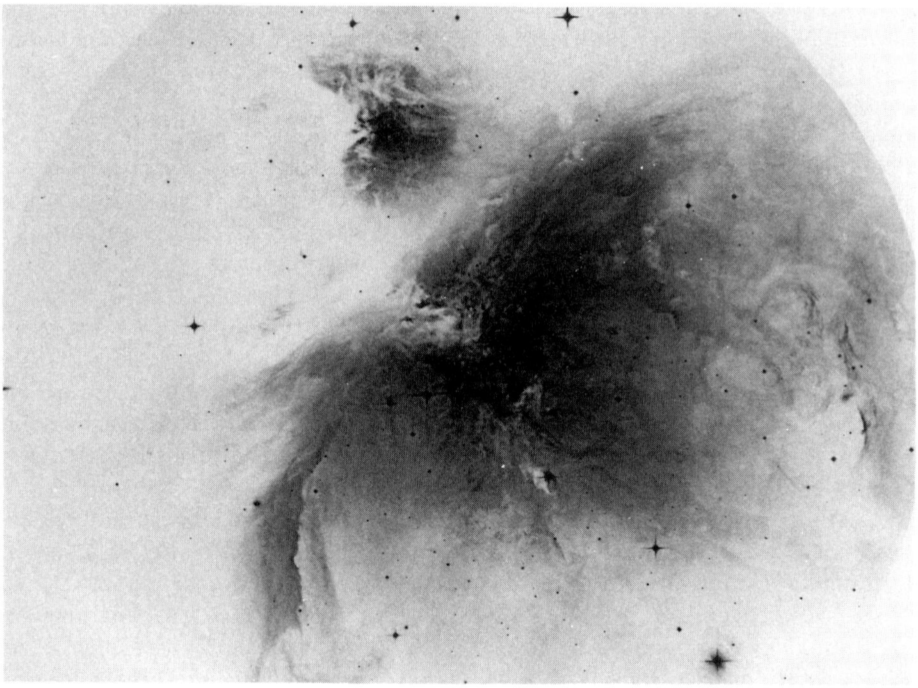

Figure 1b

In principle one would like to have all the brightness distribution information combined into a single image, at each frequency. In practice this turns out to be almost impossible. The basic problem is that the EM dynamic range from the brightest compact component to the faintest diffuse emission is too large to be recorded with all strucures easily visible on any single image. This is true for optical photographs, CCD images, and radio images. For sources like the Orion Nebula, where a very large range of EM are present, the combination of two or more adjacent VLA configurations is always dominated by the lower resolution configuration. Although we have attempted to do this in a few instances, here we will present only the maps obtained from single VLA configurations. Also, due to the limited space and to the fact that some maps are still being reduced, we will present only a sample of the information that can be extracted from these data.

At the lowest resolution the most obvious features are the roundish structureless shape of M43, the bright bar and central bright core in M42. Figure 1a shows our 20 cm C-configuration map (resolution 20"). The *s indicate the positions of reference stars to be used in the comparison with the Hα photograph of figure 1b (from ESO). Both images have the same angular scale. In general the two images are in very good agreement, implying that, on a large scale, foreground dust absorption plays a minor role. This is consistent with the widely accepted idea that the Trapezium stars are eroding a molecular cloud in the backgroud and that the ionized gas is flowing toward the observer. We also call attention to the very sharp gradient along the NE boundary of M42 which probably indicates dense neutral gas at this ionization-bounded front, although part of this effect could also be due to the negative depression produced by the interferometer response to M42 and M43. The mean EM in the map is 1.3 10^7 pc cm^{-6}.

We will now concentrate on M42. At a resolution of 7" the diffuse extended structure begins to fade away and only the bright bar and the compact core stand out clearly. Figure 2a shows the 20 cm B-configuration image and figure 2b shows the Hα short exposure photograph of G. Munch (Munch and Wilson, 1962), reproduced with the same scale as the radio image. The structure of the ionization front and the small scale components in the central core compare very well in both images. Two "jets" of emission perpendicular to the bright bar are evident in the radio map that are not clearly visible on the optical photo. In the dark bay area, NE of the Trapezium stars, there is a deficiency of ionized gas. The optical dark bay is apparently not caused by foreground extinction but by a deficiency of high EM ionized gas.

For higher resolutions, we now shift to the 2 cm maps. Due to the limited field of view (3') we can now see only the central core. The bright bar is beyond the HPW of the primary beam. Figure 3 shows the 2 cm D-configuration map (resolution 4"), overlayed on Munch's Hα photograph. The agreement between optical and radio structures is impressive. Only some areas in the outer NE part of the core may be affected by foreground obscuration. For the rest of the structures there is almost a one to one correspondence. The Trapezium stars are located slightly to the south of the center of the backward C-shaped ionized core. The core is rather sharply bounded to the west but is more irregularly bounded to the east. $\Theta^1 OriC$, the earliest spectral type star of the cluster, is located in a region of lower emission in the core, suggesting that either radiation pressure and/or a stellar wind might have evacuated a region around the star

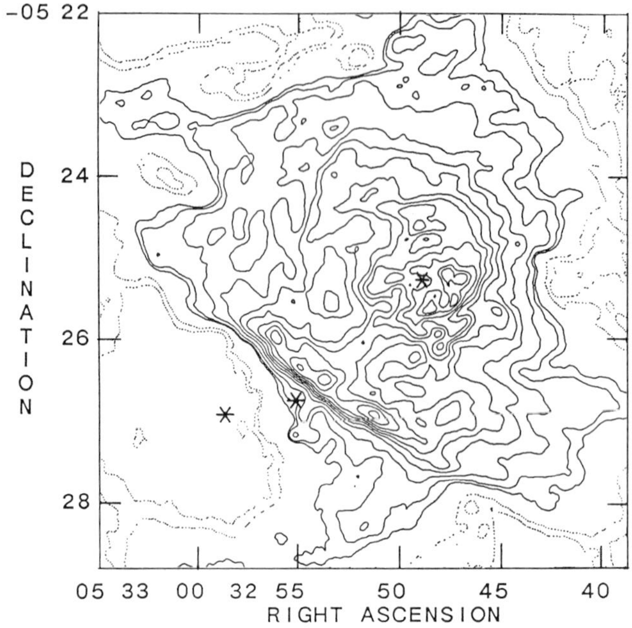

Figure 2a

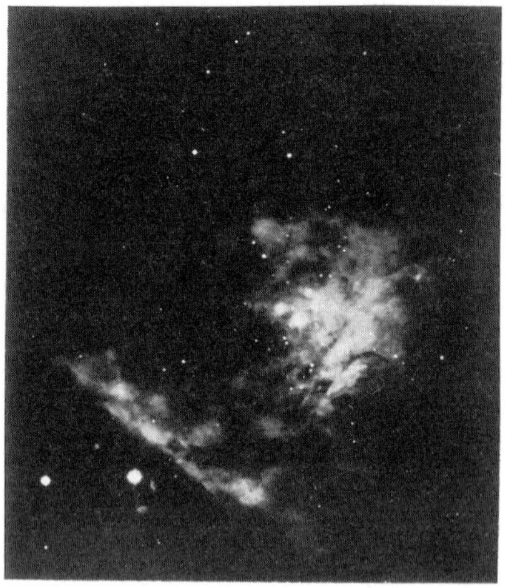

Figure 2b

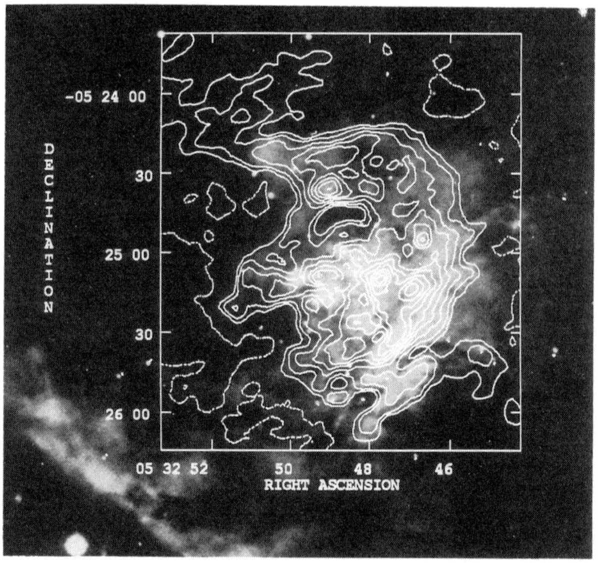

Figure 3

(Barvainis and Wootten, 1987). The structured appearance of the core on this scale size (approximately 10^{-2} pc) must originate from neutral blobs which are embedded in the ionized nebula and have thin ionized envelopes (i.e PIGs). If they were fully ionized density enhancements, their lifetimes against expansion would be too short to understand the large number observed.

Of the many comparisons that can be made between figure 3 and other types of emission in the Orion core, we show in figure 4 a skematic comparison of the continuum emission with the $^{12}C^{18}O$ (1-0) map of Wilson et al (1986) (which shows the morphology of the high density, $\sim 10^6$ cm^{-3}, molecular gas) and with the 20 μm map of the BN-KL region (Downes et al, 1981). It is obvious that the ionized core is bounded to the W by a dense molecular cloud, which probably determines the structure of the ionized core. The BN-KL region is located at the interface between the ionized and molecular gas. No radio continuum emission from BN or IRc2 is apparent at this resolution; this map is still dominated by diffuse emission. A similar result holds for the bright bar, in which the ionized gas is bounded to the south by a CO molecular cloud (Ohashi et al, 1989).

At even higher resolutions (1"), a new and unexpected aspect of the surface brightness distribution begins to appear. In figure 5, the 2 cm C-configuration is shown overlayed on Munch's photograph. While the radio emission of the bright extended blobs predominant in figure 3, they begin to fade away due to increased resolution, and a new class of small or unresolved components begins to stand out. They were not apparent in the lower resolution maps and they do not necessarely coincide with the peaks of those maps. A cluster of these components is clearly evident in the Trapezium region, one of which coincides with $\theta^1 OriA$, while others are located in regions not connected with bright extended nebulosities. At this resolution, both BN and IRc2 are detected as well as a host of other compact sources. These sources represent a wide variety of objects and emission mechanisms, most of which are not yet understood. Several have been discussed in some detail by Laques and Vidal (1987); Garay, Moran, and Reid (1985, 1987); and Churchwell et al (1987). These objects have been referred to as the "Orion Zoo" by Garay (1987) and they will be discussed further by him later in this meeting. In the spirit of comparing the structure at different resolutions, we simply point out that at 2cm in the A-configuration, all the diffuse emission is gone and only the compact sources with EM $> 10^8$ pc cm^{-6} and angular diameters $< 0.5''$ are left. The total flux density contained in this map is about 180 mJy, less that three orders of magnitude smaller than the total flux density of the nebula.

The coincidence between the radio source and the star $\theta^1 OriA$ can be established with an accuracy of $\sim 0.2''$ (this limit is imposed by the optical position of the star). However from VLA data alone the interpretation of the radio emission was unclear. While thermal emission from an ionized wind or an ultracompact HII region seemed the most obvious possibilities, these conflicted with the brightness temperature derived from the VLA size (T_b >54,000K) and the fact that, during a period of ~5 years, the radio emission varied from a minimum of 3 mJy to a maximum of 74 mJy (see figure 7). To obtain a better estimate of the angular diameter, and therefore of the brightness temperature, we have used the European VLBI Network at 6 cm to observe this source. The visibility curve is consistent with an angular diameter < 4 mas (i.e. 1.8 AU) and a flux density of 13 mJy (equal to that measured simultaneously with the VLA). These data imply a brightness temperature $> 4\times10^7$ K on 28 Sept. 1987, which unambiguously

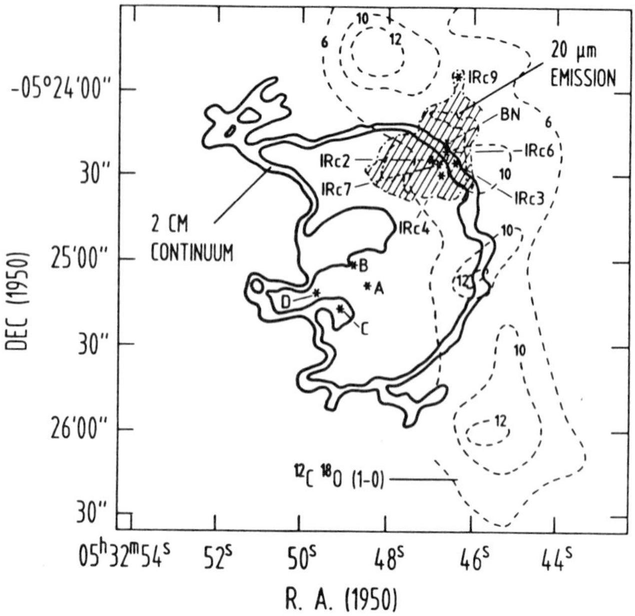

Figure 4

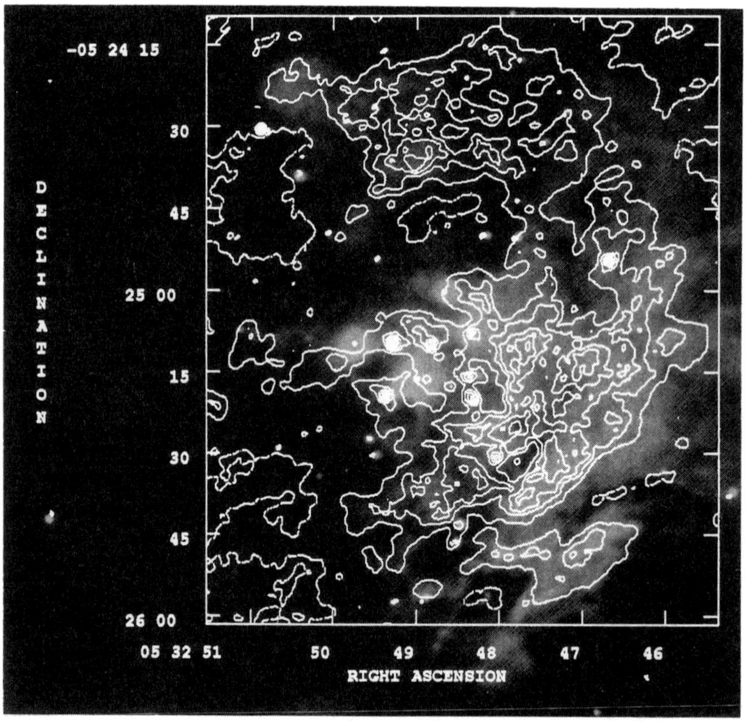

Figure 5

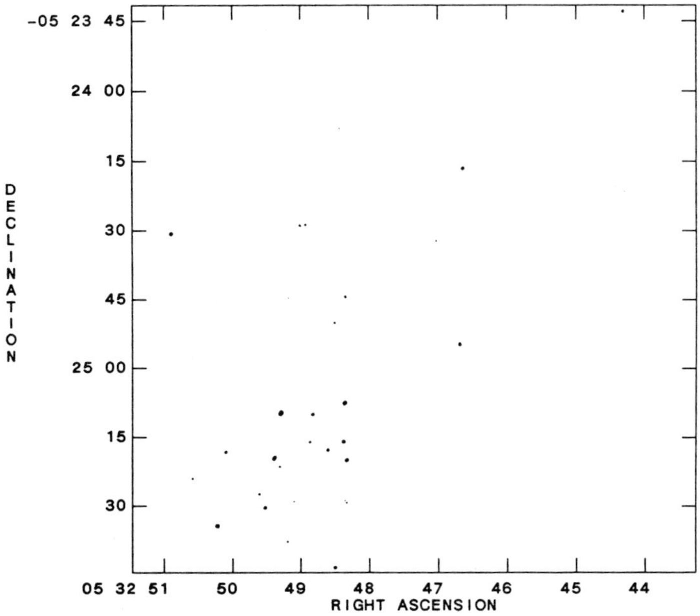

Figure 6

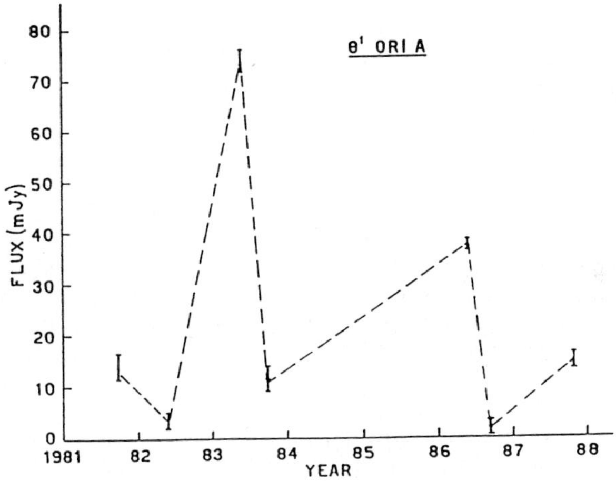

Figure 7

<1.8 AU) and a flux density of 13 mJy (equal to that measured simultaneously with the VLA). These data imply a brightness temperature $> 4\ 10^7$ K on 28 Sept. 1987, which unambiguously proves that the emissiom mechanism must be non-thermal (Felli, Massi and Churchwell, 1989). $\Theta^1 OriA$ is a binary system composed of a B3III-V star and a possible T Tauri companion whose mean separation is 0.71 AU. While it is too early to establish what the non-thermal mechanism is in the star, one possibility may be syncrotron emission from relativistic electrons in large magnetic loops within the binary system energized by flare events on the T Tauri star.

References

Barvainis, R., Wootten, A., 1987, A. J., **93**, 168

Churchwell, E., Felli, M., Wood, D.O.S., Massi, M., 1987, Ap. J., **321**, 516

Downes, D., Genzel, R., Becklin, E.E., Wynn-Williams, C.G., 1981, Ap. J., **244**, 869

Dyson, J.E., 1968, Ap. Space Sci.,**1**, 388

Felli, M., Massi, M., Churchwell, E., 1989, Ast. Ap. , in press

Garay, G., Reid, M.J., Genzel, R., Ho, P.T.P., 1982, Symposium on the Orion Nebula to Honor Henry Draper, Glassgold, A.E., Huggins, P.J., Schucking, E.L., Eds., The New York Academy of Sciences, 204.

Garay, G., Moran, J.M., Reid, M.J., 1985, in Radio Stars, ed R.M. Hjellming and D.M. Gibson, Dordrecht, Reidel, p131

Garay, G., Moran, J.M., Reid, M.J., 1987, Ap. J., **314**, 535

Garay, G., 1987, Rev. Mex. Astron. Astrophys., **14**, 489

Johnston,K.J., Palmer, P., Wilson, T.L., Bieging, J.H., 1983, Ap. J., **271**, L89

Laques, P., Vidal, J. L. 1978, Ast. Ap., **73**, 97

Martin, A.H.M., Gull, S.F., 1976, Mon. Not. R. A. S., **175**, 235

Moran, J.M., Garay, G., Reid, M.J., Genzel, R., Wright, M.C.H., Plambeck, R.L., 1983, Ap.J., **271**, L31

Munch, G., Wilson, O.C., 1962, Z. Astrophys., **56**, 127

Ohashi, N., Mizuno, A., Tatematsu, K., Sugitani, K., Kasugu, T., Fukui, Y., 1989, Ap. J., in press

Righini-Cohen, G., Simon, M., Felli, M., 1981, Sky and Telescope, **62**, 225

Wilson, T.L., Pauls, T., 1984, Ast. Ap., **138**, 225

Wilson, T.L., Serabyn, E., Henkel, C., Walmsley, C.M., 1986, Ast. Ap. **158**, L1

HIGHLY EXCITED MOLECULAR HYDROGEN IN ORION

Hans H. Hippelein, Guido Münch
MPI für Astronomie, Heidelberg

INTRODUCTION

Observations of H_2 lines in the IR have been mostly restricted to those with upper levels of low energy, which can be excited either collisionally in shocks or radiatively by UV starlight. In order to discriminate between the two excitation mechanisms we have measured in 1 μm range lines of the v=2-0 band arising from high rotational levels J$\leq$13. Their intensities, together with those of the IR lines, allow an estimate of the line of sight effective extinction and a determination of the rotational temperature measuring their joint degree of excitation. The latter parameter provides information about the energy state of the molecules at their formation and ejection from grain surfaces and thus constrains the hypothetical models for H_2 molecule formation.

M 42

With a Fabry-Perot interferometer using an instrumental resolution of 11 km s^{-1} we scanned the lines (2,0)S(10)λ10516, S(11)λ10523, S(7)λ10639 and S(5)λ10848 at various positions in M42. Towards the BN-object the S(5) and S(7) lines show distinct blue wings similar to that observed in the infrared (1,0)S(1) line, explained by a shock of ~20 km s^{-1} that excites the H_2 gas and leads to an expansion of the BN gas bubble. The relative line intensities of the blue wings determine the foreground extinction to be A_V ~4.5 (Münch et al., 1984).

In the S(10) and S(11) lines the wings are not seen, since the contribution from shock excitation is much too small to be detectable at these high energy transitions. The observed lines, therefore, must arise from regions excited by resonance absorption of UV starlight and subsequent radiative cascade (Hippelein and Münch, 1989).

The intensity distribution in M42 for the highest excited line S(11), based on 33 line scans through diaphragms between 40" and 80" over an area of ~8 arcmin, appears rather flat with a maximum at the Trapezium and nearly centrisymmetric to it. The intensity map shows no enhancement at the cluster of IR-sources and does not reflect any of the dominant structures of the HII region, except the dark bay in the East produced by foreground extinction.

The radial velocities over the whole nebula have an rms deviation of only $\pm$2 km s^{-1} around the mean, without any systematic trend with position in the nebula. After coaddition of the single scans we derived a mean S(11) profile which can be fitted by a Gaussian with a dispersion of σ=7.1 km s^{-1}. This leads, after instrumental profile deconvolution to a velocity dispersion along the line of sight of σ_V=4.7 km s^{-1}, somewhat larger than the widths of other molecular lines emitted from OMC.

Towards the Orion Bright Bar where Hayashi et al. (1985) have measured for the probably partly shock excited (2,1)S(1) line an intensity of

$I(1) = 40\ 10^{-6}$ erg s^{-1} cm^{-2} sr^{-1},

we derived intensities for the high J transitions in the v=2-0 band of

I(5) = 13.5 10^{-6},

I(7) = 8 10^{-6},

I(10) = 4 10^{-6} and

I(11) = 8 10^{-6} in the same units. From these numbers, under assumption of a foreground axtinction of A_V =3.2 mag, we can directly determine the column densities N(J') for the upper transition levels.

In Fig. 1a the excitation parameters ln[N(J')/(2J'+1)g] for the various J' using an para/ortho ratio of 1/2.2 (corresponding to an H_2 formation on grains with a temperature of 50 K) are plotted versus their relative energy in excitation temperature units. From the slope of the curves, we find for the levels lower than J'=7 a rotational temperature of ~2200 K, whereas for J'>7 the temperature is as high as ~11000 K.

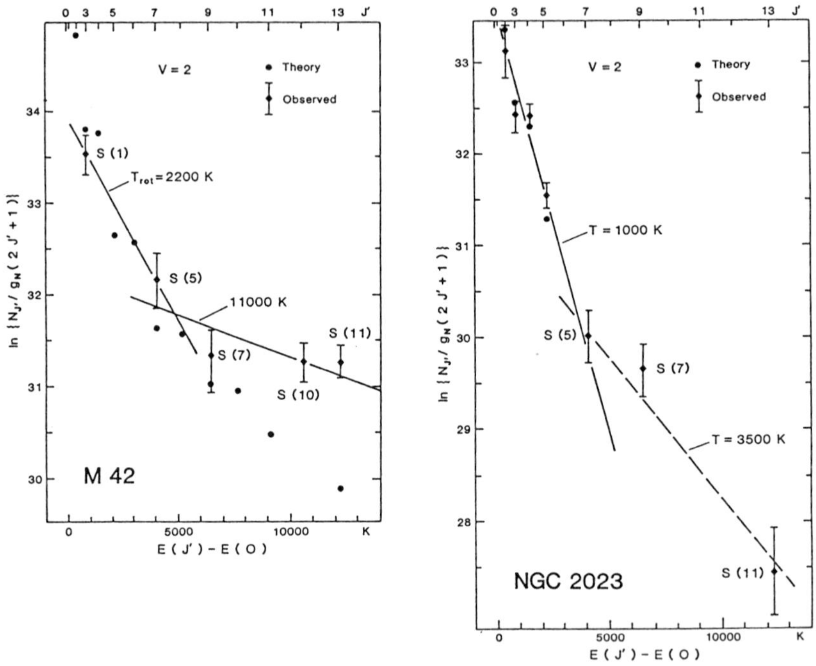

Fig 1a and b. Fit of a Boltzmann distribution to the populations of the various rotational levels in the (2,0) band for lines measured in M 42 (left) and in NGC 2023 (right).

COMPARISON WITH THEORY

From extensive theoretical model calculations done by Black and van Dishoeck (1987) it became obvious, that a significant population of the high J levels can not be realized by fluorescent excitation alone, but only in combination with the process of H_2 formation on grains. Three different formation processes were included in these models: In the first one, 1/3 of the 4.5 eV binding energy is distributed among the (v,J) states according to a Boltzmann law (Black and Dalgarno, 1976); in the two others, the nascent H_2 molecules escape from the grains at high v leves but J<2. It was found, that the total line emission is rather sensitive to paramaters such as the intensity of the radiation field of the exciting source, the gas density and the properties of the grains, whilst the relative line intensities change only slightly with these parameters. The intensities for the high rotational transitions, however, vary by orders of magnitude from one formation model to the other.

In order to compare the predicted and observed populations of the various rotational levels in v=2, we have entered in Fig. 1 the excitation parameters corresponding to the line intensities as determined by Black and v. Dishoeck for their model Nr. 14, normalized to the total line flux of model Nr. 69, which they suggest to be as representative for the Orion Bright Bar we are considering. The theoretical curve matches quite well the population of the levels up to J=7, for the higher rotational states, however, it is too low by a factor up to ~4. As the model was already calculated for the H_2 formation model providing the highest populations for the high J levels the other two formation models can be ruled out completely.

A possible explanation for the discrepancy at high J' could be the increasing importance of collisions at densities in the order of 10^5 cm^{-3} reached in M42, interchanging the energy between vibrational and rotational modes. Elastic collisions may also induce a dipole in the highly excited molecules, increasing their transition probabilities and thus simulating a higher excitation temperature. Finally, the formation model used may be unrealistic. Hunter and Watson (1978) for example proposed a formation model, where the molecules escape from the grains in a mean rotational state of J~10.

NGC 2023

We observed high rotational transitions also in M17, NGC 2023, NGC 7023 and NGC 7027, where infrared line emission has been reported. Here, we will discuss only the results for the reflection nebula NGC 2023, where we recently succeeded to observe S(11) at a very low count rate.

In Fig. 1b the excitation parameters determined for the various J' levels using a ratio 1/2 for the g values and corrected for a foreground extinction of A_V =1.6, are entered against the excitation energies for NGC 2023. The data for J'<6 are taken from Hasegawa et al. (1987) who have measured the intensities for a number of infrared H_2 lines in the brightest region 80" south from the central star. The dots stand for the predictions of the model H of Black and van Dishoeck (1987), for fluorescent excitation deriving from the star HD 37903. The predicted rotational temperature (1000 K), considerably lower than that in M 42, fits well the intensities of lines with J≤7. For lines with J≥7 a rotational temperature near 3500 K is indicated, again considerably lower than that in Orion. Model calculations for such high level lines are unfortunately not available. It is thus not possible to find out whether the discrapancy between theory and observations for the high level lines, found in M 42, is also present in NGC 2023.

REFERENCES

Black, J.H., Dalgarno, A.: 1976, Astrophys. J. 203, 132

Black, J.H., van Dishoeck, E.F.: 1987, Astrophys. J. 322, 412

Hasegawa, T., Gatley, I., Garden, R.P., Brand, P.W.J.L., Ohishi, M., Hayashi, M., Kaifu, N.: 1987, Astrophys. J. 318, L77

Hayashi, M., Hasegawa, T., Gatley, I., Garden, R., Kaifu, N.: 1985, M.N.R.A.S. 215, 31P

Hippelein, H.H., Münch, G.: 1989, Astron. Astrophys. 213, 323

Hunter, D.A., Watson, W.D.: 1978, Astrophys. J. 226, 477

Münch, G., Hippelein, H., Pitz, E.: 1984, Astron. Astrophys. 135, L11

Discussion:

PECKER: How well does one know the transition probabilities and the collisional cross sections? And are not the higher levels likely to be "safer" than the low levels used in the theoretical interpretation you mention?

DALGARNO (to Pecker): I computed these quantities, they are good, and as good for low levels as they are for high levels.

HASEGAWA (Comment): As can be seen in the classic case of the Orion Bright Bar, the excitation of H_2 emission (e.g. vibrational temperature) changes over the nebula (Hayashi et al. 1985, MNRAS, 215, 31P). This can explain the bimodal excitation apparent in your population diagram.

DALGARNO: It may be possible to explain the high J population by fluorescence pumping into high V, low J levels followed by collisional transfer into low V high, J levels. Collisions of this kind in which the energy transfer is small should be rapid.

Aperture Synthesis Observations of NH_3 and CS in Orion-KL

Y. MURATA[1], R. KAWABE[2], M. ISHIGURO[2], T. HASEGAWA[3], T. TAKANO[2], T. KASUGA[2], K.-I. MORITA[2] AND M. HAYASHI[1]

[1] Department of Astronomy, Faculty of Science, The University of Tokyo

[2] Nobeyama Radio Observatory, National Astronomical Observatory

[3] Institute of Astronomy, Faculty of Science, The University of Tokyo

ABSTRACT. We have made aperture synthesis maps of Orion-KL in NH_3(1,1),(2,2) and CS(J=1-0) emissions using the Nobeyama Millimeter Array. Both NH_3 and CS maps show new detailed structures that have not been recognized before. Two NH_3 filaments in the size of $\sim$0.45 pc $\times$0.04 pc are found extending to the northwest direction of Orion-KL. These filaments are associated with the HH objects and the finger-like H_2 emissions; they are probably formed through the interaction with high velocity, highly channeled winds from the KL region. The CS maps show, on the other hand, shell structures around IRc2 as well as the well-known rotating disk. These shells coincide with the two lobes of the shock-excited H_2 emission, being interpreted as the shock-compressed shells of ambient molecular gas interacting with the outflow from IRc2.

1. Introduction

Orion-KL is often taken to be a prototype of massive star-forming regions. There are at least three self-luminous sources in the region. IRc2 (L $\sim 10^5$ $L_\odot$), one of the three, is confirmed to be a source of essentially all the luminosity in this region (Downes *et al.* 1981; Wynn-Williams *et al.* 1984).

A massive molecular disk and a bipolar molecular outflow have been found around IRc2. The disk is elongated to the NE-SW direction and is rotating differentially (Plambeck *et al.*1982; Hasegawa *et al.* 1984; Vogel *et al.* 1985). The bipolar molecular outflow has been recognized by high resolution observations. Its axis is parallel to the rotating axis of the disk on the sky (Genzel *et al.* 1981; Erickson *et al.* 1982; Wright *et al.* 1983; Masson *et al.* 1987). The outflow is confined in a small area in spite of its high velocity reaching $\sim$100 km s^{-1}; its dynamical timescale is 10^3 years. This timescale is one or two orders of magnitude less than those of the other outflows (Bally and Lada 1983). Orion-KL deserves detailed studies in order to understand the evolution of molecular outflows and their interaction with ambient material in the early stage of star formation. Here we present higher resolution interferometric maps of NH_3 and CS (J=1-0) around Orion-KL. The high resolution maps show the new structures such as interacting shells around the bipolar outflow and filamentary structures extending out from the KL region as well as the rotating disk along the OMC-1 ridge.

2. Observations

The NH_3 (1,1),(2,2) and CS (J=1-0) lines were observed with the Nobeyama Millimeter Array (NMA; Ishiguro *et al.* 1984). The obtained spatial resolutions for our

NH_3 and CS maps are 16″ and 9″, respectively. The parameters of the observations are summarized in the table below. We used Orion-KL as the phase center for the observation (*R.A.*(1950)= $5^h32^m47^s$, *decl.*(1950)= −5°24′22″). The data reductions have been carried out using the AIPS package installed in FACOM M380 and VP50 at Nobeyama Radio Observatory. No primary beam corrections were made for the maps.

TABLE: OBSERVATIONAL PARAMETERS.

Transitions	NH_3(1,1),(2,2)	CS (J=1-0)
Frequency (GHz)	23.694 and 23.722	48.991
Date	Dec, 1986 ~May, 1987	March ~June, 1988
Mapping Area	5′×6′	2′×2′
Frontends	Cooled HEMT Receiver*	SIS Receiver**
T_{sys}(SSB)	200 K	300 K
Backends	Digital spectro-correlator*** (1024 ch., B = 80 MHz)	
Spatial resolution	16″	9″
Velocity resolution	1.0 km s^{-1}	0.5 km s^{-1}
Velocity coverage	1000 km s^{-1}	500 km s^{-1}
Bandpass calibrator	3C84	
Visibility calibrator	0528+134 (S = 2.0 Jy)	0605−085 (S = 3.2 Jy)
Field of view	300″	150″

*Kasuga *et al.* 1986, **Kawabe *et al.* 1989, ***FX; Chikada *et al.* 1987

3. Results and discussion

3.1 NH_3 filaments

Figure 1*a* and 1*b* show the spatial distributions of the NH_3 (1,1) emission in the LSR velocity ranges of 6.9 ~8.9 and 8.9 ~10.9 km s^{-1}, respectively. Three main features are prominent in Figure 1: (*1*) a strong compact source at the center, corresponding to the "hot core", (*2*) the OMC-1 ridge extending to the north at V_{LSR}= 9.9 km s^{-1} (Figure 1*b*) and to the south at V_{LSR}= 7.9 km s^{-1} (Figure 1*a*), and (*3*) two clumpy filaments extending toward the north at 7.9 km s^{-1} (Figure 1*a*), which are newly found in our observations. The length and width of the filaments are 0.45 pc and less than 0.04 pc, respectively. If we assume that the NH_3 clouds are optically thin, the NH_3(1,1-2,2) rotational temperature is 20 ~30 K and the total mass of the filaments is about 10 $M_\odot$, when the fractional abundance of NH_3 is ~10^{-8}.

These filamentary structures are seen in other observations. The west-side filament (W-filament) also appeared in the NH_3(1,1) map in Batrla *et al.* (1983). The southern part of the east-side filament (E-filament) can be identified in the high resolution CS map (Mundy *et al.* 1988) and was pointed out as the HCO^+ ridge (Olofsson *et al.* 1982).

HH objects and finger-like H_2 emissions (H_2 fingers) have been observed along these filaments as shown in Figure 2 (Axon and Taylor 1984; Taylor *et al.* 1984). The HH objects are located on the NH_3 filaments: HH9 and HH1 are at their southern edges. The H_2 fingers seem to extend from the southern edge of the E-filament. From these positional coincidences, the HH objects and the H_2 fingers may be physically associated with the NH_3 filaments. Similar case of high velocity HH objects associated with low velocity cloudlets has been observed in the HH7-11 region (Rudolph and Welch 1988). These observational results are consistent with the idea that the filamentary NH_3 cloudlets are condensed through the interaction of the high velocity winds with the ambient gas, and that the strong shocks causing the HH objects are created at the

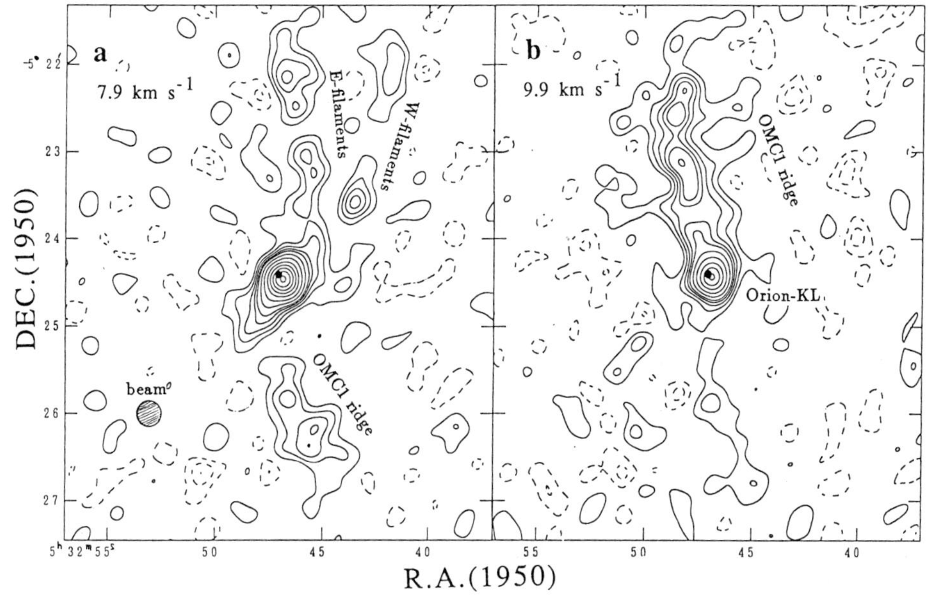

FIGURE 1—a. NH_3 (1,1) map with the V_{LSR}= 6.9 ~8.9 km s^{-1}. The central velocity are shown at the top-left corner. Filled-circle shows the position of IRc2. **—b.** NH_3 (1,1) map with the V_{LSR}= 8.9 ~10.9 km s^{-1}.

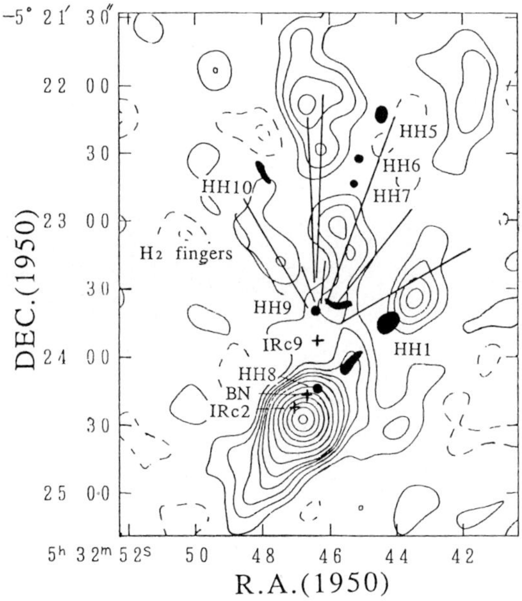

FIGURE 2—NH_3 (1,1) filaments with the HH objects (Axon and Taylor 1984) and H_2 fingers (Taylor *et al.* 1984).

front surfaces of these cloudlets. The tangential velocities of the winds are estimated to be about 200 km s^{-1} from the measurement of proper motions of the HH objects (Jones and Walker 1985), which gives the dynamical timescale of $\sim 10^3$ years consistent with the timescale of the molecular outflow from IRc2. This suggests that the wind originates from the Orion-KL region as is discussed below.

3.2 CS disk

Figure 3 shows the spatial distributions of CS (J=1-0) emission in the LSR velocity ranges of 7.2 $\sim$8.7 and 8.7 $\sim$10.2 km s^{-1}, respectively. Three main features are seen in the CS maps: (*1*) a strong peak located at the south of IRc2, which correspond to the "hot core" showing broad line emissions, (*2*) the CS disk, seen as a ridge structures from SW to NE with P.A. = 30°, and (*3*) the CS shells shown by the thick lines in Figure 3, which has been newly found in our observations.

A thin edge-on disk structure appears in our maps. The thickness and diameter of the disk are 0.03 pc and 0.3 pc, respectively. A position-velocity map of the disk is shown in Figure 4 along the line A – A' in Figure 3. This map shows two features of the disk. One is the broad line (ΔV $\sim$10 km s^{-1}), hot core component at 10″ south of IRc2. This component is considered to be an active part of the inner edge of the disk. The other is the main body of the disk which extends to the north at V_{LSR}= 10 km s^{-1} and to the south at V_{LSR}= 8 km s^{-1}. The rotational velocity at the emission peak (r $\sim$0.07 pc) is $\sim$ 1 km s^{-1}, implying the rotation period of 5 $\times 10^5$ years. The higher contours in Figure 4 indicate that the rotation is not a simple rigid motion, but rather a differential rotation as has been suggested by Vogel *et al.* (1985).

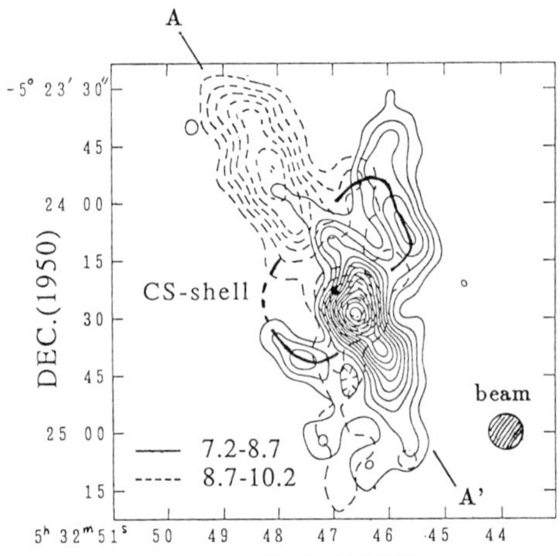

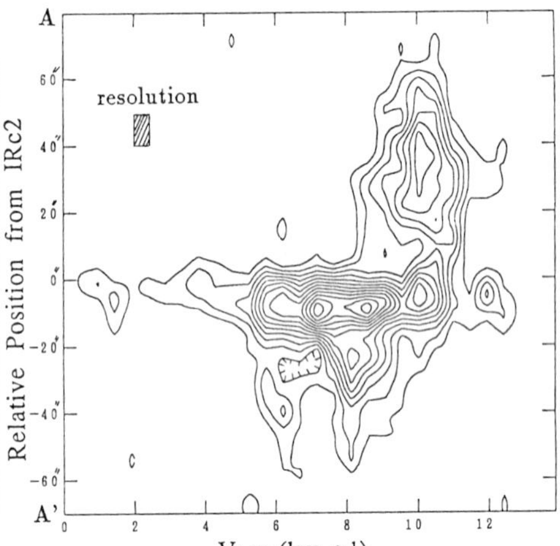

FIGURE 3(*top-left*)— CS (1-0) map with the V_{LSR}= 7.2 $\sim$8.7 km s^{-1}(*thin solid lines*) and with V_{LSR}= 8.7 $\sim$10.2 km s^{-1}(*broken lines*). Thick solid lines show the CS shell structures. Filled-circle shows the position of IRc2.

FIGURE 4(*top-right*)— Position-velocity map along the line A – A' in Figure 3. The spatial and velocity resolutions are shown at the top-left corner.

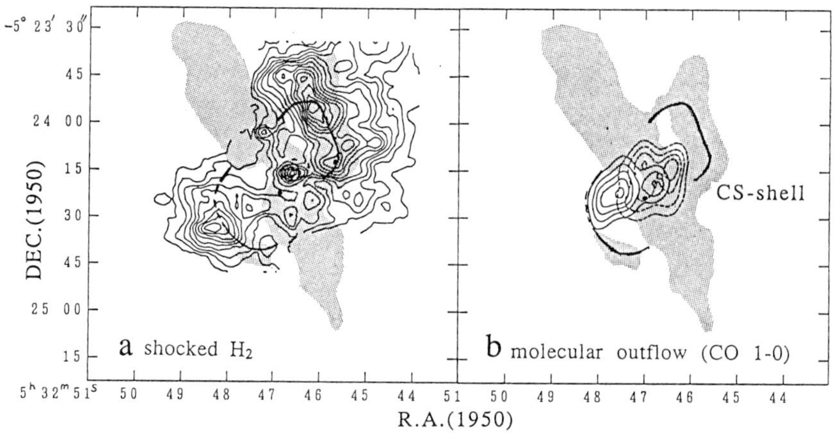

FIGURE 5—**a.** The shocked H_2 emissions (Hasegawa *et al.*1989) and **b.** the high velocity molecular outflow (Masson *et al.*1987) superposed on the CS emission presented in Figure 3. The CS shells are shown with solid lines.

3.3 CS shells

The shell structures are seen in Figure 3 as faint extensions from the disk component, with IRc2 being located at their center. They are elongated perpendicular to the disk with the size of 0.08 pc $\times$0.13 pc. The thickness of the shell have not been resolved yet by our 9″ beam and is less than 0.02 pc. Figure 5*a* and 5*b* shows the comparison of these shell structures with the shock excited H_2 emission (Beckwith *et al.* 1978; Hasegawa *et al.* 1989) and with the high velocity molecular outflow (Erickson *et al.* 1982; Wright *et al.* 1983; Masson *et al.* 1988). There is a good positional coincidence between the CS shells and the two lobes of the shocked H_2 emission. The H_2 emission is weak at the center where the massive disk is located. The molecular outflow is confined inside the shell structures. These results support the hydrodynamical models of molecular outflows that high velocity winds accelerate neutral gas which in turn sweeps up the ambient material to form a shock-compressed shell (Königl 1982, Okuda and Ikeuchi 1986). It seems that the NH_3 filaments are located on the extension of the blue-shifted lobe of the molecular outflow. We can suppose that the high velocity winds expand out through the CS shell toward the directions along which the NH_3 filaments and the HH objects are formed, as discussed before. Although the density of the shocked CS shell cannot be inferred only from our data, detection of the CS emission from the shell implies the gas density higher than 10^4 cm^{-3}, which gives the mass of the shell of $\sim$0.1 $M_\odot$, if we assume the thickness of the shell is 0.02 pc.

References

Axon,D.J., and Taylor,K. 1984, *Monthly Notices R.Astron.Soc.*, **270**, 241.

Bally,J. and Lada,C.,J. 1983, *Astrophys.J.*, **265**, 824.

Batrla,W., Wilson,T.L., Bastien,P., and Ruf,K. 1983, *Astron.Astrophys.*, **128** ,279.

Beckwith,S., Persson,S.E., Neugebauer,G.,and Becklin,E.E. 1978, *Astrophys.J.*, **223**, 464.

Chikada,Y. *et al.* 1987, *Proc. of the IEEE*, **75**, No.9, 1203.
Downes,D., Genzel,R., Becklin,E.E., and Wynn-Williams,C.G. 1981, *Astrophys.J.*, **244**, 869.
Erickson,N.R., Goldsmith P.F., Snell,R.L. Berson,R.L. Huguenin,G.R.,Ulich,B.L. and Lada,C.J. 1982, *Astrophys. J.(Letters)* , **261**, L103.
Genzel,R., Reid,M.J., Moran,J.M., and Dowens,D. 1981, *Astrophys.J.*, **244**, 884.
Hasegawa,T. *et al.* 1984, *Astrophys.J.*, **283**, 117.
Hasegawa,T. *et al.* 1989, in preparation
Ishiguro,M. *et al.* 1984, in the *proceedings Int. Symp. Milli-meter and Submillimeter Wave Radio Astronomy*, Granada, 75
Jones,B.F., and Walker,M.F. 1985, *Astron.J.*, **90**, 1320.
Kasuga,T., Kawabe,R., Ishiguro,M., Yamada,K., Kurihara,H., Niori,M., and Hirachi,Y. 1987, *Rev. Sci. Instrum.*, **58**, 379
Kawabe,R., *et al.* 1989, in preparation
Königl,A. 1982, *Astrophys.J.*, **261**, 155.
Masson,C.R., Lo, K.Y., Phillips, Sargent, A.I., Scoville,N.Z., and Woody,D.P. 1987, *Astrophys.J.*, **319**, 446.
Mundy,L.G., Cornwell,T.J., Masson,C.R., Scoville,N.Z., Baath,L.B., and Johanssen,E.B. 1988, *Astrophys.J.*, **325**, 382.
Okuda,T. and Ikeuchi,S. 1986 *Publ.Astron.Soc.Japan* , **38**, 199.
Olofsson,J., Ellder,A.Hjalmarson, and G. Rydbeck, 1982, *Astron.Astrophys.*, **113**, L18.
Plambeck,R.L., Wright,M.C.H., Welch,W.J., Bieging,J.H., Baud,B., Ho,P.T.P., and Vogel,S.N. 1982, *Astrophys. J.(Letters)* , **259**, 617.
Rudolph,A. and Welch,W.J. 1988, *Astrophys. J.(Letters)* , **326**, L31.
Taylor,K.N.R., Storey,J.W.V., Sandell,G., Williams,P.M., and Zealey,W.J. 1984, *Nature*, **311**, 236.
Vogel,S.N., Bieging,J.H., Plambeck,R.L., Welch,W.J., Wright,M.C.H. 1985, *Astrophys.J.*, **296**, 600.
Wright,M.C.H., Plambeck,R.L., Vogel,S.N., Ho,P.T.P.,and Welch,W.J. 1983, *Astrophys. J.(Letters)* , **267**, L41.
Wynn-Williams,C.G., Genzel,R., Becklin, E.E., and Downes,D. 1984, *Astrophys.J.*, **281**, 172.

Discussion:

DYSON (Comment): Your model is similar to the one we (Axon, Dyson, Tylor & Hughes) proposed for the Orion HH objects based on their optical spectroscopy. We also invoked channels in the neutral cloud to direct stellar wind onto denser blobs.

The Trapezium Radio Cluster of the Orion Nebula

Guido Garay

Departamento de Astronomía, Universidad de Chile

ABSTRACT. We review the characteristics and discuss the nature of a dense group of compact radio sources found projected toward the Trapezium cluster of the Orion nebula. There are twenty-six radio sources, with flux densities greater than 2 mJy, clustered within a region of 35" radius around Θ^1C Orionis, the most luminous star of the Trapezium. The density of radio objects, of 1.4×10^4 pc^{-3}, is extraordinarily high, about a thousand times greater than the density of stars in typical galactic clusters.

Most sources show flat, or slowly rising, spectra between 5 and 15 GHz, brightness temperatures smaller than 10^4 K, and flux densities that are constant on time scales of days to a few months. These characteristics suggest that the radio emission is free-free radiation from an HII region. The ionization must be external however, most likely produced by the UV radiation from Θ^1C. Possible models for these objects are: (1) neutral condensations surrounded by ionized envelopes; and (2) low mass stars surrounded by evaporating protostellar neutral accretion disks. The ionized gas flowing out from either the globules or the stellar disks and expanding into the surrounding medium is likely to produce the turbulence observed in the central region of the Orion nebula.

I. INTRODUCTION

The term "Trapezium radio cluster" is used here to refer to a dense group of compact radio sources, first discovered by Moran et al. (1982), found projected toward the Trapezium stars Θ^1 Orionis. From several VLA observations, made at different frequencies, spatial resolutions and sensitivities (Garay, Moran, and Reid 1987; Churchwell, Felli, Wood, and Massi 1987; Garay 1987), a total of 26 compact radio sources with flux densities greater than 2 mJy have been detected in a region of $\sim$ 35" radius centered near Θ^1C, the brightest Trapezium star. A map of the radio sources at 5 GHz is shown in Figure 1. Assuming that the three-dimensional region is a sphere, with a radius of 0.076 pc (35" at 450 pc), the density of compact radio sources is $\sim 1.4\times10^4$ pc^{-3}. This density is extraordinarily high, about seven times higher than the density of Trapezium stars brighter than $M(I_c)$ = 6.0 (Herbig and Terndrup 1986) and about a thousand times greater than the density of stars in Galactic clusters.

All but one of the Trapezium radio sources have optical and near infrared counterparts. Twenty are clearly appreciated in a multifrequency (1.6, 2.2, and 3.8 μ) near infrared image of the central region of the Orion nebula (Allen et al. 1984); eighteen are seen in short exposure plates taken in deep red (0.7-0.9 nm) light (Herbig 1982); and seven are associated with optically visible 'nebular condensations' observed by Laques and Vidal (1979) and Vidal (1982) (see Table 2 of Garay 1987, for a summary).

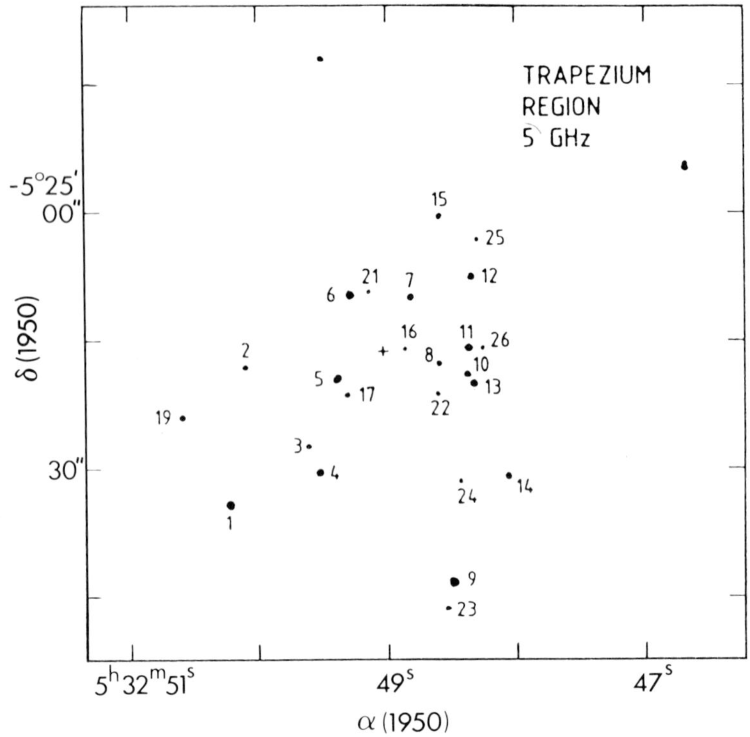

Figure 1. Map of the Trapezium compact radio sources at 5 GHz. The cross indicates the position of the Trapezium star Θ^1C.

Table 1. Observational parameters

Date 1985	Config.	Frequency (GHz)	Bandwidth (MHz)	HPBW (")	rms noise (mJy/beam)
Jan 18....	A	4.86	100.	0.43x0.37	0.45
Jan 19....	A	4.86	100.	0.44x0.37	0.27
		14.94	100.	0.13x0.12	0.20
Feb 2....	A	4.86	100.	0.44x0.39	0.61
		14.94	100.	0.31x0.23	0.47
Feb 16....	A	4.86	100.	0.44x0.39	0.42
		14.94	100.	0.15x0.15	0.30
Mar 1....	A	4.86	100.	0.52x0.50	0.51
		14.94	100.	0.20x0.15	0.26
Mar 6....	A	4.86	100.	0.42x0.36	0.29
		14.94	100.	0.15x0.12	0.22
Mar 8....	A	4.86	100.	0.40x0.35	0.31
		14.94	100.	0.13x0.12	0.21
Mar 15....	A	4.86	100.	0.57x0.49	0.59
		14.94	100.	0.15x0.12	0.37

II. CHARACTERISTICS OF THE RADIO EMISSION

From the data of GMR (Tables 2 and 3) and CFWM (Table 2) it appears that several of the compact radio sources show variability in their flux density on time scales of a year. It is not clear if these variations, which are typically of 40% about the mean, are intrinsic to the sources or extrinsic, such as for instance due to calibration errors, the use of different array configurations (hence different angular resolutions and sensitivities to extended structure) and/or the use of different image processing.

We report here observations that were designed to look for variability in the flux density of the Trapezium compact radio sources on time scales ranging from days to a few months. The data were obtained with the VLA of the NRAO[1] during the first quarter of 1985, a period in which the array was in the A configuration. Table 1 gives the observing dates and instrumental parameters.

In order to study variability, and to take into account the slight difference in the size of the synthesized beams between the different epochs, we decided to compare brightness temperatures. Assuming that the brightness temperature, T_b, is uniform over the synthesized gaussian beam, then $T_b = 1.22{\times}10^3\ \nu^{-2}\ \Theta_b^{-2}\ S_\nu$, where ν is the frequency in GHz, Θ_b is the HPBW in arcsec, and S_ν is the observed peak flux density per beam in mJy. The derived brightness temperature at 6 cm of the sixteen strongest radio sources within the Trapezium radio cluster are given in Table 2. Most of these sources appear to be constant (1 σ rms variations about the mean < 20%). Source No. 12 is clearly variable, while sources 1, 9, and 14 are probably variable. Among the weaker sources, Nos 22, 23, and 25 are probably variable. We will assume that the rest of the Trapezium radio sources are constant. The variability observed on time scales of years will be assumed, until demonstrated otherwise, as due to extrinsic reasons.

The observational parameters of the Trapezium radio sources are listed in Table 3. The data used to derive these parameters were obtained by combining, in the (u,v)-plane, all data from the individual observations. The deconvolved angular diameters of the sources, given in columns 6 and 7, are typically between 0"1 and 0"4. The sizes derived from the observations at 5 and 15 GHz are in good agreement, in spite of the fact that the synthesized beam at 5 GHz (0"39) is about three times larger than that at 15 GHz (0"12). There is a significant difference between the deconvolved angular diameters at 15 GHz given here and those reported by CFWM, even though they were derived from observations with similar angular resolutions. Several of the sources not resolved by CFWM appear extended to us. The peak brightness temperatures observed by us are consistent with the sources being resolved.

The spectra, between 5 and 15 GHz, of most of the Trapezium radio sources are either flat or rising toward the higher frequency. In addition, their brightness temperatures are smaller than 10^4 K (mean average of 2000

---

[1] The National Radio Astronomy Observatory is operated by Associated Universities Inc., under cooperative agreement with the National Science Foundation.

Table 2. 6 cm brightness temperature (×10^3 K)

Date	Source															
1985	1	2	3	4	5	6	7	8	9	10	11	12	13	14	15	19
Jan 18	2.4	1.1	1.3	2.1	2.4	3.6	1.8	1.3	1.9	1.3	2.2	2.9	1.7	1.0	0.9	1.3
Jan 19	1.8	1.2	1.2	2.0	2.4	3.6	2.0	1.3	1.7	1.3	2.2	6.7	2.1	1.2	1.1	1.1
Feb 2	0.9	1.4	1.3	2.1	2.1	4.4	1.7	---	---	---	2.5	4.6	1.3	---	---	1.5
Feb 16	1.7	1.3	1.0	2.0	2.6	3.4	2.4	1.3	1.7	1.2	2.1	4.4	2.3	1.1	1.4	1.2
Mar 1	1.6	1.3	0.8	1.9	2.4	3.1	2.3	1.0	2.5	1.3	2.4	4.6	1.8	2.2	1.5	1.2
Mar 6	1.7	1.1	1.1	1.9	2.6	3.7	2.0	1.0	1.5	1.4	2.4	2.7	2.0	1.3	1.1	1.2
Mar 8	1.7	1.3	1.1	1.9	2.5	3.5	2.1	1.3	1.9	1.4	2.5	6.0	1.8	1.3	1.2	1.1
Mar 15	1.9	---	1.0	2.1	2.8	3.4	1.2	0.8	2.5	1.5	2.0	2.5	2.2	1.4	1.2	0.9

Table 3. Trapezium radio cluster

Source	Coordinates(1950)		Flux density (mJy)		Deconvolved size[c] (")	
	R.A.	Dec.	5 GHz	15 GHz	5 GHz	15 GHz
	$5^h\ 32^m$	$-5°\ 25'$				
1	$50^s.21$	$34''.1$	12.9	7.7	0.48	0.36
2	50.10	18.2	3.8	4.8	0.16	0.16
3	49.61	27.3	4.4	4.8	0.20	0.15
4	49.52	30.3	9.9	9.0	0.34	0.30
5	49.39	19.6	12.0	15.0	0.30	0.16
6	49.29	9.8	16.6	26.3	0.30	0.21
7	48.82	10.0	8.1	10.6	0.21	0.13
8	48.60	17.7	4.0	6.3	0.14	0.10
9	48.49	43.1	14.8	10.8	0.56	---
10	48.39	18.9	6.2	5.4	0.31	0.36
11	48.37	15.9	10.5	11.8	0.31	0.15
12	48.35	7.5	---a	---a	u	u
13	48.33	20.0	9.1	10.0	0.31	0.33
14	48.06	30.9	6.7	5.0	0.40	0.36
15	48.60	0.5	4.1	3.6	0.18	0.15
16	48.87	16.0	2.1	4.0	u	u
17	49.31	21.3	3.6	4.4	0.20	u
19	50.58	24.0	5.3	4.7	0.30	0.27
20	49.38	19.3	---b	4.4	---	0.12
21	49.15	9.4	1.4	2.0	u	u
22	48.61	21.2	2.3	---	0.23	---
23	48.54	46.4	4.0	---	0.41	---
24	48.44	31.4	1.4	4.4	u	0.18
25	48.30	3.3	1.4	1.4	u	u
26	48.27	15.9	1.8	2.6	u	0.13

[a] variable flux density; [b] blended with source No. 5; [c] u means unresolved (HPBW of 0."39 and 0."12 at 5 and 15 GHz, respectively)

and 300 K, at 5 and 15 GHz, respectively). These characteristics suggest that the radio emission is free-free radiation arising from a compact HII region.

The number of Lyman continuum photons needed to ionize each source is $\sim 4\times10^{44}$ s^{-1}, which could be supplied by B2 ZAMS stars. However, the presence of such stars is doubtful, owing to: (1) the weakness of the optical continuum associated with the radio sources; and (2) the absence of strong near-infrared emission, which argues against their being embedded in the cloud.

III. NATURE OF THE COMPACT RADIO SOURCES

Because of its proximity, at a distance of 450 pc, the Orion region has become the prototype of a star forming site in our Galaxy. The discovery of the compact radio sources embedded in the Orion nebula should add new information for studying the process of star formation; therefore to establish their nature should be of considerable interest. Since the high density of radio sources is found only in the vicinity of Θ^1C (no similar cluster was found within a 9'×9' region of the Orion nebula), the existence of the Trapezium radio cluster must be intimately related to the presence of such a luminous star. We suggest that the radio emission from most of the Trapezium radio sources is thermal bremsstrahlung from ionized gas that is externally excited by the UV radiation from Θ^1C. The Trapezium radio cluster represents, thus, an aspect of the early interaction of a young, luminous star with its surroundings.

The Trapezium compact radio sources can be modeled as neutral condensations surrounded by ionized envelopes (GMR). If the temperature of the ionized envelope is 10^4 K and if its electron density decreases as the square of the distance from the core center, then a typical neutral condensation has a radius of 7×10^{14} cm and an electron density just beyond that radius of 5×10^5 cm^{-3}. The mass loss rate by ionization is $\sim 10^{-7}$ $M_\odot$/yr. In order to have lifetimes greater than the age of the cluster, the initial mass of the globules would have to be > 0.02 $M_\odot$. A detailed discussion of the structure and stability of neutral globules immersed in an ionized medium has been given by Dyson (1968).

Some of the Trapezium radio sources are associated with optical objects with apparent visual magnitude of ~ 15 mag (see GMR, Table 4). If the optical luminosity is due to the presence of an optically visible low mass star, as suggested by CFWM, the neutral condensation model encounters problems since the star should be obscured by the globule. CFWM proposed that these radio sources might correspond to evaporating protostellar accretion disks surrounding low mass stars. The physical properties derived from the observations, such as density, total mass and mass loss rate of the ionized gas, are about the same for the protostellar disk and neutral condensation models.

Not all the Trapezium radio sources are likely to be globules or disks, however. The radio emission from source No. 12, which is associated with the binary system Θ^1 Ori A, shows strong variability on a time scale of days, suggesting that the radiation arises in flare events. The non-thermal origin of this emission was conclusively established by Felli et al. (1989), who derived a brightness temperature > 4×10^7 K and suggested that the emission probably arises from the T Tauri component of the binary system. Source No. 25 is associated with the star Θ^1 Ori E which is a strong and variable source of X-ray emission, hence its unresolved and probably variable radio emission could well be attributed to non-thermal emission. Sources 9 and 23 are associated with young pre-main sequence stars which show variable optical and X-ray emission. A pure non-thermal interpretation for the origin of their radio emission is problematic however, since both radio sources appear to be extended. A possible interpretation might be that the radio emission arises from both a central T Tauri star and a surrounding accretion disk.

Several observations toward the Orion nebula have shown the presence of large turbulent motions (Münch 1958; Wilson et al. 1959; Castañeda and O'Dell 1987). Castañeda and O'Dell (1987) concluded that the input scale of the turbulence is smaller than ~ 0.3 pc and is probably introduced at several sizes. Since the decay of supersonic turbulence is rapid, energy sources are needed to maintain it. Wilson et al. (1959) suggested that the eddies containing energy are secondary ionization fronts arising from the non-uniform growth of the HII zone into the surrounding molecular cloud.

The globules and stellar disks, which act as sources of ionized gas flowing out into, and stirring up, the surrounding nebula, are natural sources of turbulence. We thus propose that the Trapezium compact radio sources are a major source of turbulence within the Trapezium region. It was Dyson (1968) who first proposed, based on theoretical work, that partially ionized globules were the source of turbulence in the central region of the Orion nebula. The parameters of the (smaller) condensations predicted by him are in good agreement with the ones derived from the observations.

I gratefully acknowledge support by the Director General of the European Southern Observatory, Prof. H. van der Laan.

REFERENCES

Allen, D., Bailey, J., and Hyland, A.R. 1984, Sky Teles., **67**, 222.
Castañeda, H.O., and O'Dell, C.R. 1987, Ap. J. (Letters), **315**, L55.
Churchwell, E., Felli, M., Wood, D.O.S., and Massi, M. 1987, Ap. J., **321**, 516.
Dyson, J.E. 1968, Ap. Space Sci, **1**, 388.
Felli, M., Massi, M., and Churchwell, E. 1989, Astr. Ap., in press.
Garay, G. 1987, Rev. Mex. Astron. Astrof., **14**, 489.
Garay, G., Moran, J.M., and Reid, M.J. 1987, Ap. J., **314**, 535.
Herbig, G.H. 1982, Symposium on the Orion Nebula to Honor Henry Draper, eds. A.E. Glassgold et al. (New York: New York Academy of Science), p.64.
Herbig, G.H., and Terndrup, D.M. 1986, Ap. J., **307**, 609.
Lacques, P., and Vidal, J.L. 1979, Astr. Ap., **73**, 97.
Moran, J.M., Garay, G., Reid, M.J., Genzel, R., and Ho, P.T.P. 1982, Symposium on the Orion Nebula to Honor Henry Draper, eds. A.E. Glassgold et al. (New York: New York Academy of Science), p.204.
Münch, G. 1958, Rev. Mod. Phys., **30**, 1035.
Vidal, J.L. 1982, Symposium on the Orion Nebula to Honor Henry Draper, eds. A.E. Glassgold et al. (New York: New York Academy of Science), p.176.
Wilson, O.C., Münch, G., Flather, E.M., and Coffeen, M.F. 1959, Ap. J. Suppl., 4, 199.

TURBULENCE IN THE ORION NEBULA

H.O. Castañeda[1] and C.R. O'Dell[2]

[1] Inst. de Astrofisica de Canarias, Spain

[2] Rice University, USA

We have examined the radial and turbulent velocities of the Orion Nebula at the seeing limit, observing the [OIII]λ5007 line and the [OII]λ3727 doublet with the coudé feed system of the Kitt Peak National Observatory, at very high spatial and spectral resolution.

We show evidence for the existence of multiple components in the nebular profile of the different lines, representing different elements of the gas flow. We find at least four distinct flows, including an highly blueshifted [OII] component over the entire observed area. A model to explain the velocity structure is advanced, as the interaction of a "champagne" flow with a stellar wind bubble and secondary streams from inhomogeneities in the nebula.

The structure function has been calculated and compared with the Kolmogorov theory of turbulence. There is evidence for a correlation between random velocities and distance across the face of the nebula, but the observation are in disagreement with the classical theory for nebular turbulence.

HIGH-RESOLUTION MOLECULAR LINE OBSERVATIONS OF THE CORE AND OUTFLOW IN ORION B

J.S. Richer, R.E. Hills and R. Padman

Mullard Radio Astronomy Observatory, Cambridge, England

High-resolution CO $J \rightarrow 1 \rightarrow 2$ observations of the Orion B molecular outflow show that the outflow is unipolar, and that there is evidence of acceleration of molecular gas at up to 0.5pc from the driving star. The highest-velocity material, as well as being furthest from the source, seems to lie close to the central axis of the flow, and is presumably being accelerated by entrainment in the flow or jet emanating from the star. We have also mapped the HCO^+ $J = 3 \rightarrow 2$ emission at 19-arcsec resolution. We derive an excitation temperature of around 25 K in the cloud core, and a core mass of about 75 $M_\odot$, this estimate is in accord with a model in which the core has a kinetic temperature of 30-50 K, with no molecular depletion on to grains. This is in contrast to the recent suggestion that the core contains cold isothermal protostars.

THE EXTRAGALACTIC ISM

STAR FORMATION IN GALACTIC NUCLEI

Roberto Terlevich
Royal Greenwich Observatory,
Herstmonceux, BN27 1RP,
U.K.

Summary. The term AGN or Active Galactic Nucleus, although with rather ill-defined boundaries, signifies extragalactic systems with non-stellar nuclear activity. Non-stellar activity indicators are: high luminosity, presence of broad emission lines, variability, nuclear UV excess, x-ray emission, radio emission, variable polarization, jets, etc.

The list of AGNs includes LINERs, Seyfert galaxies, QSOs (optically selected), Quasars (radio selected), narrow-line and broad-line radio galaxies, optically violent variable Quasars and BL Lac type objects. The large luminosity of AGNs, is believed to derive from gravitational energy and is usually associated with an accretion disk circling a super-massive blackhole. It is also customary to assume that the size of the emitting region is limited by its variability to a few light weeks and that the very broad emission lines (FWHI$\sim$5000 km s^{-1}) of the so called Broad Line Region (BLR) are consistent with photoionization of high density filaments orbiting the inner few parsecs of the AGN.

Here I will discuss the possibility that the energy source in at least some AGNs is a burst of star formation. In this scenario the BLR is the product of SN remnants evolving in high density medium. I will show that the observed variability of the broad line region in Seyferts type 1 and QSOs is consistent with the predictions of the starburst scenario, and that the total luminosity of even the most luminous QSOs is well inside the expectations of dissipative galaxy formation scenarios.

I. Overview of the scenario. There is a fundamental dichotomy, first reported by Heckmann *et al.* (1980), in the distribution of Hubble types of galaxies with emission line nuclei. While most galaxies with active nuclei have Hubble types *earlier* than Sbc, those with HII region type nuclei have Hubble types *later* than Sbc.

Independently of the origin of the bi-modal Hubble type distribution, (see Terlevich *et al.* 1987) the fact remains that is only in luminous bulges that metal rich stellar populations are found (Cowley *et al.* 1982), and when those bulges show nuclear emission lines, they are systematically classified as 'active' as opposed to HII region type.

For the past few years a large part of my work has been devoted to the study of regions with active star formation. In particular the study of the intrinsic properties of giant extragalactic HII regions like 30-Doradus in the LMC and HII galaxies like IIZw40 and IZw18. HII galaxies have small dimensions, spheroidal shape, a young (very hot) stellar population and spectroscopic properties indistinguishable from those of giant extragalactic HII regions. With Jorge Melnick, Mariano Moles and Josefa Masegosa, I finished a large scale spectrophotometric survey of about 500 HII galaxies selected from objective prism Schmidt surveys. A totally unexpected result from that study was the relatively narrow metallicity distribution of galaxies (FWHM~0.7 Dex). We found only few galaxies showing oxygen abundance less than 0.1 solar and no galaxy with oxygen abundance larger than 1.5 times solar (Terlevich 1988).

Where are then the luminous metal rich bursts of star formation? With Jorge Melnick, Mariano Moles, Angeles Díaz and Elena Terlevich I started to estimate the properties of bursts of star formation at high metallicities. We found that unlike the low metallicity case, metal rich massive stars have their evolution fundamentally affected by mass-loss in the form of stellar winds. Without exception all evolutionary computations find the same differences between conservative, $\dot{M}=0$, and mass-losing models found initially by Tanaka (1966); namely a change in the H and He burning scale and an evolution towards high temperatures after hydrogen exhaustion. During the blueward evolution the star reaches effective temperatures well in excess of those typical of the ZAMS. Wolf-Rayet stars with massive progenitors ($M>60\,M_\odot$) are believed to be in the blueward evolutionary stage (Conti 1976, Chiosi 1981, Maeder 1983). The Wolf-Rayet stage is reached after a highly unstable phase, corresponding to the minimum effective temperature ever reached by the star (~15,000K) and roughly before the onset of helium burning. Luminous blue variables, like the Hubble-Sandage variables, η Carina and P-Cygni type stars are believed to be in this unstable phase. The most recent evolutionary star models for solar composition and incorporating mass loss and overshooting, indicates that by the end of the helium burning phase the effective temperature reaches up to 200,000K and their bolometric luminosity could be up to a factor of 2 larger than at the ZAMS. We call these extremely hot and luminous Wolf-Rayet stars WARMERS.

WARMERS do exist. Barlow and Cohen (1982) have indicated the existence of a new class of extremely hot Wolf-Rayet stars (WO stars), characterized by the presence of high ionization emission lines of oxygen and carbon (OIV, CIV) in their optical and UV spectra. Davidson and Kinman (1982) reported observations of an HII region in the dwarf galaxy IC1613 ionized by a Wolf-Rayet star with broad OVI and CIV in emission probably an extreme WC5 or even WC4. Using the spectrum of the HII region they determine that most of the luminosity is emitted at an energy of 37eV similar to the peak of a 110,000K blackbody spectrum. The estimated bolometric luminosity is $1.3\text{x}10^5\ L_\odot$.

With Jorge Melnick I investigated the changes that a population of WARMERS will introduce in the emitted spectrum of a young metal rich cluster (Terlevich and Melnick 1985; hereafter TM85). Using Maeder's isochrones, we studied the evolution of a burst of star formation using our computer code IMF. This program computes from the theoretical isochrones at stellar atmospheres the integrated spectrum of a star cluster as a function of age; then calculates the emission line spectrum of the associated HII region using the photoionization code CLOUDY developed by Gary Ferland.

We found that the emitted spectrum of a metal rich HII region suffers a qualitative change after about 3Myr of evolution, when the most massive stars reach the WARMER phase. In a very short time the ionizing spectrum of the cluster is fundamentally modified by the appearance of a luminous and hot component, the WARMER component. Consequently the emission line spectrum is transformed from that of a typical low excitation HII region into a high excitation Seyfert type 2. Following the evolution still further, shows that after 5Myr as the ionizing flux decreases and therefore the ionization parameter also decreases, the Seyfert type 2 nucleus becomes a Liner.

In brief, TM85 have shown that the 'traditional' method of using the Baldwin, Phillips and Terlevich (1981) diagnostic diagrams to classify emission line regions should be used with care. Power law type ionizing continuum can be the result of a few WARMERs modifying the emitted UV spectrum of a young cluster. The best evolutionary star models predict this behaviour for solar and over solar abundances.

This result raised the possibility of a direct relation between star formation and some forms of nuclear activity. This possibility has been discussed in the literature by several authors (Pronik 1973, Adams and Weedman 1975, Harwit and Pacini 1975, Osterbrock 1978, Weedman 1983, Perry and Dyson 1985). Some researchers have preferred the view that after the star formation event, the resulting cluster of compact remnants will eventually coalesce and form a massive blackhole (Rees 1978, 1984), while others suggest that the gas lost during the evolution of the massive stars is able to fall into the gravitational potential of the bulge, cool and form a blackhole (Norman and Scoville 1988). All these scenarios have a common point, they have to start the evolution with extremely compact clusters with large velocity dispersions to allow the formation of a blackhole.

Another possibility originally suggested by Shklovskii (1960), and further developed by Field (1964) and McCrea (1976) and more recently by Terlevich *et al.* (1987), Terlevich and Melnick (1987) and Terlevich (1989) is that no blackhole forms and that *all* the observed properties of at least some AGNs are the *direct* consequence of a nuclear starburst.

The observed variability of the Broad Line Region (BLR) in Seyferts type 1 and QSOs, the line profiles and line ratios, total luminosity and small size of the BLR, poses difficulties for the starburst scenario. In this paper I will discuss these points.

II. The BLR in the starburst scenario. With Melnick and Moles (1987; hereafter TMM87), I extended the work of TM85 to include the supernova phase in the evolution of the nuclear starburst. In that paper we postulated that the observed properties of Seyfert type 1 nuclei may also be related to nuclear starburst activity. We distinguished two different supernova activity phases,

1 - a SN type Ib at the end of the life-time of the most massive stars ($M>25\,M_{\odot}$). These SNe have Wolf-Rayet/WARMERS progenitors and they are predicted to be optically dim and radio loud. This stage is associated with the Seyfert 2 phase.

2 - a SN type II phase at the end of the life-time of intermediate mass stars ($8 < M < 25\,M_{\odot}$). These SNe have red supergiant progenitors. The SN ejecta after leaving the atmosphere of the star will presumably interact with dense ($n\sim10^7$)

circumstellar/interstellar medium. This interaction will produce a hot and luminous remnant with a life time of about 2 years. The predicted velocities, dimensions, temperature, density and luminosity of these remnants are similar to those of the canonical BLR. The emitted spectrum of a fast shock ($V=1000\,km\,s^{-1}$) in a high density ($n=10^7$) environment has been studied by Daltabuit *et al.* (1978). They found that the emitted spectrum of the post shock cool dense shell, photoionized by the radiation of the shock, resembled that of a QSO. Thus in all aspects, we expect this rapidly evolving remnants to closely resemble the BLR.

Terlevich and Melnick (1988) provided evidence that this may be the case. They showed that the reported flare in the Seyfert type 1 NGC 5548, may have been the first detection of a SN in the nucleus of a galaxy, rather than an accretion event (Peterson and Ferland 1986). The spectrum of the flare looked very similar to that of SN 1983K the only supernova known with a probable Wolf-Rayet progenitor (Niemela *et al.* 1985). Also the luminosity and duration of the flare were similar to those of SN 1983K.

Perhaps the most compelling evidence comes from some recent observations of supernovae. Filippenko (1989) has reported the discovery of a supernova in an HII region in one of the spiral arms of the SBc galaxy NGC 4615. After maximum the spectrum was dominated by broad permitted emission lines of hydrogen, FeII and CaII; it has a striking resemblance to the spectrum of a Seyfert type 1 or a QSO. A second supernova reported in the same paper, SN 1988I, showed a similar spectrum. Even the luminosities of the SNe were comparable to those of Seyfert 1 nuclei.

The previous discussion supports the idea that at least some low luminosity AGNs can be associated with nuclear star formation. To extend the model in order to include the high luminosity AGNs, it is necessary to discuss two fundamental observational constraints:

1 - the amplitude of the optical variability in QSOs, and

2 - the required star formation rate to explain the luminosity of the brightest QSOs.

III. Optical variability in QSOs. Long term monitoring of radio-selected quasars has shown that many are variable on time scales of years and with variable amplitudes (Tritton and Selmes 1971, Barbieri and Romano 1981, Pica and Smith 1983; PS83), and some may not vary at all. A small and perhaps unrepresentative subset, the Optically Violently Variable (OVV) are the most variable ones. Work on optically selected QSOs, has indicated that most of them show only small amplitude (~0.2mag) variability (Bonoli *et al.* 1979, PS83). Perhaps the largest dataset is that obtained at the Rosemary Hill Observatory (PS83). It includes 13 years of continuous monitoring of 130 AGNs and OVVs representing an average of 45 epochs per source. The average rms error of the photometry is 0.13 mag. PS83 found a trend with luminosity in the variability amplitude. This is illustrated in Figure 1 were I plotted the observed variability amplitude versus the absolute blue luminosity estimated for $H_o=50\,km\,s^{-1}\,Mpc^{-1}$ and $q_o=0$. As can be seen from the figure luminous QSOs are somehow less variable than low luminosity QSOs and Seyfert nuclei. From least-squares fit to the data,

PS83 estimated the change in peak-to-peak amplitude with luminosity as 0.12 mag±0.02mag per magnitude for all sources and 0.18±0.03mag per magnitude for the radio quasars only. This trend is expected if the variability is produced by random flares, but Pica and Smith did not favour this explanation because the slope of the observed relation is much more shallow than the theoretically expected. The variability amplitude should decrease roughly as the inverse of the square root of the luminosity. In their simple analysis Pica and Smith did not include the observational errors; this would tend to flatten the relation.

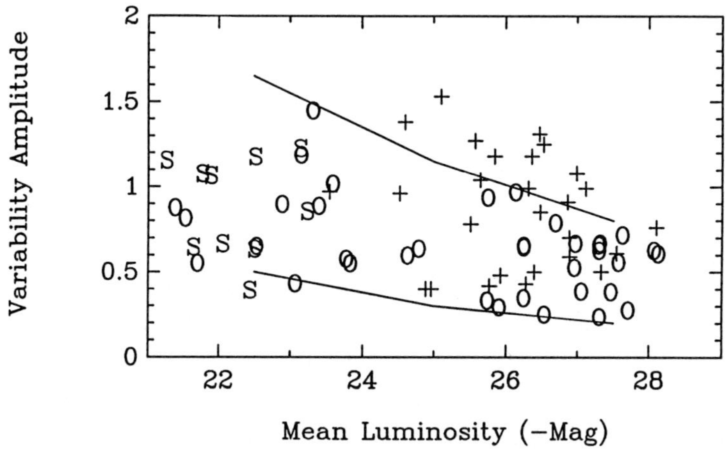

Figure 1 - Mean blue luminosity ($H_o = 50$, $q_o = 0$) versus peak-to-peak variability amplitude for all non-OVV sources from the survey by PS83. S = Seyfert galaxies; O = optically selected QSOs; + = radio loud Quasars. The lines represent the expected range of values for a starburst.

To check this effect, and in collaboration with Alejandro Terlevich, I estimated using Montecarlo techniques, the variability of a Starburst during the supernova stage. We used the predicted light curve for a supernova and its remnant evolving in a n=10^7 density medium (Terlevich 1989). The light curve consists of the SN event lasting two months and reaching M_b=-19. The SN remnant reaches a similar maximum luminosity 240 days after that initial explosion and then decays as $(t/240days)^{-11/7}$. We constructed the light curve by superposition of random events with random amplitude between $M_b = -18$ and $M_b = -20$. Gaussian noise with an rms equal to the estimated observational errors by PS83 (σ=0.13mg), was then added to this light curve. The resulting curve was then randomly sampled 45 times to reproduce the average number of epochs in the data set. With each light curve the sampling was repeated 100 times to estimate the rms change in the determination of the peak-to-peak amplitude. The lines in Figure 1 represent the result of the simulations. They correspond to the average variability amplitude plus and minus 2σ fluctuations. They represent the range of values that we expect the observer to measure 95% of the time. We can see that most of the QSOs lie inside the predicted region. Only some radio selected quasars have variability slightly over the predicted values.

We are lead to conclude that the variability observed in optically selected QSOs and in many radio selected quasars is well inside the expected in the starburst scenario. Therefore, the variability time scale is *not* constraining the size of the BLR; in our scenario the BLR of a luminous QSO, can be and probably is distributed over few kpc and still will produce the required variability.

If monitoring is performed with accuracy of 0.01 magnitudes then we predict variability amplitudes of

0.10 magnitudes for a M_b=-27.5 *QSO*
0.30 magnitudes for a M_b=-25.0 *QSO*
0.85 magnitudes for a M_b=-22.5 *QSO*

IV. Supernova rates in young galaxies and starbursts. A typical L* elliptical galaxy has a mass of about 10^{12} $M_\odot$ and an effective radius (half mass radius) of 5 kpc and its present absolute blue luminosity is M_b=-22. The supernova rate (SNR) depends on the star-formation rate (SFR), and the details of the initial mass function (IMF). In the numerical examples that follow I assumed a Salpeter IMF. For this type of IMF one supernova explodes every 30-50 $M_\odot$ of new stars, depending on the adopted lower and upper mass limits. Following Larson's dissipative models for the formation of elliptical galaxies, during the formation of the inner kiloparsec at the end of the collapse, the SFR for a 10^{12} $M_\odot$ elliptical reaches a peak of about 3000 $M_\odot$/year corresponding to about 100 SN/year. If the blue luminosity of the SN and SN remnant is $M_b \sim 19.5$, and the remnants last for about 1 year, it is simple to estimate that the integrated blue luminosity of such a galaxy during the SN phase is about -25. A very massive elliptical has a total mass of 10^{13} $M_\odot$ and during the formation of the inner part will reach SN rates of about 1000 SN/year and M_b of -27 or -28.

A different approach is to estimate the supernova rate for a burst of star formation that represents 5% of the total mass of the galaxy. Norman and Scoville (1988) have recently estimated the flux of stars off the main sequence for a Salpeter IMF cluster of stars following stellar evolution models by Renzini and Buzzoni (1986) and Iben and Renzini (1986). They find that the flux of stars off the main sequence is between 20 and 10 stars/year for a cluster of 10^{10} $M_\odot$ and for stellar masses larger than 7 $M_\odot$, i.e. inside the lower limit for type II SNe. The SN rate is almost equal to the flux of stars off the main sequence, therefore, a burst of strength 5% in a 10^{12} $M_\odot$ galaxy will also produce about 100 SN/yr. This type of burst may be triggered by dynamical interactions, and presumably they are more common at high redshifts when galaxies had a larger gas content.

V. The size of the BLR. The size of the BLR will depend on the SN rate as a function of radius in a young elliptical galaxy. In Larson's model, the SN rate is at the maximum value at the end of the collapse and inside the effective radius. Assuming that the SN distribution follows the mass ditribution of the galaxy, we can then estimate its half intensity diameter or $2xR_c$. From the comparison of the $r^{1/4}$ law and the Hubble law of brightness distribution Kerr (1957) found that $R_h = R_e/11$ and as the R_c in Hubble's law is $0.4142xR_h$ then in a typical elliptical $2xR_c \sim R_e/13$. In our typical L* elliptical this corresponds to less than 400pc or 0.014 arcsec at a redshift of 1.

A different estimate can be done using the scaling laws of bursts of star formation. Terlevich and Melnick (1981) and Melnick *et al.* (1987) studied the scaling laws for giant HII regions and HII galaxies. They found relations of the form $L \propto R^2$ are valid over more than an order of magnitude in radius. Applying the scaling to the determination of the core radius of 30-Doradus by Moffat *et al.* (1985) (0.26pc), and using a $M_b = -14.5$, the predicted R_c for the $M_b = -25$ burst of the L* elliptical is less than 100pc. I have to mention that the scaling law used implies constant surface density and the volume density is therefore smaller in the largest systems.

Putting both estimates together the size of the BLR in the starburts scenario should be,

$$0.10\, arcsec 10^{-0.2*(m_b-14.5)} \leq FWHM \leq 0.40\, arcsec 10^{-0.2*(m_b-14.5)}$$

where m_b is the apparent blue magnitude. The expression is valid for objects more luminous than $M_b = -21$. The limit is to ensure that the BLR is not dominated by a single SN remnant.

VI. Concluding remarks. The ideas presented here demonstrate that the starburst model based in extremely simple assumptions is potentially able to reproduce the main observed properties of the BLR in Seyfert galaxies and QSOs.

The fact that the luminosities and variability of QSOs can be explained with the expected SN rates during the formation of a spheroidal galaxy leads naturally to the suggestion that perhaps most of the optically selected QSOs, representing the majority of the high redshift luminous objects, are young galaxies in the process of formation.

One important difference between the starburst and the blackhole scenario is the size of the BLR. The fact that the BLR should be tens of parsecs FWHM in luminous Seyferts, constitutes a potential test for the scenario. Based in simple scaling laws I predict that the HST may be able to resolve the BLR of some of the nearest luminous Seyfert galaxies.

A second test is based in the predicted relation between variability amplitude and luminosity. This second test requires 0.01 mag photometry of few luminous QSOs during 3 or 4 years.

I would like to thank Elena Terlevich for her comments that greatly improved this paper.

References

Adams, T.F. and Weedman, D.W., 1975. *Astrophys. J.* , **199,** 19.

Baldwin, J.A., Phillips, M.M. and Terlevich, R., 1981. *Publ. astr. Soc. Pacif.* , **93,** 5.

Barbieri, C. and Romano, G., 1981. *Astr. Astrophys. Suppl.* , **44,** 159.

Barlow, M. and Cohen, M., 1982. *in: Wolf-Rayet Stars: Observations, Physics, Evolution, p. 387, eds de Loore,C. and Willis,A. Reidel, Dordrecht, Holland.*

Bonoli, F., Braccesi, A., Federici, L., Zitelli, V. and Formiggini, L., 1979. *Astr. Astrophys. Suppl.* , **35,** 391.

Chiosi, C., 1981. *in: The most massive stars, p.261, eds D'Odorico, S.Baade and Kjar,K. (Proceedings ESO workshop)*

Conti, P.S., 1976. *Mem.Soc.R.Sci.Liege*, **6eme Serie 9,** 193.

Cowley, A.P., Crampton, D. and McClure, R.D., 1982. *Astrophys. J.* , **263,** 1.

Daltabuit, E., MacAlpine, G. and Cox, D., 1978. *Astrophys. J.* , **219,** 372.

Davidson, K. and Kinman, T.D., 1982. *Publ. astr. Soc. Pacif.* , **94,** 634.

Field, G.B., 1964. *Astrophys. J.* , **140,** 1434.

Filippenko, A.V., 1989. *Astron. J.* , **97,** 726.

Harwit, M. and Pacini, F., 1975. *Astrophys. J. Lett.* , **200,** 1127.

Heckmann, T.M., Balick, B. and Crane, P.C., 1980. *Astr. Astrophys. Suppl.* , **40,** 295.

Iben, I. and Renzini, A., 1986. *Ann. Rev. Astr. Astrophys.* , **21,** 271.

Kerr, F.J., 1957. *Astron. J.* , **62,** 93.

Larson, R.B., 1974. *Mon. Not. R. astr. Soc.* , **166,** 585.

Maeder, A., 1983. *Astr. Astrophys.* , **120,** 113.

McCrea, W.H., 1976. *in: The galaxy and the local group, eds Dickens, R.J. and Perry, J.E. (RGO Bull 182).*

Melnick, J., Moles, M.,Terlevich, R. and Garcia-Pelayo, J-M, 1987. *Mon. Not. R. astr. Soc.* , **226,** 849.

Moffat, J.W., Seggewiss, W. and Shara, M.M., 1985. *Astrophys. J.* , **295,** 109.

Norman, C. and Scoville, N., 1988. *Astrophys. J.* , **332,** 124

Osterbrock, D.E., 1978. *Phys. Scripta*, **17,** 285.

Perry, J.J. and Dyson, J.E., 1985. *Mon. Not. R. astr. Soc.* , **213,** 665.

Peterson, B.M. and Ferland, G., 1986. *Nature*, **324,** 345.

Pica, A.J. and Smith, A.G., 1983. *Astrophys. J.* , **272,** 11; (PS83).

Pronik, I.I., 1974. *Soviet astr.*, **16,** 628.

Rees, M.J., 1978. *Ann. N.Y. Acad. Sci.*, **320,** 613.

Rees, M.J., 1984. *Ann. Rev. Astr. Astrophys.* , **22,** 471.

Renzini, A. and Buzzoni, A., 1986. *in: Spectral evolution in Galaxies, eds. C.Chiosi and A. Renzini, p. 195, Reidel.*

Shklovskii, I.S., 1960. *Soviet astr.*, **4,** 885.

Tanaka, S., 1966. *Publ. astr. Soc. Japan* , **18,** 47.

Terlevich, R., 1988. *in The post recombination Universe, eds. N. Kaiser and A.N. Lasenby, Kluwer.*

Terlevich, R., 1989. *in Evolutionary phenomena in Galaxies, eds. J.E. Beckman and B.E.J. Pagel, Cambridge Univ. Press.*

Terlevich, R., Melnick, J. and Moles, M., 1987. *in Observational evidence for activity in galaxies, eds. E.Ye. Khachikyan, K.J. Fricke and J. Melnick, Reidel, Dordrecht; (TMM87).*

Terlevich, R and Melnick J., 1987. *in Starburst and galaxy evolution eds. T.X. Thuan, T. Montmerle and J. Tran Than Van, Frontiere.*

Terlevich, R. and Melnick, J., 1988. *Nature*, **333,** 239.

Terlevich, R. and Melnick, J., 1981. *Mon. Not. R. astr. Soc.* , **195,** 839.

Terlevich, R. and Melnick, J., 1985. *Mon. Not. R. astr. Soc.* , **213,** 841; (TM85).

Terlevich, R. and Terlevich, A., 1990. *in preparation.*

Tritton, K.P. and Selmes, R.A., 1971. *Mon. Not. R. astr. Soc.* , **153,** 453.

Weedman, D.W., 1983. *Astrophys. J.* , **266,** 479.

Discussion

PAGEL: What SN rate do you need to account for the variability of the most luminous AGNs?

TERLEVICH: Three SN a day keep the blackholes away.

OSTERBROCK: It is very important to explore all possible models for interpreting AGNs, such as this one which is quite different from the more conventional blackhole picture. The basic feature of AGN spectra is photoionization by a spectrum that extends to high energies; clearly if very high temperature stars actually exist, assembling enough of them will give almost any desired photoionizing spectrum. Detailed predictions of the profiles and variations should help to distinguish between proposed models and are clearly important. The difficulty I had with such proposals is that the best models I know, and the observational data I know, do not suggest that Wolf-Rayet stars emit high-energy photoionizing spectra.

TERLEVICH: Wolf-Rayet stars cover a large range of temperatures, most of them are relatively cool stars, but for a minority mainly composed of extreme WC and WO stars the evidence suggests very high effective temperature. Probably the best case is reported by Davidson and Kinman (1982). They obtained observations of an HII region ionized by a WC5 or WC4. From the analysis of the emission lines they comcluded that the ionizing radiation peaked at 37ev similar to the peak of a 110,000K blackbody. The luminosity of this WC star is 1.3×10^5 $L_\odot$.

WAMPLER: You have presented an interesting alternative to the 'standard' model. The standard model is not troubled by observations of high polarization in the continuum. I wonder if detailed polarization measurements might be able to distinguish your 'thermal' model from the standard non-thermal model.

TERLEVICH: I believe it can give some discrimination between thermal vs. non-thermal models specially if variable polarization is detected. My difficulty is that I do not believe that the canonical accretion disk + blackhole model is fundamentally non-thermal. I'm shure it may be possible to create scenarios in which most of the luminosity is emitted by thermal processes.

LARSON: I would worry about the stupendously high star formation rates required in your picture. You require 1,000 SNe per year, which in turn require a star formation rate of order 100,000 $M_\odot$ per year. Even allowing 10^{10} $M_\odot$ of material to participate in this phenomenon, it can only last for 10^5 years, which makes it a very short-lived and therefore very rare phenomenon.

TERLEVICH: You have picked the most extreme case. It corresponds to the most luminous QSOs. For this case I used a very massive elliptical galaxy with total mass 10^{13} $M_\odot$ and your models for galaxy formation. The peak star formation rate is, depending on the model, between 50,000 $M_\odot$/yr and 20,000 $M_\odot$/yr for about 10^8 years. This gives a SN rate between 2,000 SN/year and 200 SN/year depending on the choice of IMF parameters. This rate and life-time are exactly what is needed to scale the space density of present day luminous elliptical galaxies to the space density in co-moving coordinates of redshift 2 luminous QSOs.

DYSON: Have you taken into account the fact that the SN remnants are evolving in a very high, presumably non-thermal radiation field, which would affect the physics via, for example, Compton cooling?

TERLEVICH: The radiation field is not expected to be intense. These SN remnants are not concentrated in a small volume around the active nucleus like in your scenario. They are spread over kiloparsecs, they can even cover most of the galaxy. Compton cooling should be effective only in the early stages of the evolution of the remnant.

DULTZIN-HAYAN: I believe that you are leaving out of the scenario not only OVV and BL Lacs but also radio galaxies. How can you possible collimate radio-jets in an 'only SN' scenario?

TERLEVICH: I agree, it is very difficult to explain the collimation of objects like Cygnus A. On the other hand most objects classified as AGNs either do not have radio emission or if they do, the ejecta is not highly collimated.

IONIZED GAS AND STELLAR CONTENT IN A SAMPLE OF HII GALAXIES

J.M. Vilchez, J. Cepa, C. Esteban
Instituto de Astrofisica de Canarias
38200-La Laguna, Tenerife, Spain

Abstract

We present the first results of an extensive study of the stellar content and the ionized gas structure for a sample of BCGs. Deep broad (BVRI Johnson) and narrow (H_α) CCD images have been obtained for about 25 galaxies. For a subset of objects, selected on the basis of their apparent structure, long-slit spectroscopy has also been obtained in order to investigate the interaction between the underlying stellar population and the ionized gas.

1. Introduction

Blue Compact Galaxies (BCG) are dwarf systems which present vigorous star formation giving rise, for the most recent burts, to several giant HII regions. These giant HII regions generally dominate the optical light of the galaxy, and therefore their emission line spectrum can be easily measurable with intermediate size telescopes. BCG's with such giant HII regions, most of them being rich in HI gas and oxygen poor, form a substantial subset of the class of objects known as *HII Galaxies* (Hazard 1985).

The environment where bursts of star formation take place in these galaxies is a matter of considerable interest, because of the importance for the understanding of the evolution of galaxies in general. HII galaxies offer an opportunity to study the very first moments of the evolution of galaxies. In this context, the study of the underlying (young/old) stellar population and its interaction with the observable (ionized) gas, could provide useful information about how star formation proceeds. On the other hand, the knowledge of the nature of the stellar content allows to decide wether there are genuinely young galaxies (in contrast with metal rich *but* fresh starbursts) and whether the elements present in the ionized gas result from nucleosynthesis in previous generations of stars or just come from the one responsible of the observable HII regions. In order to answer some main questions related to the stellar content in these dwarfs and its interaction with the ionized gas, we carried out a program of combined deep CCD imaging and long-slit spectroscopy for a *selected* sample of HII galaxies. In the present paper we report some recent results on the photometric properties of the, up to date, studied galaxies; also we present combined spectroscopy and imaging work for several well known dwarfs, and pay especial attention to I Zw 18 (the lowest abundance known HII galaxy).

Several studies of the underlying population in BCG's have been carried out in the last few years making use of CCD cameras capable of reaching very low surface brightness; notably by Loose and Thuan (1985) using B and R filters and by Kunth et al. (1985, 1988) using Geneva B_1, B_2, Gunn R and Gunn-Mould I filters. The results appear to show that BCG's present a large variety of morphological types, ranging from more or less smooth circular or elliptical structures easily detectable out to a few Kpc, to cases where nothing is visible beyond the HII region itself. Our CCD observations combine broad (Johnson B, V, R, I) and narrow (mainly H_α) band filters, and have been performed for about 25 galaxies. In addition, long-slit spectroscopy in the range $\lambda\lambda$ 3500-7400 Å has been obtained for a subset of $\approx$ 15 galaxies of the sample, *selected* mainly on the basis of their observed structure. Since typical dimensions of Giant HII complexes in galaxies are of the order of some 200 pc (e.g. Diaz et al. 1987), our imaging has been performed with good spatial resolution (seeing and pixel sampling) in order to resolve individual star forming areas (1 arcsec is equivalent to some 60 pc at 10 Mpc). The requiered pixel sampling was such that allowed the Point Spread Function to be well resolved and sampled by a significative number of pixels; this fact is crucial for a correct continuum substraction in the narrow band images. Finally, good seeing conditions, necessary to optimize our instrumental set up, were achieved in most cases. Table I is a journal of the CCD observations, including the telescopes, instrumentation and observing conditions of the program.

The spectroscopic observations were performed using the IDS + IPCS attached to the 2.5m INT at the Observatorio del Roque de los Muchachos (La Palma). A dispersion of 2 Å/pixel was used giving a spectral resolution of $\approx$ 3.5 Å from 3500-7400 Å. For each object, a total of 110 spectra were recorded in the spatial direction, one every 1.5 arcsec along the slit. The seeing was oscillating between 0.9 and 1.3 arcsec. The spectral resolution attained was sufficient to resolve underlying absorption features present in some of the galaxies studied (notably in the cases of Mkn 314 and II Zw 33 discussed below), while the spatial sampling allowed to localize these apparently more evolved HII complexes within the galaxies.

The sample of galaxies was selected from the master lists of Thuan and Martin (1981), Thuan (1985) and Longmore et al. (1982), being the more restrictive condition the existence of accurate data of HI/total masses. Some of our objects are in common with the work of Loose and Thuan (1985), though they have been observed in more bands and with reasonably photometric conditions. A complete description of the galaxy sample and derived properties will be published elsewhere (Vilchez, Cepa, Esteban 1989b). The present paper is intended to present a progress report of some of the basic structural parameters derived for the sample, and the methodology applied, as well as to review some of the results obtained for three particularly interesting galaxies: Mkn 314, II Zw 33 and, especially, I Zw 18.

TABLE I.- Journal of CCD Observations

Telescope	Instrument	Resolution	Mean Seeing
WHT 4.2m	Cass. TaurusII Cam.	0.26 "/pix	0.7-1.2 "
CAHA 3.5m	Prime CCD Cam.	0.50 "/pix	2 "
INT 2.5m	Prime CCD Cam.	0.70 "/pix	1-1.6 "
JKT 1.0m	Cass. CCD Cam.	0.50 "/pix	1-1.4 "

2. Global Structural Parameters

Total fluxes and colours have been derived for the program galaxies. Total fluxes were obtained integrating the flux in all the pixels with signal above the 3 σ level, generally corresponding to 26.5 mag/arcsec^2, (previously, the CCD frames were masked to avoid stars in the field, hot spots, etc.). As an inmediate consequence the derived total luminosities are far more reliable than those obtained by means of aperture photometry and/or magnitude limited observations. Figure 1 (a) shows the correlation between the derived absolute magnitudes in the blue M_B, (H_o=75 assumed), and the abundance of oxygen (12 + Log O/H) for the galaxies in the program with known O/H from the literature and/or our spectroscopy. In Figure 1 (b) the relationship between HI mass and the derived M_B is presented. HI masses have been taken from Thuan and Martin (1981), Brinks and Klein (1985) and Combes (1985). A strong correlation is present in both figures in the sense that lower abundance galaxies seem to have larger blue magnitudes, whilst larger HI masses are associated to brighter galaxies. The brightest galaxy studied appears to be III Zw 102, with an absolute magnitude more adecuated to that of a giant galaxy (unless there is something wrong with the distance modulus quoted in the literature (Thuan and Martin, 1981). This suggestion could be consistent with the very high dust content found for this galaxy (the galaxy presents very clear dust lanes in our colour images), a feature which appears to be not very common among low abundance compact HII galaxies; in addition, the high O/H value that is indicated by its IPCS spectra points to the same direction. Overall, the relationship between M_B and oxygen abundance found by Skillman et al. (1989) for his compilation of Dwarf Irregulars, 12 + Log O/H = -0.15 M_B + 5.5, appears to be very well followed by the program galaxies, from the "metallicity" of I Zw 18 (M_B= -14) up to the one of the giant III Zw 102.

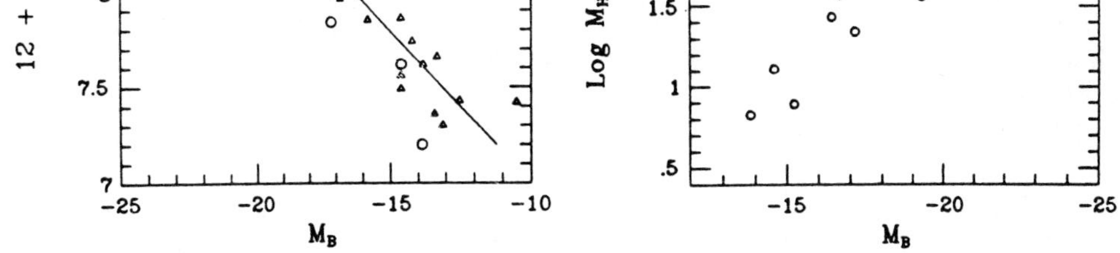

Figure 1.- *The relationship between the oxygen abundance, a), and HI mass, b), with the derived M_B magnitude. In a) the triangles are objects from Skillman et al. (1989).*

In order to make a systematic analysis of structural parameters of the galaxies, independently of their particular apparent morphology, we are investigating the extension of the formalism developed for elliptical/spiral galaxies to these type of galaxies; these show such a rich variety of morphologies that systematical studies of their structure are very rare. We have extended the concept of *"effective radius"* to this type of galaxies by means of the definition of an *"equivalent effective area"*, which is the galactic area enclosed within the isophote that includes the half of the total luminosity of the galaxy. Also, an *"equivalent effective radius"* can then be defined, formally, as the one of a circle with area equal to the *"equivalent effective area"*; likewise, an *"equivalent surface brightness"* profile can be derived plotting the surface brightness at every isophote versus its corresponding *"equivalent radius"*. The basic equations used to extend this formalism can be expressed as follows:

$$\frac{1}{2} F_{Tot}^{\lambda} = \int_{SB_{eff}}^{SB_{max}} f_{\lambda}(x,y) \quad dx \; dy$$

where F_{Tot}^{λ} is the total flux of the galaxy in the band λ, $f_{\lambda}(x,y)$ is the measured flux in the pixel (x,y) in the band λ and SB_{eff} and SB_{max} are the surface brightnesses of the isophote enclosing the *"equivalent effective area"* and the maximun surface brightness observed in the galaxy, respectively, both referred to the band λ. And similarly, the *"equivalent effective area"* can then be defined by:

$$A_{eff} = \int_{SB_{eff}}^{SB_{max}} \quad dx \; dy = \pi \; R_{eff}^2$$

Following this method it should be possible to distinguish , for instance, between Low Surface Brightness and Very Compact Galaxies, since they should present different relative fractions of the total surface per unit interval of surface brightness. A detailed application to the program galaxies is in course.

Selected Examples: the cases of II Zw 33, Mkn 314 and I Zw 18

In the following we present, as an example, some preliminary results for three individual galaxies choosed on the basis of their particular structure. First, the galaxies Mkn 314 and II Zw 33 show a striking morphology in our broad band frames, characterized by the existence of twisted isophotes and the presence of aligned knots of star formation. This morphology has been interpreted, in the case of Mkn 314 (e.g. Loose and Thuan 1985), as the possible relic of a merging process; although the HI velocity pattern for Mkn 314 is highly turbulent, in contrast with the ordered motions observed in the merging BCG II Zw 40 (Brinks and Klein, 1985, 1988).

On the other hand, on the basis of CCD imaging, it has been claimed that II Zw 33 should be a barred spiral galaxy (Loose and Thuan 1985), a suggestion consistent with recent findings indicating that II Zw 33 is an old object (Vilchez, Cepa, Esteban

1989a). Our spectroscopy shows that certainly the giant HII complexes in II Zw 33 seem to be more chemically evolved, as a whole, than the ones in Mkn 314, which presents an oxygen abundance at least a factor three lower (12 + Log O/H = 8.15). On the other hand, II Zw 33 shows clear evolutionary differences between the starburst in the centre of the galaxy, and the HII knots apparent also in broad band images; the central regions are more evolved as judged from their equivalent width of Hβ, and the presence of strong (broad) absorption features in the underlying continuum. Figure 2 illustrates the relative behaviour of the equivalent width of Hβ in II Zw 33, from the comparison of the spatial profiles of Hβ and the Continuum along the slit.

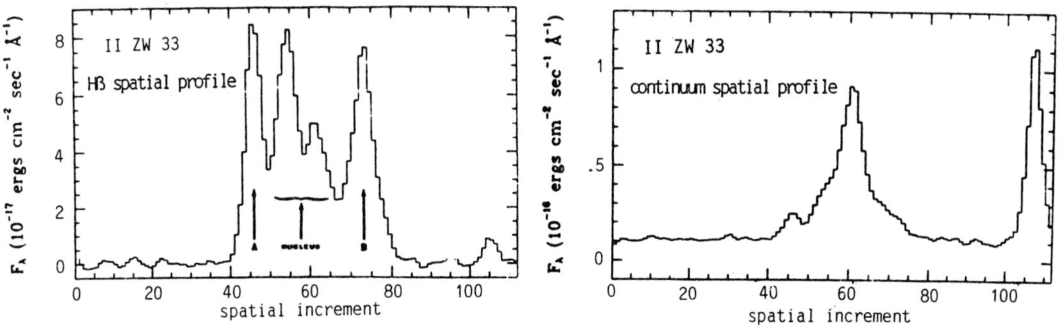

Figure 2 *Relative Behaviour of the Hβ (left) and Continuum (right) profiles in II Zw 33*

Perhaps the most interesting case among BCG's is the one of I Zw 18, because of its extreme abundance, O/H =1/50 solar. This characteristic has lead to the suggestion that the observed starburts in the galaxy constitute the very first star formation in I Zw 18, and the observed abundance ratios have been interpreted in terms of nucleosynthesis in the current burst (Kunth and Sargent 1986). Our observations have revealed the presence of a substantial amount of stellar population having colours typical of disc stars of some 10^9 years; suggesting that the current burst is perhaps the first important burst of star formation *but* definitively not the first star formation in the galaxy (Vilchez, Cepa, Pagel 1989). Therefore I Zw 18 is not a genuinely young galaxy (in the sense pointed out in section 1 above). In addition, a set of around 20 new non stellar sources in the field of I Zw 18, have been catalogued and measured. All the sources are well above the 3 σ level in the R frame, and present non stellar well resolved profiles. Most of them we believe are actually related to I Zw 18, since the search was performed within an area roughly defined by the HI map of Lequeux and Viallefond (1980). Figure 3 shows the colour-colour diagram for the measured sources, with colours ranging from those typical of OB to KM stars. Finally, the photometry obtained for I Zw 18 has allowed us to compute an approximate mass in stars which in turns, when compared to the total mass of HI (e.g. Viallefond et al. 1987), has revealed that the prediction of the Simple Model of chemical evolution is not far from the locus of I Zw 18. Therefore, the outstanding difference between "dynamical" and "chemical" masses can be telling us that a substantial amount of dark matter is probably present.

Figure 3.- *Colour-Colour diagram of the set of non-stellar sources detected in the field of I Zw 18.*

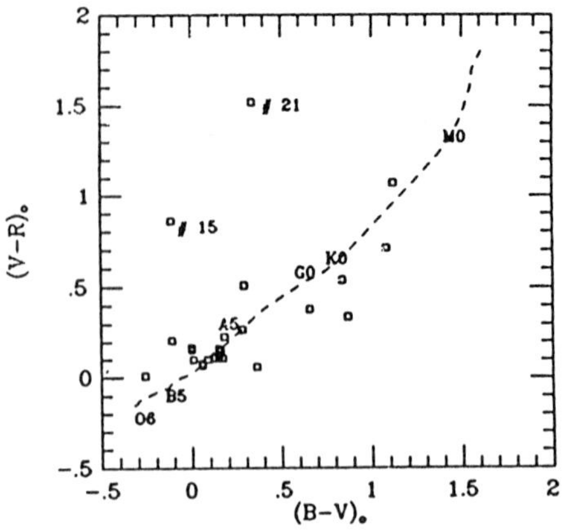

References

Brinks, E., Klein, U., 1985. in *Star-Forming Dwarf Galaxies and Related Objects.* D. Kunth, T.X. Thuan, J.T. Thanh Van, eds., p. 281. Frontieres. Paris

——————, 1988. Mon. Not Roy. astr. Soc., 231, 63P.

Combes, F., 1985. in *Star-Forming Dwarf Galaxies and Related Objects.* D. Kunth, T.X. Thuan, J.T. Thanh Van, eds., p. 307. Frontieres. Paris

Díaz, A., Terlevich, E., Pagel, B.E.J., Vilchez, J.M., Edmunds, M.G.,1987. Mon. Not. Roy. astr. Soc., 226, 19.

Hazard, C., 1985. in *Star-Forming Dwarf Galaxies and Related Objects.* D. Kunth, T.X. Thuan, J.T. Thanh Van, eds., p. 9. Frontieres. Paris.

Kunth, D., Martin, J.M., Maurogordato, S., Vigroux, L., 1985. in *Star-Forming Dwarf Galaxies and Related Objects.* D. Kunth, T.X. Thuan, J.T. Thanh Van, eds., p. 89. Frontieres. Paris.

Kunth, D., Maurogordato, S., Vigroux, L., 1988. Astron. Astrophys., 204, 10.

Kunth, D., Sargent, W.L.W., 1986. Astrophys. J., 300, 496.

lequeux, J., Viallefond, F., 1980. Astron. Astrophys., 91, 269.

Longmore, A.J., Hawarden, T.G., Goss, W.M., Mebold, U., Webster, B..L., 1982. Mon. Not Roy. astr. Soc., 200, 325.

Loose, H.H., Thuan, T.X., 1985. in *Star-Forming Dwarf Galaxies and Related Objects.* D. Kunth, T.X. Thuan, J.T. Thanh Van, eds., p. 73. Frontieres. Paris

Skillman, E.D., Kennicut, R.C., Hodge, P., 1989. Preprint.

Thuan, T.X., 1985. Astrophys. J., 299, 881.

Thuan, T.X., Martin, G.E., 1981. Astrophys. J., 247, 823.

Viallefond, F., Lequeux, J., Comte, G., 1987. in *Starbursts and Galaxy Evolution.* T. Montmerle, J.T.T. Van, eds., p. 139. Frontieres. Paris.

Vilchez, J.M., Cepa, J., Esteban, C., 1989a. in 2nd *Tex-Mex Conference on Astrophysics.* In press.

Vilchez, J.M., Cepa, J., Esteban, C., 1989b. In prep.

Vilchez, J.M., Cepa, J., Pagel, B.E.J., 1989. Preprint.

Discussion:

PAGEL (Comment): I am delighted to see your relation between abundance and ratio of mass of gas to mass of gas plus stars. The discrepancies with total mass based on the virial theorem may raise from dark matter which takes no part in the chemical evolution of visible material.

THE SHAPING OF THE OPTICAL JET OF THE GALAXY NGC 4258

P. Martin, J.-R. Roy, and L. Noreau
Département de physique, Université Laval
Québec, Qc, Canada

and

K.-Y. Lo
Department of Astronomy, University of Illinois
349 Astronomy Building
Urbana, IL 61801, USA

SUMMARY. Hα + [NII] and red continuum CCD images as well as high resolution aperture synthesis CO maps were obtained in order to study the optical jet of the barred spiral galaxy NGC 4258. The CO observations show two clouds near the center of the galaxy; these clouds outline a channel and the Hα jet follows this channel. The observations are consistent with the jet being in or making a small angle with the galaxy plane. It is concluded that the interstellar medium may play an important role in making jets detectable optically and in shaping their forms.

I. OBSERVATIONS

We have obtained narrow-band images at Hα + [NII] of the remarkable galaxy NGC 4258 (M 106) with the Mont Mégantic 1.6 m telescope using a f/8 to f/3.5 focal reducer and a RCA CCD camera ($\approx$ 1.1 arcsec/pixel). The stellar continuum derived from images at 7020 Å ($\Delta\lambda$= 200 Å) was subtracted to produce a pure emission line image of the galaxy at Hα + [NII]; CCD frames were obtained at several redshifts to cover a velocity range of about 800 km/s centered on the systemic velocity of the galaxy.

Figure 1 shows the Hα image of NGC 4258 thus obtained. The most striking feature is an elongated and S-shaped feature centered on the nucleus and detectable only as line emission. This structure discovered by Courtès and Cruvellier (1961) has been called "anomalous arms", and coincides with a radio jet identified by van der Kruit, Oort, and Mathewson (1972). Figure 2 shows our Hα image superposed with a 21-cm radio contour map obtained by van Albada and van der Hulst (1982). In the following, we refer to this unusual optical structure as the "optical jet" of NGC 4258, in the sense that it is due to emission from material shocked and excited by the energetic particles which are also responsible for the radio jet.

Although the jet bends in the same direction as the spiral arms, its structure is quite different, being more continuous and less knotty. Moreover the spectrum of the jet (Figure 3) displays strong [OI], [NII] and [SII] lines relative to Hα. Despite the difficulty of estimating the strength of the underlying stellar Hα absorption, the spectrum is clearly typical of shock-excited emission. In this article, we focus on the interaction between the jet and the interstellar medium.

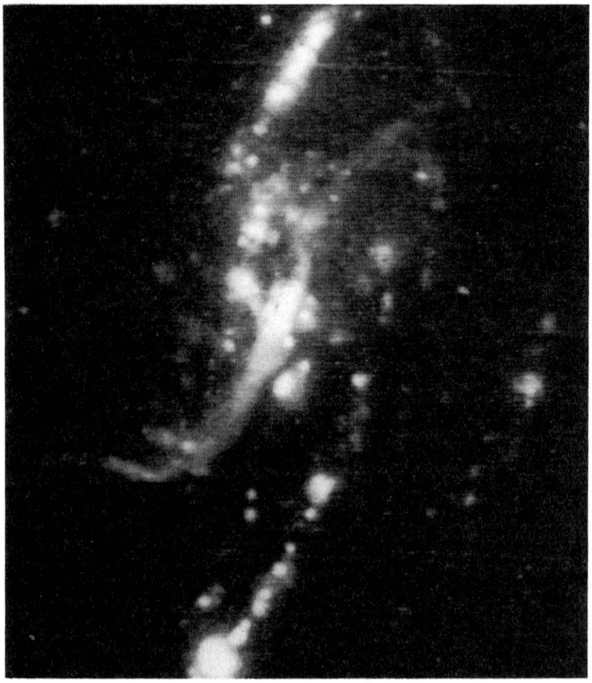

Figure 1. Image at Hα + [NII] of the giant barred spiral galaxy NGC 4258. The field shown is about $5' \times 5'$. North is at the top and east at left.

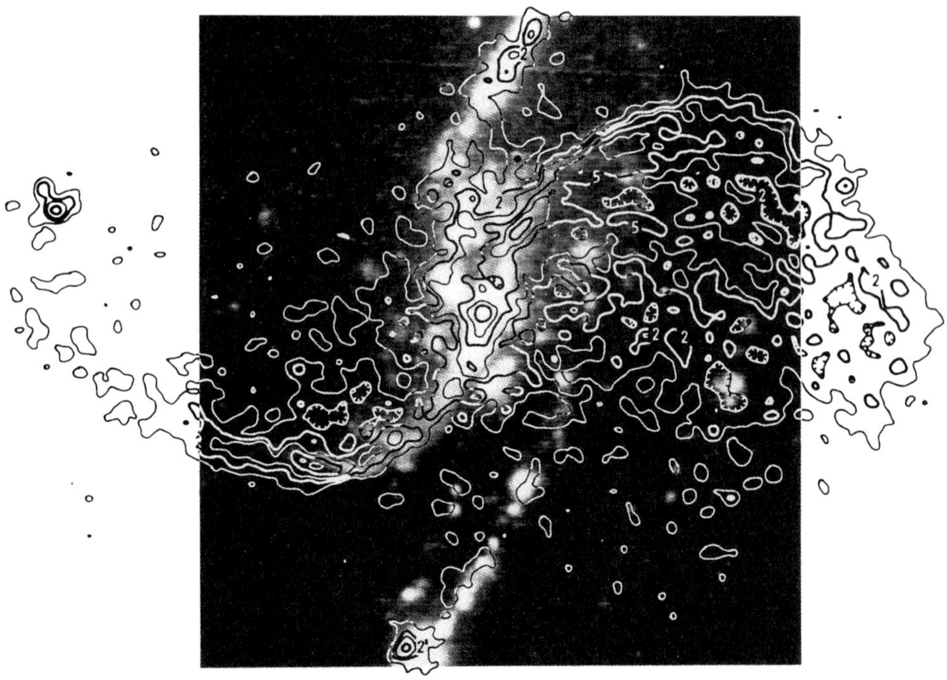

Figure 2. Superposition of the 21-cm contour map of van Albada and van der Hulst (1982) on the Hα image of NGC 4258; the radio map has a resolution of $6.5'' \times 6.5''$.

II. THE NATURE OF THE OPTICAL JET OF NGC 4258

A jet of material expulsed from the center of a galaxy would be bent by the ram pressure of the interstellar medium and would normally appear as a "leading" structure (Wilson and Ulvestad 1982). The jet of NGC 4258 appears to be "trailing". In agreement with previous investigators (see in van Albada and van der Hulst 1982), we present evidences that the jet is in the plane of the galaxy and is indeed "trailing", a behavior that we will explain in section III.

Several characteristics of the jet betrays the role of the interstellar medium in shaping the jet, and demonstrate that the jet is not a pair of spiral arms.

1. Where Hα + [NII] is bright, 21 cm continuum emission is always strong (Figures 1 and 2). The converse is not true however: the radio jet remains visible much further away from the nucleus than the optical jet. Nevertheless this emphasizes that optical line emission is most probable in the inner parts of the galaxy where the density of the interstellar medium is high.

2. Splitting and bending of the optical jet at its SE and NW ends take place at the co-rotation radius as determined from HI observations (van Albada 1980).

3. A weakening in surface brightness and broadening of the SE portion of the optical jet (just before the splitting) is associated with a region of low density HI (van Albada 1978).

4. An integrated ^{12}CO (2.6 mm) intensity map of the central $1' \times 1'$ of NGC 4258 (Figure 4) obtained with the Owens Valley interferometer reveals molecular clouds outlining a well-defined channel. The optical-radio jet falls like a stream in this elongated gap. Moreover the optical jet closely espouses the shape of the CO channel; to the north, the jets bends eastward for a short distance alike the wall formed by the CO western CO clouds. The Hα brightness of the jet is high there; then the optical jet appears to dim in intensity and the radio jet fans out much more in the NW quadrant (Figure 2) when compared with the SE quadrant. The close match between the jet and molecular cloud morphology strongly suggests that the energetic particles responsible for the jet have bored a channel (or a tunnel) in the interstellar medium, and that the particles are maintaining the tunnel by a snow-plough action.

5. The optical jet has a smooth morphology indicating continuous interaction between energetic particles and the diffuse interstellar medium of NGC 4258. The optical jet is also relatively narrow, suggesting that the interstellar medium has a confining effect on the jet. The role played by interstellar magnetic fields may be important in confining such a structure.

III. DISCUSSION

Observational evidences are that the jet of NGC 4258 is flowing within the plane of the galaxy and that it is strongly interacting with the interstellar medium. We propose that the optical jet is made of interstellar gas which has been shocked and entrained

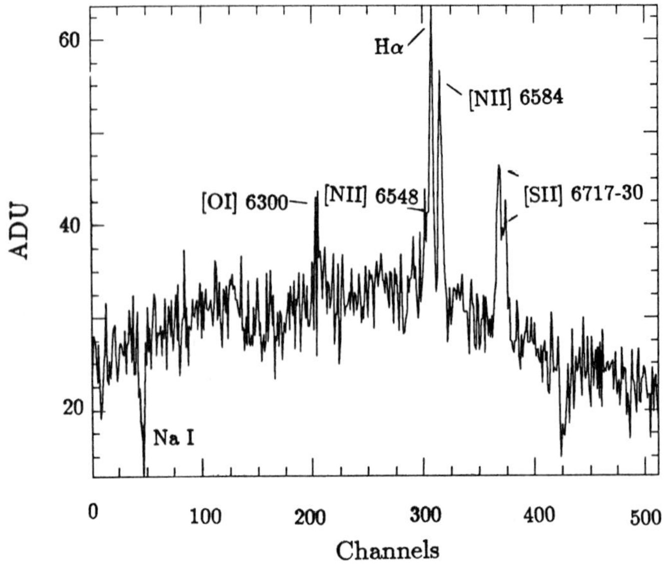

Figure 3. Spectrum in the red of the optical jet of NGC 4258 in the central region. The spectrum has not been corrected for atmospheric extinction nor for the detector spectral response, but sky background has been subtracted.

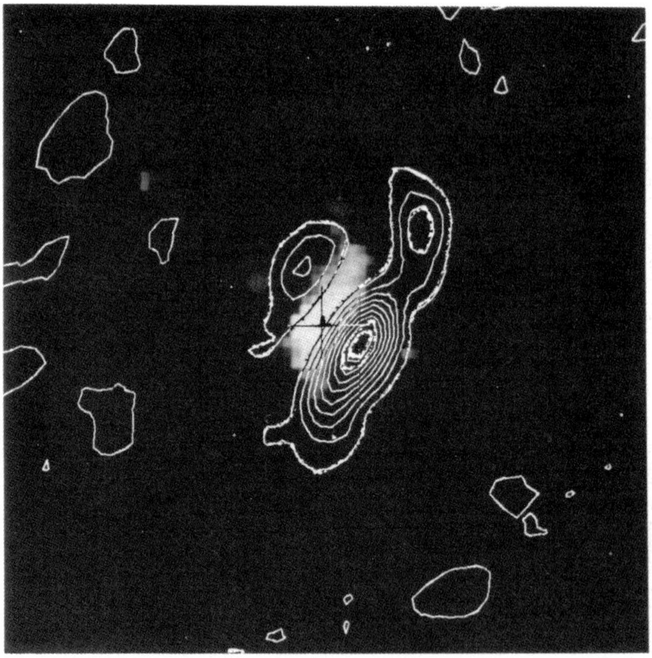

Figure 4. An integrated CO intensity contour map of the central $1' \times 1'$ of NGC 4258 obtained with the Owens Valley interferometer superposed on the Hα image of NGC 4258 selecting only the brightest levels of optical emission.

by the high-energy particles responsible for the radio jet (Martin *et al.* 1989); such entrainment has been discussed by de Young (1986). It is unlikely that the material visible optically originates from the nucleus; although it is not known how far interstellar gas can be entrained by a jet of energetic particles, one can say the optical emission is probably due to gas being excited locally. Detailed velocity maps of the optical jet are needed.

The particles originating in the nucleus have bored a tunnel through the interstellar medium. This tunnel offers a path of least resistance to the beam of particles. However this tunnel is subjected to the rotational motion of the galaxy interstellar medium forcing the jet to "trail". Through its action on the tunnel, the interstellar medium shapes the morphology of the jet in the inner parts of NGC 4258. The jet in turn affects the interstellar medium of NGC 4258. It has hollowed out molecular clouds near the galaxy center and has evacuated a channel at larger galactocentric distances. Even at the edges of the galaxy, the disk emission seems to be disturbed in the regions intersected by the jet (van Albada and van der Hulst 1982).

REFERENCES

Courtès, G., and Cruvellier, P.: 1961, *Comp. Rend. Acad. Sci. Paris*, **253**, 218.

de Young, D. S.: 1986, *Astrophys. J.*, **357**, 62.

Martin, P., Roy, J.-R., Noreau, L., and Lo, K.-Y.: 1989, *Astrophys. J.* in press.

van Albada, G. D.: 1978, *Ph. D. Dissertation*, University of Leiden.

van Albada, G. D.: 1980, *Astron. Astrophys.*, **90**, 123.

van Albada, G. D., and van der Hulst, J. M.: 1982, *Astron. Astrophys.*, **115**, 263.

van der Kruit, P. C., Oort, J. H., and Mathewson, D. S.: 1982, *Astron. Astrophys.*, **21**, 169.

Wilson, A. S., and Ulvestad, J. S.: 1982: *Astrophys. J.*, **263**, 576.

A PLASMON DRIVEN BOWSHOCK MODEL FOR THE NARROW LINE REGION OF NGC 5929

D. Taylor, J. E. Dyson
Department of Astronomy,
University of Manchester,
Manchester M13 9PL, U.K.

D. J. Axon, A. Pedlar
Nuffield Radio Astronomy Observatories,
Jodrell Bank,
Macclesfield,
Cheshire SK11 9DL, U.K.

Close links appear to exist between the optical narrow line region (NLR) in Seyfert galaxies and the nuclear non-thermal radio emission. Both occur on similar scales ($\sim 10^2 - 10^3$ pc), and correlations have been found between the 21 cm radio power and both the [OIII] $\lambda 5007 Å$ luminosity and linewidth (de Bruyn & Wilson 1978; Whittle 1985). Approximate balance appears to exist between the thermal pressure of the NLR gas and the relativistic and magnetic pressure of the radio components (de Bruyn & Wilson 1978; Unger *et al.* 1985; Pedlar, Dyson & Unger 1985). There are also several examples of a direct spatial association between the optical and radio components, for example Mkn 78 (Pedlar *et al.* 1988).

A particularly good example for modelling this association is NGC 5929, a Type 2 Seyfert with a redshift of 2625 km s^{-1} (Whittle *et al.* 1986), corresponding to a distance of 53 Mpc for $H_0 = 50$ km s^{-1} Mpc^{-1}. This galaxy has a simple two component structure in its nuclear region at both optical and radio wavelengths. Narrow band [OIII] $\lambda 5007 Å$ imaging (Whittle *et al.* 1986) shows two core components with a separation of $\sim$ 1.7 arcsec ($\sim$190 pc) along PA $63 \pm 2°$, on either side of the stellar nucleus. They are of unequal brightness, the stronger being to the south-west. Long-slit spectroscopic observations (Whittle *et al.* 1986) reveal two distinct [OIII] $\lambda 5007$ velocity components which, with the slit aligned along PA 60°, are also separated spatially.The components have heliocentric peak velocities of 2310 ± 10 km s^{-1} (NE) and 2625 ± 25 km s^{-1} (SW). By simply partitioning the line profiles into two portions about the median velocity of the two components, Whittle *et al.* (1986) derived their spatial separation to be $\sim$0.95 arcsec ($\sim$240 pc).

A 5 GHz VLA map of the nuclear region (Ulvestad & Wilson 1984) shows a two lobe radio structure; the components are unresolved and have a separation of $\sim$1.3 arcsec in PA $\sim$60°. A VLBI observation (Whittle *et al.* 1986) gives an estimate of 10 pc for the scale size of the lobes.

We have developed a model for the NLR in which the linear motions of radio-emitting bubbles of plasma ('plasmons') ejected from the Seyfert nucleus drive bowshocks into the ambient nuclear medium. The cooled shocked gas, accelerated by the plasmons and photoionised by the UV nuclear continuum, produces the optical NLR emission. This provides a natural explanation for the relationship between the radio continuum emission and the NLR, and avoids the severe problems of confinement inherent in the standard picture of the NLR as an assembly of clouds.

Our model consists of a steady state axisymmetric bowshock driven ahead of a plasmon which is moving at constant velocity through a homogeneous medium. The ambient gas is completely photoionised by continuum radiation from the Seyfert nucleus. This gas is compressed and heated on entering the shock, and then flows along the bowshock. As it does so it gradually cools into thermal equilibrium with the nuclear radiation field and finally merges back into the ambient medium. The deceleration of the plasmon by ram pressure is ignored and the flow pattern and plasmon shape hence assumed time independent. At any point on the bowshock surface each gas parcel is assumed to have a pressure equal to the ram pressure at that point and a velocity determined by global conservation of mass and momentum. Individual parcels of gas are assumed to cool independently, both radiatively and adiabatically. In this model, therefore, gas at a given distance from the apex of the bowshock comprises many individual packets each with a unique temperature, density and ionisation structure, but with the same pressure and velocity. The [OIII] emission from each point on the surface is calculated by summing the contributions from these elements. The radiative cooling law used in the calculation of the temperatures of the gas parcels is a numerical approximation to the non-equilibrium cooling curve of Falle (1975). To calculate the ionisation states of the gas elements, ionisation equilibrium is assumed at all times. This is justified for the purposes of calculating the optical emission as this is predominantly produced by cooled, photoionised gas.

The ionising UV radiation field of the galactic nucleus is modelled as a simple power law with spectral index $\alpha = 1.4$, (defined so that the energy flux $f_\nu \propto \nu^{1-\alpha}$), which is typical of the Seyfert 2 galaxies observed by Ferland & Osterbrock (1986).

The radio components in NGC 5929 are not resolved, so their shapes are unknown. However, one case of a Seyfert 2 galaxy where radio components are resolved is NGC 1068, which has a similar two component structure to NGC 5929, although on a larger scale. For want of a better guide to the bowshock shape to use in modelling NGC 5929, a Chebyshev fit was made to the outline of the NE radio lobe of NGC 1068 from the VLA map of Wilson & Ulvestad (1983), and scaled down to the appropriate size.

We have modelled the NLR of NGC 5929 as a pair of radio plasmons of radius $\sim$10 pc travelling in opposite directions away from the galactic nucleus, each producing a bowshock. We invoke an outflow in the ambient nuclear gas to produce the velocity separation between the two optical components, although this could equally be the result of galactic rotation.

A representative (but not unique) solution is sketched in Figure 1. The plasmon velocities are 550 $\mathrm{km\,s^{-1}}$ and 1020 $\mathrm{km\,s^{-1}}$ and the ambient medium has a density of 5 $\mathrm{cm^{-3}}$ and radial velocity 120 $\mathrm{km\,s^{-1}}$, resulting in shock velocities (at the stagnation points) of 430 $\mathrm{km\,s^{-1}}$ and 900 $\mathrm{km\,s^{-1}}$ respectively.

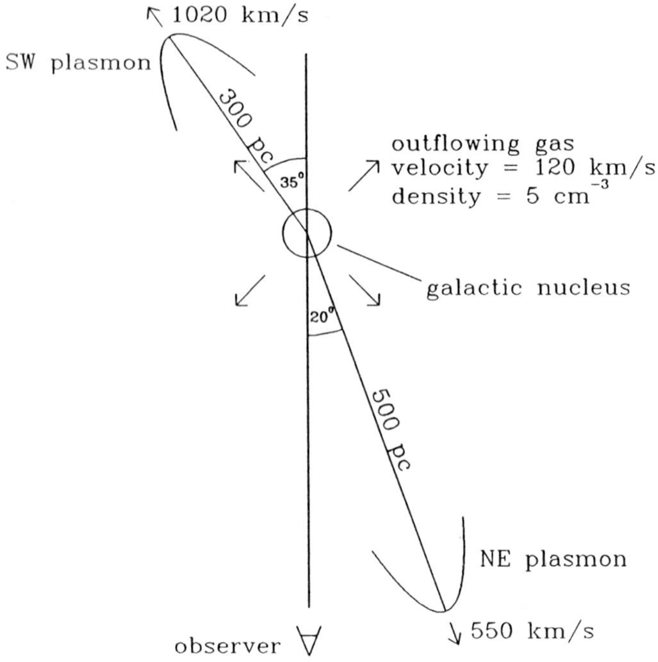

FIGURE 1: Sketch of the plasmon model for NGC 5929.

Figure 2 shows the sequence of [OIII] λ5007 line profiles from Whittle *et al.* (1986) with the slit aligned along PA 60°, and the equivalent profiles produced by the model. To obtain these, the bowshock emission was convolved with a gaussian of FWHM 400 pc to simulate the effects of seeing, and the profiles convolved with a 45 km s^{-1} FWHM gaussian to account for instrumental broadening. Figure 3 shows the [OIII] component intensity distribution from Whittle *et al.* (1986) along with that produced here. Table 1 lists the observable parameters of the model, and the corresponding values from Whittle *et al.* (1986).

TABLE 1: [OIII] component properties produced by the plasmon model and the corresponding observational values (from Whittle *et al.* 1986).

		Observational	**Theoretical**
Separation (arcsec)		0.95 ± 0.11	1.19
Velocity difference (km/s)		315 ± 18	325
FWHM (km/s)	**NE**	80 ± 10	96
	SW	185 ± 20	184
$\mathbf{L_{5007}}$ **(erg/s)**	**NE**	1.4×10^{40}	9.8×10^{39}
	SW	3.0×10^{40}	3.2×10^{40}

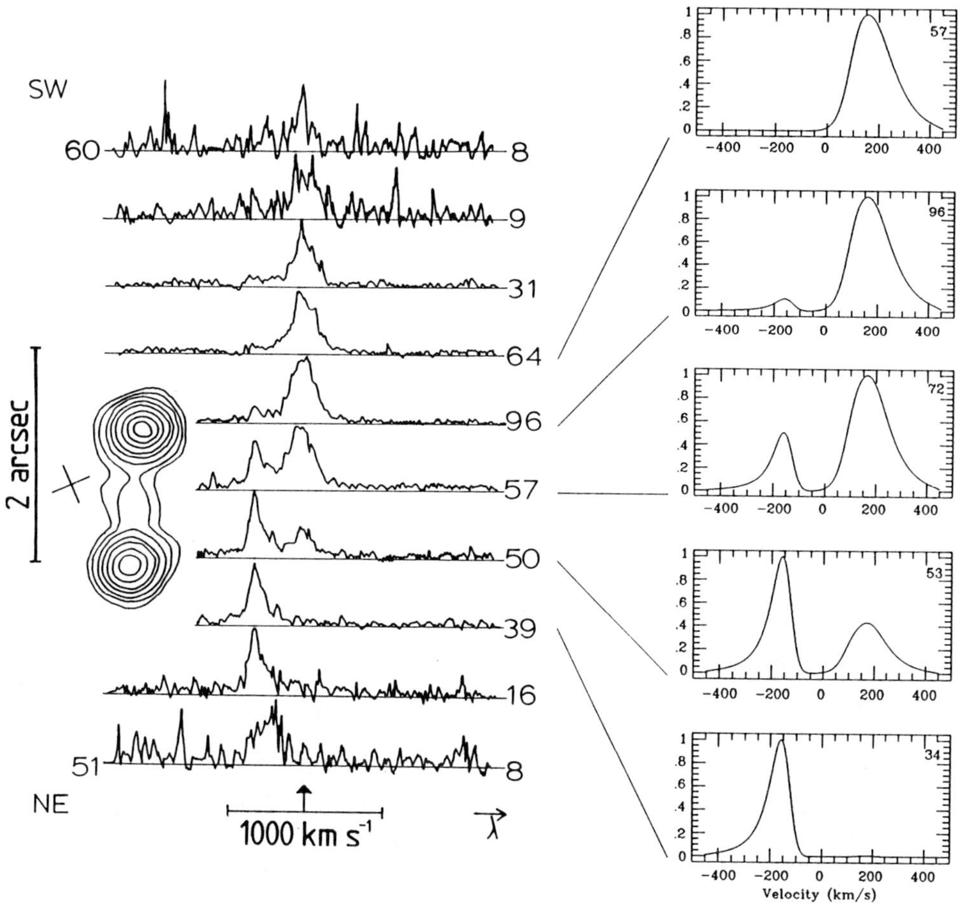

FIGURE 2: Observational and theoretical [OIII] $\lambda 5007$ line profiles for NGC 5929. The observations are from the long-slit data of Whittle *et al.* (1986). Spatial increments along the slit are 0.63 arcsec long and the increment numbers run from 51 (NE) to 60 (SW). The profiles are autoscaled, a relative intensity scale is given by the numbers to the right of each spectrum, which are proportional to the peak intensity of that spectrum. The vertical arrow at the bottom indicates a reference velocity of 2625 $\mathrm{km\,s^{-1}}$ (the galactocentric redshift of NGC 5929). Superimposed at the same scale and orientation is a radio map, taken from Ulvestad and Wilson (1984). The cross marks the position of the stellar nucleus. The profiles generated by the plasmon model correspond to slit increments 53 to 57. The velocities are relative to the galactic nucleus. The profiles are autoscaled, with the relative intensities of each spectrum given by the number in the top right hand corner.

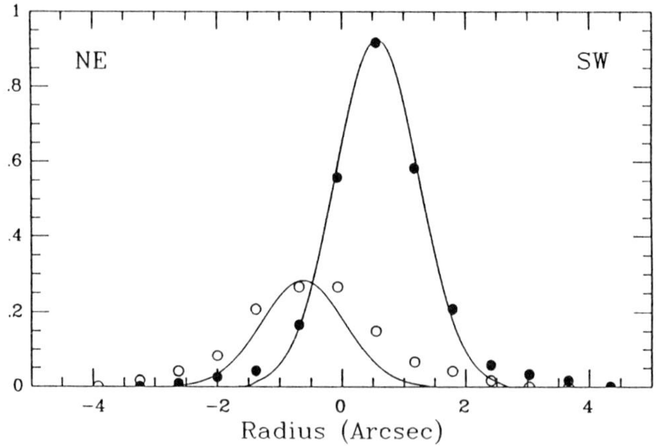

FIGURE 3: [OIII] component intensity distribution of the plasmon model. The points represent the Gaussian fits to long-slit data made by Whittle *et al.* (1986), with the open circles corresponding to the NE component and the filled circles to the SW component. The solid lines show the distributions produced by the model. The horizontal axis represents the distance from the position of the galactic nucleus in the model.

REFERENCES

De Bruyn, A. J. & Wilson, A. S., 1978. *Astron. Astrophys.*, **64**, 433.

Falle, S. A. E. G., 1975. *Astron. Astrophys.*, **42**, 323.

Ferland, G. J. & Osterbrock, D. E., 1986. *Astrophys. J.*, **300**, 658.

Pedlar, A., Dyson, J. E. & Unger, S. W., 1985. *Mon. Not. R. astr. Soc.*, **214**, 463.

Pedlar, A., Meaburn, J., Axon, D. J., Unger, S. W., Guerrine, N., Whittle, M., Meurs, E. & Ward, M., 1989. *Mon. Not. R. astr. Soc.*, **238**, 863.

Ulvestad, J. S. & Wilson, A. S., 1984. *Astrophys. J.*, **285**, 439.

Unger, S. W., Pedlar. A., Booler, R. V. & Harrison, B. A., 1985. *Mon. Not. R. astr. Soc.*, **219**, 387.

Whittle, M., 1985. *Mon. Not. R. astr. Soc.*, **213**, 33.

Whittle, M., Haniff, C. F., Ward, M. J., Meurs, E. J. A., Pedlar, A., Unger, S. W., Axon, D. J. & Harrison, B. A., 1986. *Mon. Not. R. astr. Soc.*, **222**, 189.

Wilson, A. S. & Ulvestad, J. S., 1983. *Astrophys. J.*, **275**, 8.

Discussion:

OSTERBROCK: You have presented this model for NGC 5929. Do you consider it as more generally valid to many or all Seyfert 2 galaxies?

TAYLOR: Yes. Optical NLR emission appears to be closely related to the radio emission in Seyfert galaxies, and could be a result of the interaction between radio components and the interstellar medium. Detailed modelling, however, is only possible in the very few cases where the optical structure can be resolved.

AGE EFFECTS IN GIANT EXTRAGALACTIC HII REGIONS

J. Masegosa and M. Moles

Instituto de Astrofísica de Andalucía, Granada. Spain

Spatial inhomogeneities for a sample of Giant Extragalactic HII Regions (GEHR) have been analyzed. The results indicate that, whitin the observational errors, the O/H and He/H abundance ratios are constant for a given GEHR. The Ne/O ratio on the other hand shows a trend with the evolution of the ionizing stellar clusters.

Spectral changes between different zones cannot be explained in terms of effective temperature variations. We found out that they could be accounted for as differences in the ionization parameter. This parameter, governing the spatial peculiarities in GEHR, is probably related to the age of the ionizing clusters, a fact not yet taken into account in photoionization models.

EVOLUTION OF CLUMPY GAS IN GALAXIES

R. Kunze, H.W. Yorke, R. Spurzem

Institut für Astronomie und Astrophysik, Universität Würzburg

The basic equations for treating a multicomponent medium: stars, gas clouds and an ambient gas are discussed elsewhere (Yorke, et al., this meeting). Here we apply the 2-D multicomponent code to the problem of explosions in a clumpy medium (cool "clouds" embedded in a warm $T \sim 10^4\ K$ or hot $T \sim 10^6\ K$ gas). The evolution of supernova remnants and supershells in media of varying degrees of clumpiness are compared and contrasted to the results of earlier calculations.

INTERSTELLAR MATTER AT HIGH GALACTIC LATITUDES

H_α-EMISSION IN DIRECTIONS TOWARD HIGH VELOCITY 21cm CLOUDS

Guido Münch and Eckhart Pitz
Max-Planck-Institut für Astronomie
Königstuhl, 6900 Heidelberg, FRG

Summary

Measurements of H_α-emission in fields known to emit 21cm radiation at intermediate and high velocities are presented. The observations are partial results of a survey being carried out at the Calar Alto Observatory with a dedicated Fabry-Perot spectrometer of 15cm aperture, providing velocity and angular resolutions of 11km s^{-1} and 0.5 deg^2. The limited material available indicates that the H_α-emission, when detected at a radial velocity nearly the same as that measured in 21cm, appears more extended in the sky than the HI-radiation.

1. **Introduction**

The presence of extended galactic H_α-emission, from almost all over the entire sky, has become known through the pioneering observations carried out with the 2-etalon PEPSIOS spectrometer of the University of Wisconsin (Reynolds, Roesler and Scherb, 1974 and 1977; Reynolds, 1980, 1981 and 1983). The H_α intensities measured by Reynolds (1984) in 9 fields with galactic latitude $|b|>45°$ are very low, corresponding to emission rates $0.5<ER<1.7R$ (1R = 1 Rayleigh = 2.47(-7) $erg.cm^{-2}s^{-1}sr^{-1}$ at H_α), but imply an average hydrogen recombination rate of 2 to $7x10^6$ s^{-1} per cm^2 of the galactic disk, large enough to raise an important unsolved problem in regard to the nature of the ionization agents. The small number of sample fields in the Reynolds (1984) survey, chosen in directions to known pulsars, clearly cannot provide information about the porosity of the H_α emitting medium or about its likely geometrical relation to the 21cm features. In order to answer such questions and provide theoreticians with an idea of how unrealistic their modeling of the thermal phases of the interstellar medium may be (Shull, 1987), many thousands of observing hours with large thruput spectrometers, such as PEPSIOS and field widened Michelsons, will be required. In consideration of the fact that at present only the Wisconsin PEPSIOS is operational, under non-optimal conditions of atmospheric transparency, we undertook the construction of a spectrometer similar to PEPSIOS, to be used at the Calar Alto Observatory.

In this communication we shall present, in a condensed form, some results of our H_α measurements in fields where 21cm line emission at intermediate and high velocities has been mapped. Although our survey is far from completed, it suffices to show that, at least in some cases, there exists a positional and kinematical correspondance between the HI and HII distributions.

2. Instrumentation

The essential difference between our instrument and the PEPSIOS is that the high resolution master etalon of 15cm aperture, manufactured by Queensgate Instruments Ltd., is piezoelectrically scanned and servo-controlled for plate parallelism. The etalon gap was chosen to deliver a resolving power of 30,000 and can be varied as to provide a continuous scan range of 850km s^{-1}, around a manually set nominal zero. As it will become apparent below, such a large scan range is of great value to fix the background "continuum" against which the faint galactic emission lines have to be discriminated. The off-order suppressing slave etalon, also piezoelectrically scanned and controlled for parallelism, has an aperture of only 50mm and, correspondingly, it has been lens coupled to the master with a gap ratio 1:8.12. The field stop, generally chosen to be 0°8 in diameter, is placed at the intermediate focus between etalons, while the interference prefilter, of FWHM=15A, is located in the parallel beam entering the slave etalon. The price we have paid for the high reproducibility and stability in plate parallelism provided by the piezoelectric system, is a relatively low finesse (no more than 14), due to imperfect dielectric coatings and master flats deformation at the feet of the cemented piezoelectric transducers. The effective resolution, nevertheless, is near 11km s^{-1}, as it can be seen in Fig. 1 by the profile of the Th I 6554 line from a hollow cathode discharge.

The instrument has a vertical optical axis and is enclosed in a 6m long sea-going container, admitting sky-light from a polar siderostat through a trap side door. Light and heat tight partitions separate the spectrometer from the "room" containing observer, control electronics and system microcomputer. We use Ga:As photomultipliers as detectors for the scanned signal and for a co-aligned fixed wide band signal from the sky, which serves only as monitor of sky conditions.

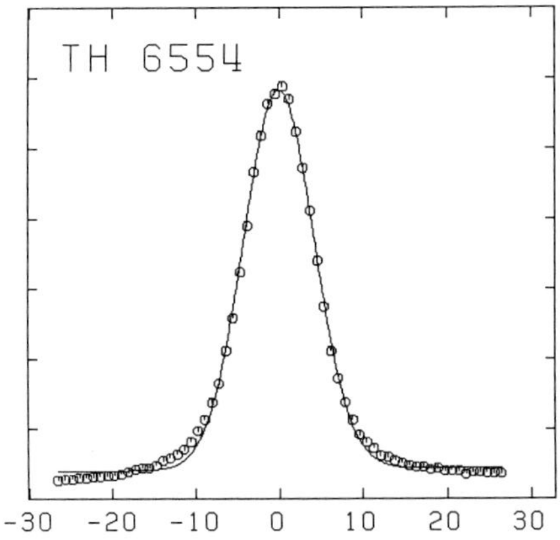

Fig. 1. Intensity profile of the Th I 6554.15A line with abscissae in km s^{-1}. The Gaussian fit drawn as a continuous curve, with a dispersion of 4.23km s^{-1} fits well the core of the line. At distances from line centre $|\Delta V|>10$km s^{-1} the extended wings of the 2-fold Airy profile become apparent. The slight slope of the baseline is due to curvature in the pre-filter transmission.

3. **Procedures**

The main difficulty encountered in the measurement of galactic H_α-lines with emission rates ER at the 1R level arises from the presence of the much stronger geocoronal H_α (GCH_α) and of its first order ghosts shifted ± 136km s^{-1}, which the slave etalon suppresses only to an intensity of 4.8% with respect to the parent. The airglow OH lines $P_1(3)6553.66$ and $P_2(4)6568.78$ of the (6,1) band, only partially suppressed by the pre-filter and slave etalon, also have ghosts in the range of interest. By choosing a master spectral free range of 136km s^{-1} we managed to place the ghosts of those OH lines near the H_α ones, in such a way that relatively extended spectral ranges became available for the search for faint galactic H_α features.

The dependence on time and position in the sky of the GCH_α and the OH line intensities severely constrains the time available to scan a chosen range with a S/N-ratio sufficient to allow recognition of faint galactic emission. Even under optimum "photometric" conditions of atmospheric transparency, it is impossible with a single beam instrument to "subtract" completely from a scan the night sky features by comparison with another scan not obtained simultaneously on a neighbouring field. On the basis of these considerations, as a general strategy in our survey for faint galactic H_α emission with shifts $|\Delta V|<100$ km s^{-1}, we have set a limit of 30 minutes for the duration of one scan, sampling a range not larger than 300km s^{-1} with about 50 points. The implied 30sec integration time per sample provides then $S/N\approx20$, due to photon arrival statistics for a single sample in the "continuum" of a typical scan at $|b|>30°$. A line with height above the continuum equal to the r.m.s. PM- dark noise of $2s^{-1}$ can then still be recognised in a 3σ-sense.

4. **Results**

a) *OYM 190, a cloud at intermediate latitude*

As an example of a fairly strong displaced galactic H_α emission we show in Fig. 2a the spectrum of the 21cm cloud OYM190 of Meng and Krauss (1970) with V_{LSR}=-50km s^{-1}, which happens to be near the star HD215733 (l=85°.2, b=-36°.4), long known (Münch and Zirin, 1961) to show 4 interstellar CaII components with LSR shifts of -57, -44, -26 and -11km s^{-1}. The H_α spectrum of Fig. 2a is a coaddition of 2 scans obtained in the same night, to which 3 Gaussians with free parameters have been fitted. The subtraction from Fig. 2a of the Gauss fitted GCH_α and of its ghost, taken with fixed height, position and dispersion, respect to the parent, gives the spectrum of Fig. 2b. While "removing" the GCH at V_{LSR}=+15.7km s^{-1}, it is very likely that some of the low velocity galactic H_α has also been subtracted. The subtraction of the GCH_α ghost has also not removed the faint nearby OH ghosts. Nevertheless, the complex character of the residual galactic spectrum is apparent in Fig. 2b, which is confirmed by other scans secured at the same position. The Gaussian fit shown in Fig. 2b has two components with LSR peaks at -43.7 and -0.4km s^{-1} and FWHM-widths of 24 and 36km s^{-1}, respectively. Adopting as photometric standard an emission rate of 850R, as determined by Scherb (1981) and independently confirmed by our own measurements, we find for the two galactic components emission rates of 0.6 and 2.3 R.

The scans of Fig. 2 were obtained at the centroid of the OYM190 cloud as drawn by Meng and Krauss (1970). A few other scans secured beyond the 21cm boundaries definitely show galactic H_α extending to -60km s^{-1} and more, but somewhat weaker and with a different shape from those of Fig. 2. Pending a detailed comparison of the 21cm and H_α profiles within the OYM190 cloud and in its extended neighbourhood, however, the apparent enhancement of the -44km s^{-1} component in Fig. 2b cannot prove the side by side coexistence of the HI and HII emitting gases. In this regard it is of interest to remark that the high resolution 21cm profile obtained by Lockman, Hobbs and Shull (1986) at HD215733 shows a broad displaced component peaking at -47km s^{-1}, but weaker than and well separated from the near zero velocity component. The H_α profile we observe has a component at -44km s^{-1}, but it is not well separated from the nearby zero velocity component. We believe this is an effect not due to the smaller beam width (21 arcmin) of the 21cm profile or to the larger thermal width of the HII emission, but rather to the relatively greater extension of the ionized emitting volume.

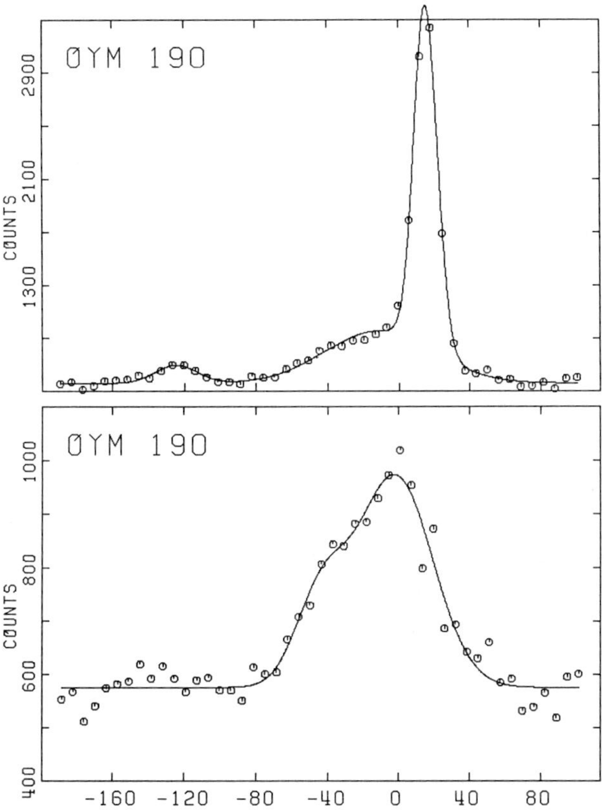

Fig. 2. H_α spectrum of the 21 cm cloud OYM190 (104°,-29°) from two scans, each with 30s integrations. Above (a) is shown the raw data fitted with 3 Gaussian components. Below (b) is shown the spectrum obtained upon subtracting from (a) the Gaussian fitted geocoronal H_α and its ghost, with a two component Gaussian fit for the residual galactic emission. Notice that the count scale below is x4 times greater than above.

b) *OMM360 and OLH393, high latitude clouds*

The OMM360 field of Meng and Krauss (1970), with V_{LSR}=-50km s^{-1}, attracted our attention because of its large extension in the sky and its neighbourhood to other closed contours in HI column density at similar "intermediate" velocity. One of such nearby clouds, OLM351 with V_{LSR} between -45 and -66km s^{-1}, covers completely the "high" velocity cloud OLH393 (185°, +65°) with V_{LSR}=-86km s^{-1}, also called MII by Hulsbosch (1986). It was thus thought that the detection in the OMM360 field of an H_α component at about -50km s^{-1} and at the position of OLH393 of components at -85 and -50km s^{-1} would by itself offer strong evidence for the close relationship between the HII and HII emitting gases. We have scans in 10 fields in and around the OMM360 complex, all of which unmistakably show galactic H_α in the range -60 to -40km s^{-1}. Shown in Fig. 3 is the average of two scans obtained on the brightest condensation (201°, +62°) appearing in the high resolution map of Giovanelli, Verschuur and Cramm (1972), after subtraction of the GCH_α and its ghost in the manner explained for Fig. 2. The residual galactic spectrum clearly shows evidence of complexity, again amply confirmed by many other scans. The Gaussian fit of 2 components, shown in Fig. 3, has peaks at -54 and -14km s^{-1}, widths of 26 and 34km s^{-1}, and emission rates of 0.35 and 0.85R, respectively. Spectra obtained in fields outside the 21cm boundaries of OMM360 and the other nearby clouds also show the shifted galactic H_α, somewhat but not much weaker than within. Especially, the western boundary of OMM360 in the map of Meng and Krauss (1970) appears at RA=$10^h 0^m$, Dec=+30°, but shifted galactic H_α begins to be noticeable in scans taken at $9^h 20^m$ and the same declination.

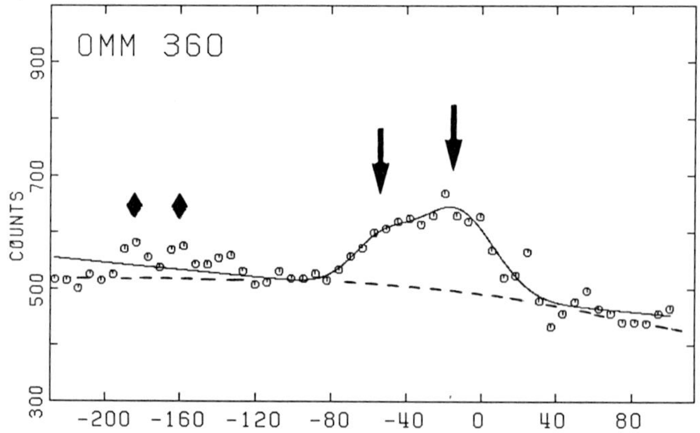

Fig. 3. Spectrum of the galactic H_α emission in OMM360 from two coadded scans, each with 30s integration time per sample. The geocorona H_α at 15.8km s^{-1} (LSR) and its ghost have been subtracted from the raw spectrum.

Unlike OMM360, the OLH393 or MII cloud is relatively compact and small (4°x8°) (Giovanelli, Verschuur and Cramm, 1972), although it probably belongs, together with MI=OMH425, to an elongated "chain" of clouds extending to the North East. Two spectra of OLH393, obtained in Jan. 1989 and Dec. 1988, each representing the coaddition of 2 single scans are shown in Fig. 4, a and b, with the GCH and ghost removed. The comparison of Fig. 4a with Fig. 3 immediately shows the existence of radiation in the range -80 to -40km s^{-1}, while hardly any can be seen around zero velocity. In the spectrum of Fig. 4b, obtained at larger zenith distance and angle from the antisolar point, the subtraction of the GCH_{α} has not been as complete as in Fig. 4a, but the existence of line emission in the same range can also be noticed. The shape of the emission feature, with a secondary minimum around -65km s^{-1} appears to be similar in the two spectra, suggesting the reality of two components, one at -80 and the other at -50km s^{-1}, which we ascribe to OLH393 and OLM351, respectively.

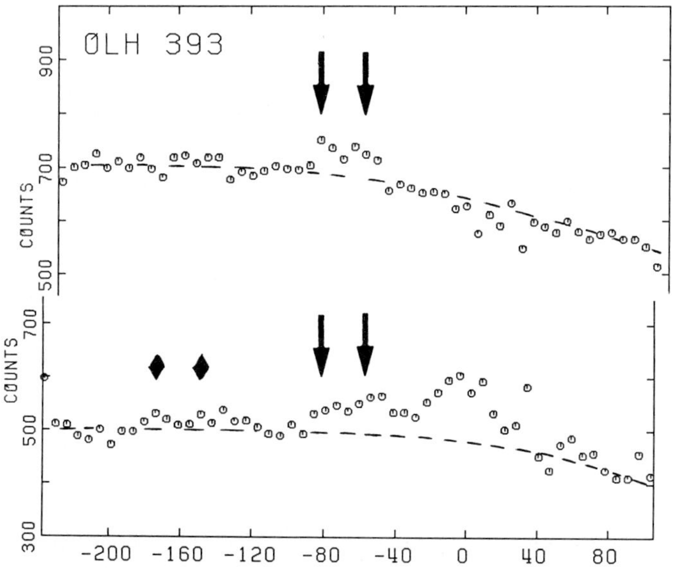

Fig. 4. Two spectra of OLH393, each obtained from the coaddition of 2 scans, with the geocoronal H_{α} and its ghost subtracted, and so aligned as to have common abscissae. Above is shown the spectrum (a) obtained in Jan. 89, while below appears the one (b) secured one month earlier. The baselines of the two spectra (dashes) are not identical because of differing contributions of atmospheric scattering (from stars, zodiacal light and artificial illumination). The emission peaks marked with a filled diamond in (b) are due to OH. The features highlighted by arrows are assigned to galactic H_{α} emission at LSR-velocities of -80 and -50km s^{-1}, to which correspond emission rates of 0.15 and 0.2R.

In this context we should mention that Reynolds (1987) has unsuccessfully searched for H_{α} associated with the cloud OMH323 (177°.0, +67°.0) with V_{LSR} (21cm)=-82km s^{-1}, to the North-East of and very close to OLH393. We do not yet have scans at OMH323, but we believe that the

S/N ratio of the scan shown by Reynolds (Fig. 1, 1987) would not suffice to detect a shifted galactic H_α as weak as that we see in OLH393.

Our selection of the two clouds OLH393 and OLM351 for study was to some extent conditioned by the presence of the O9V type star HD93521 (183°.2, 62°.2) just at the 21cm boundaries of both clouds. The star has an interstellar CaII absorption component at -56km s^{-1} (Münch 1952) and the 21cm profile at the star also has a well defined broad component peaking at -55km s^{-1} (Lockman, Hobbs and Shull, 1986), with no significant signal around -80km s^{-1}. The few scans we have at the position of HD93521 show unmistakably signal of H_α radiation only around -50km s^{-1}. We believe the negative result found by Reynolds and Ogden (1982) for the same star must again be due to insufficient S/N. The conclusion to be drawn from the presence of a CaII absorption component at the velocity of the 21cm and $H\alpha$ emissions is not profound or new, namely, that the height above the galactic plane of the absorbing/emitting mass is smaller than that (1800pc) of the star (Hobbs, Morgan and Ebert, 1982). Nevertheless, it serves as another indicator for the spatial association between the HI and HII gases.

5. **Concluding remarks**

The positional and kinematical correspondance between the H_α and 21cm emissions in the few clouds with V_{LSR}>-100km s^{-1} we have studied demonstrates the possibility of studying the interstellar gas at large heights above the galactic plane through H_α observations. On the basis of the observations available, we cannot draw general conclusions about the state of ionization, relative spatial distribution between HI and HII gas, etc. Rather, we can only draw a program for further studies. The experience we have gained in regard to the observing time and S/N needed to gather reliable information about line emission with ER<0.2R leads us to estimate that in two years of observations at most we could survey the Galactic Polar caps with about 200 samples, quite insufficient for a complete coverage. The need for other dedicated installations, especially in the Southern Hemisphere, to participate in an extensive survey seems to us obviously urgent. The recent announcement by Kutyrev and Reynolds (1989) of the H_α detection in a -300km s^{-1} 21cm cloud in Cetus, also speaks for the important results waiting to be gained by an enlarged group of observers.

The possibility of improving the instrumentation is also open. Most valuable would be to build an instrument with two (or more) independent beams, mounted equatorially, which could be achieved with piezoelectrically servoed etalons.

Acknowledgements We are deeply indebted for the collaboration of Günther Hille in the design, construction and installation of the spectrometer. Our thanks are due to Walter Rauh and Bernhard Grimm for the system software and electronics. The help of Hans Hippelein in the adaptation of his data reduction programs to our needs is also gratefully acknowledged.

References

Giovanelli, R., Verschuur, G.L., Cram, T.R.: 1973, *Astron. Astrophys. Suppl.* **12**, 209
Hobbs, L.M., Morgan, W.W., Ebert, C.E.: 1982, Astrophys. J. **263**, 690
Hulsbosch, A.N.M.: 1968, *Bull. Astr. Inst. Netherlands* **18**, 413
Kutyrev, A., Reynolds, J.R.: 1989, *Bull. Am. Astron. Soc.* (In Press)
Lockman, F.J., Hobbs, L.M., Shull, J.M.: 1986, *Astrophys. J.* **301**, 380
Meng, S.Y., Krauss, S.D.: 1970, *Astron. J.* **75**, 535
Münch, G.: 1952, *Pub. Astr. Soc. Pac.*, **64**, 312
Münch, G., Zirin, H.: 1961, *Astrophys. J.* **133**, 11.
Reynolds, J.R., Roesler, F., Scherb, F.: 1977, *Astrophys. J.* **211**, 215
Reynolds, J.R.: 1980, *Astrophys. J.* **226**, 153
Reynolds, J.R.: 1983, *Astrophys. J.* **268**, 698
Reynolds, J.R.: 1984, *Astrophys. J.* **282**, 191
Reynolds, J.R.: 1987, *Astrophys. J.* **323**, 553
Scherb, F.: 1981, *Astrophys. J.* 243, **644**
Shull, J.M.: 1987, *Interstellar Processes*, pp. 225-243, D.J. Hollenbach and H. Thronson, Edit. (Dordrecht: Reidel)

Discussion:

DANLY: I would like to comment that although your Ca components in HD93521 are seen only out to -55 kms^{-1} the more sensitive ultraviolet absorption lines extended all the way to -80 kms^{-1}. Perhaps the gas giving rise to emission at -80 kms^{-1} has too small a column density to be seen in CaII, but is nonetheless present in front of the star, as seen in CII ot SiII, for example.

MÜNCH: You are quite right. As I said, the absence of a high velocity line in absorption does not necessarity imply that the star is nearest to us than the cloud, because of the small "filling factor" or porosity of the material in CaII. The gas in CII and SiII, however, may have less porosity as the ionization potencials for C and Si are higher than of Ca.

COURTES: The reason of the difficulty to obtain an optical identification of the HVCs is likely due to the fact that the structure of the emission clouds is mainly filamentary (as it is in the diffuse Hα emission of M33 for example, 6 m telescope observations (Courtes et al. 1987). The large aperture etalon technics integrates structures that are maybe hundred times brighter than the mean background, killing the lest hope of contrast. The best should be to use the same interference technics but with a high resolution imagery permited by the new CCD detectors and, of course a minimum sized telescope.

MÜNCH: The surface brightness I am meaning - say, 0.1 Rayleigh - are at least one order of magnitude lower than those you can reach with imaging techniques at small pixel size because of the solid angle factor which can not be gained back by "binning" n pixels because while doing so you increase the r.m.s. readout noise as $\sqrt{n}$. For the low level of detection we have to pay the price of low angular resolution, for a given etalon size. An increase in collecting area indeed would bring higher angular resolution, which could be achieved by using a mosaic of parallel servo-controlled etalons, say 6 to 8 of them, each with 10 cm diameter mounted in front of a fast all reflecting steerable "light bucket" with $\sim$40 to 50 cm aperture.

HIGH-VELOCITY ABSORPTION COMPONENTS TOWARD THE LMC

G. Vladilo[1], P. Molaro[1], S. Monai[2], and M. Centurion[1]

[1] Osservatorio Astronomico di Trieste
[2] Dipartimento di Astronomia - Università di Trieste

ABSTRACT

High-resolution ($\lambda/\delta\lambda \simeq 6 \times 10^4$) spectra of R136 (the central object of 30 Dor) and of 2 LMC supergiants obtained at ESO with the 1.4-m telescope have been analysed in order to study the high-velocity absorption components at $v_{LSR} \simeq +60$ km s^{-1} and $\simeq +130$ km s^{-1} first seen in satellite ultraviolet spectra in the general direction of the LMC. Our data have been integrated with previous literature results of SN1987a and of other 4 LMC supergiants obtained from spectra of similar spectral resolution.

At least 6 (possibly 9) interstellar components are found in the range $+40$ km s^{-1} $\leq v_{LSR} \leq$ $+150$ km s^{-1}, suggesting that the IUE features at $\simeq +60$ km s^{-1} and $\simeq +130$ km s^{-1} are in fact splitted in several components when observed at higher resolution. Two components originally discovered toward SN1987a ($v_{LSR} \simeq +42$ km s^{-1} and $+107$ km s^{-1}), are detected for the first time in the CaII interstellar spectrum of R136. A component at $v_{LSR} \simeq +49$ km s^{-1} shows significant variations both in CaII and in NaI along the different lines of sight. A relatively broad component at $v_{LSR} \simeq +58$ km s^{-1} is present in CaII toward R136, but is generally undetected in NaI in the other lines of sight. Also the components in the range $+100$ km s^{-1} $\leq v_{LSR} \leq +150$ km s^{-1} are detected in CaII, but not in NaI. Two weak absorptions, detected at $+125$ km s^{-1} and $+141$ km s^{-1} toward R136, may be associated with the $\simeq +130$ km s^{-1} component characteristic of the ultraviolet spectra.

The almost perfect velocity coincidence of the $+107$ km s^{-1}, $+42$ km s^{-1}, $+49$ km s^{-1} and $+57$ km s^{-1} components along different lines of sight suggests a common origin of the absorbing gas and a location close to our Galaxy. In spite of the low detection limit in our spectra ($W_\lambda \simeq 5$ mÅ), the lack of detection of NaI is compatible with the presence of foreground gas clouds with hydrogen column density N(HI) $\leq 2 \times 10^{19}$ cm^{-2}, or even greater. The fact that we can detect more easily CaII absorptions than the corresponding NaI absorptions indicates a very low value of CaII depletion.

INTRODUCTION

During the last decade the interstellar medium in the direction of the Large Magellanic Cloud (LMC) has been investigated by means of absorption line spectra in the ultraviolet (Savage and de Boer, 1979, 1981; FWHM $\simeq$ 20-25 km s^{-1}) and in the optical (Blades and Meaburn, 1980; Songaila, 1981; Songaila et al., 1986 and ref.s therein; FWHM $\simeq$ 8-15 km s^{-1}). The IUE data lead to the discovery of absorption features near $+60$ km s^{-1} and $+130$ km s^{-1}, interpreted by Savage and de Boer (1981) as produced in an extended gaseous halo, co-rotating with the disk of the Galaxy. Songaila at al. (1986), from the analysis of CaII spectra of 30 stars in the LMC, suggested that these features may arise inside the LMC.

In the past few years, improvements in the ground-based technology have allowed to perform observations at very high spectral resolution (FWHM $\simeq$ 5 km s^{-1}) and signal-to-noise (S/N $\simeq$ 50) toward some LMC supergiants (Ferlet et al., 1985a; Molaro et al., 1989). The explosion of SN1987a has provided a unique background source toward which an extremely rich interstellar spectrum has been observed in great detail (Vidal-Madjar et al., 1987; FWHM $\simeq$ 3 km s^{-1}; S/N $\geq$ 100). In this communication we present new CaII spectra of R136 (the central object of the 30 Dor nebula) and NaI spectra of two LMC supergiants (Sk-67°44 and Sk-67°201). Our data are compared with previous optical observations towards other lines of sight characterized by a similar quality (FWHM $\simeq$ 5 km s^{-1}; S/N $\simeq$ 50), in order to study the high-velocity components detected by Savage and de Boer in the ultraviolet. The relevant data for the target stars are shown in Table I.

OBSERVATIONS AND DATA REDUCTION

The observations were performed in April 1988 with the Coudè Echelle Spectrograph (CES) fed by the 1.4-m telescope at the European Southern Observatory (ESO), La Silla. The spectra were obtained using a CCD detector with the short camera of the CES at $\lambda/\delta\lambda \simeq 6 \times 10^4$. Due to the fall of sensitivity of the CCD toward the blue spectral region, only the brightest object, R136, was observed in CaII. Sk-67°44 and Sk-67°201 were observed in NaI. A template star (αPic) was used as a reference for correction of the telluric contamination in the NaI spectral region. The spectra were calibrated using standard commands of IHAP, the ESO image processing system. The resulting minimum detectable equivalent width in our spectra (3 σ level) is of about 5 mÅ. After normalization of the spectra to the stellar continuum, the interstellar absorptions were measured using multiple component gaussian fits. The wavelength scale was converted to LSR radial velocity scale ($v_{LSR} = v_{hel} - 15.2$ km s^{-1}), with a final error lower than ± 1.0 km s^{-1}. In Fig.s 1 and 2 we show the CaII spectrum of R136 and the NaI spectrum of Sk-67°44, respectively.

HIGH-VELOCITY ABSORPTION COMPONENTS

In Table II we list a summary of the data in the range +40 km s$^{-1} \leq v_{LSR} \leq$ +150 km s^{-1}. For each component detected in at least one line of sight we give the LSR radial velocity, v_{LSR} (km s^{-1}), and the equivalent width, W_λ (mÅ). Only optical spectra with spectral resolution $\lambda/\delta\lambda \geq 5 \times 10^4$ have been considered. Data from previous literature works are indicated in the Table. The comparison in the NaI region has been performed along 7 lines of sight, while in the CaII region only between R136 and SN1987a.

In the range +40 km s$^{-1} \leq v_{LSR} \leq$ +60 km s^{-1} there are at least 3 interstellar components detected in CaII toward the supernova and in our spectrum of R136. This suggests that the absorption originally discovered in the ultraviolet near +60 km s^{-1} is in fact the result of a blend of different components.

A weak CaII absorption at +42 km s^{-1} is detected in R136, which could be associated to a CaII feature at the same velocity observed in the supernova. This component is not observed in NaI in any line of sight.

The component around +49 km s^{-1} is detected in 4 (possibly 5) of the 7 lines of sight investigated in NaI. The strong CaII feature visible at the same velocity in Sn1987a is almost absent in R136. This component shows strong variations along the different lines of sight, both in NaI and in CaII. An indication for the location of the +49 km s^{-1} component comes from the NaI spectra of Sk-69°220 and Sk-69°221. These stars are LMC members, but do not show any significant NaI absorption at the LMC systemic velocity, suggesting that they are situated in front of the bulk of the LMC gas. Since the +49 km s^{-1} feature is present in their spectra, the corresponding cloud cannot lie behind the bulk of the LMC gas, and should be associated to the halo of our Galaxy, or to a halo surrounding the LMC.

A relatively broad CaII component is seen near $v_{LSR} \simeq$ +58 km s^{-1} toward SN1987a and R136. In the NaI region the same component is marginally detected in the supernova, is present

TABLE I

LMC bright sources observed at high spectral resolution (FWHM $\leq$ 5 km s^{-1}) in the optical

Star	m_V	Spectral type	R.A. (1950)	Decl. (1950)
SN1987a			5^h36^m	-69°19'
R136	9.50	WR	5^h39^m	-69°07'
Sk-67° 44	9.21	A4Ia	5^h06^m	-67°57'
Sk-67°201	9.90	A0Iae	5^h34^m	-67°03'
Sk-69°203	12.29	B0.5	5^h36^m	-69°16'
Sk-69°211	10.36	B8 I	5^h37^m	-69°26'
Sk-69°220	11.40	Bep	5^h37^m	-69°32'
Sk-69°221	10.63	B1Iab	5^h37^m	-69°30'

TABLE II

High-velocity interstellar components toward the targets of Table I

NaI D2

SN1987a[1]		Sk-67°44[2]		Sk-67°201[2]		Sk-69°203[3]		Sk-69°211[3]		Sk-69°220[4]		Sk-69°221[4]	
v_{LSR}	W_λ	v_{LSR}	W_λ	v_{LSR}	W_λ	v_{LSR}	W_λ	v_{LSR}	W_λ	v_{LSR}	W_λ	v_{LSR}	W_λ
49.4	14		< 4		< 5	49.2	4:	48.8	42	46.7	20	47.6	32
≃58	2		< 4		< 5	58.5	14		< 5		< 6		< 6

CaII K

SN1987a[1]		R136[2]	
v_{LSR}	W_λ	v_{LSR}	W_λ
41.9	10[5]	42.5	11
49.4	60[5]		< 5
56.4	20	57:	20
60.2			≤ 5
107.4	24[5]	106.5	12
113.6	4[5]		
		125.4	6
		141.3	6
≃150	18		< 5

[1] Vidal-Madjar et al. (1987)
[2] This work
[3] Molaro et al. (1989)
[4] Ferlet et al. (1985)
[5] Blend solved from the data of Vidal-Madjar et al. (1987)

as a broad, shallow absorption in Sk-69°203, and is not detected in the other lines of sight.

In the range $+100$ km s^{-1} $\leq v_{LSR} \leq +150$ km s^{-1} absorptions are detected in CaII, but not in NaI. One component at $+107$ km s^{-1} is observed both in the supernova and, for the first time, in R136. The perfect velocity coincidence of this component along the two lines of sight suggests a common origin of the absorbing gas. If this gas were located in the LMC this would imply the existence of a coherently moving structure with linear dimensions of about 300 pc, i.e., the projected linear size corresponding to the angular separation between R136 and SN1987a at a distance of $\simeq$ 50 kpc. This argument favours a local origin (within a few kpc) of the absorbing gas. A similar conclusion can be reached for the components at $+42$ km s^{-1}, $+49$ km s^{-1} and $+57$ km s^{-1}, for which a good agreement in radial velocities along the different lines of sight is also found.

Two CaII components at $+125$ km s^{-1} and $+141$ km s^{-1}, close to the detection limit, are visible toward R136 but not toward SN1987a. These features are likely associated to the absorption observed by Savage and de Boer (1981) around $v_{LSR} \simeq +130$ km s^{-1} and again suggest a complex structure of the gas originally detected at lower resolution in the ultraviolet. Finally, a broad, shallow CaII absorption at $+150$ km s^{-1} is present in the supernova, but not in R136.

DISCUSSION AND FUTURE WORK

In all the cases discussed above the CaII components are stronger than the corresponding NaI features. In many cases there are components detected in CaII, but not in NaI. These facts can be used to constrain the HI column density and the calcium to hydrogen ratio of the foreground gas, if present. The lack of detection in NaI implies $N(NaI) < 2.5 \times 10^{10}$ cm^{-2} and this in turn translates in an upper limit $N(HI) < 2 \times 10^{19}$ cm^{-2}, when the ratio $N(NaI)/N(HI) \simeq 1.3 \times 10^{-9}$ given by Ferlet et al. (1985b) for the Galactic gas is used. If in the LMC the sodium is underabundant with respect to the solar value, the upper limit in the HI column density would be even higher. A cloud with hydrogen column density in the range 10^{18} cm^{-2} $< N(HI) \leq 2 \times 10^{19}$ cm^{-2} would escape NaI detection, but not CaII detection, provided CaII depletion is very low. The same cloud would be easily detected in the ultraviolet thanks to the presence of the strongest lines of the most abundant interstellar ions in that spectral region. In fact, Savage and de Boer (1980) estimated $N(HI) > 10^{18}$ cm^{-2} from their analysis of high-velocity clouds. We conclude that the lack of NaI detection, even at the very low level of the data presented in this communication, is not in contraddiction with the presence of foreground halo gas of the type suggested by Savage and de Boer (1981) on the basis of their IUE data. Of course, in those directions where NaI is detected, the cloud density must be higher. The clouds must be very patchy in order to explain the strong variations of the absorptions, even between lines of sight separated only by a few arcminutes (Molaro et al. , 1989).

Recent 21-cm observations reported by de Boer et al. (1989a) indicate that high-velocity gas in the range of radial velocities of interest is present in the general direction of the LMC, but well outside the LMC boundary, in agreement with a Galactic halo origin of the observed clouds. The low HI column densities measured by de Boer et al. (1989a) — $N(HI) \leq 6 \times 10^{18}$ cm^{-2} — are consistent with the above mentioned contraints derived on the basis of our NaI upper limits.

A preliminary study of the kinematics and the stratification of the LMC gas is clearly important in order to address the problem of the origin of the high-velocity absorptions in that direction of the sky. Observations by means of the IUE satellite are inadequate to reach this goal because the S/N and spectral resolution are insufficient to solve the interstellar components. A large number of lines of sight must be investigated in the optical at high spectral resolution and with sufficient S/N in order to detect NaI and CaII absorptions. Perhaps the best way of tackling this problem is to survey selected regions of the Magellanic Clouds and to sample the depth structure of the gas by looking at stars distributed at different distances within each region. We plan to perform

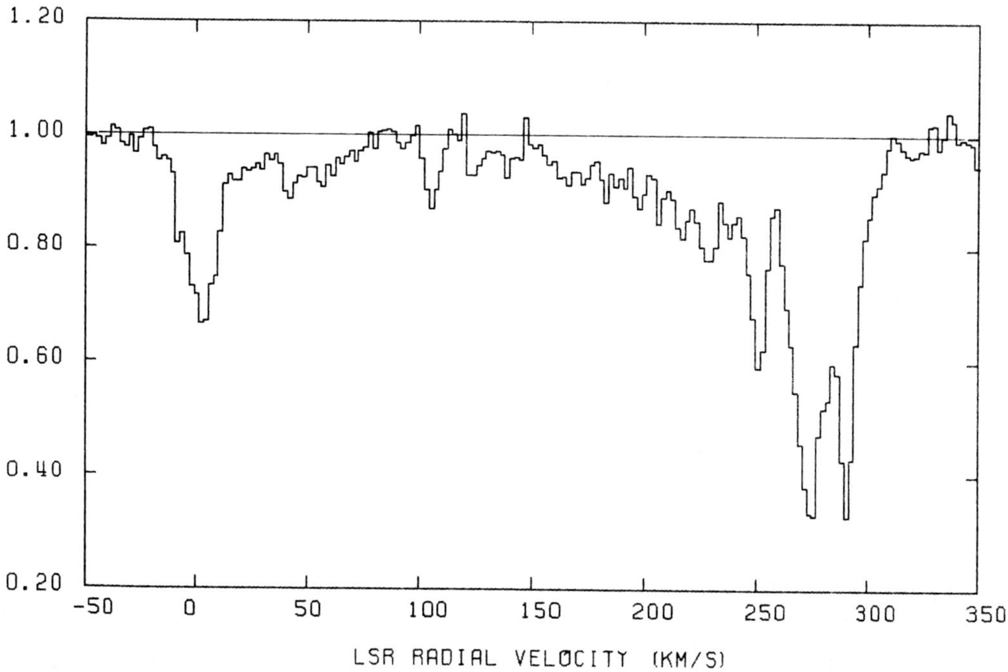

Fig. 1 - CaII K interstellar spectrum toward R136.

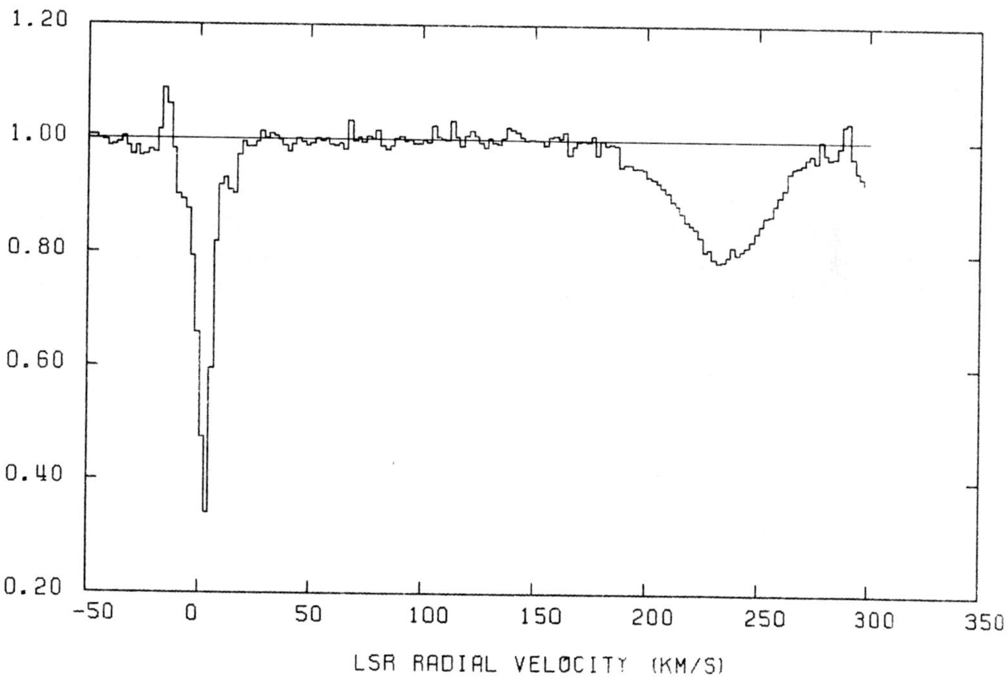

Fig. 2 - NaI D2 spectrum toward Sk-67°44 after correction for telluric contamination. The broad absorption close to $v_{LSR} \simeq 235$ km s^{-1} is due to stellar NaI.

these kind of observations during the next few years in the frame of a coordinated investigation of selected regions in the Magellanic Clouds (de Boer et al., 1989b). The use of a fiber link to connect the 3.6-m ESO telescope with the CES (Avila and D'Odorico, 1988) will allow to obtain CaII and NaI spectra at the required spectral resolution for a significant number of targets in the Magellanic Clouds.

REFERENCES

Avila, G., and D'Odorico,S.: 1988, *Proc.of the Very Large Telescope Symposium*, ed. M.H.Ulrich, (in press).

Blades, J.C., and Meaburn, J.: 1980, *M.N.R.A.S.*, **190**, 59P.

de Boer, K.S., Azzopardi, M., Baschek, B., Dennefeld, M., Israel, F.P., Molaro, P., Seggewiss, W., Spite, F., and Westerlund, B.D.: 1989b, to be pubblished on the *Messenger*.

de Boer, K.S., Morras, R., and Bajaja: 1989a, submitted for pubblication on *Astron. Astrophys.*

Ferlet, R., Dennefeld, M., and Maurice, E.: 1985a, *Astron. Astrophys.*, **152**, 151.

Ferlet, R., Vidal-Madjar, A., and Gry, C.: 1985b, *Astrophys. J.*, **298**, 838.

Molaro, P., Vladilo, G., Avila, G. and D'Odorico, S.: 1989, *Astrophys. J. (Letters)*, **339**, L63.

Savage, B.D., and de Boer, K.S.: 1979, *Astrophys. J. (Letters)*, **230**, L77.

Savage, B.D., and de Boer, K.S.: 1981, *Astrophys. J.*, **243**, 460.

Songaila, A.: 1981, *Astrophys. J.*, **248**, 945.

Songaila, A., Blades, J.C., and Hu, E.M.: 1986, *Astrophys. J.*, **303**, 198.

Vidal-Madjar, A., Andreani, P., Cristiani, S., Ferlet, R., Lanz, T., and Vladilo, G.: 1987, *Astron. Astrophys.*, **177**, L17.

Discussion:

VAN WOERDEN (Comment): From the strong variations in line strength over distances of a few arcminutes, you conclude that the lines around +50 kms^{-1} velocity arise in or near the Large Magellanic Cloud. However, such strong variations over small angular separations do also occur in high-velocity clouds (see following paper by B.P. Wakker) and even in low-velocity gas in the disk (as shown in Westerbok studies by Schwarz, Kalberla and Goss).

de BOER (Comment): As Hugo van Woerden just mentioned, also the high velocity gas is patchy. The problem of association of the -60 and -130 kms^{-1} components with the LMC is resolved. Together with Bajaja and Morras we observed in 21 cm directions outside the boundaries of LMC. We have found the -60 and -130 kms^{-1} gas outside those boundaries, so I claim these clouds belong to the Milky Way and not to the LMC.

DISTANCE AND CHEMICAL COMPOSITION OF HIGH-VELOCITY CLOUDS

Hugo van Woerden, Ulrich J. Schwarz and Bart P. Wakker
Kapteyn Astronomical Institute
Postbus 800, 9700AV Groningen, The Netherlands

Abstract

This contribution reports results of recent absorption-line studies on La Palma and reviews earlier work done elsewhere. CaII has been found in absorption in five HVCs, mostly against extragalactic background sources. The CaII/HI ratios differ strongly, from 0.002 to 0.1 times solar, suggesting different origins of HVCs. Nondetections of CaII absorption in stellar spectra may provide lower limits to HVC distances, or be caused by small-scale structure or a low CaII/HI ratio in the HVC. Proper interpretation requires comparison of CaII/HI ratios in stellar and extragalactic probes, and use of high-resolution HI maps. No solid direct distance determinations of HVCs are available yet. The distance of Complex C is controversial.

1. Introduction

The cloud structure and associated motions of the interstellar medium were discovered by Beals (1936) and Adams (1943, 1949) through high-resolution observations of CaII absorption lines. Guido Münch (1953, 1957) was the first to measure the large-scale distribution of interstellar calcium and sodium clouds, confirming the spiral structure found in the distributions of emission nebulae (Morgan, Sharpless and Osterbrock 1952) and in the 21-cm neutral hydrogen maps (Van de Hulst 1953; Van de Hulst, Muller and Oort 1954).

Following these studies at low galactic latitudes, Münch and Zirin (1961) used the CaII and NaI lines in a first exploration of interstellar clouds at large distances from the plane. A five-year search for HI in the galactic halo led to the discovery of high-velocity clouds (Muller, Oort and Raimond 1963). In view of the wide distribution of HI at velocities up to $|V| > 70\,\mathrm{km\,s^{-1}}$ (Van Woerden, Takakubo and Braes 1962; Blaauw and Tolbert 1966), the Dwingeloo studies defined high-velocity clouds (HVCs) as HI clouds at $|b| > 10°$ and $|V| > 70 - 80\,\mathrm{km\,s^{-1}}$ a velocity range significantly higher than that in earlier optical studies. (Velocities V throughout this paper are relative to the local standard of rest.)

A recent review (Van Woerden, Schwarz and Hulsbosch 1985) summarizes the HVC phenomena; cf. also Mirabel's paper in this volume. After 25 years of study, nature and origin of HVCs remain enigmatic. The key to better understanding lies in the determination of the distances of HVCs; estimates implied by various hypotheses concerning their origin range between 0.1 kpc (nearby supernova shells) and 500 kpc (protogalaxies in the Local Group). The distance also is an important parameter in determinations of physical properties of HVCs: sizes, masses, densities, velocity gradients, and time scales vary as distance to the power $+1, +2, -1, -1$ and $+1$; column densities, kinetic and spin temperatures are distance-independent.

The chemical composition of HVCs is another important item in discussions of their nature and origin. For instance, in a nearby supernova shell a high calcium abundance would be expected, while protogalaxies in the Local Group might be devoid of metals. Also, the composition plays a key role in attempts to determine the distance of HVCs, as will be shown below.

The present paper discusses the results of our recent work on the distance and composition of HVCs, carried out with the Isaac Newton Telescope (La Palma) and the Anglo-Australian Telescope, in collaboration with Aad Hulsbosch (Nijmegen) and Gordon Robertson (AAO). For perspective, we also summarize the results of other workers in this field, and we critically review methods, problems and prospects. We restrict the discussion to HVCs detected in the 21-cm line, leaving aside observations of high-velocity halo gas which have not been correlated with hydrogen HVCs.

2. The method and its problems

Many indirect methods have been followed to estimate the distance of HVCs, but their results differ by several orders of magnitude. Since no physical associations of HVCs with objects at well-known distances have been found, the only direct method available is through measurement of interstellar absorption lines in the spectra of stars lying at known distances projected onto HVCs.

The probe stars must have clean spectra so that the HVC absorption is not confused with or obliterated by stellar absorption lines. Early-type stars would be best, but these are extremely rare at distances >1 kpc in latitudes $|b| > 10^{\circ}$. RR Lyrae stars have been well catalogued, and have reasonable magnitudes: +11 to +16 at distances of 1 to 10 kpc, but their stellar lines are in general too strong, except possibly for a brief interval around maximum brightness. The best probes are blue horizontal-branch (BHB), sdO and sdB stars; unfortunately, searches for these objects have so far been sparse, and the number known to lie within an HVC boundary remains small. Since absorption by HVCs at $|V| \gtrsim 100\,\mathrm{km\,s^{-1}}$ must be distinguished from that by gas at low and intermediate velocities, a velocity resolution of order $30\,\mathrm{km\,s^{-1}}$ is required. At magnitudes >12, spectra of $30\,\mathrm{km\,s^{-1}}$ resolution require a large telescope.

Detection of a CaII or NaI absorption line at the HVC velocity shows the cloud to be in front of the star, and hence places an upper limit on the HVC's distance. Nondetection of an HVC absorption line, however, may have several causes. The HVC may be behind the star; or it may be in front, but contain too few Ca^+ ions on the line of sight to the star. A lack of Ca^+ ions may be due to: 1) a preponderance of neutral or doubly ionized calcium, 2) a low calcium abundance, or 3) a low column density of hydrogen (and other) gas on the stellar sight-line. A test for causes 1) and 2) is provided by observation of an extragalactic probe, preferably one with a bright nucleus, such as a quasar or Seyfert galaxy; combination of the HVC CaII line strength in this spectrum with a 21-cm line profile measured in the same direction supplies a N(CaII)/N(HI) ratio, which one may then reasonably assume to be valid elsewhere in the same HVC. The chance that cause 3) applies is great: as shown elsewhere in this volume (Wakker 1989), HVCs are very rich in small-scale structure, and HI column densities may vary by factors of order 10 on scales of a few arcminutes. High-resolution maps, preferably obtained with a synthesis telescope (Westerbork or VLA), are required to test for this situation, and to provide a more reliable ratio N(CaII)/N(HI) in the direction of the extragalactic probe, for comparison with a — similarly measured! — upper limit derived from nondetection on a star.

The preceding arguments, summarized in Table 1, are of key importance in interpreting nondetections of metal absorptions in HVCs, as well as in deriving metal abundances in HVCs from detections. Only after the effects of causes B 1) 2) 3) have been assessed by the appropriate observations, can a nondetection yield a reliable lower limit to the distance of a HVC.

Table 1 Possible causes of non-detection of an HVC absorption line toward a star.

A. The HVC lies behind the star
B. The sight-line to the star contains too few Ca^+ ions
(1) low ratio N(CaII)/N(Ca) [ionisation equilibrium shifted: Ca predominantly in form of CaIII or CaI]
(2) low ratio N(Ca)/N(H) [i.e. low calcium abundance]
(3) low column density of hydrogen (and other atoms) [small-scale structure in HVC]
Checks for B1 and B2: Measure N(CaII)/N(HI) on extragalactic background source.
Check for B3: High-resolution HI map with synthesis telescope.

3. Calcium abundances in high-velocity clouds

Table 2 summarises the ratios of ionized calcium to neutral hydrogen measured in high-velocity clouds. The velocities v_{lsr} are from single-dish 21-cm observations at Dwingeloo (D), Parkes (P) or Villa Elisa, Argentina (VE). Except for one detection against an RR Lyrae star, only results on extragalactic probes have been listed. In view of the unknown ionisation equilibrium, the ratios N(CaII)/N(HI) have not been corrected to abundances N(Ca)/N(H). For comparison, the table also gives the ratios measured in various clouds at low and intermediate velocities, and the calcium abundance measured in the Sun.

So far, CaII detections have been reported for 5 different HVCs, all lying far apart on the sky. [In the spectrum of stars in the Magellanic Clouds many absorption lines at velocities between 0 and $+310\,\mathrm{km\,s^{-1}}$ have been observed. The relation of these lines with galactic HVCs remains uncertain, however; hence we have not listed these detections.] Clearly, the 5 HVCs detected do not consist of primordial gas, but are enriched in heavy elements. In several HVCs, the CaII/HI ratio is much higher than in low — or intermediate — velocity gas, showing that in these HVCs the calcium is not depleted by being locked up in dust grains. Alternatively, the dust may have been affected by the acceleration process that caused the HVCs, releasing the metals.

Table 2 Calcium II absorptions measured in high-velocity clouds.

Name of HVC	Probe Name, Type	l	b	v_{lsr} $\mathrm{km\,s^{-1}}$	N_{CaII}/N_{HI} 10^{-9}	Reference
Magell. Stream	Fairall 9, Sy	294	−58	+195 P	11	Songaila 1981
287+22+240	NGC 3783, Sy	287.5	+22.5	+240 VE	4.6	West *et al.* 1985
237+17+105	0837-120, Q	237.2	+17.4	+105 P	160	Robertson *et al.* 1989
						Van Woerden *et al.* 1986
Complex C	BT Dra, RR	99.4	+51.2	−154 D	120	Songaila *et al.* 1988
				−94 D	50	Songaila *et al.* 1988
String A	MK 106, Q	161.1	+42.9	−156 D	30	Schwarz *et al.* 1989
Various	, Q, Sy				?	Pettini 1986
For comparison:						
low-velocity gas	stars			<20	0.5-6	Pottasch 1985
IVCs	stars			20–60	6–30	Albert 1983
Sun					2200	Pottasch 1985

The CaII/HI ratio varies strongly from cloud to cloud. This indicates that different HVCs may have different origins, or different ionization conditions, and that it is important to measure the ratio in various clouds; for many clouds, no information is available yet. The variation in the CaII/HI ratio further indicates that, in interpreting CaII nondetections in terms of HVC distances, one cannot use CaII/HI ratios measured in other HVCs; it is necessary to compare upper limits obtained on stellar probes with ratios measured on extragalactic probes behind the same HVC.

Does the CaII/HI ratio vary within one HVC? At present we have no unequivocal answer to this question. In Complex C, there is apparent disagreement between the ratio measured against a stellar probe and the upper limits found on two more distant probes; however, as discussed in section 4, this disagreement may not be significant.

In view of the strong small-scale structure present in many HVCs (cf. the papers by Wakker and by Danly in this volume), the CaII/HI ratios listed in Table 2 must be considered with caution. Observations at high angular resolution are required to obtain more reliable ratios.

4. Distances of high-velocity clouds

From the above it will be clear that, in order to have any handle on the distances of HVCs, we must have either 1) a CaII (or any other suitable metal ion) absorption detected against a star; or 2) a CaII (etc.) detection against an extragalactic probe, accompanied by a nondetection against a star at a significantly lower level of CaII/HI. The first three HVCs listed in Table 2 do not fulfil this requirement: they show absorption against extragalactic probes, but no useful measurements on stellar probes have been obtained. Hence, for these three HVCs the only distance information is an upper limit of many megaparsecs. For the other two HVCs, String A and Complex C, both stellar and extragalactic probes have been measured. We discuss these objects in some detail.

4.1 String A

For this object, the first HVC discovered, a distance of a few kpc has often been assumed (e.g. Oort and Hulsbosch 1978). Pettini (1986, private communication), on the basis of a nondetection of CaII against a BHB star, derived a lower limit of 15 kpc. However, while a 1973 Green Bank map at ~10 arcmin resolution had this star lying within the boundary of Cloud A IV, our recent Westerbork map (Wakker and Schwarz 1989) shows no measurable HI in the direction of the BHB star. Hence, the nondetection of CaII is not significant and the 15 kpc distance limit spurious. Songaila *et al.* (1988), from a comparison of several stellar spectra, have tentatively suggested a distance of 1-2 kpc for String A, but the evidence is not strong.

Figure 1 illustrates two spectra obtained by us with the Isaac Newton Telescope on La Palma (Schwarz *et al.* 1989). The quasar MK 106, projected on cloud A VI, shows CaII absorption near the HVC velocity of $-156\,\mathrm{km\,s^{-1}}$; the derived ratio N(CaII)/N(HI) is $(30 \pm 9)\times10^{-9}$. The star 291-257, projected on cloud AIII, shows no absorption by the HVC; the upper limit on the CaII/HI ratio is 7×10^{-9}. Assuming this star to be on the BHB at absolute magnitude $\sim+1$, we estimate its distance as 4 kpc. The nondetection of the HVC in the stellar spectrum is confirmed by an independent observation by Brown *et al.* (1989) who, from detailed analysis, find a spectral type B 1V for the star, and a distance of 31 kpc!

At face value, these results would indicate that cloud A III in chain A is beyond 4, or even 31, kpc distance. The latter limit especially is surprising, since chain A is rich in small-scale structure, suggestive of shocks and instabilities, and hence possibly of some kind of interaction. However, the lower distance limits found are no more than tentative, for several reasons. 1) The HVC absorption in the quasar spectrum is only just above 3σ and requires confirmation. 2) The small-scale structure in various clouds of String A is particularly strong; hence high-resolution HI maps are required before the CaII/HI ratios

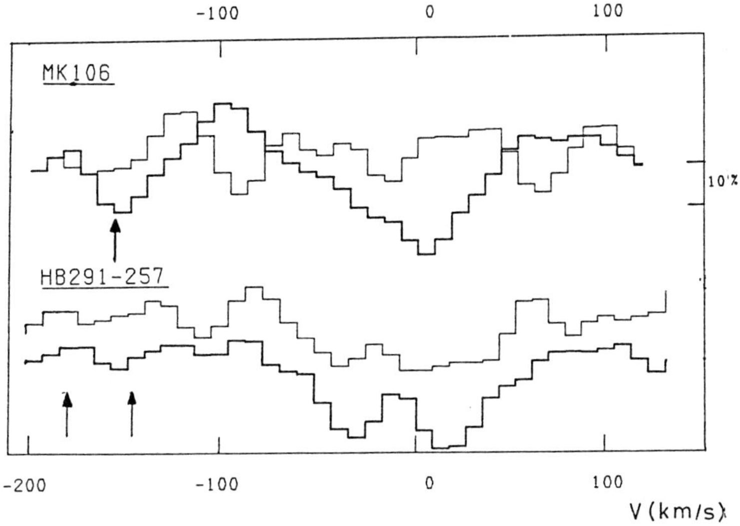

Figure 1. Spectrum portions around the CaII K line (fat line) and H line (thin line), measured at the Isaac Newton Telescope on La Palma, for 2 probes projected on string A. Spectra have been smoothed to obtain optimum signal/noise ratios. A 10% absorption scale is indicated. Arrows show HVC velocities measured at the probe positions. MK 106 is a quasar, 291-257 a blue horizontal-branch (or main-sequence) star at 4 (or 31) kpc distance. The star shows no HVC absorption, the quasar does. Absorption at low and intermediate velocities is seen in both spectra. Velocities are relative to the local standard of rest.

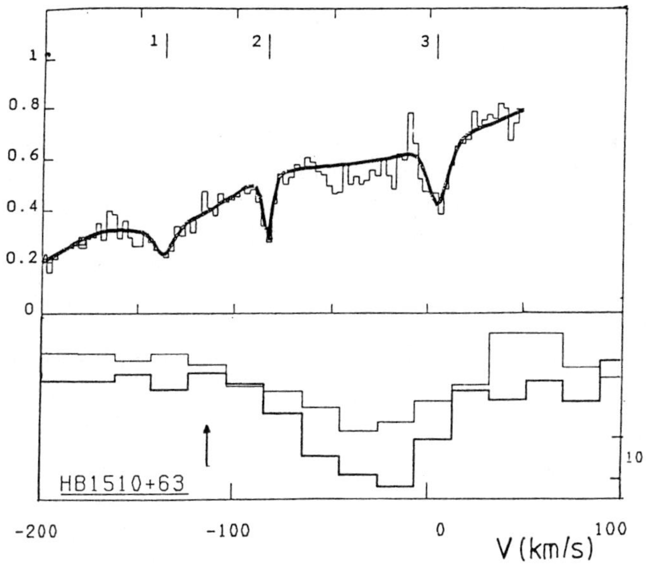

Figure 2. Top: Spectrum of the RR Lyrae star BT Dra at CaII K (histogram) and model fit (thick line) obtained by Songaila et al. (1988). Bottom: Spectrum of the horizontal-branch star 1510+63 at K (fat line) and H (thin line), observed with the Isaac Newton Telescope on La Palma (Schwarz et al. 1989). The stars lie 4° apart, both projected on Complex C. Arrows indicate HVC velocities.

can be trusted. We have observed both fields at Westerbork and the analysis is in progress. 3) The quasar and the star are 13° apart and lie projected on different clouds in String A, which is highly fragmented; the CaII/HI ratio may well vary between these clouds, and it would be desirable to measure more extragalactic and stellar probes, closer together.

We conclude that more detailed observations, both optical and at 21 cm, are required before the distance of String A can be pinned down.

4.2 Complex C

This is the largest HVC, measuring about 70° ×20°; it may be connected with the "Outer Arm" of the Galaxy, which lies in a strongly warped portion of the disk.

Songaila, Cowie and Weaver (1988) have found absorption at $V = -136\,\mathrm{km\,s^{-1}}$ in the RR Lyrae star BT Dra at 2.1 kpc distance, a possible absorption at a similar velocity in SW Dra at 1.2 kpc, and no absorption in several stars at distances $\lesssim 0.8$ kpc. They conclude that Complex C must be at 1–2 kpc distance. [Songaila *et al.* (1985, 1988) also report CaII and NaI absorption at $V = -83\,\mathrm{km\,s^{-1}}$ in BT Dra. However, it is not clear whether gas at this velocity belongs to Complex C, or may rather be connected with the large complex at intermediate negative velocities ($-90 < V < -30\,\mathrm{km\,s^{-1}}$) described by Wesselius and Fejes (1973).]

Together with the Songaila *et al.* (1988) spectrum of BT Dra, Figure 2 shows the spectrum obtained by us with the Isaac Newton Telescope on the BHB star 1510+63 at 3.7 kpc distance, 5° from BT Dra. Combining the 3σ upper limit of 50 mÅ with N(HI) = $12\times10^{18}\,\mathrm{cm}^{-2}$ obtained from special measurements at Dwingeloo by R.J. Habing, we find N(CaII)/N(HI) $\leq 50\times10^{-9}$, significantly lower than the 120×10^{-9} derived from the K-line detection by Songaila *et al.*. A noisier spectrum on the quasar 0.087, at 8° from BT Dra, also gives a significantly lower CaII/HI ratio.

Our upper limits for the CaII/HI ratio on a more distant star and on an extragalactic probe would appear inconsistent with the detection in BT Dra reported by Songaila *et al.* (1988). Contamination with stellar lines in the spectrum of the RR Lyr-type star might be a possible cause of the discrepancy. However, other possible causes include noise in the spectra, small-scale HI structure, and true variations in the CaII/HI ratio. Again, further observations will be required to settle the issue and firmly measure the distance of Complex C.

5. Concluding remarks

In several high-velocity clouds, CaII (and sometimes also NaI) has now reliably been detected. The ratios N(CaII)/N(HI) vary over a wide range, suggesting that different HVCs may have different origins.

Almost all CaII detections in HVCs have been made against extragalactic probes. This fact might suggest that most HVCs are at great distances (see also Wakker 1989), but that conclusion would be premature: firm distance limits based on stellar observations are lacking. As a consequence, the nature and origin of HVCs and the values of their physical properties remain open questions.

Progress in this field will require further optical spectra of high quality, for several probes per HVC. Detections and non-detections of HVC absorption are both useful, for distance as well as chemical composition of the clouds. In both cases, HI synthesis maps are required to obtain reliable abundances and distance estimates. These quantities are of great importance to theories of the origin of HVCs and their role in the evolution of our Galaxy. Finally, searches for suitable probes - blue horizontal branch stars, post AGB stars, quasars and seyferts - are badly needed.

Acknowledgements The Isaac Newton Telescope at the Roque de los Muchachos observatory is operated by the Royal Greenwich Observatory and the Instituto de Astrofísica de Canarias. The Dwingeloo and Westerbork Radio Observatories are operated by the Netherlands Foundation for Research in Astronomy (ASTRON/NFRA) with financial support from NWO. Wakker has been supported by ASTRON. We thank Rolf Jan Habing for his share in the Dwingeloo observations.

References

Adams W.S., 1943, *Astrophys. J.* **97**, 105

Adams W.S., 1949, *Astrophys. J.* **109**, 354

Albert C.E. 1983, *Astrophys. J.* **272**, 509

Beals C.S., 1936, *Monthly Notices Roy. Astr. Soc.* **96**, 661

Blaauw A., Tolbert C.R., 1966, *Bull. Astr. Inst. Neth.* **18**, 405

Brown P.J.F., Dufton P.L., Keenan F.P., Boksenberg A., King D.L., Pettini M., 1989, *Astrophys. J.* **339**, 397

van de Hulst H.C., 1953, Observatory **73**, 129

van de Hulst H.C., Muller C.A., Oort J.H., 1954, *Bull. Astr. Inst. Neth.* **12**, 177

Morgan W.W., Sharpless S., Osterbrock D.S., 1952, *Astron. J.* **57**, 3

Münch G., 1953, *Publ. Astron. Soc. Pacific* **65**, 179

Münch G., 1957, *Astrophys. J.* **125**, 42

Münch G., and Zirin 1961, *Astrophys. J.* **133**, 11

Muller C.A., Oort J.H., Raimond E., 1963, C.R. Acad. Sci. Paris **257**, 166

Oort J.H. and Hulsbosch A.N.M. 1978, in "Astronomical papers dedicated to Bengt Strömgren" (eds. A. Reiz and T. Andersen), Copenhagen Univ. Observatory, p. 409

Pettini M. 1986, private communication

Pottasch S.R., 1985, in "The Milky Way Galaxy", (eds. H. van Woerden, R.J. Allen,. W.B. Burton), IAU Symposium **106**, p.575

Robertson J.G., Schwarz U.J., Van Woerden H., Murray J.D., Morton D.C., Hulsbosch A.N.M., 1989, in preparation

Schwarz U.J., Wakker B.P., Van Woerden H. 1989, in preparation

Songaila A., York D.G., Cowie L.L., Blades J.C, 1985, *Astrophys. J.* **293**, L15

Songaila A., Cowie L.L., Weaver H.F., 1988, *Astrophys. J.* **329**, 588

Songaila A., 1981, *Astrophys. J.* **243**, L19

Van Woerden H., Takakubo K., Braes L.L.E. 1962, *Bull. Astr. Inst. Neth.* **16**, 321

Van Woerden H., Schwarz U.J., Hulsbosch A.N.M., 1985, in "The Milky Way Galaxy" (eds. H. van Woerden, R.J. Allen, W.B. Burton), IAU Symposium **106**, 387

Van Woerden, H., Schwarz U.J., Robertson J.G., Hulsbosch A.N.M. 1986, in "Gaseous Haloes of Galaxies" (eds. J.N.Bregman & F.J.Lockman), Workshop no. 12, NRAO, p126

Wakker B.P., 1989, this volume

Wakker B.P., Schwarz, U.J., 1989 in preparation

West K.A., Pettini M., Penston M.V., Blades J.C., Morton D.C., 1985, *Monthly Notices Roy. Astr. Soc.* **215**, 482

Wesselius P.R., Fejes I., 1973, *Astron. Astrophys.* **24**, 15

Discussion:

MÜNCH (Comment): In my opinion, to the two reasons you listed for an optical IS line, expected from a foreground cloud, not to appear on the spectrum of a distant star, the porosity of the HI medium should be added.

INFLOW OF NEUTRAL GAS TOWARD THE GALACTIC DISK

I.F. Mirabel
Department of Physics. University of Puerto Rico. Box AT. San Juan. Puerto Rico 00931.

Summary: I highlight the evidence for inflow of neutral gas toward the disk of the Galaxy. The Milky Way is accreting 0.2-0.5 $M_\odot$/yr of extragalactic atomic hydrogen at very high velocities. The interaction of infalling clouds with galactic material produces large-scale disturbances in the interstellar medium. Although the injection of energy into the galactic disk by infalling neutral gas is only 1% of the energy from supernovae, the impinging of high velocity neutral gas may be a relatively important source of energy in localized regions of the outer Galaxy.
In the solar neighborhood the downfall rate of HI at intermediate velocities is 2.72×10^{-8} $z(kpc)^{-1}$ $M_\odot$ pc^{-2} yr^{-1}, which if representative of the whole galactic disk, is at least 10 times more massive than the estimated accretion rate of extragalactic HI at very high velocities. This implies that most of the neutral gas that is infalling in the solar vicinity has originated in the galactic disk. It is concluded that in the Milky Way galaxy there is a moderate inflow of extragalactic neutral gas on top of a more intensive disk-halo circulation.

I. INTRODUCTION

The infall of gas toward the galactic disk is of outmost importance for our understanding of the structure and dynamics of the interstellar medium, and here I concentrate in the following questions: Is there an inflow of neutral gas toward the Galaxy?. What is the mass inflow rate?. What is the energy injected into the disk by infall of high velocity neutral gas and how it compares with other sources of energy?. What is the downfall rate of HI at intermediate velocities in the solar neighborhood, and how it compares with the accretion of neutral gas at very high velocities?

The studies on high velocity clouds of neutral hydrogen (HVC's) could provide clues to answer these questions. However, as pointed out by several authors (see the review by van Woerden, Schwarz, and Hulsbosch 1985, and references therein) the distance to these objects is unknown. Furthermore, it is not known how the dynamics of the system of high velocity clouds are related to the rotation of the galactic disk.

Despite these difficulties there are several indications that a subsample of HVC's is of extragalactic origin. Mathewson, Cleary, and Murray (1974) discovered a complex of clouds (Magellanic Stream) that extends more than 180° in the sky and is physically connected to the Magellanic galaxies. Besides, in the galactic anticenter there is a complex of clouds approaching the Milky Way with very high velocities (VHVC's; [V]>140 km/s). This gas must be extragalactic, since using current mass models for the Galaxy (e.g.: Clutton-Brock, Innanen, and Papp, 1977; Rohlfs and Kreitschmann, 1981), I find that for these clouds to be accelerated to their extreme galactocentric velocities of 250 km/s, they should have started their descent from distances greater than 30 kpc from the galactic plane. At last, there is a galactic hemispheric asymmetry in the distribution of VHVC's, since most VHVC's are in the Southern Hemisphere. Giovanelli (1981) has argued that this asymmetry invalidates the possibility of a galactic fountain origin for most of the VHVC's.

II. HIGH VELOCITY INFLOW OF NEUTRAL GAS.

Because the line-of-sight component of the solar motion in the galactic center and anticenter directions is small, surveys in these two regions of the sky offer favorable conditions for the detection and unbaised study of gas with high anomalous motions. In the galactic anticenter, Hulsbosch and Wakker (1988, and references therein) find a population of clouds having negative velocities, up to -250 km/s with respect to the galactic standard of rest. In Figure 1a I plotted the distribution of clouds with [V]>140 km/s using data from Hulsbosch and Wakker (1988). More than 99% of the column density of the VHVC's in this region of the sky has negative velocities.
Figure 1b shows the distribution of clouds in the opposite direction of the sky, namely, in the inner Galaxy, using the results of Mirabel and Morras (1984) complemented by data recently reported by Hulsbosch and Wakker (1988) and Bajaja et al. (1985).

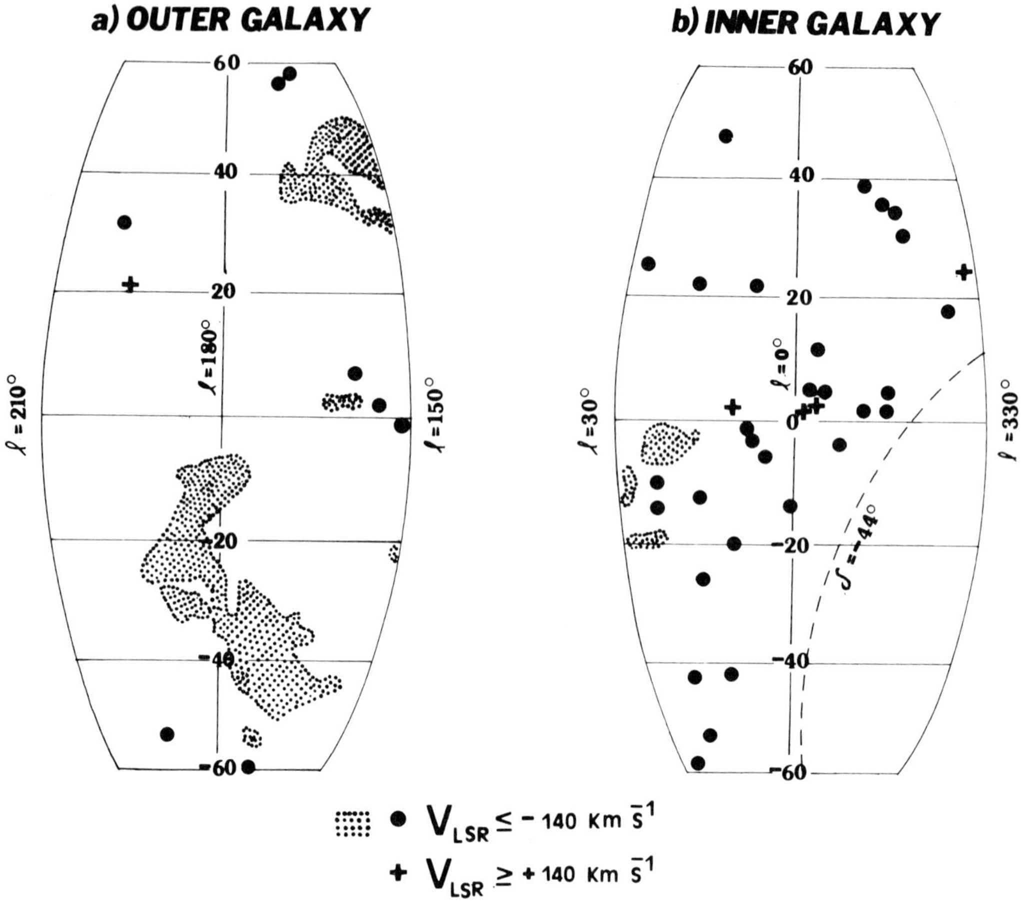

Figure 1: Distribution of atomic hydrogen with very high velocities in the outer and inner Galaxy. The preponderance of inward motions in these two regions of the sky is the best evidence for a high velocity inflow of gas toward the Milky Way. Data are from Mirabel and Morras (1984), Bajaja et al. (1985), and Hulsbosch and Wakker (1988).

The region south of declination -44°, the southern limit of the survey conducted from Green Bank by Mirabel and Morras (1984), was surveyed by Bajaja et al. (1985) using a less sensitive receiver system. Figure 1b shows that in the inner Galaxy there is also a clear preponderance of individual clouds (84%) and overall column densities (95%) having very high negative velocities.

The striking preponderance of clouds with the same extreme inward motions in the center and anticenter regions of the sky shown in Figure 1, is the strongest evidence for a high velocity inflow of neutral gas toward the Galaxy. In the most simple infall scenario it is likely that the clouds will move preferencially toward the central regions of the Galaxy. Hence, for an observer in the solar system, most of the infalling gas in the anticenter should have radial motions toward the observer. In the direction of the inner galaxy, there must be a preponderance of clouds with negative velocities. However, some infalling objects located between the Sun and the galactic center will be receeding from the observer. This is what shows Figure 1 where 99% of the gas with very high velocities in the anticenter has negative velocities, whereas in the inner galaxy about 5% of the gas is detected at positive radial velocities.

In the context of the inflow hypothesis, most clouds with very high negative velocities observed in the direction of the inner Galaxy are likely to be at distances greater than 10 kpc from the Sun, beyond the galactic center. Hence, for a given physical size, the clouds detected in the inner Galaxy should appear smaller than their counterparts in the anticenter. Maps of several clouds carried out by Mirabel and Morras (1984) confirm this prediction, since clouds with high negative velocities in the inner Galaxy typically have angular sizes that are about 5 times smaller than their counterparts in the anticenter.

The computation of the net influx of gas toward the Milky Way depends on several unknown factors, such as the distance to the clouds, the motion of the objects in the plane of the sky, and the actual distribution of infalling gas over the whole sky. To estimate the influx of neutral gas toward the Galaxy I have used the results summarized in Figure 1. In other words, I have used data from the 30% of the sky that offer the most favorable conditions for an unbiased detection of neutral gas with truly large anomalous motions ([V]>140 km/s).

The HI with V_{lsr} <-140 km/s in the 7200 squares degrees of the anticenter shown in Figure 1a has a mass $M(HI) = 2.63 \times 10^4 D^2(kpc) M_\odot$, after correcting for the partial sky coverage of the survey by Hulsbosch and Wakker (1988). From the search by Kulkarni and Mathieu (1986) for interstellar optical absorption lines in the clouds of the anticenter we can assume that the very high velocity gas in this region is at distances greater than 2 kpc from the Sun. An upper distance of 10 kpc seems reasonable, given the galactocentric velocities reached by the infalling gas. Hence, assuming that the gas is at a mean distance of 6 kpc from the Sun, and that it is moving toward the inner Galaxy with a space velocity of 250 km/s, I obtain an infall rate of 0.03 $M_{\odot}$/yr for the anticenter, which extrapolated to the whole sky implies 0.18 $M_{\odot}$/yr.

The HI with V_{lsr} <-140 km/s in the direction of the inner Galaxy that is shown in Figure 1b has a mass M(HI) = 7.23 x 10^3 D^2(kpc) $M_{\odot}$, after correcting for the partial sky coverage of the surveys. Using the current mass models for the Galaxy, I infer that the clouds with such high infall velocities shoud be located beyond the galactic center at distances from the Sun of 10-30 kpc. Assuming that the VHVC's are at a mean distance of 20 kpc from the Sun, and a space velocity toward the galactic center region of 250 km/s, I obtain an infall rate of 0.04 $M_{\odot}$/yr in that region, which extrapolated to the whole sky implies 0.22 $M_{\odot}$/yr.

The results from VHVC's can be summarized as follow: from unbiased surveys that cover 30% of the sky we conclude that 0.20 $M_{\odot}$/yr of neutral gas is being acreeted by the Milky Way at very high velocities.

III. HVC's-MILKY WAY INTERACTIONS: A "SPLATTER" IN THE OUTER GALAXY.

A natural consequence of the high velocity infall of neutral gas must be the interaction with the interstellar gas of the Milky Way. Oort (1970) proposed that most high velocity streams of HI clouds with more moderate velocities ([V_{lsr}]<140 km/s) are galactic clouds which have been set in motion by gas of extragalactic origin that has penetrated into the galactic halo. However, no direct observational evidence for the physical interaction between extragalactic VHVCs and galactic material had been found until recently.

Mirabel (1982) and Kulkarni, Dickey and Heiles (1985) found evidences for such an interaction in the anticenter region of the sky. Figure 2 shows the tip of the anticenter complex of very high velocity clouds together with a remarkable supershell of HI at V = - 71 km/s identified by Heiles (1984). Kulkarni, Dickey and Heiles (1985) find that this supershell consists of post-shock gas that is cooling down. They give several reasons to interpret this supershell as the result of a "cosmic

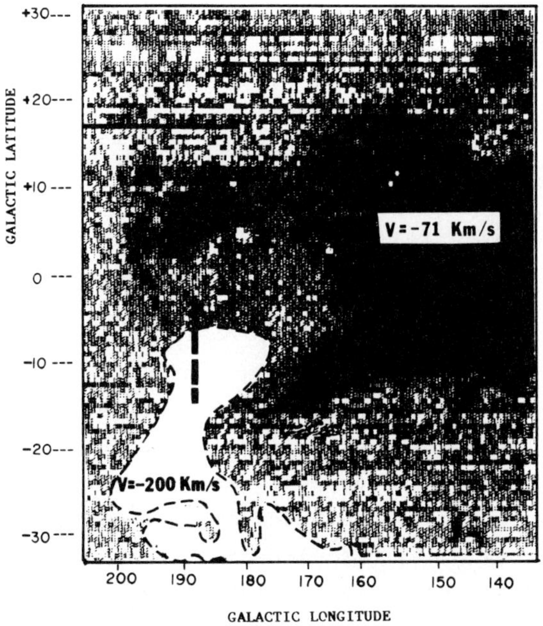

Figure 2: Colliding very high velocity cloud ACI (v = - 200 km/s) represented in white, superimposed on a grey-scale representation of the "anticenter shell" centered at v = -71 km/s (Heiles, 1984). The broken lines represent the extent of the position-velocity map at the edge of ACI that is presented in Figure 3.

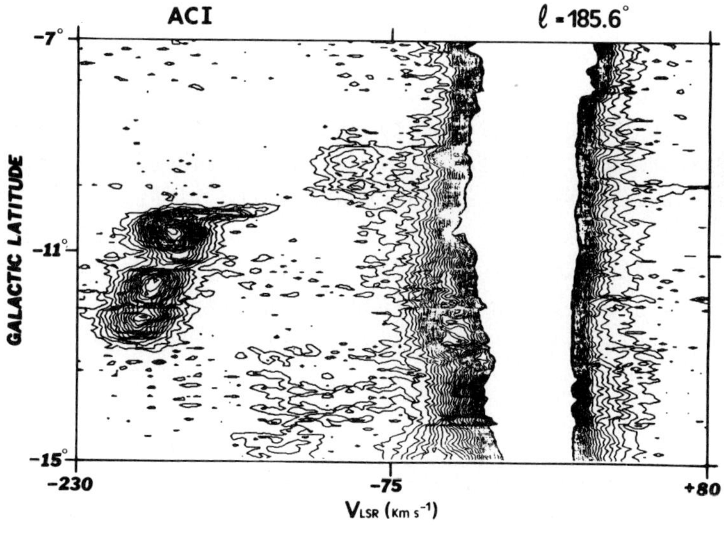

Figure 3: High resolution position-velocity map by Morras and Mirabel (1989) along the broken lines shown in Figure 2. The interaction of ACI with the Milky Way is revealed by the steep edge and decceleration of the very high velocity cloud at the same positions where are found kinematic disturbances spanning over 200 km/s.

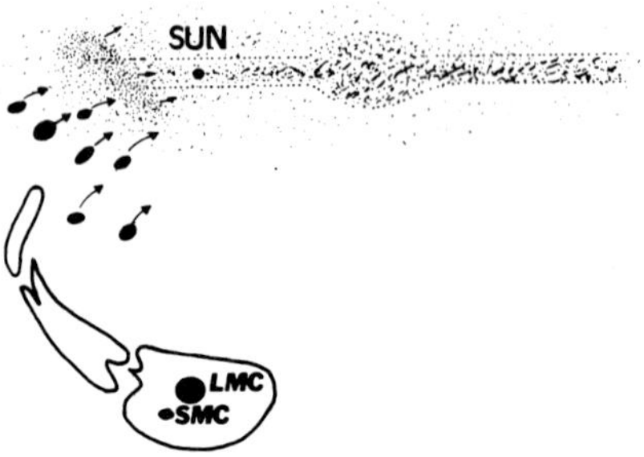

Figure 4: Sketch of the model proposed by Mirabel (1981) for the impinging stream of very high velocity clouds producing the "galactic splatter" in the anticenter of the Galaxy.

splatter" produced by impinging HVC's, rather than by supernovae explosions.

The complementary distribution on the sky between the gas at -200 km/s and the gas at -71 km/s shown in Figure 2, supports that hipothesis. This type of distribution is expected as a result of an interaction, since at the position where the extragalactic clouds have been slowed down, the galactic halo gas has already been accelerated to -71 km/s. At greater distances, where the gas infalling with - 200 km/s has not yet reached the Milky Way gas, no prominent features at -71 km/s have been produced.

The anticenter complex of very high velocity clouds extends more than 50° in the southern galactic sky, but strikingly, it does not "cross" toward the northern hemisphere, suggesting a break in the motion of the -200 km/s stream at its closer angular distance from the galactic disk. Using the Arecibo telescope, Mirabel (1982) studied the region where the VHV gas breaks down, and found direct evidence for the physical connection of gas with velocities in the range of - 200 km/s to + 20 km/s. Figure 3 shows a high sensitivity, position-velocity map by Morras and Mirabel (1989) along the broken line in Figure 2, which cuts the northern edge of the very high velocity cloud ACI. The jump in velocity and the "shock" structure at the edge of the ACI, occur at the same position of disturbances observed in the gas with lower velocities, and this argues for an interaction with high-z galactic material.

IV. THE INFLOW OF MASS AND THE ENERGY BUDGET IN THE GALACTIC DISK.

Mirabel (1981, 1982) has proposed that the anticenter complex of VHVC's is a fragment of the Magellanic Stream that is impinging into the Milky Way. As sketched in Figure 4, the northern clouds of the anticenter complex are falling obliquely toward the galactic disk, and have already reached the high-z gas of the Milky Way, producing the "galactic splatter" studied by Kulkarni, Dickey and Heiles (1985).

The remarkable anticenter shell noted by Heiles (1984) has an energy of 10^{53} erg (Kulkarni, Dickey, and Heiles 1985), which is difficult to explain on the basis of supernovae explosions. On the contrary, this energy is comparable to the kinetic energy of the VHVC's in that region of the sky. If the VHVC "ACI" in Figures 2 and 3 is at distance of 5 kpc from the Sun (Kulkarni and Mathieu, 1986), its mass would be 4 x 10^{5} $M_{\odot}$, with a kinetic energy of 2.5 x 10^{53} erg, enough to produce the most energetic HI supershells found by Heiles (1984).

The Magellanic Stream has a total mass of 2.8 x 10^{5} D^{2} (kpc) including the HI corona in which the Magellanic galaxies are embedded. The long streamer with negative velocities that is approaching the Milky Way has one third of the column density of the whole Magellanic Stream (Mathewson, Cleary and Murray, 1974), and assuming that it is at a mean distance of 40 kpc, one finds 1.5 x 10^{8} $M_{\odot}$ for the northern streamer. The ballistic downfall of the northern streamer from a distance of 40 kpc will take 3 x 10^{8} years, producing a mass inflow of 0.5 $M_{\odot}$/yr. This is comparabable with the present infall rate infered from the VHVCs found in the region of the sky shown in Figure 1.

Taking the Galaxy as a whole, the energy injected by HVC's will be about 2 x 10^{47} erg/yr, only 1% the energy contributed by supernovae of 10^{51} erg exploding at a rate of 1/50 yr^{-1}. However, outside the solar circle, the energy due to the infall rate of neutral gas may be relatively important, since the supernovae rate decreases with galactocentric distance. As Heiles (1989) stated, comparativelly few molecular gas and massive stars populate the outer regions of the Galaxy, and in that region, supernovae explosions may not dominate completely the energy balance of the interstellar medium. It remains an open question if the downfall energy of HVC's dominates over supernovae in the outskirts of the Galaxy.

Franco et al. (1988) followed Tenorio-Tagle's (1980) idea and propose a model in which the molecular cloud complexes in Orion and Monoceros are the result of interaction of HVC's with Milky Way material. In their model, cloud-galaxy collisions are able to generate massive shocked layers and self-gravity can then provide the conditions for the transformation of these layers into molecular clouds.

Ostriker has pointed out at several meetings that if the gas is stopped abruptly when hitting the disk, this energy would have to be radiated in the x-rays, violating the observed background. The temperatures of the collisionally heated gas would be in the range of 10^6-10^7 K for velocities in the range of 100-300 km/s, which would correspond to radiation in the range of 0.25-2.5 kev. In this respect, I wish to point out the following: the decceleration in the upper halo of clouds with densities smaller than 0.1 HI cm^{-3} must be gradual, and the kinetic energy converted into thermal energy would be radiated in the soft x-rays as the clouds cool down. The mean free path for photons radiated by a plasma of 5 x 10^5 K is 10^{19} HI/cm^{-2} (Bloch et al. 1986), implying that they must be completely absorbed after travelling 3 pc inside a typical HI cloud of the disk with a density of 1 HI atom cm^{-3}. The anticenter complex of HVC's is likely to be at distances 3 orders of magnitude larger than the mean free path of the soft x-rays, and therefore, there is no chance for detecting the collisionally ionized gas. Furthermore, Stern and Bowyer (1979) find a high latitude excess in the 0.1 kev flux whose origin is still unclear. This plasma could partially be due to gas heated as a result of the downfall of neutral gas.

V. DOWNFALL OF ATOMIC HYDROGEN IN THE SOLAR NEIGHBORHOOD.

The galactic polar caps offer favorable conditions for the study of motions perpendicular to the galactic plane. 25 years ago, Dieter (1964) proposed that a large fraction of the HI detected at high latitudes could be condensations descending from the halo. The downflow of atomic hydrogen in the solar neighborhood was clearly established by the work of Wesselius and Fejes (1973) and Weaver (1974). The best illustration of the downfall phenomenon has been presented in Figure 9 of Weaver (1974). Despite the large amounts of observations, to my knowledge, there has been no quantitative estimation of the mass infall rate.

To make a quantitative estimate of the inflow of neutral gas in the solar neighborhood, Mirabel and Morras (1989) have recently used the 21 cm survey by Stark et al. (1983), which was conducted with the horn antenna at Crawdford Hill. We find a clear asymmetry in the high velocity wings of the gas in the galactic polar caps: the negative velocity wings extend to higher velocities than the positive velocity wings in the majority of the directions. This phenomenon is

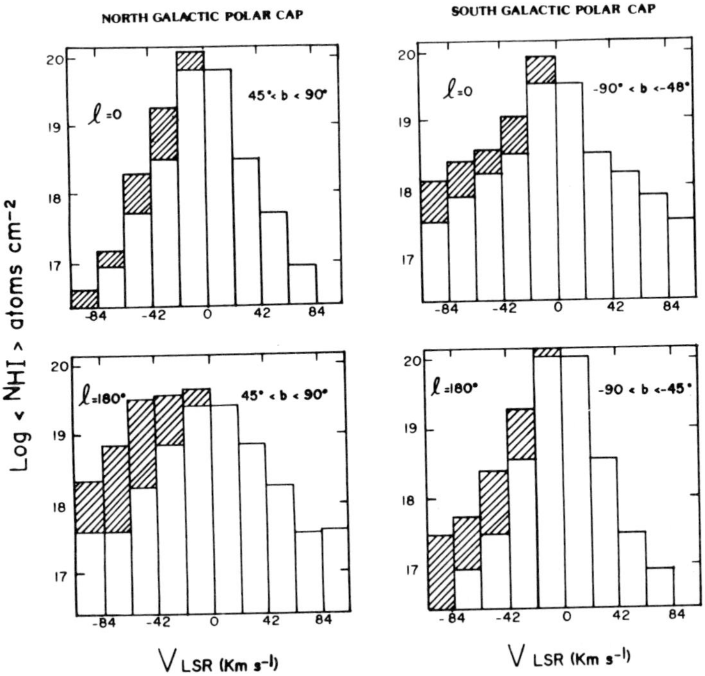

Figure 5: Histograms for $l = 180^{o}$ and $l = 0^{o}$ of the HI column densities in the galactic polar caps. The excess of blueshifted HI relative to the redshifted HI is indicated by hatched areas. The column densities were derived by Mirabel and Morras (1989) from the 21 cm survey by Stark et al. (1983).

particularly striking in the directions of $l = 180^{o}$ and $l = 0^{o}$, where no effects due to the solar motion about the galactic center are expected.

Figure 5 shows histograms of the mean column density $\langle N_{HI} \rangle$ in the galactic polar caps for $l = 180^{o}$ and $l = 0^{o}$, computed from the averages of the individual profiles, in intervals of 21 km/s (4 channels). The histograms in Figure 5 show a clear asymmetry with considerable more blueshifted gas than redshifted. The excesses of blueshifted column densities relative to the redshifted ones are indicated in the histograms by the hatched areas.

Kulkarni and Fich (1985) have shown that a large fraction of the HI in any direction of the sky is fast moving, with large positive and negative velocities. Therefore, to estimate the actual mass infall rate of atomic hydrogen in the solar neighborhood we use only the **excesses** in column densities of the blueshifted gas relative to the redshifted gas. We have taken the excesses in the polar caps for $l = 0^o, 90^o, 180^o$ and 270°. The mass infall rates were computed with the equation:

$$\dot{M} = 8.1 \times 10^{-12} \sum \langle N_{HI} \rangle_i (10^{18}\ HI/cm^2)\ V_i (km/s)\ z_i^{-1} (kpc)$$

where $\dot{M}$ is in $M_\odot\ pc^{-2}\ yr^{-1}$. We find $\dot{M} = 2.72 \times 10^{-8}\ z^{-1}(kpc)\ M_\odot\ pc^{-2}\ yr^{-1}$, where z is the height of the infalling gas above the galactic plane. The gas with 0 > V > - 21 km/s contributes 30% of the total amount, and the downfall from the Northern Hemisphere is 60% of the total inflow.

If the solar neighborhood were representative of an area of the galactic disk of total radius r(kpc), the infall rate would be $\dot{M} = 8.6 \times 10^{-2}\ r^2 (kpc)\ z^{-1} (kpc)\ M_\odot\ yr^{-1}$. For z = 3 kpc (Kahn, 1989) and r = 10 kpc, we obtain $\dot{M} = 3\ M_\odot\ yr^{-1}$, comparable to the output rates of matter toward the halo due to supernovae (Heiles, 1989). This infall rate is about one order of magnitude higher than the infall rate from VHVC's. It is interesting that the mechanical momentum of the VHVC's is about the same as the mechanical momentum of the infalling gas at intermediate velocities.

Since the infall rate at intermediate velocities in the solar neighborhood is more than 10 times higher than the accretion rate of extragalactic neutral gas at very high velocities, most of the infalling matter in the solar vicinity must have originated in the disk, implying an active circulation flow of gas from the disk to the halo, and from the halo to the disk.

Acknowledgements: I thank Ricardo Morras for permission to present in this review unpublished results from our work. I thank Jesus Tharrats for his help in computing the infall velocities of VHVC's using various mass models for the Galaxy. I also wish to thank Bart Wakker for information on HI high velocity data, and Laura Danly for information on unpublished research using interstellar ultraviolet lines. I also thank Dietmar Nieves for his help with drawing the figures.

REFERENCES

Bajaja, E., Cappa de Nicolau, C.E., Cersosimo, J.C., Loiseau, N., Martin, M.C., Morras, R., Olano, C.A., Poppel, W.G.L. 1985, Astrophys. J. Supp. Ser. **58**, 143.

Bloch, J.J., Jahoda, J., Juda, M., McCammon, D., Sanders, W.T., Snowden, S.L. 1986, Astrophys. J. **308**, L59.

Clutton-Brock, M., Innanen, K.A., and Papp, K.A. 1977, Astrophys. Ap. **47**, 299.

Dieter, N.H. 1964, A. J. **69**, **288**.

Giovanelli, R. 1980, Astron. J. **85**, 1155.

Heiles, C. 1984, Astrophys. J. Supp. Ser. **55**, 585.

Heiles, C. 1989, Structure and Dynamics of the Interstellar Medium. Eds. Tenorio-Tagle, G., Moles, M., and Melnick, J. in press.

Franco, J., Tenorio-Tagle, G., Bodenheimer, P., Rozyczka, M., and Mirabel, I.F. 1988, Astrophys. J. **333**, 826.

Hulsbosch, A.N.M. and Wakker, B.P. 1988, Astron. Astrophys. Suppl. ser. **75**, 191.

Kahn, F.D. 1989, Structure and Dynamics of the Interstellar Medium. Eds. Tenorio-Tagle, G., Moles, M., and Melnick, J. in press.

Kulkarni, S.R., Dickey, J.M., Heiles, C. 1985, Astrophys. J. **291**, 716.

Kulkarni, S.R., and Fich, M. 1985, Astrophys. J. **289**, 792.

Mathewson, D.S., Cleary, M.N., and Murray, J.D. 1974, Astrophys. J. **190**, 291.

Mirabel, I.F. 1981, Astrophys. J. **250**, 528.

Mirabel, I.F. 1982, Astrophys. J. 256, 112.

Mirabel, I.F., and Morras, R. 1984, Astrphys. J. **279**, 86.

Mirabel, I.F., and Morras, R. 1989, in preparation.

Morras, R., and Mirabel, I.F. 1989, in preparation.

Oort, J.H. 1970, Astr. Ap., **7**, 381.

Rohlfs, K. and Kreitschmann, J. 1981, Astrophys. Space Science **79**, 289.

Stark, A.A., Heiles, C., Bally, J., and Linke, R. 1989, in preparation.

Stern, R. and Bowyer, S. 1979, Astrophys. J. **230**, 755.

Tenorio-Tagle, G. 1980 , Astron. Ap. **88**, 61.

van Woerden, H., Schwarz, U.J., and Hulsbosch, A.N.M. 1985, in "The Milky Way Galaxy" (eds. H. van Woerden, R.J. Allen, and W.B. Burton, IAU Symp. **106**, 387.

Weaver, H. 1973, in Highlights of Astronomy **3**, ed. G. Contopoulos, (Dodrecht: Reidel), p. 423.

Wesselius, P.R. and Fejes, I. 1973, Astron. Ap. **24**, 15.

ULTRAVIOLET OBSERVATIONS OF HALO CLOUDS

Laura Danly and Chris Blades
Space Telescope Science Institute
3700 San Martin Drive, Baltimore MD 21218

I. Introduction

Since the earliest optical absorption line studies of Munch and Zirin (1961) identified clouds of gas located at large distances from the galactic plane, considerable effort has gone into trying to understand the origin and nature of Milky Way halo gas. Subsequent high resolution optical absorption studies (Albert 1981; Blades et al 1989) have expanded on the early results, demonstrating clearly that (1) halo clouds are more likely to have velocities outside the range allowed by galactic rotation and (2) halo clouds show smaller depletion of refractory elements compared to their disk counterparts (i.e. the Spitzer-Routley effect).

Further insight into the nature of halo gas has been provided by absorption studies in the near ultraviolet region of the spectrum (Savage and deBoer 1979, 1981; Pettini and West 1983; Savage and Massa 1987; Danly 1987, 1989). Despite the lower (25 km/s) resolution, the wider range of ionization states available for study in the ultraviolet and the greater sensitivity provided by higher f-value lines of more abundant species have permitted several additional pieces of observational information to be uncovered. These include: (1) the nearly ubiquitous presence of C IV and Si IV in the halo, (2) the higher scale height of the high ionization species, (3) the very low column density, low ionization gas that pervades the halo, (4) the strong bias toward infall of gas in the northern galactic hemisphere and (5) the strong asymmetry in the characteristics of halo gas toward the NGP vs that toward the SGP.

Comprehensive absorption line and 21-cm surveys have provided insight into the gross characteristics of halo gas and have lead to several plausible theoretical explanations for its origin. However, no definitive conclusions have been reached as to the nature of its support or ionization, or to the origin of diffuse halo gas or of the high velocity clouds (HVCs). Differentiation between competing theories for halo gas will only be achieved by in-depth studies of physical characteristics on a component-by-component basis. This is the hope and expectation for GHRS (the Goddard High Resolution Spectrometer on HST): to observe and analyze ultraviolet species at ground based resolution.

In order to make the most efficient use of data available in advance of the launch of HST, we have been combining high-resolution, ground based optical data with lower resolution IUE data on several important species observed toward high latitude halo stars. The selected lines of sight are those known from previous work to have complex, multi-component spectra where at least one cloud lies outside the bulk of the HI layer (e.g. $z>250$ pc). Here we present results toward one of our stars, HD 203664.

II. The line of sight toward HD 203664 -- Qualitative Description

HD 203664 lies in the direction of l=61°, b=-28° at z=-650 pc. Within four degrees lie two nearer stars, HD 203699 (z=-320 pc) and HD 204862 (z=-50 pc). All three of these stars were observed at Ca II by Blades et al (1989, which provided the data in advance of publication; see also Albert et al, this volume). Only HD 203664 has been observed with IUE; time has been allocated to observe the other stars in the ultraviolet during the coming (13th) IUE episode. The data are discussed below.

optical: The bottom three spectra in Figure 1 show the Ca II profiles toward the three stars in order of increasing distance from the plane. The data have a resolution of 5 km/s. The data clearly show an increase in the complexity of the interstellar spectra with increasing distance from the plane. The interstellar spectrum toward HD 204862 (-50 pc) reveals only a single, weak primary component with broadened wings suggestive of additional component structure. At -320 pc, HD 203699 shows a broadening of the primary component out to -35 km/s, and an additional component at roughly -60 km/s. Between -320 pc and -650 pc one finds in the spectrum of HD 203664 the onset of a strong feature at +70 km/s. Interestingly, the *negative* velocity absorption shows its central minimum at about -40 km/s, compared to the -60 km/s for the *nearer* HD 203699. The lack of -60 km/s absorption in the more distant HD 203664 is very likely due to structure in the intermediate velocity gas at scales < 4°.

ultraviolet: The top three spectra (nos. 4-6) in Figure 1 show the absorption profiles of Si II λ1526, Al II λ1670, and C IV λ1548, respectively, *all in the direction of HD 203664*. The data have a resolution of 25 km/s. The Si II and Al II profiles are very similar in overall shape to the Ca II K profile, although the UV lines are stronger. Note the strongest UV line does show absorption beyond -60 km/s, suggesting that the -60 km/s cloud in front of HD 203699 does extend in front of HD 203664, but perhaps has too little column density to be detected at Ca II K.

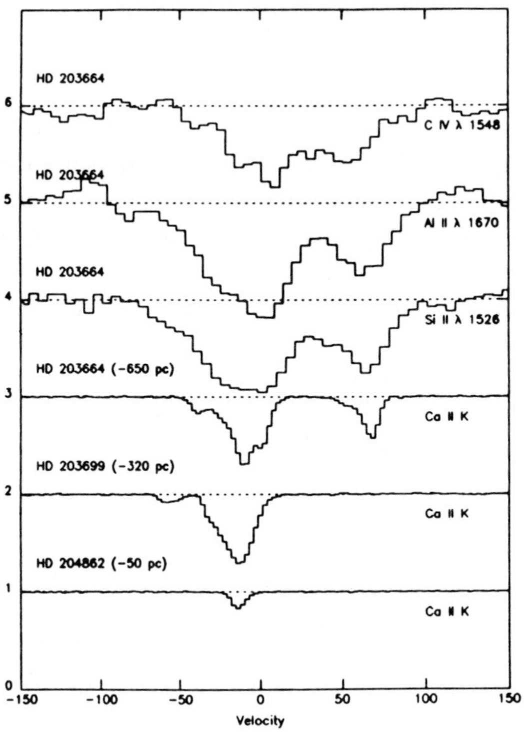

The C IV profile reveals a markedly different structure from the low ionization profiles. The most noticeable difference is the continuity in the absorption between the 0 km/s and the +70 km/s features. While the low ionization species show very weak (or non-existent in the case of Ca II) absorption near 40 km/s, the C IV absorption is strong at these velocities -- nearly as strong as the 0 km/s and +70 km/s features. The data suggest that the gas near 40 km/s range is in a higher ionization state than the strong 0 km/s and +70 km/s components. The data also suggest that the component structure in the high ionization gas may be *different* from that in the low ionization gas. For example, the C IV profile does not show the strong feature at +70 km/s seen in the low ions. Instead the C IV shows a local minimum near ~50 km/s, and the line has nearly returned to the continuum by +70 km/s.

III. Quantitative Discussion -- the Method for Determining Column Densities

In order to accurately model the cloud structure along any line of sight through the interstellar medium one must know the number of components, their central velocities, their b-values, and their column densities. The lower the resolution, the more difficult it is to determine any of these quantities. For those species with the doublet transitions of 2s ^2S--2p $^2P^0$ or 3s ^2S--3p $^2P^0$, such as N V, C IV, and Si IV, limits on the column densities can be set using the doublet ratio method. But generally speaking, cloud b-values and column densities are found simultaneously using a curve-of-growth (c.-o.-g.) analysis by fitting the measured equivalent widths of several lines of the same species to theoretical curves. Frequently, interstellar column densities are determined from the *total* measured equivalent width which is independent of component structure. Results from such a "single-component c.-o.-g." analysis can have large uncertainties and give little insight into the physical conditions of gas along lines of sight with complex structure.

For example, in their IUE survey of 261 lines of sight, Van Steenberg and Shull (1988) found a column density for Si II toward HD 203664 based on a single-component c.-o.-g. analysis of $14.87 < \log N_{\rm Si\,II} < 17.86$ and a range in b-values of 0.05 km/s $< b < 60$ km/s! Although their best fit gives a total $\log N_{\rm Si\,II} = 15.37$ and b=15.23 for the entire line, the values still provide no information on the relative abundance, depletion or ionization state of different clouds along the line of sight. An examination of the profile shape for Si II in Figure 1 immediately shows that a one-component model is not an accurate representation of the physical conditions of the gas toward this star. The higher resolution optical data reveal that *at least* six components are needed to account for the Ca II K profile shape toward this star (Figure 2) .

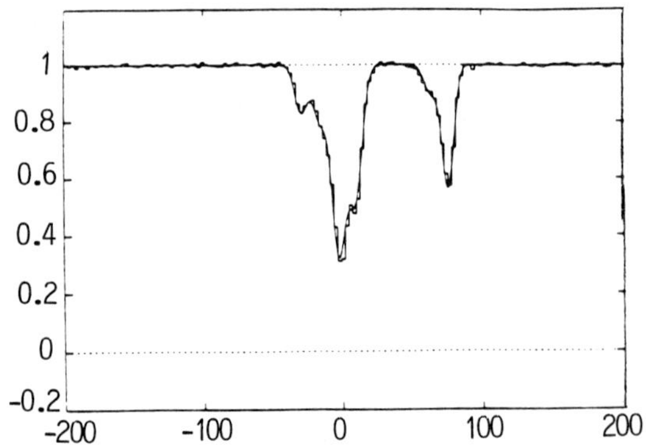

Six component fit to Ca II K line

Cloud Model:

1	4.0000	−29.000	3.00000E+11
2	5.0000	−15.000	4.00000E+11
3	4.0000	−2.0000	2.00000E+12
4	5.0000	10.000	1.20000E+12
5	5.0000	65.000	2.00000E+11
6	3.0000	77.000	9.60000E+11

The velocity model provided by the high resolution optical data fixes most of the free parameters of the fit to a much greater accuracy than on the basis of the lower resolution UV data alone, and can be applied to other species along the sightline. For each of the six observed Ca II clouds, the central velocity and b-value is determined and the values remain fixed for any species (such as Si II, Al II, Fe II, etc.) that also arise under the same interstellar conditions as Ca II. The only remaining free parameter is the column density for each component which can be adjusted until the best fit to the observed profile is found. The most accurate fits are achieved for those species (such as Si II; see Figure 3) where two or more lines are observable with IUE, spanning a wide range in f-value.

IV. Limitations

There are several limitations to this method of determining column densities, including:

• *Ca II cannot detect weakest components.* There may be components along the line of sight which are not detected at Ca II. We have already shown that toward HD 203664, the Si II and Al II profiles show absorption at -60 km/s and +40 km/s while the Ca II profile does not. Therefore at least two additional (bringing the total to eight) components are required to model the interstellar gas along the line of sight to HD 203664.

There is very little that can be done to get around this problem. However, with the bulk of the ultraviolet absorption accounted for by the "known" components, the range of possible values for the column densities and b-values of the weak lines are more restricted than they would be without the benefit of the optical data. Furthermore, the non-detection of Ca II puts an upper limit on how much of any given ion, X, can be present, by assuming "reasonable" values for the Ca II/X ratio.

• *Ca II velocity model is not (necessarily) applicable to the high ionization species.* As discussed in §II, the profiles of the high ionization lines suggest different velocity structure than that seen in the Ca II and other low ionization species. As C IV and Si IV arise under different physical condition than the low ions, this is not unexpected. Indeed, one of the aims of this study is to investigate how well (or how poorly) the velocity structure seen in the low ion lines relates to that seen in the high ions.

• *The Ca II lines are probably not fully resolved.* Even at 5 km/s resolution, the optical data probably does not adequately resolve the true component structure along the line of sight. The apparent components in Figure 2 may actually be made up of several narrower components. Observations by Hobbs (1969a,b) at a resolution of ~1 km/s show that clouds have b-values that are typically 1.5 - 2.5 km/s. Even lower velocity dispersion clouds may exist, though they would have to be *extremely* narrow (<0.2 km/s) or very weak to be undetected in Hobbs' data.

Fortunately, the signal-to-noise ratio in the Ca II data is so high that even components with b-values ~1 km/s would be detected if they contain comparable or greater column density than the 5 km/s component. Therefore, we claim that our Ca II column densities are good to a factor of 2, and that there are no hidden narrow components of significant additional column density. The only exception to this result is for the very core of the near-zero velocity component where even the Ca II components are fairly heavily saturated. The determination of the column density near 0 km/s is therefore only a lower limit. We return to this in the next section.

The applicability of the 5 km/s components is supported by the work of Jenkins (1986) who showed analytically that if, for a given ensemble of lines, the range of b-values and optical depths are smoothly distributed (i.e. not bi-modal), a single component analysis provides a remarkably accurate determination of the column density. The result holds for a wide range of b-values and/or optical depths, though it breaks down at levels of high saturation.

As a final comment, we note that *any* work done with GHRS on the abundances, ionization and depletion of interstellar clouds will necessarily be limited by these same assumptions. Although the resolution is improved over the optical data (3 km/s vs. 5 km/s), the detection limit for narrow components will likely be comparable because of a reduction in signal-to-noise ratio.

V. Results for HD 203664

In this section we illustrate our method by showing profile fits of two species, Si II and C IV, along the line of sight to HD 203664. Si II is an ideal species for study since it has several lines observable with IUE which span 2.5 orders of magnitude in f-value (values taken from Morton and Smith, 1973). Of the high ionization lines, the C IV lines are the strongest and the best exposed. Figure 3a shows the "best fit" to the Si II data. The listed column densities are good to a factor of two for all but the components near 0 km/s which are lower limits. An upper limit to the amount of column density that could be hidden in a ≤1 km/s wide component near 0 km/s is set by the absence of radiation damping wings which would become evident at column densities on the order of 10^{16} cm^{-2}. Interestingly, the *total* column density of Si II along this line of sight is about 1.6×10^{15} cm^{-2}, within a factor of two of the best estimate given by Van Steenberg and Shull (1988) for the sightline. The reasonable agreement between the two different methods supports Jenkins' (1986) view that a one component c.-o.-g. gives a reasonably reliable estimate of total column density.

Comparing to the fit of the Ca II K line in Figure 2, one sees that the Ca II/Si II ratio varies from component to component. Even more significant variations among components are seen in the Si II/H I ratio, though this may be due to beam size effects in the 21-cm data (the profile and discussion of beam effects can be found in Danly et al. 1989). In particular, the +70 km/s component is very weak, or even non-existent in the HI data. Interferometric observations of the HI in this direction will be made in September, 1989, in the hope of answering this question in greater detail.

By far the most difficult data to fit is in the region of the +40 km/s gas. While the fit appears a bit too strong in the λ1304 and λ1526 lines, it is a bit too weak for the λ1260 line. Some of the error may be due to uncertainties in the f-values, though errors in the atomic parameters should be uniform over the entire profile. Our best fit is achieved with a component having a significantly larger b-value of 20 km/s. Components with smaller b-values result in λ1304 and λ1526 lines which are even stronger and a λ1260 line that is even weaker! This can be understood in view of the fact that narrower lines saturate "quicker" (i.e. at lower column density) and thus there is *less* of a difference between the resulting λ1304 and λ1526 profiles and the λ1260 profile. A single narrow component does not have a large enough equivalent width to make the λ1260 line uniformly black, while adding additional narrow components makes the λ1304 and λ1526 lines unreasonably strong.

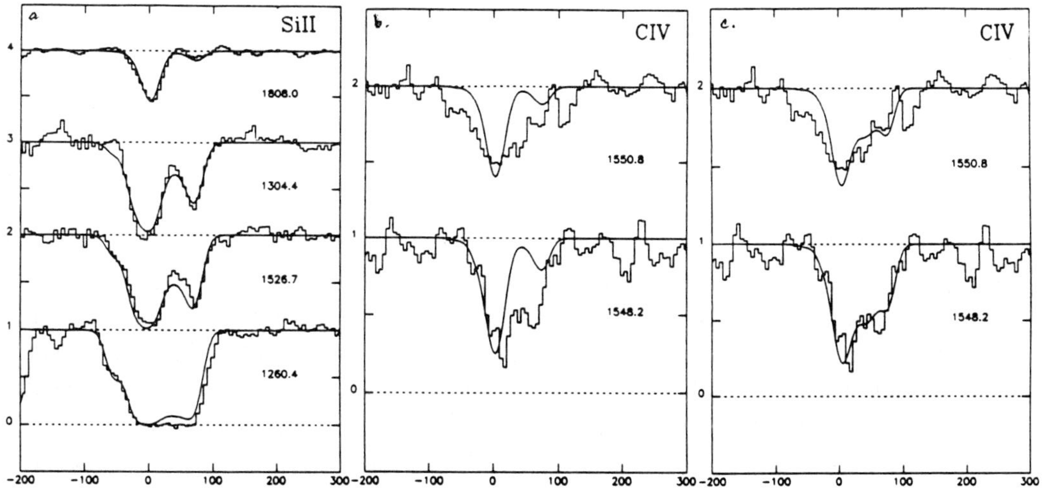

Cloud Model:

#	b	v	N
1	7.00	−55.00	8.00E+12
2	4.00	−29.00	4.00E+13
3	5.00	−15.00	8.00E+13
4	4.00	−2.00	8.00E+14
5	5.00	10.00	4.80E+14
6	20.00	40.00	3.00E+13
7	5.00	65.00	4.00E+13
8	3.00	77.00	9.00E+13

Cloud Model:

#	b	v	N
1	7.00	−55.00	6.40E+11
2	4.00	−29.00	3.20E+12
3	5.00	−15.00	6.40E+12
4	4.00	−2.00	6.40E+13
5	5.00	10.00	3.84E+13
6	20.00	40.00	2.40E+12
7	5.00	65.00	3.20E+12
8	3.00	77.00	7.20E+12

Cloud Model:

#	b	v	N
1	7.00	−55.00	6.40E+11
2	4.00	−29.00	3.20E+12
3	5.00	−15.00	6.40E+12
4	4.00	−2.00	6.40E+13
5	5.00	10.00	3.84E+13
6	20.00	40.00	4.00E+13
7	5.00	65.00	3.20E+12
8	3.00	77.00	3.00E+13

Figure 3b shows the C IV λ1548 profile, along with the best Si II model which has been scaled in order to bring the central optical depth of the 0 km/s gas into the best agreement possible with the data. The C IV/Si II scaling is a factor of 0.08. In this figure one clearly sees that a uniform scaling of all the components is not appropriate: significant component-to-component variations are found in the C IV/Si II ratio. A better fit is seen in Figure 3c, where the appropriate scaling for the +70 km/s gas is more like 0.33, while the +40 km/s feature has *more* column density in C IV than in Si II! There is *at least* a factor of 170 increase in the C IV/Si II ratio in the +40 km/s gas compared to the 0 km/s gas. The difference in the ratio could be even greater since the Si II column density in the 0 km/s gas is a lower limit.

VI. Discussion

From the models shown in §V and even from the profiles in Figure 1 it is evident that significant variations in physical conditions are likely to exist from one component to the next in gas along the line of sight toward halo stars. The data indicate moderate differences in the ion ratios between the +70 km/s halo cloud and disk gas, and even more extreme variations in the +40 km/s gas. Lack of space prevents more data from being shown, but this trend is seen in other species as well: both the O I (which is found only in H I regions) and the C II* (indicative of higher density) lines show strong 0 km/s and +70 km/s absorption features while they are very weak or non-existent near +40 km/s. The gas in this region shows higher ionization, lower column density, lower number density and higher velocity dispersion than the gas in the stronger "cloud" components. These are just the characteristics that one might expect to see in an inter-cloud medium and it is tempting to speculate that we are detecting such inter-cloud gas in our observations.

In addition to providing a comparison of conditions among various components, the method shown here has reduced the uncertainty in column density by two orders of magnitude and has placed more meaningful limits on b-values than those provided by a one component analysis. Further improvements can be made through better signal-to-noise in the UV data and better atomic data. The uncertainties introduced by possibile hidden narrow components can be reduced by studying this line of sight at very high (~ 1 km/s) resolution from the ground and/or by observing very weak lines that are probably not saturated (e.g. Cr II and Zn II) using HST. Finally, only GHRS observations of the UV species can provide the final word on whether such a cloud model is appropriate. Similarly, only GHRS can reveal the differences in velocity structure between the high ionization species and the low ionization species which are suggested in our data.

We thank Elise Albert for helpful discussions and Duncan Forbes for assistance with the data reduction. This work is supported by NASA under grant NAG-5-999.

References

Albert, C.E. 1983, Ap.J., **272**, 509

Blades, J.C., Albert, C.E., Morton, D.C., and Proulx, M. 1989, in preparation

Danly, L. 1987, *PhD Thesis*, University of Wisconsin--Madison

Danly, L. 1989, Ap.J., **342**, 785

Danly, L., Savage, B., Lockman, F. J., and Meade, M. R. 1989, submitted

Hobbs, L.M. 1969a, Ap.J., **157**, 135

Hobbs, L.M. 1969b, Ap.J., **157**, 165

Jenkins, E.B. 1986, Ap.J., **304**, 739

Morton, D.C. and Smith, W.H. 1973, Ap.J. Suppl., **26**, 333

Pettini, M., and West, K. A. 1982, Ap. J., **260**, 561

Savage, B. D., and deBoer, K. S. 1979, Ap. J. (Letters), **230**, L77

Savage, B. D., and deBoer, K. S. 1981, Ap. J., **243**, 460

Savage, B. D., and Massa, D. M. 1987, Ap. J., **314**, 380

Van Steenberg, M.E. and Shull, J.M., 1988, Ap.J., **330**, 942

Discussion:

MÜNCH (Comment): The difficulties you find in fitting the observed profiles (strong components) are only a consequence of the known fact (W.S. Adams, G. Münch, L. Hobbs) that as the resolving power of the spectrograph is increased the strong components unfold into a number of weaker partially overlapping components. Even with the highest resolving power - say 500000 - there is no hope of resolving all of them. Now, it is also known that the velocity distribution, macroscopically observable as line multiplicity and microscopically as line with, depends on the optical depth of the cloud. The net result is that the strongest lines in abundant ions have apparently larger Doppler parameters than the same lines in less abundant ions.

DANLY: By all means, uncertainties in obtaining column densities from unresolved lines are well documented. As described in our paper, we investigated the magnitude of the uncertainty in our analysis by attempting to "hide" both narrow and very broad components in the model, and them evaluate their contribution to the resultant profile. We claim our column densities are accurate to a factor of two, with the exception of the core of the gas near 0 km s^{-1} for which our analysis gives a lower limit (and then only if the hidden gas is in a component narrower than $\sim$ 1 km s^{-1}).

PECKER (Comment): I found your data very impresive and rich. But don't you take too much out of them? I observed CIV, SiII, UV lines, etc. in the solar UV spectrum: they are definitely formed in different regions. Then what is the meaning of their ratios? Secondly, in order to obtain, from line intensities, the column density, you insisted (correctly) that one important source of error is the uncertainty in the f-values. True! But not less worrying is the bad knowledge of all radiative and collisional transitions are involved in ionization and excitation of the atomic -ionic species you have observed; therefore I would take your column densities with extreme caution!

DANLY: I agree that the Si II and CIV lines are probably produced in different regions. Indeed the velocity structure differences between these lines suggest different component structure along the lines of sight. The "meaning" of our ion ratios is to suggest that differences in physical conditions exist in different regions along the line of sight. For example, gas near +40 km s^{-1} toward HD 203664 is almost certainly more highly ionized and lower density than the +70 km s^{-1} or 0 km s^{-1} gas.

Regarding your second point, a lack of complete knowledge of transition rates, non-LTE effects, etc., introduces uncertainties not in the derived column densities, but in their interpretation. One of our aims in embarking on this observational study in the pre-HST era is to challenge the theorists: are current models sufficiently detailed and sufficiently predictive to make use of the wealth of data of this sort which will become available with HST?

ANALYSIS OF LOW- AND HIGH-RESOLUTION OBSERVATIONS OF HIGH-VELOCITY CLOUDS

Bart P. Wakker
Kapteyn Laboratorium, Rijks Universiteit Groningen
Postbus 800, 9700AV Groningen, The Netherlands

1. Introduction

For almost three decades neutral hydrogen moving at velocities unexplicable by galactic rotation has been observed. These so-called high-velocity clouds (HVCs) have been invoked as evidence for infall of neutral gas to the galaxy, as manifestations of a galactic fountain, as energy source for the formation of supershells, etc. No general consensus about their origin has presently been reached. However, it is becoming clear that no single model will suffice to explain all HVCs. A number of clouds may consist of material streaming toward the galactic center, as Mirabel (this conference) has advocated for several years, though their origin still remains unclear. A better understanding is mainly hampered by the fact that the distance remains unknown. An overview of the current status of the distance problem is given by van Woerden elsewhere in this volume.

Recently a deep and complete survey of the HV gas has been completed (Hulsbosch & Wakker 1988, Bajaja *et al.* 1985). Older surveys, notably the one by Giovanelli (1980) provided much insight into the phenomenon but suffered from incompleteness. In the new surveys the whole sky north of declination $-18°$ is covered on a $1°\times1°$ grid with a $36'$ beam and a detection limit of 0.05 K (corresponding to about $10^{18}\,\mathrm{cm}^{-2}$), while south of $-18°$ the grid is $2°\times2°$ with detection limit 0.08 K. In Section 2 one of the new results based on these data is presented, which is related to the absorption-line studies of galactic halo gas. The survey data also allow one to constrain models for the origin of HVCs, as is described in Section 3.

More detailed maps of HVCs have been made since 1973 (Giovanelli, Cram & Verschuur). These data revealed the presence of spatial and velocity fine structure within the clouds. The smallest concentrations were unresolved with the $12'$ to $20'$ beams used. Giovanelli & Haynes (1976) discovered that at this resolution the velocity profile often showed a sharp (FWHM 7 $\mathrm{km\,s^{-1}}$) peak on top of a broad (FWHM 23 $\mathrm{km\,s^{-1}}$) plateau. From 1979 onward data at $1'$ resolution have been taken with the Westerbork synthesis telescope. These are described in Section 4 and show that the intrinsic linewidths are narrower (FWHM 3 $\mathrm{km\,s^{-1}}$) and that the broader profiles are likely to be due to beam smearing.

2. Sky coverage

Figure 1 shows a number of curves $C(T>T_0)$, defined as the percentage of sky area at which HVCs are found with brightness temperature above a certain limit T_0

for a given velocity limit. At the survey limit of 0.05 K 9% of the sky is covered by gas having $|v_{lsr}| > 100\,\mathrm{km\,s^{-1}}$, an increase of a factor two when compared to older surveys (with detection limits of the order of 0.2 K).

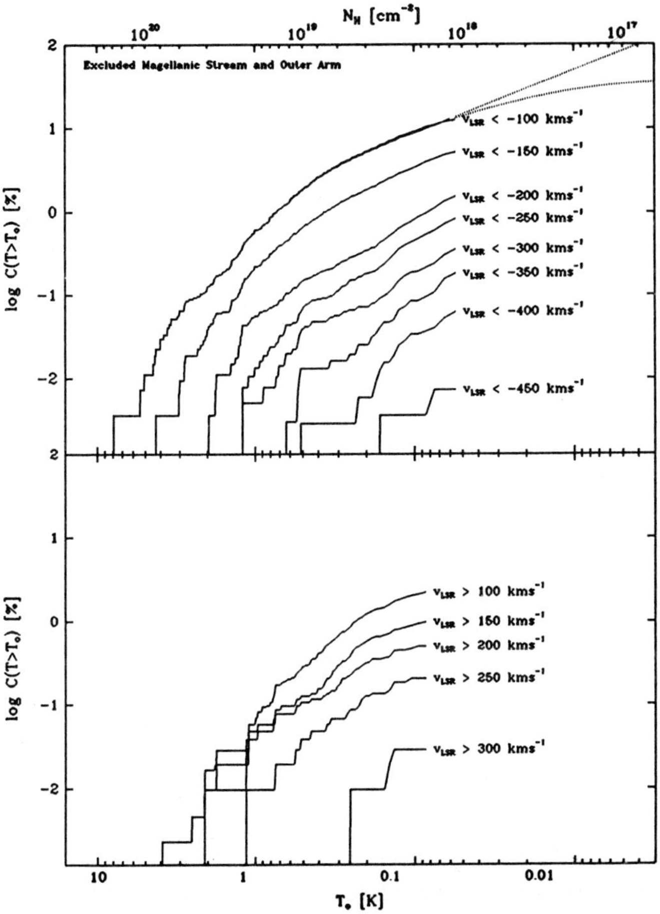

Figure 1. Each curve shows the percentage of sky area covered by high-velocity gas with velocity above the indicated limit as function of detection limit. The lower x-scale is in degrees Kelvin brightness temperature, the upper scale in column density, assuming a line width of 20 $\mathrm{kms^{-1}}$. The dotted lines show the extrapolation for a velocity limit of $-100\ \mathrm{kms^{-1}}$. A linear and a parabolic least-squares fit were made to the corrected observational curves in the range between 0.06 and 0.20 K.

Using UV absorption lines and assuming an element abundance one can probe much lower hydrogen column densities than in the 21-cm emission line. For instance, of the lines that are in the wavelength band of the IUE satellite, SiII $\lambda1260$ allows to detect material where the hydrogen column is $2\times10^{17}\ \mathrm{cm^{-2}}$. Therefore (at a velocity limit of $-100\ \mathrm{km\,s^{-1}}$) the relation between sky coverage and detection limit has been extrapolated from the survey limit of $10^{18}\ \mathrm{cm^{-2}}$ down to $10^{17}\ \mathrm{cm^{-2}}$ as shown by the dotted lines in Fig. 1.

To find the extrapolation a correction of the order of 10% was made for incomplete sky coverage and for the overestimation of sky area in the case of single-point and two-point clouds, as will be described in Wakker 1989a. The extrapolation predicts that at the level of $2\times10^{17}\ \mathrm{cm^{-2}}$ between 25 and 60% of the sky is covered by gas having $v_{lsr} < -100\,\mathrm{km\,s^{-1}}$. However, in Danly's (1989) sample there are 19 stars more distant than 1.5 kpc and having $l < 180°$, but in none absorption at velocities more negative than than $-100\,\mathrm{km\,s^{-1}}$ is detected. Unless the HVC's do not contain heavy elements this implies that they, at least statistically, are farther away than 1.5 kpc.

3. Modeling

In order to fully understand the implications of the survey data for the understanding of the origin of HVCs, modeling is necessary. Models have been constructed that are based on the following scheme:

1) Choose a galactic potential.
2) Inject clouds into this potential according to some prescription. A prescription consists of drawing from a given distribution a random radius, azimuth, height, velocity vector and mass at the time of formation.
3) Follow the clouds in their ballistic orbit and in their evolution until the clouds either disperses or is destroyed, for instance when it hits the galactic disk.
4) At each timestep determine the observables (longitude, latitude, LSR velocity, flux, area).

The galactic potential that was chosen consists of a bulge, halo and disk with parameters for the bulge and halo as given by van der Kruit (1986):

$$\rho_{bulge}(R) = \rho_{0,bulge}\left(\frac{R}{b}\right)^{-\frac{3}{2}}\left(1+\frac{R^2}{b}\right)^{-1}, \qquad \rho_{halo}(R) = \rho_{0,halo}\left(1+\frac{R^2}{a}\right)^{-1},$$

with b=300 pc, $\rho_{\odot,bulge}$=0.0002 $\mathrm{M_\odot\,pc^{-2}}$, a=2700 pc and $\rho_{\odot,halo}$=0.008 $\mathrm{M_\odot\,pc^{-2}}$. The halo is cut off at 60 kpc radius. The disk density is given by:

$$\rho_{disk}(r,z) = \rho_{0,disk}\exp-\frac{r}{h}\exp-\frac{z}{s},$$

with radial and vertical scalelengths h=5 kpc and s=325 pc. In order to get a rotation velocity of 220 $\mathrm{km\,s^{-1}}$ at the solar radius (8.5 kpc) the total disk surface density at the sun must be 75 $\mathrm{M_\odot\,pc^{-2}}$.

Cloud evolution is modelled in a very simple way: they are assumed to be pressure confined throughout their lifetime and therefore always stay the same size. Only when they hit the gaseous disk are they assumed to be destroyed. At formation a mass is given to each cloud, chosen from a power-law spectrum with slope α. Assuming the same average density for all clouds makes it possible to calculate the cloud radius. Together with the cloud distance this allows to calculate the cloud flux, brightness temperature and area. By comparing the observed distributions of flux and area with the model distributions it follows that $\alpha = -1.5$ and that the density, averaged across the whole cloud, must be about 0.01 $\mathrm{cm^{-3}}$.

The different types of models are characterized by different presciptions for the injection scheme. Various possibilities exist, three of which will be described here.

1) The cannonball model: Clouds form in the galactic disk and are corotating. They also get a vertical velocity of 50 $km\,s^{-1}$. Such a model describes e.g. gas directly ejected from the disk by means of supernova explosions.

2) The galactic fountain model: As suggested by Bregman (1980) the HVCs with velocities up to 200 $km\,s^{-1}$ may consist of cooled halo gas. The prescription used here to parametrize this fountain model is the following: the radial density distribution is exponential with scalelength 5 kpc (like the galactic disk) and radius 15 kpc); condensation takes place at heights between 2 and 6 kpc above the disk; the vertical velocity at formation is a random value between -75 and $+75$ $km\,s^{-1}$; the component in the direction of galactic rotation is determined from an estimate of the radius at which the hot gas rose up by assuming that it moved upward and outward by one scalelength in both r and z ($h_r = \frac{kT}{\mu g_r(r_{form})}, h_z = \frac{kT}{\mu g_z(r_{form})}$) and assuming conservation of angular momentum.

3) The infall model: Clouds form outside the galactic disk at distances up to 150 kpc, having a r^{-2} density distribution. The velocity distribution is found by taking the first and second moments of the radial velocities of all dwarf galaxies in the list of Richter *et al.* (1987) nearer than 300 kpc. Assuming that the azimuthal dispersions are equal to the radial dispersion, this gives the clouds a velocity dispersion of 50 $km\,s^{-1}$ in all directions and an infall component of -20 $km\,s^{-1}$.

In Figure 2 the observed distribution of HVCs in longitude and LSR velocity is shown, together with the predictions obtained from the first runs of the three described models. From the comparison of the observed distribution with the model pictures it is immediately clear that none of the models can completely explain the data. The fact that at longitudes below 180° mostly negative velocities are observed and above 180° mostly positive velocities is clearly seen in the models, however. This is due to the reflection of the rotation of the sun around the galactic center. In the infall model clouds with velocities up to 400 $km\,s^{-1}$ are seen, but on the whole the density distribution of points is not what is observed. The cannonball and fountain models show the concentration of clouds at velocities between +/−100 and +/−200 $km\,s^{-1}$, but are unable to explain the existence of clouds with higher velocities and of the HVCs in the anticenter region. A more complete description of models and conclusions will be presented by Wakker & Bregman (1989).

4. High-resolution observations

Since Schwarz & Oort (1981) published the first high-resolution map of a HVC, a large number of new fields has been observed at Westerbork, with integration times of $2x12^h$ and $4x12^h$. Six of these fields have been completely reduced. A full description of the datataking, reduction, analysis and the data themselves will be presented by Wakker (1989b).

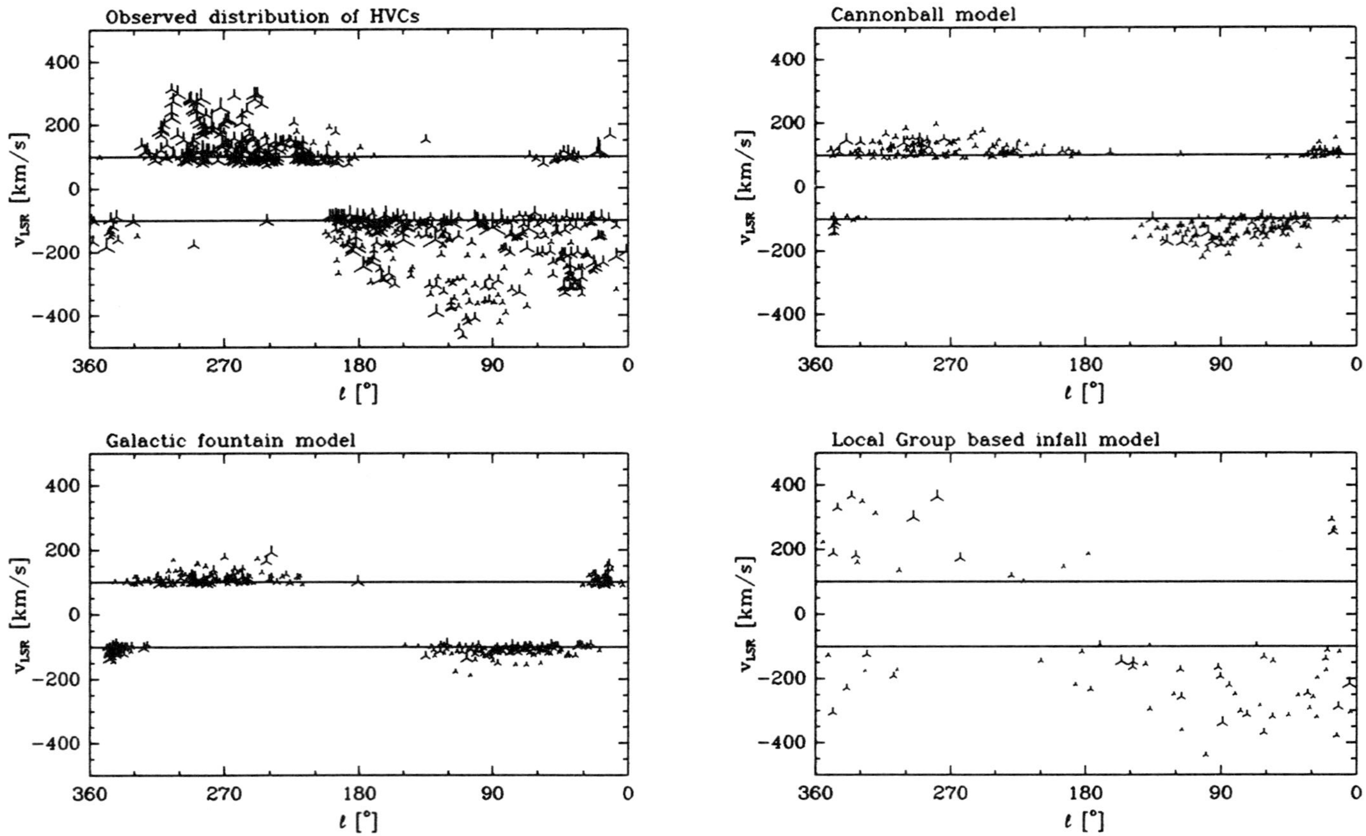

Fig. 2. Observed and predicted distribution in longitude and LSR velocity of HVCs. The model parameters are described in the text. The size of the symbols is proportional to the cloud flux.

4.1 Fine structure

In Fig. 3 a representative total-hydrogen map of a HVC is presented. Immediately apparent is the existence of much fine structure. It can be shown that even with a beam of 1′ the smallest concentrations are not completely resolved. This means that there is structure in the gas at linear scales less than 0.3 D(kpc) parsec (with D the cloud distance).

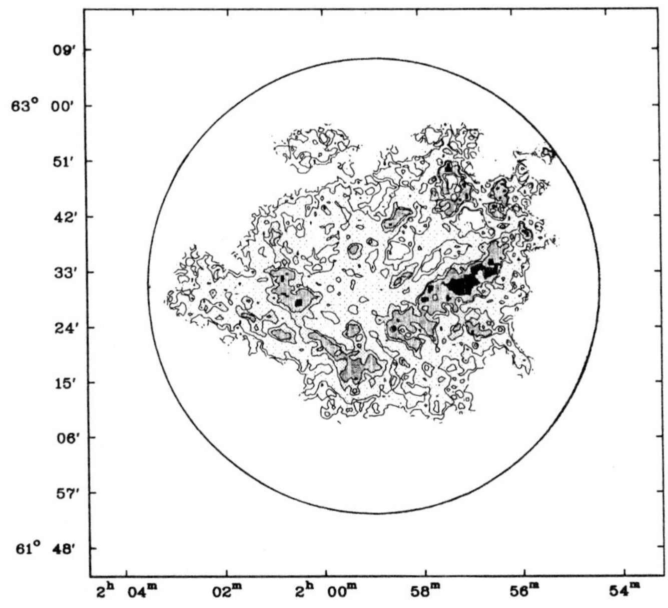

Figure 3. Total hydrogen map of HVC131+1-200 as observed by the WSRT. Contours and gray scales are at the level of 5, 10, 15, 20 and 25×10^{18} cm^{-2}. The beam is 1′. The circle indicates the radius where the primary beam attenuation is a factor 2.

The density contrast can be as high as 5 on scales of 3′. For the absorption-line studies of galactic halo gas this introduces a considerable uncertainty in the derivation of abundances. A small feature which lies right along the line of sight to a halo star may provide a strong absorption component but go undetected in a single-dish beam due to beam-dilution effects. Alternatively, a cloud seen in HI but not in absorption could in fact lie in front of the probe star, but sufficiently filamentary or patchy that the bulk of the gas lies just off the line of sight from the pencil-beam to the star.

4.2 Line widths and spin temperature of the gas

At each of the positions in the WSRT data cubes a gaussian fit was made to the spectrum of the clouds. Usually only one component is present, but profiles with 2 or even 3 components are not rare. From the fit one obtains the velocity field of the cloud and the line width of its profiles.

The two VHVCs that have been observed (HVC114-10-440 and HVC110-7-465) give a surprising result. They have very different line widths (FWHM 2.1 and 17.5 $\mathrm{km\,s^{-1}}$ respectively), although they are only 5 degrees apart on the sky and have similar spatial structure (both at high resolution and in the single-dish observations by Cohen & Mirabel (1979)). The single-dish data already suggested this: for HVC110-7-465 the profiles had an FWHM of 23 $\mathrm{km\,s^{-1}}$, while HVC114-10-440 showed an envelope with FWHM 23 $\mathrm{km\,s^{-1}}$ and a core with FWHM 7 $\mathrm{km\,s^{-1}}$. At high-resolution HVC110-7-465 still does not show narrow profiles, but for HVC114-10-440 only extremely narrow ones are seen close to the velocity resolution of the data. It can be shown that because there exists a large velocity gradient across the field, the wider lines are due to beam-smearing. A possible explanation for the difference between the two clouds may be that they are both in a cooling stage, but HVC110-7-465 has not yet cooled as far as the other one. This would provide evidence for a hot origin of the VHVCs. On the other hand, both objects may have been intergalactic gas clouds being accelerated by the Galaxy and now are in the process of being disrupted by friction with halo gas. In this picture HVC114-10-440 is still in its original cool state, while in the other HVC a substantial amount of heat has already been generated.

In the other four fields the fitted linewidths are rather small: 3 $\mathrm{km\,s^{-1}}$. Such narrow widths imply that the kinetic temperature of the gas is below 200 K. A lower limit of 30 K is provided by the peak brightness temperature, thereby considerably narrowing the previously known limits to the temperature. No evidence is found for broad lines, indicative of hot gas, although these might be hidden in the smooth background which has been filtered out by the interferometer. However the broad lines seen with single-dish instruments are likely to be due to beam-smearing effects.

References

Bajaja E., de Cappa Nicolau C.E., Cersosimo J.C., Loiseau N., Martin M.C., Morras R., Olano C.A., Poppel W.G.L., 1985, *Astrophys. J. Suppl. Ser.* **58**, 143

Bregman J.N., 1980, *Astrophys. J.* **236**, 577

Cohen R.J., Mirabel I.F., 1979, *Monthly Notices Roy. Astr. Soc.* **186**, 217

Danly L., 1989, *Astrophys. J.* **342**, ..

Giovanelli R., 1980, *Astron. J.* **85**, 1155

Giovanelli R., Haynes M.P., 1976, *Monthly Notices Roy. Astr. Soc.* **177**, 525

Giovanelli R., Verschuur G.L., Cram T.R., 1973, *Astron. Astrophys.* **12**, 209

Hulsbosch A.N.M., Wakker B.P., 1988, *Astron. Astrophys. Suppl. Ser.* **71**, 191

van der Kruit, P.C., 1986, in "The Galaxy", NATO ASI series, ed. G.Gilmore and B.Carswell, p27

Richter O.G., Tammann G.A., Huchtmeier W.K., 1987, *Astron. Astrophys.* **171**, 33

Schwarz U.J., Oort J.H., 1981, *Astron. Astrophys.* **101**, 305

Wakker B.P., 1989a, in preparation

Wakker B.P., 1989b, in preparation

Wakker B.P., Bregman J.N., 1989, in preparation

Discussion:

VAN WOERDEN: Do your small velocity dispersions (1-2 $\mathrm{kms^{-1}}$) and consequent low temperatures (~200K) apply only to the dense concentrations, or also to the thinner gas in between?

WAKKER: In the less dense regions the fitted dispersions are also low, but the accuracy with which they can be measured is less because the signal to noise is less.

MEBOLD: Calculating your temperature limits, did you have the zero-spacing HI emission added to your data? It has been found that the spatially more extended HI emission tends to have the larger velocity dispersion.

WAKKER: I did not add in the zero and short-spacing information. And from comparing with single-dish data it is clear that a significant amount of flux is missing. Without really going through the process of adding short-spacing to the maps it is impossible to know exactly the linewidths of this "missing" flux. However, because the linewidths measured from the WSRT maps increase when you smooth them and because the interferometer filters out only low spatial frecuencies, not low spectral frecuencies, I think that it is not likely that the smooth background has broad lines. The two-component structure seen in single-dish maps at 10' resolution could be the result of beam-smearing, because there are large velocity gradients and jumps in the cloud. In one particular case (HVC 114-10-440) I could reproduce the 2-component shape of the single-dish profile by adding all the profiles fitted at 1' resolution, all of which had an FWHM of only about 1 kms^{-1}.

HEILES: Did you try to do a 21cm absorption experiment using radio continuum sources?

WAKKER: I once correlated a catalogue of 21-cm radio sources with the Dwingeloo survey of HVCs. Unfortunately none of the strong continuum sources fell on top of an appreciable HVC component. And if you try to make a prediction of the absorption strength for the correlating sources, only a few are expected to give an absorption of more than 3 sigma. Clearly it is necessary to get a more complete catalogue of 21cm sources and to do the correlation again, and also to use long integration times on the candidate sources.

de BOER : In particular the WSRT data that you showed are very nice and extremely important. But by gost, I wished you had not shown it since it makes clear how difficult a task the halo-absorption line people have!

WAKKER: That certainly is true. There are large variations of integrated column density on scales of just a few arcminutes. But of course, in the WSRT maps the short-spacing information is missing and it is known that the integrated flux calculated from the WSRT maps is between 50 and 80% of the single-dish flux so that there must be a smooth background present in addition to all the fine structure. This might help to get a measurable column density toward the star.

DANLY: I cannot determine the contrast scale in your slide (of the high spatial resolution HVC observations). Is there sufficient structure over scales of 2 arcmin to account for the differences seen in the 50 kms^{-1} Na components toward SN1987a and Sk-69 203, as shown in the previous talk?

WAKKER: The contrast in the WSRT maps is quite high. Certainly on scales of 5 arcmin the intensity can drop from a high value to alm ost nothing. On 2 arcmin scales the contrast is about a factor 3. So, I think this can easily explain the difference in the Na lines seen toward the LMC.

Molecules at the Interface of an HVC and a High-z HI Filament

U. Mebold, U. Herbstmeier, P.W.M. Kalberla, I. Souvatzis
Radioastronomisches Institut der Universität Bonn
Auf dem Hügel 71, D 5300 Bonn 1, F.R.G.

1. Introduction

The Draco nebula, the faint reflection nebulosity LBN 406/412/415 (Lynds 1965) at $l \approx 91°$, $b \approx 38°$, $z \gtrsim 500$ pc (Goerigk and Mebold, 1986), contains delicate filaments at its low-l, low-b (SW) boundary (Goerigk et al. 1983, Mebold et al. 1985). The density in the boundary is characterized by the presence of molecules like CO and H_2CO (Mebold et al. 1985) and NH_3 (Mebold et al. 1987). Johnson (1986) found an excess of unidentified IRAS point sources in the area of the Draco nebula, a large fraction of which is located close to the SW boundary.

The velocity field at the boundary of the Draco nebula is characterized by a gradient of about $+5 kms^{-1}/5'$ on top of a systemic velocity of $V=-22 kms^{-1}$ (velocities refer to the LSR). The direction of the gradient is towards the galactic position angle GPA=230° measured from galactic north (N) through the direction of increasing galactic longitude (E). The gradient along most of this boundary was derived from HI observations made with the 100-m telescope (Goerigk et al. 1983). An association of the Draco nebula, in particular its SW boundary, with HI high velocity clouds (HVCs) has been discussed by Mebold et al. (1985) in the context of the origin of one of the filaments visible at prints of the Palomar Observatory Sky Survey (POSS) by Rohlfs et al. (1989a).

In the present paper we present a complete map of the CO distribution in the Draco nebula, review the evidence for an interaction of HVCs with the Draco nebula and the dust and HI cloud IVC86.0+38.5-44 located close to the Draco nebula. Finally we will discuss the origin of the molecules and the dust in these two nebulae in relation to their possible origin.

2. Data and Results

Fig.1 shows the 100μm emission of the Draco nebula (cf. Johnson l.c.) observed by IRAS. The distribution of the integrated ^{12}CO line, WCO, is presented in Fig.2. The CO data were obtained with the 3-m KOSMA mm telescope at Gornergrat. The beam size is 3.9 arcmin, the velocity resolution is $\delta V = 0.08 kms^{-1}$. The intensity scale is calibrated in K (T_R^*) and the 3rms temperature uncertainty is $\delta T = 0.8$K. More details of the data will be published elsewhere.

We see that the CO molecules are observed nearly exclusively at the low-l, low-b (SW) boundary of the Draco nebula. Comparing the WCO distribution with that of the 100μm emission, we find that the molecules are concentrated even more than the dust at a rather narrow strip along the fore-mentioned boundary. The intensity of the 100μm emission drops off very steeply to the SW and rather slowly to the NE of the nebula. This head-tail morphology is also present - but much less pronounced - in the WCO-distribution.

The head-tail morphology of the dust- and the WCO-distribution, the velocity gradient at the SW side of the HI distribution (s. section 1) and the filamentary morphology of the dust boundary at the POSS suggest that the Draco nebula is moving towards the SW (galactic position angle GPA = 225^{0}) and that shock fronts and instabilities exist at this boundary.

We have therefore studied the HI and the CO emission of this boundary with high angular resolution using the WSRT and VLA aperture synthesis telescopes for the HI- and the Onsala 20-m and the Kitt Peak 12-m single dish telescopes for the CO-observations. Reports about these observations have been published (Rohlfs et al. 1989a) or will be published (Rohlfs et al. 1989b). Here we will discuss high resolution HI results relating to the velocity gradient and the narrow dust filaments at the SW boundary of the nebula and review high resolution CO results relating to two suspected areas of interaction between HVCs and molecular gas.

Fig.3 shows the velocity gradient observed perpendicular to the SW-boundary in one of its undisturbed parts. The location is indicated by a heavy line in Fig.5. The gradient increases from a value $\approx$ 0.0 $kms^{-1}arcmin^{-1}$ inside the nebula monotoneously to a maximum value of $\approx$+1.5 $kms^{-1}arcmin^{-1}$ at the outer edge of the boundary. We have compared the location of that edge to that of the dust visible at the POSS. We find that the edge of the dust distribution (indicated by the vertical line in Fig.3) is coincident within the positional accuracy of our VLA 21-cm line data ($\approx$30arcsec) with the location of the positive most HI gas at $V\approx-12\ kms^{-1}$.

Fig.4 shows a schematic presentation of the results found by Rohlfs et al. (1989a) for the dust filament which we will call DRA91.3+37.0-22. The frame of Fig.4 is indicated by its four corners in Fig.1. Rohlfs et al. suggest a model for the evolution of the filament during the last 10^{6}yr. At about that time an HVC consisting of a dense core ($n_{H}\approx100\ cm^{-3}$) and a less dense halo ($n_{H}\approx0.6cm^{-3}$) which was located close to the upper left hand corner outside Fig.4, was moving along the direction of the vectors in that Figure and had a radial velocity of V=-180kms^{-1}. It started colliding with the boundary layer of the Draco nebula ($n_{H}\approx10^{3}cm^{-3}$) which had a radial velocity of -18kms^{-1}. During the inelastic collision a cloud, called the molecular clump in Fig.4, was formed from the combined mass of the HVC-core and the boundary layer. After the collision the clump left behind a turbulent wake which we see now as the filament connecting the clump and the Draco nebula. The presence of the wake indicates a large mass loss rate. Hence the clump appears to be a rather transient phenomenon. This conclusion is supported by the fact that the gravitational potential of the clump is not adequate to bind it. The authors can not give an estimate of the density of the medium into which the clump is moving now.

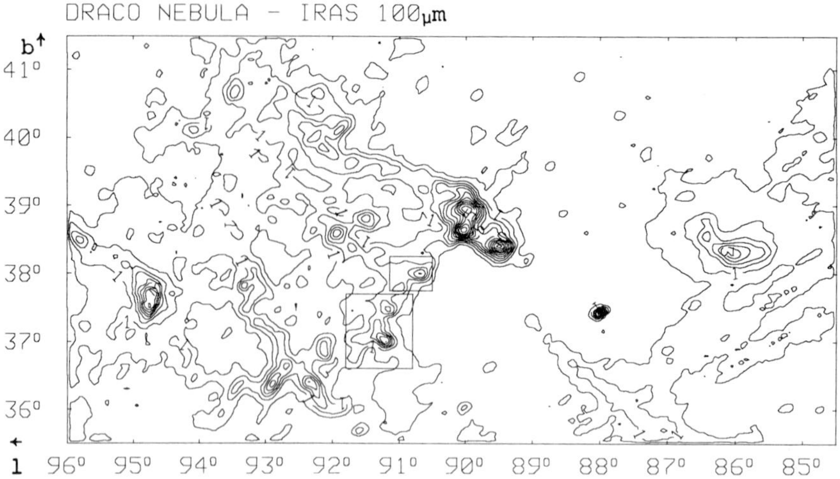

Fig.1: IRAS 100μm image of the Draco nebula and IVC86+38.5-44. The lowest contour is at 0.5 MJy/sr and the contour intervals are 0.5 MJy/sr. The frames of Figures 4 and 5 are indicated by rectangles.

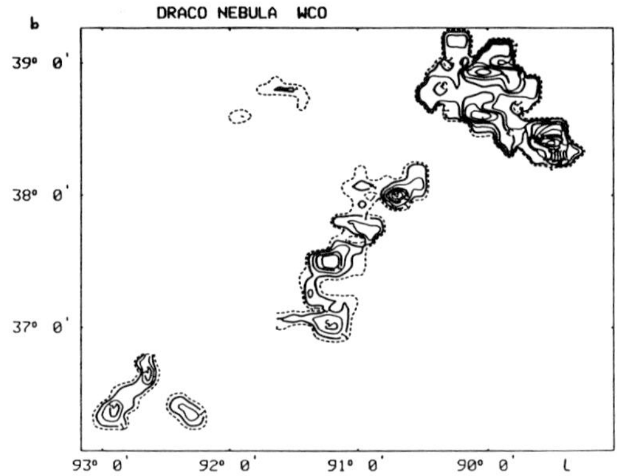

Fig.2: The ^{12}CO line integral of the Draco nebula observed with the KOSMA telescope and the mm-telescope at Bordeaux. The dashed contour corresponds to 0.5 Kkm/s, the first full contour is at 1 Kkm/s. The contour interval is 1 Kkm/s.

Fig.3: Position-velocity plot of HI brightness temperatures along the axis indicated in Figure 5. The lowest contour is at 2K (Fig.3a) and 1K (Fig.3b) and the contour interval is 0.8K. The mark indicates the position of the dust edge at the POSS plates.

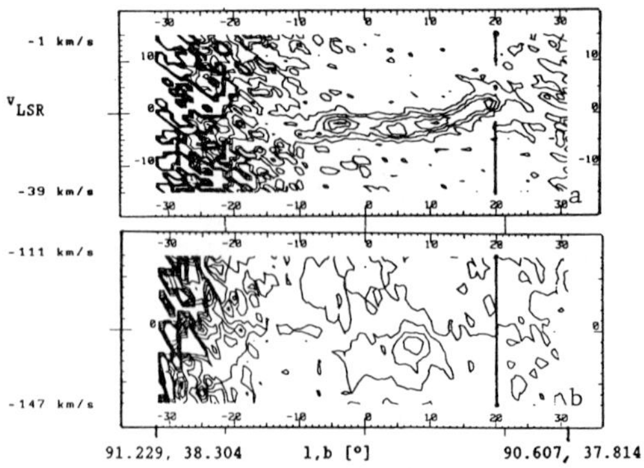

The quoted densities were derived from line intensities of ^{12}CO, ^{13}CO and H_2CO molecules and standard factors for the conversion of the corresponding molecular column densities into column densities of molecular hydrogen or hydrogen nuclei. It is likely that these conversion factors give too large densities (s. section 3).

Fig.5 shows a schematic presentation of the results found by Rohlfs et al. (1989b) for the complex of filaments which we call DRA90.7+38.0-20. The frame of Fig.5 is indicated by its four corners in Fig.1. The model proposed for the features in Fig.5 is similar to that just described with the main difference beeing that the collision of the boundary layer of the Draco nebula with HVCs, in particular the dense cores like HVC90.7+38.0-126, is much less evolved. The HVC clumps at velocities negative of $-133 kms^{-1}$ are probably unaffected by the interaction with the Draco nebula and therefore indicate the initial velocity of the HVCs. The clump at $V=-126 kms^{-1}$ appears to be running into a molecular cloud at $V=-24 kms^{-1}$ associated with the Draco nebula. This cloud has been accelerated towards the observer by $\approx 5 kms^{-1}$. At the same time the HVC clump appears to have been decellerated by $\approx 10 kms^{-1}$ compared to the velocity of the undisturbed HVC clumps. So, in the case presented in Fig.5 the collision has just started, while in Fig.4 it apparently has already come to an end.

Fig.6 shows the results of 21-cm line observations with the 100-m telescope of the third cloud which we believe to be the product of a collision of an HVC and gas at low velocities (Herbstmeier 1989). It is called IVC86+38.5-44, is located close to the Draco nebula and shows strong 100μm emission (see Fig.1). The cloud has also been observed with the KOSMA telescope for ^{12}CO lines. No lines were detected to a limit of $T_R^* \approx 0.4K$. We see the cloud at the SW end of an HVC filament. This HVC filament appears as an extension of the head-tail elongation of IVC86+38.5-44. It can also be traced for $\approx 5^0$ along a faint dust tail starting at the core of the IVC and extending up to $l \approx 90°$, $b \approx 44°$ (Herbstmeier 1989). The head-tail structure of the IVC and the HVC may be regarded as evidence for an interaction of the IVC with the surrounding medium. Such an interaction is also suggested by rather wide 21-cm lines just at the SW edge of IVC86+38.5-44. The rather large dust colour temperature ($\approx 25K$) derived from the 60μm and 100μm IRAS emission may or be not related to a possible interaction.

3. Discussion and Conclusion

An interesting aspect of these high-z dust and molecular clouds is their internal pressure. While Wakker (this volume) gives rather low values for the pressure in his HVCs, Souvatzis (1989) finds rather large values ($n*T \approx 10^5 Kcm^{-3}$) for the HVC clumps in Fig.5. Here n*T has been estimated from the HI column density the linear size and the observed velocity dispersion. The n*T values for the HI gas in the associated features of the Draco nebula are close to $10^4 Kkms^{-1}$, i.e. a factor of ≈ 10 smaller (Souvatzis 1989). If the partial pressure of the molecular constituent of the Draco nebula is added, the total pressure may be close to that in the associated HVCs. Although it is difficult to compare the n*T values for our high-z clouds with those ($\approx 10^4 Kkms^{-1}$) derived for galactic plane clouds, the former values are probably clearly

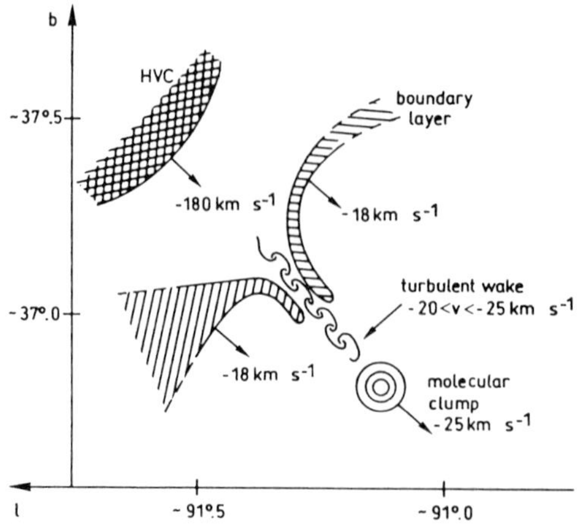

Fig.4: Sketch of the collision scenario DRA91.3+37.0-22. The velocities are relative to the LSR. For details see text.

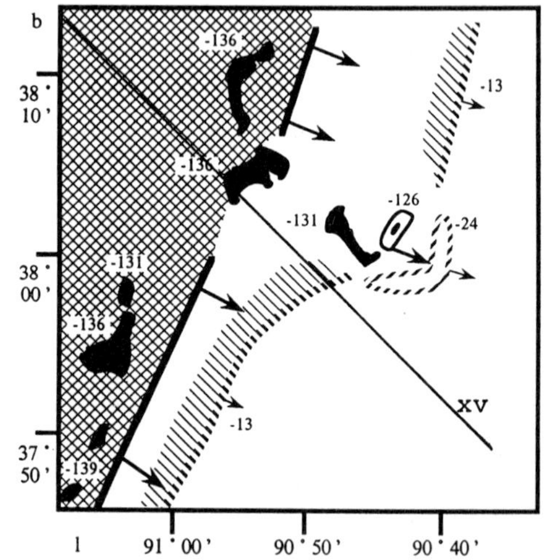

Fig.5: Sketch of the collision scenario around HVC90.7+38.0-126 indicated by "⊙". The velocities are relative to the LSR. For details see text. The location of the position velocity plot in Fig. 3 is indicated by "XV".

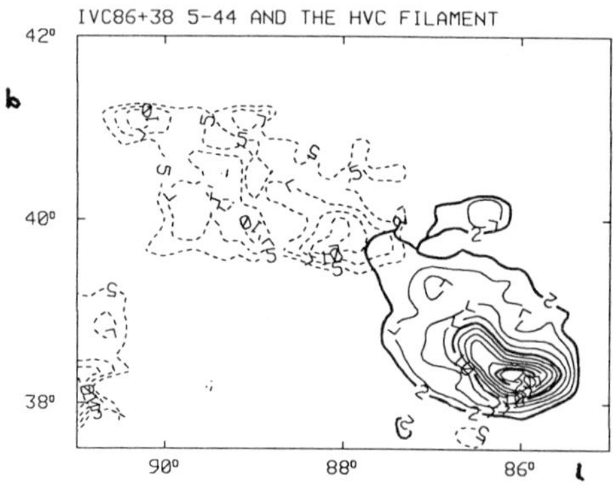

Fig.6: HI column density distribution of IVC86+38.5-44 (full lines) and the HVC filament (dashed lines). The contours are labelled in $10^{19}\,cm^{-2}$.

larger: For similar values of the temperature (10^3 K) the densities in our HVC clumps are probably larger by a factor of ≈10.

The rather large pressures found in the HVCs and the boundary layer of the Draco nebula is additional evidence for non-steady-state conditions, e.g. some kind of collisional interfaces, in this area (cf F.Kahn, this volume). From this evidence and the results given in section 2 we obviously can identify two different kinds of collision fronts:
1. There is the high velocity ($\delta V \approx 100 kms^{-1}$) collision between the HVCs and the Draco nebula. This kind of collision is observed in the area displayed in Fig.5, has occured - according to the model of Rohlfs et al. (1989a) - some 10^5 yr ago in the area displayed in Fig.4 and has possibly produced the cloud IVC86+38.5-44 (Fig.6).
2. The other kind of collision is the low velocity ($\delta V \approx 20 kms^{-1}$) collision between the Draco nebula and the gas at zero velocties, i.e. the medium at the SW side of the Draco nebula.

We propose to interpret the Draco nebula as well as IVC86+38.5-44 as the results of collisions between HVCs and an HI filament which has a velocity close to zero and is reaching out of the galactic plane to $b \approx 40^0$ (Heiles 1984). We hypothesize that it extends to an elevation of $z \gtrsim 500$pc.

The cloud IVC86+38.5-44, as the least complicated structure of those discussed here, may be considered as the region between the down-stream and the up-stream shocks which have developped after a collision of a HVC at a velocity of $V_{HVC}=-110 kms^{-1}$ versus the zero-velocity filament. Here the direction of the motion is assumed to be parallel to the tail in Fig.6. Using the relation

$$V_{CD} = V_{HVC}/(1+(n_{HVC}/n_0)^{-1/2})$$

for the velocity of the contact discontinuity, V_{CD}, as a function of V_{HVC} and the densities of the HVC, n_{HVC}, and the zero-velocity filament, n_0, we can reproduce the velocity of our IVC, $V_{IVC}=-44 kms^{-1}$, if we adopt a density contrast of $n_{HVC}/n_0=0.3$ and identify V_{IVC} with V_{CD}. The adopted density contrast is realistic for the kind of high-z HI filament considered here. The latter condition may be fullfilled if the material cools rapidly behind the shock. In this picture the SW edge of the IVC is identified with the down-stream shock, its NE end is identified with the up-stream shock and cooled material is the HI gas which we observed.

A similar explanation may be given for the Draco nebula. We can reconcile its velocity of $-22 kms^{-1}$ by a collision of HVCs with velocities between -180 and $-130 kms^{-1}$ if the density contrast is much larger than that in the case of IVC86.0+38.5-44: n_{HVC}/n_0 = 0.02 to 0.04, respectively. Here the relative velocity of the up-stream shock is between 160 and $100 kms^{-1}$ while that of the down-stream shock is only close to $22 kms^{-1}$. It is suggestive to identify the up-stream shock with the SW boundaries of the HVCs in Fig.4 and Fig.5. The down-stream shock is then at the SW edge of the Draco nebula, in the region were the velocity gradient - displayed in Fig.3 - is found. As predicted by our model, a velocity gradient similar to that in Fig.3 is found for most of the SW boundary of the Draco nebula.

In this model the up-stream shock is a fast shock which probably dissociates molecules and possibly even destroys dust particles. The down-stream shock is a slow shock which allows for rapid cooling and the building up of the molecules which we have observed in the post-shock region of the Draco nebula. The 100μm emission of the dust, which is seen in this region too, has been swept up behind the slow shock and in front of the contact discontinuity from the material into which the shock is moving. The contrast of the gas and dust density behind the slow shock and that in the general field in front of the slow shock may therefore be used as an indicator of the length of the path of the HVCs through our Galaxy.

Summing up, the proposed collision of HVCs versus a high-z HI filament qualitatively explains most of the observed features of the Draco nebula and IVC86+38.5-44. However, a large number of details have to be worked out before this model is completely acceptable. In this context and finally in this paper, we will discuss the conversion of molecular line intensities, in particular WCO, into column densities of molecular hydrogen or hydrogen nuclei. An analysis of the data of the Draco nebula indicates that molecules, in particular CO, are overabundant compared to typical clouds in the galactic plane (Herbstmeier et al. 1989). Similar results were also found for Cirrus clouds (Heithausen and Mebold 1989). An explanation of the apparently large molecular abundances may either be related to the reduced depletions of the heavy elements found towards stars in high galactic latitudes or related to shocks - similar to those discussed in the present paper - which may produce an enhancement of molecules in the Cirrus (cf e.g. the Discussion in Heithausen and Mebold 1989). Although this problem does not affect the rather qualitative interpretation given above, it is of major importance, if these or similar clouds are used to study phenomena like shock fronts or are used as probes for the astrophysical conditions in the halo of our Galaxy

References

Goerigk, W., Mebold, U., Reif, K., Kalberla, P.M.W., Velden. L.: 1983, Astron. Astrophys. **120**, 63

Goerigk, W., Mebold,U.,:1986,Astron. Astrophys. **162**, 279

Heiles, C.: 1984, Astrophys. J. Suppl. **55**, 585

Heithausen,A., Mebold, U.: 1989, Astron. Astrophys. **214**, 347

Herbstmeier, U., Rohlfs, R., Mebold, U.: 1989 , in: "The Physics and Chemistry of Interstellar Molecular Clouds", ed. G. Winnewisser and J.T. Armstrong, Springer Verlag.

Herbstmeier,U.:1989, in prep.

Johnson, H.M.: 1986, Astrophys. J., **309**, 321

Lynds, B.T.: 1965, Astrophys. J. Suppl. , **12**, 163

Mebold, U., Cernicharo, J., Velden, L., Reif, K., Crezelius, C., Goerigk, W.: 1985, Astron. Astrophys. **151**, 427

Mebold, U., Heithausen, A., Reif, K.: 1987, Astron. Astrophys. **180**, 213

Rohlfs,R., Herbstmeier, U., Mebold, U., Winnberg, A.: 1989a, Astron. Astrophys. **211**, 402

Rohlfs,R., Herbstmeier,U., Mebold,U., Fink, U.:1989b, Astron. Astrophys, subm.

Souvatzis,I.1989: Diploma Thesis, University of Bonn

Discussion:

WAKKER: Do you expect that the high velocity gas becomes ionized at the interface between low and high-velocity gas that you claim to see. If so, what about Hα emission from that region? Is that expected there?

MEBOLD: We do expect ionized gas at this interface and I know that G. Münch has tried to find it. I am not sure about his results, but obviously any possible emission is not strong.

OBSERVATIONS OF THE GALACTIC HALO

Klaas S. de Boer
Sternwarte, Universty of Bonn, F.R. Germany

Summary. After reviewing the history of the early research on gas in the Milky way halo, observational methods to investigate the halo are described. The meaning of the concepts Halo and HVC is discussed and selectional bias in databases is pointed out. The essentials of the interpretation of observed velocities are given. The scanty information on distances of HVCs is reviewed and an outlook is given on the possibilities to firmly settle the distance of at least a few HVCs.

General History

The concept of a halo around the disk of the Milky Way dates from the studies of stellar distributions at the beginning of the century. The globular stars clusters appeared to exist in a roughly spherical volume around the disk, a space called halo. Spectroscopy then showed that stars at high galactic latitudes have, on average, a lower metal content than stars in the solar vicinity. The nature of these stars tends towards the properties of the globular cluster stars and Baade named these field objects stars of the halo-population. The astronomical use of the word halo was taken from the meaning "halos" of saints, but it originally derives from the greek word for "(round) threshing ground".

The threshing ground aspect of the word halo remained meaningless to me until three years ago. During an observing stay at the Calar Alto Observatory looking for halo gas, I spent (after a misty night) a pleasant afternoon with my colleagues touring the valleys of the Sierra de los Filabres. At a very scenic spot we noticed an abandoned village and decided to take a look. At an elevated part, away from the few crumbling walls of foregone houses, we found such a "halos", a circular well-paved area. I imagined how the people of the hamlet would have threshed their grains using the convective airflow up into their valley and past their village to produce clouds of small and light particulate stuff high into the air.

The studies of the interstellar gas in the late 1930's had revealed that many a line of sight showed multiplicity in the interstellar absorption spectra. Adams (1949) reports 2 stars at $b > 20^{\circ}$ having multiplicity in CaII K. To follow up on these findings Guido Münch started to systematically observe stars at high galactic latitude (Münch 1952). The full data appeared much later (Münch and Zirin 1961). Several absorption structures were found with velocities substantially different from the expected 0 km s^{-1} for local galactic gas. These structures then represented High Velocity

Clouds (HVCs), at that time defined as clouds with velocities differing by more than 20 km s^{-1} from the LSR. Based on the distances of the stars used as background light sources we know roughly how far from the Milky Way disk the clouds detected by Münch in the visual are.

The radio 21-cm measurements at high galactic latitudes showed the presence of HI gas with surprisingly large velocities. Isolated features were found, but also larger complexes were discovered, like Complex C in the northern sky. Most clouds seemed the move towards the Solar Neigbourhood (see review by van Woerden et al., 1985). For none of these structures detected at 21 cm could distances be derived; the presence of these clouds could not be confirmed through absorption-line measurements in stellar spectra. This by itself implied rather large distances from the Milky Way disk. Later systematic surveys in both the northern and southern hemisphere (Giovanelli 1980; Hulsbosch 1985; Bajaja et al 1985; Hulsbosch and Wakker 1988) demonstrated that about 1/3 of all directions at high latitudes show the presence of HVCs.

The likely existence of large volumes of ionzed gas was most elegantly advocated from theoretical considerations by Spitzer (1956). This proposition was based on the early results of the observations of Guido Münch, of which Spitzer was informed during frequent visits from Princeton to the Mount Wilson Observatory. The very presence of neutral clouds had to mean that there was a confining medium as well, and since there was no observational evidence for other forms of cool or neutral gas, Spitzer postulated that the space outside the disk was filled with hot and well ionised gas possibly out to 10 kpc and with a temperature of the order of a million degrees.

In the following, methods to study halo gas are reviewed and results are described. A number of reviews has appeared on the topic of gas in the halo such as by York (1982), de Boer (1985), Bergeron, Savage, Green (1987).

Ultraviolet, X-ray, Infrared and H-alpha Data

With the opening of the UV-part of the spectrum in 1972 resonance transitions of highly ionised states of interstellar atoms finally (Spitzer 1956) became accessible. The first unambigous proof for the existence of well ionised gas outside the disk of the Milky Way came with the IUE from the high dispersion spectra of stars in the Large Magellanic Cloud (Savage and de Boer 1979, 1981). Absorption by CIV and SiIV was detected at velocities up to 140 km s^{-1} LSR, velocities almost certainly pertaining to gas of the Milky Way. This gas clearly is unaffected by the ionisation environment of the background light sources. Earlier studies on the possible existence of hot and ionised gas generally distributed in the Milky Way disk (see Jenkins 1978) always had to consider the possibility of ionisation by the radiation field of the actual star observed.

The measurements of the absorption lines in the ultraviolet have a definite advantage over those in the visual. In the UV one finds the transitions of the elements abundant in the gas phase in neutral clouds such as CII and MgII and also the numerous lines of SiII and FeII. These have on average an optical depth more than a factor 100 larger than the lines in the visual, thus making the detection of small amounts of gas, in particular at exceptional velocities, relatively easy. However, the UV instrumentation available up to now is attached to the rather small 45 cm IUE telescope. Also the gas-phase abundances of depleted elements (such as TiII) may be a function of distance from the galactic plane (Albert 1983; Edgar and Savage 1989).

The metal content of HVCs seen at high galactic latitudes is not too far from the solar composition. The very fact that absorption lines from metals, such as from CaII, FeII, MgII, CII, etc., are seen (toward globular clusters and toward the LMC, see below) shows that HVCs consist largely of processed material. CaII has been found in HVCs using a number of extragalactic background light sources, but the Ca abundance is uncertain due to possible depletion effects. The abundances of FeII, MgII, etc., could be determined for the HVCs on the lines of sight to the LMC (Savage and de Boer 1981) to be within 5 and 25% of solar. Having recorded HI 21-cm emission profiles showing these HVCs, McGee, Newton, and Morton (1983) and McGee and Newton (1986) could fix the metal content of the HVCs at about 40% of solar while the high signal-to-noise data toward SN1987A showed this to be the case for gas in three HVCs on that line of sight (Blades et al 1988). Suggestions that the -60 and -130 km s^{-1} clouds are associated with the LMC are negated by observations by de Boer, Morras, and Bajaja (1989).

During the 1970's, Wisconsin rocket soft X-ray scans of the almost complete sky showed that there is a fair amount of soft X-ray background intensity of which it was speculated that it might come in part from the Milky Way halo (the full account of the data is given by McGammon et al 1983). The spatial structure in the background appears to be anti-correlated with the location of HI clouds as known from 21-cm measurements and the background flux reduces nowhere on the sky to zero level. The explanation is that there is a local source of soft X-ray emission (local hot bubble) and neutral clouds absorb the soft X-rays coming from a more distant source, thus producing "cloud-shadows". Recent reanalyses of the all-sky soft X-ray data (Hirth et al. 1989, Snowden and Sanders, these proceedings) show that the observed intensities fit to a 10^6 K medium whose source function increases with z-distance of about half a kpc and in which neutral clouds are embedded.

The IRAS satellite revealed the widespread presence of thin veils of IR emitting dust at high galactic latitudes which was named infrared cirrus (Low et al 1984). Essentially all filaments have been identified with HI 21-cm gas, and known high latitude diffuse reflection nebulae coincide with IRAS cirrus as well. For one of these features the distance could be derived using the gas density distribution towards stars in almost the same directions as derived from the increase in strength of absorption lines (Hobbs, Blitz, Magnani 1986). Also studies using star counts (Magnani and de Vries 1986) and generalised surveys of the CO associated with the IRAS cirrus (Magnani, Lada, and Blitz 1986) led to distances < 200 pc.

For most of the individual cirrus the distance is still unknown.

H-alpha emission from diffuse galactic gas has intensively been investigated by Reynolds since 1978 with a Fabry-Perot device allowing a velocity resolution of 12 km s^{-1}. Among the multitude of directions and objects to look at also a few known HVCs have been measured. In those directions Reynolds (1987) sets firm upper limits to the intensity of H-alpha radiation. This relative lack of H-alpha emission indicates that the ionised gas surrounding the HVCs has a density $n(e) < 0.1$ cm^{-3} or a temperature $T > 10^{5}$ K, or both. Also Guido Münch (these proceedings) has looked at H-alpha from HVCs and he emphasized the great precission required in such measurements because of the faintness of the signal and the sources of contamination.

Interlude on Definitions

Where does the halo begin, when looking from within the Milky Way disk, and how far does the halo extend? It seems relevant to sort out a few phrases. The early definition of halo referred to the older stars, to lower metallicity, and to spatial extent. In later radio and infrared studies the word halo often has been used to just mean "at high galactic latitudes". That the halo extends far out is, I feel, a notion common among the workers in the field. But where then does the halo start? If we believe that the gas outside the Milky Way disk is largely ionised, then the halo may be thought to start at a z height equal to the scale height of neutral gas.

The velocities need some important comments. High-Velocity originally meant gas at speeds differing by more than 20 km s^{-1} from the LSR, a definition based on the reach of optical spectroscopy. Later, with 21-cm measurements of the entire galaxy, the definition of HVC became "cloud with velocity not in agreement with galactic rotation models". This means that in practice the limit in velocity of HVCs is a function of galactic direction: only such "HVCs" are recognizable which do not blend in the normal HI 21-cm emission velocities.

Is high-velocity gas in the halo? It may be but may be it is not. There are numerous examples of high velocity absorption features present in spectra of disk stars. It is likewise possible that high-velocity gas seen at high galactic latitudes is near the disk.

May zero-velocity gas be in the halo? Yes, why not, but one cannot find such gas that easily since it will blend in velocity with any gas present locally in the Milky Way disk.

Distances to the Halo Gas Components

The location of the base of the halo may, for practical purposes, be taken as the same as the end of the disk. Of course, since gas is a diffuse matter there is no borderline between disk and halo so one may use the notion of the scale height of the gas. Lockmann, Hobbs, and Shull (1986) investigated the vertical extent of the neutral gas of the Milky Way. They collected Lyman-alpha data for 40 stars and could for 25 derive reliable HI column densities from the interstellar absorption line. Comparison with HI 21-cm emission in the same directions showed that at most 15% of the neutral gas is further than 1 kpc from the Milky Way disk. This demonstrates unambiguously the limited thickness of the neutral layer.

The well ionised gas appears to start showing up in absorption lines as of vertical distances of about 1 kpc. The investigations of the lines of sight to the Magellanic Clouds (Savage and de Boer 1981) showed that the, in those directions normally strong, SiIV and CIV lines (these being proof of well ionised gas) were absent in the spectra of two foreground stars with distances of about 1 kpc. In a subsequent study, Pettini and West (1982) showed that along their 30 sight lines into the halo the absorption by SiIV and CIV started showing up only outside 1 kpc, their most distant object being 3 kpc below the Milky Way disk. Related but different is the positive detection of large amounts of SiIV, CIV and NV in the spectra of stars with z about 2 kpc in the direction of the central parts of the Milky Way (Savage and Massa 1987). They detected well ionised gas outside a layer of 1 kpc, but the directions probed sample parts of the Milky Way where the disk may be thicker than near the Sun. Very accurate data have become available using SN1987A as background light source. This line of sight shows essentially that what was known toward the LMC in general, but in particular the lines from SiIV and CIV, which are very smooth in velocity, show that large amounts of well ionised gas are present in the halo (Savage et al., 1989).

The distribution of free electrons also defines the location of ionised gas in the galaxy and outside the Milky Way disk. Again globular clusters appear to be important in this respect. The newly discovered pulsars in globular clusters allow to add very essential dispersion measures to the data base available. Reynolds (1989) showed that also the electrons may have a spatial distribution with a scale height of 1 to 2.5 kpc, very similar to the scale height of the ions SiIV and CIV.

The quest to find distances to HVCs has not ended although in some cases distance limits have been claimed. The number of pertinent studies is rather limited. The first comes from Münch (1952) with 2 cases of absorption at high velocity, but few studies followed. The possibility to observe the dominant ionic stages of the interstellar elements through UV absorption lines, and the detection of two HVCs in front of the LMC (Savage and de Boer, 1979, 1981), increased interest in the halo. Many individual lines of sight were investigated. Pettini and West (1982) observed with the IUE many stars at high latitudes and with z-distances up to 3 kpc but found no high velocity absorption. Towards two globular clusters CII was seen in absorption at velocities in the HVC range of up to -100 km s^{-1} (de Boer and Savage, 1983, 1984). Very nice results are being obtained by Danly (1989) who has analysed IUE spectra of high-latitude stars and she finds in some cases absorption at velocities up to -90 km s^{-1}. In the visual a very

Table 1. Sight lines with detected high-velocity gas in spectra of objects with z > 0.7 kpc and with gas velocities > /50/ km s^{-1}

Object	l	b	type	d	z	v(rad) abs.	21-cm	ref
M10	15	+23	G.C.	4.3	1.7	-78	-85	KK72
vZ1128 M3	42	+79	G.C.	10	10	-70		BS84
B29 M13	59	+41	G.C.	6.3	4.1	-80	-80	BS83
HD215733	85	-36	B1II	2.5	-1.5	-50	-50	A83
BT Dra	99	+51	RR Lyr	2.1	1.6	-136	-140	SCW88
HD121800	113	+50	B1.5V	2.2	1.7	-113		D89
HZ 25	153	+80	B2:	3.3:	3.3:	-91		D89
HDE 233622	168	+44	B1.5:	3.5:	2.4:	-72		D89
BD+38 2182	182	+62	B2:	4.6:	4.1:	-96		D89
HD 93521	183	+62	O9V	2.1	1.8	-53	-53	M52, A83[a])
HD100340	259	+61	B1IV	5.7	5.0	-70		D89
LMC	280	-33	OB*	55	30	-60	-60	SB81
						-120	-120	SB81
HD119608	320	+43	B1Ib	4.3:	3.0:	-77		D89

[a]) PW82 quote d = 0.8 kpc and find no HV absorption in FeII

refs.: A83= Albert 1983 (CaII), BS84= de Boer and Savage 1984 (CII), BS83= de Boer and Savage 1983 (CII), D89= Danley 1989 (CII), KK72= Kerr and Knapp 1972 (CaII), M52= Münch 1952 (CaII), PW82= Pettini and West 1982 (FeII), SB81= Savage and de Boer 1981 (FeII), SCW88= Songaila, Cowie and Weaver 1988 (CaII)

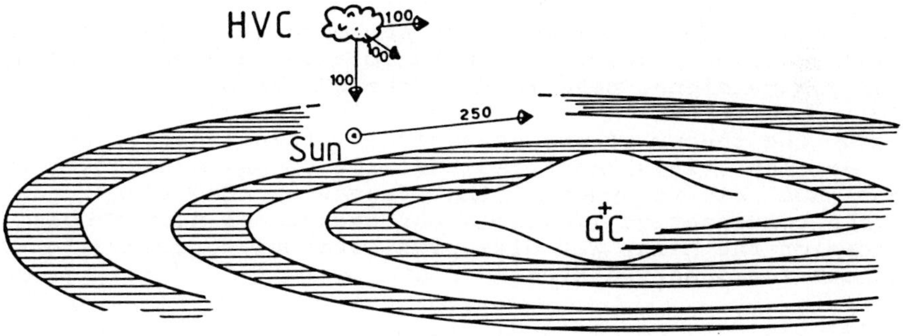

Fig. 1. The observed velocities of almost all HVCs are consistent with the following simplified characteristics: z distances between 2 and 5 kpc and velocity components of the order of 100 km s^{-1} (Kaelble et al 1985), as indicated. Figure from de Boer (1985).

elegant study was carried out by Albert (1983) who measured TiII, CaII and NaI in pairs of stars with very similar directions but with very different distances. Two cases were found with absorption at velocities differing by more than 40 km s^{-1} from the LSR. This work is being continued by Blades et al (see these proceedings). Keenan et al. (1988) completed a programme on CaII observations of halo stars without finding HV absorption. Songaila, Cowie, and Weaver (1988) published CaII data on stars in the direction of HV-gas complexes of the northern hemisphere and they see CaII absorption in spectra of an RR Lyr star at velocities related to those of the Complex C. The intensive investigations (star counts and radio lines) of the Draco Cloud (see Rohlfs et al 1989, Mebold, these proceedings) show that a HVC collides with Milky Way gas at a z-height of over 0.5 kpc.

From just HI 21-cm measurements (one observes after all only one velocity component with a cloud direction) it is not possible to derive the distance of individual HVCs. However, if one assumes that all high-latitude HVCs belong to one category it may be possible to arrive at rough values for their kinematic behaviour. Using the HVC-survey of Giovanelli (1980) as database and including the limits on cloud motions derived from the UV absorption lines seen in globular clusters stars (de Boer 1982, de Boer and Savage 1983, 1984), a consistency analysis was carried out by Kaelble, de Boer and Grewing (1985). They showed that most of the observed velocities fitted to the following description (see Fig.1):

* a velocity of up to 100 km s^{-1} in the direction of the disk,
* a velocity toward the rotation axis of the Milky Way of up to 100 km s^{-1},
* a velocity in the sense of the galactic rotation of about 100 km s^{-1},
* z-distances of 2 to 5 kpc.

It was noted that forcing the HVC sample to be at a distance of e.g. only 1 kpc from the disk resulted in bizarre dynamic behaviour in the sense that it would require some form of speedy flow (with respect to the LSR) of gas from the first to the third galactic quadrant. Also, the corotation hypothesis for halo-gas, used in most publications as a first order approach to understanding the HVC phenomenon, appeared to be untenable because of resulting very peculiar 3-dimensional behaviour of the HVC sample.

Based on the theory of the galactic fountain (Shapiro and Field, 1976), Bregman (1980) modelled the flow of gas condensing out of the hot and rising part of the fountain. He found that for reasonable condensation locations the pockets of neutral gas would rain down on the Milky Way disk with velocities rather similar to those known for observed HVCs.

All available data giving indications about the distances of HVCs has been collected and positive detections are assembled in Table 1. The selection criteria for inclusion were that the stars had to be at $z > 0.7$ kpc and the velocities of the gas had to differ by more than 50 km s^{-1} from the LSR. However, many negative results can be found in the literature (see in Albert 1983, de Boer et al 1989, Keenan et al 1988, Münch 1952, Pettini and West 1982, Songaila et al 1988, Danly 1989; also Blades et al, these proceedings), and it turns out that most of these stars are not in the direction of known HV gas. The list of negative results is, of course, very inhomogeneous: in the visual one has measured the CaII line region of the spectra, while in the UV mostly the FeII lines were used. This all by itself results in a very different limit in terms of equivalent HI optical depth. Be as it may, HVCs are seen only in spectra of objects at distances larger than 1.5 to 2 kpc from the Milky Way disk, while in objects with z-distances less than 1.5 kpc no HV gas is found. The observational result agrees with the consistency analysis of the 21-cm HVC sample by Kaelble, de Boer and Grewing (1985) as discussed above.

Observations of the Future

We have a lot of information on gas in the galactic halo, but only few hard facts. Neutral gas is abundant in the disk and well ionised gas starts showing up outside z = 1 kpc. This in fact defines the "disk" and gives the onset of the "halo". In the halo one finds the high-velocity clouds, in general not nearer than 1 kpc, although there are a few clear cases suggesting collisions of HVCs with disk material. Ionised gas can be observed but not the hottest component which would have been detectable through OVI.

New and deeper HI 21-cm surveys looking for HI at high galactic latitudes will probably show that there is HV gas everywhere, i.e., in all directions but 21-cm data will not bring the answer of distances any nearer than it is now. Radio-synthesis maps of individual clouds are needed to verify the usefulness of the stellar probes (Wakker, these proceedings).

Absorption line studies are the most promising but one essential problem here is that one needs a background light source of the proper kind. Extragalactic sources are of limited value; they usually are diffuse and provide limited numbers of photons and their distances are irrelevant for the questions related with the location of halo gas. Thus the best objects for absorption-line studies are stars. For studies in the visual the stellar spectrum has to be devoid of structure at the location of the interstellar absorption lines (CaII or possibly NaI). This means that the stars used either must have wide and shallow lines (fast rotators) or hardly any lines at all (hot stars). Both requirements are difficult to meet. The further away in the halo, the older the stars are (assuming no star formation in the halo) and such stars are cool and slow rotators (Petersson et al., 1983 and further papers). That leaves us with hot and old stars, which are of the subdwarf kind, or generally speaking, horizontal-branch like stars. These stars form an effective temperature continuum which runs from 7000 K for RR Lyr stars via HBA, HBB and sdB stars to 10^5 K for sdO stars. There is a difficult trade-off to be made here: the cooler stars are the more luminous ones thus can be seen and used over larger distances but they in general have a complex intrinsic continuum; the hotter stars are less luminous and thus in general are more nearby but they have usually a clean background continuum. This implies that the HBA and HBB stars can be very useful for studies in the visual, while the sdB and sdO stars are rather useful for measurements in the UV. In fact, the sdB stars seem to suffer least from the various drawbacks and might be regarded as an optimum choice as background light source. One special characteristic sets RR Lyr stars apart: they have shallow lines at their maximum brightness and seem to be useable then, but clearly the moment of maximum brightness must be known.

York et al (1986) have listed the known RR Lyr type stars in the direction of known HV complexes and that project eventually resulted in the measurements reported by Songaila, Cowie, and Weaver (1988). Also Schwarz and van Woerden and collaborators have used RR Lyr stars to look for HVC absorption. Keenan et al (1988) have searched in spectra of normal stars at high latitude; the "normality" remains a decisive factor in the interpretation of their results. Recognizing the value of sdB stars, de Boer, Heber and Richtler (1988) announced their search for sdB stars and a large number of the stars of approximately that kind present in the Palomar-Green catalogue of blue objects (Green et al 1986) have been investigated to verify their nature and to determine the luminosity and thus the distance (Möhler 1988, Möhler, de Boer et al, in preparation).

The technique of recording high dispersion spectra requires as large an amount of photons possible and sensitive detectors. The first requirement can in the visual be met by using big telescopes, but in the UV that is difficult (IUE 45 cm, HST 1 m). In the visual, the second requirement will be met by using CCDs with low readout noise and/or photon-counting systems and both of the detector types are available today. The spectral resolution required at all wavelengths is of the order of $1.5\ 10^4$ or equivalent to 20 km s^{-1}. UV absorption lines have the highest optical depth available for the detection of neutral gas.

Further H-alpha data on HVCs would be very helpful, even if only upper limits can be obtained. A big improvement here could be achieved by using a bigger telescope (Reynolds works with 15 cm). The problem here is the surface-brightness/aperture-field ratio, which can be made substantially larger than the one used by Reynolds when employing focal reducers at larger telescopes.

It is a pleasure to acknowledge the discussions I had with Guido Münch on the history of the early halo gas studies and the organisers of the meeting for their financial support.

References

Adams, W.S.: 1948, Astrophys. J. 109, 354
Albert, C.A.: 1983, Astrophys. J. 272, 509
Bajaja, E., Cappa de Nicolau, C.E., Cerosime, J.C., Louiseau, N., Martin, M.C., Morras, R., Olano, C.A., Pöppel, W.G.L.: 1985, Astrophys. J. Suppl. 58, 143
Bergeron, J., Savage, B.D., Green, R.F.: 1987, in "Exploring the Universe with the IUE Satellite", Ed. Y. Kondo, Reidel, p 703
Blades, J.C., Wheatley, J.M., Panagia, N., Grewing, M., Pettini, M., Wamsteker, W.: 1988, Astrophys. J. 332, L 75
Bregman, J.N.: 1980, Astrophys. J. 236, 577
Danley, L.: 1989, Astrophys. J. in press
de Boer, K.S.: 1982, in IAU Highlights of Astronomy, 6, 657, Reidel
de Boer, K.S.: 1985, Mitt. Astron. Ges. 63, 21

de Boer, K.S., Savage, B.D.: 1983, Astrophys. J. 265, 210
de Boer, K.S., Savage, B.D.: 1984, Astron. Astrophys. 136, L 7
de Boer, K.S., Heber, U., Richtler, T.: 1988, Astron. Astrophys. 202, 113
de Boer, K.S., Morras, R., Bajaja, E.: 1989, Astron. Astrophys. submitted
de Boer, K.S., Decker, F., Preussner, P.-R., Schwarz, U.J., Tobin, W., van Woerden, H.: 1989, Astron. Astrophys. submitted
Edgar, R.J., Savage, B.D.: 1989, Astrophys. J. in press
Giovanelli, R.: 1980, Astron. J. 85, 1155
Hirth, W., Dahlem, M., Mebold, U., Müller, P.: 1989, Astron. Astrophys. subm.
Hobbs, L.M., Blitz, L., Magnani, L.: 1986, Astrophys. J. 306, L 109
Hulsbosch, A.N.M.: 1985, in "The Milky Way Galaxy", IAU Symp. 106, Eds H. van Woerden, R.J. Allen, W.B. Burton; Reidel, p 409
Hulsbosch, A.N.M., Wakker, B.P.: 1988, Astron. Astrophys. Suppl. Ser. 75, 191
Keenan, F.P., Conlon, E.S., Brown, P.J.F., Dufton, P.L.: 1988, Astron. Astrophys. 192, 295
Kerr, F.J., Knapp, G.R.: 1972, Astron. J. 77, 354
Lockmann, F.J., Hobbs, L.M., Shull, J.M.: 1986, Astrophys. J. 301, 380
Low, F.J., et 13 al.: 1984, Astrophys. J. 278, L 19
Magnani, L., de Vries, C.P.: 1986, Astron. Astrophys. 168, 271
Magnani, L., Lada, E.A., Blitz, L.: 1986, Astrophys. J. 301, 395
McGammon, D., Burrows, D.N., Sanders, W.T., Kraushaar, W.L.: 1983, Astrophys. J. 269, 107
McGee, R.X., Newton, L.M., Morton, D.C.: 1983, Mon. Not. Roy. Astron. Soc. 205, 1191
McGee, R.X., Newton, L.M.: 1986, Proc. Astron. Soc. Australia 6, 358
Mebold, U., Cernicharo, J., Velden, L., Reif, K., Crezelius, C., Goerigk, W.: 1985, Astron. Astrophys. 151, 427
Möhler, S.: 1988, Diploma Thesis Univ. Bonn
Münch, G.: 1952, Pub. Astron. Soc. Pac. 64, 312
Münch, G.: 1957, Astrophys. J. 125, 42
Münch, G., Zirin, H.: 1961, Astrophys. J. 133, 11
Peterson, R.:, Tarbell, T.D., Carney, B.W.: 1983, Astrophys. J. 265, 972
Pettini, M., West, K.A.: 1982, Astrophys. J. 260, 561
Reynolds, R.J.: 1987, Astrophys. J. 323, 553
Reynolds, R.J.: 1989, Astrophys. J. 339, L 29
Rohlfs, R., Herbstmeier, U., Mebold, U., Winnberg, A.: 1989, Astron. Astrophys. 211, 402
Savage, B.D., de Boer, K.S.: 1979, Astrophys. J. 230, L 77
Savage, B.D., de Boer, K.S.: 1981, Astrophys. J. 243, 460
Savage, B.D., Jenkins, E.B., Joseph, C.L., de Boer, K.S.: 1989, Astrophys. J. in press
Savage, B.D., Massa, D.: 1987, Astrophys. J. 314, 380
Shapiro, P.R., Field, G.B.: 1976, Astrophys. J. 205, 762
Songaila, A., Cowie, L.L., Weaver, H.: 1988, Astrophys. J. 329, 580
Spitzer, L.: 1956, Astrophys. J. 124, 20
van Woerden, H., Schwarz, U.J., Hulsbosch, A.N.M.: 1985, in "The Milky Way Galaxy", IAU Symp. 106, Eds. H. van Woerden, R.J. Allen, W.B. Burton; Reidel, p 387
York, D.G.: 1982, Ann. Rev. Astron. Astrophys. 20, 221
York, D.G., Burks, G.S., Gibney, T.B.: 1986, Astron. J. 91, 354

Discussion:

VÖLK: How strong is the empirical evidence against corotation already at heights $|Z| = 10$ kpc? Theoretically, one should expect corotation certainly up to about 25 kpc, because up to this heights the Alfven velocity should exceed systematic fluid speeds of the halo.

de BOER: The absorption line measurements toward M13 (star Barnard e.g.) show most clearly that there is gas which does not rotate along with the disk. M13 is at l$\sim$60, b$\sim$45 and distance about 6kpc, so Z = 4kpc. Absorption is seen down to about -100 kms^{-1} whereas +20 kms^{-1} is expected with corotation. This means that the gas between us and M13 stays behind, unless it has an extreme peculiar velocity. The analysis of the 21cm HVC data base confirms in other directions that these clouds do rotate along but with velocities well below the disk 220 kms^{-1} (see Kaelble, de Boer, Grewing 1985).

A NEW HIGH-RESOLUTION OPTICAL STUDY OF HALO GAS

C. E. Albert
U.S. Naval Academy, Annapolis, MD, USA

J. C. Blades
Space Telescope Science Institute, Baltimore, MD, USA

D. C. Morton and M. Proulx
Herzberg Institute of Astrophysics, Ottawa, ON, Canada

F. J. Lockman
National Radio Astronomy Observatory, Charlottesville, VA, USA

Abstract: We present preliminary results from a new high resolution optical study of halo gas at the coudé focus of the Canada - France - Hawaii Telescope. Our work is still in progress so two general results are presented here: significant absorption is produced in interstellar gas beyond 500 pc from the galactic plane, and well-resolved halo clouds are identified.

1. Introduction.

The identification and study of interstellar clouds in the galactic halo has been a subject of interest since the pioneering work of Münch and Zirin in 1961. They measured the velocities and intensities of absorption lines of Ca II and, in some cases, Na I toward 24 high latitude OB stars ranging from 200 pc to 2 kpc beyond the plane. Their observations showed a real increase with distance in the number of Ca II components per star, implying that some of the observed gas must be located at distances up to a kiloparsec from the galactic plane. This important study prompted further work by, among others, Greenstein (1968), Rickard (1972), Richstone and Morton (1975) and Cohen and Meloy (1975).

Morton and Blades (1986) surveyed Ca II absorption lines toward a total of 25 extragalactic objects. Somewhat surprisingly, they found that, along these "infinite" path lengths, the Ca II is confined to a thick disk with a scale height of about 1 kpc. This result was consistent with Albert's (1983) study of the lower halo, where high resolution observations of Ti II, Ca II, Na I and 21 cm H I revealed 2 types of cold gas with distinct distribution, kinematics and abundance. A thick, low-velocity component appears to extend from the plane to well outside the thin disk of OB stars and, in addition, higher velocity gas is observed only beyond the foreground stars. Since singly ionized titanium is the dominant ionization stage in H I regions (Wallerstein and Goldsmith 1974), the ratio of the column densities N(Ti II)/N(H I) gives a direct lower limit to the depletion of the interstellar gas. The "thick" component is characterized by a gas phase titanium abundance averaging 6 times that of the disk and probably includes a complex mixture of cloud and intercloud regions as well as, perhaps, low velocity gas near its turning point in a galactic fountain. The higher velocity gas is even less depleted, with an average gas phase titanium abundance 15 times that of the disk.

2. Data.

Our new CFHT observations extend the earlier studies to 26 lines of sight, including 17 with one or more nearly-aligned foreground stars as well as local stars along 6 extragalactic sight lines studied by Morton and Blades. With the good ultraviolet throughput of CFHT, we were able to observe Ca II toward 64 stars and Ti II toward 40 stars, which more than doubles the number of sight lines observed at high resolution. The corresponding 21 cm emission lines of H I were obtained at the 140 foot telescope of the National Radio Astronomy Observatory in October 1987.

The goal of our study is to extend our previous studies to determine the distribution, kinematics and titanium abundances of a significant sample of halo clouds.

Figure 1 shows representative spectra along the line of sight toward a distant star, HD 119608, first studied by Münch and Zirin. Three stars are observed, at increasing distance above the galactic plane: HD 117880 (z = 420: pc), HD 123884 (z = 1000 pc) and HD 119608 (z = 2600 pc). All the continua are normalized to unity, and the length of the vertical scale from zero to one is the same for all spectra. The K line of Ca II is shown for all three stars and Ti II (λ 3384 Å) is observed toward the distant star; the instrumental profile full width at half maximum is approximately 8 km/s. We detect only one Ca II interstellar cloud toward the foreground star at z = 420 pc. At z = 1000 pc, HD 123884 shows two strong Ca II components, implying that substantial amounts of gas must be located between 420 - 1000 pc. The same structure is seen toward HD 119608 and is echoed in the strong Ti II line. All lines show additional weak unresolved components. However, the equivalent width of Ca II toward HD 123884 is stronger than that toward the more distant HD 119608 so the interstellar gas is evidently quite clumpy and relatively little material seems to exist beyond 1000 pc. These lines of sight are plotted in Figure 2: we seem to be detecting a large inhomogeneous cloud well below the star at z = 2600 pc, but above 420 pc. This example illustrates two major results of our study: an increase in integrated interstellar equivalent width and total absorption velocity range beyond the galactic plane and the identification of a set of discrete clouds with good distance limits. The latter will be a major focus of our abundance analysis.

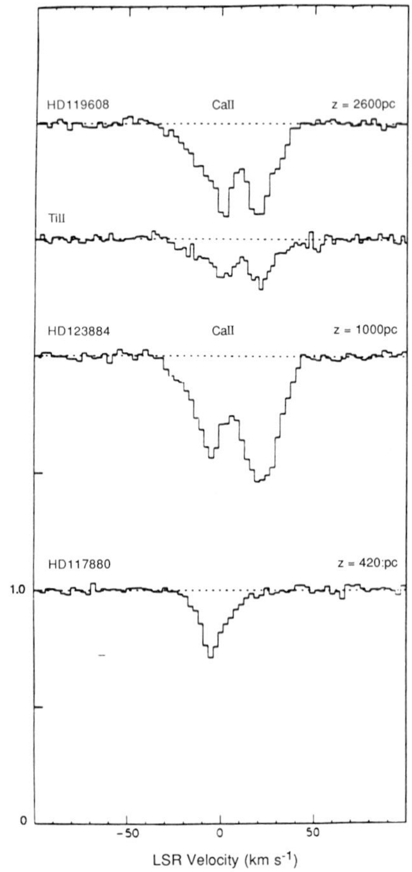

FIG. 1. - Interstellar absorption lines along the line of sight toward HD 119608 and foreground stars. All continua are normalized to unity and the length of the vertical scale from 0 to 1 is the same for all spectra.

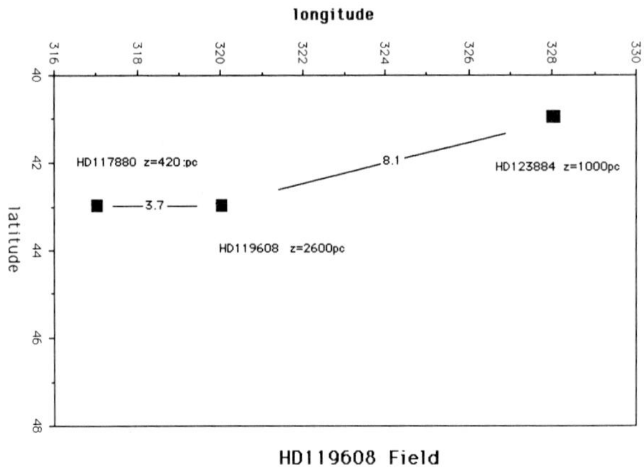

FIG. 2. - HD 119608 field. The separation between stars is shown in degrees.

3. Integrated Equivalent Widths and Absorption Velocity Ranges.

There is a clear increase in both the total equivalent widths and the absorption velocity ranges from the average distance to the foreground stars (z ~ 145 pc) to the average distance to the halo stars (z ~ 1100 pc). Prominent low-velocity absorption is observed in all directions, but higher velocity gas appears beyond the foreground stars, including several discrete high velocity clouds at both positive and negative velocities. Figure 3 depicts observations of the K line of Ca II toward all the stars, where the data are averaged in three distance ranges: z < 500 pc, 500 < z < 1000 pc and z > 1000 pc. There is substantial material between 500 - 1000 pc, but not significant gas beyond a kiloparsec. This confirms our earlier work.

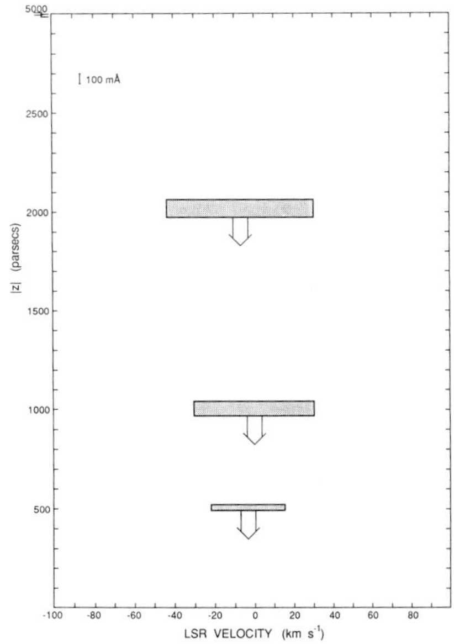

FIG. 3. - Integrated equivalent widths of the K line of Ca II toward 64 high latitude stars, averaged for 3 distance ranges: z < 500 pc, 500 < z < 1000 pc and z > 1000 pc. The vertical height of each bar is proportional to the total observed equivalent width within the indicated velocity ranges. The small symbol in the upper left corner denotes the height corresponding to an equivalent width of 100 mA.

4. Discrete Halo Clouds.

Twenty-three good cases of individual high latitude Ca II clouds have been identified, including several examples of clouds well outside the galactic plane. These are resolved clouds, chosen outside the average velocity range of absorption toward stars with z < 500 pc ($-22 \leq v_{LSR} \leq +14$ km/s). In 10 cases, the velocity exceeds ± 70 km/s and six of these are in directions showing high velocity H I clouds in Wakker's (1989) survey (also Hulsbosch and Wakker 1988). The signs of the Ca II cloud velocities are the same as those observed in H I.

The Ca II absorption lines are generally weak, with log N(Ca II) typically ranging from 10 - 11, but a few strong components are observed. Some of these Ca II clouds are detected in the 21 cm spectra while others are not. It is likely that all of these clouds have low H I column densities. For example, for cosmic abundances, even if all

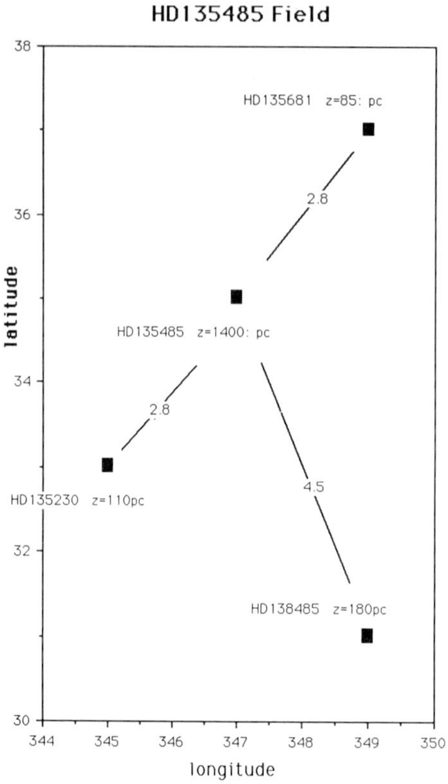

FIG. 4. - HD 135485 field. The separation between stars is shown in degrees.

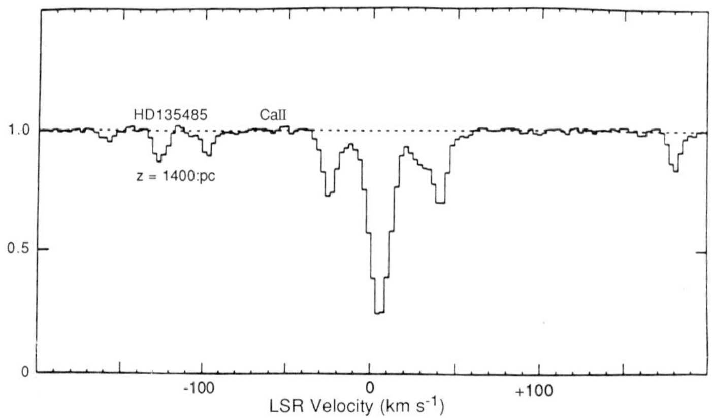

FIG. 5. - The interstellar Ca II K line along the line of sight toward HD 135485.

the calcium in the cloud toward HD 3379 (log N(Ca II) = 10.54) were singly ionized, the corresponding H I column density of log N(H I) ≃ 17 would be too low for us to detect. Thus, these observations alone do not allow us to conclude whether the clouds are neutral or ionized.

A basic conclusion of this work is that we do not find a significant population of massive high velocity clouds: there are certainly several interesting individual cases, but they seem to contain little mass. This material also appears to be quite patchy with a great deal of spatial structure.

The line of sight toward HD 135485, originally studied by Münch and Zirin, provides a good example: the observed field is plotted in Figure 4 and the Ca II K spectrum toward HD 135485 is shown in Figure 5. Only weak low velocity Ca II is detected toward the three foreground objects in this direction. However HD 135485, 1400: pc above the plane, shows complex and strong Ca II over a wide velocity range. At low velocities, the 21 cm H I emission is consistent toward all four stars, implying that a large cloud covers much of this area. In HD 135485 there are spectral lines which, if attributed to interstellar Ca II K, would have LSR velocities of -98 and -128 km/s. These are not detected toward the foreground stars in this field. The features have possible, but very weak, H I emission counterparts. For depleted disk clouds the observed log N(Ca II) $\simeq$ 11 would be easily detected at H I. These halo clouds appear to be significantly less depleted than typical disk clouds, which is consistent both with our earlier work and with Danly's (1989) ultraviolet observations toward halo stars.

REFERENCES

Albert, C. E. 1983, Ap.J., 272, 509.
Cohen, J. G., and Meloy, D.A. 1975, Ap.J., 198, 545.
Danly, L. 1989, this volume.
Greenstein, J. L. 1968, Ap.J., 152, 431.
Hulsbosch, A. N. M., and Wakker, B. P. 1988, Astr. Ap. Suppl. Ser., 75, 191.
Morton, D. C., and Blades, J. C. 1986, Mon. Not. R. astr. Soc., 220, 927.
Münch, G., and Zirin, H. 1961, Ap.J., 133, 11.
Richstone, D. O., and Morton, D. C. 1975, Ap. J., 201, 289.
Rickard, J. J. 1972, Astr. Ap., 17, 425.
Wakker, B. 1989, this volume.
Wallerstein, G., and Goldsmith, D. 1974, Ap.J., 187, 237.

Discussion:

MÜNCH (Comment): Your IS line profiles in HD119608 seem to show distinct extended wings, not apparent in the lower resolution and S/N-ratio photographic observations I made 30 years ago. Their appearance suggests their origin is a superposition of many weak unresolved lines. From the behaviour of the Ca II and Na I lines arising in high velocity clouds, Spitzer and Routly, first, and later myself, suggested a relation between the kinematics of clouds and their physical state. On basis of your observations, we should be able to establish in a more quantitative fashion that relation between kinematics (or distance to the galactic plane) and degree of ionization, electron temperature, etc.

GALACTIC WINDS

Heinrich J. Völk[1], D. Breitschwerdt[1], J.F. McKenzie[2]
[1]Max-Planck-Institut für Kernphysik, Heidelberg
[2]Max-Planck-Institut für Aeronomie, Lindau

1. Introduction

In the sequel we will be discussing that - apart from all other halo flows - it is highly probable that our Galaxy has a systematic mass loss; most other galaxies should behave in a similar way. This mass loss should occur in the form of a supersonic galactic wind provided the intergalactic pressure is low enough. We will investigate this possibility mainly for our Galaxy. We shall argue [1] that the energetic particle component of the interstellar medium, the cosmic rays (CRs), is basically responsible for driving a wind for typical gas temperatures below 10^6 K. Thus for asymptotic wind velocities of the order of the escape velocity, the mass flux is essentially determined by the observable CR energy loss from the Galaxy. Galaxies with very hot gas on the other hand hardly need the CR component and drive a wind almost exclusively through their thermal pressure.

2. Is the halo on the average static or convective?

Presumably the gas flows in the halo above the disk (defined by the cool neutral hydrogen gas layer with a half-thickness of about 100 pc) are of a three-dimensional nature [2]. Therefore an upward flow at one point above the disk does not imply that an upward flow must occur elsewhere as well. On the contrary, even with locally nonzero mass flows the halo can be static on average with zero average mass velocity. From a gas dynamic point of view such a halo is entirely compatible with a Galactic fountain picture [3,4], where upwelling parcels of gas reach finite heights above the plane to fall back afterwards. However it is not clear whether the CRs can escape through the gas to extragalactic space withough dragging part of the gas with them. At least this appears as an extreme assumption. As another possibility the halo could on average be convective. This could either be due to a mass loss, say in the form of a wind, or due to mass accretion. Mass accretion could occur in the form of infall of extragalactic clouds as advocated a long time ago by Oort [5] and suggested by the 21 cm observations of Mirabel and Morras [6] and van Woerden et al. [7]. Again, due to the three-dimensional nature of the halo flows, either wind or infall, or both simultaneously might occur, together with a system of fountains. In all cases the CRs will escape to "infinity". We consider a (stochastic) combination of wind outflow and cloud accretion as most likely.

3. Cosmic Ray Component in the Galaxy

The CRs constitute a High Energy Component of the interstellar medium (ISM) which appears to be quasi-steady with a pressure in the disk $p_{cd} \simeq 3 \cdot 10^{-13}$dyn cm$^{-2} \lesssim p_{gd} \simeq 6 \cdot 10^{-13}$dyn cm^{-2}, where p_{gd} is the pressure of the thermal disk gas. The outward CR energy flux density at the disk boundary, based on the spallation grammage of GeV particles, is $F_{cd} = 3.6 \cdot 10^{40}$ erg sec$^{-1} \cdot A^{-1}_{gal,d}$, where $A_{gal,d}$ is the surface area of the galactic disk (e.g. [8]). Actually, because the grammage decreases with increasing particle energy, F_{cd} should be multiplied by a factor of order 3. The total energy loss $\simeq 10^{41}$ erg/sec due to CRs then corresponds to about 10 percent of the energy produced by supernovae in the Galaxy (assuming one supernova of 10^{51} erg per 30 years). There is no radiative cooling of the CR component. Upon its drift through the gas it excites MHD waves, basically Alfvén waves. These waves tend to propagate outwards into the halo. CR particles scatter on the associated magnetic fluctuations and thus transfer outward momentum to the gas. The basic alternative is whether there occurs an outflow together with the gas or whether the particles just diffuse outwards leaving the gas behind.

4. Halo dynamics

In analogy to the solar corona we consider the galactic magnetic field configuration as a mixture of open and closed field lines. Open field lines should be due to the production of hot gas and CRs by OB stars and supernova remnants in the upper disk, or due to the Parker instability of a stratified medium [9]. The CRs diffuse in the frame of the outward propagating Alfvén waves towards $|z| = \infty$, perpendicular to the plane of the Galaxy (Fig. 1). This combination of Alfvénic drift and diffusion of the CRs through the gas [10] constitutes a form of "heat conduction". Ionisation of the thermal gas should be effectively maintained by upper disk OB stars, halo stars and globular cluster stars from "within", and quasars from "outside". At heights $|z| \gtrsim 1$ kpc above other (hydrodynamic) wave sources the CR pressure gradient should lead to a very strong resonant generation of waves in the fully ionized halo gas. As a result, the coupling of the thermal plasma and the CRs is strong (e.g. [11]). To lowest approximation we then can neglect CR diffusion keeping only their drift with the waves through the plasma. Together with the wave pressure gradient this implies a sizeable outward force density on the thermal gas. The neglect of CR diffusion in the halo (but not in the disk up to $|z| \simeq 1$ kpc!) is consistent with an energy dependent escape of the CRs, as it is inferred from observations.
Thus we consider an adiabatic flow of a 3-component system consisting of thermal gas, waves and CRs (cf. [12]) at heights $|z| \gtrsim$ few kpc above the galactic midplane. Neglecting forces due to the average magnetic field $\underline{B}$, we adopt a one-dimensional flux tube model with "radially" diverging flux tubes at distances exceeding the radius of the galactic disk. Such a model neglects centrifugal effects with their obvious tendency to strengthen any wind flow.

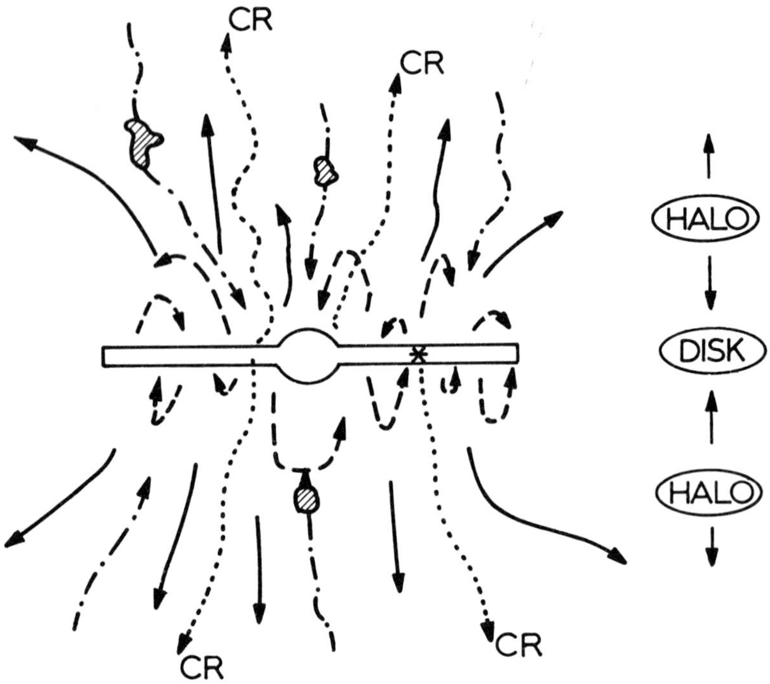

Figure 1: Cartoon of the various possible halo flows. The dashed lines correspond to a fountain flow. The dash-dot lines indicate extragalactic clouds (hatched) falling onto the Galaxy disk, indicated schematically. A galactic mass loss flow is given by the solid lines, whereas the escaping CRs correspond to the dotted flow lines.

5. Sample solutions

The hydrodynamical theory is described in a number of papers [13,14,15]. Here we will confine ourselves to a discussion some sample results. Taking the inner boundary at $\mid z_0 \mid = 1$ kpc (reference level), with a density $n_0 = 10^{-3}$ H-atoms cm^{-3} and a gas temperature $T_0 = 10^6$K, a small CR pressure p_{co}=10^{-13} dyn cm^{-2}, and B_0= 1μG, we take a gravitational potential due to a massive dark halo $(8 \cdot 10^{11} M_\odot)$ apart from a disk $(2.5 \cdot 10^{11} M_\odot)$ and a bulge $(2 \cdot 10^{10} M_\odot)$ contribution [16]. The wave pressure $p_{wo} = 10^{-2} \cdot B_0^2/8\pi$ is taken quite small at the reference level to obtain a conservative estimate.

For a magnetic flux tube at a distance R_0=10 kpc from the galactic center the resulting flow exhibits a critical point at a height of $\mid z_c \mid = 34$ kpc. This implies that the intergalactic pressure is sufficiently small. The flow speed u_0 at the reference level is about 10 km sec^{-1}, reaching an asymptotic terminal velocity u_f= 310 km sec^{-1} at $\mid z \mid \gg \mid z_c \mid$. The CR and wave pressures dominate the gas pressure for $\mid z \mid \gtrsim 10$ kpc, demonstrating that the wind is basically driven by the CRs which have a much softer equation of state than the gas (Fig. 2).

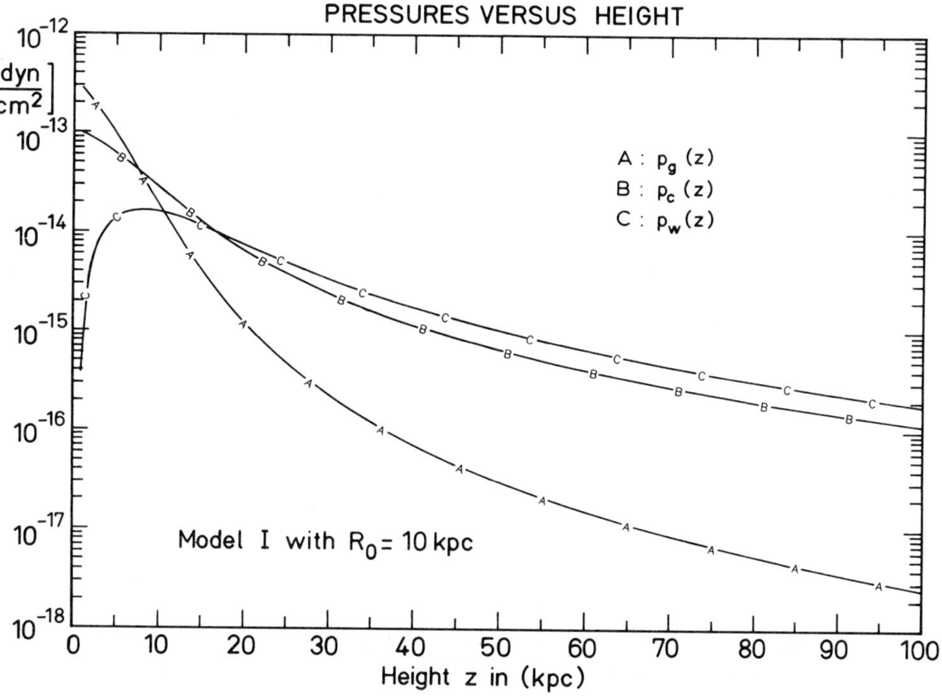

Figure 2: Results of an outflow calculation along a flux tube $| z |>| z_0 |$, starting at the radial distance R_0= 10 kpc in the disk and a reference level $| z_0 |$= 1 kpc above it. The three pressures p_g, p_c, and p_w, given by the symbols A, B, and C on the curves correspond to the pressures of the thermal plasma, the CRs, and the scattering waves.

Keeping, in a simple-minded approach, the boundary conditions at $| z_0 |$, while varying R_0, shows a monotonic increase (decrease) of $n_0 u_0 u_f$ with decreasing R_0, u_f being about twice the escape speed from the point $| z_0 |$, R_0. The resulting galactic mass loss rate is $\dot{M}_{gal} \simeq 0.3 M_\odot/yr$ for the conservative CR energy loss rate $F_{cd} \cdot A_{gal,d} = 3.6 \cdot 10^{40}$ erg $\sec^{-1}$ quoted above.
Assuming $\dot{M}_{gal} \simeq$ const. for a galactic age of the order of 10^{10}yr implies that over this period a mass of the order of the entire present ISM will have been lost from the Galaxy. This is a significant but not excessively large mass loss.
Without CR effects taken into account (p_{co}=0) one needs for a wind gas temperatures $T_{go} > 8 \cdot 10^6$ kpc at $R_0 = 1$ kpc, and $T_{go} > 4 \cdot 10^6$ kpc at $R_0 = 10$ kpc, including centrifugal effects [16,17,18]. The average hot ISM is certainly much below such temperatures. However in "supershells" temperatures of such magnitudes might exist [19]. For such regions a purely thermally driven wind might result, as generally for galaxies with sufficiently large hot ISM temperatures.

6. Conclusions

The existence of galactic winds clearly depends on the assumption of a sufficiently small intergalactic pressure. Thus in galaxy clusters with a high intracluster pressure the individual galaxies may not exhibit a wind, whereas the cluster as a whole should show a "cluster wind".

Winds do not constitute a contradiction to fountains on closed or even open field lines ("coronal arcs") in galaxies, or to cooling flows in clusters.

Observations have as yet not given X-ray halos indicative of a wind [20,21]. On the other hand we interprete the radio halos and the polarization characteristics of NGC 4631 [22], or NGC 891 (Hummel, private communication) as indicative of open magnetic field configurations. These are most likely produced by mass outflows.

References

1. F. Ipavich: Astrophys. J. 196, 107, (1975)
2. G. Münch, H. Zirin: Astrophys. J. 133, 11 (1961)
3. P.R. Shapiro, G.B. Field: Astrophys. J. 205, 762 (1976)
4. F.D. Kahn: In Investigating the Universe (D. Reidel Publ. Comp.) p. 1 (1981)
5. J.H. Oort: Bull. Astron. Inst. Netherlands 18, 421 (1966)
6. I.F. Mirabel, R. Morras: Astrophys. J. 279, 86 (1984)
7. H. van Woerden, U.J. Schwarz, A.N.M. Hulsbach: In The Milky Way Galaxy, ed. by H. van Woerden et al. (D. Reidel Publ. Comp.), p. 387 (1985)
8. V.L. Ginzburg: Proc. 20th Int. Cosmic Ray Conf. 7, 7 (1987)
9. E.N. Parker: Astrophys. J. 145, 811 (1966)
10. J. Skilling: Mon. Not. Roy. Astron. Soc. 172, 557 (1975)
11. H.J. Völk: In High Energy Astrophys., Proc. 19th Rencontre de Moriond, ed. by Tran Than Van (Editions Frontières, Gif-sur-Yvette), p. 281 (1984)
12. J.F. McKenzie, H.J. Völk: Astron. Astrophys. 116, 191 (1982)
13. D. Breitschwerdt, J.F. McKenzie, H.J. Völk: In Interstellar Magnetic Fields, ed. by R. Beck and R. Gräve (Springer-Verlag Berlin, Heidelberg), p. 131 (1987a)
14. D. Breitschwerdt, J.F. McKenzie, H.J. Völk: Proc. 20th Int. Cosmic Ray Conf. 2, 115 (1987b)
15. J.F. McKenzie, D. Breitschwerdt, H.J. Völk: Proc. 20th Int. Cosmic Ray Conf. 2, 119 (1987)
16. A. Habe, S. Ikeuchi: Progress in Theor. Phys. 64, 1995 (1980)
17. R.A. Chevalier, W.R. Oegerle: Astrophys. J. 227, 398 (1979)
18. J.N. Bregmann: Astrophys. J. 236, 577 (1980)
19. C. Heiles: this volume (1989)
20. D.P. Cox, D. McCammon: Astrophys. J. 304, 657 (1986)
21. J.N. Bregmann, A.E. Glassgold: Astrophys. J. 263, 564 (1982)
22. E. Hummel, H. Lesch, R. Wielebinski, R. Schlickeiser: Astron. Astrophys. 197, L29 (1988)

Discussion:

MEBOLD: I want to add the galaxy NGC 1808 to the list of galaxies you used as observational illustrations. Its morphology is suggestive of a poloidad magnetic field symmetrical to the center of the galaxy. Recent observations with the VLA have supported this suggestion.

VÖLK: I have not seen the NGC 1808 morphology up to now and would like to see it. A poloidal field is of course not necessarily open and thus per se does not allow a mass loss to infinity. For that it must be pushed open by an internal "gas" overpressure.

PECKER: If Heiles (and Terlevich) are right, the SN explosions are, and have been, constantly heating local holes in the galactic disk matter; therefore, UV radiation might flow more easily in the halo, and radiation pressure alone (Lyα) acting on neutral hydrogen might be an efficient "motor" for a wind. How could it compare with your model of galactic C.R. driven galactic wind?

VÖLK: I certanly believe that UV photons will escape into the halo and add to the gas scale height there as well as to the ionization level. I have not compared photon radiation pressure to CR pressure effects but I agree that it should be done.

Cycling of dust grains through the galactic halo

A. Ferrara+, J. Franco*, B. Barsella†, F.Ferrini†

+ Dipartimento di Astronomia, Università di Firenze, Firenze, Italy
* Instituto de Astronomía, UNAM, México City, México
† Istituto di Astronomia, Università di Pisa, Pisa, Italy

Abstract

The analysis of the vertical motion of dust grains inside the disk of spiral galaxies shows that grains at high latitudes ($z/H > 0.3 - 0.5$, where H is the scale height of the gaseous disk) can be removed from the gaseous disk by radiation pressure from field stars. Once the dust particles are located at the base of the galactic corona, and neglecting the presence of magnetic fields, their motion is determined by the global light and mass distributions. We follow the dynamical evolution of grains embedded in a hot galactic corona ($T \geq 10^5$ K and $n \leq 10^{-2}$ cm^{-3}), including drag and sputtering. We find that typical sputtering time scales are larger than 10^8 yrs, and, in these times scales, the grains can travel large radial and vertical distances from their original positions. This process can remove grains from the inner part of a galaxy and send them to large galactocentric distances. The evolution resembles a dusty galactic fountain which may pollute the outer galaxy with grains and heavy elements. Similarly, for some cases the dust particles can be detached from the galactic gravitational field and they are expelled to the intergalactic medium. In these cases, the chemical pollution is not restricted to the host galaxy and may reach material which is otherwise primordial.

1. Introduction

Dust grains can be pushed by the radiation pressure produced by the stellar photon field and they, in turn, can transmit their momentum to the surrounding gas. Franco (1989) has demonstrated that small dusty clouds can be expelled from the disk by radiation pressure. The process, which is called "photolevitation" and requires an efficient dust–gas coupling, depends upon the cloud column density, cloud–cloud velocity dispersions, and on the vertical distribution of disk stars. In the absence of magnetic fields, the minimum gas column density for dust–gas coupling is about $N \sim 10^{19}$ cm^{-2} and dust grains can drift inside the cloud within these column densities. The clouds get flattened during the upward acceleration process and some grains may detach from the cloud. Similarly, given the patchiness of the interstellar medium, some naked grains can escape from the disk through low density "tunnels".

Here we present, as discussed in Barsella *et al.* (1989b), the dynamical evolution of naked grains that are able to drift into the base of the galactic corona. The properties of the interstellar drag change drastically in the transition region between the disk and the halo: the interactions with localized dense gaseous structures disappear and sputtering with the hot, but substantially less dense, coronal gas becomes more prominent.

2. The model for the evolution of dust grains in the halo

The equation of motion is described in detail in Barsella *et al.* (1989b) and contains three terms: radiation pressure, gravity and gas drag. The optical grain properties are taken from Draine and Lee (1984) and the gravitational field and light distributions are simulated as in Barsella *et al.* (1989a). The effects of the halo gas, drag and sputtering, are followed with the scheme described by Draine and Salpeter (1979). Evaporation effects due to the hard radiation field expected in the halo are not considered.

For simplicity, to maintain the variation of the physical processes to a minimum, the pressure and density of the coronal gas are assumed constant. The considered values are in the range $10^{-3} - 10^{-2}$ cm^{-3} and $3 \leq \log(p/k) \leq 4$, where p is the pressure and k is the Boltzmann constant. These are commonly considered as typical values (see *e.g.* Pettini *et al.* , 1982; de Boer, 1985) and determine the thermodynamical properties of the coronal gas in our model. Note that the assumed values may represent upper limits for the halo, and the effects are rather severe for what concerns to sputtering and gas drag, allowing a good confidence in the results.

3. Results

The motion of grains has been studied for the galaxy NGC 3198. The parameters have been taken by Kent (1987). We used bare grains of astronomical silicate with an initial radius of 100 nm, typical in a galactic environment. The initial position has been chosen at a height of 100 pc above midplane and at a variety of galactocentric distances (from 0.1 to 2.0 kpc). The grain has zero initial velocity relative to the gas, and the coronal gas is assumed to rotate at the same rate of the galaxy.

The figures show the model results for NGC 3198. Fig. 1 displays the height evolution as a function of time for two different choices of pressure and density of the coronal gas: solid curves correspond to $\log(p/k) = 3$ and $n = 10^{-3}$ cm^{-3}, and the dashed ones correspond to the same pressure but $n = 10^{-2}$ cm^{-3}. Each case is shown for two different initial positions, the upper and lower curves correspond to initial galactocentric distances of 0.1, 1 kpc, respectively. The height Z is measured in units of the halo radius, which for this galaxy is $R_H = 33$ kpc.

The evolution of the grains sizes, in nm, is shown in Fig. 2; the upper dashed curve corresponds to the initial location near the galactic center, and the lower curve to an initial galactocentric distance of 1 kpc. For the lower density case, $n = 10^{-3}$ cm^{-3}, there is almost no sputtering of the grains. This plot shows clearly that dust can survive for a rather long time. We have also explored models by doubling the efficiency of sputtering given by Draine and Salpeter (1979) (i.e. to mimick the eventual effect of X==rays and cosmic rays in the hot galactic corona). While there is no appreciable change in the evolution for grains in the lower density case,for $n = 10^{-2}$ cm^{-3} the grains reach a complete desruption after 4.5×10^8 yrs.

Finally, Fig. 3 shows the radial spread of the grains starting from height $z = 100$ pc and at galactocentric distances $r = 0.1$ and 1 kpc. It appears that the radial position attained is of the order of the radius of the outer arm zone. It is clear from this plot that dust grains are removed from the internal regions of the galaxy, where the density of stars producing grains is higher.

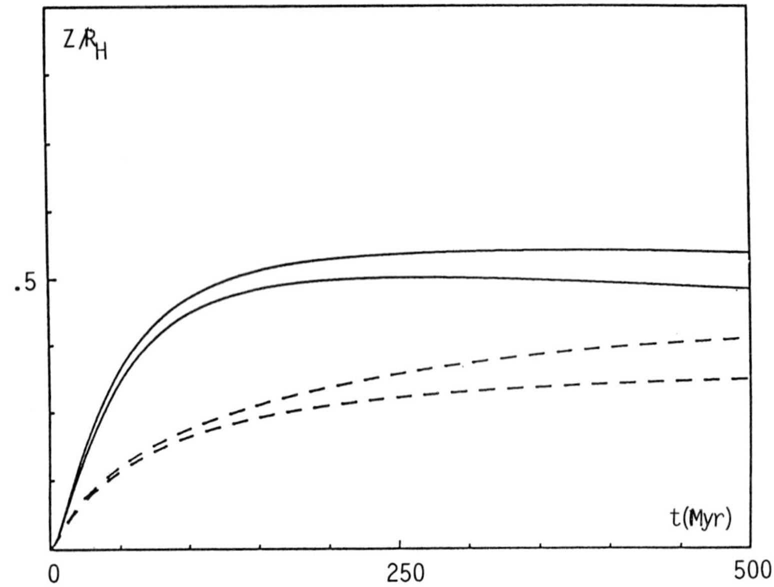

Fig. 1. Evolution of the grain height (in terms of the halo radius, R_H = 33 kpc). The solid lines correspond to $\log(p/k) = 3$ and $n = 10^{-3}$ cm^{-3}; dashed lines correspond to the same pressure but $n = 10^{-2}$ cm^{-3}. The calculation has been done for a galactocentric start distance of 0.1 and 1 kpc.

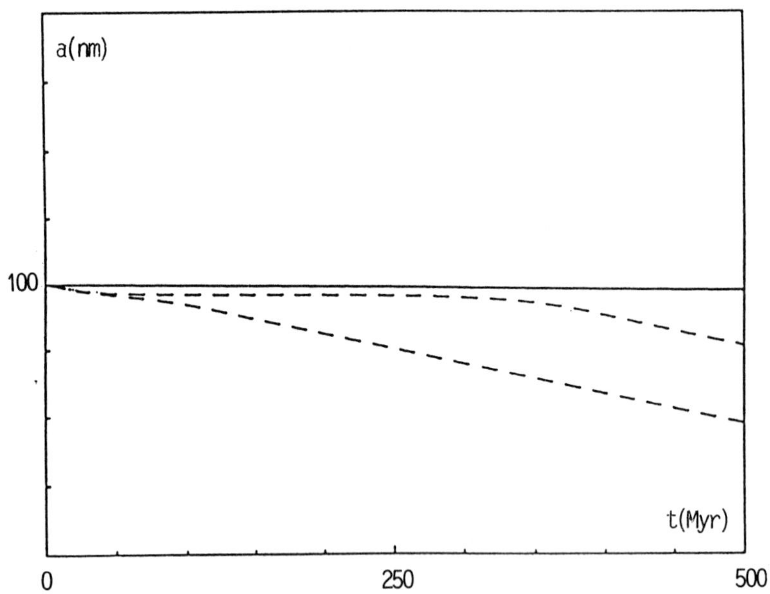

Fig. 2. Evolution of the grain radius as a function of time. The solid and dashed lines have the same meaning as in Fig. 1

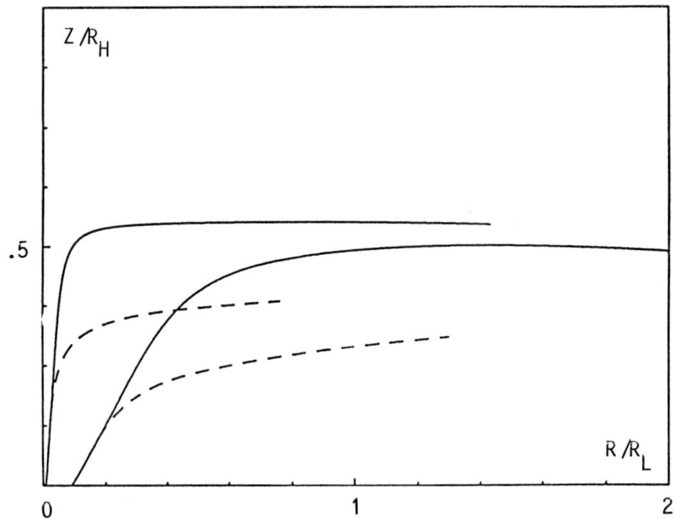

Fig. 3. $z - r$ trajectory of the grain; the ordinate is measured in units of the halo radius, the ascissa in units of the luminosity radius $R_L = 10.8$ kpc. The solid and dashed lines have the same meaning as in Fig. 1

An interesting consequence of the large-scale motion and of the erosion suffered by dust grains during their flight in the galactic halo, is the pollution of heavy elements as well as of small grains in regions that are otherwise poor in these interstellar components. The future work will concerns an extention to other galaxies, following as a guideline the static analysis of the forces by Ferrara *et al.* (1989), which revealed a continuity of behaviours in the galaxies, ranging from a almost complete capture of grains in the disk to their efficient expulsion out the gravitational field of the whole galaxy.

References

Barsella, B., Ferrini, F., Greenberg, J.M., and Aiello, S. 1989a, *Astron. Astrophys.*, 209, 349

Barsella, B., Franco, J., Ferrini, F., and Ferrara, A. 1989b, in preparation

de Boer, K. S. 1985, in "The Milky Way Galaxy", IAU Symp. 106, 415, H. van Woerden *et al.* (eds)

Draine, B.T., and Lee, H.M. 1984, *Ap. J.*, 285, 89

Draine, B.T., and Salpeter, E.E. 1979, *Ap. J.*, 231, 77

Ferrara, F., Ferrini, F., Barsella, B., and Aiello, S. 1989, submitted to *Astron. Astrophys.*

Franco, J. 1989, preprint

Kent, S.M. 1987, *Astron. J.*, 93, 816

Pettini, M., West, K., 1982, *Ap. J.*, 260, 561

Discussion:

VÖLK: Did you consider grain charging in the context of grain-gas coupling?.

FERRARA: No, the project is in the early phases and we are analysing the simplest cases (i.e. no B-fields, constant halo pressure and density, etc.). We will address this issue, and others, in the near future.

THE INTERMEDIATE VELOCITY CLOUD IVC86+38.5-45, RELATED TO HIGH VELOCITY CLOUDS?

U. Herbstmeier

Radioastronomisches Institut der Universität, Bonn

Examining the high velocity cloud complex C and its possible interaction with gas of the galactic disk we became aware of a dust cloud located at galactic longitude $l = 86°$ and latitude $b = 38\overset{\circ}{.}5$ seen in the $100\mu m$ image of the IRAS. This cloud consists of a cucumber shaped core surounded by a horseshoe shaped halo opening towards larger galactic coordinates. At very low $100\mu m$ intensities dust lanes starting at the core or the ends of the horseshoe can be traced for about 5° towards high galactic latitude and longitude.

HI observations with the 100m telescope in Effelsberg reveal that these lanes run parallel to an HI spur at velocities of about $-120\ km\ s^{-1}$. This high velocity cloud has its tip at the end of the dust core which shows up in the HI spectra at velocities of about $-45\ km\ s^{-1}$. The spatial anticoincidences mentioned, the similarity in direction, and the horseshoe shaped structure of the intermediate velocity cloud are evidence for an interaction between high velocity clouds and the surrounding material. The rather high ratios of the IRAS $60\mu m$ to $100\mu m$ emission and between the $100\mu m$ intensities and HI column densities respectively suggest that the dust is probably heated up by this collision.

THE NaI INTERSTELLAR SPECTRUM OF HVC 287.5 + 225 + 240

P. Molaro *et al.*

Osservatorio Astronomico di Trieste, Italy

We present high resolution (FWHM≈7 km/sec) and high signal-to-noise observations of the bright Seyfert galaxy NGC3783 obtained with the CES spectrograph linked by fibres to the 3.6m telescope at La Silla, Chile.

Three NaI components have been detected in the Galactic range of radial velocities: one at $V_{HEL} = -45\ km/s$ and two, which represent the main Galactic absorption, at $V_{HEL} = -10\ km/s$ and $V_{HEL} = -5\ km/s$. For the -45 km/s component the ratio CaII/NaI is about 9, pointing for a halo origin of this cloud.

No traces of NaI absorptions are found at +40 km/s and +64 km/s, where CaII components have been found by West et al. (1985, M.N.R.A.S. 215,481). The resulting CaII/NaI ratios are thus even greater of what found in this previous work, providing a further indication of a halo origin for these clouds, where Ca is dominantly in the gaseous form.

We have a detected a NaI component at $V_{HEL} = 240\ km/s$, related to the HVC 287.5+22.5+240 (Morras and Bajaja, 1983, Astron. Astrophys. Suppl. Ser. 51, 131). Combining this result with the CaII from West *et al.* we derive CaII/NaI≃1.7. This very low value of the CaII/NaI ratio gives support to the suggestion of West *et al.* that the high velocity observed is merely the effect of the Galactic rotation and that the cloud is located outside our Galaxy.

COLLISIONS BETWEEN HIGH LATITUDE CLOUDS: THEORY MEETS OBSERVATIONS

J.C. Lattanzio[1] and E.R. Keto[2]

[1]Institute of Geophysics and Planetary Physics

[2]Lawrence Livermore National Laboratory

We report on fully three-dimensional hydrodynamic and radiative transfer simulations of collisions between high latitude clouds. Our model uses the smoothed particle hydrodynamics code described by Lattanzio and Henriksen (1988) to compute the velocity, temperature, and density fields in the impacted clouds and a recently developed radiative transfer code to compute ^{13}CO line radiation from the simulated source. By including the instrumental effects involved in a particular observations we can make detailed comparisons with the observations. The model shows that: 1) The previously unexplained energy source for the broad CO line wings reported by Blitz, Magnani, and Wandel (1988) derives from the collisions. 2) Collisions can induce rapid gravitational collapse and star formation in these clouds which are otherwise supported against gravitational contraction via their internal energy content. 3) The external pressure due to intercloud HI, firts proposed for these objects by Keto and Myers (1986), plays a significant role in the stability and evolution of the high latitude clouds.

THE STRUCTURE OF THE INTERSTELLAR MEDIUM

MULTIPLE-SUPERNOVA REMNANTS

M. Różyczka
Warsaw University Observatory
Al. Ujazdowskie 4, PL-00-478 Warszawa, Poland

ABSTRACT

The present observational and theoretical status of multiple-supernova remnants is briefly reviewed, and evolutionary models of the remnants around OB associations are presented. The remnants are followed for up to almost 30 Myr in various galactic environments (Gaussian or composite disks combined with cold or hot halos). It is found that the presence of a hot halo profoundly modifies the evolution of the remnants, promoting the evaporation of their shells and leading to noneruptive merging of the ejecta with the halo gas. The results also suggest an explanation for the observed deficiency of evolved multiple-supernova remnants.

I. INTRODUCTION

Multiple-supernova remnants (hereafter MSR) is a fairly complicated subject. Both theory and observations of MSRs provide us with more controversies than ultimate conclusions, and in principle every controversy should be given its share of author's attention. As for the present status of the theory, one has to divide the available time and space between things that have been done and things that should be done, and a fair reviewer would probably do it on a fifty-fifty basis. But what makes the situation really complicated, is the multitude of proven or expected connections leading from MSRs to such issues of the galactic astronomy like circulation of chemically enriched gases, galactic winds, or structure and state of the galactic halo. With similar connections pointing towards propagating star formation, the MSRs seem to be one of the vital links in the still mysterious chain of events responsible for the chemical evolution of galaxies.

To get a MSR, one needs an OB association and a galactic disk. An average association contains 20-40 stars of spectral type B3 or earlier whose masses are greater than 8 $M_\odot$, and which can explode as type II supernovae (the number of stars in "record" associations can be 10-20 times larger). One-tenth of those stars are massive ($M > 30\ M_\odot$), rapidly evolving O stars, which are powerful sources of ionizing radiation and stellar winds. It can be estimated, however, that more than 80% of the energy deposited by the association in the disk is provided by supernovae originating from less massive and more slowly evolving B stars. The least massive supernova progenitors explode after $\sim 3 \times 10^7$ yr, during which the association expands at an average rate of ~5 km/s. Thus, the size of a dying association is on the order of 150 pc. MSRs can easily grow to sizes appreciably larger than that, which means that from their point of wiev the associations can be regarded as essentially point-like sources of multiple explosions.

Gaseous galactic disks receiving the energy liberated by OB associations are highly nonuniform, differentially rotating structures permeated by magnetic fields and enveloped in hot halos. To make the things worse, they are very likely turbulent. On the average, their density decreases rapidly as one moves away from the equatorial (or symmetry) plane. The best fits to Ly-α and 21 cm data reflecting this decrease contain Gausssian as well as exponential components. The density of a halo is thought to be on the order of 0.003 atoms/cm^3, and the temperature of a halo - on the order of 1-3$\times 10^6$ K. The Galactic halo is loosely described as "a disk with $\sim$5 kpc thickness (Heiles 1987).

An evolving MSR sweeps the disk gas and stores it in a cool and dense shell which in most Galactic cases is the only observationally detectable part of the whole remnant (in other galaxies the swept volume itself can be detected as a "hole" in the disk). Thus, from the observational point of view the Galactic MSRs belong to the class of large-scale HI structures (among these latter, ring-like or arc-like features prevail outside of the solar circle, while in the inner Galaxy tightly packed, roughly cylindrical "worms" are almost exclusively found). In the Milky Way there is no one-to-one correspondence between OB associations and radio features. On the other hand, the HI holes in M31 (especially those with diameters $\leq$ 300 pc), as well as the HI shells in Magellanic Clouds, seem to be rather well correlated with OB associations and HII regions (Tenorio-Tagle and Bodenheimer 1988 and references therein).

It has been suggested (Heiles 1987) that the observed volume of the Galactic disk occupied by MSRs is much smaller than the arguments based on the total number and mean lifetimes of associations would suggest (similar discrepancies have also been found in other galaxies). Moreover, the expected mass input rate to the Galactic halo due to supernovae from OB associations seems to significantly exceed the one implied by the diffuse X-ray data. Coupled with the above controversies, the widespread belief that the sequential supernovae in OB associations dominate the dynamics and morphology of the interstellar medium caused the evolution of MSRs to become an increasingly attractive theoretical challenge.

The first suggestion that large HI structures (or at least some of them) may be identified with MSRs was due to Cowie et al. (1979). The theory of MSRs has been developed since then by Bruhweiler et al. (1980), Tomisaka et al. (1981), Tomisaka and Ikeuchi (1986), Ikeuchi (1987), McCray and Kafatos (1987), Tenorio-Tagle et al. (1987, hereafter Paper I), Mac Low and Mc Cray (1988), Mac Low et al. (1989, hereafter MLMCN) and Igumentshchev et al. (1989). The methods applied by those authors ranged from analytical estimates based on the classical theory of interstellar bubbles, through one-dimensional hydrodynamical simulations of spherically symmetric MSRs and simplified two- dimensional hydrodynamics (snowplough and Kompaneyets approximations), to full two-dimensional simulations.

In the theoretical investigations, the associations are approximated as point sources of either "superwinds" or repeating explosions. In the first scenario the mechanical luminosity of a superwind is defined as the energy of all supernovae which explode in the association divided by the lifetime of the association. Based on the initial mass function and stellar lifetimes, the time intervals ΔT_{sn} between consecutive explosions in the second scenario can be shown to differ very little from each other (Tomisaka and Ikeuchi 1986, Mac Low and Mc Cray 1988), and the commonly adopted value of ΔT_{sn} is 2×10^5 yr. The disks are represented by smooth density distributions in hydrostatic equilibrium, stratified in the direction perpendicular to the equatorial plane. Self-gravity, magnetic fields and rotation are usually neglected (see, however, Umemura et al. 1988, and Palouš, this volume). The microphysical processes taken into account in theoretical papers on MSRs include optically thin cooling (in all cases), heat conduction (in one case) and various approximation to the evaporation from the shell (in nearly all cases).

Some theoretical results seem to be already fairly well established. Mac Low and Mc Cray (1988) found in agreement with Paper I that in an evolved MSR the blast waves from subsequent supernovae decay into sound waves before they hit the outer shell. Further, based on the Kompaneyets approximation, they formulated a criterion predicting whether an MSR would break out of a stratified gaseous disk of a galaxy. MLMCN and Igumentshchev et al. (1989), who modelled the breakout phenomenon with the help of 2-D hydrodynamical codes, found in agreement with that criterion that the breakout occurs readily in a Gaussian disk but may be entirely prevented in an exponential disk with a scale height of a few hundred parsecs.

However, even 2-D simulations have not yet led to a fully consistent theoretical description the evolution of MSRs. A good example of a disputed issue is shell instabilities. All authors agree that the shell of a MSR evolving in a stratified disk becomes appreciably irregular due to the Rayleigh-Taylor instability which begins to operate as soon as the breakout commences (i.e. as the velocity of the upper part of the shell starts to increase). However, according to our results from Paper I the shell can become irregular even without the presence of a density gradient in the ambient medium. We attributed this effect to either the Rayleigh-Taylor instability induced by supernova shocks decaying into sound waves before they hit the shell, or a more general, thin shell instability discussed e.g. by Różyczka (1985) or Bertschinger (1986). It must be stressed that the early presence of shell irregularities may significantly increase the diffusion rate through the shell/ejecta interface, and influence the MSR evolution due to strongly enhanced cooling of the interior still prior to the breakout.

Also, entirely open problems connected with MSRs exist in addition to the unsettled ones. The results of MLMCN and Igumentshchev et al. (1989), who estimated that during the breakout 5-10% of the outer shell and a large fraction of the ejecta accelerate into the halo, indicate that the MSRs are likely to play the major rôle in the processes of mixing between galactic disks and halos, and that MSR modelling may provide important informations about the chemical evolution of spiral galaxies (see e.g. Ikeuchi 1987). However, if the exchange of matter between the disk and the halo is to be followed, the calculations should be extended into much later evolutionary phases of the MSRs than it has been done so far. Also, various gas distributions should be examined in order to clarify their influence on MSR evolution and/or circulation of supernova ejecta.

Another open problem is the very possibility of the breakout in a composite disk which constitutes the best fit to the Lyman-α and 21 cm observations of HI (Lockman et al. 1986, Bloemen 1987). According to Mac Low and Mc Cray (1988), in such a medium the breakout may not occur, and as a result giant, coherent HI arcs or shells would be formed. It can be calculated that more than 2000 structures of this type should be seen in the Galaxy, which is clearly not the case.

To look at these questions, Tenorio-Tagle, Różyczka and Bodenheimer (1989, hereafter MRB) obtained a broad sample of evolutionary models of MSRs, some of which were followed for over 17 Myr (up to almost 30 Myr in one of the cases). Their models evolve in various environments (Gaussian, exponential or composite disks combined with hot or warm halos), and in several cases the diffusive effects are included in a simplified way. A brief description of their numerical methods and input physics is given in Section II. Section III presents some of their models. Finally, the results of MRB and their implications are discussed in Section IV.

II. THE CALCULATIONS

a) Numerical methods and physical assumptions

The equations of hydrodynamics are solved on a cylindrical grid by means of an explicit, second order scheme described by Różyczka (1985). Due to symmetry constraints only one quadrant of the remnant is actually followed. The (r,z) grid consists of 100×200 - 300×600 equally spaced points and provides a resolution of $\Delta r = \Delta z =$ 5.0 - 2.5 pc.

Radiative, optically thin cooling operates everywhere in the grid (except in the undisturbed gas, in order to keep the latter in the dynamical equilibrium during the calculations). Magnetic fields and self-gravity are neglected. The models are subject, however, to an external gravitational field defined in such a way that the gravitational acceleration balances pressure gradients in the undisturbed gas at the beginning of the evolutionary run. Only the z-component of that field is different from zero, and its value is lower than the Galactic one by factors of 2 to 50 at low and high latitudes, respectively. The deceleration of MSR shells is therefore uderestimated. However, since at the end of each run the characteristic time for the real deceleration is always longer than, or equal to the evolutionary time, the errors introduced in this way should not be too significant.

b) Initial and boundary conditions

A pulsed energy input is employed, i.e. all consecutive supernova explosions are modelled separately. Every explosion is induced by an instantaneous release of 2.5^{50} ergs of thermal energy within a small volume representing a quarter of a sphere centered at $z = r = 0$. Following the estimates discussed by Tenorio-Tagle and Bodenheimer (1988), a value of 2×10^5 yr is chosen for the interval between explosions (1×10^6 yr in one of the cases). The mass of the ejecta is limited to 8 $M_\odot$ per single explosion.

In several cases the diffusive effects are simulated by adding an extra mass to the ejecta of every supernova while leaving the mass of the shell unchanged. For the extra mass a value of 50 $M_\odot$ is adopted, resulting from the formula given by Mac Low and McCray (1987) for temperatures slightly above 10^6 K. The total mass injected into "diffusive" MSRs amounts to about 6000 $M_\odot$, while the total mass of the evolved shell is one order of magnitude larger. Admittedly, this is a very crude description of the diffusion. However, since the diffusion affects the evolution of a MSR mainly by enhanced cooling of the interior, such a procedure is justified at least as a first approximation to the problem. To check its validity, in one of the "diffusive" models the mass was gradually removed from the shell: no qualitative differences in the flow pattern were observed.

In the present calculations two types of density distributions of the gaseous component of a galactic disk are considered. The scale height H of the pure Gaussian disk $\rho = \rho_0 \exp(-(z/H)^2)$ is equal to 100 pc, whereas in the composite disk $\rho = \rho_e \exp(-(z/H_e) + \rho_g \exp(-(z/H_g)^2)$ the scale heights H_e and H_g amount to 300 and 100 pc (Fuchs and Thielheim 1979). The parameters ρ_0, ρ_e and ρ_g are equal to 1.67×10^{-24}, 1.25×10^{-24} and 0.42×10^{-24} g/cm^3, respectively. In all models the pressure P_0 in the symmetry plane of the disk is close to the canonical value of 1×10^{-12} dyn/cm^2 (Bloemen 1987).

The disks are either isothermal ($T_D = 3 \times 10^3$ K), or the temperature rises in them from 7.5×10^3 K in the symmetry plane up to several hundred thousand K far away from it. As a result, in nonisothermal disks the pressure gradient is much gentler than the density gradient. According to Bloemen (1987), such a situation is observed in the Galaxy when the effective (i.e. turbulent and magnetic) pressure is considered instead of the purely thermal one.

The Gaussian models and most of the composite ones have a uniform halo which extends above 280 pc from the galactic plane and remains in pressure equilibrium with the stratified part of the disk. The density of the halo ρ_H is in all models equal to 5×10^{-27} g/cm^3, while its pressure P_H ranges from 3×10^{-15} to 7×10^{-13} dyn/cm^2, resulting in halo temperatures T_H of 7.5×10^3 - 1×10^6 K.

III. RESULTS OF CALCULATIONS.

All MRB cases discussed here are listed in Table 1, in which "evp" denotes the model in which the gas was artificially removed from the shell. The remaining, self-explanatory abbreviations refer to the information contained in Section II.

Cases A-B.

These models (Fig.1) illustrate the possibility of the breakout in composite and Gaussian disks. Case A (Fig. 1a) was run at a resolution of 3.3 pc for 2.91×10^7 yr. At that time several dense tongues resulting from the shell instability separate from the main shell and plunge into the MSR. However, the breakout does not yet take place, and the shell does not fragment. The model corroborates the statement of MLMCN that the MSRs are confined by the large scale height of the Fuchs and Thielheim (1979) density distribution (it should be noted that in terms of scale heights climbed by the shell Fig. 1a should be compared to Fig. 2b of MLMCN). A fully developed breakout can be observed in Fig. 1b which shows a Gaussian model with a cold halo at an age of 5.57×10^6 yr. In agreement with MLMCN it is found that only a small percentage of the shell mass is accelerated directly upwards, and that a secondary main shock wave sweeping through the halo forms, after the shell has fragmented. Interior to that wave the halo gas is rapidly mixed with the SN ejecta and the disk gas which diffused from the shell into the MSR. The secondary wave also sweeps through the upper layers of the disk and heats them up to several hundred thousands K.

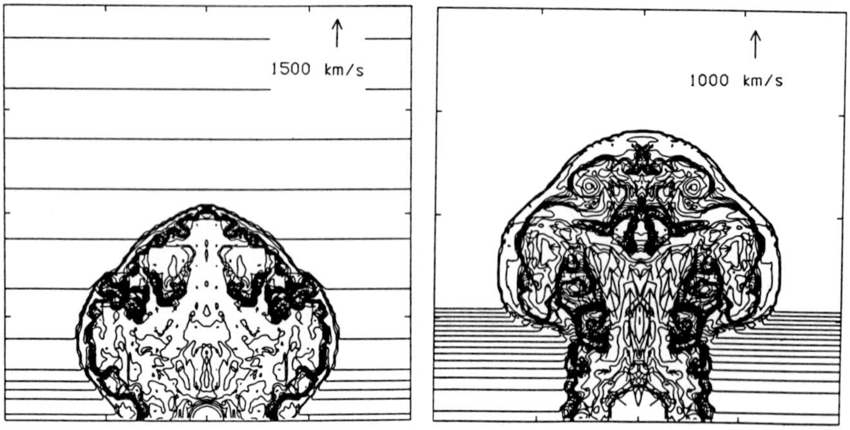

Fig.1. Density distributions and velocity fields of models A and B (frames a and b taken at an evolutionary age of 2.91×10^7 and 5.57×10^6 yr, respectively). The contours are logarithmically spaced with $\Delta \log(\rho) = 0.2$. The distance betwen tick marks is 500 pc in frames a, and 250 pc in frame b.

TABLE 1

Parameters of the Models

case	disk					halo			supernovae		remarks
	P_O	ρ_O	T_O	distr ρ	distr T	P_H	ρ_H	T_H	rate	mass	
A	4_{10}^{-13}	1.7_{10}^{-24}	3_{10}^{3}	composite	unif	-			std	std	large grid
B	4_{10}^{-13}	1.7_{10}^{-24}	3_{10}^{3}	Gaussian	unif	3_{10}^{-15}	5_{10}^{-27}	8_{10}^{3}	std	std	-
C	4_{10}^{-13}	1.7_{10}^{-24}	3_{10}^{3}	composite	unif	4_{10}^{-14}	5_{10}^{-27}	5_{10}^{5}	std	std	-
D	4_{10}^{-13}	1.7_{10}^{-24}	3_{10}^{3}	composite	unif	4_{10}^{-14}	5_{10}^{-27}	5_{10}^{5}	low	std	low sn rate
E	4_{10}^{-13}	1.7_{10}^{-24}	3_{10}^{3}	Gaussian	unif	3_{10}^{-15}	5_{10}^{-27}	8_{10}^{3}	std	high	low resolution
F	1_{10}^{-12}	1.7_{10}^{-24}	8_{10}^{3}	Gaussian	var	7_{10}^{-13}	5_{10}^{-27}	1_{10}^{6}	std	high	low resolution
G	1_{10}^{-12}	1.7_{10}^{-24}	8_{10}^{3}	composite	var	7_{10}^{-13}	5_{10}^{-27}	1_{10}^{6}	std	high	low resolution
H	1_{10}^{-12}	1.7_{10}^{-24}	8_{10}^{3}	composite	var	7_{10}^{-13}	5_{10}^{-27}	1_{10}^{6}	std	high	low res evap

Cases C-D.

In these models a hot (5×10^5 K) halo was placed atop the composite disk. Case C (Fig.2a) is shown at 1.32×10^7 yr, when the upper one-third of the MSR extends into the halo and drives a sound wave into it (the velocity of the shell has already fallen below the sound velocity of the hot halo gas, so that the propagation of a shock wave through the halo is inhibited). The diameter of the MSR measured in the disk plane amounts to about 500 pc. To reach a comparable size in Case D evolving at a low SN rate, 1.94×10^7 yr are needed (Fig.2b). However, even at that time the internal instability is not so well developed as in Case E (in the latter case more wavelenghts seem to be excited and the average length of the dense tongues is larger; some of them have even separated from the shell). Both cases show once again that the growth rate of the internal instabilities depends on the way the energy is injected into the MSR. In both cases mixing of halo and ejecta is impossible. To enable it, the upper part of the shell would have to be either broken or washed out by the diffusion.

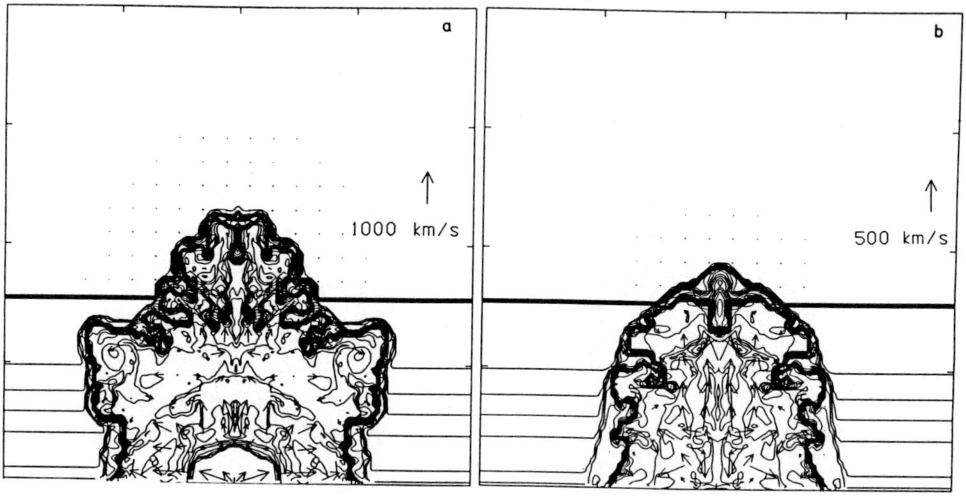

Fig.2. Density distributions and velocity fields of models C and D (frames a and b taken at an evolutionary age of 1.32×10^7 and 1.94×10^7 yr, respectively). Contour spacing like in Fig.1. The distance between tick marks is 250 pc in both frames.

Cases E-F.

These models illustrate the influence of the halo temperature on the late evolutionary stages of MSRs breaking out of a Gaussian disk, and are shown at approximately the same age of 1.2×10^7 yr (Fig. 3). An extra mass is added to the SN ejecta in both models, and the resolution is decreased to 5pc (the amount of information lost due to poorer resolution is insignificant as far as the global properties of the models are considered). In Case E (cold halo, Fig. 3a) essentially the same phenomena as in Case B are observed, the only difference being the size of the MSR. The flow within the MSR is very complicated: freely expanding ejecta are first shocked in an almost spherical primary wave, then they expand again in a nozzle formed by the broken shell to reach higly supersonic velocities, and finally they are re-shocked in an almost horizontal secondary wave. While spreading above the secondary wave, they are directed sideways and downwards, so that they reach the disk again.

In Case F (hot halo, Fig. 3b) the flow pattern is entirely different: instead of multiple shocks a gentle, sonic expansion prevails, directed predominantly upwards. Mixing between the interior gases and the halo mainly occurs in a series of vortices resembling puffs of smoke from a broad chimney. It should also be noticed that unlike in Case E, the whole halo has already been set in motion (the motions of the halo are now induced by sound waves, whose speed is higher than the speed of the shock wave in Case E due to the high temperature of the halo). Furthermore, the sound waves from the halo induce sound and weak shock waves in the upper layers of the disk. Typical velocities in a large vortex range from 100 to 200 km/s. Velocities in the halo far away from the vortices do not exceed 50 km/s, and those in the upper disk are smaler than 10 km/s.

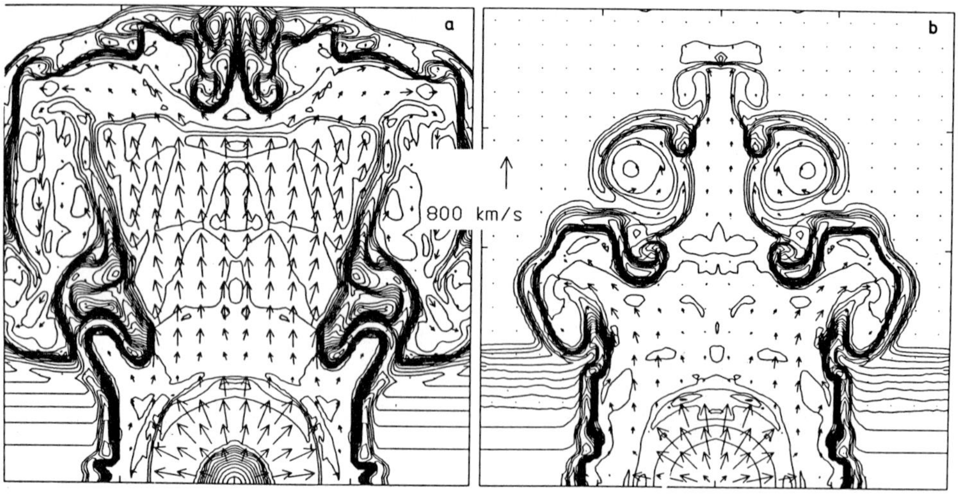

Fig.3. Density distributions and velocity fields of models E and F (frames a and b taken at an evolutionary age of 1.18×10^7 and 1.21×10^7 yr respectively). Contour spacing like in Fig.1. The distance between tick marks is 230 pc in both frames.

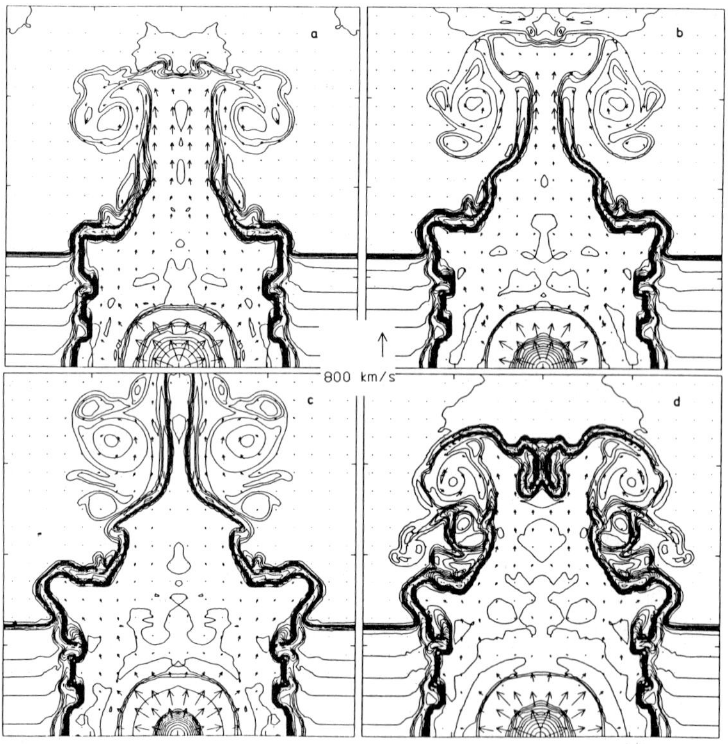

Fig.4. Density distributions and velocity fields of models G (frames a,b and c taken at an evolutionary age of 1.45×10^7, 1.58×10^7 and 1.71×10^7 respecively), and H (shown at an evolutionary age of 1.72×10^7 yr in frame d). Contour spacing like in Fig.1. The distance between tick marks is 230 pc in all frames.

Cases G-H

In these calculations the late evolutionary stages of MSRs developing in a composite disk with a hot halo are examined. An extra mass is added to the SN ejecta in both models, and in Case G the mass is gradually removed from the shell on a time scale of 1×10^7 yr (only the part of the shell extending into the halo is affected by this process). Both runs demonstrate the possibility of the breakout. The overall flow pattern is similar to the one observed in Case F: the whole halo is set in motion, and mixing proceeds in a number of vortices. The outflow from the interior of the MSR into the halo can be very efficient, and occasionally velocities up to 400 km/s are reached in the "chimney". On the other hand, the vortices may occasionally squeeze the chimney and quench the outflow (shortly afterwards a new vortex begins to develop). The last models obtained in both cases may be compared in Figs. 4c and 4d: in the latter figure the chimney is appreciably broader due to the inhibited squeezing action of the vortices which are hampered by dense fragments of the shell. The fragments of the shell in the halo move at a speed of 30 km/s.

IV. DISCUSSION.

The calculations described in the present paper generally agree with the results of MLMCN and Iumentshchev et al. (1989). In particular, it is found that long-lived, fairly regular MSRs form in composite disks without halos (Case A). Such structures are able to reach radii of 1 kpc and ages in excess of 3×10^7 yr (still later evolutionary stages may be dominated by the gravity of the stellar component of the galactic disk which was not taken into account in the models). The SN ejecta are contained in them and cool rather in place than spread throughout the disk.

On the other hand, the MSRs break readily out of Gaussian disks with cold halos (Cases B and E). Evolutionary phases preceding the breakout are very short ($t<3$ Myr). The shell fragments upon breaking out, and releases the SN ejecta into the halo. However, a secondary shell composed of halo gas is formed immediately afterwards and direct mixing between the ejecta and the halo is inhibited (it is only possible due to the diffusion at the interface between the secondary shell and the ejecta). Compared to the composite case, a large fraction of the MSR energy goes into heating the halo rather than increasing the kinetic and potential energy of the shell. MSRs assume mushroom shapes and spread sideways above the disk, so that it is possible for them to merge and to form extended, hot regions ($T> 10^6$ K) far above the symmetry plane of the disk. Thus, it is conceivable that even a thermally uniform galactic environment will evolve towards a thermally stratified one, with a hot halo atop of a rather cool disk.

In the presence of a hot halo, independently of the assumed stratification of the disk, chimneys are formed rather than mushrooms (Cases F-H). Due to sound waves excited at the outer edge of the shell during the breakout, and subsequently due to vortices above the already formed chimney, extended turbulent r egions develop around every MSR. Excited primarily in the halo, the turbulence spreads into the upper layers of the disk via sound and/or weak, randomly oriented shock waves. Upon the breakout the shell is not screened from the halo by a shock wave, so that the diffusion may act on its both sides, leading to a rapid evaporation. As a result, shortly after the MSR has reached the base of the halo, an efficient dynamical mixing between the ejecta and the halo gas is enabled, and the metal-enriched gas may propagate to large distances from the center of the bubble, also permeating the upper disk.

The time scale for the formation of a chimney is of the order of 10-15 Myr, whereas the average lifetime of an OB association is at least two times longer. To find the expected number of MSRs in a given sample of associations, however, an estimate of the lifetime of MSRs in the form of observable HI features is needed, which, due to the diffusion, may be rather short. Indeed, instabilities increasing the surface of the shell, and the fact that the diffusion operates on both sides of the shell, may bring the diffusion time scale down to a value much smaller than $\sim$ 100 Myr characteristic of a bubble extending to 250 pc in the Galactic plane (Mac Low and Mc Cray (1988). Since short wavelegth modes of the shell instabilities grow faster (Różyczka 1985 and references therein), a tenfold decrease of the diffusion time is entirely conceivable if one assumes that the rate of the diffusion from the shell is directly proportional the surface of the latter. In particular, the lifetime of an observable MSR might turn out to be shorter than the lifetime of its parent association. Unfortunately, the precise nature of the instabilities operating in the shell has not been revealed by the present calculations: as before, Rayleigh-Taylor and/or general thin shell effects may be in action. Very high resolution models of small regions of the shell would probably solve the problem.

The hot halo models provide a straightforward explanation to the missing coherent HI structures. They also suggest that qualitative corrections to the galactic circulation scenario proposed by Ikeuchi (1987) may be necessary, accentuating the importance of turbulent mixing. Moreover, it is likely that the circulation time (basically, the time the ejecta spend in the halo before getting incorporated into the disk again) may be much shorter than it was estimated on the basis of cooling rates (Heiles 1987, Ikeuchi 1987). Additional computational effort is needed to verify this hypothesis.

ACKNOWLEDGMENTS.

This work was partly supported by the grant CPBP 01.11 from the Polish Academy of Sciences.

REFERENCES

Bertschinger, E. 1986, *Astrophys. J.* **304**, 154

Bloemen, J.B.M.G. 1987, *Astrophys. J.* **322**, 694

Bruhweiler, F.C., Gull, T. Kafatos, M. and Sofia, S. 1980, *Astrophys. J. Letters* **238**, L27

Cowie, L.L., Songaila, A., and York, D.G. 1979, *Astrophys. J.* **230**, 469

Fuchs, B., and Thielheim, K.O. 1979, *Astrophys. J.* **227**, 801

Heiles, C. 1984, *Astrophys. J. Suppl.* **55**, 585

Heiles, C. 1987, *Astrophys. J.* **315**, 555

Igumentshchev, I.V, Shustov, B.M., and Tutukov, A.V.: 1989, *Astron.Astrophys.* (submitted)

Ikeuchi, S. 1987, in *Starbursts and Galaxy Evolution*, ed. T.X. Thuan, T. Montmerle, J. Tran Thanh Van, p.27, Gif sur Yvette: Ed. Front

Lockman, F.J., Hobbs, L.M., Shull, J.M. 1986, *Astrophys. J* **301**, 380

McCray, R. and Kafatos, M. 1987, *Astrophys. J.* **317**, 190

Mac Low, M.-M. and McCray, R. 1988, *Astrophys. J.* **324**, 776

Mac Low, M.-M., McCray, R., and Norman, M.L. 1989, *Astrophys. J.* **337**, 141

Różyczka, M. 1985, *Astron. Astrophys.* **143**, 59

Tenorio-Tagle, G. and Bodenheimer, P. 1988, *Ann. Rev. Astron. Astrophys.* **26**, 145

Tenorio-Tagle, G., Bodenheimer, P., and Różyczka, M. 1987, *Astron. Astrophys.* **182**, 120

Tenorio-Tagle, G., Różyczka, M., and Bodenheimer, P. 1989, *Astron. Astrophys.*, submitted

Tomisaka, K., Habe, H., and Ikeuchi, S. 1981, *Astrophys. Sp. Sci.* **78**, 273

Tomisaka, K., and Ikeuchi, S. 1986, *Publ. Astr. Soc. Japan* **38**, 697

Umemura, S., Kazuo, I., and Kazunari, S. 1988, *Publ. Astr. Soc. Japan* **40**, 25

DISCUSSION

J.DYSON: What is the efficiency of conversion of the energy of a typical multiple-supernova into kinetic energy of the shell compared to the case of a single supernova in the ISM?

M.RÓŻYCZKA: As long as there is a continued energy input in the center of the MSR, and the nonspherical effects are negligible, the kinetic energy of the shell stays close to 0.1 of the total energy that has been delivered by supernovae. The largest conversion factors found for single explosion remnants amount to 0.08, but in most cases they are by half an order of magnitude smaller. It is also known that until late evolutionary stages the radii of nonspherical MSRs, as measured in the symmetry plane of the galactic disk, grow proportionally to t^{β}, where β is appreciably larger than the snowplough value of 0.25 (even in the early breakout $\beta \sim 0.44$). Thus, the MSRs which have not yet "discharged into the halo" convert the supernova energy into the kinetic energy of the shell more efficiently than their classical counterparts.

J.WAMPLER: Is the neglection of rotational effects in your models justified?

M.RÓŻYCZKA: At the solar circle the galactic shear does not cause appreciable effects for the first 30 Myr of the MSR's evolution (at that time the MSR's cross-section in the equatorial plane becomes an ellipse whose axis ratio is rougly 2/3).

GALACTIC FOUNTAINS

F. D. Kahn
Department of Astronomy, The University,
Manchester M13 9PL, England

Summary

The intercloud medium (ICM) is kept hot and supplied with gas by supernova remnants of Type II. The rate at which mass is added balances the rate at which gas flows away from the ICM into the galactic halo. A fountain is thus formed, since the gas only has enough energy to escape from the disk, but not from the Galaxy as a whole. The upward flow must be followed by the descent of the gas, in the form of a shocked sheet: the return flow seems to be observed in the form of intermediate velocity clouds.

These conclusions will be modified if cosmic ray pressure is important, or if there is a significant inflow of extragalactic gas into the halo.

1. Heating the Inter-Cloud Medium

The Population I component of the Galaxy lies within a disk extending some 100 parsecs above and below the galactic plane. It includes the brightest and most active stars, notably O and B stars, which emit powerful winds, and which eventually evolve into the progenitors of Type II supernovae. These stars consequently provide a massive input of kinetic and thermal energy into the interstellar medium close to the galactic plane.

Most of the interstellar mass is contained in molecular clouds, where the gas is cold ($T \sim 10$ K) and in atomic clouds, where it is cool ($T \sim 100$ K). A small fraction of the mass is contained in HII regions, which are warm ($T \sim 10^4$ K); however the major fraction of the space within the disk is occupied by the inter-cloud medium (ICM) in which the density is low, with $n \sim 10^{-2}$ cm^{-3} and the temperature is high ($T \sim 10^6$ K). The gas here gains its thermal energy from shocks in supernova remnants. A more detailed description of the evolution of a remnant in the actual interstellar medium is given by Shull et al (1985).

The sudden injection of energy E_0 by a supernova explosion drives a blast wave into the surrounding medium, here modelled as being homogeneous, with density ρ_0. According to the Sedov solution the shock heading the blast has radius

$$r = \left(\frac{2E_0}{\rho_0}\right)^{1/5} t^{2/5} \tag{1}$$

at time t after the explosion. The adiabatic parameter immediately behind the shock is

$$\kappa \equiv P/\rho^{5/3} = 0.016\, E_0^{2/5}\, \rho_0^{-16/15}\, t^{-6/5}\,, \tag{2}$$

and the mass enclosed by the shock is

$$M_s = 6.4 \rho_0^{2/5} E_0^{3/5} t^{6/5}. \tag{3}$$

The passage of the shock heats a mass

$$M(> \kappa) = 0.10\, E_0\, \rho_0^{-2/3}\, \kappa^{-1} \tag{4}$$

so that its adiabatic parameter exceeds κ. The inner parts of a remnant experience the fastest moving shocks, and the gas there is heated so intensely that it can stay hot long enough to flow to the edge of the galactic disk before cooling. (The fate of hot gas left behind by a SNR is described by Kahn 1976.) To fix ideas, a parcel of gas in which κ exceeds $\kappa_* \equiv 5 \times 10^{30}$ ($\mathrm{cm}^4\ \mathrm{gm}^{-2/3}\ \mathrm{s}^{-2}$) will not cool for 10^6 years and will then still be hot enough to escape from the disk. With values 3×10^{51} erg and 2×10^{-24} gm cm^{-3}, for E_0 and ρ_0 respectively,

$$M(> \kappa_*) = 4 \times 10^{35} \mathrm{gm} \equiv 200 M_\odot\ . \tag{5}$$

A reasonable estimate is that each supernova of Type II makes 200 $M_\odot$ of gas sufficiently hot that it can escape from the disk. Such supernovae occur typically three times a century in a galaxy like ours. A given supernova provides a source of hot gas for the fountain that lasts about a million years, and so the active remnants, in this sense, are typically spaced 60 parsecs apart, with our assumed supernova rate. It seems reasonable to make the approximation that the injection of hot gas is smoothly distributed within the galactic disk, with the rate being largest close to the plane, and tapering off on either side on a length scale of order 100 parsecs.

2. Starting the Flow into the Fountain

The gas in the ICM has so large a sound speed that it will not be confined by gravity to the galactic disk. In fact little error is made by ignoring gravity in a description of the transition from subsonic flow, within the disk, to supersonic well outside. A steady plane parallel flow satisfies the three conservation conditions

$$\text{mass} \quad \rho u = \Phi(z) \tag{6}$$

$$\text{momentum} \quad P + \rho u^2 = \Pi \tag{7}$$

$$\text{energy} \quad \frac{1}{2} u^2 + \frac{5}{2} \frac{P}{\rho} = \epsilon\ ; \tag{8}$$

here P = pressure, ρ = density, u = velocity, $\Phi(z)$ = mass flux at height z, Π = momentum flux.

The stagnation enthalpy ϵ remains equal to the injection energy ϵ_* if there is no radiative heat loss. The mass flux increases with height, starting from zero at the galactic plane and reaching some value $\Phi_{\max}$ at greater heights.

It follows from equations (6), (7) and (8) that

$$\epsilon = \frac{5}{2} \frac{\Pi u}{\Phi} - 2u^2\ ; \tag{9}$$

regarded as a function of u the value of ϵ reaches its maximum

$$\epsilon = \epsilon_{\rm max} = \frac{25}{32}\frac{\Pi^2}{\Phi^2} \tag{10}$$

when $u = 5\Pi/8\Phi$.

Solutions for u exist for an uncooled flow provided $\epsilon_{\rm max}$ exceeds or is equal to ϵ, or if

$$\Phi \leq \frac{5\sqrt{2}}{8}\frac{\Pi}{\sqrt{\epsilon}} . \tag{11}$$

The sonic point is then reached at the outer limit of the region where hot gas is injected, and $\Phi = \Phi_{\rm max}$; this condition also fixes Π and thereby the pressure in the intercloud medium, see Figure 1. A better description allows for ϵ to decrease with z, since the gas at greater heights has had more time to lose energy by radiation. The gas can then pass through a sonic point, and flow away supersonically from the outer layers of the disk; see Figure 2.

3. The Gravitational Field in the Halo

The flow into the fountain leaves the disk with supersonic speed and its later motion is controlled by the gravitational field in the halo. A convenient model has half the gravitating mass of the Galaxy in each of Populations I and II (Sellwood and Sanders 1988), and so distributed that the rotation speed v_0 of the galactic disk is uniform. The gravitational field then has components

$$F_\varpi \equiv \frac{\partial \mathcal{V}}{\partial \varpi} = -\frac{v_0^2}{2\varpi}\left\{1 - \frac{z}{(\varpi^2+z^2)^{1/2}}\right\} - \frac{v_0^2\varpi}{2(\varpi^2+z^2)} \tag{12}$$

and

$$F_z \equiv \frac{\partial \mathcal{V}}{\partial z} = -\frac{v_0^2}{2(\varpi^2+z^2)^{1/2}} - \frac{v_0^2 z}{2(\varpi^2+z^2)} \tag{13}$$

respectively in the ϖ-direction, perpendicular to, and the z-direction, parallel to the galactic axis of rotation. The leading terms on the right, in both expressions, represent the field due to the mass of Population I, and the other terms that due to Population II. When z is small

$$\frac{\partial \mathcal{V}}{\partial \varpi} = -\frac{v_0^2}{\varpi} + \frac{v_0^2 z}{2\varpi^2} + \cdots , \tag{14}$$

and the first term here is balanced by the centrifugal force. The second term only becomes relatively important with greater height z, and will be ignored here. The results of this calculation can therefore only be used in the part of the halo where $|z|$ is rather smaller than ϖ. In the same approximation

$$\frac{\partial \mathcal{V}}{\partial z} \sim -\frac{v_0^2}{2\varpi} \equiv -g , \tag{15}$$

typically 10^{-8} cm s^{-2}, with $v_0 = 250$ km s^{-1} and $\varpi = 10$ kpc. Provided that the fountain does not rise too high, the gravitational acceleration in the ϖ-direction will everywhere be in approximate balance with the centrifugal force, and the z-component will be uniform above the galactic disk. The present treatment will not serve, though, to describe a galactic wind in which the gas inevitably is carried to large heights z.

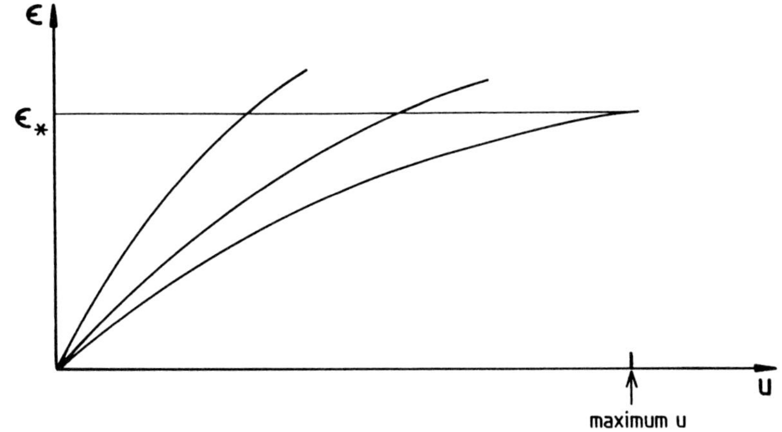

Figure 1: The successive curves show the dependence of ϵ, the stagnation enthalpy, on the velocity u, for various values of the mass flux Φ, starting on the left with small Φ. The actual flow speed is determined in each case by the intersection of the curve with the horizontal line $\epsilon = \epsilon_*$. The last curve corresponds to the maximum value that Φ can reach. In these examples the gas is not allowed to cool radiatively. The flow is everywhere subsonic—and in the limiting case exactly sonic.

4. The Rising Stream in the Fountain

A simple ballistic pattern describes the upward flow in the fountain. The mass flux Φ is taken to be constant, and the upward velocity to be u_0 at the galactic plane, $z = 0$. The speed of sound is negligible compared with the upward velocity, which is

$$u = (u_0^2 - 2gz)^{1/2} \qquad (16)$$

at height z; the density is

$$\rho = \Phi(u_0^2 - 2gz)^{-1/2} \qquad (17)$$

there. During its rising phase the fountain has an upper boundary at $z = Z(t)$, say, and

$$\frac{dZ}{dt} = u \qquad (18)$$

so that the flow is still rising towards its maximum height

$$Z_{\max} = \frac{u_0^2}{2g}, \qquad (19)$$

here taken to be 3 kpc, or 10^{22} cm.

There are no effective pressure forces in the gas so that no expansion wave propagates back towards the galactic plane from the free boundary further up. An initial speed $u_0 = 140$ km s^{-1} corresponds to the $Z_{\max}$ chosen here, and the stagnation enthalpy in the flow is

$$\epsilon_0 = \frac{1}{2}u_0^2 = 10^{14}\mathrm{cm}^2\,\mathrm{s}^{-2}\,.$$

This value is also a good first estimate for the enthalpy in the hot ICM, although there will have been some radiative heat loss in the flow; the corresponding temperature is $T_0 = 3 \times 10^5$ K. If the pressure in the ICM is 10^{-12} dyne cm^{-2}, then the electron density is $n_{e,0} = 1.25 \times 10^{-2}$ cm^{-3} and the density $\rho_0 = 2.5 \times 10^{-26}$ gm cm^{-3}. The adiabatic parameter of the ICM is

$$\kappa = P_0 \rho_0^{-5/3} = 4.7 \times 10^{30} \, (\text{cm}^4 \, \text{gm}^{-2/3} \, \text{s}^{-2}) \, ,$$

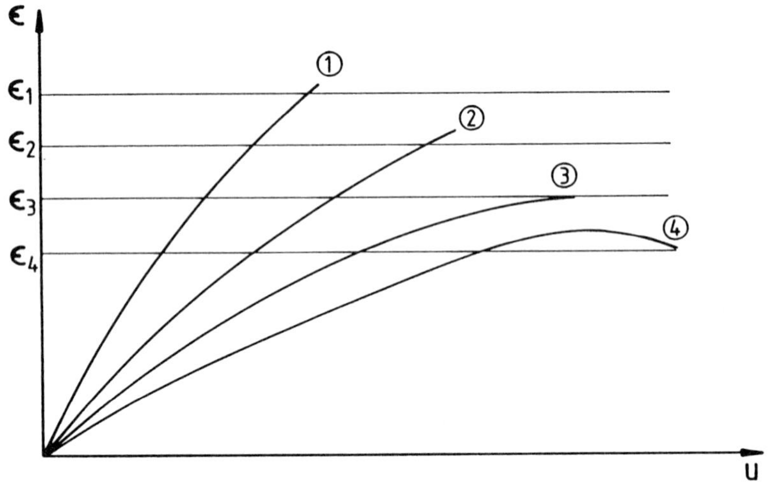

Figure 2: As in Figure 1, but with radiative cooling allowed. As a result the stagnation enthalpy decreases with height, in the order 1, 2, 3, 4. The contact of curve (3) with $\epsilon = \epsilon_3$ depicts the sonic solution, but the flow can continue to become supersonic (intersection of (4) with $\epsilon = \epsilon_4$).

in good agreement with the earlier estimate for its likely value on injection from supernova remnants. The mass flux into the halo, on each side of the disk, is

$$\Phi = \Pi / u_0 = P_0 / u_0 = 7 \times 10^{-20} \, \text{gm} \, \text{cm}^{-2} \, \text{s}^{-1} \, ,$$

and the total rate of mass loss into the fountain, from a galactic disk of radius 10 kpc, is

$$\dot{\mathcal{M}} = 4 \times 10^{26} \, \text{gm s}^{-1} = 6 \, M_\odot \, \text{year}^{-1} \, ,$$

again in good agreement with the earlier statement that some three supernovae of Type II are expected to occur, per century, and that each will inject some 200 $M_\odot$ of hot gas into the ICM.

5. Descent of the Shocked Sheet

Once the uppermost part of the flow has reached height Z_{max} the gas will start to fall back towards the galactic plane. A well-cooled shock separates the ascending and descending parts of the fountain. The shocked gas cools rapidly because the shock speed is relatively low, and it forms a thin sheet in hydrostatic equilibrium in a rest frame moving downwards with speed w.

It is convenient to use u, the velocity of the upward flow, as the independent variable in describing the descent of the sheet of shocked gas. With

$$\frac{d}{dt} \equiv -w\frac{d}{dz} = \frac{wg}{u}\frac{d}{du} \tag{20}$$

the equation of conservation of mass becomes

$$u\frac{d\Sigma}{dt} = wg\frac{d\Sigma}{du} = \Phi(u+w)\,, \tag{21}$$

where Σ is the surface density in the descending sheet.

The equation of motion becomes

$$\frac{d}{dt}\Sigma w = \Sigma g - u\frac{d\Sigma}{dt} \tag{22}$$

or

$$w\frac{d}{du}\Sigma(u+w) = \Sigma(u+w)\,, \tag{23}$$

after some algebra. From (21) and (23)

$$\frac{d}{d\Sigma}\Sigma(u+w) = \frac{\Sigma g}{\Phi}$$

so that

$$\Sigma = \frac{2\Phi}{g}(u+w)\,, \tag{24}$$

and on substituting from this relation in (21) it follows that

$$2w\frac{d}{du}(u+w) = u+w$$

or

$$w = \frac{1}{2}u\,, \tag{25}$$

since $u = w = 0$ when $z = Z_{\max}$. The falling sheet of shocked gas has just half the speed of the upward stream, at any level. Its downward acceleration is

$$-w\frac{dw}{dz} = -\frac{u}{4}\frac{du}{dz} = \frac{g}{4} \tag{26}$$

and the layer above the shock experiences an effective gravitational acceleration $3g/4$. It also follows from (21) and (25) that

$$\frac{d\Sigma}{dt} = \frac{3}{2}\Phi$$

and since $w = gt/4$

$$\Sigma = 6\Phi\, w/g\,. \tag{27}$$

The pressure at the base of the shocked layer is then

$$P_b = 3\Sigma g/4 = 9\Phi\, w/2\,. \tag{28}$$

The descending gas is kept partly or fully ionized by Ly-c photons coming up from the galactic disk. This radiation originates on early type stars that are only partly shielded by their attendant HII regions. To fix ideas, a Ly-c flux of 10^7 photon cm^{-2} s^{-1} will be produced, on each side of the galactic plane, by a population of naked O stars, each with luminosity $10^5\,L_\odot$, having a surface density of one star per 20 000 pc^2.

A descending sheet, as described earlier, reaches the vicinity of the galactic plane with a base pressure $P_b = 2.2 \times 10^{-12}$ dyne cm^{-2}; the flux of Ly c photons needed to keep it ionized is

$$S_{\mathrm{req}} = \frac{b}{m_a^2}\int_0^\infty \rho^2\,dh = \frac{b\Sigma_0^2}{2m_a^2 H} \tag{29}$$

where $b = 2 \times 10^{-13}$ $\mathrm{cm}^3\,\mathrm{s}^{-1}$ is the effective recombination coefficient, $\Sigma_0 = 3\Phi u_0/g = 3\times 10^{-4}$ gm cm^{-2} is the surface density in the sheet, h measures height from the bottom of the sheet, and $H = 4c_i^2/3g = 1.3 \times 10^{20}$ cm is the scale height within the sheet. With these parameters

$$S_{\mathrm{req}} = 1.7 \times 10^7 \text{ photons cm}^{-2}\,\mathrm{s}^{-1}\ ;$$

the shocked sheet will therefore be partly neutral if it descends on a region of the galactic disk where the surface density of unshielded O stars is about equal to the standard value assumed earlier. If so the neutral hydrogen it contains will be observable by means of the 21 cm line, and the object would be classified as an intermediate velocity cloud. Mirabel (1989) gives an up-to-date discussion of relevant observations.

The gas returning in the fountain will never descend fast enough to be identified with the high velocity clouds, whose speeds are closer to 150 km s^{-1}. A descending sheet will only move that fast if pushed from behind, perhaps by matter falling into the Galaxy. Doubling the downward speed raises the post-shock pressure by a factor $(4/3)^2 = 1.78$, and the flux of Ly c photons needed for full ionization goes up to, typically, 5.4×10^7 photons cm^{-2} s^{-1}.

To sum up, a shocked sheet descends to the galactic plane from the fountain with a typical speed of 70 km s^{-1}. It is likely to be only partly ionized on returning to the galactic disk. The ionizing radiation comes from below; the neutral gas, which is cool, consequently overlies the ionized gas, which is warm ($T \sim 10^4$ K). This stratification is Rayleigh-Taylor unstable; if an intermediate or high-velocity cloud can be observed with the 21cm line it must be partly neutral, and should therefore show an internal velocity structure, with typical speeds of 10 km s^{-1}, resulting from the instability.

6. Cosmic Rays in the Fountain

Cosmic rays contribute appreciably to the energy content of the interstellar medium; some authors even argue that they provide the major part of the interstellar pressure. If this is so then the flow into the halo must be driven by cosmic rays rather than by hot gas. Substantial physical differences follow, the most important being that the cosmic ray gas cannot cool radiatively and so the energy content of the flow is less easily lost (Breitschwerdt et al 1987). But here a less extreme position will be taken.

Suppose that the flow starts into the halo with the specifications given in Section 4, that is mass flux Φ, negligible thermal energy content and upward velocity u_0, at $z = 0$. Now add cosmic ray gas with pressure $\mathcal{P}_*$ at level $z = 0$, and postulate a backgound magnetic field with uniform strength B and lines of force perpendicular to the plane $z = 0$. In the presence of the field the cosmic ray plasma is coupled to the thermal plasma; it can diffuse down a density gradient of cosmic rays but the diffusion speed is generally restricted to be not much larger than the Alfvén speed v_A. The coupling between the thermal and relativistic plasmas is provided by Alfvén waves, which are excited whenever the speed of relative streaming exceeds the Alfvén speed and which then resonantly scatter the relativistic particles. The wave field only carries a relatively small fraction of the energy involved, and its contribution to the momentum flow Π or to the rate of working W will be ignored (Skilling 1970).

For the cosmic-ray assisted flow in a steady state

$$\Pi = \Phi u + \mathcal{P} \tag{30}$$

and

$$W = \frac{1}{2}\Phi u^2 + 3\mathcal{P}(u - v_A) + \mathcal{P}u , \tag{31}$$

and here $\mathcal{P}$ is the cosmic ray pressure, $3\mathcal{P}(u - v_A)$ gives the rate per unit area at which cosmic ray energy is convected and $\mathcal{P}u$ the rate, per unit area, at which cosmic ray pressure does work. W is the total rate of working, per unit area. The gas flow slows down with height, the gas density and the cosmic ray density increase upwards, and therefore the cosmic rays diffuse downwards relative to the thermal plasma. This explains why u and v_A occur in the combination $u - v_A$ in relation (31).

Gravity has to be allowed for here and consequently Π and W become functions of z and satisfy the equations

$$\frac{d\Pi}{dz} = -\rho g , \quad \frac{dW}{dz} = -\Phi g , \tag{32}$$

so that

$$\frac{dW}{d\Pi} = \frac{\Phi}{\rho} = u . \tag{33}$$

Finally the Alfvén speed is

$$v_A = \frac{B\, u^{1/2}}{\sqrt{4\pi\Phi}} \equiv \beta\, u^{1/2} . \tag{34}$$

It is now convenient to use dimensionless variables ϖ and V, defined by

$$v_A = \beta^2 V \text{ (or equivalently } u = \beta^2 V^2) \text{ and } \mathcal{P} = \mathcal{P}_* \varpi . \tag{35}$$

A differential equation for ϖ in terms of V then follows from using relations (30), (31) and (35) in equation (33). The cosmic ray pressure is found from the solution to be

$$\mathcal{P} = \mathcal{P}_* \frac{V_0(V_0 - 1)^{5/3}}{V(V-1)^{5/3}} , \tag{36}$$

and increases with decreasing V, or increasing z. The first of equations (32) can be expressed in terms of V and becomes

$$\left\{ \frac{B^2}{2\pi} V - \frac{\mathcal{P}_*(8V-3)V_0(V_0-1)^{5/3}}{V^2(V-1)^{8/3}} \right\} \frac{dV}{dz} = -\frac{4\pi\Phi^2 g}{B^2 V^2} . \tag{37}$$

The term in braces starts positive at the plane $z = 0$, where $V = V_0$, and then decreases with increasing height. It will certainly reach zero while V still exceeds unity and the flow is still moving upwards, but if so dV/dz becomes infinite and that would be unphysical. Here then is a crucial difference which distinguishes between fountains with or without cosmic rays. The table below shows the maximum value of $\mathcal{R}_*$, the ratio of cosmic ray to magnetic energy density at $z = 0$, which is allowable if the flow is to be able to slow down smoothly from the initial V_0 to V without encountering an infinity in dV/dz. The last column gives the speed corresponding to V, in km s^{-1}. In the calculation it is assumed that $B = 10^{-6}$ gauss, and that $\Phi = 7 \times 10^{-20}$ gm cm^{-2} s^{-1} and $u_0 = 1.4 \times 10^7$ cm s^{-1} as before, so that

$$V_0 = \sqrt{\frac{4\pi\Phi u_0}{B^2}} = 3.54 \,.$$

The term in braces, in equation (37), is positive provided that

$$\mathcal{R}_* \equiv \frac{24\pi\mathcal{P}_*}{B^2} < \frac{36V^3(V-1)^{8/3}}{(8V-3)V_0(V_0-1)^{5/3}} \,. \tag{38}$$

Small values of $\mathcal{R}_*$ correspond to a low energy density for the cosmic rays: if $\mathcal{R}_*$ is as small as 0.13 then the upward flow can continue almost to a complete standstill before meeting the singularity in the velocity gradient. For a more realistic value of $\mathcal{R}_*$, say 5.82, the energy density due to cosmic rays would be 2.3×10^{-13} erg cm^{-3}, the fountain could flow without serious hindrance until its speed had halved and it had

V	Max^m $\mathcal{R}_*$	u (cm s^{-1})
1.5	0.13	2.56 $\times 10^6$
2.0	1.32	4.55 $\times 10^6$
2.5	5.82	7.11 $\times 10^6$
3.0	21.7	1.02 $\times 10^7$
3.5	42.5	1.39 $\times 10^7$

risen to about three quarters of height $Z_{\max}$. For still larger $\mathcal{R}_*$ (say 21.7) the cosmic ray energy density would be 8.6 $\times 10^{-13}$ erg cm^{-3}, rather larger than the kinetic energy density, and the flow could only rise to height $\frac{1}{2}Z_{\max}$, approximately. Clearly even a small contribution of cosmic ray energy can make a noticeable difference to the flow: a large contribution will radically change it. But cases where this contribution is large call for a special treatment of the dynamics of the transition zone from subsonic flow in the galactic disk to supersonic flow outside. Without such a calculation it is not possible to define $\mathcal{P}_*$ in terms of the cosmic ray energy density in the galactic disk. This problem will have to be left unsolved for the present.

7. Discussion

The flow into the galactic fountain is caused by the high temperature, and possibly the considerable cosmic ray pressure, in the inter-cloud medium. The gravitational field perpendicular to the galactic plane is much too small to confine the ICM within the disk. The hot gas, and the cosmic rays, therefore stream away into the halo, but there is insufficient energy available to allow the gas to escape from the Galaxy, according to this model. Consequently a return flow is set up, very probably to be identified with the intermediate velocity clouds that are observed with the 21 cm line. There are strong similarities between the description that has been given here and that in an earlier account (Kahn 1981). For a view with a somewhat different flavour see Bregman (1980). The treatment presented here cannot account for the observed high velocity clouds, unless there is an additional push given to the descending gas by the infall of material from outside the Galaxy. A possible configuration is shown in Figure 3.

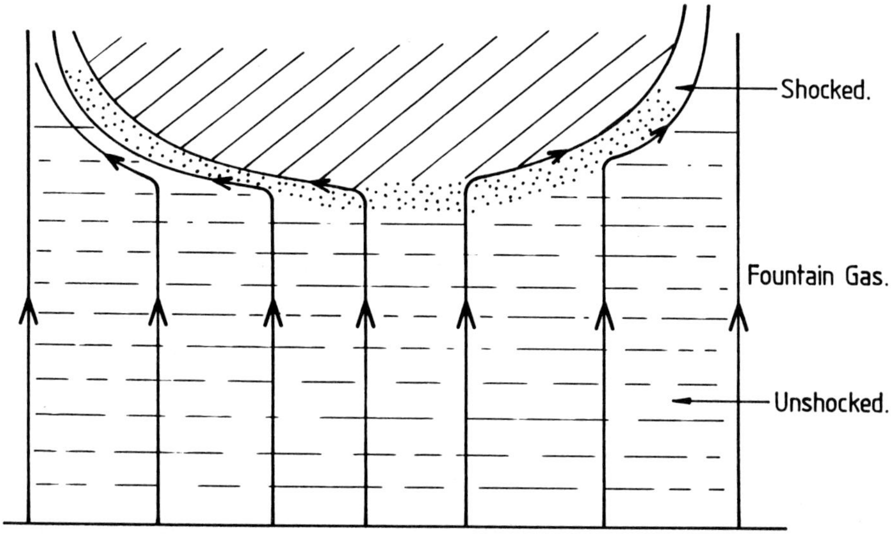

Figure 3: An infall of material (sloping hatched lines) from outside the Galaxy pushes down on the layer of shocked gas that is descending towards the galactic plane. At the same time it distorts the magnetic lines of force (shown with arrows) and compresses them. There is a possibility that such an interaction will increase cosmic ray confinement because of the creation of magnetic mirrors at the locations where the field is most compressed.

References

Bregman, J. N. 1980, Astrophys. J. **236**, 577.

Breitschwerdt, D., McKenzie, J. F. and Völk, H. J. 1987, "Magnetic Field Structure of the Galaxy" in *Interstellar Magnetic Fields* (ed. R. Beck and R. Gräve), Springer, p. 131.

Kahn, F. D. 1976, Astron. Astrophys., **50**, 145.

Kahn, F. D. 1981, "Dynamics of the Galactic Fountain" in *Investigating the Universe* (ed. F. D. Kahn), Reidel, p. 1.

Mirabel, J. F. 1989, "Inflow of Neutral Gas Toward the Galactic Disk" in IAU Colloquium 120 *Structure and Dynamics of the Interstellar Medium* (eds. G. Tenorio-Tagle, M. Moles and J. Melnick), Springer.

Sellwood, J. A. and Sanders, R. H. 1988, Mon. Not. R. astr. Soc., **233**, 611.

Shull, P., Dyson, J. E., Kahn, F. D. and West, K. A. 1985, Mon. Not. R. astr. Soc., **212**, 799.

Skilling, J. 1970, Mon. Not R. astr. Soc., **147**, 1.

CLUSTERED SUPERNOVAE
vs.
THE GASEOUS DISK AND HALO: A Rematch

CARL HEILES

University of California, Berkeley

ABSTRACT. Recent developments, both observational and theoretical, require a reevaluation of the effects of clustered supernovae on the two-dimensional porosity parameter Q_{2D} and the rates of mass injection into the halo $\dot{M}$ of both cold and hot gas. Clustered supernovae produce two types of bubble. Most clusters produce breakthrough bubbles, which do no more than break through the dense gas disk. But large clusters produce enough energy to make blowout bubbles, which blow gas up into the halo. We calculate area filling factors and mass injection rates into the halo for different types of galaxy. We relate our calculations to two observables, the area covered by H I 'holes' and the area covered by giant H II regions. We also reiterate the difficulty of producing the very largest supershells by clusered supernovae.

1. INTRODUCTION

In a spiral galaxy the gas and new stars are concentrated into a relatively thin disk. The stars are formed in clusters. The massive stars become supernovae and these explosions are correlated in space and time. These correlated supernovae produce one large bubble instead of many small ones. If the bubble is large enough, it becomes larger than the thickness of the disk and 'breaks through'. We made calculations on these matters in Paper I (Heiles 1987), which provided unsatisfactory agreement with observation. New theoretical and observational developments cause us to perform new calculations, which are presented in detail in Heiles (1989) and briefly summarized here.

Mac Low and McCray (1988, hereafter MM) and Mac Low, McCray, and Norman (1989, hereafter MMN) have made detailed calculations of this process and made a very important point: just because a bubble breaks through the 'classical' dense gas disk does not mean that it 'blows out' into the halo. This is because of the extensive, low-density 'Lockman' (1984) $|z|$ extension of disk gas. Communication with the halo requires that the shell break through these components and open out into the halo, which requires much more energy. We define *'breakthrough'* bubbles as those that break out of the dense, relatively low-scale-height part of the disk; and *'blowout'* bubbles as those that actually break through all disk gas and communicate with the halo. Both types of bubble can be observed as HI shells and supershells in our own Galaxy and as H I 'holes' in external galaxies.

Crucial to our calculation is knowledge of the fraction of clusters that are large enough to produce breakout or blowout. Recently, we have been blessed with a remarkable paper that allows us to calculate the fraction of clusters of each type. Kennicutt, Edgar, and Hodge (1989; KEH) have derived the frequency and spatial distribution of Hα luminosities $L(obs)$ of bright H II regions in external galaxies, as functions of galactic type. These bright H II regions are produced by precisely those clusters that produce superbubbles. We can use KEH to obtain the formation rates per unit area on the disk of clusters as a function of N.

2. FORMATION RATES OF CLUSTERED SUPERNOVAE

2.1. The relation between $L(obs)$ and N.

To use KEH's observations of $L(obs)$ to calculate the influence of supernovae on the interstellar medium, we must know the relation between $L(obs)$ and the number of supernovae N. This is not a simple matter, because it is thought that all stars more massive than 8 $M_\odot$ become supernovae. Since the initial mass function (IMF) favors less massive stars, most of the supernovae come from stars with masses just above this value. However, most of the ionizing photons that produce the H II regions come from the very massive stars. Thus, both the slope and upper mass cutoff of the assumed IMF affect the desired relation.

Fortunately, careful calculations have been done by Lequeux *et al.* (1981), Melnick, Terlevich, and Eggleton (1985), and McKee (1989), and all agree quite well in predicting about 4.8×10^{62} Ly-continuum photons per supernova. McKee (1989) has evaluated the reliability of this relation by using it to predict the overall Galactic Type II supernova rate from the measurements of Galactic thermal emission of Güsten and Mezger (1982) and comparing the result with the rate derived from pulsar statistics and from rates in other galaxies. Agreement is quite satisfactory.

However, performing the same comparison for external galaxies using the integrated Hα luminosities provides unsatisfactory agreement. The extragalactic Hα luminosities imply a much smaller ionizing flux, and thus a much smaller supernova rate, than the Galactic thermal radio emission. It is unlikely that our Galaxy has many more ionizing photons than otherwise comparable external galaxies, and we do not understand the source of discrepancy. To recover the expected supernova rates in external galaxies, we must adjust the relation between number of supernovae in a cluster and the cluster's Hα luminosity upwards by a factor of about five. Our calculations use this adjusted rate.

For a cluster, $L(obs)$ is equal to the total Hα energy emitted divided by the lifetime of the cluster's H II region τ_{HII}. With $\tau_{HII,7}$ being τ_{HII} in units of 10^7 yr and $L(obs)_{38}$ the Hα luminosity in units of 10^{38} erg s^{-1}, the adjusted rate is

$$N = 222\ L(obs)_{38}\ \tau_{HII,7}\ . \tag{1}$$

For τ_{HII} we adopt 20 Myr because there exists ample observational evidence for star formation occuring over a substantial time interval. In our Galaxy, individual clusters such as Orion and Scorpius/Ophiuchus have undergone sequential star formation over intervals of some 15 Myr (Blaauw 1964).

2.2. Summary of results from KEH.

KEH give the distribution in $L(obs)$ of extragalactic H II regions in the form of a power law. KEH also give the observed surface density of H II regions on the disks of external galaxies, which depend on galactic type. In our calculations, we shall require weighted averages of the form $\langle L^y \rangle \Sigma$, where Σ is the formation rate per kpc^2 on the disk of clusters having L in the range L_{Min} to L_{Max}. L_{Min} is the luminosity of the smallest cluster of interest, for example the smallest whose supernovae will produce a breakthrough bubble. L_{Max} is the upper L cutoff in KEH's observed power-law distributions, which depends on galactic type. For Sb galaxies, $L(obs)_{Max,38} \approx 3$, where L_{38} has units of 10^{38} erg s^{-1}; this corresponds to a cluster than contains about 1350 supernovae.

It is convenient to express these weighted averages in terms of Σ alone. Skipping the details, we obtain

$$\langle L(obs)_{38}^y \rangle \Sigma \approx \frac{7.8 \times 10^{-3}}{1.14 - y}\ \tau_{HII,7}^{-1} L(obs)_{Min,38}^{-(1.14-y)}\ \text{kpc}^{-2}\ \text{Myr}^{-1}\ . \tag{2}$$

The steepness of the $L(obs)$ distribution guarantees that the clusters with fewer supernovae, which are much more numerous, dominate the interaction with the interstellar medium.

3. ISM PARAMETERS

In Paper I we used numerical parameters for the ISM as follows: the 'intercloud' gas density $n_0 = 0.24$ cm^{-3}; the 'scale height' $h_{100} = 1.85$ (see below for definition), where the subscript indicates units of 100 pc; the pressure $P_{04} = 0.40$, where the subscript indicates units of $nT = 10^4$ cm^{-3} K; and the rms velocity $v_{rms} = 9.9$ km s^{-1}.

However, the $|z|$ structure of the ISM is not the classical, simple thin disk. Instead, there is also an extended component, the Lockman component. We follow Lockman, Hobbs, and Shull (1986; LHS) and approximate the $|z|$ distribution as

$$n(|z|) = n_c \exp(-|z|^2/z_c^2) \; + \; n_L \exp(-|z|/z_L) \; . \tag{3}$$

This equation lumps the 'classical' CNM and WNM, which have different scale heights, together into the first term; the new Lockman component is represented by the second term. We adopt $z_c \approx 190$ pc, $z_L \approx 500$ pc, $n_c = 0.316$ cm^{-2}, and $n_L = 0.107$ cm^{-2}. n_0, which we use in various equations below, is the total density at $|z| = 0$, equal to 0.422 cm^{-2}.

Below we use an artificial scale height h. In our approximate theory we assume that $n(|z|) = const.$ for $|z| < h$ and $n(|z|) = 0$ for $|z| > h$. We relate h to z_c or z_L by requiring that the column densities to $|z| = \infty$ be correct. Thus, if we are discussing the classical component, we take $h_{100} = (\sqrt{\pi}/2) z_c \approx 1.7$; if we are discussing the Lockman component, we take $h_{100} = z_L = 5.0$.

4. BREAKTHROUGH: Q_{2D} FROM SUPERNOVAE

4.1. Rederivation of Q_{2D}.

The two-dimensional porosity parameter, roughly equal to the fraction of the disk area occupied by breakthrough bubbles, is denoted by Q_{2D}. Q_{2D} is given by $f \Sigma \pi R_f^2 \tau_{SN}$, where Σ is the formation rate per unit disk area of clusters that produce breakthrough bubbles; R_f is the final bubble radius in the disk, somewhat larger than h, the disk thickness; τ_{SN} is the persistence time of the bubble; and f is a factor, not too far from unity, that accounts for the details of breakthrough bubble dynamics. Paper I set τ_{SN} equal to R_f/v_{rms}, the time required for the ambient gas to repenetrate the bubble with its typical random velocity v_{rms}. This is not quite conceptually correct (Koo 1989), although a conceptually correct expression provides essentially the same result.

We assume that bubble dynamics are those outlined in Paper I, namely we assume that the energetic winds and sequential explosive impulses of the N supernovae in the cluster act as a 'superwind' and produce bubble dynamics equal to that of a continuous stellar wind in the manner described by Weaver *et al.* (1977). The mechanical luminosity of the superwind, $L(wind)$, is equal to the total energy released divided by the time interval in which it is released, τ_{SN}. Paper I assumed that the energy superwind blows for 30 Myr. However, McCray and Kafatos (1987) use a better value, 50 Myr, and in this paper we will increase this to 60 Myr to account for the fact that not all the stars in a cluster are formed simultaneously. Keeping τ_{SN} as a free parameter, we have

$$L(wind)_{38} = 3.72 \times 10^{-2} \tau_{SN,7}^{-1} N. \tag{4}$$

Applying all to the theory given in Paper I, we derive

$$Q_{2D} \approx 340 \, \langle L(wind)_{38}^{1/2} \rangle \Sigma v_{rms}^{-5/2} h_{100}^{2} n_0^{-1/2} \tag{5}$$

4.2. Breakthrough Dynamics.

Paper I argued that at least 12 SN are required for breakthrough. This corresponds to $L(wind)_{38} > 0.15$. MM's more recent detailed calculations show that breakthrough occurs when their parameter D (their equation [29]),

$$D \approx 940 \; L(wind)_{38} h_{100}^{-2} P_{04}^{-3/2} n_0^{1/2} \; , \tag{6}$$

exceeds a value somewhat smaller than 100. For our adopted ISM parameters (section 3), this occurs when $L(wind)_{38} > 0.12$, or $L(obs)_{38} > 0.0145\tau_{SN}/\tau_{HII}$. The near equality of this more rigorous limit with Paper I's limit is purely fortuitous.

4.3. Evaluation of Q_{2D}.

Combining equations (4) and (5) and using equation (2) together with $L(obs)_{Min,38} = 0.0145\tau_{SN}/\tau_{HII}$, we obtain $Q_{2D} = 2.6\tau_{HII,7}^{0.14}\tau_{SN,7}^{-1.14} = 0.37$. This corresponds to a an area filling factor for the hot bubbles of $Q/(1+Q) = 0.27$. Our value $Q_{2D} = 0.37$ is ~ 90 times smaller than the values derived in Paper I. This is fortunate, because the extremely large values of Paper I are not supported by observational data.

In comparing this prediction with observation, we recall that our derived area filling factor is a Galaxy-wide average because we used the average value of Σ. The average should apply roughly to the Solar neighborhood, and Q_{2D} should increase toward the interior. Our predicted area filling factor of 0.27 is about twice as large as the volume filling factor of large H I holes in the Solar neighborhood (Heiles 1980). For M31, another Sb galaxy, the observed area filling factor can be derived from Figures 21 and 22 of Brinks and Bajaja (1986). It peaks at ~ 0.09, corresponding to $Q_{2D} = 0.10$, for galactocentric radius $\sim$ 10 kpc; presumably the M31-wide average $Q_{2D} \approx 0.05$. Our calculated value should also apply roughly to M31, and it is again too large. M33 is an Sc galaxy and has an observed area filling factor < 0.4 (Deul and den Hartog 1989), corresponding to $Q_{2D} < 0.67$. KEH's data cause us to predict that Sc galaxies have 3 to 5 times higher values of Q_{2D} than do Sb's, and this again suggests that our predicted value is too high. We conclude that our predicted values of Q_{2D} are systematically too high by factors of ~ 3.

5. Q_{2D}^{HII} FOR H II REGIONS

The basic theory of the Stromgren sphere (see Spitzer 1978), together with the approximate ratio of $L(obs)$ to $L(uv)$, allows us to write

$$R_{HII} < 0.12 \; L(obs)_{38}^{1/3} n_0^{-2/3} \text{ kpc} \; . \tag{7}$$

The inequality results from clumping, which should be minimal (McKee *et al.* 1984). Q_{2D}^{HII} is equal to $\Sigma^{HII}\pi R_{HII}^2 \tau_{HII}$, or

$$Q_{2D}^{HII} < 0.45 \langle L(obs)_{38}^{2/3} \rangle \Sigma^{HII} n_0^{-4/3} \tau_{HII,7} \ , \tag{8}$$

where Σ^{HII} is the formation rate per Myr per unit area of disk of clusters that produce the H II regions of interest.

We assume that the only H II regions to produce observable holes are those that attain breakthrough, *i.e.* $R_{HII} > h$. For the ISM parameters values of section 3, and assuming equality in equation (7), this requires $L(obs)_{38} > 0.51$. Applying equation (2) yields $Q_{2D}^{HII} = 0.032$, about 8 times smaller than Q_{2D}. The main reason is that the H II region produced by a cluster's massive stars is smaller than the bubble produced by its supernovae.

The partial correlation of observed H I holes with OB associations and H II regions in M31 found by Brinks and Bajaja (1986) and in M33 by Deul and den Hartog (1989) may be consistent with our results. The H II regions are not easily visible on broad-band optical photographs because the emission measures are small. The emission measure of an H II region that breaks through is $2hn_0^2$, or 68 cm^{-6} pc for the ISM parameters of section 3.

6. BLOWOUT: Q_{2D} AND $\dot{M}$

Blowout requires a more stringent condition on D in equation (6), because we must use the scale height of the Lockman component (and, according to MM, the density at $z = 0$). With $h_{100} = 5$, $D > 100$ requires $L(wind)_{38} > 1.036$. This corresponds to 167 supernovae, or $L(obs)_{38} > 0.125\tau_{SN}/\tau_{HII} = 0.375$. We assume that equation (5) applies for these bubbles, but use $h_{100} = h_{c,100} = 1.7$ instead of $h_{100} = h_{L,100} = 5$ because the dynamics of the classical dense disk gas should not be affected much by the evolution of the bubble after breakthrough.

We obtain $Q_{2D} = 0.65\tau_{HII,7}^{0.14}\tau_{SN,7}^{-1.14} = 0.0931$. Thus, about 1/3 of the H I hole area is occupied by blowout bubbles. Associated with these blowout shells are two forms of mass injected into the halo, cold shell fragments and hot gas.

6.1. $\dot{M}_{cold}$: cold shell fragments.

The supernovae drive a radiative shock into the ambient ISM. The matter in this cold radiative shell moves up in $|z|$ through the negative density gradient, accelerating and undergoing Rayleigh-Taylor instability, which makes it break up into fragments of cold neutral gas (McCray and Kafatos 1987). If the supernovae are infrequent and do not approximate a continuous wind, there may perhaps be further instabilities (Tenorio-Tagle, Bodenheimer, and Różyczka 1987). If these clouds pursue ballistic trajectories beginning at $|z| = 4z_L \approx 2$ kpc with a of 100 km/s, they would rise to $|z| \approx 3.7$ kpc and fall back to $|z| = 0$ after about 80 Myr.

For the particular models treated numerically by MMN, about 0.075 of the total mass of ambient ISM in the cylinder of height h and radius R_f was injected into the halo as cold clouds. If this applies generally, then the rate of injected mass of cold cloud fragments is

$$\dot{M}_{cold} = 5.0 \times 10^5 \ \langle L(wind)_{38}^{1/3} \rangle \Sigma h_{100}^{7/3} n_0^{2/3} v_{rms}^{-1} \ \mathrm{M_\odot \ kpc^{-2} \ Myr^{-1}} \ . \tag{9}$$

For the Galactic parameters in section 3, this is $10300\tau_{HII,7}^{0.14}\tau_{SN,7}^{-1.14}$ $M_\odot$ kpc^{-2} Myr^{-1}. If this occurs uniformly over a disk of radius 10 kpc, it becomes $\dot{M}_{cold} = 0.46$ M$_\odot$ yr^{-1}. This is split equally between the 'northern' and 'southern' halo hemispheres. If clouds remain well-defined and follow ballistic trajectories, the amount of gas resident in the halo in the form of these clouds is $\dot{M}_{cold}$ multiplied by the residence time for a ballistic trajectory, or about 3.7×10^7 M$_\odot$.

6.2. $\dot{M}_{hot}$: hot, diffuse gas.

Hot gas is produced by evaporation of gas from clouds and the inside of the cold shell. This gas is important for the halo, because it is injected at high temperature, can travel to high $|z|$, and can spread out into a large volume. Using equations (5) and (6) from Paper I and equation (4) above, we obtain

$$\dot{M}_{hot} = 9.9 \times 10^3 \ \langle L(wind)_{38}^{8/21} \rangle \Sigma n_0^{1/3} h_{100}^{41/21} + 1.65 \times 10^4 \ \langle L(wind)_{38}^{5/7} \rangle \Sigma h_{100}^{2/7} \tau_{SN,7} \ \mathrm{M_\odot \ kpc^{-2} \ Myr^{-1}}. \quad (10)$$

For h we use $h_{c,100} = 1.7$, because after the bubble breaks through the thin dense disk the interior hot gas expands very rapidly, so evaporation becomes nearly as ineffective as it would be if the bubble had blown out (see Figure 8 of Mac Low and McCray 1988).

With our adopted parameters, we obtain $\dot{M}_{hot} = 3650$ $\mathrm{M_\odot \ kpc^{-2} \ Myr^{-1}}$. If this occurs uniformly over a disk of radius 10 kpc, it becomes $\dot{M}_{hot} = 1.15$ $\mathrm{M_\odot \ yr^{-1}}$. Again, this is split equally between the 'northern' and 'southern' halo hemispheres.

What is the fate of the diffuse hot gas that is injected into the halo? It is injected at a high temperature, and is heated further by the Type I supernovae. In Paper I we took the scale height of Type I supernovae, h_{SNI}, to be 325 pc, larger than the scale height of the gas, h. Thus the Type I supernovae were very effective in heating the diffuse halo gas. However, Lockman's disk component has $h_L = 500$ pc, which is larger than h_{SNI}. If all of the Type I supernova energy is absorbed by Lockman's gas, then the Type I supernovae will not be an effective agent for the diffuse halo gas.

The question is very important. Paper I showed that, for negligible radiative cooling and $\dot{M}_{hot} \lesssim 2.1$ $\mathrm{M_\odot \ yr^{-1}}$, the energy input from the Type I supernovae would heat the gas so much that it would exit as a wind. With $h_{SNI} < h_L$, a smaller fraction of the Type I supernova energy will heat the halo gas, and the gas might survive without leaving as a wind. A significant fraction of $\dot{M}_{hot}$ might exit the Galaxy as a wind, depending on several details. If the gas exits as a wind it should do so with a velocity of order 200 km/s, which would make its residence time of order 50 Myr. If the gas does not exit as a wind, it will fall to the Galactic plane after it cools. The cooling time should be smaller than this value. Thus the amount of hot gas in the halo, which is equal to $\dot{M}_{hot}$ multiplied by the residence time, should be $\lesssim 6 \times 10^7$ $\mathrm{M_\odot}$.

7. THE LARGEST SHELLS

Detailed properties of Galactic shells were given by Heiles (1979). There were errors for two shells in Table 2 of that paper. For GS123+07–127 and GS139–03–69, the values of $\log R_{sh}$ should be 3.0 and 1.9, smaller by factors of 1.6 and 2.5 than the values given; listed values for $\log n_0$, $\log M$, and $\log E_k$ are also erroneous.

Shell radii in the Galaxy range up to 1300 pc, if we include only those shells with maximum confidence. Can such large shells be produced by the largest clusters? As originally emphasized by Tenorio-Tagle(1981) it is difficult for clustered supernovae to produce a shell radius very much larger than h, because once blowout occurs most of the additional wind energy is dissipated into the halo and does not produce much extra expansion in the disk. In addition there are other observational reasons for suspecting a different mechanism might operate (Mirabel 1982, and his paper presented at this meeting).

We must appeal to other possibilities. One is that there are fluctuations in h; a large cluster, located in a region where h happens to be large (perhaps because of a previous cluster's supernovae), will be 'lucky' and make a bigger splash than usual. We might have 'clusters of clusters': the effects of multiple large clusters, located nearby in space and time, can be additive, not only in terms of $L(wind)$ but also, probably more importantly, in terms of the earlier clusters modifying the ambient ISM for later ones, for example by increasing the local value of h. Occasionally, several strategically located clusters might create neighboring holes that look like one large hole.

Apart from such effects, the most likely mechanism is completely different. Tenorio-Tagle(1981)and Tenorio-Tagle *et al.* (1987) have suggested that infalling high-velocity clouds can impart large energies to the disk ISM and cause the very largest supershells; they have also suggested ways by which this process can be observationally distinguished from clustered supernovae. There is direct observational evidence for this in our own Galaxy, as discussed by Mirabel at this meeting, and in external galaxies (Brinks 1989). In our Galaxy, some large shells are morphologically associated with high-velocity gas. The only external galaxies that exhibit very large holes have been observed are those with high-velocity gas. These include our Galaxy (Heiles 1984) and M101 (van der Hulst and Sancisi 1988). M31 contains no very large holes, and high-velocity gas at the level seen in our Galaxy is absent in M31 (Brinks 1989).

8. Discussion

Comparison of our predictions with observational data yield significant discrepencies, although they are much smaller than in Paper I and not unreasonable given the uncertainties and approximations in the approach. Different types of galaxy have different rates of clustered supernovae: rates increase in later-type galaxies, so that Sb, Sc, and Irregular galaxies have progressively higher rates.

There is observational evidence in support of our fundamental approach. The evidence is the H I supershells observed in our Galaxy (Heiles 1984) and H I holes observed in M31 (Brinks and Bajaja 1986) and M33 (Deul and den Hartog 1989). Further, in external galaxies the beautiful Hα photographs of M33 by Courtès *et al.* (1987) and of the LMC by Meaburn (1980) exhibit many large ringlike H II regions. The large rings can hardly be anything else but evolved shells produced by the superwinds of the central clusters. It takes about 5 Myr for shells to expand to the typical size of these rings, and because stars in a cluster form over a longer period of time a fraction of the shells should be ionized.

Central to our ideas is the fact that an observationally-derived quantity, Q_{2D}, and a desired but observationally elusive quantity, $\dot{M}$ (the mass injection rate into the halo), are inextricably related, although the quantitative details depend on the ISM parameters. Mass cannot be injected into the halo unless there is a corresponding Q_{2D}. Q_{2D} can be observationally derived, either relatively directly by observing the H I holes, or much less directly (and with more uncertainty) by observing the large H II regions and applying equations (5) and (8).

It is a pleasure to acknowledge discussions with Elias Brinks, Robert Kennicutt, Bon-Chul Koo, Mordecai-Mark Mac Low, Richard McCray, Chris McKee, Guellermo Tenorio-Tagle and Rene Walterbos. This material is based upon work supported by the National Science Foundation under Award No. AST-8818544.

REFERENCES

Blaauw, A. 1964, *Ann. Rev. Astron. Ap.*, **2**, 213.

Brinks, E. 1989, personal communication.

Brinks, E. and Bajaja, E. 1986, *Astron. Ap.*, **169**, 14.

Courtès, G., Petit, H., Sivan, J.-P., Dodonov, S., and Petit. M. 1987, *Astron. Ap.*, **174**, 28.

Deul, E.R., and den Hartog, R.G. 1989, *Astron. Ap.*, in press.

Güsten, R., and Mezger, P.G. 1982, *Vistas in Astron.*, **26**, 159.

Heiles, C. 1979, *Ap. J.*, **229**, 533.

Heiles, C. 1980, *Ap. J.*, **235**, 833.

Heiles, C. 1984, *Ap. J. Suppl.*, **55**, 585.

Heiles, C. 1987, *Ap. J.*, **315**, 555 (Paper I).

Heiles, C. 1989, *Ap. J.*, submitted.

Kennicutt, R.C. Jr., Edgar, B.K., and Hodge, P.W. 1989, *Ap. J.*, **337**, 761.

Koo, B-C 1989, private communication.

Lequeux, J., Maucherat-Joubert, M., Deharveng, J.M., and Kunth, D. 1981, *Astron. Ap.*, **103**, 305.

Lockman, F.J. 1984, *Ap. J.*, **283**, 90.

Lockman, F.J., Hobbs, L.M., and Shull, J.M. 1986, *Ap. J.*, **301**, 380.

Mac Low, M. and McCray, R. 1988, *Ap. J.*, **324**, 776.

Mac Low, M., McCray, R., and Norman, M.L. 1989, *Ap. J.*, **337**, 141.

McCray, R. and Kafatos, M. 1987, *Ap. J.*, **317**, 190.

McKee, C.F. 1989, *Ap. J.*, in press.

McKee, C.F., van Buren, D., and Lazareff, B. 1984, *Ap. J.*, **278**, L115.

Meaburn, J. 1980, *Mon. Not. R. Astron. Soc.*, **192**, 365.

Melnick. J., Terlevich, R., and Eggleton, P.P. 1985, *Mon. Not. R. Astron. Soc.*, **216**, 255.

Mirabel, F. 1982, *Ap. J.*, **256**, 112.

Spitzer, L. 1978, *Physical Processes in the Interstellar Medium* (Wiley: New York).

Tenorio-Tagle, G. 1981, *Astron. Ap.*, **94**, 338.

Tenorio-Tagle, G., Bodenheimer, P., and Różyczka, M. 1987, *Astron. Ap.*, **182**, 120.

Tenorio-Tagle, G., Franco, J., Bodenheimer, P., and Różyczka, M. 1987, *Astron. Ap.*, **179**, 219.

van der Hulst, T., and Sancisi, R. 1988, *Astron. J.*, **95**, 1354.

Weaver, R., McCray, R., Castor, J., Shapiro, P., and Moore, R., 1977, *Ap. J.*, **218**, 377.

Discussion:

FRANCO: Your presentation stressed the multiple SN scenario to create large structures and at the end you just gave a hint that cloud-galaxy collisions may also be operative to provide the large scale structures. The HVC-galaxy model has been worked out in some detail and there are predictions that may help to differentiate (i.e. X-ray emission in the case of the multi-supernova model) the origin of the energetic superstructures.

HEILES: Forgive me for neglecting to properly acknowlege your contribution. The paper by Tenorio-Tagle et al.(A.A. 179, 219, 1987) presents calculations of the HVC-galaxy collisions, and certainly must apply for many of the largest supershells. I was simply trying to point out that it is conceivable that multiple supernovae might produce a (probably small) portion of the very largest supershells.

PECKER: I presume that each individual injection of matter in the halo, from SN explosions in a cluster, would enrich the halo in "processed" matter. This would certanly affect the chemical composition of the young halo stars: this is certanly something that can be tested. Does it confirm or deny your figures about the rate of mass injection in the halo?

HEILES: I don't now. I haven't estimated what fraction of mass injected into the halo comes directly from the exploding supernovae.

KHAN: Have you allowed for radiative heat loss from the gas at large heights?

HEILES: No. I've set $T = \text{const}\ \dot{E}/\dot{M}$, where $\dot{E}$ comes from the supernovae that exploded at high Z and thus might heat the halo gas directly. If one wants to do better, one must include not only cooling, but other heating processes.

DANLY: I am wondering about the observational effects of the strong gradients with galactocentric radius in the SN rate and SN correlation which you alluded to. Do you also therefore expect a variation with galactocentric radius in the characteristics, such as ionization, of galactic halo gas? Savage & Massa looked for such variations in their survey toward the galactic center and found none. Do you feel their observations do not constrain your model?

NORMAN & HEILES: Pressure should be higher and ionizing photon flux should be higher. Not clear how this will affect observed UV lines. However, total recombination rate should be larger, so $H\alpha$ emission (Reynolds type) should be brighter.

BLADES: Is it meaningful to derive a scale-height for the material that is blown out into the halo? Ultraviolet-optical observations seem to indicate that considerable gas is situated between 0.5 and 1 kpc, and the high ions (C IV, SiIV) seem to appear at about $Z = 1$kpc.

HEILES: In this ballistic-type model you would not expect to find an exponential-like scale height. However, I would expect to see this material concentrated at Z-distances of 2-3kpc.

TENORIO-TAGLE: I do not understand how a model of sequential cluster formation could lead to a supershell. By definition (see Heiles 1979, 1984) a supershell presents a large amount of kinetic energy E $>10^{53}$ erg. You have shown here how easy it is to exceed the dimensions of the disk and how energy from further explosions leaks out into the halo and does not provide the remaining shell with further energy. This clearly will also be the situation in the sequencial cluster scenario. Once the edge of the disk is reached, all or most of the further deposited energy will go into the halo. The remaining remnant in the galactic disk will be large but it will not present the energy of a supershell. The above point is also related to the question of Prof. Pecker most of the mass ejected into the halo is in fact supernova matter and not original cold disk material. Could you comment on this point?

HEILES: First, point 1. I was only saying that, occassionally, you might have a shell of radius that could be produced by clusters of multiple supernova. But I agree that most are produced by infalling HVC's, as your theoretical models have shown. Point 2: This is a good idea! I will be intented in making an estimate.

TERLEVICH, R. (Comment): The models you have used to compute the number of SN per unit Hα luminosity are for ZAMS stars and do not include mass loss. These two effects, reduce the total Hα luminosity per unit mass of a cluster. The HII region population of a spiral or irregular galaxy disk, samples young clusters with a distribution of ages between zero and 4/5 Myr whith a median of perhaps 3 Myr. Thus the integrated Hα flux will on average correspond to a cluster of such age and will be about a factor of 2 lower than at ZAMS. Another important ingredient is stellar winds. Massive star evolution is dominated by the effects of stellar winds. The presence of the wind reduces the U.V. flux by as much as a factor of 2. Finally, the slope of the IMF you have used ($\alpha = 2.5$) seems to be a bit flatter than the present best value for massive stars (i.e., $8<M<100\ M_{\odot}$). Massive stars IMF slope in the solar neighbourhood is more likely between 2.7 and 3.0. Changing to this value will further reduce the Hα flux per unit cluster mass by another factor of 2. All in all I believe that your estimates may be off by something between a factor of 3 to 10. Also I will suggest to use the models by Melnick, Terlevich and Eggleton (1985). They estimate many others parameters: the Hβ flux per unit mass for young clusters of stars with a range of IMFs, mass loss rates and abundances, using a self-consistent set of interior and atmosphere theoretical models.

Gamma rays from violent interstellar events

J.B.G.M. Bloemen
Leiden Observatory and Laboratory for Space Research Leiden
P.O. Box 9513, 2300 RA Leiden, The Netherlands

Summary

As an illustration of what the next generation of γ-ray telescopes may show us, an up-to-date COS-B 'finding chart' of potential γ-ray point sources and unexplained extended γ-ray features is presented. The latter, in particular, may be related to energetic phenomena in the interstellar medium, capable of enhancing the local cosmic-ray density. As an example, a prominent feature, extending over at least $10° - 15°$ almost perpendicular to the Galactic disk, is discussed in some detail, linking it to the giant HII region S54 and Complex C of high-velocity clouds.

1. Introduction

Energetic events in the interstellar medium associated with regions of relatively high gas density, such as, supernova-cloud interactions and the collision of high-velocity clouds (HVC's) with the Galactic disk, may be visible as regions of enhanced γ-ray emission (see, e.g., the review by Morfill and Tenorio-Tagle, 1983). The basic requirement is that the energy supplied is converted into the acceleration of particles with a reasonable efficiency; the interstellar matter in the region of enhanced cosmic-ray (CR) density converts part of the CR energy into γ rays. This occurs through collisional nuclear excitation (leading to γ-ray lines at MeV energies), inelastic nuclear interactions with the subsequent production of $\pi°$ mesons (leading to $\pi°$-decay γ-ray emission), and Coulomb scattering of CR electrons on the nuclei and electrons of the interstellar gas (leading to bremsstrahlung γ-ray emission). Also, γ rays may be produced by inverse-Compton scattering of high-energy CR electrons on the interstellar radiation field. The relative importance of the emission processes is a strong function of γ-ray energy (for details, see e.g. Stecker, 1971), but in first order approximation the continuum γ-ray flux from a certain volume is a measure of the product of the average CR density and the gas (or photon) density. Hence, regions of high gas densities (such as molecular clouds), regions with strong radiation fields (such as the immediate surroundings of OB associations in their early stages), and regions with high CR densities (such as probably the evironments of energetic events like supernovae, stellar winds, and violent cloud collisions) have high γ-ray production rates per unit of volume. The latter, 'active' regions of γ-ray emission are most pronounced when they coincide with the former two.

The SAS-2 and COS-B γ-ray surveys ($E_\gamma \gtrsim 50$ MeV) show that the appearance of the γ-ray Milky Way is not dominated by such active regions of γ-ray emission. A fairly homogeneous distribution of cosmic rays throughout the Galaxy can account in good detail for the observed γ-ray intenstity distribution, with most of the structure attributable to variations in the total gas column density (see review by Bloemen, 1989). Detailed modelling of this diffuse emission, however, reveals several features that are not accounted for.

2. Unexplained γ-ray excesses in the COS-B data

The search for unexplained γ-ray excesses is closely related to the search for γ-ray point sources. After some early claims based on balloon experiments, firm detection of γ-ray point sources was first achieved with the SAS-2 satellite, in particular that of the Crab and Vela pulsars and an hitherto puzzling source, now known as 'Geminga'. The COS-B observations show evidence for several additional sources, which in early stages were simply defined as statistically significant peaks in the γ-ray intensity distribution, with an angular shape consistent with the COS-B point-spread function (see Hermsen, 1983). This lead to the so called 2CG catalogue (Swanenburg *et al.*, 1981), containing 25 γ-ray sources. Crab and Vela are unambiguously identified by their timing signatures, but all other identifications have to be based solely on positional coincidence. Given the limited angular resolution (above 100 MeV $\sim 2.5°$ FWHM) it is difficult to differentiate between intrinsically compact objects and objects that might be up to a few degrees in size, such as CR-irradiated clouds. With the availability of CO surveys (in addition to already existing HI 21 cm data) it became feasible to take this diffuse emission into account in the search for sources, which were then defined as excesses, consistent with the COS-B point-spread function, superimposed on a model of the diffuse emission based on HI and CO surveys (Pollock *et al.*, 1985ab; Simpson and Mayer-Hasselwander, 1987, 1989). Hence, those original sources that find an explanation in terms of the same CR-matter interactions as the underlying diffuse emission (about 30% - 50%) are excluded. Remaining sources may either be intrinsically compact objects or still be of diffuse origin, indicating a high CR density (our point of interest here) or an enhanced radiation field somewhere along the line of sight.

Evidently, the detection of extended γ-ray excesses, not particularly aimed at in the analyses of Pollock *et al.* and Simpson and Mayer-Hasselwander, is of interest for our purposes as well. Also, it is advantageous to combine the γ-ray data for all energies, which was not done by these authors. The energy range 100 MeV – 5 GeV is a good compromise between counting statistics and angular resolution (which degrades with decreasing energy). Further, allowing more flexibilty in the predicted diffuse emission seems appropriate, because the CR distribution adopted for the modelling (a function of Galactocentric radius only) is probably a gross approximation. The following describes a first attempt to take these points into account in a simple manner.

Fig. 1 presents a 'finding chart' of potential γ-ray sources and extended features for the 100 MeV – 5 GeV range. Basically, this map represents the γ-ray excesses that are left when the expected diffuse emission (from Strong *et al.*, 1988) is subtracted from the observed γ-ray intensity distribution. It is not a pure subtraction. In order to correct for deviations between model and observations on scales of $10° - 20°$ in longitude direction, the modelled intensities were first re-adjusted to the observed intensities on this scale: for a given position, the predictions in a strip of 21 pixels in longitude and 1 pixel in latitude (pixel size is $0.5° \times 0.5°$), centered on this position, were scaled upwards or

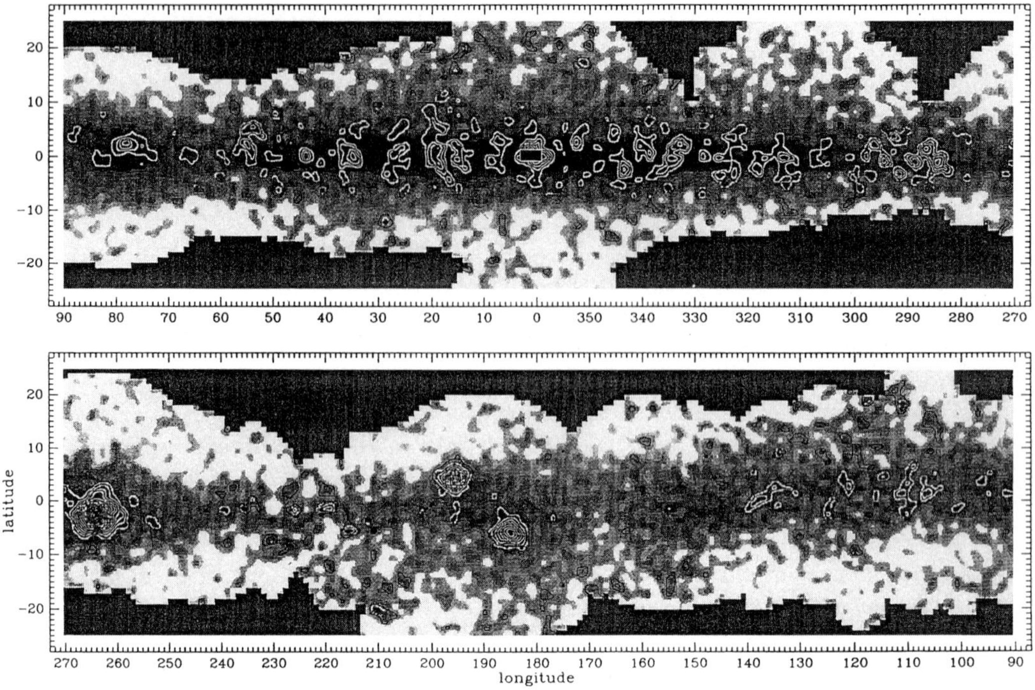

Fig. 1. 'Finding chart' of potential γ-ray point sources and extended excesses, derived from the COS-B γ-ray survey (100 MeV – 5 GeV). For details, see Section 2. The underlying gray-scale map represents the observed γ-ray sky. The small box in the Galactic-centre direction indicates a region that was excluded from the analysis; the feature surrounding it is probably an artifact. The strong excesses near $\ell = 185°$, $195°$, and $263°$ correspond to the Crab, Geminga, and Vela γ-ray sources. Note the extended features visible in the map. Contour values: 8, 13, 18, ...$\times 10^{-5}$ photon cm^{-2} s^{-1} sr^{-1}.

downwards such that for only 20% of the pixels the observed intensities are below the predicted intensities. This approach was chosen, instead of a pure normalization over a certain area, because it will reveal all indications for excesses in the data (the value of 20%, however, is a rather arbitrary choice). Clearly, no formal statistical significance can be assigned to the excesses; 'finding chart' is probably the best denomination for the resultant map. To give the reader some indication of the significance, it may be useful to point out, on the basis of detailed studies of some selected regions, that even excesses with only two contours can be considered of real interest. Less attention should be paid to the precise shape of the excesses. Interestingly, however, some features are clearly extended, most pronounced being the ones near $\ell \approx 15° - 20°$ and $\ell \approx 330° - 340°$.

3. An example: the excess near $\ell \approx 15° - 20°$

With a flux of $\sim 2 \times 10^{-6}$ photon cm^{-2} s^{-1} (100 MeV – 5 GeV), this excess would be among the most intense sources in the 2CG catalogue, if not extended. It coincides with two unique phenomena: 'Stockert's chimney', a (thermal) radio-continuum spur

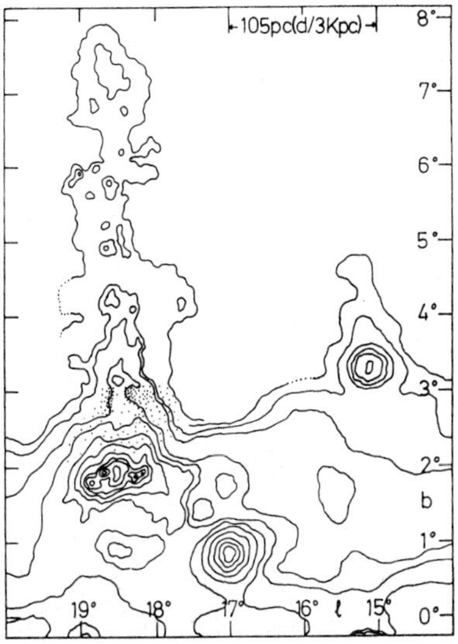

Fig.2. Sketch of 'Stockert's chimney' at 2.7 GHz, with S54 located at $\ell \approx 18.5°$, $b \approx 2°$ (Müller *et al.*, 1987; Kundt and Müller, 1987).

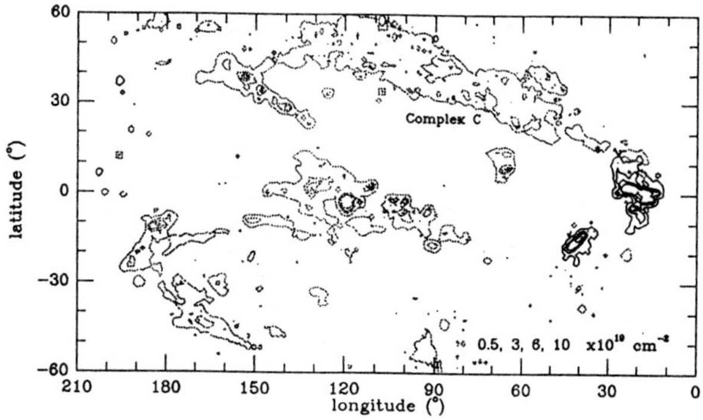

Fig.3. HI column-density map of the HVC's in the catalogue of Hulsbosch and Wakker (1988), illustrating the possible connection between Complex C and the γ-ray excess near $\ell \approx 15° - 20°$. Dotted: negative velocities. Full: positive velocities. The so-called Outer Arm complex has been excluded.

associated with the giant HII region S54 (Müller *et al.*, 1987; Kundt and Müller, 1987; see Fig. 2) and the crossing of the extension of the so called Complex C of HVC's with the Galactic disk (Fig. 3). At a kinematic distance of $\sim$ 3 kpc (the molecular cloud associated with S54 is visible in Fig. 5 at $V \approx 30$ km/s), S54 is located at $z \approx 100$ pc and the radio spur extends up to $z \approx 400$ pc. Kundt and Müller argued that the spur is an outflow phenomenon driven by the HII region, consisting of a light relativistic pair plasma (with an admixture of partially ionized hydrogen and cosmic rays), particularly for reasons of energy and absence of recoil and explosively receding fragments. Apart from the energy requirement ($E_k \approx 10^{54}$ erg), however, these arguments may not apply (i.e., the chimney may consist of 'ordinary' interstellar matter), if the HVC's and the intermediate-velocity gas seen in this region of the sky are to be associated with S54, as suggested by Fig. 4. Shock acceleration of particles in the surroundings of S54 is then certainly plausible, whatever the precise energy source may be. A quite different interpretation of the findings may be the interaction of infalling HVC's with the Galactic disk. It has been suggested that shock ionization, formation of giant HII regions, and shock acceleration accompany such phenomena (e.g. Tenorio-Tagle, 1981; Morfill and Tenorio-Tagle, 1983). At a distance of 3 kpc, the CR enhancement (relative to the CR density of $\sim$ 1 eV cm^{-3} in the solar vicinity) that is required to explain the excess γ-ray flux is about $10^7/M$, where M (in units of $M_\odot$) is the gas mass involved. If, for instance, the observed excess flux would originate largely from the HVC's with positive velocities surrounding S54 (Fig. 4c), with a mass of $\sim 5 \times 10^4 M_\odot$ at 3 kpc, then the average CR density would be enhanced by more than two orders of magnitude, and the total CR energy in this volume would exceed 10^{53} erg. A detailed analysis will be presented elsewhere.

It may be clear that with the launch of the Soviet-French γ-ray telescope GAMMA-1 and NASA's Gamma Ray Observatory, γ-ray astronomy can soon be expected to give exciting new insight into energetic interstellar phenomena.

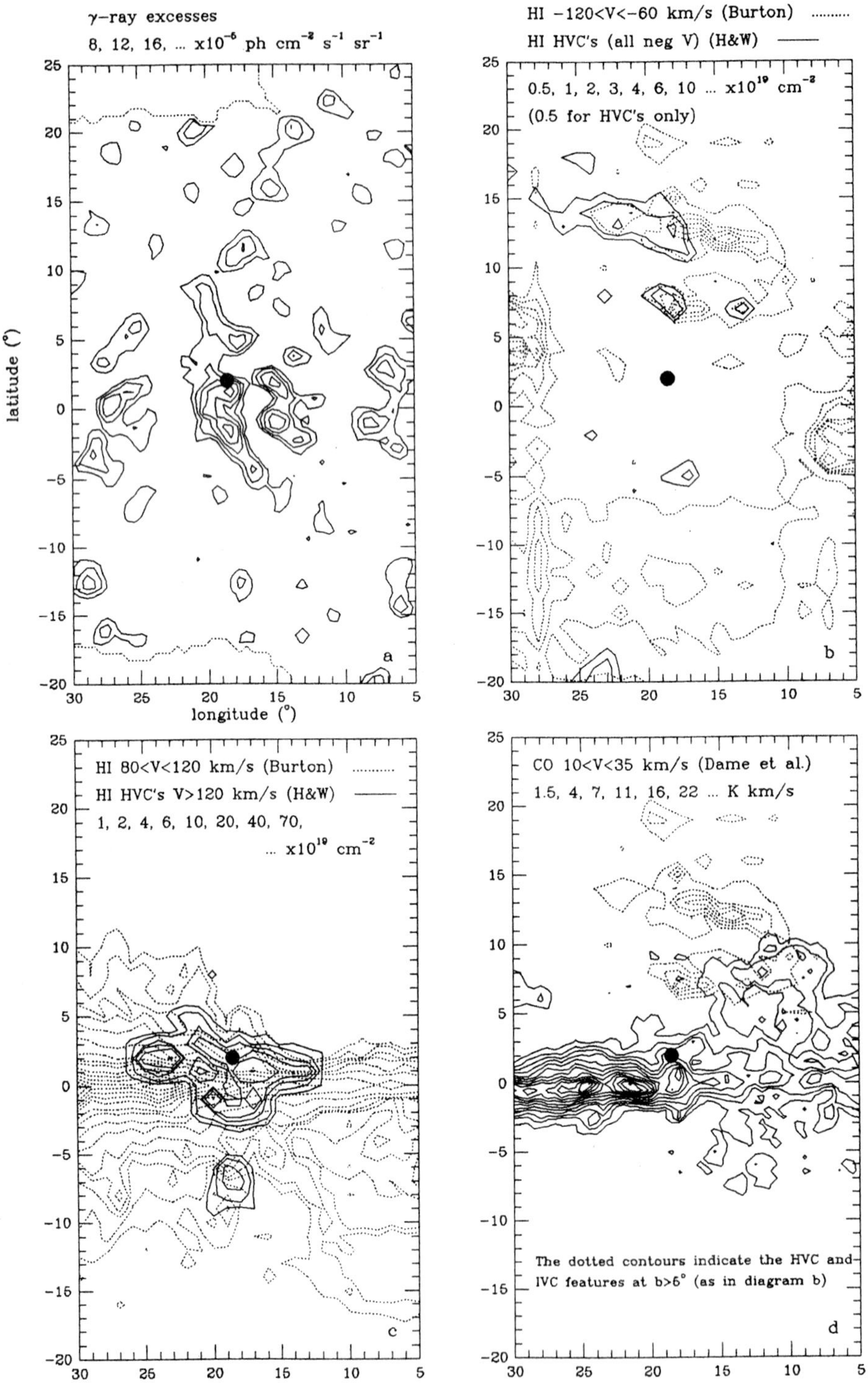

Fig. 4. A compilation of selected HI (b and c) and CO (d) observations around the extended γ-ray excess near $\ell \approx 15° - 20°$ (a). The black dot indicates the position of S54. In Fig. 4b, the extended low-intensity feature seen at $b < -5°$ is probably spurious (see Burton and te Lintel Hekkert, 1985). Also, in Fig. 4c, the feature seen at $b < -10°$ (extending to $\ell \approx 350°$) needs confirmation.

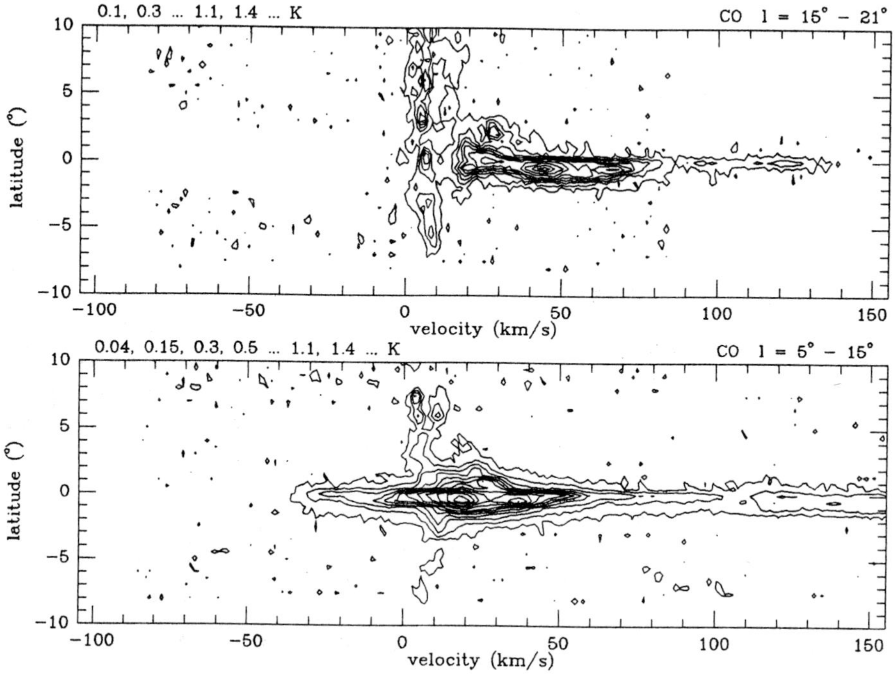

Fig. 5. Latitude, velocity maps of the CO line, averaged over $5^\circ \lesssim \ell \lesssim 15^\circ$ (lower) and $15^\circ \lesssim \ell \lesssim 21^\circ$ (upper), derived from the Columbia – Cerro Tololo CO survey (Dame *et al.*, 1987, and references therein). The upper diagram includes the molecular cloud adjacent to S54, visible at $V \approx 30$ km/s, $b \approx 2^\circ$. Note the presence of high-latitude gas at $V > 10$ km/s.

Acknowledgements It is a pleasure to thank B.P. Wakker for informative discussions on high-velocity clouds. I acknowledge the receipt of a Fellowship from the Royal Netherlands Academy of Arts and Sciences (KNAW).

References

Bloemen, J.B.G.M. 1989. Ann. Rev. Astr. Ap. 27 (in press)
Burton, W.B. 1985. Astr. Ap. Suppl. 62:365
Burton, W.B., te Lintel Hekkert, P. 1985. Astr. Ap. Suppl. 62:645
Dame, T.M., Ungerechts, H., Cohen, R.S., de Geus, E. et al. 1987. Ap. J. 322:706
Hermsen, W. 1983. Space Sci. Rev. 36:61
Hulsbosch, A.N.M., Wakker, B.P. 1988. Astr. Ap. Suppl. 75:191
Kundt, W., Müller, P. 1987. Astrophys. Space Sci. 136:281
Mayer-Hasselwander, H.A., Simpson, G. 1989. Adv. in Space Research (in press)
Morfill, G.E., Tenorio-Tagle, G. 1983. Space Sci. Rev. 36:93
Müller, P., Reif, K., Reich, W. 1987. Astr. Ap. 183:327
Stecker, F.W. 1971. 'Cosmic Gamma Rays', Mono Book Corp.
Pollock, A.M.T., Bennett, K., Bignami, G.F., Bloemen, J.B.G.M. et al. 1985. Astr. Ap. 146:352
Pollock, A.M.T., Bennett, K., Bignami, G.F. et al. 1985. Proc. 19th Int. Cosmic Ray Conf. 1:338
Simpson, G., Mayer-Hasselwander, H.A. 1987. Proc. 20th Int. Cosmic Ray Conf. 1:89
Strong, A.W., Bloemen, J.B.G.M., Dame, T.M., Grenier, I. et al. 1988. Astr. Ap. 207:1
Swanenburg, B.N., Bennett, K., Bignami, G.F., Buccheri, R. et al. 1981. Ap. J. Lett. 243:L69
Tenorio-Tagle, G. 1981. Astr. Ap. 94:338
van Woerden, H., Schwarz, U.J., Hulsbosch, A.N.M. 1985. in 'The Milky Way Galaxy', eds. H. van Woerden, R.J. Allen, W.B. Burton, Reidel, Dordrecht, p. 387

STRUCTURE OF THE DIFFUSE INTERSTELLAR MEDIUM

Donald P. Cox
Department of Physics, University of Wisconsin-Madison
1150 University Ave., Madison, WI, 53706, USA

ABSTRACT

The interstellar medium has a thick disk structure ($|z| \lesssim 1500$ pc, the ECL or extracloud layer) with a thin zone of cloud contamination at $|z| \lesssim 100$ pc. The properties of all components other than clouds (e.g. cosmic rays, magnetic field, extracloud matter, pressure) drop slowly with $|z|$ across the thick distribution, giving this layer a very important influence in the evolution of superbubbles. It seems likely to quench blowout or breakout and virtually all fountain activity. The weight of this layer stabilizes the clouds at low z; its high pressure and low density provide a cushion for impacts of infalling clouds.

The intercloud (or more properly "extracloud") medium is stable against supernova disruption because of the incompressibility and elasticity contributed by its magnetic field. Coronal ions in the galactic disk probably derive in large part from quiescent bubbles of hot gas generated by Type I supernovae. Most of the diffuse ionization close to the midplane is probably caused by O star radiation leaking out of the HII regions, but Type I supernovae could maintain enough partial ionization in the bulk of the extracloud medium to contribute significantly to $\bar{n}_e$. At high z, the increasing degree of ionization is almost certainly due to cones of ionizing radiation emerging from OB associations. The pattern is like sunlight rays through a partially cloudy sky, the neutrals lying in shadows of low z dense material, and between cones.

Above the extracloud layer, there could be a hot, interesting, dynamic corona, and between the two a vigorous transition layer or chromosphere. If so, their study (except for the high z high ions, high velocity clouds, and upper limits to the x-ray emission) lies almost entirely in the future.

I have tried to highlight what I regard as key unsolved problems in this field, should the reader wish to be helpful.

I. To Guido's Health, Happiness, and Continued Good Science.

When I asked among my friends in Wisconsin for stories I could tell on Guido, Dan McCammon told me he was a good guy to go drinking with, Art Code had promised his stories to Don Osterbrock, Ron Reynolds gave me a paper to take to Guido and asked me to bring careful notes on his talk, and Chris Anderson told me there were no stories that my tender ears would be prepared to hear, but anyway to take Guido his fond greetings. It's a sorry lot altogether and quite the pity. I'd hoped to begin my talk, in this last session of the meeting, with a bit of humor.

In any case, Guido, Salud!

II. Conclusions

Carl Heiles set a good precedent, starting with the conclusions.

The picture which I have been exploring is orthogonal to that of most modelers of galactic fountains, winds, disk-halo interactions, high velocity clouds, superbubbles, and interstellar media dominated by supernovae.

It tends to ignore OB associations, stellar winds, and Type II supernovae, imagining them to be locally contained inflamations in an otherwise fairly docile environment. (At high latitude, the neglect of OB associations' ionizing flux has been an error on my part.)

It ignores interstellar clouds of all types, noticing that even in the midplane their volume occupation is negligible, that their weight contributes only a little to the vertical confinement of the magnetic field, and that (except under the most extreme circumstances) their magnetic field configurations probably adjust to shield them from appreciable interaction with their environment.

In this view, Type I supernovae are an important source of disturbance. But, when the ambient magnetic field is considered, it is quite likely that the medium behaves elastically to first order (Cox 1986, 1988; Cox and Slavin 1989). An old remnant probably consists of a moderate sized hot bubble within a much larger region of shocked but only very slightly compressed material. The latter "shell" of the remnant is indistinguishable from the surroundings, its inner edge rebounds to fill the bubble in a few million years. These conclusions need confirmation via 2D magnetohydrodynamic calculations so that the complex behavior of oblique shocks can be secured. (See Cox, 1988; Spitzer 1989.)

There are several important consequences of such an evolution for isolated remnants.

(1) Viability of Warm Intercloud Medium. The supernovae do not disrupt the intercloud medium into a froth of hot gas and dense shells or clouds -- a warm intercloud medium of density 0.1 to 0.2 cm^{-3} is possible owing to the lack of compressibility of $B^2/8\pi$.

(2) Coronal Ions. Preliminary calculations of the ion content of the quiescent hot bubbles show that their population can provide the observed interstellar mean densities of O VI, N V, C IV, and Si IV (Cox and Slavin, 1989). (Although much work remains to be done on this topic, the picture is already at least as successful as any other for providing these ions.)

(3) SN Energy Dispersal. This evolution drastically alters one's perception of the ultimate distribution of the supernova energy. After the onset of the radiative phase, much of the energy is stored in magnetic field compression, rather than being radiated in EUV. At later times it is recovered, driving the re-expansion of the shell material. The energy is ultimately dissipated over a very large volume in quite weak shocks. The prime functions of such shocks may be: (A) the partial ionization of hydrogen along with an appreciable production of Lyman α by collisional excitation*, and, (B) occasional reheating of the intercloud gas to roughly 10^4 K, from which it cools only very slowly.

(4) Survivability of Interstellar Dust. The significant change in the character of these shocks compared to the highly compressive radiative shocks of earlier models may result in much lower dust destruction rates.

(5) The Cowie and York Test. One must similarly reevaluate the expected frequency of radiative shock signatures, before the significance of the Cowie and York (1978) limit is known.

In this picture, the intercloud medium, the high z medium and the interarm medium are all the same thing, with parameters varying gradually with increasing height above the plane, z.

A three layer description seems to be useful for the diffuse galactic environment.

1. The Cloud Zone. At the galactic midplane is the cloud contaminated zone, a thin disk with $|z| \lesssim 100$ pc. The dominant interstellar beasts of this neighborhood include dark clouds, diffuse clouds, lower density cloud envelopes, the intercloud medium (in neutral and ionized forms), hot bubbles, cosmic rays, and a strong magnetic field (with a significant component parallel to the disk). There are also the superactive subzones around sites of present and former OB associations, considered extensively by other speakers at this conference.

Characteristic densities, temperatures, pressures, filling factors, and power requirements associated with thin disk components are shown in Table 1. The notes mention some of the major issues. I differ from most other authors, in discarding the assumption of thermal pressure balance between components (noting that $B^2/8\pi$ is large enough to make up any discrepancies), and thus in my conclusion that the neutral intercloud medium occupies a large fraction of the total volume. As mentioned previously, the magnetic field stabilizes this phase against disruption by supernovae.

There are three aspects of the thin disk parameters about which I remain uneasy: (A) I don't believe we have yet located the cosmic ray acceleration sites; (B) the maintenance of widespread ionization in the intercloud medium requires more power than any source other than O stars can comfortably provide, but O stars should

---

*John Mathis points out that this could depend somewhat sensitively on the extent of preshock ionization.

Table 1

Midplane Characteristics of the Interstellar Medium[a]

		Power	Pressure	Density	Temperature	Volume
		(vol. avgd.) $[10^{-26}\text{erg cm}^{-3}\text{ s}^{-1}]$	(p/k) $[\text{cm}^{-3}\text{ K}]$ (thermal)	(local) $[\text{cm}^{-3}]$	[K]	[%]
Clouds	dark	?	moderate	large	15	small
	diffuse	10^{b}	3000	40	80	$\lesssim 2$
	envelopes	$\lesssim 1$	3000	$\lesssim 8$	400	$\lesssim 10$
Intercloud	neutral	$\lesssim 1$	1000^{c}	0.2	5000	$\gtrsim 60^{c}$
	ionized	7^{d}	3000	0.2	8000	15
Type I SNR	bubbles	(at 10^6 years)	1000	0.003	3×10^{5}	12
	power	2^{e}	(S ~ 4×10^{-14} pc^{-3} yr^{-1} at 5×10^{50} ergs)			
Ionizing UV	O stars	110	from Abbott, 1982			
	B stars	2.2				
	Other	0.3	From Panagia and Terzian, 1984			
			(non thermal)			
Cosmic Rays		1^{e}	6000			
Magnetic Field		small[g]	6000-10,000[h]			

Notes:

[a]These are best guess schematic results. Power neglects dust absorption of light and production of FIR.

[b]10^{-25} egs/atom-sec to balance gas cooling

[c]Low thermal pressure of neutral intercloud medium is commonly used as evidence that the local density is higher and filling factor lower, leaving room for a hot component, but see h below.

[d]Power to keep the diffuse ionized gas ionized is very large. (See Reynolds, 1984)

[e]The cosmic ray power requirement is a significant fraction of the Type I SN output. (Both are assumed to have 300 pc scale ht. in this chart.)

[f]The large UV power from O stars does not leak well into the intercloud component far from OB associations if the neutral hydrogen is randomly distributed.

[g]The field replacement timescale from windup or radial inflow is long. SN driven dynamo could be significant.

[h]Large magnetic pressure makes medium difficult to compress, the compression is elastic, thermal pressure balance is not enforced between components except over timescales much longer than those of major disturbances.

have difficulty delivering their photons where needed (but see section IV); and (C) the diffuse cloud heating mechanism seems accidently to result in minimum cloud thermal pressures of an allowed magnitude, somewhat below the weight of the interstellar medium (but see discussion of FGH model below).

2. The Extracloud Layer. Surrounding the cloud-contaminated zone is a much thicker disk, having $|z| \lesssim 1500$ pc. I have come to regard this zone with great respect; it is not the location of just a few trace amounts of this and that, of no particular consequence. Rather it is a zone in which the magnetic field is only slightly less than that at midplane; it is the trapping volume for the cosmic rays; it is home to a distribution of gas whose heating and ionization require an extraordinary amount of power and whose weight defines the midplane pressure. It is the atmosphere, the overburden, the thermostat, the buffer for the interstellar system, dictating the conditions within the cloud environment at low z.

The thick distribution of cosmic rays and magnetic field is displayed by the galactic synchrotron emission, the low mean density within the cosmic ray trapping volume, and the galactic contribution to Faraday rotation of extragalactic sources (see references in Boulares and Cox, 1989). The γ-ray emission of the Galaxy also contains hints of it (Bloemen, private communication). These components drop slowly, to $|z| \sim 1500$ pc, from their midplane values. The presence of "intercloud" material at high z has shown up in 21 cm studies (e.g. Lockman, Hobbs, and Shull, 1986), in studies of trace ions such as Ti II (Edgar and Savage, 1989, and references therein), and in the distribution of pulsar dispersion measures (Reynolds, 1989a). The neutral component of the intercloud medium, the Lockman Layer, has a midplane mean density $\bar{n}_{HI}(0) \sim 0.1$ cm^{-3} and scale height of roughly 400 to 500 pc. The ionized component, the Reynolds Layer, has a midplane mean density $\bar{n}_e(0) \approx 0.025$ cm^{-3} and scale height of about 1500 pc. The masses in the two are comparable, and similar to that of the clouds. Because they are at high z, their weight exceeds that of the clouds.

Considering the Lockman and Reynolds Layers as the neutral and ionized portions of the same material distribution, Ron Reynolds and I would like to propose subsuming both, along with the "intercloud medium", the "interarm medium", the magnetic field, and the cosmic rays into one general term, the "extracloud layer", ECL*. Then ECM (extracloud matter, medium, or material) is appropriate to refer to the material itself. The material in the Lockman and Reynolds layers can then be NECM and IECM, or one can continue to use the WNM and WIM designations.

There are several outstanding problems regarding the thick disk. These include (A) determination of the high z supernova rate, (B) understanding cosmic ray transport and escape, (C) determination of the magnetic field configuration, (D)

---

*"Extra" connotes outside of, as well as above. "Layer" calls to mind the overall distribution in a thick disk; in addition, Reynolds points out that it was the term used by Hoyle and Ellis (1963) when they originally proposed the existence of what I now call the Reynolds Layer.

measuring the degree of intermixture of the neutral and ionized components, (E) determining the origin of the ionization (which as in the midplane, requires an extraordinarily large amount of power--see Reynolds, 1984, 1989b), (F) understanding the binding of the large field energy to the Galaxy and the associated hydrostatics of the extracloud layer, and (G) understanding the layer's effect on the formation of superbubbles, chimneys, and fountains. Models of the latter which do not include the interaction with 10^{-12} dyn cm^{-2} of nonthermal pressure over kpc scales, as well as with the material in the ECL, cannot hope to be realistic. The ECL, for example, could make possible the growth of superbubbles to much larger dimensions in the plane than previous calculations have found, while quenching blowout and fountain activity.

Of possibly more than historical interest, the marriage of zones 1 and 2 into a comprehensive ISM picture was an essential feature of the Field, Goldsmith, and Habing (1969, FGH) model. The view was rather like the atmosphere-ocean phase equilibrium. Without the overburden of an atmosphere, the ocean would evaporate explosively until the atmospheric pressure exceeded the vapor pressure of the water, suppressing boiling. In the FGH picture, there was just enough intercloud material for its weight to provide the critical pressure required for the existence of clouds in the vicinity of the galactic midplane.

In my view, (Cox, 1988) the clouds have a powerful heating mechanism (possibly photoelectric ejection off grains by starlight) which provides them with a certain equilibrium temperature (and therefore pressure) at each density. That pressure has a very flat minimum around $p/k \sim 3000$ cm^{-3} K. This value is higher than what I regard as typical of the thermal pressure in the extracloud medium ($p/k \sim 1000$ to 3000 cm^{-3} K from Table 1), but is considerably lower than the weight of the ECM ($p/k \sim 2$ to 3×10^4 cm^{-3} K), much of which is borne by the cosmic rays and magnetic field. Thus it is possible yet that the system forces material into the extracloud component until there is sufficient weight to confine the clouds, using the magnetic pressure as the intermediary. Clouds would then be possible in regions of slightly lower than typical field or higher than typical external thermal pressure.

There are at least two bugs in this prescription. One is that the cloud and extracloud components have very similar amounts of mass (in contrast to the ocean and atmosphere). This is a very awkward situation for the clouds, having used up a substantial amount of their material to stabilize themselves. A second peculiarity is that there are other models of the ISM in which supernova explosions define the background pressure dynamically (McKee and Ostriker, 1977; Cox, 1981). One necessarily wonders at the coincidence of the results from two apparently independent mechanisms.

3. The Galactic Corona. Beyond the thick disk there may well be a qualitatively different environment, the galactic corona. In this region I imagine that the density is extremely low, the temperature is possibly very high ($\sim 2 \times 10^6$ K), the galactic material and cosmic rays outflowing, and extragalactic material infalling. Almost every aspect of this region is unknown or known poorly

and represents an outstanding problem. An important constraint on its state distribution is that the X-ray emission is very weak.

Between the thick ECL and the galactic corona there could also be a very interesting transition layer, something like the galactic chromosphere suggested by Sciama (1972), in which downward energy flow lights up the higher density gas. This boundary could be very irregular; in a conversation, Laura Danly and I imagined it might provide an excess population of high ions between $|z| = 1$ and 2 kpc. An important energy source for the ECL might be the stoppage of infalling clouds. For the very high velocity clouds, high temperatures (and their tell-tale but unobserved X-rays) might be avoided if the HVC motions had low Alfven Mach number in a high B low n environment above 1.5 kpc. (At 1 kpc with $B \sim 2\ \mu G$ and $n \sim 0.01\ cm^{-3}$, the Alfven speed is about 40 km s^{-1}. Continuation of B to much lower densities in the corona would probably be required for this to work.)

III. The Problems of Hydrostatics in the ECL

Two recent papers, at least, have discussed the hydrostatics of the thick extracloud layer (Bloemen, 1987; Boulares and Cox 1989), addressing the same difficulty with different methods. For some time the essential problem had been thought to be that, although the tracers of the nonthermal components all point to a thick layer, the weight of the known material was concentrated at lower z. Under the normal assumption that weight is what restrains the field and cosmic rays from free escape, one concluded that the pressure gradient is concentrated lower than the pressure itself. The pressure almost surely decreases somewhat at moderate z, by an amount needed to support the material, but what then holds down the field and cosmic rays at greater heights? What does their residual pressure gradient do?

In the context of the 1D analysis, four solutions to this problem seemed to be possible.

(1) A possibility explored by Badhwar and Stevens (1977) and Bloemen (1987) is that there is unseen low density material at high z, too cold to be supported by its thermal pressure gradient.

(2) A second but perhaps unlikely possibility is that the Galaxy is surrounded by a significant external pressure.

(3) Hydrostatics may be the wrong picture. Perhaps the gradient in the nonthermal pressure at high z accelerates a wind from the Galaxy, somewhat as described here by Heinz Volk.

(4) A solution proposed by Cox (1988) and explored in greater detail by Boulares and Cox (1989) is that magnetic tension is important. At high z the net magnetic force could even be downward, restraining the cosmic rays, coupling them to the anchoring weight below by tension in the field. This configuration works wonderfully for suspension bridges.

Solution (4) has the advantage that it seems to be a natural consequence of instabilities in the system (e.g. Mouschovias, 1974). It is also consistent with data on rotation measures of extragalactic objects (Simard-Normandin and Kronberg 1980). The idea has been criticized on the grounds that the particles and fields may not actually be bound, but I regard this as a subtle point. Tension can restrain expansion; it does so in the field around a straight current-carrying wire. The wire itself experiences a net inward force. Similarly, the magnetic force on a current carrying loop is toward the plane of the loop (attempting to squash the wire--parallel currents attract) but radially outward (attempting to stretch the wire into a larger loop). It is quite possible that the galactic field on the larger scale is restrained in part by the radial gravitational force on interstellar material.

An additional sobering truth is that the configuration we observe should have questionable stability. The cosmic rays do, after all, escape. Very likely some of the field does also. I regard the suggestion by Kraushaar (1963) that the field is strained to near breaking by the cosmic ray pressure as almost certainly correct. As such, I don't think it surprising that we cannot easily demonstrate that the configuration is stable.

But recently there has been a shift in the nature of the hydrostatics problem. The Reynolds layer is intermixed rather well with the cosmic rays and magnetic field to high z and has sufficient weight to provide their restraint, making solution (1) above potentially adequate within the uncertainties, at least to $z \sim 1$ kpc. But when looked at in detail, the layer's weight is so great for the commonly assumed gravity distributions that the total ISM weight exceeds any current estimate of total midplane pressure. Cox and Snowden (1986) and Cox (1988) used this as further evidence for an rms midplane magnetic field of 5μG. Spitzer (1989, private communication) and Boulares and Cox (1989) find typical required fields even larger. Spitzer has remarked that the ECM may have a dispersion velocity at high z of 30 km s^{-1}, aiding its support, but this does not help the difficulty with the midplane pressure unless, perhaps, the high velocity dispersion extends down to $z = 0$. Boulares and Cox favored the use of a weaker gravity (less weight yields lower midplane pressure), consistent with the observed galactic matter distribution (no dark matter) as proposed recently by Bienaymé, Robin, and Crézé (1987). But there is yet considerable room for debate and need for observational clarification.

IV. The Ionization Power Problem

For some time Ron Reynolds has been emphasizing the importance of the diffuse ionization problem. As shown in Table 1, only O stars among the known sources have sufficient power to provide the observed diffuse Hα. (Adding Type II makes SN a marginal contender as well.) But ionization exists far from O stars and it has been difficult to see how the ionizing photons can find clear paths over great distances.

(Note, for example, that in the McKee and Ostriker, 1977, model, the mean free path between cloud envelope intersections was only 12 pc.) The WNM or neutral extracloud medium gets in the way. (But see Harrington and Bregman, 1986)

In preparation for this meeting I tried to solve the problem using Type I supernovae and their new late time evolution that can pump a substantial fraction of their energy, perhaps as much as 1/3, into ionization. Cox and Slavin (1989) found that supernovae with energies of 5×10^{50} ergs need a rate $S \sim 4 \times 10^{-14}$ pc^{-3} yr^{-1} to explain the high ions (O VI, N V, C IV, Si IV) in the galactic plane as arising from their bubbles. That result is tentative and could be low by as much as a factor of 2. Taking the larger rate, 1/3 of the total energy, and assuming 30 eV is expended per ionization, the average ionization power would be 1.4×10^{-26} erg cm^{-3} s^{-1}, the ionization rate 3×10^{-16} cm^{-3} s^{-1}, the sustainable mean square density 10^{-3} cm^{-6} (at 8000 K) and the rms electron density 0.03 cm^{-3}. I was encouraged by the fact that the rms density is comparable to the observed mean density $\bar{n}_e$ (z = 0) = 0.025 cm^{-3} (see references in Reynolds 1989a). The supernovae could provide the diffuse electrons seen via pulsar dispersion measures, but that uniform distribution would miss providing the Hα by a factor of 7.

I next wondered whether the supernovae might at least provide the bulk of the ionization found at high z, the distribution being (Reynolds, 1989a) $\bar{n}_e (z) = 0.025\ e^{-z/1500\ pc}$ cm^{-3}. I decided first to compare the supernova ionization rate possible in a 1 cm^2 column toward the pole. Following Heiles (1987) I took the Type I scale height to be 300 pc, yielding a maximum achievable emission measure of (10^{-3} cm^{-6}) (300 pc) = 0.3 cm^{-6} pc. This is again a factor of about 7 lower than the characteristic emission measure "toward the pole" as inferred by Reynolds (1984). I moaned about this fact at the meeting, causing Guido to respond that he had observed emission measures as low as 0.5 cm^{-6} pc. Something was clearly wrong with my simplistic view of the Hα background.

Since the meeting I have discussed Guido's remark and several other aspects of the diffuse Hα background with Ron Reynolds. The results of those discussions will require a paper all their own (Cox and Reynolds, in preparation). The tentative conclusions are as follows:

(A) The distribution of electrons <u>in the galactic midplane</u> (not counting those in bonafide HII regions) requires at least 3 components for even a rudimentary description:

1. A dense component ($n_e \sim 1$ cm^{-3}) occupying very little of the interstellar volume, contributes modestly to $\bar{n}_e$ and appreciably to $\overline{n_e^2}$;

2. A fully ionized portion of the ECM ($n_e \sim 0.2$ cm^{-3}) occupying a few percent of the volume, provides roughly half of both $\bar{n}_e$ and $\overline{n_e^2}$.

3. Partial ionization in the rest of the ECM ($n \sim 0.2$ cm^{-3}, $n_e \sim 0.01$ cm^{-3}) occupying 75% of interstellar space, provides about a third of $\bar{n}_e$ but very little of $\overline{n_e^2}$.

(B) Type I supernovae are capable of providing the partial ionization in the mostly neutral regions of the ECM as well as making a modest contribution to the fully ionized ECM; but most of the fully ionized material (in the ECM and dense regions) is due to O star radiation.

(C) Although the Type I supernovae are plausibly capable of providing an emission measure contribution of 0.3 cm^{-6} pc toward the pole, their probable contribution is less than that unless the supernova rate has been underestimated.

(D) OB associations are fully capable of drilling high latitude ionized rays through the neutrals in the ECM except where their radiation is significantly shadowed by denser foreground clouds (the resulting rays resembling the pattern of sunlight through partial cloud cover). As in the paper presented at this meeting by José Franco, the ionized zone is in a cone above the association. The overlapping of those cones at z ~ 1 kpc results in a nearly fully ionized ECM above that point.

(E) Lower latitude HI in the ECM survives because of shadowing by clouds (in shells?) and self shadowing, resulting in the finite opening angles of the associations' ionization beams. Due to shifting source and cloud patterns with time, as well as Type I supernovae, the neutral regions have a modest partial ionization, contributing significantly to $\bar{n}_e$.

(F) The low emission measures at high latitude (typically 1 cm^{-6} pc at $|b| > 50°$ but as low as half that in some directions, compared to the expected 2.3 cm^{-6} pc found by Reynolds (1984) from a best fit, assuming a csc $|b|$ dependence, to the b variation at $|b| \lesssim 50°$) result from our not being located directly below the illumination cone of an OB association.

In short, the Reynolds Layer receives most of its ionization from direct rays from OB associations. The neutrals lie between the rays in shadows of clouds, are in no way randomly distributed or intermixed, and do not get in the way of the ionizing photons. (When they do, they are rapidly ionized.)

V. Summary

(A) The things we know are not well known, at least not to me.

(B) The thick extracloud layer changes everything.

(C) Shadows of vapor, as usual, are important in the interstellar medium.

(D) There is much interesting work left to do.

This research was supported, in part, by the National Science Foundation under grant number AST-8643609 and by the National Aeronautic and Space Administration under grant number NAG5-629.

As usual, the reader should be aware that I consistently overstate my certainty in interpretations, in order to present as clear a picture as possible.

I would like to acknowledge useful conversations with Ron Reynolds, Jon Slavin, Ahmed Boulares, Guido Münch, Pepe Franco, Dick Edgar, Laura Danly, Charlie Goebel, Hans Bloemen, John Mathis, Blair Savage, and Lyman Spitzer. I am particularly grateful to Ron for patiently teaching me about the diffuse Hα and for a critical reading of the manuscript, and to Lyman for his constructive criticism regarding the hydrostatic problem.

REFERENCES

Abbott, D. C. 1982, Ap. J., 263, 723.
Badhwar, G. D., and Stevens, S. A., 1977, Ap. J., 212, 494.
Bienayme, O., Robin, A. C., and Creze, M. 1987, Astron. Astrophys., 180, 94.
Bloemen, J. B. G. M. 1987, Ap. J., 322, 694.
Boulares, A. and Cox, D. P. 1989, in preparation.
Cowie, L. L. and York, D. G. 1978, Ap. J., 223, 876.
Cox, D. P. 1981, Ap. J., 245, 534.
Cox, D. P. 1986, in Workshop on Model Nebulae, Observ. de Meudon; ed: D. Pequinot (Paris Observ.) p. 11.
Cox, D. P. 1988, in IAU Colloq. 101, Interaction of Supernova Remnants with the Interstellar Medium, eds. T. L. Landecker and R. S. Roger (Cambridge: Cambridge University Press) p. 73.
Cox, D. P. and Snowden, S. L. 1986, Adv. Space Res. 6, 97.
Cox, D. P. and Slavin, J. D. 1989, in EUV Astronomy, eds. R. F. Malina and S. Bowyer (Pergamon), in press.
Edgar, R. J. and Savage, B. D. 1989, Ap. J., 340, 762.
Field, G. B., Goldsmith, D. W., and Habing, H. J. 1969, Ap. J. (Letters), 155, L149.
Harrington, J. P. and Bregman, J. N. 1986, Ap. J., 309, 833.
Heiles, C. 1987, Ap. J., 315, 555.
Hoyle, F. and Ellis, G. R. A. 1963, Australian J. Physics, 16, 1.
Kraushaar, W. L. 1963, Proc. Int. Conf. on Cosmic Rays at Jaipur, 3, 379.
Lockman, F. J., Hobbs, L. M., and Shull, J. M. 1986, Ap. J., 301, 380.
McKee, C. F. and Ostriker, J. P. 1977, Ap. J., 218, 148.
Mouschovias, T. Ch. 1974, Ap. J., 192, 37.
Panagia, N. and Terzian, Y. 1984, Ap. J., 287, 315.
Reynolds, R. J. 1984, Ap. J., 282, 191.
Reynolds, R. J. 1989a, Ap. J., 345, in press (Oct. 15).
Reynolds, R. J. 1989b, preprint.
Reynolds, R. J. 1989c, Proc. of IAU Symposium 139, Galactic and Extragalactic Background Radiation, eds: Bowyer, Leinert (Dordrecht:Reidel) in press.
Sciama, D. W. 1972, Nature, 240, 456.
Simard-Normandin, M., and Kronberg, P. P. 1980, Ap. J., 242, 74.
Spitzer, L. 1989, Mat. Fys. Medd. Dan. Vid. Solsk., in press.

Discussion:

VÖLK (Comment): Assuming strong, large scale B-fields up to high $|Z|$, then these force free fields should behave similar to the solar case, random fluid motions resulting from stellar mass loss moving the feet of the field lines around. As argued by Parker, this should lead to magnetic field dissipation and corresponding gas heating. Indeed the high $|Z|$ highly ionized species like CIV, NV, etc., which have been discussed by K. de Boer and L. Danly at this meeting might very well be produced by this nonthermal heat source.

The Mass Spectrum of Interstellar Clouds

John M. Dickey
Sterrewacht Leiden and University of Minnesota
and
R. W. Garwood
University of Pittsburgh

Abstract

The abundance of 21-cm absorption lines seen in surveys at high latitudes can be translated into a line of sight abundance of clouds vs. column density using an empirical relationship between temperature and optical depth. As VLA surveys of 21-cm absorption at low latitudes are now becoming available, it is possible to study the variation of this function with galactic radius. It is interesting to compare the abundance of these diffuse atomic clouds (with temperatures of 50 to 100 K and masses of 1 to 10 $M_\odot$) to the abundance of molecular clouds. To do the latter we must make assumptions about cloud cross-sections in order to convert the line of sight abundance of diffuse clouds into a number per unit volume, and to convert from cloud column density to mass. The spectrum of diffuse clouds matches fairly well the spectrum of molecular clouds, although observationally there is a gap of several orders of magnitude in cloud mass. Optical absorption studies also agree well with the 21-cm results for clouds of column density a few times 10^{20} $M_\odot$.

I. Background

The 21-cm line is seen in absorption from cool atomic gas; the kinetic temperature in this medium is typically 50 to 100 K. In contrast 21-cm emission comes from atomic hydrogen at all temperatures; more than half of the atomic gas is warm (~6000 K) so that emission spectra consist of a blend of warm and cool material which is often impossible to separate. Because of this blending with warm gas, line profiles in 21-cm emission are blends of various inseparable features. Thus it is not possible to inventory the population of diffuse clouds using 21-cm emission surveys, as has been done for molecular clouds using CO surveys. On the other hand, absorption profiles are made up of distinct lines which can be fitted quite well by Gaussians. Recent surveys at low and high latitudes give us a good idea of the abundance of these lines (which we empirically identify as clouds) as a function of optical depth and position in the galaxy.

Physical conditions in these diffuse clouds are quite different from those in molecular clouds, even the "diffuse" clouds seen in molecular absorption. Densities are typically 20 to 100 cm^{-3}, and the hydrogen is mostly atomic (HI). The warm HI may be partially associated with these clouds as well, perhaps as a halo, but at least some of the warm gas is independent of clouds (Payne *et al.* 1983, Kulkarni and Heiles, 1988). Deep absorption at 21-cm is also seen from molecular clouds where the bulk of the hydrogen is molecular, because their lower temperatures make the absorption coefficient per atom much greater, but these clouds are relatively rare. In the inner galaxy, however, the atomic and molecular clouds are highly correlated; there apparently most cool atomic gas is associated spatially with the molecular cloud regions, although it is not necessarily mixed with the molecular gas on a small scale.

II. $\Phi(N)$– the Line of Sight Abundance vs. Column Density

Many lines of sight have been observed at high latitudes, so for the region within one or two kpc of the sun we have good statistics on the number of clouds per unit distance as a function of optical depth, $\phi(\tau)$, which is shown on figure 1. Also shown are estimates from

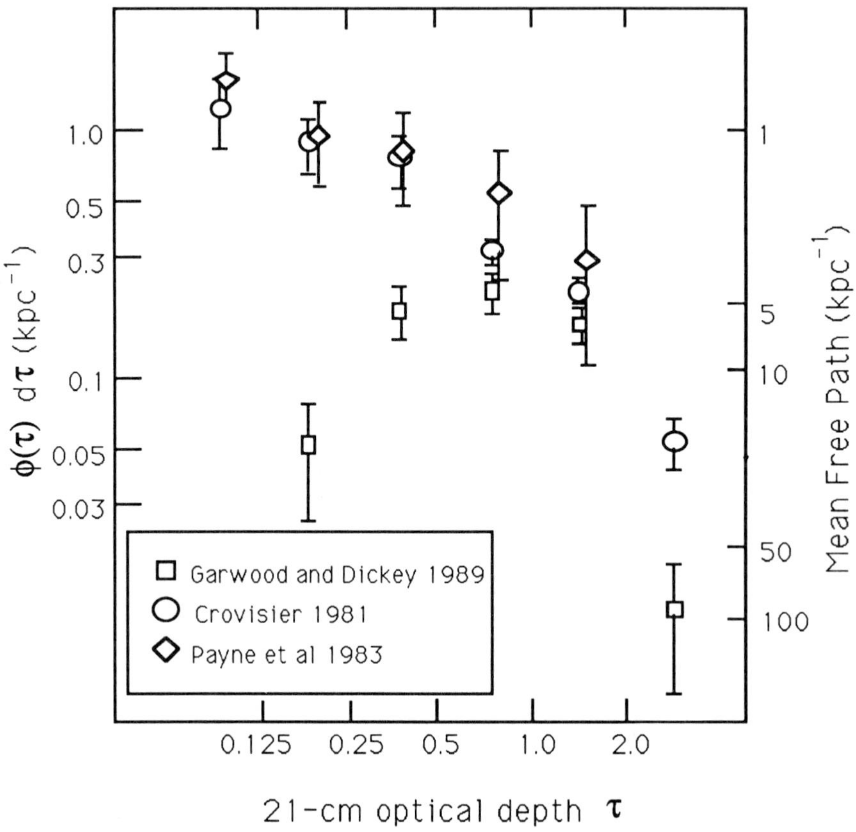

Figure 1. Observational Results for $\phi(\tau)$, the abundance of absorption lines as a function of optical depth. The Garwood and Dickey survey is incomplete below $\tau = 0.5$ because of line blending.

a recent low latitude absorption survey for $\phi(\tau)$ in the inner galaxy (Garwood and Dickey 1989). Apparently $\phi(\tau)$ decreases with decreasing galactic radius (which is the opposite of what we expected before doing this survey). For the solar neighborhood $\phi(\tau)$ is fitted adequately over the range $0.1 < \tau < 2$ by a simple power law : $\phi(\tau)\ d\tau = 0.35\ \tau^{-1.6}\ d\tau$ kpc^{-1} (Crovisier 1981, Payne et al 1983).

The function $\phi(\tau)$ can be translated into $\Phi(N)$, the number of clouds per unit line of sight distance as a function of HI column density N, if the spin temperature, T_{sp}, of the gas is known, since the optical depth measures the ratio N / T_{sp}. Empirically there is a correlation between T_{sp} and τ, which may reflect the relative column density of warm and cool gas in clouds (Liszt 1983, Mebold et al 1982). Whatever its origin, the T_{sp}-τ relation allows us to translate 21-cm optical depth to HI column density, so that we can construct $\Phi(N)$(Kulkarni and Heiles 1986). For very optically thick clouds the T_{sp}-τ relation must break down, as the HI will not get cooler than the molecular gas in dense clouds, which is typically in the range 5 to 20 K. It may be appropriate for clouds with optical depths above 2 to discard the T_{sp}-τ relation and simply assume that the HI has temperature about 20 K on average. These two approaches give roughly similar results for $\Phi(N)$, illustrated on figure 2 (Dickey and Garwood 1989). The actual range of N over which these are measured

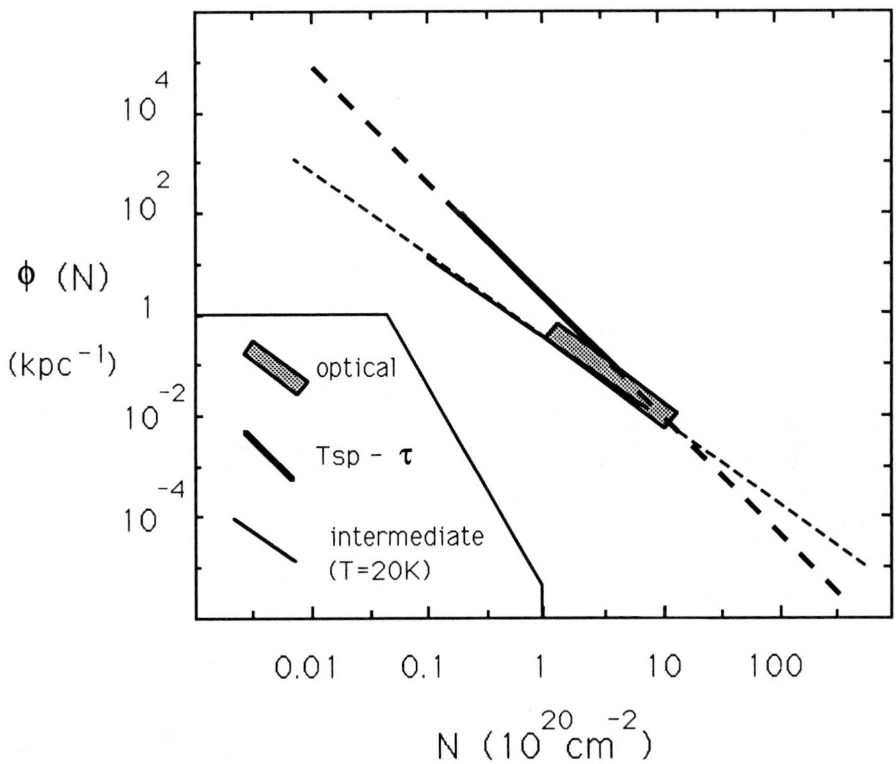

Figure 2. The function $\Phi(N)$deduced from $\phi(\tau)$ using two possible spin temperature assumptions. The corresponding abundance deduced from optical absorption lines is shown.

is only about 0.5 to 2.5 in units of 10^{20} atoms cm^{-2}. Figure 2 also shows the result from surveys of optical absorption lines (Hobbs 1974), which is in good agreement with both the slope and the absolute normalization of the 21-cm result.

III. The Cloud Cross-Section Determines $\rho(m)$

Absorption observations at all wavelengths typically measure the properties of gas only along a narrow column in front of the background continuum source. Cloud sizes perpendicular to the line of sight are usually not measured directly. To compare with emission surveys we need to estimate the cloud cross-section, σ, which may be defined as the ratio of the cloud mass, m, to the mean column density, $< N >$, ie.

$$\sigma = \frac{m}{< N > \cdot m_H} = 1.3\mathrm{pc}^2 \cdot (\frac{m}{m_\odot}) \cdot (\frac{< N >}{10^{20}\mathrm{cm}^{-2}})^{-1} \qquad 1.$$

where $< N >$ is the column density averaged over the projected area of the cloud,

$$< N > = \frac{1}{\Omega_{cl}} \int N(\theta, \phi) d\Omega \qquad 2.$$

where Ω_{cl} is the solid angle of the cloud.

For a homogeneous sphere the internal density is simply

$$n = 38.6 \frac{< N >^{\frac{3}{2}}}{\sqrt{m}} \ \mathrm{cm}^{-3} \qquad 3.$$

(with $< N >$ in units of 10^{20} cm^{-2} and m in $M_\odot$), but diffuse clouds are well known to have highly elongated shapes (eg. Heiles 1967), so it is better to model their shapes with ellipsoids with fairly high axis ratios. The ratio of $< N >$ to that for a sphere with the same mass goes only as axis ratio to the 1/3 power, so even for axis ratios of 100 or so the actual internal density is within a factor 5 of the value given by equation 3.

Although the interstellar medium is by no means at constant pressure (eg. Jenkins *et al.* 1983), it is probably a reasonable approximation to say that the gas pressure in diffuse clouds is generally roughly 3000 cm^{-3} K. Using this and the spin temperature as discussed above we can fix the density, and solve for the mass as a function of $< N >$. The same geometrical assumptions give us the cloud cross-section,

$$\rho = \frac{\phi}{\sigma} \qquad 4.$$

where ϕ is the line of sight density as above, and ρ is the number of clouds per unit volume. Now we can convert from $\Phi(N)$to $\rho(m)$, the mass spectrum of interstellar clouds, which is shown on figure 3 (from Dickey and Garwood 1989). This function we can compare directly with the result for molecular clouds (Scoville and Sanders 1987, Terebey *et al.* 1987, Elmegreen 1987), which is also shown on figure 3. The molecular cloud spectrum is known only for clouds with masses in the range 10^5 to 10^6 $m_\odot$for which the CO surveys are complete, and for which virial mass estimates are reliable. In between the giant molecular clouds and the diffuse clouds (with masses of a few solar masses) there is a gap of several orders of magnitude which is not well sampled; these intermediate mass clouds may correspond to the high latitude molecular clouds seen in recent surveys (eg. Magnani *et al.* 1988). It is also interesting that the slope of the mass spectrum of diffuse clouds is near the critical value, $\rho(m) \alpha \ m^{-2}$, for which equal mass contributions are made by equal mass intervals, and both the maximum and minimum mass are critical to the total mass density of cloud material :

$$\rho = \int m \cdot \rho(m) dm \qquad 5.$$

We can check the normalization of the atomic portion of this spectrum, since we know that in the solar neighborhood the cool HI makes up about half of the total atomic hydrogen (eg. Colgan *et al.* 1988), whose surface density is 5 to 6 $M_\odot pc^{-2}$ at the solar circle (Burton 1987 and references therein). The diffuse cloud spectrum shown on figure 3 gives this value if the limits of integration in equation 5 are taken to be about 0.1 and 1000 $M_\odot$.

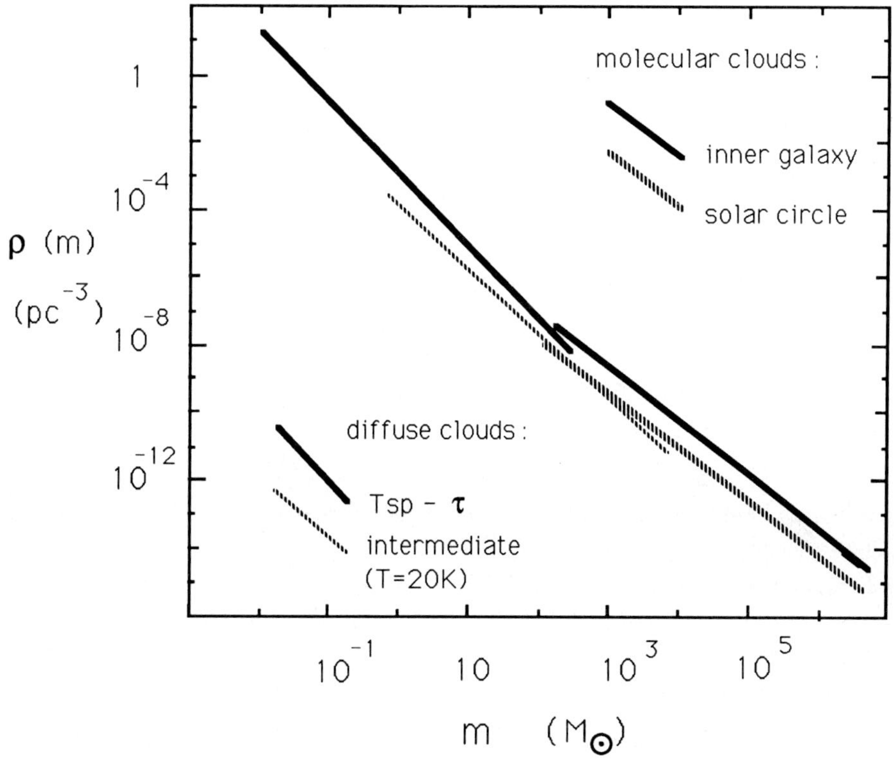

Figure 3. The cloud mass spectrum derived from the 21cm absorption survey results compared with the molecular cloud mass spectrum deduced from CO surveys.

It is important to distinguish between the approach taken here and other derivations of the "cloud spectrum" based on measurements of cloud sizes. Not only are cloud edges difficult to determine observationally due to spatial blending by the telescope beam in emission surveys, but more importantly the relation between cloud size and mass is very steep. For the two alternatives of constant temperature clouds or the T_{sp}-τ relation $< N >$ goes as radius to the

2nd or 3rd power, whereas for molecular clouds which are self-gravitating and therefore centrally condensed the central column density is an even steeper function of the radius. Thus it is unwise to construct a cloud mass spectrum starting from a cloud size spectrum derived indirectly from observations.

Acknowledgements

This research was supported in part by NSF grant AST-872290 to the University of Minnesota. JMD is grateful to the Sterrewacht Leiden for hospitality during part of this work.

References :

Burton, W.B., 1988, in *Galactic and Extragalactic Radio Astronomy*, 2nd Ed., G.L. Verschuur and K.I. Kellerman (Heidelberg: Springer Verlag).

Colgan, S.W.J, Salpeter, E.E., and Terzian, Y., 1988, *Ap. J.* **328**,275.

Crovisier, J., 1981, *Astron. Astrop.* **94**, 162.

Elmegreen, B., 1987, in *Interstellar Processes*, ed. D. Hollenbach and H. Thronson, (Dordrecht : Reidel) p. 259.

Garwood, R.W., and Dickey, J.M., 1989, *Ap. J.* **338**, 841.

Heiles, C., 1967, *Ap. J. Supp.* **15**, 97.

Hobbs, L., 1974, *Ap. J.* **191**, 395.

Jenkins, E.B., Jura, M., and Loewenstein, M., 1983, *Ap. J.* **270**, 88.

Kulkarni, S.R., and Heiles, C.E., 1987, in *Interstellar Processes*, ed. D. Hollenbach and H. Thronson, (Dordrecht : Reidel) p. 87.

Kulkarni, S.R., and Heiles, C.E., 1988, in *Galactic and Extragalactic Radio Astronomy*, 2nd Ed., G.L. Verschuur and K.I. Kellerman, (Heidelberg: Springer Verlag) pp. 95-153.

Liszt, H.S., 1983, *Ap. J.* **275**, 163.

Magnani, L., Blitz, L., and Wouterloot, J.G.A., 1988, *Ap. J.* **326**, 909.

Mebold, U., Winnberg, A., Kalberla, P.M.W., and Goss, W.M., 1982, *Astron. Astrop.* **115**, 223.

Payne, H.E., Salpeter, E.E., and Terzian, Y., 1983, *Ap. J.* **272**, 540.

Scoville, N.Z. and Sanders, D.B., 1987, in *Interstellar Processes*, ed. D. Hollenbach and H. Thronson, (Dordrecht : Reidel), p. 21.

Terebey, S., Fich, M., Blitz, L., and Henkel, C.,1987, *Ap. J.* **308**, 357.

Discussion:

MÜNCH (Comment): Your success in establishing that the estimate of the first moment in the mass distribution of IS "clouds" (their mean number per kpc) is the same when derived from 21cm absorption data as from optical IS lines multiplicity, is most gratifying. But we should recall that the same number is obtained from extinction data (stellar color excesses) and from brightness fluctuations of the Milky Way. This overall agreement shows that the concept of "clouds" as one dimensional structures of zero measure has a physical meaning. We owe to the intuition of Ambartsumian this abstraction of an inmensely complicated physical situation. Chandrasekhar and I worked out in detail the mathematics of the stochastic processes involved, as Chandra mentioned during the Banquet of this meeting. After my "divergence" to Caltech I studied further the questions on the basis of "infinitely divisible distributions" and applied the results to various sub-sets of measured stellar color excesses. My efforts to extend the analysis to joint distributions of optical depths and velocities of IS absorption lines met only partial sucess. With the recent "explosion" of observational data, however, new theoretical tools for interpretation need to be elaborated. I fully agree with John Dickey's remark that to ask for the "shape" of an interstellar cloud is a meaningless proposition. In this context I should further add that the "pictures" of molecular clouds under higher and higher angular resolution, as shown by J. Bally, are a beautiful visualization of the concept of "infinite divisibility", that is to say of the topological invariance of their 2D distribution under varying angular resolution.

SCWARZ: In deriving the optical depth distribution as function of distance from the center, do you take "shadowing" (you cannot see through optically thick clouds) effects into account?

DICKEY: Yes, we take account of line blanketing which leads to incompleteness in the cloud spectrum $\phi(\tau)$. There are very few velocities where the total optical depth saturates so as to difficult to measure (optical depths add linearly, in any case).

DETERMINISTIC SELF-PROPAGATING STAR FORMATION

Jan Palouš
Astronomical Institute of the Czechoslovak Academy of Sciences
Budečská 6, 120 23 Prague 2, Czechoslovakia

Abstract

The evolution of large scale expanding structures in differentially rotating disks is studied. High column densities in some places may eventually lead to molecular cloud formation and initiate also star-formation. After some time, multi-structured arms evolve, where regions of intensive star-formation are separated from each other by regions of atomic gas or molecular clouds. This is due to the deterministic nature and to the coherence of this process. A simple model of galactic evolution is introduced and the different behaviour of Sa, Sb, and Sc galaxies is shown.

1. Introduction

Star formation in disks of spiral galaxies operates coherently on scales of a kiloparsec or more: expanding superassociations /Ambartsumian, 1947/, complexes of star-forming regions /Efremov, 1985/, coeval clusters /Lyngå and Wramdemark, 1984/ and stellar superclusters or stellar streams /Eggen, 1989, Palouš and Hauck, 1987/ were found. Clusters of HII regions and OB associations reside at separations of 1 to 4 kpc in arms of many normal spirals /Elmegreen and Elmegreen, 1983/. This superstructure of star formation in galaxies may be associated with superclouds composed of 10^6 to $4\ 10^7\ M_o$ of material partly in the form of molecules /Elmegreen and Elmegreen, 1987/. The fraction of molecules in superclouds decreases from 70% to 5% with increasing galactocentric distance.

The formation of stars on such large scales could be interpreted as due to a process propagating in the galactic disk. What may be the mechanism of propagation? Groups of massive stars or OB associations release during the short lifetime of their members a large amount of energy in the form of ionising photons, stellar winds and supernovae. The total of some 10^{53} erg is injected over 10^7 years to the vicinity of an association. It pushes the ambient medium and it forms an expanding supershell. The supershells are considered in the present paper as being responsible of completing the star formation cycle (see Tenorio-Tagle and Bodenheimer 1988).

We focus on the evolution of supershells in differentially rotating disks. Molecular clouds can be formed on the tips of these elongated superstructures. They may eventually become the sites of next generation star formation. We assume that this is the mechanism of self-propagating star formation, which is deterministic in nature. Starting with the limited number of clouds, all the galaxy may be after some time populated by clouds. We also discuss a simple model of galactic evolution and show the differences between Sa, Sb, and Sc galaxies.

2. The supershells

10^{52}- 10^{54} erg released by the massive stars from an OB association in a relatively small volume in a galaxy modify over 10^7 years substantially the enviroment of the stellar cluster. An expanding supershell leaves a giant hole in the ISM. The ambient medium accumulates in the shock and the shock decelerates due to the conservation of the total amount of momentum.

The evolution of supershells was studied numerically. We use the strong shock approximation developed by Sedov /1946/, and assume that the thickness of the shock front is much smaller than the radius of the cavity. Then the completely 2D treatment of gas dynamics may be substituted by a simpler " 1.5D " approach, where the thickness of the shock is disregarded. This is sometimes also called the snow-plough model.

A more detailed description of our model may be found in our paper I /Tenorio-Tagle, and Palouš, 1987/. The evolution of a supershell in the differentially rotating disk of a galaxy is shown in Fig. 1. The shocked, swept up matter accumulates in the shock and at the same time it slides along the front to the tips. Large column densities near the two tips result in formation of molecules as soon as the critical column density $N_{crit} = 10^{21} Z_o/Z$ cm^{-2}, required to shield

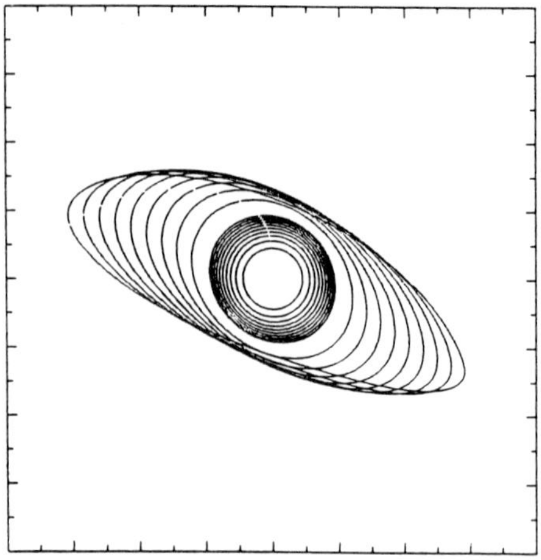

Figure 1.

Time evolution of a supershell. The first ten contours were drawn at an interval equal to 10^6 years, while for the other ones it was set equal to 10^7 years.

the external UV radiation field /Franco, and Cox, 1986/, is exceeded.

The evolution of the size, shape, inclination, density, and of the total mass have been expressed with the approximation formulas in our Paper II /Palouš, et al., 1989/. The rate of superstructure evolution, which varies from a galaxy to another, is proportional to the reference time, $t_{ref} = \pi / \alpha \kappa \ 10^9$ years, where κ is the epicyclic frequency in km s^{-1} kpc^{-1}. $\alpha = 0.74 \ \sigma_g^{0.1}$, where σ_g is the gas surface density in the galactic disk in $M_o pc^{-2}$.

In this communication we use the approximate formulas from Paper II and discuss the propagation of star formation. We assume that the lifetime of a molecular cloud is 2 to 3 10^7 years. After this time, the stars are formed in its interiors. They release over 10^7 years about 10^{53} erg disrupting the parent cloud.

3. After 500 Myr

We assume that the star formation propagates if two conditions are met:

1. the column density in the supershell is higher than the critical value $N_{crit} = 10^{21} \ Z_o/Z \ cm^{-2}$,
2. the total mass accumulated in the supershell is larger than the critical value M_{crit}. The propagation of star formation is connected with massive stars and we assume that they are formed in massive clouds only. We take

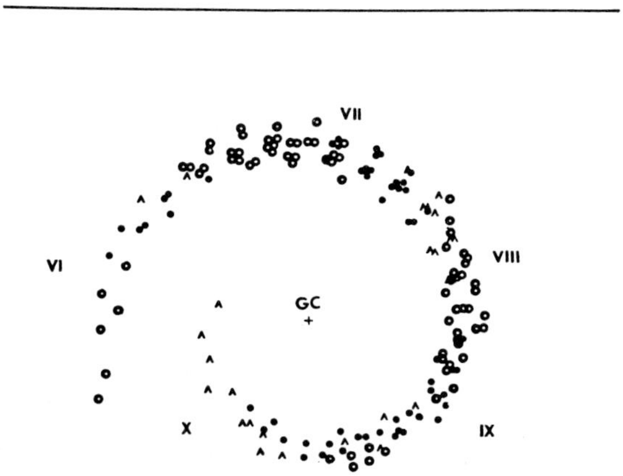

Figure 2.

The distribution of atomic /o/ and molecular clouds /•/ and OB associations /^/ after 5 10^8 years. The solid line is 32 kpc long.

$$M_{crit} \simeq 10^5 M_o.$$

Low mass clouds and clouds with low column densities, where also some star formation occurs, are, according to our assumptions, not able to propagate the star formation.

The time t_{crit} after which the column density N_{crit} is reached may be expressed from our formulas in Paper II:

$$t_{crit} = t_{ref}/f, \quad f = 1 + 0.25 \left[\frac{n_o}{1 cm^{-3}}\right]^{0.52} \left[\frac{E_o}{10^{53} erg}\right]^{0.15} \left[\frac{Z}{Z_o}\right]^{0.67} , \qquad /1/$$

where n_o is the density of ambient medium and E_o is the initial energy released. The total mass accumulated in the supershell is for $t > 0.25\ t_{ref}$

$$M_{tot} = 5.2 \left[\frac{h_d}{100pc}\right]^{1.5} \left[\frac{n_o}{1\ cm^{-3}}\right]^{0.54} \left[\frac{E_o}{10^{53} erg}\right]^{0.44} 10^5\ M_o , \qquad /2/$$

where h_d is the half-thickness of the gaseous disk. With our assumption 2 we may introduce the critical density n_{crit}

$$n_{crit} = 0.05 \left[\frac{h_d}{100pc}\right]^{-2.8} \left[\frac{E_o}{10^{53} erg}\right]^{0.8} . \qquad /3/$$

If the density of ambient medium n_o is less than n_{crit} the star formation do not propagate.

Let us start with one star-forming region in the galactic disk. We use the flat rotation curve with $\Theta = 220$ km s^{-1} representing Sb galaxies /Rubin et al., 1985/. We also introduced the gradients of ambient medium density, metalicity, and disk thickness: n_o and Z are larger and h_d smaller inside the diks than outside.

After 5 10^8 years /Fig.2/ several generations of clouds are created. As a consequence of shorter t_{ref}, the propagation is faster inside the disk than outside. The Xth generation of clouds and stars at a small galactocentric radius evolves now when only a few generations have occurred outside /Fig.2/.

Due to the galactic differential rotation, clouds and young stellar groups form a spiral arm. It is a multistructured arm composed of several regions of 1 - 4 kpc in size, which are preferentially atomic, molecular, or star-forming. This is a consequence of the deterministic nature of propagation, and of the coherence of the process, which started from one star-forming region.

A higher fraction of gas is in the form of molecules inside rather than outside the disk. This is the result of:a. larger metalicity inside implying lower critical column density for molecule formation, and b. higher density of ambient medium inside making higher the column density in the supershell.

How galaxies differ from each other? The rate of propagation is proportional to t_{ref}, which is related to the rotation curve. Higher rotation speeds imply smaller t_{ref}. With shorter t_{ref} clouds multiply quicker, thus after certain time more rotation yields more clouds. Higher rotation yields also higher wind up of the spiral arm.

4. A simple model of galactic evolution

Starting with many randomly distributed star-forming regions, the galactic plane will be populated after some time by a huge number of clouds and hardly any spiral arm could be distinguished. In our model, the number of clouds N is a function of time. $N_{old}/t/$ randomly distributed clouds will create $N_{new}/t + \Delta t/$ clouds. The correlation in cloud positions, as described in the preceding chapter, is a consequence of limited number of clouds. For large number of clouds the correlation will be smeared out and we assume that the distribution of new clouds is a random one.

We investigate a simple model of the galactic disk evolution and express how the number of clouds change in the subsequent time-steps. The disk is represented by three components: diffuse gas, clouds and stars. Star-forming clouds are formed in the evolution of supershells from the diffuse component. But the lifetime of clouds is limited. They are disrupted as a consequence of the formation of stars in their interiors after about 3 10^7 years. The gas recycles and 90% goes back to the diffuse component. 10% does not return, it is locked up in low mass stars and in remnants after the evolution of high mass stars. The

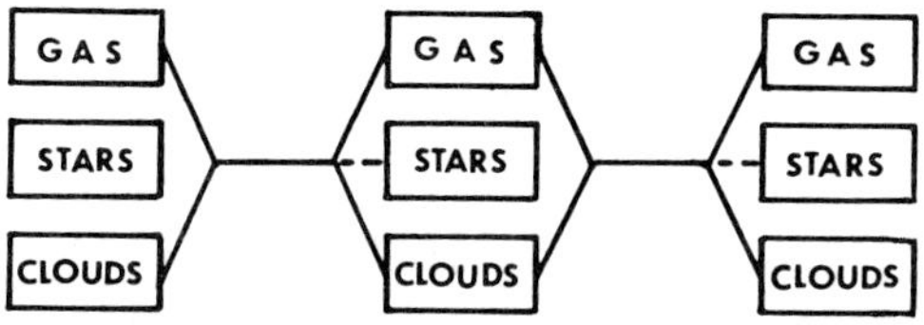

total mass in all three components is, in this simple model, kept constant.

Our model introduce a time-delay: the old ambient gas density and the old number of clouds influence the new number of clouds and the new ambient gas density. This feedback is highly nonlinear, which is inherently connected with the deterministic nature of our model of self-propagating star formation.

The evolution starts with all mass in the diffuse gas, but few initial star-forming clouds. In early phases the number of clouds grows rapidly. This is connected with high density of the diffuse component. If the number of clouds is related to the star formation rate, the initial burst of cloud formation is followed by a burst of star formation. Subsequent decrease of n_o reduces the number of clouds. Finally, after some time, when the number of clouds is reduced to a few thousands, the evolution is very slow at almost constant density of the diffuse component of gas. t_{crit} is short in rapidly rotating Sa galaxies. This results in the steeper increase of N causing stronger burst of star formation and larger gas exhaustion in Sa than in Sc galaxies.

The above model of galactic evolution can be generalized, the halo-disk interaction introduced, as well as the chemical evolution, etc. Coalescence of small clouds, or large-scale gravitational instabilities are also relevant for star formation in galaxies. They cooperate with our mechanism of cloud formation. Evolution of galaxies results from several processes. In our opinion, the deterministic self-propagating star formation contributes to a more complex scenario.

References

Ambartsumian, V.A.:1947, Stellar Evolution and Astrophysics,Arm.A.Sci.
Efremov,Yu.N.:1985,Soviet Astron.Letts. 11, 169
Eggen,O.J.:1989,Fund.Cosm.Phys. 13, Nos. 1-2
Elmegreen,B.G.,Elmegreen,D.M.:1983,Mon.Not.R.astr.Soc. 203, 31
Elmegreen,B.G.,Elmegreen,D.M.:1987,The Astrophys.J.320, 182
Franco,J.,Cox,D.P.:1986,Publ.Astron.Soc.Pacific, 98, 1076
Lyngå,G.,Wramdemark,S.:1984,Astron.Astrophys. 132, 58
Palouš,J.,Hauck,B.:1987, Astron.Astrophys. 162, 54
Palouš,J.,Franco,J.,Tenorio-Tagle,G.:1989,Astron.Astrophys.,submitted
Tenorio-Tagle,G.,Bodenheimer,P.:1988,Ann.Rev.Astron.Astrophys.26,145
Tenorio-Tagle,G.,Palouš,J.:1987,Astron.Astrophys.186,287
Sedov,L.I.:1946,Reports of the Acad.Sci.USSR 52,No.1.

THE VIOLENT INTERSTELLAR MEDIUM IN MESSIER 31

Elias Brinks[1], Robert Braun[2], Stephen W. Unger[1]

[1] *Royal Greenwich Observatory, Herstmonceux Castle, Hailsham BN27 1RP, England*

[2] *National Radio Astronomy Observatory, Very Large Array, Socorro, New Mexico 87801-0387, U.S.A.*

ABSTRACT: TAURUS observations in the line of Hα and VLA HI mapping of the HII complex No. 722 in M31, reveal what seems to be a spherical cavity 330 pc in diameter blown out by a stellar association of over 20×10^6 year old. Evidence of induced star formation which was initiated less than 5×10^6 years ago is present in the form of bright HII emission and numerous O, B and Wolf-Rayet stars which are found within the shell surrounding the cavity. The energy necessary to create the HI shell is estimated to be about 5×10^{51} erg.

INTRODUCTION

Vast amounts of energy are pumped into the interstellar medium (ISM) by an OB association, firstly in the form of photons and stellar winds by stars more massive than about 25 $M_\odot$, and later by supernova explosions of all stars heavier than about 8 $M_\odot$. One of the effects that this has is the creation of large cavities in the disks of spiral galaxies which are filled with hot gas and which are surrounded by expanding shells of ionised and neutral hydrogen (HI) gas. Clear evidence for this has been found by Heiles (1979, 1984) in our Galaxy and by Meaburn (1980) and Dopita *et al.* (1985) in the Large Magellanic Cloud. Recently Brinks and Bajaja (1986) have found similar features, HI holes, in the disk of the Andromeda galaxy in the HI survey of M31 with the Westerbork Synthesis Radio Telescope (WSRT). Very similar results have been derived for M33 on the basis of an HI survey with the WSRT by Deul (1988).

In recent years the standard two phase model of the ISM first proposed by Field, Goldsmith and Habing (1969) has been extended to describe a supernova dominated medium (Cox and Smith 1974; McKee and Ostriker 1977). These models consider three phases, hot, warm and cool gas, each roughly in pressure equilibrium. Later, several modifications have been suggested to account for the fact that the energy is not deposited uniformly throughout the disk but that the SN explode within the boundaries of their parent OB association (Heiles 1987; Norman and Ikeuchi 1989). As was mentioned by Heiles (this conference) there is considerable debate on the actual number of Type I and Type II supernovae, the number of OB associations, their distribution in the Galaxy, etc., which results in wildly varying predictions for the structure of the ISM and the filling factors of the various phases. Due to our unfavourable position within our own system it is unlikely that this situation will improve.

In order to complement the detailed but necessarily incomplete view which we have of our Galaxy we have undertaken a study at high linear resolution of M31 with the aim of providing better constraints for the input parameters of the different models. In addition, in M31 we can look at the structure of the ISM as a function of galactocentric distance. A further important justification to undertake this study is that we can inspect the interaction of an evolving OB association with its surroundings and test the various descriptions which have been put forward (see Tenorio-Tagle and Bodenheimer (1988) for a comprehensive review).

Below we will describe a set of observations obtained with a scanning Fabry-Perot interferometer in the line of Hα, mapping the warm ionised component of the ISM. Our aim is to study the morphology of the Hα emission and its kinematics, notably the structure at the interface of the hot and ionised phase and the interaction between the ionised and cool components. Also, we will look for evidence of sites of secondary or induced star formation triggered by the violent events.

OBSERVATIONS

The observations that we report here were done with the TAURUS-I scanning Fabry-Perot interferometer at the Cassegrain focus of the 2.5-m Isaac Newton Telescope of the La Palma Observatory. They were conducted on 11/12 October 1987. For a description of the instrument and detector and information on the calibration and data reduction aspects the reader is referred to Bland *et al.* (1987). Eight fields in M31 were covered. A full account of the observations will be given elsewhere (Brinks *et al.*, in preparation). Here we will highlight some results from one area. This field is located at a galactocentric distance of 9.7 kpc in the northeastern part of the galaxy which is particularly rich in terms of HII regions and OB associations. It is centred at 34.2N and 7.8E distance from the nucleus of M31 as measured in arcminutes along the North major and East minor axis. It covers the area of HII region P722, object No. 722 from the list of Pellet *et al.* (1978). The field of view of TAURUS on the INT is about 5 arcmin which at the assumed distance of M31 of 690 kpc corresponds to 1000 pc. We mapped the line of Hα at a resolution of $\sim$ 8 km s^{-1}. Our exposure time was $\sim$ 80 min.

Figure 1 presents the sequence of channel maps which show in each frame the distribution of the Hα flux on the sky at a particular velocity. The central heliocentric velocity corresponding to each channel is indicated on top of each frame. The maps were smoothed to a resolution of 5 arcsec to increase the signal to noise ratio and to bring the resolution more in line with that of surveys of HI (see below). The equatorial (1950.0) coordinates of the maps were determined by measuring the positions of a set of foreground stars and are accurate to better than 1 arcsec. Channel to channel variations in the sky background due to terrestial emission lines such as OH at λ = 655.37 and 656.90 nm have been removed by determining the background in areas free of line emission. A map representing the optical continuum emission has been constructed on the basis of those channels which do not contain any line emission. This map has been subtracted from the remaining channels.

RESULTS AND DISCUSSION

The Hα maps compare well with the photographs of Pellet *et al.* (1978) and shell structure corresponding to P722 can be readily identified. The entire area has a line width (FWHM) of 44 km s^{-1} and an integrated flux density of 2.5×10^{-12} erg cm^{-2} s^{-1} (Walterbos and Braun, private communication). In Hα the shell is three quarters

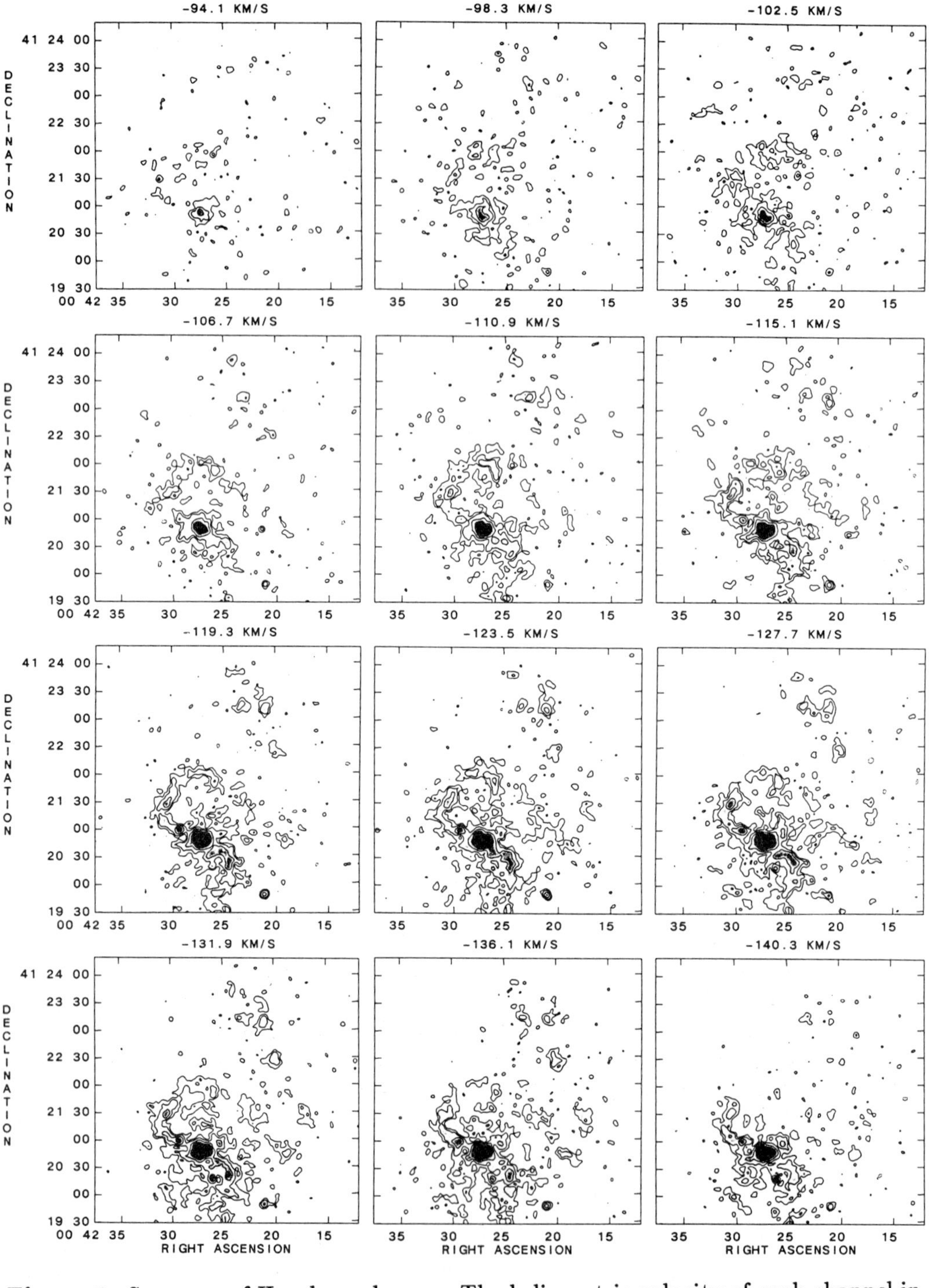

Figure 1: Sequence of Hα channel maps. The heliocentric velocity of each channel in km s^{-1} is shown above each frame. Contour levels are at -1, 1, 2, 4, 8 and 16 in units of 2.14×10^{-17} erg cm^{-2} s^{-1} $arcsec^{-2}$ per channel. The resolution is 5 arcsec, the channel separation 4.2 km s^{-1}.

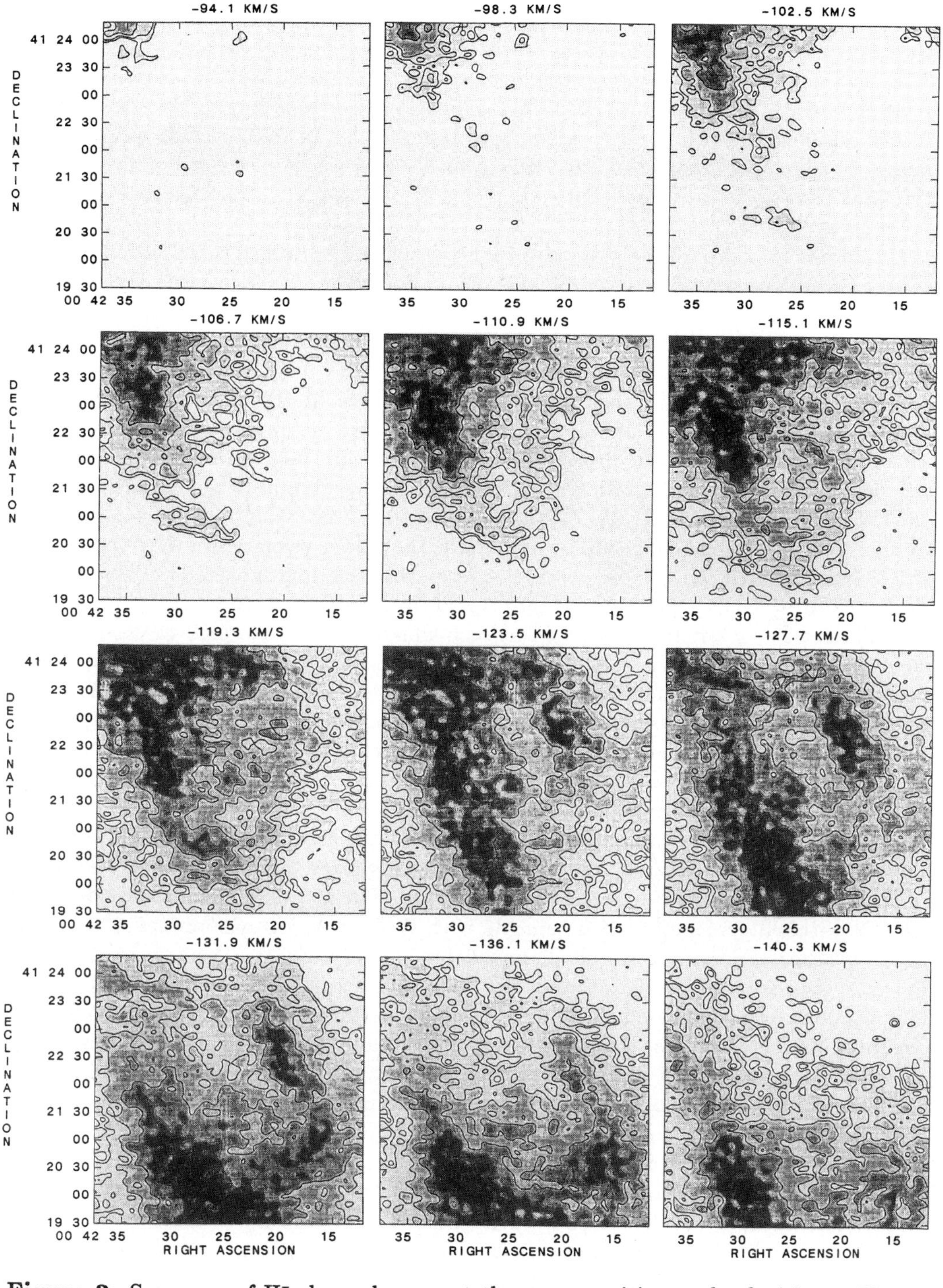

Figure 2: Sequence of HI channel maps at the same position and velocities as Figure 1. Contour levels are at -1, 1, 1.4, 2, 2.8, 4, 5.6, and 8 in units of 19.8 Kelvin. The rms noise level is 6.6 K. The beam is elongated in the East-West direction and measures 11 × 8.7 arcsec.

complete and measures $\sim$ 250 pc in diameter. On a set of higher resolution maps the shell is resolved and has a thickness of order 20 pc. No signs of expansion are evident. The shell contains many clumps which correspond to unresolved HII regions such as P710, P716, P719, P724 and P727. They extend over some 40 km s^{-1} in velocity. The bright resolved HII region P721 which is also embedded in P722 stands out from the maps. It measures about 20 $\times$ 22.5 arcsec or 65 $\times$ 75 pc. An integrated spectrum of this area shows a broad profile with a FWHM of $\sim$ 36 km s^{-1}. The Hα flux density is 9.7×10^{-13} erg cm^{-2} s^{-1}.

In order to compare the Hα maps with the neutral hydrogen distribution we have selected from the new HI survey of M31 made by Braun (1989) with the Very Large Array (VLA) the same area. Figure 2 shows a sequence of HI channel maps on exactly the same coordinate grid and sampled at the same velocities as the Hα data cube. The resolution of the HI maps is $\Delta\alpha \times \Delta\delta \times \Delta V = 11.0'' \times 8.7'' \times 5.1$ km s^{-1}. The Hα shell partly coincides with an HI shell whose outline can best be traced in the channel map at -115.1 km s^{-1}. The Hα emission is located such that it overlaps with the HI in the southern and eastern rim of the shell. The northern arc falls within the cavity. The HI shell is slightly larger, practically complete and near circular in appearance. The HI shell was not seen in the WSRT HI maps due to the lower resolution and sensitivity of that survey. Because of confusion by abundant HI at lower velocities in the northeastern spiral arm, the shell can only be observed over part of the range covered in velocity. The shell measures $\sim$ 330 pc in diameter and is marginally resolved. As in Hα, no change of shell diameter as a function of velocity is seen which could be ascribed to expansion. If we use the same method as employed by Brinks and Bajaja (1986) we derive an upper limit for the age of the shell of 20×10^6 year and an energy requirement of 5×10^{51} erg. Because the shell is stationary we have assumed a value of 8 km s^{-1} for its expansion velocity which is equivalent to the velocity dispersion in the ISM.

The northeastern arm is particularly rich in OB associations and our field coincides with OB48 from the list of van den Bergh (1964). Efremov *et al.* (1987) have reclassified van den Bergh's objects and redrawn the borders of this association. Their photographs show that the HI and Hα shell coincide with star complexes 48-3 and 48-4. At the centre of P722 there is a relative absence of young stars. The OB associations a-d which are located within star complex 48-4 coincide with P727, 724, 721, 719 and 716 which lie in the southeastern part of the shell. In a detailed study of OB48 Massey *et al.* (1986) describe the region as younger than 5×10^6 year. They base their estimate on colour-magnitude maps and on the fact that they find seven Wolf-Rayet stars. These WR stars generally coincide with the OB associations of Efremov *et al.* According to Massey *et al.* OB48 seems to consist of several groups, each measuring about 200 pc in diameter. From their Figure 3 it can be easily seen that the blue objects tend to be concentrated near the perimeter of the shell.

In an attempt to synthesize the various observations we arrive at the following picture. The oldest feature is the HI shell. Some 20×10^6 year ago an evolving OB association has most likely blown a cavity in the ISM. The most massive stars from this association have all evolved, the main sequence turn-off being located around B0 or at slightly later spectral types. Therefore no remnant association is seen by Massey *et al.* whose CCD photometry cuts off at stars older than B0. The HI shell, after having swept up a considerable amount of mass has stalled. Within the shell densities have reached high enough values to allow secondary star formation to commence. This star formation shows up as prominent HII regions. This explains why the Hα emission, notably on the southeastern side coincides with the HI shell. The Hα feature forming part of the northern rim lies within the HI shell. This can be understood if this HII region is part of the near or far side of the neutral shell such that its projection on the sky falls within the HI hole. The arguments given by Massey *et al.* allow us to put an age to the secondary star formation event of less than 5×10^6 year. Most of the Hα emission stems from this new event. Narrow band CCD direct imaging in Hα and in the line of [SII] corroborates this interpretation showing that although shock ionisation is present, this region is dominated by photoionisation (Walterbos and Braun, private communication).

These first results illustrate that it is feasible with current instruments to study the violent ISM in nearby galaxies at linear resolutions which could hitherto only be achieved in our Galaxy. Further analysis of this and the other fields is currently in progress.

ACKNOWLEDGEMENTS

The Isaac Newton Telescope is operated on the island of La Palma by the Royal Greenwich Observatory in the Spanish Observatorio del Roque de los Muchachos of the Instituto de Astrofisica de Canarias. The National Radio Astronomy Observatory is operated by Associated Universities, Inc., under contract with the National Science Foundation.

REFERENCES

Bland, J., Taylor, K., Atherton, P.D.: 1987, *Monthly Notices Roy. Astron. Soc.* **228**, 595

Braun, R.: 1989, *Astrophys. J. Supp.* (in preparation)

Brinks, E., Bajaja, E.: 1986, *Astron. Astrophys.* **169**, 14

Brinks, E., Burton, W.B.: 1984, *Astron. Astrophys.* **141**, 195

Cox, D.P, Smith, B.W.: 1974, *Astrophys. J. Lett.* **189**, L105

Deul, E.: 1988, *Ph. D. Thesis* (Leiden Observatory)

Dopita, M.A, Mathewson, D.S., Ford, V.L.: 1985, *Astrophys. J.* **297**, 599

Efremov, Yu.N., Ivanov, G.R., Nikolov, N.S.: 1987, *Astrophys. Space Sci.* **135**, 119

Field, G.B., Goldsmith, D.W., Habing, H.J.: 1969, *Astrophys. J. Letters* **155**, L149

Heiles, C.: 1979, *Astrophys. J.* **229**, 533

Heiles, C.: 1984, *Astrophys. J. Suppl.* **55**, 585

Heiles, C.: 1987, *Astrophys. J.* **315**, 555

Massey, P., Armandroff, T.E., Conti, P.S.: 1986, *Astron. J.* **92**, 1303

McKee, C.F., Ostriker, J.P.: 1977, *Astrophys. J.* **218**, 148

Meaburn, J.: 1980, *Monthly Notices Roy. Astron. Soc.* **192**, 365

Norman, C.A., Ikeuchi, S.: 1989, *Astrophys. J.* (submitted)

Pellet, A., Astier, N., Viale, A., Courtès, G., Maucherat, A., Monnet, G., Simien, F.: 1978, *Astron. Astrophys. Suppl.* **31**, 439

Tenorio-Tagle, G., Bodenheimer, P.: 1988, *Ann. Rev. Astron. Astrophys.* **26**, 145

van den Bergh, S.: 1964, *Astrophys. J. Suppl.* **9**, 65

CONTRIBUTIONS OF SUPERNOVAE TO THE CHEMICAL AND DYNAMICAL EVOLUTION OF THE ISM

Virginia Trimble
Department of Physics, University of California, Irvine CA 92717 USA
and
Astronomy Program, University of Maryland, College Park MD 20742 USA

ABSTRACT: Supernovae of Types Ia, Ib, and II contribute, on average, 10^{51} ergs of turbulent energy to the interstellar medium for each 2 M_o of new heavy elements. This permits a dynamical extension of the homogeneous one-zone model (with constant IMF and instantaneous recycling) that is familiar from studies of galactic chemical evolution. Chemically possible scenarios predict current kinetic energy inputs ranging from 10^{39} to 10^{40} erg yr^{-1} per solar mass of interstellar gas. Dynamical studies might narrow this.

I. THE R.M.S. SUPERNOVAE

Widely-believed current theory (Woosley and Weaver 1986) attributes Type II supernovae to core collapse in massive stars and classical Type Ia's to deflagration of carbon and oxygen to iron (etc.) in accreting degenerate dwarfs or merging degenerate dwarf pairs. The slightly fainter Type IIb's (distinguished by minor spectral differences and probable association with youngish stars) have been variously blamed on variants of both of the preceeding mechanisms (Tornambè and Matteucci 1987; Iben et al. 1987; Branch and Nomoto 1987; Uomoto 1986; Begelman and Sarazin 1986; Gaskell et al. 1986; Filippenko and Sargent 1986).

For the core collapse case, heavy element production depends primarily on initial stellar mass, $M_z = -7.5 + 0.5\ M^{MS}$, in solar masses (Fig. 1, derived from calculations by Arnett, 1978). The average over a stellar population with Salpeter initial mass function ($x = -2$), minimum mass for core collapse = 8 M_o, and maximum effective main sequence mass = 50 M_o (Schramm 1989; stars of still higher initial mass shed so rapidly that their nucleosynthesis mimics that of lower mass stars) is 2.4 M_o per event. An average kinetic energy of 10^{51} ergs in the ejected envelope is derived both from model light curves and from observed line widths, attributed to 5-10 M_o moving at 3000-4000 km s^{-1}. Apparently the shock doesn't much care what is outside the bouncing core.

The average values apply also to several otherwise deviant SNIIs. The explosion that gave rise to Cas A either was very faint or went undetected (Ashworth 1980), but studies of the remnant show sizable excesses of heavy elements and an energetic shock moving out into the ISM. Since the remnant also contains hydrogen, we assume that the spectrum, if recorded, would have shown the defining lines of a Type II and that the energy source was core collapse, possibly to a black hole, since no X-rays attributable to a young, hot neutron star have been seen. Similarly for 1987A, models

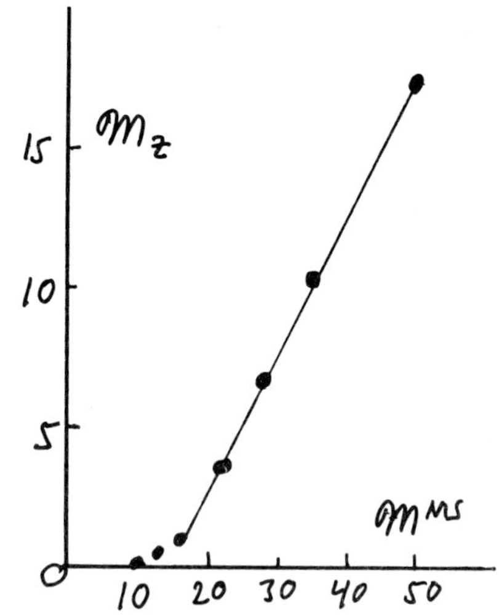

Figure 1: Mass of new heavy elements ejected by Type II, core collapse, supernovae, as a function of main sequence mass. Units are solar masses; models are those of Arnett (1978).

directed explicitly toward its progenitor indicate the synthesis of about 2.5 M_o of heavy elements (Hashimoto et al. 1989; Woosley et al. 1989) and an ejected envelope mass of 5-10 M_o. Since the time development of the line profiles showed the base of the hydrogen-rich envelope to lie at about 4000 km s^{-1}, the total kinetic energy (confirmed by models of the light curve) is the canonical 10^{51} ergs, despite the moderate faintness of the event.

Calculations directed toward SN 1987A have also made clear that moderate changes in initial composition and in nuclear reactions rates (especially the critical and uncertain $C^{12}(\alpha, \gamma)O^{16}$ rate) do not much affect the total amount of metal production, though they do modify the relative amounts of the major nucleosynthetic products (C,O, Ne,Mg,Si,S,Fe).

The supernova of 1954 is a curious exception to the pattern. The observed expansion energy of the remnant is at very most 10^{50} erg (1-2 M_o of line-emitting gas moving at less than 2000 km s^{-1}). And its contribution of new heavy elements is also negligible! (Davidson et al. 1982).

For the carbon detonation case, roughly a Chandrasekhar mass of heavy elements is necessarily kicked out into the world. Some of this will be newly-synthesized iron, some unburned CO, and some elements in between (the ratios impacting directly on the value of H_o found from SN Ia's and so not to be mentioned in polite company). Models of the light curves and confirming evidence from young supernova remnants tell us that the ejected kinetic energy is again about 10^{51} erg. If the current ratio of core collapse to deflagration supernovae is about two to one (van den Bergh et al. 1987), then the average over the supernova ensemble is 10^{51} ergs for each 2 M_o of new heavy elements.

II. THE HOMOGENEOUS, ONE-ZONE MODEL REVISITED

The first detailed models of galactic chemical evolution were numerical ones derived by Tinsley (1968). Shortly thereafter, Searle (1973) showed that some of the results could be arrived at analytically in the case where a galaxy constitutes a homogeneous, closed system, the birthrate of stars is a separable function of time and mass, and stellar nucleosynthesis occurs on a time scale short compared to anything else in the problem. This is called the homogeneous, one-zone model, with constant initial mass function and instantaneous recycling. Let u be the fraction of material in the form of gas, Z be the fractional metal abundance in the gas, and y be the yield (the fraction of material going into stars that comes out as heavy elements). Then chemical evolution is described as:

$$\frac{d(uZ)}{dt} = Z\frac{du}{dt} - y\frac{du}{dt}(1-Z) \qquad (1)$$

where the first term on the right represents loss of metals going into stars and the second term represents the gain from supernovae (notice that du/dt is intrinsically negative). After rearranging terms and assuming $Z \ll 1$, eqn. (1) has the solution:

$$Z = y \ln(1/u) \qquad (2)$$

which turns out to represent average galaxies, from irregulars to ellipticals, rather well. A little more manipulation leads to the predicted numbers of stars of various metal abundances as (where Z_o and u_o are the present values):

$$\frac{N(Z)}{N(Z_o)} = \frac{1 - u_o^{(Z/Z_o)}}{1 - u_o} \qquad (3)$$

Eqn. (3) presents the theoretical side of the classic G dwarf problem (meaning that we see far fewer old, metal-poor stars in the solar neighborhood than the model predicts). Other populations, including globular clusters and old giants in Baade's window are, on the other hand, well fit by the prediction (Pagel 1987).

If, instead of simplifying eqn. (1), we multiply it by the mass of the galaxy,

$$\frac{d(M_gZ)}{dt} = Z\frac{dM_g}{dt} - y\frac{dM_g}{dt} = Z\frac{dM_g}{dt} + k\frac{dE}{dt} \qquad (4)$$

is the result. The second term on the right can then be replaced by +k dE/dt, where dE/dt is the rate of input of kinetic energy into the gas and k is the 2 M_o per 10^{51} ergs produced by our rms supernova. Here and now, with a galactic gas mass of 7.35 X 10^9 M_o (8% of a disc with R = 8.5 kpc and V_c = 220 km/s) and $\Delta Z/\Delta t = 0.016/8\times10^9$ yr, we predict an input of 7 X 10^{48} erg yr^{-1} or 2.5 X 10^{47} erg per solar mass of gas per galactic rotation period. These numbers are not obviously in conflict with anything in particular. Closer examination of chemical enrichment history and its implications, however, reveals a wide margin of uncertainty, which dynamical considerations may be able to constrain.

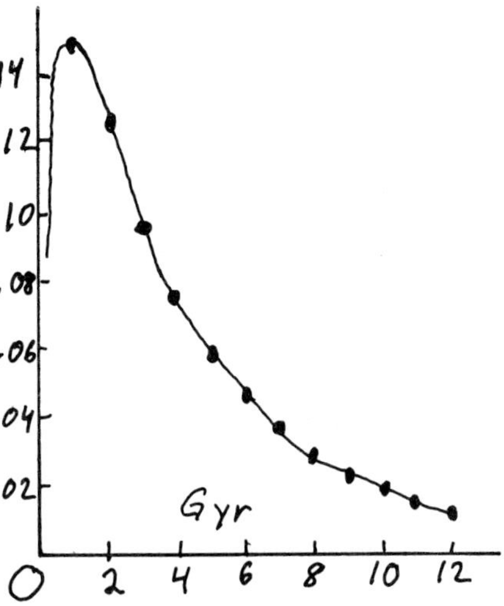

Figure 2: Rate of Type II SNe, in events per year, vs. age of the galactic disc in Gyr, from a model by Matteuci and Tornambè (1988)

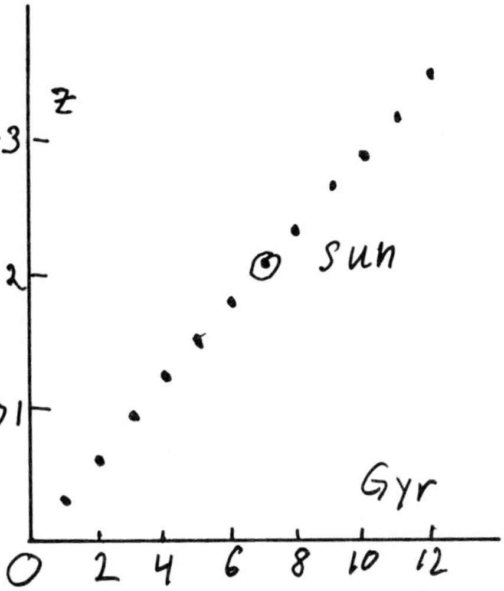

Figure 3: Change of gas metal abundance, Z, vs. age of the galactic disc in Gyr implied by the SN rates of Fig. 2. The exponential decline (t_e = 4.3 Gyr) is balanced by exponentially declining gas mass to yield a linear Z(t).

III. PAST AND PRESENT SN RATES AND THEIR IMPLICATIONS

A typical, successful, model of Milky Way chemical evolution (Matteucci and Tornambè 1988), incorporates SNII rates vs. time as shown in Fig. 2. Heavy element abundance in the residual gas is then as shown in Fig. 3. The value for the sun is tolerable, but the present Z_{gas} = 0.035 is clearly too high (compare Z = 0.018 for Orion, Peimbert et al. 1988, and [Fe/H] = -0.08 to +0.125 for six young clusters, Boesgaard 1989), though the present SNII rate is in good accord with that expected observationally (van den Bergh et al. 1987). In addition, our mental picture of galactic chemical evolution is really more like Fig. 4 or 5 than Fig. 3 (another way of saying that the G dwarf problem is peculiar to the solar neighborhood).

Consider the three components separately and quantitatively. For the halo, initial metallicity Z_i = 0 and final metallicity Z_f = 0.005, achieved over a time interval of at most 3 Gyr. Thus dZ/dt $\gtrsim$ 0.002/Gyr. For the disc component, maximum metallicity rises to Z = 0.01 in 3-4 Gyr, for a dZ/dt of about 0.002/Gyr. Finally, in the thin disc, Z rises by perhaps another 0.016 in about 8 Gyr, yielding dZ/dt = 0.002/Gyr yet again.

This would seem to correspond to a kinetic energy input of 10^{39} erg yr^{-1} per M_o of gas over all of galactic history and a supernova rate of only 1/136 yr at the present time. But what are the real uncertainties?

Recent observational work (Capellaro and Turatti 1988, van den Bergh et al. 1987, Tammann 1982, Richter and Rosa 1989; see Trimble 1989 for further references) comes rather close to recovering Zwicky's (1938) supernova rate of (in modern units and

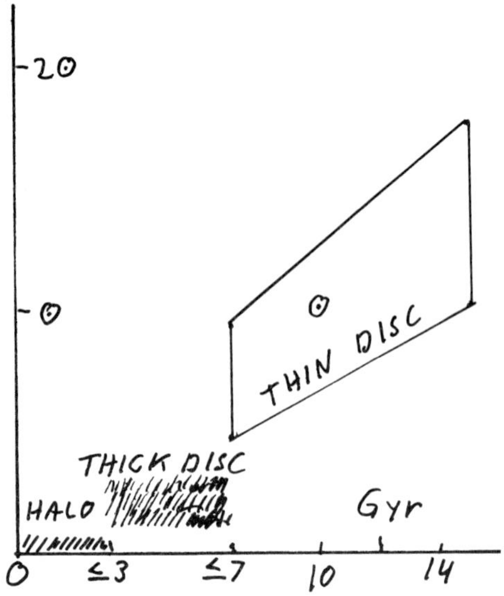

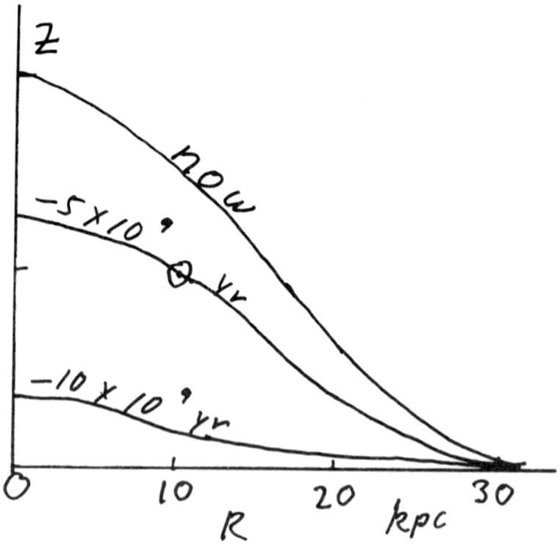

Figure 4: Scenario for galactic chemical evolution with three discrete stellar populations. Recent work on stars with large proper motions (Carney 1988) suggests such discreteness at least in the solar neighborhood.

Figure 5: Scenario for galactic chemical evolution with continuous distributions in both space and time. Norris and Green (1989) find that this may be a more suitable picture than one with separable components.

distance scales) 1.7 h^2 SNe per century per 10^{10} $L_{\odot}^{B}$, or about 3.4 h^2 per century in the Milky Way, implying a kinetic energy input rate of 5 X 10^{39} erg yr^{-1} per $M_{\odot}$ of gas. Finally, Schramm (1989) has advocated one nucleosynthetic, core-collapse supernova every 10 years, which would feed 10^{40} erg yr^{-1} per $M_{\odot}$ of interstellar gas into the system. Thus I would like to ask the question: Can dynamical studies help to narrow down this order-of-magnitude uncertainty?

ACKNOWLEDGEMENTS

Guido Münch began the singularly difficult task of guiding me through a PhD dissertation some 25 years ago, for which I shall always be enormously grateful to him. I wish him a long life, filled with salud, amor, y pesetas, in whichever order he would like them.

REFERENCES

Ashworth, W. 1980. J. Hist. Astron. 11, 1

Begelman, M.C. & C.L. Sarazin 1986. Astrophys. J. 302, L59

van den Bergh, S., R.D. McClure, & R. Evans 1987. Astrophys. J. 323, 44

Boesgaard, A.M. 1989. Astrophys. J. 336, 798

Branch, D. & K. Nomoto 1986. Astron. Astrophys. 164, 113

Cappelaro, E. & M. Turatti 1988. Astron. Astrophys. 190, 1

Carney, B.A. 1988. Private communication

Davidson, K. et al. 1982. Astrophys. J. 253, 696

Filippenko, A.V. & W.L.W. Sargent 1986. Astron. J. 91, 691

Hashimoto, M. , K. Nomoto & T. Shigeyama 1989. Astron. Astrophys. 210, L5

Iben, I, K. Nomoto, A. Tornambe, & A. Tutukov 1987. Astrophys. J. 317, 717

Matteucci, F. & A. Tornambè 1988. Comm. Astrophys. 12, 245

Norris, J. & E.M. Green 1989. Astrophys. J. 332, 272

Pagel, B.E.J. 1987. in G. Gilmore & R.F. Carswell (eds.) The Galaxy (Dordrecht: Reidel; ASI Inst.) p. 341

Peimbert, M. et al. 1988. Publ. Astron. Soc. Japan 40, 581

Schramm, D.N. 1989. Talk at UCLA Workshop, The Next Supernova

Searle, L. 1973. in G. Cayrel de Strobel & A.M. Deplace (eds.) Stellar Ages (Meudon, Obs. de Paris) LII, 1

Richter, O.-G. & M. Rosa 1989. Astron. Astrophys. 206, 219

Tammann, G.A. 1982. in M.R. Rees & R.J. Stoneham (eds.) Supernovae: A Survey of Current Research (Dordrect: Reidel, ASI Inst.) p. 371

Tinsley, B.M. 1968. Astrophys. J. 151, 547

Tornambè, A. & F. Matteucci 1987. Astrophys. J. 318, L25

Trimble, V. 1989. in D.S. Hayes & R.M. Genet (eds.) Remote Access Automatic Telescopes Fairborn Press, in press

Uomoto, A. 1986. Astrophys. J. 310, L35

Woosley, S.E. & T.A. Weaver 1986. Ann. Rev. Astron. Astrophys. 24, 205

Woosley, S.E., P.A. Pinto & T.A. Weaver 1989. Proc. Astron. Soc. Australia. in press

Zwicky, F. 1938. Astrophys. J. 88, 529

Discussion:

MÜNCH (Comment): We have heard during this Meeting about the crucial role played by SN in determining the structure and physical state of the ISM, but little about their influence in the chemical composition of the local ISM. In this respect I should mention that last Summer Gerry Wasserbug exposed to me existing problems in our understanding of the high abundance of radionuclides Al^{26} in meteorites. Wasserburg believes that Astronomers are not fully aware of the implications of the isotopic anomalies, and for this reason promised me he would attend our Meeting. Since he could not make it, I can only repeat his plea to us for paying attention to the message contained in the early solar system material.

1/4 Ke V DIFFUSE BACKGROUND AND THE LOCAL INTERSTELLAR MEDIUM

S.L. Snowden

Max-Planck-Institute for Extraterrestrial Physics

The 1/4 keV diffuse X-ray background (SXRB) is discussed in relation to the local interstellar medium (LISM). The most likely source for these soft X-rays is thermal emission from a hot diffuse plasma. The existence of a non-zero flux from all directions and the short ISM mean free path of these X-rays ($10^{20} HI\ cm^{-2}$), coupled with ISM pressure constraints, imply that the plasma has a local component and that it must, at least locally (nearest hundred parsecs), have a large filling factor. Our understanding of the geometry and physical parameters of the LISM is therefore directly tied to our understanding of the SXRB.

Two extreme SXRB source models are: 1) The displacement model; the source plasma is contained within a cavity in the galactic HI surrounding the Sun and the observed negative correlation between soft X-ray flux and HI is produced entirely by a competition for space. 2) The interspersed model; the emitting plasma and HI are intermixed with the plasma having no effect on the HI distribution. The observed negative correlation is produced entirely by absorption, or by a combination of absorption and latitude effects due to different scale heights for the hot and cold gas. Observational consequences of these models are discussed, with the results favoring something much closer to 1).

With the assumption that the displacement model provides an accurate representation of solar neighborhood, the structure of the LISM is discussed. The local plasma-filled cavity is irregulary shaped and is distended out of the galactic plane with an aspect ratio of ~ 2.5 with a mean radius of ~ 100 pc. Some parameters of the plasma are: electron density, $\sim 0.005\ cm^{-3}$, pressure, (p/k) $\sim 9000\ cm^{-3}K$, and temperature, $10^{6.0}K$. Finally, the displacement model is shown to be consistent with diverse observations of the LISM.

LARGE-SCALE IRREGULARITIES IN THE INTERSTELLAR MEDIUM

Hari Om Vats

Physical Research Laboratory, India

The radio flux measurement of the compact extragalactic objects are known to show two types of variablity i.e. intrinsic and extrinsic. Some of the unusual minima in the flux measurement seem to involve large structures in the interstellar medium. In this article we outline a simple ray approach for simulation of these minima in the flux measurement. Simulation will help in estimating the physical parameters of these large scale irregularities or structures in the interstellar medium. Some preliminary results of the simulation will also be described.

THE MILKY WAY DISK WARP

E. Florido[1], E. Battaner[1], E. Alfaro[2] and M.L. Sanchez-Saavedra[1]

[1]Universidad de Granada, Spain

[2] Instituto de Astrofísica de Andalucía, Spain

A warped disk in our own galaxy is evident by means of HI, HII, γ-rays and dust observations, but unexistent when star distributions are considered, specially those of late type stars. This fact is in disagreement with the theories which assume a gravitational origin of warps, for instance a tidal interaction with the Magellanic Clouds. We tried to find the z-distribution of open clusters of different ages, for which a warp distribution was neither found nor rejected. Assuming an intergalactic magnetic field origin of the warp, we obtain a direction of the field in the Milky Way neighborhood given by (b,1) = (45°, 74°).

Lecture Notes in Mathematics

Vol. 1236: Stochastic Partial Differential Equations and Applications. Proceedings, 1985. Edited by G. Da Prato and L. Tubaro. V, 257 pages. 1987.

Vol. 1237: Rational Approximation and its Applications in Mathematics and Physics. Proceedings, 1985. Edited by J. Gilewicz, M. Pindor and W. Siemaszko. XII, 350 pages. 1987.

Vol. 1250: Stochastic Processes – Mathematics and Physics II. Proceedings 1985. Edited by S. Albeverio, Ph. Blanchard and L. Streit. VI, 359 pages. 1987.

Vol. 1251: Differential Geometric Methods in Mathematical Physics. Proceedings, 1985. Edited by P.L. García and A. Pérez-Rendón. VII, 300 pages. 1987.

Vol. 1255: Differential Geometry and Differential Equations. Proceedings, 1985. Edited by C. Gu, M. Berger and R.L. Bryant. XII, 243 pages. 1987.

Vol. 1256: Pseudo-Differential Operators. Proceedings, 1986. Edited by H.O. Cordes, B. Gramsch and H. Widom. X, 479 pages. 1987.

Vol. 1258: J. Weidmann, Spectral Theory of Ordinary Differential Operators. VI, 303 pages. 1987.

Vol. 1260: N.H. Pavel, Nonlinear Evolution Operators and Semigroups. VI, 285 pages. 1987.

Vol. 1263: V.L. Hansen (Ed.), Differential Geometry. Proceedings, 1985. XI, 288 pages. 1987.

Vol. 1265: W. Van Assche, Asymptotics for Orthogonal Polynomials. VI, 201 pages. 1987.

Vol. 1267: J. Lindenstrauss, V.D. Milman (Eds.), Geometrical Aspects of Functional Analysis. Seminar. VII, 212 pages. 1987.

Vol. 1269: M. Shiota, Nash Manifolds. VI, 223 pages. 1987.

Vol. 1270: C. Carasso, P.-A. Raviart, D. Serre (Eds.), Nonlinear Hyperbolic Problems. Proceedings, 1986. XV, 341 pages. 1987.

Vol. 1272: M.S. Livšic, L.L. Waksman, Commuting Nonselfadjoint Operators in Hilbert Space. III, 115 pages. 1987.

Vol. 1273: G.-M. Greuel, G. Trautmann (Eds.), Singularities, Representation of Algebras, and Vector Bundles. Proceedings, 1985. XIV, 383 pages. 1987.

Vol. 1275: C.A. Berenstein (Ed.), Complex Analysis I. Proceedings, 1985–86. XV, 331 pages. 1987.

Vol. 1276: C.A. Berenstein (Ed.), Complex Analysis II. Proceedings, 1985–86. IX, 320 pages. 1987.

Vol. 1277: C.A. Berenstein (Ed.), Complex Analysis III. Proceedings, 1985–86. X, 350 pages. 1987.

Vol. 1283: S. Mardešić, J. Segal (Eds.), Geometric Topology and Shape Theory. Proceedings, 1986. V, 261 pages. 1987.

Vol. 1285: I.W. Knowles, Y. Saitō (Eds.), Differential Equations and Mathematical Physics. Proceedings, 1986. XVI, 499 pages. 1987.

Vol. 1287: E.B. Saff (Ed.), Approximation Theory, Tampa. Proceedings, 1985–1986. V, 228 pages. 1987.

Vol. 1288: Yu. L. Rodin, Generalized Analytic Functions on Riemann Surfaces. V, 128 pages, 1987.

Vol. 1294: M. Queffélec, Substitution Dynamical Systems – Spectral Analysis. XIII, 240 pages. 1987.

Vol. 1299: S. Watanabe, Yu.V. Prokhorov (Eds.), Probability Theory and Mathematical Statistics. Proceedings, 1986. VIII, 589 pages. 1988.

Vol. 1300: G.B. Seligman, Constructions of Lie Algebras and their Modules. VI, 190 pages. 1988.

Vol. 1302: M. Cwikel, J. Peetre, Y. Sagher, H. Wallin (Eds.), Function Spaces and Applications. Proceedings, 1986. VI, 445 pages. 1988.

Vol. 1303: L. Accardi, W. von Waldenfels (Eds.), Quantum Probability and Applications III. Proceedings, 1987. VI, 373 pages. 1988.

Lecture Notes in Physics

Vol. 325: E. Brändas, N. Elander (Eds.), Resonances, Proceedings, 1987. XVIII, 564 pages. 1989.

Vol. 326: A. Grauel, Feldtheoretische Beschreibung der Thermodynamik für Grenzflächen. IX, 317 Seiten. 1989.

Vol. 327: K. Meisenheimer, H.-J. Röser (Eds.), Hot Spots in Extragalactic Radio Source. Proceedings, 1988, XII, 301 pages. 1989.

Vol. 328: G. Wegner (Ed.), White Dwarfs. Proceedings, 1988. XIV, 524 pages. 1989.

Vol. 329: A Heck, F. Murtagh (Eds.), Knowledge Based Systems in Astronomy. IV, 280 pages. 1989.

Vol. 330: J.M. Moran, J.N. Hewitt, K.Y. Lo (Eds.), Gravitational Lenses. Proceedings, 1988. XIV, 238 pages. 1989.

Vol. 331: G. Winnewisser, J.T. Armstrong (Eds.), The Physics and Chemistry of Interstellar Molecular Clouds mm and Sub-mm Observations in Astrophysics. Proceedings, 1988. XVIII, 463 pages. 1989.

Vol. 332: P. Flin, H.W. Duerbeck (Eds.), Morphological Cosmology. Proceedings, 1988. VII, 438 pages. 1989.

Vol. 333: I. Appenzeller, H.J. Habing, P. Léna (Eds.), Evolution of Galaxies – Astronomical Observations. Proceedings, 1988. X, 391 pages. 1989.

Vol. 334: L. Maraschi, T. Maccacaro, M.-H. Ulrich (Eds.), BL Lac Objects. XIII, 497 pages. 1989.

Vol. 335: A. Lakhtakia, V.K. Varadan, V.V. Varadan, Time-Harmonic Electromagnetic Fields in Chiral Media. VII, 121 pages. 1989.

Vol. 336: M. Müller, Consistent Classical Supergravity Theories. VI, 125 pages. 1989.

Vol. 337: A.P. Maclin, T.L. Gill, W.W. Zachary (Eds.), Magnetic Phenomena. Proceedings, 1988. VI, 142 pages. 1989.

Vol. 338: V. Privman, N.M. Švrakić, Directed Models of Polymers, Interfaces, and Clusters: Scaling and Finite-Size Properties. VI, 120 pages. 1989.

Vol. 339: F. Ehlotzky (Ed.), Fundamentals of Laser Interactions II. Proceedings, 1989. XI, 317 pages. 1989.

Vol. 340: M. Peshkin, A. Tonomura, The Aharonov-Bohm Effect. VI, 152 pages. 1989.

Vol. 341: E.F. Milone (Ed.), Infrared Extinction and Standardization. Proceedings, 1988. III, 79 pages. 1989.

Vol. 343: M. Month, S. Turner (Eds.), Frontiers of Particle Beams; Observation, Diagnosis and Correction. Proceedings, 1988. IX, 509 pages. 1989.

Vol. 344: M. Rascle, D. Serre, M. Slemrod (Eds.), PDEs and Continuum Models of Phase Transitions. Proceedings, 1988. VII, 229 pages. 1989.

Vol. 345: H. Holden, A. Jensen (Eds.), Schrödinger Operators. Proceedings, 1988. V, 458 pages. 1989.

Vol. 346: D. Lüst, S. Theisen, Lectures on String Theory. VII, 346 pages. 1989.

Vol. 348: A. Bohm, M. Gadella, Dirac Kets, Gamow Vectors and Gel'fand Triplets. VII, 119 pages. 1989.

Vol. 349: Y.A. Kubyshin, J.M. Mourão, G. Rudolph, I.P. Volobujev, Dimensional Reduction of Gauge Theories, Spontaneous Compactification and Model Building. X, 80 pages. 1989.

Vol. 350: G. Tenorio-Tagle, M. Moles, J. Melnick (Eds.), Structure and Dynamics of the Interstellar Medium. Proceedings, 1989. XXI, 537 pages. 1989.